W9-BNZ-942

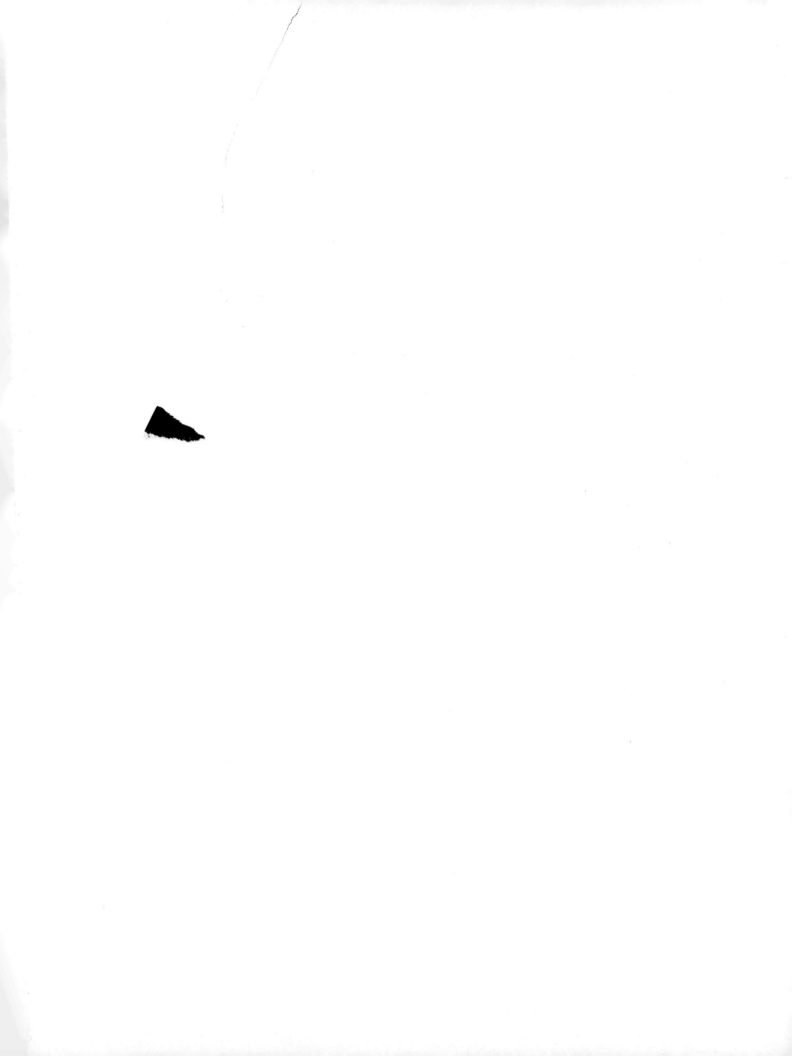

ABOUT ISLAND PRESS

Island Press is the only nonprofit organization in the United States whose principal purpose is the publication of books on environmental issues and natural resource management. We provide solutions-oriented information to professionals, public officials, business and community leaders, and concerned citizens who are shaping responses to environmental problems.

In 1994, Island Press celebrated its tenth anniversary as the leading provider of timely and practical books that take a multidisciplinary approach to critical environmental concerns. Our growing list of titles reflects our commitment to bringing the best of an expanding body of literature to the environmental community throughout North America and the world.

Support for Island Press is provided by The Geraldine R. Dodge Foundation, The Energy Foundation, The Ford Foundation, William and Flora Hewlett Foundation, The James Irvine Foundation, The John D. and Catherine T. MacArthur Foundation, The Andrew W. Mellon Foundation, The Pew Charitable Trusts, The Rockefeller Brothers Fund, The Tides Foundation, Turner Foundation, Inc., The Rockefeller Philanthropic Collaborative, Inc., and individual donors.

ABOUT THE CRAIGHEAD WILDLIFE–WILDLANDS INSTITUTE

The Craighead Wildlife–Wildlands Institute is an independent, nonprofit, conservation organization devoted to long-term ecological research and scientific activism in the Rocky Mountain West. Founded in 1980 by Dr. John J. Craighead, and headquartered in Missoula, Montana, the Institute has won national recognition for its scientific accomplishments in the field of conservation biology. Although a privately supported organization, the Institute maintains close links with its colleagues at the University of Montana, national and local environmental organizations, government agencies, and the general public in order to fulfill its mission. The Craighead Wildlife–Wildlands Institute encourages the application of ecological principles and scientific discoveries to promote wise use of the nation's public lands.

THE GRIZZLY BEARS OF YELLOWSTONE

The Grizzly Bears of Yellowstone

THEIR ECOLOGY IN THE YELLOWSTONE ECOSYSTEM, 1959–1992

John J. Craighead, Jay S. Sumner, and John A. Mitchell
CRAIGHEAD WILDLIFE–WILDLANDS INSTITUTE

ISLAND PRESS
Washington, D.C. • *Covelo, California*

BOWLING GREEN STATE
UNIVERSITY LIBRARIES

Copyright © 1995 by John J. Craighead and the Craighead Wildlife–Wildlands Institute

All photographs were reproduced with permission for one-time use only from the photographers noted in the photo captions.

All rights reserved under International and Pan-American Copyright Conventions. No part of this book may be reproduced in any form or by any means without permission in writing from the publisher: Island Press, 1718 Connecticut Avenue, N.W., Suite 300, Washington, DC 20009.

ISLAND PRESS is a trademark of The Center for Resource Economics.

Library of Congress Cataloging-in-Publication Data

Craighead, John Johnson, 1916
    The grizzly bears of Yellowstone : their ecology in the Yellowstone
ecosystem, 1959-1992 / John J. Craighead, Jay S. Sumner, and John A.
Mitchell
        p.   cm.
    Includes bibliographical references and index.
    ISBN 1-55963-456-1 (cloth)
    1. Grizzly bear—Yellowstone National Park. 2. Grizzly bear—
Ecology—Yellowstone National Park. 3. Mammal populations—
Yellowstone National Park. 4. Wildlife conservation—Yellowstone
National Park. I. Sumner, Jay S. II. Mitchell, John A.
    III. Title.
QL737.C27C72   1995
599.74'446—dc20   95-22680
                                                          CIP

This book is printed on paper manufactured using pulp from sustainably harvested forests. The paper was generously donated to the Craighead Wildlife–Wildlands Institute by Crescent Cardboard, a long-time supporter of conservation efforts.

Manufactured in the United States of America

10 9 8 7 6 5 4 3 2 1

To the memory of Charles Engelhard,
consummate businessman, philanthropist, outdoorsman
and practicing conservationist, whose outlook on nature
is expressed most aptly through the poetic words of
William Wordsworth. . . "A lover of the meadows and the woods,
and mountains; and of all that we behold from this green earth."

To Seward Prosser Mellon,
whose concern for the future environment of mankind
has preserved for posterity pristine landscapes
with their magnificent diversity of life.

Contents

CHAPTER 7. REPRODUCTIVE BIOLOGY

CHAPTER 8. SURVIVORSHIP AND WEANING OF OFFSPRING

CHAPTER 9. DENSITY-DEPENDENT
INFLUENCES ON POPULATION DYNAMICS

CHAPTER 10. FOOD HABITS AND
FEEDING BEHAVIOR OF THE GRIZZLY BEAR POPULATION

Section II: Population Comparisons
Chapter 14. 1974-1992: The Era of Decreed Research

Chapter 15. Grizzly Bear Distribution and Movements Following Catastrophic Events

Chapter 16. Comparative Food Habits and Feeding Behavior Following Ecocenter Closures

Preface

This monograph first describes and interprets the social behavior, movements, reproductive and population biology, demography, food habits, and habitat requirements of grizzly bears that frequented the open-pit dump at Trout Creek in Yellowstone National Park during the years 1959-70, a site exemplary of what we term "ecocenters"—sites that predictably attract annual aggregations of bears, in response to massive food resources. When appropriate, information derived from a study of this population subunit is compared with parameters of the larger ecosystem population. A major focus is the complex social interrelationships that develop between grizzly bears when they congregate to feed and breed. Study results reveal fundamental aspects of bear biology that may apply in varying degrees to populations of grizzly/brown bears throughout their range in North America, Asia, and Europe.

In Section I, we describe demographic parameters of the Trout Creek and ecosystem-wide bear populations. We then examine the social, feeding, and reproductive behaviors exhibited when bears gathered at the ecocenters. Section II compares biological parameters of the Yellowstone ecosystem-wide population before closure of the ecocenters with those of the postclosure population; it concludes with some thoughts on the dynamics of the changes observed and their management implications. A plan for the recovery of the threatened grizzly bear population of Yellowstone is proposed, based on biological analyses and their relevance to ecosystem resource management concepts.

We believe that understanding of current public land resource problems can be achieved only through long-term investigations that contribute to a holistic approach to natural resource management. The research reported here has extended over a time span of more than 33 years.

This monograph is directed to the scientific reader and, specifically, to bear biologists and resource managers. To make the contents more readily available to resource administrators, policy makers, politicians, and the lay reader, we have briefly summarized chapters and have prepared informative legends for the figures and plates. A general, but comprehensive understanding of the contents can be obtained by sequentially reading chapter summaries and the figure and plate captions pertaining to each chapter.

The major conclusion of our study is that resource management agencies must focus on the broad problem of preserving and managing habitat in a resource exploitive society where politics and economic policies thwart sustainable resource management. For government agencies to practice ecosystem and bioregional resource management in the future will require input from the private scientific sector, as well as from the general public. We believe that working together, private research organizations and government resource agencies can create concepts of land-use economics that can be integrated with the principles of ecosystem ecology. This monograph reveals the necessity of working cooperatively to achieve this in the Northern Rocky Mountain bioregion. The grizzly bear is the umbrella species . . . how fares the grizzly bear, so fares the ecology of the region.

Acknowledgments

For the senior author, work on this monograph has spanned a period of 36 consecutive years, with support and cooperation from many people and organizations.

The long span of time this work covers complicates the task of appropriately recognizing assistance and support. Nevertheless, we shall attempt to credit all and ask for understanding where we fail.

All contributions have been essential to completion of this work, but none more so than family love and solidarity. Collecting, collating, analyzing, and interpreting the data have required a continuity of effort that has impinged on the family and professional lives of all three authors. We owe special appreciation to Margaret Craighead, Donna Mitchell, and Janna Sumner, who gallantly accepted lost weekends and the frequent preoccupation of their husbands as evils necessary to the authoring of this monograph. Our work has been a labor of love, not without sacrifice.

For the senior author, gratitude to family begins a generation back, to a scientist, Dr. Frank C. Craighead, Sr., and his wife Carolyn Craighead, who nourished in their offspring the seeds of inquisitiveness and instilled in their children, John, Frank, and Jean, a love and concern for the natural world. It is to them the senior author owes the best part of a lifetime intimate with the wild and with the magnificent subject of this monograph, the grizzly bear.

The senior author's family accompanied him on much of the field work. His eldest son, Derek, was introduced to the study of grizzly bears in Yellowstone as a grade-schooler. He assisted in the early years of the Scapegoat Study as a graduate student at the University of Montana, then later worked on the Alaskan project as a field biologist with the Alaska Department of Fish and Game, and finally, has contributed as an independent bear researcher. The senior author's daughter, Karen, received her indoctrination to grizzlies and their habitat in Yellowstone while a teenager. Karen's great love of the outdoors kept her at her father's side throughout the summers of her college career, and she was indispensable on the grizzly bear habitat-mapping project in the Scapegoat. John Willis, the youngest of the senior author's children, radio-tracked bears from his father's backpack before he could read, and after graduating from the University of Montana, he assisted whenever a free summer allowed. His assistance with botanical aspects of the various habitat studies is especially appreciated. Wife Margaret, a biologist and outdoor enthusiast in her own right, kept the family and household intact, yet frequently joined the researchers in the field. Here, as a family, we enjoyed the unique work and adventure that only wilderness living offers.

In the Yellowstone Study, the senior author's brother, Frank C. Craighead, Jr., played an absolutely essential role, participating in nearly all aspects of the research. In collaboration with Hoke Franciscus, Joel Varney, Richard Davies, and the senior author, he developed the first radio collars and radio-location system used to track large carnivores. Frank has published his account of the Yellowstone grizzly bear research in the 1979 Sierra Club publication, *Track of the Grizzly*. At that time, the senior author and his brother agreed to separate courses in the publication of their work, as set forth in the preface to his book. This monograph is the product of the senior author and his colleagues, Jay Sumner and John Mitchell, and may be viewed as a technical companion to Frank's description of our field efforts in Yellowstone.

In this monograph, we report on the scientific aspects of three separate field studies—the Yellowstone Study, the Scapegoat Wilderness Study (Montana), and the Kobuk River Study (Alaska). Basically, the field teams consisted of professionals and students who made substantial contributions to the research effort.

We are grateful to Maurice Hornocker, who, as a graduate student, helped initiate the

Yellowstone project and set a high standard for field observation, and to Robert Ruff, who added important dimensions to the quantitative recording and collating of the field data. Harry V. Reynolds, III, first as a high school senior, and then as a University of Montana undergraduate, served as an able trapper and observer through a number of field seasons.

Others who assisted at one time or another in the field or laboratory work in Yellowstone are Gerry Atwell, Fritz Bell, Peter Bixler, Jim Claar, Lance Craighead, Henry McCutchen, Jerome McGahan, Bart O'Gara, Bert Pfeifer, Charles Ridenour, Jack Seidensticker, Mike Stevens, Henry Tomingas, James Wiley, Wes Woodgerd, Vince Yannon, and Dick Ellis.

Team members on the senior author's Scapegoat Study in western Montana were Jay Sumner, Gordon Scaggs, John Mitchell, Derek, John, and Karen Craighead. Chris Servheen, Steve Ford, and Richard Brown participated during one field season.

Team members on the senior author's Alaskan Kobuk River project were Derek Craighead, Lance Craighead, Harry Reynolds, III, John Hecktel, and John Mitchell.

The cooperation and the integrity of Yellowstone District Ranger Harry Reynolds, Jr., and Yellowstone Superintendent Lemuel Garrison were instrumental in ameliorating the professional and administrative conflicts that invariably arise in a long-term cooperative project. They are the unsung heroes of the Yellowstone grizzly bear controversy. For support in that bureaucratic struggle, we are most grateful to Jim Nakamura, Frank Sogandares, and Richard Fevold of the University of Montana; Melvin Payne, Chairman of the Board, National Geographic Society; Vincent Schaeffer, Director, Atmospheric Sciences Research Center, State University of New York at Albany; Lawrence Gould, President, Carlton College; Jerry Stout, Pennsylvania State University; and Dan Leedy, Rollin Sparrowe, Robert Streeter, and Charles Stone of the U. S. Fish and Wildlife Service.

An intensive research effort of long duration is possible only with adequate and continuing financial aid. We feel fortunate for and grateful to our financial supporters. Aid was of two types: funds to conduct the field studies and funds to produce an integrated account of those studies. Organizations and foundations contributing major financing for the research effort were the R.K. Mellon Foundation, the Engelhard Foundation, the Murdock Charitable Trust, the National Geographic Society, the Atomic Energy Commission, and the National Science Foundation.

Major personal contributions of time or funding came from Vincent Schaeffer, Gilbert Grosvenor, Bud Ozmun, Scott Ozmun, Charles Pihl, Jean Vollum, George Hixon, and William Gallagher. Many of these contributors joined the senior author in the field. The personal interest they took in the research and the camaraderie they engendered were especially gratifying.

For major grants to finance the writing and facilitate publication, we are indebted to the R.K. Mellon Foundation, the Engelhard Foundation, the Murdock Charitable Trust, Sophie Engelhard Craighead, and Earl Morgenroth. Others who contributed significantly with both funds and wise counsel were Loren Kreck, Spencer Beebe, Jim Castles, Meri Jaye, and Craighead Wildlife-Wildlands Institute (CWWI) board members Yvon Chouinard, John Castles, Richard Gooding, Thomas McGuane, and William Zader.

The Boone & Crocket Club, the National Rifle Association, the American Museum of Natural History, and the New York Zoological Society provided one-year financial contributions to the work. The U. S. Fish and Wildlife Service and Montana Cooperative Wildlife Research Unit contributed to the senior author's salary and transportation during the 12 years of research in Yellowstone. Administrative support and facilities were provided by the University of Montana, the National Park Service (NPS), the Yellowstone Park Company, the Wildlife Management Institute, the Montana Department of Fish, Wildlife and Parks, and the Fish & Game Departments of Idaho and Wyoming. The late Donald Comstock, Engineer, U. S. Forest Service, and his staff deserve special credit for providing photogrammetry expertise that greatly facilitated our vegetation-mapping projects.

The authors cannot adequately express their thanks and admiration to John Hogg, CWWI Senior Biologist, whose contributions throughout the monograph, always generously offered, were both fundamental and unique. He added significantly to its analytical and interpretive quality. We thank Brian Steele for performing the more complex statistical analyses; Noel Weaver, for her computer expertise in preparing tables and graphs, carefully checking each database and co-directing publication; Roly Redmond, for assisting in various analyses; Greg Tollefson and Terry McEneaney, for assembling some of the historical data; Martin Onishuk and Jennifer O'Laughlin, for editing assistance; John W. Craighead, for organizing illustrative materials and photographic layouts and co-directing publication; and Derek Craighead, for his vital contributions to the technology for satellite-based animal location. For the essential tasks of word-processing text and tables, we are indebted to Sharon Gaughan, Susan Sindt, Nadine Frances, Jan MacFarland, Alvina Barclay, Marian Jenkins, Ginger Schwarz, and Rae Dabbert. We offer sincerest thanks, too, to Bernadette Bannister, Hope Johnston, Bryony Schwan, and Deana Ross—members of the CWWI staff who, although not directly involved in the production of this publication, always stood ready to provide assistance, as needed.

Richard Knight, Bonnie Blanchard, and David Mattson of the Interagency Grizzly Bear Study Team generously provided data crucial to making many of the between-population comparisons. Kerry Gunther, NPS biologist, provided a wide range of NPS data that were helpful in evaluating bear movements following closure of the open-pit dumps. We are also grateful to Jon Robinett, manager of the Diamond G Ranch, Wyoming, for invaluable information on bear sightings and movements in the Brent Creek area of Wyoming.

The scope and volume of data presented in this monograph, many of them for the first time, made critical review of the entire manuscript by single individuals both impractical and burdensome. Accordingly, we requested reviews for specific chapters from colleagues selected for their specialized expertise. We are indebted to Morgan Berthrong, Stewart Brandborg, Erick Green, Charles Jonkel, John Hogg, David Mattson, Lee Metzgar, Sterling Miller, Harry Reynolds, Jr., Harry Reynolds, III, and John Weaver for critiques of one or more chapters. All performed their tasks beyond the call of duty, and therefore, their comments and recommendations have contributed significantly to the quality of this publication. We know the costs of dropping personal work to assist others, and we offer these individuals our heartfelt appreciation.

James Wiley, editor for the Western Foundation of Vertebrate Zoology, has read the entire manuscript critically, made valuable suggestions, and masterfully edited the monograph for publication. We have greatly appreciated his assistance.

Fig. 1.1. The 899,000-hectare (2.2-million-acre) Yellowstone National Park.

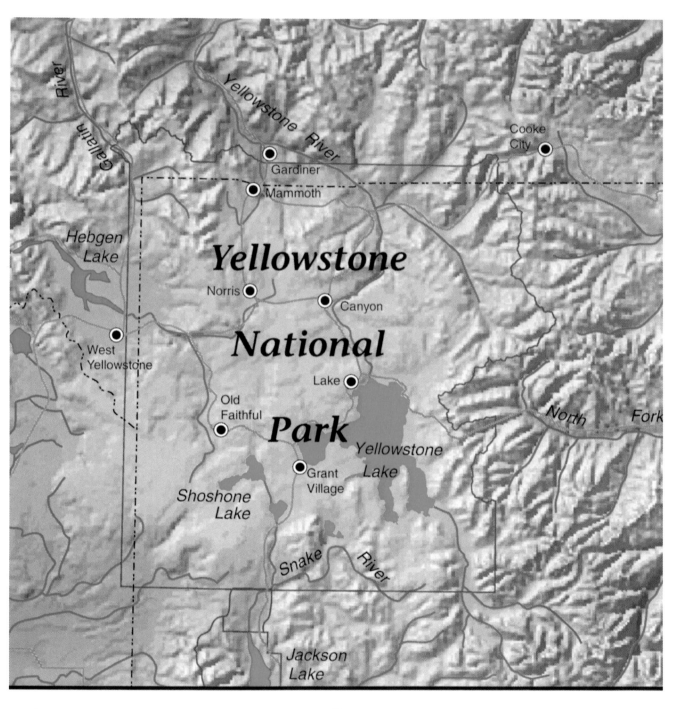

Albers Equal Area Projection

SECTION I: THE STUDY (1959-70)

1
Historical Prelude

INTRODUCTION

In 1959, the senior author and his brother, Frank C. Craighead, Jr., began a long-term study of the grizzly bear (*Ursus arctos horribilis*) in Yellowstone National Park (Fig. 1.1) and the surrounding national forests (Fig. 1.2). At that time, relatively little was known about the ecosystem-wide bear habitat or the grizzly bear population inhabiting it. We designed the research to obtain basic biological information that could be used to develop a management program. The project was enthusiastically embraced by the Park Superintendent, Lemuel Garrison, and his staff. A Memorandum of Agreement defined the respective roles and responsibilities of our research team and the Yellowstone Park administration. Funding was obtained, with minor exceptions, from the private sector.

Working relations were smooth and productive until 1968, when a change in park administration and policy caused frictions that could not be easily resolved. Working relationships deteriorated rapidly. By 1970 (after 12 years of research), we found it impossible to continue our work and, reluctantly, terminated the project in 1971. Following this event, a controversy developed over National Park Service bear management and its effects on the integrity of the bear population. Politics and biology soon became intertwined, preventing a scientific resolution in spite of a wealth of biological information and the advice of numerous investigative committees.

The political aspects of the controversy are nearly as complex and fascinating as the biological ones. This monograph, however, is confined to the biology and management of the grizzly bear, with only brief reference to politics to place our research project in political perspective.

We completed the 1970 field season under unacceptable Park Service restrictions and were permitted only a limited field effort in 1971. During this time, the Park Service mounted its own independent research effort in Yellowstone under the aegis of Glen F. Cole, a Supervisory Research Biologist. Cole believed, incorrectly, that park grizzly bears were found in two dynamically distinct groups, a "dump" population and a "backcountry" one, and he supervised bear research activities accordingly (Cole 1970). During his tenure (1970-74), marking of bears in the Park was prohibited. The considerable outside criticisms of Cole's research methodology and bear management decisions were confirmed by the "National Academy of Sciences (NAS) Committee Report on the Yellowstone Grizzlies" (Cowan *et al.* 1974). A year after that report, marking of bears was again permitted (1975), and data reliable for comparative analyses finally began to emerge.

The NAS Committee Report supported our research findings and concluded that the research program carried out by the National Park Service from 1970-74 had been inadequate to provide the data essential for devising sound grizzly bear management policies (Cowan *et al.* 1974). The legacy of that NPS program is a four-year break (1971-74) in the availability of valid quantitative data documenting the far-reaching effects of dump closure on the Yellowstone grizzlies. It is widely acknowledged that the population underwent drastic readjustments involving, at the very least, an episode of extremely high bear mortality. The "data-gap" has severely complicated interpretation of much of the biological data secured in later years.

As a result of the NAS Committee's recommendations (Cowan *et al.* 1974), control of grizzly bear research in the Park was transferred from the Park Service to a newly formed Interagency Grizzly Bear Study Team (IGBST) under the direction of Dr. Richard Knight. In 1975, the IGBST was permitted to trap and mark bears and to fit bears with radio-transmitter collars. Permission to radio-track bears from airplanes, earlier denied to us, greatly improved data-gathering procedures and fostered development of a reliable life history database. The IGBST also perpetuated a National Park Service practice of identifying so-called "unduplicated females with cubs-of-the-year." These annual counts of largely

An ecocenter is a site with a large, dependable source of high-calorie food that seasonally attracts and holds large aggregations of bears for periods of two to three months. *(Photo by John J. Craighead)*

unmarked animals continue to be portrayed, to this day, as accurate data appropriate for use in projecting population estimates for purposes of the recovery program.

Despite being allowed some latitude in its research, the IGBST was constrained by park administration from full interagency cooperation and from using the data from, or even citing, the many publications documenting our study. With the exception of the 1978-79 field season report (Knight *et al.* 1980), IGBST annual reports neither mentioned nor utilized any pre-1970 data. Our research was simply acknowledged in most of these annual reports as "Earlier research on grizzlies . . . (Craighead *et al.* 1974)." This restriction not only flouted basic scientific processes, but more practically, it also prevented development of an IGBST research design that would provide data that could be compared with our data from earlier research.

Although there is now little disagreement that the dump closures had a significant short-term impact on the Yellowstone grizzly bear population, the long-term effects are less clear. Also lost in the

controversy is the related question of what the Yellowstone situation can teach about the role of "ecocenters" in grizzly ecology and evolution. An ecocenter is a site with a large, dependable source of high-calorie food that seasonally attracts and holds large aggregations of bears for periods of two to three months (Craighead and Mitchell 1982). The open-pit dumps in Yellowstone prior to 1971 were ecocenters in this sense. It is true that they were "artificial," if we consider only the kinds and conditions of food available. But if humans (past and present populations) are considered as part of the biota, the term "artificial" as opposed to "natural" has no significance. An ecocenter formed at a bison jump is as natural or artificial as one formed at garbage disposal sites. Ecologically, they are equivalent. In other words, the Yellowstone bears responded to the food in open-pit dumps just as bears in other populations now respond, and probably have responded for thousands of years, to "natural" ecocenters such as salmon spawning runs or the carcasses of large mammals such as bison and whales.

Because more than 20 years have elapsed since the Yellowstone field work was terminated, the study area has changed significantly, as have administrative policies, research philosophies, and research endeavors. What was current perspective in the decade of the 1960s is, three decades later, historical perspective. An awareness of more recent historical events is as important to understanding the ecology of the grizzly bear in Yellowstone as the earlier historic events. The area has experienced some large-scale construction and a cataclysmic fire that affected biological processes and systems, as well as the population of grizzly bears and its cohabitants. The concepts of "natural regulation" and "let it burn" fire policy that emerged from the Leopold Report of 1963 have had major effects on park in-house research and on wildlife management, as well as on resource philosophy in general.

Our early published research on both elk (*Cervus elaphus*) (Craighead et al. 1972b, 1973a) and grizzly bears, as well as that of other researchers, has contributed substantially, if slowly, to administrative perception of research and its application both park-wide and ecosystem-wide. The evolving concept of ecosystem-based management (Craighead 1982, Craighead and Mitchell 1982, Noss 1990, Craighead 1991, Craighead and Craighead 1991, Bader 1992, Monnig and Byler 1992, Grumbine 1994) has also altered perspectives on management of natural resources, including the grizzly bear. Recent events in the long history of Yellowstone have altered the study area and the research climate in ways that cannot yet be fully evaluated.

For nearly 100 years Yellowstone National Park managers practiced fire suppression. This permitted heavy fuel loads to accumulate and led to the devastating fires of 1988.
(Photo by Derek Craighead)

The point here is that events since 1970 have changed the study area and the research climate just as importantly as the events we describe later in historical perspective (Chapter 2). The area and its inhabitants that we studied from 1959 to 1970 were different in many significant respects from those of the 1974-92 period. We refer to and discuss some of these modifying events as we encounter them in our analyses. Heraclitus expressed it succinctly when he stated, "You can't step into the same river twice." Francis Thompson was more encompassing, and expressive of ecological change, when he wrote:

> All things by Almighty Power
> Near and far
> Hiddenly to each other connected are
> That thou canst not stir a flower
> Without the troubling of a star.

MAJOR PHYSICAL, BIOLOGICAL, AND POLITICAL FEATURES

The Park

The roughly 899,000 hectares (2,222,000 acres) that make up Yellowstone National Park (Fig. 1.1) form the core of the study area (Fig. 1.2). The Park is defined within more or less arbitrary boundaries, encompassing a series of high volcanic plateaus that straddle the Continental Divide where the northwest corner of Wyoming meets Montana and Idaho. Ranging in altitude from just over 1615 meters (5300 ft) where the Yellowstone River flows northward out of the Park at Gardiner, Montana, to 3462 meter (11,358-ft) Eagle Peak along the southeast boundary, the area's average elevation is slightly more than 2134 meters (7000 ft).

The broad range in elevation is reflected in climatic variation throughout the Park. The mean annual temperature at park headquarters (Mammoth) is just over 4°C (39.2°F), but the mean temperature decreases with increased elevation, leaving most of the Park somewhat cooler. Annual precipitation ranges from 51 centimeters (20 in) near the North Entrance to as much as 102 centimeters (40 in) elsewhere in the Park, and averages about 60 centimeters (24 in) (Houston 1982).

Guarding the northern and eastern flanks of the Park is the Absaroka Range, running from the Yellowstone River near the North Entrance at Gardiner all the way to the Wind River to the south. With headwaters in the southernmost reaches, the Yellowstone River flows north into the Park, feeding Yellowstone Lake, then carves its way out through the spectacular Grand Canyon of the Yellowstone, before turning gradually north then east toward its eventual confluence with the Missouri.

OSPREY

PEREGRINE FALCON

TRUMPETER SWAN

MULE DEER

COYOTE

AMERICAN BISON

PRONGHORN ANTELOPE

ROCKY MOUNTAIN ELK

BIG HORN SHEEP

GREAT GRAY OWL

MOOSE

Wildlife is abundant in Yellowstone National Park. *(Photos by John J. Craighead)*

Fig. 1.2. The 72,850-km^2 (28,125-mi^2) Yellowstone Ecosystem (The Greater Yellowstone Coalition 1991).

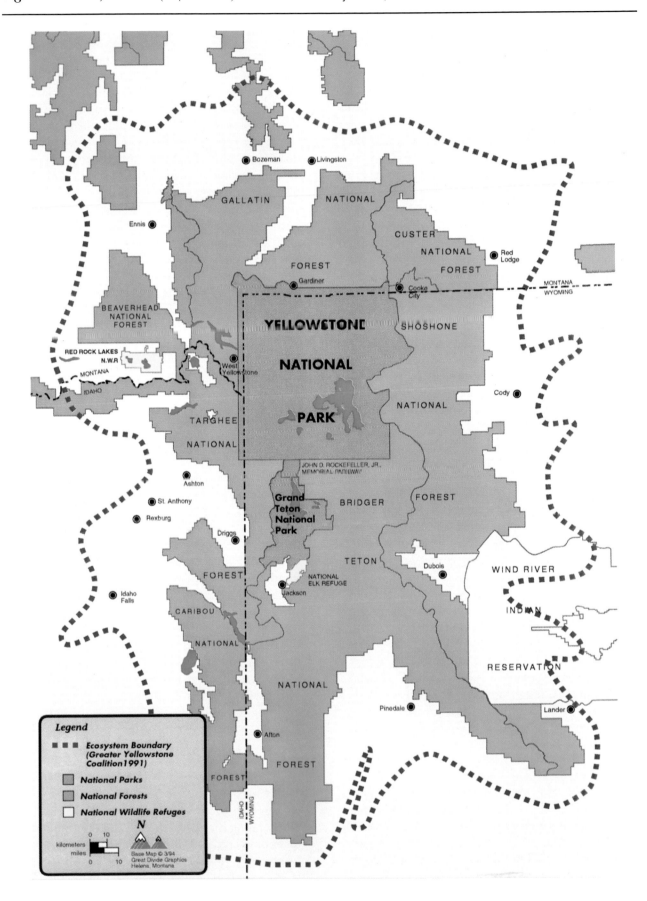

The Continental Divide enters from near the southeast corner and meanders erratically across the southern half of the Park to its exit some 19 kilometers (12 mi) south of West Yellowstone. West of the Two Ocean Plateau, from the headwaters of the Yellowstone River, the Snake River forms and flows southward out of the Park. The Snake is joined by the Lewis River near the South Entrance. It then flows south through Teton National Park and back westward and northward to its confluence with the Columbia.

The Madison River also has its headwaters in Yellowstone, formed by the confluence of the Firehole and Gibbon Rivers in the west-central area of the Park. The Madison exits Yellowstone a short distance north of the West Entrance.

The Gallatin Range provides a final mountain buffer, spanning the boundary from the northwestern corner of the Park to the Yellowstone River at Gardiner. From this range flows the Gallatin River, which eventually joins the Madison and the Jefferson to form the Missouri River near Three Forks, Montana.

The majority of Yellowstone is forested. Although the Park supports patches of Engelmann spruce (*Picea engelmannii*), Douglas-fir (*Pseudotsuga menziesii*) climax forest, and aspen (*Populus tremuloides*), it is dominated by fire-successional lodgepole pine (*Pinus contorta*) forest, which accounted for nearly 75-77% of the entire forest complex before the fires of 1988. When Yellowstone was established in 1872, well over half of the Park was covered with young stands of lodgepole pine, a successional stage projected to be eventually replaced by spruce, subalpine fir (*Abies lasiocarpa*), Douglas-fir, and whitebark pine (*Pinus albicaulis*), the climax forest vegetation for the prevailing climate. Development of plant communities composed of one or more of these species depends on soil factors, moisture, seeding conditions, and, in particular, altitude. Succession can occur naturally or with the aid of active fire suppression (Craighead 1991). The other important plant complexes include the sagebrush-grassland found in areas below 2286 meters (7500 ft), notably in the northern reaches of the Park, the alpine and subalpine meadows, just below and above 3048 meters (10,000 ft), and aquatic plant communities.

Although climatic and altitudinal differences account for much of the vegetation composition, fire has also been extremely important as an ecological factor. After 1872, however, park managers reduced, then virtually eliminated fire. Policy changes after 1963 introduced the concept of controlled burning and "let it burn," in an attempt to restore a "natural" balance. Yellowstone's lodgepole forests were thus protected from fire for nearly 100 years. The success of that fire suppression is evident in the area. Lodgepole forests represent a fire climax—that is, a forest destined to burn periodically—that seldom, if ever, attains the climatic climax condition over extensive areas (Craighead 1991).

In Yellowstone, large area burns have occurred in a 300- to 400-year cycle (Romme 1982). The subalpine forests are characterized by long-term, cyclic changes in landscape composition and diversity. Early, mid, and late forest successional stages provide landscape diversity throughout the entire Park and greater Yellowstone Ecosystem. However, both biotic and landscape diversity decline as lodgepole forests mature. These mature or over-mature forests contain heavy loads of both dead and live fuel. During the summer of 1988, wildfires drastically altered the entire Ecosystem in a matter of weeks, creating conditions that will take years to evaluate fully (Christensen *et al.* 1989, Despain *et al.* 1989, Greater Yellowstone Coordinating Committee 1989).

Perhaps most famous for its thermal activity, Yellowstone includes more than 5000 thermal features, including an estimated 150 geysers, and numerous hot springs, pools, and steam vents. It was this unique aspect that fostered the scientific and commercial interest which culminated in the creation of the Park (Haines 1977, Vol. 1).

Wildlife is abundant throughout Yellowstone. With the exceptions of wolves (*Canis lupus*), white-tailed deer (*Odocoileus virginianus*), and perhaps fishers (*Martes pennanti*) and wolverines (*Gulo gulo*), all mammals present within the Park at its inception remain in at least representative populations. Elk are the most numerous ungulates and have served as the focal point for much of the controversy about wildlife management in the Park over the years. Bison (*Bison bison*), mule deer (*Odocoileus hemionus*), moose (*Alces alces*), pronghorn (*Antilocapra americana*), bighorn sheep (*Ovis canadensis*), American black bear (*Ursus americanus*), and grizzly bear are residents, along with coyote (*Canis latrans*), fox (*Vulpes* spp.), lynx (*Lynx canadensis*), bobcat (*L. rufus*), marten (*Martes americana*), beaver (*Castor canadensis*), mountain lion (*Felis concolor*), and numerous smaller mammals.

More than 200 species of birds are known to inhabit Yellowstone at least part of the year. Notable among these because they are considered rare or endangered are the trumpeter swan (*Cygnus buccinator*), the bald eagle (*Haliaeetus leucocephalus*), and the peregrine falcon (*Falco peregrinus*).

Many park waters, especially in the upper-Madison area, were barren of fish before the arrival of white explorers and settlers. In those places where fish did occur, they were abundant; early explorers of Yellowstone Lake reported that fish could be "forked up by the boatload." Those fish were Yellowstone cutthroat trout (*Oncorhynchus clarki bouvieri*), the main game fish species in the Park. Other native game fishes are the arctic grayling (*Thymallus arcticus*) and the mountain whitefish (*Prosopium williamsoni*). Today several introduced species are also plentiful, including rainbow trout (*Oncorhynchus mykiss*), brown trout (*Salmo trutta*), brook trout (*Salvelinus fontinalis*), and mackinaw or lake trout (*Salvelinus namaycush*). Many of the formerly barren streams and lakes now support healthy populations of these introduced species.

Of the Park's 8992 square kilometers (3472 mi²), 611 square kilometers (236 mi²) are in Montana, 93 square kilometers (36 mi²) are in Idaho and the remainder are in Wyoming (Bartlett 1974). The Park is accessible from five entrances: the North Entrance at Gardiner, Montana; a northeastern entrance by way of Cooke City, Montana; the East Entrance, about 80 kilometers (50 mi) west of Cody, Wyoming; the South Entrance, about 96 kilometers (60 mi) north of Jackson, Wyoming; and the West Entrance, adjacent to the town of West Yellowstone, Montana.

The Ecosystem

Yellowstone National Park is the core of the Greater Yellowstone Ecosystem, which encompasses more than 7.3 million hectares (18 million acres) of public and private lands in Wyoming, Idaho, and Montana (Fig. 1.2; The Greater Yellowstone Coalition 1991). It includes all of Yellowstone and Grand Teton National Parks; portions of the Bridger-Teton, Shoshone, Targhee, Gallatin, Custer, and Beaverhead National Forests; some areas administered by the U. S. Bureau of Land Management; three national wildlife refuges (National Elk Refuge, Red Rock Lake, and Grays Lake); as well as substantial state and private lands (Fig. 1.2). National Park wilderness zones and Forest Service wilderness areas within the Ecosystem total approximately 2.5 million hectares (6 million acres). Non-wilderness park and national forest lands also total approximately 2.5 million hectares (6 million acres).

The Park Service manages its lands both for preservation of their natural values and for public recreational use. The Forest Service manages its wilderness lands under a preservationist mandate and its non-wilderness holdings under a multiple-use mandate. The former attempts to minimize

Early explorers of Yellowstone reported that fish were abundant in Yellowstone Lake and they could be "forked up by the boatload." Those fish were the Yellowstone cutthroat. Today several non-indigenous species compete for habitat with native species or inhabit formerly barren waters.
(Photo by Margaret Smith Craighead)

intrusive human activity, whereas the latter permits and encourages logging, mining, grazing, and motorized recreational activities. Land in state and private ownership is not subject to federal regulation and is open to development. Thus, the land ownership and the resource management mandates are complex and difficult to coordinate throughout the more than 7-million-hectare (18-million-acre) Ecosystem. When the Park was surveyed 118 years ago, there were no ecologists, and the concept of an ecosystem was unknown; thus, in spite of the best intentions, the boundaries were arbitrarily drawn and have been only minimally revised.

That Yellowstone National Park boundaries do not delineate a self-contained, independent ecosystem has long been recognized. As early as 1882, General Phil Sheridan suggested that the Yellowstone preserve was not nearly large enough to provide adequate protection for its wildlife. He proposed

(Photo by Derek Craighead)

(Photo by Charlie Craighead)

Until 1935, intense predator control was pursued by park managers. In 1926, the last remaining wolf of the Yellowstone Ecosystem was eliminated. The Superintendent's Report of 1907 states, "The mountain lions have almost been exterminated."

Yellowstone National Park boundaries do not delineate a complete ecosystem. In 1882, General Phil Sheridan proposed doubling the size of the Park to include land used by migrating ungulates and to provide adequate protection for other wildlife. *(Photo by John J. Craighead)*

doubling the size of the Park to include land used by migrating ungulates. The areas he proposed eventually became part of the National Forest System. Efforts to incorporate adjacent lands into a sound management entity became lost in bureaucratic struggle.

The concept emerged again in 1918, when an addition of 3276 square kilometers (1265 mi²) east and south of the Yellowstone boundary was proposed. This addition later became part of Teton National Park and Teton National Forest. Again, the original intention of creating a single biological entity was lost. Nevertheless, the concept of a "Greater Yellowstone" persists, and the soundness of the biological concept has been revealed by this and subsequent research.

The flora and fauna of the Ecosystem are essentially identical to the Park's. A delineation of the Yellowstone Ecosystem is shown in Fig. 1.2 (Greater Yellowstone Coalition 1991; a recent revision [Harting and Glick 1994] showed only minor boundary changes). The relationship of ecosystem boundaries to critical habitat for the grizzly bear also has been described (Craighead 1980b, USFWS 1982). The National Forest's portion is managed under congressional directives very different from those of the National Park portion. Some areas are managed as wilderness and some for multiple use, with priority given to extractive

uses. The missions of the two agencies often conflict, making cooperative bear management and other resource management difficult. Further compounding the complexities of ecosystem management are private inholdings and small population centers such as Gardiner, West Yellowstone, Cooke City, and Silver Gate, Montana, and Pahaska Tepee and Flagg Ranch, Wyoming. Human activities and impacts strongly affected the ecology of the grizzly bear in the past, did so during the course of this study, and will continue to do so in the future. Although people are part of the Ecosystem's biota, until recently they have not been adequately recognized as such in agency wildlife management planning.

Scientific Research

The Park Service has been responsible for management of its wild animal populations, but has had no research arm of its own. Congress had directed that such work be done by other federal agencies. Thus, prior to our study, the Park Service, and the Yellowstone administration in particular, lacked the scientific staff necessary for obtaining information for use in formulating policy and management actions. The need to anticipate and resolve resource management problems led to biological studies by several scientists, some government sponsored and some privately funded.

Some of the earliest work was done by A. W. Nelson and Henry Graves, who were called in to evaluate the elk situation in 1916 (Graves and Nelson 1919). Milton Skinner was appointed Chief Naturalist of Yellowstone in 1919. He considered the high densities of wintering elk unnatural and worked without pay, conducting research on a variety of biological problems (Skinner 1928). W. M. Rush conducted studies of the elk range beginning in 1928 (Rush 1932). George M. Wright investigated predator control activities beginning in 1929. He also investigated range conditions of the Northern Elk Herd (1934). Wright, B. H. Thompson, and Joseph S. Dixon, at the direction of Horace Albright, conducted a wildlife survey which was published as a two-volume series in *Fauna of the National Parks of the United States* (Wright *et al.* 1933, 1935). They reported on the overused ranges, the poor resource management, and the unfavorable wildlife conditions. Even at that early period, they ascribed the problems to "the insufficiency of park areas as self-contained biological units." Rudolf Grimm conducted range studies for nearly a decade (Grimm 1935-39 and 1943-47).

In 1937, Adolph Murie began his classic work on the ecology of the coyote. His publication, *Ecol-*

ogy of the Coyote in Yellowstone (1940), marked the first long-term mammal study in the area. His findings did not jibe with the predator control policy, so both the author and his work were virtually ignored by Park Service officials, and his publication was banned in the Park for many years (Adolph Murie, pers. comm.). Olaus Murie prepared a report recommending that the open-pit garbage dumps be closed to wean bears of their unnatural feeding habits (Murie 1944b). Walt Kittams carried out range and elk population studies through the 1940s and 1950s (Kittams 1953, 1959), and R. J. Jonas began his work on beavers in 1952 (Jonas 1955). Ken Greer, utilizing elk killed in control measures during the 1960s, investigated reproduction dynamics in the Northern Elk Herd (Greer 1966a). These and other researchers contributed significantly to an understanding of the ecology of Yellowstone. Nevertheless, wildlife management decisions were more often based on public opinion and political concerns than on research findings. To correct this situation, advisory committees were appointed to evaluate all available data on the status of the Park's flora and fauna and to recommend management and policy changes.

"Blue ribbon" committees were formed to address park management problems stemming from public indignation at the killing of elk within Yellowstone National Park to reduce the Northern Elk Herd.
(Photo by John J. Craighead)

The policies formulated by these committees were based largely on historical data predating the beginning of the grizzly bear study. As such, they were an integral part of the research environment, although not formulated into official policy until the bear research was underway.

Advisory Committees

The first of the "blue ribbon" committees was formed to address public indignation over the killing of elk within the Park to reduce the Northern Elk Herd. The interpretation of this and later reports by the National Park Service had a tremendous impact on the implementation of scientific findings into park management programs.

In the early 1960s, Secretary of the Interior Stewart Udall appointed an independent advisory committee to evaluate wildlife management in Yellowstone. This group, under the leadership of A. Starker Leopold, published its findings in 1963 in a document that became known as the "Leopold Report." The report generally supported park policies being followed at the time, specifically elk reduction and no public hunting (Leopold *et al.* 1963).

But the report went further and suggested that park managers take on a new goal—to maintain or recreate the conditions that existed before the arrival of the European explorers. To accomplish this, the committee urged that wildlife research be intensified to determine what those pre-existing conditions were and how to approximate them. The committee also urged active and purposeful management of park resources using valid, peer-reviewed research techniques and the hiring of appropriately trained resource management personnel. The committee's views on management were "borrowed" directly from a report of the First World Conference on National Parks (Bourliere *et al.* 1962). The most controversial management tool advocated was the use of fire. The implications of

using both natural and controlled fires to advance management goals were enormous.

On 2 May 1963, Secretary Udall directed the Park Service to implement the committee's recommendations, thus formalizing a policy that was misinterpreted and controversial from the outset. The controversy began almost immediately, when the "Robbins Report" was issued by a follow-up committee directed to examine the existing scientific research and to make recommendations for ongoing programs (Robbins *et al.* 1963). The Robbins committee warned that the Park could not be left to run itself "naturally" and that existing research was inadequate to support an active management strategy. Research was underfunded and misdirected, it said. Again, the Robbins Report underscored the need for, and lack of, good scientific data on which to base management decisions.

In response to the problems of managing bears, a committee chaired by A. Starker Leopold issued a report entitled "A Bear Management Policy and Program for Yellowstone National Park" (1969). Implementation of recommendations from this report would have enormous impact on black bear and grizzly bear populations throughout the entire Yellowstone Ecosystem. (The results of the report's recommendations and the Yellowstone Park's attempt to implement them are examined in more detail in Section II of this monograph.)

As the recommendations of the various committees were implemented, at least in part, it became apparent that "keeping primitive" a system already drastically and irrevocably altered by the presence of humans might prove to be impossible. In the meantime, however, that became Park Service policy and the basis for wildlife management decisions. The policy would have far-reaching effects, not only for elk and bison, but for other wildlife as well, particularly grizzly bears.

2
Historical Perspective

INTRODUCTION

The history of Yellowstone National Park is intricately connected to the status of the ecosystem-wide population of grizzly bears. Policies, activities, and management actions established during the first 100 years of the Park's existence have affected the grizzly population in numerous ways They also provide some historical perspective for the present study, as well as contribute to an understanding of the biological, cultural, scientific, and administrative conditions that prevailed during the 12-year field study and, in certain cases, prevail today.

The historical literature and records on the Yellowstone area are enormous. We confine our review to those subjects most directly related to major changes in the park environment and to the early accounts of wildlife and the documentations of black bears and grizzly bears from the early 1800s to the beginning of our research project in 1959.

Pre-Columbian Yellowstone was a pristine wilderness. Animals were abundant, and native Americans were important members of the biota. Blackfeet, Nez Perce, Shoshone, and Crow Indians visited the area, but only a small band of Shoshones, or "sheep-eaters," lived there year-round (Chittenden 1924; Haines 1977, Vol. 1). In the early 1800s, Yellowstone's landscape was diverse as a result of large fires in the 1700s, some probably due to deliberate burning by Native Americans. This diversity of landscape was reflected in the diversity of plant communities, which formed a vegetational mosaic consisting of lowland coniferous forests in all stages of plant succession, sagebrush parks, aspen woodlands, subalpine meadows, and upland coniferous forests, with alpine or subalpine meadows, spruce "snow mats," and boulder fields at the higher elevations.

A full complement of animal populations occupied the Yellowstone habitat in pre-Columbian times, their numbers regulated by natural processes. Elk summered on the high plateaus and migrated to distant winter ranges. Wolves and cougars preyed on the elk and on the other large mammals—deer, moose, bighorn sheep, and bison. Lynx, marten, wolverines, and coyotes were common carnivores that preyed on the small mammals. Beaver were relatively abundant. Both black bears and grizzly bears were well distributed throughout the Ecosystem, feeding on a wide range of plants and animals. The species diversity of the avifauna was probably not significantly different from that found today, but some species no doubt occurred in greater numbers; e.g., the peregrine falcon, great gray owl (*Strix nebulosa*), harlequin duck (*Histrionicus histrionicus*), bald eagle, trumpeter swan, whooping crane (*Grus americana*), and various species of woodpeckers.

Since the creation of Yellowstone National Park, its plant and animal communities have experienced substantial changes, some natural, others man induced. Plants and animals have been particularly hard hit by policies formulated to meet the conflicting mandates of preservation versus use, policies that have also left an administrative legacy of unresolved resource and management problems. The examples abound: a predator control policy, operational from 1916 to 1935, eliminated the wolf and brought the cougar close to extirpation; a policy of overprotecting large ungulates, intermittently applied, has resulted in some drastic changes in plant succession and seed production (Kay 1995); a road construction program designed to encourage human visitation has altered or denied wildlife access to much of the Park's limited riparian habitat; a wildfire policy that initially sought to suppress fires, then to encourage them, has confused the goal of forest management; park operational policy has failed to integrate humans as ecological components of the biota and has left their ecologic role in Yellowstone poorly defined; an archaic program for garbage disposal prevailed for nearly 100 years; and finally, the most critical legacy of all, Yellowstone's artificial boundary did not and does not recognize biotic boundaries or entities.

EARLY HISTORY

A hundred years ago, Yellowstone was still largely unexplored, but its scientific uniqueness was already recognized.

> As a country for sight-seers, it is without parallel; as a field for scientific research, it promises great results; in the branches of geology, mineralogy, botany, zoology and ornithology, it is probably the greatest laboratory that nature furnishes on the surface of the globe.

So wrote Lieutenant Gustavus C. Doane in his report to the Secretary of War on the 1870 Washburn Expedition's pioneering exploration of what is now Yellowstone National Park (*The Pioneer* 1872, St. Paul MN).

President Ulysses Grant signed the legislation creating Yellowstone National Park on 1 March 1872. Before its creation, the world's first national park had been visited, explored, and surveyed by both official and unofficial parties.

Native Americans

Recent human residents of the Yellowstone area were limited to small bands of Sheepeater Indians, a subgroup of the Shoshone. The Sheepeaters eked out a marginal existence over a relatively small area, subsisting on, among other things, the bighorn sheep from which they got their name. Tribes such as the Blackfeet, Nez Perce, Bannock, and Crow Indians, as well as other Shoshone groups, used the resources of the Yellowstone area as transients. Residents or transients, early Native Americans were very much a part of the Yellowstone fauna, and their hunting and foraging practices affected the ecology of the area (Chittenden 1924, Haines 1977, Kay 1994).

Hunters and Trappers

John Colter is generally credited with being the first Caucasian to visit portions of what is now Yellowstone. As a reward for his singular contribution to the success of the Lewis and Clark Expedition, he was allowed to resign from the expeditionary force when it was homeward bound. Instead of returning to civilization, he joined a group of trappers and in 1806 traveled west again (Lavender 1988). Although historians differ on Colter's route, there is agreement that during the winter of 1807-1808, he observed thermal features somewhere along the eastern perimeter of what was to become Yellowstone National Park (Chittenden 1924).

Colter's observations were expanded by those of trappers and wanderers over the next 40 years. Much of Yellowstone was visited, and each explorer brought back new facts and myths that aroused public interest in the area.

With the demise of the fur trade after 1840, Yellowstone drew little attention for more than 20 years. Gold-seekers from Montana's diggings visited the area in the late 1850s. They found little ore, but their accumulated tales of geologic oddities and thermal wonders generated renewed interest in this vast unsettled area (Chittenden 1924; Haines 1977, Vol. 1).

Exploration and Creation of Yellowstone National Park

In 1860, Captain W. F. Raynolds led the first government-sponsored attempt at exploration of the Yellowstone region. He sought a transportation route from the sources of the Wind River in Wyoming to the headwaters of the Missouri in Montana. He determined there was no easily accessible route to the upper Yellowstone drainage and did not penetrate the interior of what is now Yellowstone National Park. Further government exploration was delayed by the Civil War (Chittenden 1924; Haines 1977, Vol. 1).

In 1869, a private group of three Montanans, David E. Folsom, Charles W. Cook, and William Peterson, completed a circuit from Bozeman, Montana, up the Yellowstone River as far as Yellowstone Lake, west to Shoshone Lake, down the Firehole River to the Madison and from there on down to the Three Forks of the Missouri. Folsom is generally credited with intensifying interest in the wonders of the Yellowstone area and with being one of the first to suggest that it be set aside for the enjoyment of the public (Haines 1977, Vol. 1).

The next exploration effort, the Washburn Expedition, took place in 1870. Although a private group, the expedition had a military escort under Lt. Gustavus C. Doane. This group generally retraced Folsom's route and also explored the east side and southern extremities of Yellowstone Lake. Several members wrote graphic accounts of the journey. Expedition members are credited with furthering the concept of preserving the area (Chittenden 1924; Barber 1987).

One of the most remarkable experiences in the exploration of Yellowstone occurred when Truman Everts, a member of the Washburn party, was separated from his companions during September and October, 1870. In a near-tragic mishap, his horse threw him and he lost his mount and broke his spectacles. Without a horse, his glasses, supplies, or equipment of any kind except a hand lens, he managed to survive the cold by using the lens and the sun to start a fire. Purely by accident, he found that the starchy root of a thistle (*Cirsium foliosum*) was edible. This plant nourished him for 37 days

Enthusiastic reports from the Washburn expedition of 1870 helped spur congressional interest in the Yellowstone area. The Hayden expedition (pictured) was funded as the official United States Geological Survey exploration of the Yellowstone region. By 1 March 1872, Yellowstone National Park was reserved as "a public park and pleasuring ground for the benefit and enjoyment of the people."
(Courtesy of Colorado History Museum)

until he was found by companions in the extreme northern section of what would later be the Park (Chittenden 1924).

Such stories as this, and the enthusiastic reports from other members of the Washburn Expedition, spurred Congress to fund an official exploratory survey under the leadership of Dr. Ferdinand V. Hayden. Accompanying the Hayden Expedition were scientists, engineers, and field assistants, including photographer William H. Jackson and artist Thomas Moran. Their work helped convey the uniqueness and the magnificence of Yellowstone to Congress and to the public (Chittenden 1924; Haines 1977, Vol. 1). At the same time, a U. S. Army "reconnaissance" party known as the Barlow-Heap Expedition traversed Yellowstone. The combined data, new maps, specimens, sketches, and photographs from the two parties provided final proof that Yellowstone was indeed a wonderland worthy of preservation (Chittenden 1924; Haines 1977, Vol. 1).

Interest in the area exploded and soon spawned a movement to set aside the Yellowstone country for the benefit of the nation. On 1 March 1872, Yellowstone National Park was reserved as "a public park and pleasuring ground for the benefit and enjoyment of the people."

Early Civilian Administration

The Organic Act establishing Yellowstone did not provide funding for its administration and management. Nathaniel P. Langford was appointed Superintendent and served without pay. It became evident during his tenure that public funds and action were needed to protect wildlife and the unique features of the Park (Haines 1977, Vol. 1).

Amid growing discontent with the way Yellowstone was being managed, Philetus W. Norris was appointed Superintendent in 1877. He was already familiar with Yellowstone and had definite ideas about how to protect it and make it

The arrival of the Army in 1886 brought order to 13 years of chaotic civilian administration of the preserve. In 1916, the Army relinquished military control of Yellowstone National Park to the newly created National Park Service. *(Courtesy of Haynes Foundation Collection Montana Historical Society)*

more accessible to the public. Norris secured the first congressional appropriations for the Park, supervised construction of 246 kilometers (153 mi) of road and over 322 kilometers (200 mi) of trail, took the first steps to protect the wildlife and natural features, and constructed the first government buildings. His protective stance alienated individuals and officials who wanted to exploit the wildlife and scenic wonders of Yellowstone for profit. This led to his departure and the arrival of Patrick H. Conger in 1882 (Chittenden 1924; Haines 1977, Vol. 1).

Historians credit Conger with commencing exploitive development that led to the near-demise of Yellowstone. In 1883, the Northern Pacific Railroad completed its Yellowstone Park branch. In that year, the Yellowstone Park Improvement Company built the National Hotel at Mammoth Hot Springs. Other entrepreneurs also got into the act, providing transportation and guide services (Haines 1977, Vol. 1; Bartlett 1985). Many of Yellowstone's vast resources were exploited by concessionaires during this period. Workers ate park elk and bison, and the same fare appeared on hotel menus. Park timber was harvested for construction and for firewood by settlers north of the Park.

Conger's tenure ended in 1884, and two more civilians supervised Yellowstone in quick succes-

sion. Robert E. Carpenter served for one year and was removed for scheming to take private possession of park lands. David W. Wear took over in 1885 at a time of growing demand for tourist services, specifically to halt abuses and depredations to the Park. Unfortunately, a year later Congress failed to appropriate funds, and the Secretary of the Interior requested that the military take over management. In August 1886, the U. S. Army began what was to be a 30-year term of administering Yellowstone (Chittenden 1924; Hampton 1971; Haines 1977, Vol. 1).

Military Administration

The arrival of the Army brought order to the chaotic administration of the preserve. Soldiers enforced the laws, suppressed wildfires, and served a public relations role. Stationed at strategic locations throughout the Park, they often provided visitors with limited information on the natural phenomena, as well as directions and assistance of all kinds, and conducted regular patrols at all seasons of the year. In one recorded incident, two of these soldiers lost their lives while conducting a ski patrol between Lake and West Thumb stations.

Meanwhile, annual visitation expanded, as did service facilities. Roads and bridges were built; others were relocated or improved. Hotels were established at Canyon, Lake, Fountain Flats, and Old Faithful. A series of private camps (e.g., Indian Creek, Norris Junction, Spring Creek, Nez Perce Creek) was established as an alternative to hotel accommodations. During this period, attempts to change the original north boundary of the Park and to secure a railroad right-of-way within the Park were thwarted, as was an attempt by the Northern Pacific Railroad to monopolize concessions (Haines 1977, Vol. 2).

During the first years of military overrule, soldiers had little statutory authority to regulate visitors' actions. There were no laws addressing specific offenses and no provisions for punishment, other than the authority to expel people from the Park. Until the passage of the Lacey Act in 1894, the military used the threat of expulsion to control miscreants. The Lacey Act, however, gave the military authority to arrest individuals for specific offenses. Theoretically, the act gave absolute protection to Yellowstone's wildlife, finally providing leverage against the poachers who had frequented the Park for years (Haines 1977, Vol. 2; Bartlett 1985; Chase 1986).

Creation of the National Park Service

The National Park Service Act was signed into law by President Wilson on 25 August 1916, although the U. S. Department of Interior had a functional National Park Service as early as 1913. In 1916, the Army relinquished control of the Park to civilian administration. There was, however, a lengthy transitional period between military and civilian administration of Yellowstone. Ex-soldiers and civilian scouts formed the first ranger force; e.g., the chief scout, James McBride—selected by the military in May 1914—was named Chief Ranger in October 1919.

The military had administered Yellowstone in a rigid, uninspired, but generally predictable way. Operations under the military were not strongly directed toward preservation of the natural features or the wildlife. A letter from William H. Wright to Superintendent S. B. M. Young (Army Records. Letters Received, Sep-Oct, 1908. Vol. VII, pp. 341-349) documents some of the problems and concerns that arose under military jurisdiction.

> For nearly fifteen years, I have been in the habit of visiting the Yellowstone National park. Sometimes conducting parties, who wished to view this Wonderland and sometimes to personally explore the most remote parts of the park, to study and become more thoroughly acquainted with its animal life, particularly the Grizzly bear, which now is, I am sorry to say, rapidly becoming extinct.

> This last summer, as you will know, I have, with a party of small boys spent ten weeks in this great play ground, and owing to your kindness in furnishing me with a letter to the scouts, officers and patrols, in charge of the park, we were able and permitted to enjoy certain privileges which the ordinary travelers and campers do not enjoy. These privileges, we most heartily and thoroughly appreciated as we often wandered far from the regular beaten road and trails which mark the route of the less fortunate travelers and tourists. In our rambles, we were able to become better acquainted with the different animals in their natural surroundings and also of the methods of patrolling and protecting this natural gameland and it is of this that I am now addressing you.

> For more than a quarter of a century, I have made wild animal life a study, have seen their once extensive domain cut down to a few square miles in comparison to what it formerly was and their once vast numbers diminish to a few straggling bunches which are now to be found and seen only in a few localities, where they are not even now safe from the ruthless slaughter of the fur, tooth and head hunter, and the only place where the wild animal is, or should be immune from their ever present evil is the Yellowstone National Park.

> The advantage of your letter to the commanders of the various posts throughout the park, threw us constantly in contact with the patrols and troopers assigned to the duties of policing the Territory. We met with many courtesies and in few instances, with individuals who seemed to take some interest in their surroundings and who would, no doubt, under proper training, become efficient in their duties of patrolling and protecting the natural beauties as well as the wild life to be found within the borders of the park.

> Our experiences, however, from first to last, have forced upon us the fact that as a body, these men are not only unfitted for their task, but in more than one regard, are an actual menace to the interests they are supposed to protect. . .

> To fill this position, requires men of good sound judgement, an intimate knowledge of the country, men who are well versed in woodcraft, men who take some pride and interest in the performance of their duties.

> To whom are these duties entrusted? To the rank and file of a cavalry regiment, drafted from any and all sources and enlisted for other purposes. Good troops they make, no doubt, but if they were gifted with the qualifications their duties as patrol demanded of them, they would not be here. The great majority of them are wholly indifferent to their responsibilities . . .

> We wish to be impersonal in our criticism and purposely omit location, our complaint lies not against individuals, but against the system

and our ten weeks of constant contact with this system of protecting or pretending to protect this wonderland of the world, has filled us with disgust.

One of the greatest attractions of the park, is its animal life and is, no doubt, little thought of by the casual visitors. They take their presence for granted and expect glimpses of deer, elk, and antelope as they are making their daily drive. Then at night on the garbage heaps, near the large hotels, the guests gather to watch for the bear who are sure to come to clean up the table refuse, hauled each day to these places. Remove or exterminate these animals from the park and for many visitors its greatest attraction is gone.

We hear that these animals are on the increase that the bear are becoming so numerous and bold, that they are a menace to the camper and ought to be killed off. Just how fast the deer, elk, sheep and antelope are multiplying, I will not attempt to say, as I have not made a study to such an extent as I have the bear, particularly the grizzly, and I say without any hesitation that this animal is very rapidly disappearing from the park. What has become of them we are unable to say, but after several years of close study of them and the conditions that prevail in and around the park, we have formed a definite opinion. . . .

For a number of years, we thought that the state of Wyoming was responsible [for the observed decreases in bear numbers], in that their game laws permitted the trapping of certain animals within the state game preserve joining the park on the south, but after careful investigation, we have come to the conclusion that this cannot be the main reason. First because this state game preserve has only been created within the last few years, and up to that time, all hunters and trappers were allowed to trap and kill to their hearts content along the south border of the park. There were then many times the number of grizzly in the park that there now is and while, no doubt, a good many bear find their way into these traps, we believe the main evil lies within the borders of the park.

While in the haunts of the bear we found the troopers to a man deadly afraid of the bear, be it grizzly or black, and almost every night there were some one to half a dozen shots and some times more. Next day some of the soldiers would give us a wild and blood-curdling tale of meeting a bear who refused to give up the road or who was at the back door of a hotel or camp, cleaning up a slop barrel which had been previously left there for the bear. How they had shot to kill, but did not know if he had got him. One soldier told us of being chased on the main road by a grizzly bear at whom he emptied his revolver and the bear still came on. He loaded a second time and facing about in the saddle, took the gun in both hands and again emptied every chamber at the pursuing beast.

The childish and foolish fear that all the soldiers show towards the bear would be comic, did it not mean the destruction of this beast. At one post, the soldiers talked with us about using cyanide of potassium as a means of poisoning the bear after the park was closed. They also discussed the scheme of throwing out meat in which had been placed giant powder caps, hoping the bear would strike the cap while chewing the meat and in this way blow his head off. We asked them if they were not liable to get into trouble in thus trying to kill the bear and one remarked, he didn't care if he went to jail for life if he could but get those damn bears out of the way. One trooper told us he would have a damn fine collection of bear pelts when his time was up.

We were also told by the soldiers at the different posts that they had plenty of fresh meat during the winter months and when asked if they did not get found out, replied 'They don't dare let it get to Washington, so what's the odds?' . . .

During the winter months, it is a snap for a trapper and poacher to enter the park and under the present system, secure heads and teeth and make his escape. Judging from our talks with the troopers, I would consider it a safe undertaking to enter the park for a small expenditure of favors, get away with every bear in the park.

No wonder the wild animal life is decreasing and stages are held up, the only wonder is, that not more of this is done. We are fully convinced that the robber would be safer in the park than he would outside of it, provided, however, he knew the country and the means used in protecting it, and he undoubtedly did, or he would not have undertaken the job.

These facts and observations are from our own personal experiences, and we have felt it our duty to inform you of them. We have written as freely and frankly as we have, because we feel that you deisre [*sic*] to know the truth and to remedy the existing abuses.

In spite of the defects in the military administration of the Park, its presence was an improvement over the early civilian administration. The military protected the Park from the severe exploitation of earlier years, but it failed to consider the subtleties of managing a natural system for the enjoyment of visitors, whose very presence altered the natural values of the place. This administrative and philosophical rigidity was one of the legacies passed on to the National Park Service. One evident result has been an expanding emphasis on people management, to the detriment of natural resource management (Haines 1977, Vol. 2; Barber 1987).

Superintendent Norris began the park road system in 1878. Engineering expediency rather than ecological considerations took precedence in placing roadbeds along sensitive riparian habitats.
(Montana Historical Society, Haynes Foundation, F. Jay Haynes Photographer, 1916)

The statement of purpose in the National Park Service Act directed the newly created agency,

> . . . to conserve the scenery and the natural and historic objects and the wildlife therein and to provide for the enjoyment of the same in such a manner and by such means as will leave them unimpaired for future generations.

This mission to both use and preserve Yellowstone National Park has spawned sometimes conflicting operational policies which have had major impacts on the Park's resources.

EARLY DEVELOPMENTS THAT AFFECTED WILDLIFE

All of Yellowstone's administrators have faced the challenge of accommodating a growing influx of visitors while striving to conserve the Park's natural and scenic resources. Development and preservation have met, head-on.

Construction of roads and tourist accommodations had direct, long-term impacts on the populations of black bears, grizzly bears, and other wildlife, as did other developments associated with increased tourism, such as disposal of wastes. To understand the grizzly bear situation during the 1950s and 1960s, it is necessary to understand the historical events and trends that had modulated the area's natural conditions and dictated management policies.

Road Construction and Transportation

In 1878, Superintendent Philetus Norris began the road system that exists today. His administration completed over two-thirds of the "Loop," amounting to 167 kilometers (104 mi) of its 225-kilometer (140 mi) current length. Most of the $70,000 in Norris' budget was spent on road construction. Engineering expediency rather than ecological considerations took top priority (Bartlett 1985). Planning for roads was limited to economy of construction, visitor convenience, scenic interest, and minimum detriment to landscape. There is no evidence to suggest that ecological impacts were considered. As a result, roads generally followed

In 1915, the park roads were opened to motor vehicles, forever altering the way the Park was used by the public. *(Photo by Margaret Smith Craighead)*

water courses. Both the physical presence of roads and their human traffic disturbed scarce and ecologically sensitive riparian habitat, while creating barriers to animal use and movement to and from that habitat. Much of this prime habitat was denied to the shy grizzly bear.

Lieutenant D. Kingman, an army engineer, improved on Norris' roads and constructed new ones, eliminating steep grades by locating roads along stream beds where possible. By 1886, he had added 48 kilometers (30 mi) of new road and improved 48 kilometers of existing ones (Haines 1977, Vol. 1; Bartlett 1985).

Captain H. M. Chittenden was twice chief engineer in Yellowstone (1891-92 and 1898-1906). He completed the Loop, or figure-eight route, and constructed roads from the East Entrance at Sylvan Pass and from the South Entrance on the Snake River. By 1906, the road system was essentially complete (Fig. 2.1), with roads entering the Park at five locations, all converging on the central Loop, which allowed visitors to see all the major attractions without retracing their route (Bartlett 1985). By 1959, the basic road system in the Park had existed, and had affected wildlife, for over half a century.

Until 1915, visitation had been accomplished either on foot, horseback, or in horse-drawn conveyances. In preparation for the entry of automobiles, Congress appropriated funds in 1912 to widen roads, improve surfaces, and reconstruct bridges and culverts to make the Park safe for automobiles. On 1 August 1915, Yellowstone National Park officially opened traffic to private motor vehicles. On the first day, 50 automobiles carrying 171 persons entered the Park. Traffic control gradually became the foremost duty of nearly all rangers, so by 1959, road patrols greatly exceeded the off-road patrols that had exemplified the Park Ranger of pre-vehicular days.

Thus, 1915 brought an unprecedented increase in tourism that forever altered the way Yellowstone was used by the public and administered by the National Park Service. Enforcement of motor vehicle regulations and the management of people gradually took precedence over the management of natural resources.

Tourist Accommodations

Yellowstone's first tourists visited on their own, employing their own equipment and skills to see them safely through the Park. Soon, however, it was

Fig. 2.1. Construction of roads in Yellowstone National Park (adapted from Haines 1977); by 1905, the road system, as it exists today, had been completed.

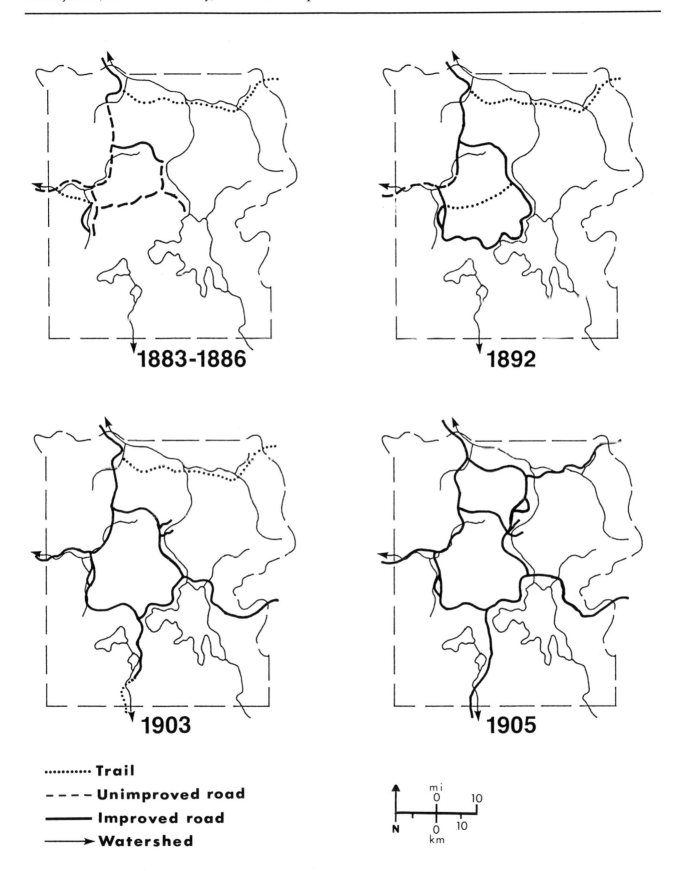

apparent that money could be made transporting and housing visitors, especially those untrained in wilderness travel.

The first commercial accommodations in the Park were at Mammoth Hot Springs, where, beginning in 1872, James McCartney ran a log hotel for people seeking the healing effects of the mineral baths. Other small commercial ventures established in those early years included several hotels, a toll bridge, and a photo studio (Haines 1977, Vol. 1; Bartlett 1985).

By 1883, the commercial possibilities of Yellowstone were evident enough to convince the Northern Pacific Railroad to build a branch line from Livingston, Montana, to a point on the Yellowstone River, 29 kilometers (18 mi) from park headquarters at Mammoth. Meanwhile, the Yellowstone Park Improvement Company, also foreseeing commercial possibilities, built the National Hotel at Mammoth without first going through official channels to secure a lease from the U. S. government. This set a precedent for preemption of choice locations and commercial rights by a single commercial enterprise, a situation that would not be rectified for many years (Bartlett 1985). The Yellowstone Park Improvement Company secured a near-monopoly on commercial operations, building tent camps at Norris Geyser Basin, the Grand Canyon, and the Upper Geyser Basin. It arranged for stage service and established lunch stations along the stage routes.

Through various financial arrangements, the Yellowstone Park Improvement Company became the Yellowstone Park Association. In 1886, the Association, a front for railroad interests, already owned six hotels. In 1889, the tent camp at the Grand Canyon of the Yellowstone was replaced by the Canyon Hotel. In response to the growing commercial potential, the Fountain and Lake hotels were constructed, and the Old Faithful Inn was added in 1904 (Chittenden 1924; Haines 1977, Vol. 2).

Accommodations for tourists were considered quite opulent for the times. For the economy-minded visitor, options were limited until 1893, when W. W. Wylie obtained a permit to establish tent camps in the Park and quickly picked up a good share of the tourist trade. Several other tent camping operations soon appeared, investing in the mounting visitation to Yellowstone by a more cost-conscious stratum of society (Haines 1977, Vol. 2; Bartlett 1985). These camps were also responsible for a growing concentration of black bears attracted to the increased number of refuse sites.

Regardless of how people traveled to see the Park, the tour was essentially the same. The geyser basins, Yellowstone Lake, and the Grand Canyon were the principal attractions. Located on the loop road, they became the sites of most facilities and developments (Bartlett 1985).

These and similar undertakings directly affected the wildlife and its habitat (Haines 1977, Vol. 2; Bartlett 1985). Early concessionaires strove to retain their monopolies, and they assumed the right to use park resources for commercial benefit. Park timber and wildlife were taken at their convenience. As late as the 1930s and 1940s, some concessionaires in the Park could still get official permission to cut timber to refurbish their facilities (Haines 1977, Vol. 2).

With the admission of automobiles to the Park, it became apparent that horse-drawn and mechanized travel could not coexist. The Yellowstone Park Transportation Company, a one-time adjunct of railroad interests, purchased 116 buses. Independent transportation companies in the Park were closed down, and facilities that had served as rest and lunch stops on the horse-drawn coach routes were abandoned, as both the business interests and the park administrators strove to consolidate the concessions (Bartlett 1985).

As a result, by 1924, Harry Childs, a man with ties to the Northern Pacific Railroad, had a virtual monopoly on transportation and accommodations in Yellowstone. Only two other concessionaires maintained significant interests in the Park after consolidation: F. J. Haynes, who operated a series of photo shops, and George Ashworth Hamilton, who held the concession for retail sales (Haines 1977, Vol. 2; Bartlett 1985).

In later years, Hamilton and Childs jointly held the concession for gasoline filling-stations in the Park. In 1968, the circle was further closed when Hamilton Stores bought out the Haynes Photo shops. The National Park Service, although allowing concessionaires to operate, reserved control over the type of service provided and the cost charged to visitors.

Despite favorable treatment, the Childs' holdings of hotels, cafeterias, and other service facilities, later known as the Yellowstone Park Company, deteriorated seriously after Childs died in 1931. Maintenance of the physical assets was erratic, and little if any modernization was done. By 1948, poor service, inadequate sanitary facilities, deteriorating structures, and high prices elicited visitor complaints and official concern (Haines 1977, Vol. 2; Bartlett 1985).

Finally, in 1966, the Yellowstone Park Company was purchased by General Host, Inc. Hailed as a savior, General Host instead managed the service

Visitation by a more cost-conscious stratum of society, who could not afford the opulent hotel services provided for early visitors, increased with automobile travel. The refuse from these and commercial tent camps attracted a growing number of black bears. *(Photos by Margaret Smith Craighead and—lower left—Haynes Foundation Collection Montana Historical Society, E. W. Hunter, Photographer)*

and housing properties with maximum attention to profit and little effort to refurbish or upgrade them. In 1979, the U. S. government bought out General Host and selected TW Services, subsidiary of TW Corporation, as the new concessionaire. After a two-year trial period, TW Services began operating the multifaceted business under a 30-year contract. The contract requires a percentage of gross revenues to be spent annually to maintain and upgrade tourist facilities (Bartlett 1985).

The arrival of the automobile combined with the growth of concessions significantly increased human impact on the natural environment. This in turn amplified activities of the National Park Service. The first free auto campgrounds were established in the summer of 1916 at Mammoth Hot Springs, Old Faithful, and Canyon. Those camps were the prototypes for the extensive campground system in the Park today. Such facilities further affected wildlife habitat and encouraged greater visitor use of the Park. The tourists increasingly required services, quickly supplied by the principal concessionaires. These included fuel and automotive repair services, groceries and dry goods stores, and restaurants and coffee shops for visitors not inclined to use hotel dining rooms. In addition, cabin camps were established for those who could not afford hotels but wanted more amenities than the campgrounds afforded (Haines 1977, Vol. 2).

Yellowstone was changing, but each new development that extended human impacts on the environment went largely unrecognized and unchallenged. Tourists, no longer confined to hotels, were competing with Yellowstone's wildlife for space. In addition to the destruction of habitat required for roads and facilities, animals were subjected to ever-increasing interaction with tourists. Humans had gradually become an integral part of the Yellowstone fauna, and motor vehicles were part of the landscape. How to integrate these elements in a comprehensive management policy has proved a knotty problem that remains largely unresolved today.

Human Visitation

During the first four years of the Park's existence, visitors relied on their own efforts and initiative to get to and traverse the area. As a result, annual visitor numbers probably did not exceed 500. But visitation began to increase soon thereafter and has risen ever since, with several exceptions due to national and international events (Table 2.1).

When commercial services and tours first became available, visitation increased only gradually. By 1883, however, with the arrival of a Northern Pacific branch line and the installment of major concessionaires, accompanied by national publicity, Yellowstone became a prestigious touring spot. Tourism rose five-fold over the previous year. And, with more facilities, visitor numbers continued to increase; 1900 was the last year the Park had fewer than 10,000 visitors. When automobiles were admitted in 1915, annual visitation rose to 50,000. With automobile travel, the Park became more accessible. Visitation dropped during World War I (1916-18), then continued to increase as Americans took to the road in the 1920s. Another drop in visitation was evident during the Depression (early 1930s) and again during World War II. Visitor numbers topped 2,000,000 for the first time in 1965 and have remained above that level since.

Until the entry of automobiles, nearly all visitors relied on transportation provided by concessionaires. After that, the number of tourists using concessionaire transportation in the Park steadily decreased. By 1960, only 14,000 of an estimated 1.5 million visitors chose public transportation (Haines 1977, Vol. 2).

Reliance on personal vehicles has led to crowded roads, less than ideal driving conditions, crowded campgrounds, and overtaxed facilities of all types. It has also led to a change in ranger duties from primarily resource inspection and management to primarily traffic regulation and tourist control. Vehicles have provided visitors with easier access to places previously viewed only from passing stagecoaches or buses.

Conflicts ensue between humans and wildlife where highways intersect wildlife travel routes or invade important habitats. Traffic management that allows tourists to stop along the Park's road system has increased chances for human-wildlife conflicts. People now visit Yellowstone year-round, but most tourism still occurs in the three summer months. In the 1990s, with nearly three million visitors annually, the summer season in the Park is an artificially imposed period of high stress for many animals. Fortunately, travel is largely confined to the loop road and to visitors who view the Park from the privacy of their vehicles along the roadside. Most of the Park is unofficially zoned as wilderness. Nevertheless, cumulative human visitation exceeded 95 million in 1991 (Table 2.1), and undoubtedly will remain an impediment to achieving the goals of preserving wilderness in the Park and throughout the entire Ecosystem.

BACKCOUNTRY VISITATION

Backcountry visitation was relatively insignificant during the first hundred years of the Park's existence, but rose dramatically in the 1970s.

Humans with their vehicles became an integral part of the Yellowstone scene as visitation rates rose annually. How best to integrate human and wildlife into a comprehensive management policy is a problem that remains largely unresolved today. *(Photos by:* Upper left, *American Heritage Center, University of Wyo. Photographer unknown;* Upper right, *Margaret Smith Craighead;* Lower right, *Haynes Foundation Collection, Montana Historical Society;* Bottom, *John J. Craighead)*

Table 2.1. Annual and cumulative human visitation to Yellowstone National Park, 1872 to 1991, in five-year increments (Haines 1977, Vol. 2; Yellowstone National Park Archives).

Year	Annual human visitation	Cumulative human visitation
1872	300	300
1875	500	1,800
1880	1,000	6,330
1885	5,000	23,330
1890	7,808	53,138
1895	5,438	82,279
1900	8,928	122,804
1905	26,188	200,086
1910	19,575	304,470
1915	51,895	447,568
1920	79,777	682,130
1925	154,282	1,298,798
1930	227,901	2,407,012
1935	317,998	3,526,595
1940	526,437	5,837,965
1945	178,296	7,039,343
1950	1,109,926	12,043,726
1955	1,368,515	18,564,665
1960	1,443,288	25,912,705
1965	2,062,476	35,222,292
1970	2,297,290	46,283,469
1975	2,246,132	56,896,220
1980	2,009,581	68,433,127
1985	2,262,455	80,313,308
1990	2,857,096	93,093,220
1991	2,957,856	95,051,076

Backcountry visitation was minimal throughout most of the 20th century, but rose dramatically in the 1970s and thereafter, as enormously-improved outdoor equipment allowed campers to remain in the backcountry for weeks at a time. In 1980, 24,000 people visited the wilderness backcountry of the Park.
(Photo by Karen Craighead Haynam)

Abetted by enormously improved outdoor equipment, hikers and campers could remain in the backcountry for weeks at a time. Official National Park Service record-keeping began in 1973, when nearly 14,000 visitors took to the backcountry trails on foot and with pack trains (Table 2.2). The impact of this increased backcountry use is expressed in the person-use nights, number of pack animals, and animal-use nights. Backcountry visitation peaked in 1981 at more than 24,000 people, representing 55,000 person-use nights. Animal-use nights have steadily increased (Table 2.2). Restrictions on visitor use have gradually reduced backcountry travel to levels more commensurate with retaining wilderness character and reducing animal-human encounters.

EARLY ESTIMATES OF WILDLIFE POPULATIONS

Soon after the creation of the National Park Service, estimates of wildlife numbers became more accurate because of the increasing use of trained biological observers and established census techniques. Beginning in 1920, the *Yellowstone Nature Notes* served as a repository for these more dependable observations of wildlife abundance. For example, a 1924 report by the Park Naturalist, Edmund J. Sawyer, estimated game abundance as follows:

Elk, including the Gallatin and Madison Herds (excluding the Jackson Hole Herd), approximately 20,000
Buffalo, (at present) 689
Buffalo, (wild herd) 85
Antelope, including about 75 kids of the year 395
Mountain Sheep 600
Deer, mule (black tailed) .... 1,500 to 1,800
Deer, white tailed (only 8 seen during past year) .... 10 to 40
Moose 450
Black Bear 200
Grizzly Bear 60

Small animals—porcupines, beaver, mink, skunk, marten, weasel, otter, and so on—are known to be plentiful in the park but no estimate as to their numbers has been attempted; beaver are particularly plentiful and otter are on the increase.

—(*Yellowstone Nature Notes* 1924)

On April 28, 1927, Samuel T. Woodring, Chief Ranger, wrote the following:

I saw one of the finest sites it has been my good fortune to witness since I first came to the Yellowstone. I arose at five o'clock and rode through the country between Mammoth and Gardiner, on both sides of the Gardiner, [*sic*] River and clear up under the side of old Mt. Sepulchre, covering a distance of approximately ten miles, I saw between five and six hundred elk, about sixty antelope, nineteen mountain sheep, ninety to a hundred mule deer, two moose, a beaver, a black bear and five coyotes . . . . I had seen all of the most important park animals, with the exception of the grizzly bear, within a radius of three miles from Mammoth Hot Springs, park headquarters, and within two hours . . .

—(*Yellowstone Nature Notes* 1927)

In 1932, George A. Baggley reported on organized game counts:

April 20 marked the completion of the animal game counts in Yellowstone National Park. The first count of elk was held on January 21 and 22. A count of antelope and deer was secured on April 12. A second elk count was accomplished on April 19 and 20 with very gratifying results.

Two or three game counts have been taken each winter for the past two or three years. This, we have found, gives a far more accurate census of wild animals than could be obtained with one attempt. One of the objects in making two or three counts is to secure mid-winter observations on the movements and activities of the wild life which are otherwise not attainable.

The following figures show the herd totals for the different game animals of the park [in 1932]:

Elk .... 10,624 (Northern Herd)	Deer .... 885 (both species)
Elk .... 2,499 (Gallatin Herd)	Antelope 668
Bear .... 180 (Grizzly)	Bear .... 500 (Black & Brown)
Buffalo 1,016	Moose 700

In quoting the number of game animals in the Yellowstone National Park, the above figures are considered official. The figure given for buffalo does not include 4 new buffalo calves which came on April 15.

The Northern Herd of elk comprises those animals which summer in the park and winter in the vicinity of Mammoth and Gardiner.

The Gallatin Herd spends a portion of the summer in the park and part on the Gallatin National Forest. They winter along the northern boundary of the park on the Gallatin watershed. The Northern and Gallatin elk herds do not intermingle to any extent.

—(*Yellowstone Nature Notes* 1932)

Table 2.2. Backcountry use, Yellowstone National Park, 1973-91 (Yellowstone National Park Archives).

Year	Number of people	Person-use nights[a]	Number of pack animals	Animal-use nights
1973	13,833	36,219	1,750	7,277
1974	16,801	41,282	1,556	6,226
1975	17,994	44,374	1,154	4,028
1976	20,077	50,580	1,258	4,121
1977	23,135	55,331	1,488	4,995
1978	24,238	52,795	1,338	5,084
1979	23,135	51,182	1,533	5,525
1980	24,010	54,874	1,808	4,694
1981	24,143	55,060	1,954	5,044
1982	21,529	49,400	1,735	6,715
1983	17,926	43,738	1,909	5,948
1984	15,737	34,936	1,625	5,527
1985	14,041	32,532	1,617	5,197
1986	13,814	31,414	2,099	7,493
1987	14,504	32,906	1,980	7,328
1988[b]	11,257	25,188	1,405	5,361
1989	14,182	32,747	2,046	7,518
1990	16,265	37,318	2,175	7,828
1991	18,585	41,476	2,403	8,399

[a] No. of people x no. of nights spent in backcountry.
[b] Backcountry areas were intermittently closed to travel due to wildland fire activity; in 1988, backcountry travel was curtailed from July through September.

In spite of improved techniques, early estimates of animal numbers were crude and are useful primarily to show general numbers and trends in numbers in relation to habitat changes. They emphasized the need for more accurate censuses, which would come with an increase in scientific investigations.

WILDLIFE PROTECTION AND ANIMAL CONTROL

The legislation that created Yellowstone National Park in 1872 placed the reserve under federal management, but included no provisions to protect its resources. Those resources were soon in danger. Tourists removed specimens and portions of thermal features, and wildlife was subjected to thoughtless and purposeful slaughter (Haines 1977, Vol. 1).

Wildlife destruction before the Park's designation may have been due, in part, to impressions of wildlife abundance created by early Yellowstone wanderers like A. Bart Henderson:

> June 2, 1870 . . . Here [Buffalo Plateau] we found all manner of wild game—buffalo, elk, blacktail deer, bear and moose. Thousands of bear, elk, buffalo and deer.

In 1875, however, three years after the Park was established, Philetus Norris commented on the slaughter of wildlife along the upper Yellowstone by market hunters (Haines 1977, Vol. 2). In the same year, William Strong was more specific:

> The terrible slaughter which has been going on since the fall of 1871 has thinned out the great bands of big game, until it is a rare thing now to see an elk, deer, or mountain sheep along the regular trail from Ellis to Yellowstone Lake . . . unless some active measures are soon taken, looking to the protection of the game there will be none left to protect. (Strong 1968)

In response to this growing concern, the Secretary of the Interior appointed Norris Superintendent in 1877. Norris' actions to control poaching and appoint a gamekeeper for the northeast corner of the Park marked the first efforts at wildlife management. It was not until 1894, however, when Congress passed the Lacey Act (U. S. Statutes at Large, Vol. 28, Chapter 72), that administrators finally had the power to control poaching in the Park. The Lacey Act provided a legal framework for protecting and managing park wildlife, and it specified mechanisms for punishing crimes in the Park. At the same time, it set the stage for contro-

The legislation creating Yellowstone National Park in 1872 included no provisions to protect its resources. Tourists enjoyed unrestricted access to delicate thermal features, removing souvenir specimens. Wildlife was subject to thoughtless slaughter. *(Photos by: Upper left, Margaret Smith Craighead; Upper right & lower, F. Jay Haynes, Haynes Foundation Collection, Montana Historical Society)*

By 1897, only 24 bison remained of the thousands of mountain-dwelling animals that formerly roamed the Park. To save the "Yellowstone bison," animals from relict plains-dwelling herds were brought to the Yellowstone Plateau. By 1912, the Park's bison were again thriving, but were no longer the pure stock native to the region. *(Photo by John J. Craighead)*

versies over official wildlife policy that continue to this day.

A major concern of early administrators was to increase the Park's populations of elk and bison because these were the animals that tourists wanted to see. The herd of mountain-dwelling bison was fast disappearing. In 1893, Superintendent George S. Anderson reported 400 bison in the Park; in 1897, Superintendent S. B. M. Young reported only 24. To save the herd, 18 animals were brought to the Park from relict herds of plains-dwelling bison, then under private ownership. Although these were conspecific with the mountain-dwelling bison, the geographic isolation of the latter, over many years, probably had conferred a degree of genetic distinction which today would be considered worth preserving. Nevertheless, the breeding program, in this case a militarily precise operation, proceeded rapidly. After the gradual release of progeny into the Park and interbreeding in the wild, bison were reportedly thriving by 1912. But these animals were

no longer the pure stock long native to the mountains of the region (Bartlett 1985).

Elk, protected from all hunting in the Park, increased dramatically in its northern portion. Few elk appear to have wintered in the Park before 1878, but part of the Northern Herd moved down the Yellowstone River drainage into the Paradise Valley every winter. The 1897 Superintendent's Report estimated 20,000 elk in the Yellowstone area, in three components: the Northern Herd (Yellowstone drainage), the Gallatin Herd (Gallatin drainage), and the Park Interior Herd. By 1907, the estimate was 30,000 to 40,000, and the need to manage these numbers had become apparent, as had the need for elk censuses (Haines 1977, Vol. 2). These increases in elk were undoubtedly related to the protection policy and to the first official coyote control program, initiated 10 years earlier in the winter of 1895-1896. The predator control effort intensified in 1904, when a pack of hounds was obtained to pursue mountain lions; 15 lions were killed that winter. Superintendent Pitcher summed up the official attitude on coyotes in his annual report:

> It is the general impression that coyotes are protected in the park, but this is far from the truth, for it is a well known fact that they are very destructive to the young game of all kinds, and we therefore use every means to get rid of them.

In 1907, Superintendent Young reported that "the mountain lions have been almost exterminated." Sixty-two were killed in the Park from 1904 to 1906, whereas no kills were reported during 1906-07. In 1908, President Roosevelt sent a personal letter to Young suggesting that no more mountain lions be killed (Haines 1977, Vol. 2).

The lions were not exterminated, but Roosevelt's request had little enduring effect on the predator control policy. The Superintendent's Report for 1914 lists 19 lions killed during the previous year; in 1918, the tally was 33 lions and 36 wolves. The last wolf was eliminated in 1926.

After the National Park Service took over administration of Yellowstone in 1916, predator control became a top priority. From 1918 to 1935, the annual kill of coyotes exceeded 100. In 1935, due to a national outcry, a ban was placed on the killing of predators in the Park. Control figures for these large predators no longer appeared in the Superintendent's Annual Reports.

During the period of intensive predator control, the elk herd expanded. Rodents also proliferated, which in turn probably hastened the deterioration of grass cover essential to the ungulates (Bartlett 1985). The development of ranching outside the Park excluded elk from using their traditional winter ranges, whereas unrestricted hunting presumably eliminated the migratory members of the herd. In 1903-04, it was assumed that the legal elk harvest was severely limited by fences erected near Gardiner to prevent elk from migrating to winter ranges outside the Park. A feeding program was started, and hay was grown in the Park for that purpose.

In response, the Northern Herd proliferated and became increasingly difficult to manage. Elk populations apparently exceeded the carrying capacities of the winter ranges. By 1912, a serious "elk problem" existed, and plans were made to live-trap and remove animals from the herd to offset the yearly increase. Trapped elk were shipped off to public zoos and to replenish or start wild herds. Some elk were also removed by legal hunting outside the Park. This control program continued into the early 1930s, but the numbers removed were far fewer than the annual increases in the herd (Haines 1977, Vol. 2).

Long-tended vegetation exclosures provided scientific evidence of retrogressive plant succession, shrub seed production, range trend, zoetic climax conditions, and natural plant succession. *(Photo by John J. Craighead)*

Because of the apparent deterioration of winter range, elk, bighorn sheep, pronghorn, and mule deer were fed during the winter. This solution was ineffective, and the situation reached a crisis in the winter of 1919-20, when more than half of the elk population died of malnutrition (Albright in Superintendent's Annual Report 1920). Although high wintering densities within the Park were viewed as unnatural (Skinner 1928), the Yellowstone administration did not decide to limit the size of the elk herd by "direct control" until 1934. From that year to 1961, rangers shot nearly 9000 elk within the Park. Legal hunting outside the Park accounted for nearly 41,000 kills; another 6000 elk were trapped and shipped elsewhere. "Direct control," practiced until the late 1960s, was effective but controversial. In 1962, removals from the Northern Herd were greatly increased, and elk numbers were reduced to about 3200 by 1968. Many of these elk were autopsied and reproductive data recorded (Greer 1965, 1966a, 1966b, 1967, 1968b). Direct control was terminated in response to increasingly adverse public opinion. Elk management continues to be a chronic problem in Yellowstone (Houston 1982, Kay 1985, Bartlett 1985, Chase 1986).

While elk populations were exploding and controversy raged over their management, white-tailed deer disappeared from Yellowstone, probably by 1924. Their willow-bottom habitat was largely eliminated by the burgeoning elk numbers (Kay 1994), combined with road and facilities construction. The destruction of that same habitat is credited with a dramatic decline in beaver numbers (Jonas 1955).

In the early 1970s, there was an attempt to dismantle long-tended vegetation exclosures, which had been established by range ecologists near Tower Junction and elsewhere in the Park. Fortunately, this move was aborted before irreparable damage was done. By excluding browsers, these fenced exclosures had dramatically demonstrated the severity of over-browsing by elk on the winter range. Fence removal would have destroyed scientific evidence of possible retrogressive plant succession, shrub seed production, range trend, zoetic climax conditions, and the natural process of plant succession in the area.

EARLY POPULATIONS OF BEARS

Not until the U. S. Army began administering Yellowstone in 1886 were bears given protection. Captain Moses Harris, the first military Superintendent, instituted a policy prohibiting the killing of any animals in the Park. Later, when predator control was reinstated, bears were not targeted; they had become major tourist attractions (Haines 1977, Vol. 2).

It is not known how the 1959 population of grizzlies in the Yellowstone area compared with earlier numbers, but available evidence indicates that the grizzly was common in Yellowstone before 1900; the black bear apparently was scarce. Early explorers frequently used the term "bear" when referring to either grizzlies or blacks, making it difficult to judge the relative abundance of the two species from historic records. Sightings and encounters with grizzlies were mentioned by N. P. Langford in 1870 and by W. E. Strong in 1876 (Strong 1968). In his "Journal of the Yellowstone Expedition of 1866 Under Captain Jeff Standifer," Bart Henderson made the following comments:

> July 4, 1870 . . . I killed several deer and three bear.
>
> July 9, 1870 . . . We returned to camp after killing five bear, and doing no prospecting.
>
> July 10, 1870 . . . killed six bear today.
>
> July 11, 1870 . . . I was chased by an old she bear today. Climbed a tree and killed under the tree.
>
> July 19, 1870 . . . Killed three bear.

After the establishment of Yellowstone National Park, references to bears in the area were more detailed than those of the early explorers, but not always explicit as to species. Grinnell (1876) noted that he had observed only one living black bear in Yellowstone and vicinity, but that grizzlies were

numerous in the Bridger Mountains and in Yellowstone Park. Superintendent W. Norris, in his annual report for 1880, stated that the grizzly ". . . is less numerous than some other varieties within the Park." This and subsequent observations by later Superintendents contradict the earlier records implying numerous grizzlies, but few if any black bears. Apparently, black bear numbers mounted rapidly as park visitation increased. In 1892, Superintendent G. S. Anderson stated in his annual report: "The bears are becoming very numerous and in some places quite troublesome, but as they are not in the least dangerous and their presence near the hotels is a source of great amusement, I do not recommend the destruction of any." (Presumably these remarks could apply only to black bears.)

In 1894, Anderson again noted, "Bears are numerous in the vicinity of all the hotels and have become very tame." In this same report, he referred to shipping "an enormous grizzly bear" to the National Zoological Park in Washington, D.C.

Hunting in the Park, with almost no regulation or supervision, was permitted until 1886, and Skinner (1928) believed this was responsible for a scarcity of bears at that time. The ban on hunting probably contributed to the increase in black bears noted by Superintendents from 1892 to 1900. By 1899, black bears were so prevalent that park officials considered control measures. Superintendent J. B. Erwin reported that year that bears were numerous and were increasing: "It will, undoubtedly, be necessary to kill some . . ." In 1907, both blacks and grizzlies were killed in control measures; in 1908, another captured grizzly bear was shipped to the National Zoological Park.

In the fall of 1908 (September-October), William H. Wright, a visitor to Yellowstone for nearly 15 years and a serious student of the grizzly bear, made what is perhaps the first crude, but comparative census of this animal in the Park. Excerpts from his report to Superintendent General S. B. M. Young (Army Records. Letters Received, Sep-Oct, 1908. Vol. VII, pp.341-349) are as follows:

> Seven years ago, there were at the Fountain and also at the Upper Basin, more grizzly bears than there are now black and grizzly combined. At the Upper Basin, there are now, as near as we were able to judge by the tracks, but one grizzly and but few black bears. We watched for several nights for the grizzly, but did not see a single one, but did on two occasions, see his tracks. At the Fountain, there were four grizzlies about two or three years old . . .
> These bear, we saw only once, although we watched for them several nights. Here there were also but few black bears, and the major-

ity of these were young bears and three were this years cubs; the old trails that were well worn several years ago, now show but little use and many of them are not traveled at all, plainly showing that the bears at this place are things of the past.

Two years ago, I spent several nights at the Lake [Yellowstone Lake], and a couple of weeks in the hills and marshes in the surrounding country, trying to photograph the grizzly by flashlight, and there were then in these feeding grounds, not less than thirty grizzlies. This year, we spent two weeks at the same place for the same purpose, and could find in all these feeding grounds but eight grizzlies and three of them were this years cubs, two were two years old, one was a three year old, and one old fellow. Two years ago, most of the bear at this place were two years old and over.

In 1906, at the Canyon, we spent nearly two weeks in the country round about and near the foot of Mount Washburn studying and trying to photograph the grizzly. In these feeding grounds, we saw night after night more than thirty grizzlies. There were two old bears with two cubs each, two with three cubs each and one with four cubs. There were in these five families, nineteen bears, then there were a large number of two or three years old, together with a goodly number of old fellows. This year we spent nearly three weeks in this same territory and saw but seventeen grizzlies, there was but one bear with two cubs, all the rest, with two exceptions, were young bears, still running in twos and threes not old enough to have families of their own. As it requires from six to eight years for a

grizzly to acquire his growth, and his life is near forty years, the question arises, what becomes of these grizzlies?

Wright goes on to describe the behavior of the soldiers entrusted with protecting park wildlife and concludes:

> We might go on and mention many more instances which came to our notice, showing why the present system of protecting the park is rotten, and why we are fully convinced that the extermination of certain of the animals is due to this means of so called protection.

Apparently Wright was unaware of the official control kill of both black and grizzly bears that occurred in 1907, a year prior to his visit to the Park and his letter of complaint to Superintendent Young. There appears to be a record of only three killed (Table 2.3), but Wright's observations suggest that those killed by soldiers and those destroyed in control measures were substantial.

On 20 August 1913, 32 grizzlies were counted on the refuse dump at Canyon. Three years later, Skinner (1925) noted 29 grizzlies—the most he had ever seen together. At about this time, he estimated a total population of 40 grizzlies in Yellowstone. In 1920, Superintendent H. M. Albright estimated there were at least 40 grizzlies in the Park and at least 100 black bears. In Albright's annual report of 1925, he estimated the black bear population at 200, of which 75 frequented points of interest along the loop road. He estimated the number of grizzly

Table 2.3. Bear control measures in Yellowstone National Park before 1931 (compiled from annual Reports of the Superintendents).

| Year | Black bears | | Grizzly bears | |
	Killed	Shipped to zoos	Killed	Shipped to zoos
1907	1		2	
1908			1	1
1911	3	2	2	10
1912				14
1913			5	5
1914				8
1915				3
1916	6		2	
1918				
1920			7	
1930	3			
Total	13	2	19	41

Note: Records of bear control were not mandatory and do not appear for all years in the Superintendent's Reports.

bears at "... not less than 75." Two years later, his annual report stated:

> Recent counts of bears of this species (grizzlies) are noted as follows: Canyon 53, Lake 11, Old Faithful 10, Total 74. This total includes 22 cubs. No recent counts have been had on the bears of this species known to frequent the more remote districts of the Park. The estimated total for the whole Park is 100.

In this report, the population of black bears was estimated at 275 (Table 2.4). Also in 1927, two woodcutters' camps at Canyon "were so infested with bears, especially grizzlies, that regular night guards were kept.... As many as fourteen bears [grizzlies] together with an undetermined number of cubs, were seen at one time about a given camp" (*Yellowstone Nature Notes* 1927).

Table 2.4. Early population estimates of black bears and grizzly bears, Yellowstone National Park, 1880-1951.

Year	Estimated number of black bears	Estimated number of grizzly bears
Before 1880		More numerous than black bear
1880		Less numerous than black bear
1892	Very numerous	
1894	Very numerous	
1895	Plentiful and increasing	
1899	Increasing	
1906	Numerous; control measures considered	Numerous; control measures considered
1907	Numerous; control measures taken	Numerous; control measures taken
1908	Numbers reduced	Numbers reduced
1913		32 counted at Canyon Dump
1920	≈100 in Park	≈40 in Park
1922		16 seen at Canyon
1925	200	75
1926	225	80
1927	275	100
1928	350	140
1929	440	150
1930	490	167
1931	465	180
1932	517	213
1933	525	260
1934 no estimate	——	——
1935	500	250
1936	500	250
1937	621	268-300
1938	520	290
1939	450	270
1940	500	300
1941	510	320
1942	550	300
1943 no estimate	——	——
1944 no estimate	——	——
1945 no estimate	——	——
1946	450	250
1947	450	200
1948 no estimate	——	——
1949	360	200
1950	360	180
1951	360	180

Note: Population figures are rough estimates made by National Park Service personnel based on sporadic and incomplete counts.

Black bears became one of the major tourist attractions of Yellowstone and as such were protected. Tourists soon learned to offer food to those that had learned to "beg" along roadsides. Bear-watching at established refuse sites such as Otter Creek became a semiofficial tourist attraction. *(Photos by:* Left, *Haynes Foundation Collection, Montana Historical Society;* Right, *John J. Craighead)*

In 1923, Park Ranger John S. McLaughlin reported 46 different bears (black and grizzly) on the dump at Canyon. On 15 August of the same year, he recorded 24 grizzly bears on the dump at one time (*Yellowstone Nature Notes* 1923).

Black bears were officially counted again in 1928. Albright listed three counts—Canyon, Lake, and Tower Falls—for a total of 227, including 52 cubs. On the basis of these counts, he estimated a population of 350 black bears in the Park. This estimate was revised to 440 in 1929, and the following year the black bear population was recorded as 490. Grizzlies were estimated at 140, 150, and 167, respectively, during these years. Estimates were continued in subsequent years, with a rather dramatic range in numbers. Black bears were estimated at a high of 621 in 1937, and grizzlies reached a peak estimate (1880-1951) of 320 in 1941.

These were crude, park-wide estimates of bear numbers, primarily valuable in showing an increasing population as visitors increased. Because estimates of both blacks and grizzlies were based on counts in areas where the two species concentrated, the figures may indicate the relative abundance of blacks to grizzlies in the more frequented portions of the Park. They probably represent relative abundance throughout the entire Ecosystem at the time and suggest the growing tendency of bears to move to and concentrate in the more protective confines

of the Park, where food in the form of garbage and tourist handouts was becoming more abundant.

Park Service estimates of black and grizzly populations were based on sporadic and incomplete counts (Table 2.4). Both species apparently increased from 1880 to 1958, but the estimates were crude, and their chief value is in showing that black bears were more numerous than grizzly bears after 1925, perhaps in a ratio of about two or three to one. In 1950, David Condon, the Chief Park Naturalist, estimated that there were 180 grizzlies in Yellowstone Park. Because of his interest in the grizzly, Condon's is perhaps the most accurate estimate before 1959, when intensive census techniques were initiated.

The black bear readily adjusted to humans and their ways and had become a frequenter of the roads, campgrounds, and refuse dumps in the Park as early as 1880. The grizzly also learned to visit dumps but, unlike the black bear, has never associated closely with man. Compared to the black bear, the grizzly has been a relatively infrequent visitor to campgrounds and only occasionally a "roadside bum."

Because of their begging proclivities, black bears became one of the major tourist attractions of Yellowstone and as such were protected. The tourist camps and hotels in and around the Park provided food, which the black bears soon learned to use, and tourists offered food to those that had

By 1916, "bear shows" at refuse sites had become part and parcel of the Yellowstone experience.
(Courtesy of Haynes foundation Collection, Montana Historical Society, E. W. Hunter, Photographer)

learned to "beg." Refuse sites near camps and hotels became favorite locations for watching both black bears and grizzly bears. The first dump to become a semiofficial tourist attraction was at the Fountain Hotel, which opened in 1891.

In 1901, Superintendent Pitcher found the situation dangerous and forbade the feeding of bears at hotels and along the roads, but he did not close refuse disposal sites. A tourist was injured by a bear during a "feeding show" at a dump in 1902, and dumps were fenced and warnings posted to protect sightseers (Haines 1977, Vol. 2). Despite this early concern, "bear shows" continued at refuse sites in the Park, usually with a ranger present. In the 1930s, bear feeding as a quasi-official activity was limited to the Canyon area. Protected by a chain-link fence, visitors saw as many as 50 bears feeding on garbage (Haines 1977, Vol. 2).

BEAR CONTROL MEASURES

By 1899, Superintendent James B. Erwin suggested that it might be necessary to "kill some [bears] to prevent [property] destruction." Not until 1907,

however, did a report record bears killed (one black and two grizzlies) because they had become troublesome (Table 2.3). Bears were also trapped and shipped to zoos, one as early as 1894. In 1908, one grizzly was shipped to the National Zoo in Washington, D. C., and another was apparently killed, as its skull and hide were sent to the Smithsonian Institution.

After 1910, bears were recognized as problems as well as attractions and were mentioned more frequently in the Superintendents' Annual Reports. Two grizzlies and three black bears were killed in 1911 when they became "dangerous to life and property." That same year, 10 grizzlies and two black bears were shipped to zoos (Table 2.3). In 1913, five more grizzlies were killed, and another five shipped to zoos. Bear numbers had apparently increased throughout the Park by then; Superintendent L. M. Brett noted that as many as 32 grizzlies were seen at the Canyon dump on a single evening.

During the 23-year period from 1907 through 1930, 15 blacks and 60 grizzlies were killed or shipped to zoos. This record is not complete, but does show

Table 2.5. Bear incidents and control measures in Yellowstone National Park, 1931-65.

Year	Personal injuries (both species)	Property damage (both species)	No. of black bears killed for control[a]	No. of grizzly bears killed for control	No. of black bears killed accidentally	Total no. of bears killed for control	Total bear mortalities	Park visitation[b]
1931	78	209	34	0		34	34	221,248
1932	34	451	55	0		55	55	157,624
1933	22	146	22	0		22	22	161,938
1934	25	38	2	0		2	2	260,775
1935	36	141	66	3		69	69	317,998
1936	42	34	10	4		14	14	432,570
1937	115	81	41	10[c]		51	51	499,242
1938	97	81	46	0		46	46	466,185
1939	78	92	23	0		23	23	486,936
1940	41	75	12	0		12	12	526,437
1941	90	102	26	0		26	26	581,761
1942	29	118	55	28		83	83	191,830
1943	5	38	14	0		14	14	64,144
1944	2	18	9	3		12	12	85,347
1945	2	7	2	0		2	2	178,296
1946	15	20	4	0		4	4	814,907
1947	40	9	8	0		8	8	932,503
1948	36	27	5	3		8	8	1,013,531
1949	47	26	11	8		19	19	1,133,516
1950	51	48	20	0		20	20	1,109,926
1951	38	75	40	4		44	44	1,166,346
1952	40	84	8	1		9	9	1,330,327
1953	43	66	10	1		11	11	1,326,858
1954	43	115	6	5		11	11	1,328,893
1955	73	121	18	5		23	23	1,368,515
1956	109	106	14	0		14	14	1,457,782
1957	91	125	20	1	18	21	39	1,595,875
1958	39	117	26	2	12	28	40	1,442,428
1959	41	269	24	6	40	30	70	1,408,667
1960	69	358	85	4	17	89	106	1,443,288
1961	58	247	58	1	12	59	71	1,524,088
1962	42	112	48	4	8	52	60	1,923,063
1963	36	278	18	2	15	20	35	1,870,644
1964	30	224	10	2	8	12	20	1,929,316
1965	39	289	9	2	7	11	18	2,062,476
Total	1,676	4,347	859	99	144	958	1,102	32,815,280
Mean	48	124	25	3	14	27	31	937,579

[a] Does not include bears accidentally killed.
[b] Annual visitation values based on Haines 1977, Vol. 2.
[c] Museum specimens.

As early as 1910, both black and grizzly bears were recognized as problems as well as attractions of the Park. Control efforts both in and out of the Park included trapping and relocating bears to zoos, but by and large, destruction of unwanted animals was most commonly practiced. *(Photo by John J. Craighead)*

that during the early years of the Park, grizzlies were considered a bigger problem than blacks.

More complete records were kept from 1931 through 1965 (Table 2.5); they show the number of bear incidents and the control exercised in an attempt to keep bear-human contacts at a level tolerable to both the public and the park administrators. The personal injury and property damage figures include both grizzlies and blacks. No guidelines were established defining what constituted a personal injury or damage to property. District Rangers made independent judgments about what should be recorded. Some recorded incidents were, no doubt, trivial. But we were unable to determine what constituted bona fide injuries or damages. If the data in Table 2.5 over-emphasize the frequency of serious bear incidents, they also reflect the judgment of the times.

Through the early 1930s, bears were controlled infrequently and only if they threatened visitors or property (Fig. 2.2). In 1935, bear control measures increased; Superintendent E. B. Rogers reported killing 66 black bears and 3 grizzlies. Prior to that,

the largest reported control was 55 bears killed in 1932, all blacks.

From 1938 through 1941, the official reports failed to mention the annual grizzly bear control actions. The 1942 Superintendent's Report lists 83 bears (55 black and 28 grizzly) killed that year (Fig. 2.2). Five of the 28 grizzlies, three females and two males, were killed at the Fishing Bridge-Lake area. These were recorded and measured by the Chief Park Naturalist, Dave Condon (*Yellowstone Nature Notes* 1950). The Superintendent's Report for 1942 also mentioned 26 black bears killed in 1941 and concluded that the average control kill of black bears for the preceding 10 years had been 33 (although data in Table 2.5 indicate an average of 31). The report did not specify grizzly bear deaths, but suggested that the "bear show" at Canyon be discontinued in an effort to return the Park to more natural conditions.

Through the late 1950s and early 1960s, the Superintendent's Reports reflect a steady killing of bears as a management tool: 23 were destroyed in 1955, 14 in 1956, and 21 in 1957 (Table 2.5 and

Table 2.6. Comparison of black bear and grizzly bear incidents in Yellowstone National Park, 1959-65.

	Black bear					Grizzly bear				
			No. bears killed					No. bears killed		
Year	Personal injury	Property damage	Control	Accidental	Total	Personal injury	Property damage	Control	Accidental	Total
1959	41	267	24	40	64	0	2	6	0	6
1960	69	346	85	17	102	0	12	3	1	4
1961	57	242	58	12	70	1	5	1	0	1
1962	40	107	48	8	56	2	5	4	0	4
1963	30	273	18	15	33	6	5	2	0	2
1964	28	220	10	8	18	2	4	2	0	2
1965	33	283	9	7	16	6	6	2	0	2
Total	298	1,738	252	107	359	17	39	20	1	21
Mean			36		51			3		3

Fig. 2.2. Annual control measures for black bears and grizzly bears, Yellowstone National Park, 1931-65.

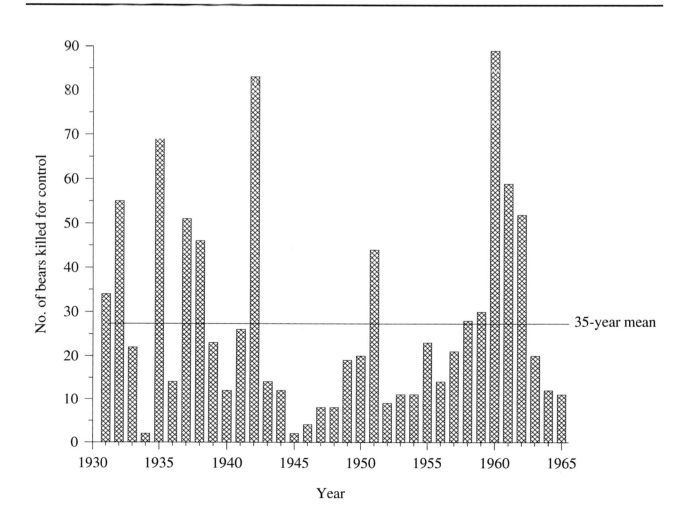

Table 2.7. Summary of bear-human incidents, Yellowstone National Park, 1931-65.

Period	Personal injuries	Property damage	No. killed for control Black bears	Grizzly bears	All bears	Black bear accidental deaths	Park visitation	Annual average park visitation	Ratio of no. of black bears killed to no. of visitors	Ratio of no. of grizzly bears killed to no. of visitors
1931-35	195	985	179	3	182		1,120,000	224,000	1:6,300	1:373,300
1936-40	373	363	132	14	146		2,411,000	482,000	1:18,300	1:172,200
1941-45	128	283	106	31	137		1,101,000	220,000	1:10,400	1:35,500
1946-50	189	130	48	11	59		5,004,000	1,001,000	1:104,300	1:454,900
1951-55	237	461	82	16	98		6,541,000	1,308,000	1:79,800	1:408,800
1956-60	349	975	169	13	182	94	7,327,000	1,465,000	1:43,400	1:563,600
1961-65	205	1,150	143	11	154	50	9,338,000	1,868,000	1:65,300	1:848,900
1931-65	1,676	4,347	859	99	958	144	32,842,000	938,000	1:38,200	1:331,700

Total mortality, black bears = 1,003
Total mortality, all bears = 1,102

Fig. 2.2). In 1956, a new category of "bears killed accidentally" was established for black bears. Why this was needed is not explained in the reports. When these "accidental kills" are added to control kills for the period 1956 through 1965, the total of black bear mortalities rises dramatically (Table 2.5).

The maximum control measures for black bears were taken in 1960, when 85 were destroyed and an additional 17 were killed unintentionally. As far back as 1932 and 1935, heavy control was practiced, with 55 and 66 blacks reportedly destroyed. Black bears were killed during each of the 35 years for which records are available (Table 2.5).

Control of the grizzly since 1931 has been far less intensive and more erratic than control of the black, with control kills reported for only 20 of the 35 years (Table 2.5). Maximum control measures were taken in 1942, when 28 grizzlies were destroyed. The ratio of grizzlies to blacks in the control kill during the 35-year period was approximately 1:8. On the other hand, the records indicate a heavier kill of grizzlies than blacks from 1907 to 1930 (Table 2.3).

PERSONAL INJURY AND PROPERTY DAMAGE

The number of personal injuries and property damage from both species of bears varied dramatically over a 35-year period from 1931 through 1965 (Table 2.5). Most of the personal injury and property damage incidents before 1959 were attributable to black bears. More precise data on personal injury and property damage, recorded by species from 1959 to 1965, show emphatically that black bears were the major culprits. They inflicted injuries and property damage, respectively, 18 and 43 times more

frequently than did grizzly bears over the same seven-year period. The ratio of property damage incidents to personal injury incidents was about two to one when grizzlies came in contact with park visitors, whereas it was about six to one in the case of the blacks (Table 2.6). Thus, grizzlies were not as destructive to property, but were considerably

Fig. 2.3. Personal injuries from black bears and grizzly bears, Yellowstone National Park, 1959-65.

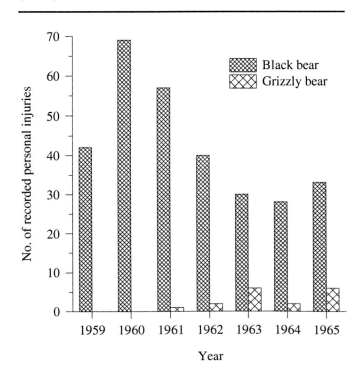

Fig. 2.4. Mean numbers of annual control kills of black bears and grizzly bears by 5-year periods, Yellowstone National Park, 1931-65.

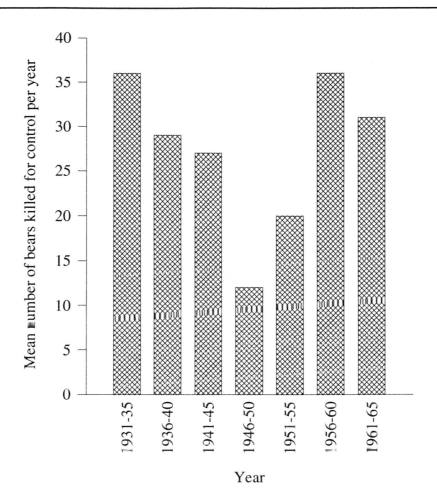

Year

more prone to attack visitors. Fortunately, grizzlies and visitors seldom met, but visitors and black bears met frequently in campgrounds and along the park roads. The number of personal injury incidents resulting from contact with grizzlies and blacks from 1959 to 1965 is illustrated in Fig. 2.3. When data for both species are lumped (Table 2.7), they show that property damage and personal injury occurred consistently and frequently over a 35-year period, with a ratio of property damage to personal injury of about three to one.

ANALYSIS OF EARLY MANAGEMENT EFFORTS

For convenience of interpretation, the information in Table 2.5 has been summarized into five-year periods (Table 2.7). These data show that the park administration reported 1676 personal injuries (most of them minor) resulting from bear-visitor contact and 4347 cases of property damage from 1931 through 1965. Park Rangers killed 859 black bears

and 99 grizzlies in control measures (accidental deaths not included), or an average of 27 bears a year. During this period, approximately 33 million visitors entered the Park, an annual average of about 938,000.

Theoretically, an increase in either numbers of bears or numbers of people should increase the opportunity for contact between the two. Many factors affect the probability of contact, however, and only some of these factors can be manipulated for management purposes. A reduction in bear numbers is the easiest and most direct way of reducing contact, but it is essentially an expedient approach, when population size or reproductive rates are unknown, as was the case for both species at the time.

Figure 2.2 shows the annual fluctuation in control kills of both species from 1931 to 1965. Figure 2.4 illustrates the annual average control kills over five-year periods. There appears to be no pattern governing the size of the annual control kill

Fig. 2.5. Incidence of personal injuries from black bears and grizzly bears, Yellowstone National Park, 1931-65.

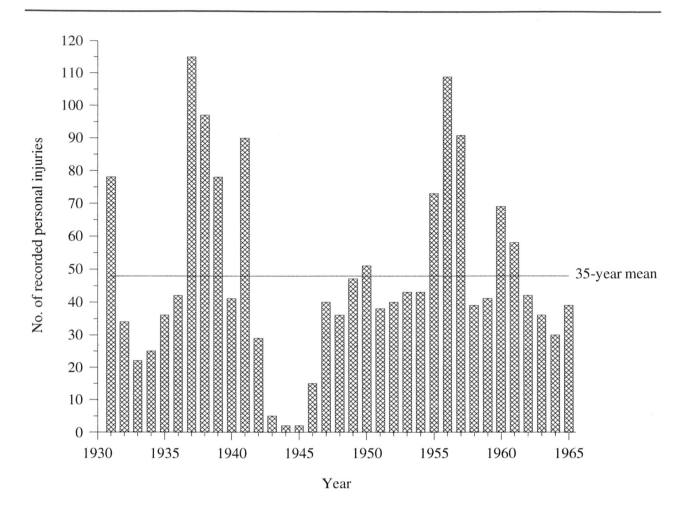

or the five-year totals until those figures are compared with the incidence of personal injuries (Fig. 2.5).

In general, the heavy control kills either coincided with or immediately followed years when personal injuries were high (Fig. 2.6). Following the heavy control kills, the incidence of personal injuries dropped, as did the control measures. Exceptions to this general trend occurred when the low incidence of personal injury in 1933-35 was followed by a heavy control kill in 1935, and when the high incidence of personal injury in 1955, 1956, and 1957 elicited only very light control until 1960, when a record number of 85 black bears was destroyed (102, if accidental deaths are included). The relationship between control kills (Fig. 2.4) and the incidence of property damage (Fig. 2.7) is not as consistent as that between control kills and personal injury, but a general pattern of increased control following increased property damage is evident (Fig. 2.6).

Personal injuries did not increase significantly from 1930 to 1965. There were 195 injuries during the five-year period from 1931 to 1935, and 205 during a similar period 25 years later (1961-65). The maximum number of injuries for a five-year interval was 349 from 1956 to 1960 (Table 2.7 and Fig. 2.5). This high incidence of injuries worried administrators and clearly indicated the need for scientific information on which to base a management program.

Incidence of property damage fluctuated widely from 1931 to 1965; 985 incidents were recorded in the first five years, and 1150 during the last five (Table 2.7 and Fig. 2.7). There were, no doubt, discrepancies in effort and commitment to record-keeping over the years. Nevertheless, there was no obvious trend in property damage that correlates with the annual increase in visitors, although park visitation had risen steadily, with the exception of the 1941-45 period that coincided with World War II.

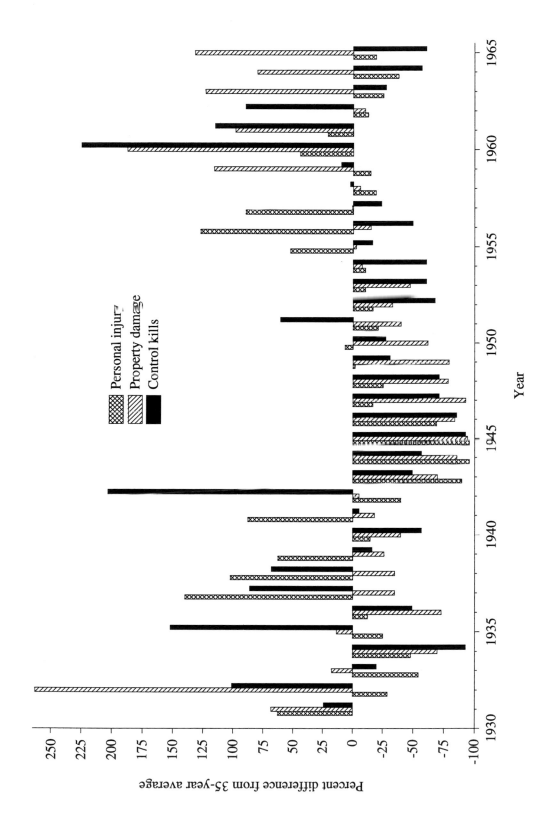

Fig. 2.6. Relationship between recorded occurrences of personal injury or property damage by black bears and grizzly bears and number of bears killed for control, Yellowstone National Park, 1931-65 (National Park Service Records). The height of each bar represents the deviation for that year, in percent, from the 35-year average. For example, "0" shows the year was average; "150" shows the number of occurrences for that year exceeded the average by 150% (i.e., totalled 2½ times the average); "-100" shows no occurrences for that year. The 35-year (1931-65) averages are: control kills, 27; personal injuries, 48; property damage incidents, 124.

Fig. 2.7. Incidence of property damage by black bears and grizzly bears, Yellowstone National Park, 1931-65.

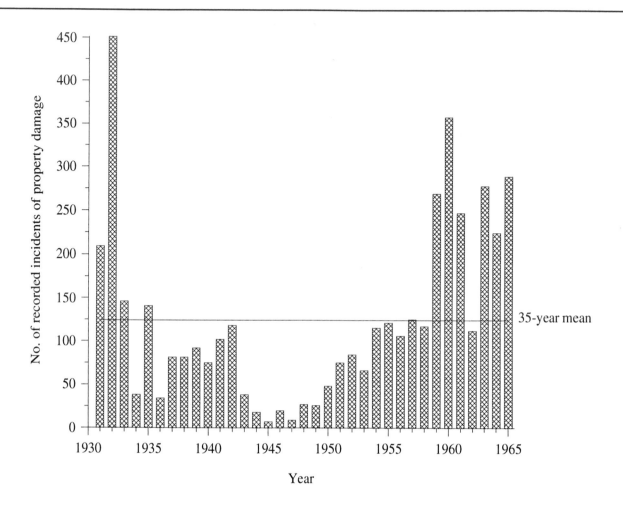

The Tort Claims Act of 1982 held the federal government responsible for personal injury resulting from gross negligence on the part of federal employees. This Act placed a new light on the potential danger from bears. Bear incidents were no longer simply a nuisance, they became career-threatening. Bears, a prime attraction for visitors, became administrative headaches. Official perspective altered; the need for scientific information became both more evident and more pressing.

Waste Disposal

The rapid growth of tourism and development of service facilities brought a new problem to the once pristine wilderness. The refuse created by transporting, housing, and feeding large numbers of visitors required disposal. The simple solution was to create dump sites. For convenience, many of the dumps were situated in close proximity with the garbage source and were relatively accessible to the public. Others, those at Trout Creek and Rab-

bit Creek, for example, were remote enough that they would not detract from the visiting public's impression of unmarred wilderness or offend their aesthetic sensibilities.

Garbage was dumped at designated sites, the most prominent being Trout Creek, Gardiner, Tower Junction, Thumb, Otter Creek, Canyon, Fishing Bridge, Rabbit Creek, and West Yellowstone. The more aggressive grizzly dominated these isolated sites, deterring all but the largest and most aggressive black bears. Only at sites situated in timber, or close to timber, did black bears and grizzlies mix; at these sites, grizzlies predominated. At open, exposed areas such as Trout Creek, black bears were rarely seen. Black bears claimed the hotels, campgrounds, roadsides, and wherever both people and food "handouts" were abundant.

Soon after the first garbage dump was established, bears quickly discovered that such places were good sources of easily obtainable food. The same behavior that in pre-Columbian times led them

At Otter Creek, near the Canyon Hotel, food wastes were dumped on a concrete platform while tourists watched the feeding bears from bleachers. Rangers provided informative lectures until 1942 when this practice was stopped. *(Courtesy of Haynes Foundation Collection, Montana Historical Society)*

The Otter Creek bear shows were closed in 1942, however, grizzly bears continued to feed at the open pit garbage dumps maintained to accommodate the waste of a continually growing human presence in the Park. *(Photos by:* Top, *Haynes Collection Foundation, Montana Historical Society;* Bottom, *Margaret Smith Craighead)*

to buffalo jumps and Indian encampments brought bears to and held them at the calorie-rich refuse sites. Tourists caught on as quickly as the bears. By 1897, garbage was being dumped on a regular evening schedule to accommodate bear-watchers near the Fountain Hotel. One early report had "from ten to sixteen bears browsing among the kitchen scraps on summer evenings" (Haines 1977, Vol. 2).

As could be expected, the proximity of visitors to bears while feeding eventually led to trouble. Signs were posted to warn visitors, and soldiers stood guard to ensure the warnings were heeded.

After a visit in 1903, Theodore Roosevelt noted the bears' behavior:

> . . . they have come to recognize the garbage heaps of the hotels as their special sources of food supply . . . . They have now come to take their place among the recognized sights of the park, and the tourists are nearly as much interested in them as in the geysers.

At Old Faithful and Otter Creek, crude bleachers were erected to accommodate the crowds, and the "bear shows" attained official status when rangers began lecturing to the tourists watching the feeding bears.

The Park Service began closing dumps to the public when bear-human conflicts became too frequent (Haines 1977, Vol. 2). By the 1930s, the Park Service had closed all but the Otter Creek dump near the Canyon Hotel. This last dump was equipped with bleachers, a chain-link fence, and a large concrete platform where the garbage was dumped and where the bears could feed. It was often difficult to find a parking space in the specially built lot, as tourists gathered to watch as many as 50 bears congregate some evenings (Haines 1977, Vol. 2). Despite its popularity with visitors, administrators believed the dangers were too great to continue operating the public feeding ground. Superintendent Edmund P. Rogers closed the Otter Creek dump to the public in 1942 (Superintendent's Report 1942).

Although the open-pit dumps were closed to public viewing, bears, especially grizzlies, continued to use them. The dumps were, by then, well-established and predictable as ecocenters for bear aggregations. Indeed, official estimates of grizzly bear populations had been based for several decades on counts taken at disposal sites. Superintendent's Reports from the late 1920s reflect some of those counts, as in the 1928 report by Superintendent Horace M. Albright: "Recent counts: Canyon 68, Lake 27, Old Faithful 8, total 103, including 31 cubs . . . . Increase shown on actual counts 39%. The total number of grizzly bears within the Park at this time is estimated as 140" (Superintendent's Report 1928).

Based on the rough estimates contained in the annual reports, it appears that the Yellowstone grizzly bear population increased from around 40 bears in 1920 to perhaps as many as 260 by 1933. Some researchers have attributed the increase primarily to the growing availability of garbage as a food source (Chase 1986). Others have attributed the apparent increase to the grizzly bear aggregations that formed at these sites, attracted from all parts of the Ecosystem. Bears in the aggregations were more easily counted.

Grizzly Bear Research

Vigorous research had not been a part of National Park strategy, but all of the earlier-detailed events strongly suggested the need for scientific evaluation of the Ecosystem. By 1960, tourists as well as park personnel were not only discouraged from visiting the dump sites, they were actually forbidden to visit them without specific permission. This could be obtained only by direct order of the Superintendent. Lemuel Garrison deserves special recognition as a visionary Superintendent, who recognized the critical need for scientific information on which to base wildlife management decisions and initiated a program to that end. Garrison effectively isolated the Trout Creek bear aggregation from tourists and park personnel; this set the stage for an in-depth research effort into grizzly bear ecology. A mutual understanding developed between us and Yellowstone National Park administrators that research would take priority over visitor and staff sightseeing. Memoranda of Understanding between our research team and the Yellowstone National Park administration were drawn up and duly signed. Responsibilities were clearly defined and a sound, cooperative relationship established. One oversight, unrecognized by either party, was the effect that evolving park policies and changes in park administrators might have on the conduct and interpretation of the research and implementation of its results.

Fig. 3.1. Location of open-pit garbage dumps and size of grizzly bear aggregations (mean of censuses, in parentheses), Yellowstone National Park, 1959-70; filled triangles denote relatively major aggregations, and open triangles, minor ones.

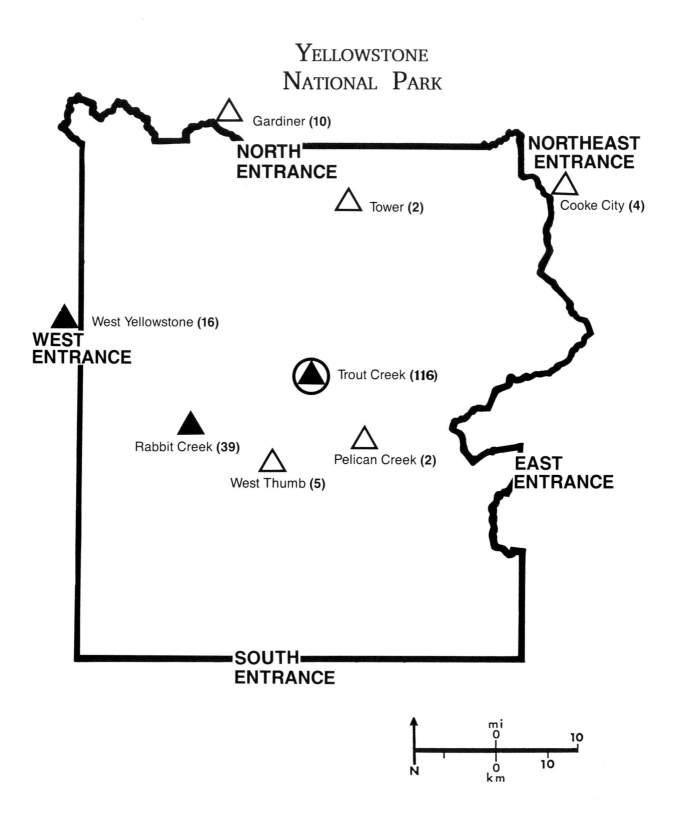

3
Study Methods

Our research encompassed the entire Yellowstone Ecosystem (Fig. 1.2), but most of our earlier population and behavioral data were obtained at the Trout Creek ecocenter (Fig. 3.1). Immobilizing, color-marking, instrumenting, radiotracking, and aging have been described by Craighead et al. (1960, 1961, 1962, 1963, 1970, and 1973b) and Craighead and Craighead (1965, 1967, and 1971). Techniques employed in population analyses, determination of home and seasonal range movements throughout the Ecosystem, and computation of death statistics have been reported by Craighead et al. (1974, 1986, 1988b), Craighead (1976, 1977), and Craighead and Craighead (1971, 1991). Methods of describing and quantifying grizzly bear habitat parameters have been described by Craighead et al. (1982, 1986, 1988a) and Craighead and Craighead (1991). Controlled experiments designed to test specific hypotheses were not feasible. Thus, our data are entirely observational. We maintained an observational effort that was as consistent from year to year and as objective as field conditions would permit.

This chapter describes some fundamental methodologies, whereas more specific methodologies are covered in contexts specific to their use. Four techniques detailed in the literature are briefly summarized here to give the reader ready access to methodologies that were both pioneering in the study of large mammals and absolutely essential to this research: immobilization, aging, radiotracking, and vegetation mapping of habitat.

CAPTURE AND IMMOBILIZATION

Much of the information we needed for a thorough study of the grizzly population necessitated handling of individual animals. The study required that the bears be individually marked with color-coded ear tags, and primary physical data taken on morphometry, breeding condition, and pathological and physiological parameters. Obviously, the first step was to develop reliable techniques for

trapping and subduing the bears with minimal injury to them (or to researchers). Immobilization was an experimental process in 1959 that underwent continuous refinement through the course of our 12-year study.

Bears were baited into steel cable foot snares or induced to enter culvert traps constructed from lengths of corrugated culvert pipe. The culvert traps were closed at one end with gates made of steel bars or grid. A trigger mechanism for gate closure was installed at the end away from the gate, and the whole affair was mounted on a trailer frame. Trapped bears were immobilized with a muscle-relaxing drug, phencyclidine hydrochloride, administered intramuscularly by means of a syringe-mounted rod ("jab stick") or by a gas-propelled syringe dart fired from a rifle (developed by Crockford et al. 1958). Later in the study, we used a CAP-CHUR gun, developed by Palmer Chemical Equipment Company, to deliver the darts. The propulsive syringe darts were also used to immobilize free-ranging bears.

Although pentobarbital sodium, a potent general anesthetic, had been used to immobilize black bears (Erickson 1957, Black 1958), the first efforts to immobilize brown bears in the wild employed ether (Troyer, pers. comm. 1960) and pentobarbital sodium for grizzly bears (Craighead et al. 1960). Because of rapid and unpredictable recovery, both anesthetics were dangerous to bear and man. We later employed the fast-acting muscle relaxant, succinylcholine chloride (Sucostrin). This drug blocks nervous transmission at the myoneural junction by competitively inhibiting acetylcholine, and is degraded by cholinesterase only very gradually. Once a bear was down, we used follow-up injections, if necessary, to keep it immobilized; however, multiple injections were difficult to control and placed the bear at greater risk, so were used only infrequently.

Over the period of study, we found that the optimal dosage for small- to medium-sized grizzly bears was about 1 mg of Sucostrin per 1.4 kilogram (3.1 lb) of body weight. This dosage prolonged

Grizzly bears were induced to enter culvert traps constructed from large-diameter corrugated culvert pipes. A baited trigger mechanism dropped a gate, securely confining the large animals. *(Photo by Frank C. Craighead, Jr.)*

immobilization in larger bears and had to be modified for age and estimated body fat (Craighead *et al.* 1960). In the middle to later years of the study, we used a second muscle-relaxant, phencyclidine hydrochloride (Sernylan), in place of Sucostrin and found that it immobilized larger bears more effectively, had a wider range of tolerance, and presented fewer hazards for the researchers. Dosages ranged from 1 mg/0.31 kg (0.69 lb) to 1 mg/1.18 kg (2.61 lbs), depending on the age and condition of the bear.

DETERMINATION OF BEAR AGE

Knowing the age of bears was essential for establishing population age distributions, constructing life tables, and examining age-specific mortality and the productivity of females. Some individuals were "known-age" because we first observed and marked them during their cub year. For others, however, we established routine methodology for accurately determining age based on modification of a technique originated by Scheffer (1950) for the study of seals, sea lions, and walruses (Craighead 1974, Craighead *et al.* 1970, 1974).

In brief, the process entails removing a tooth from an immobilized animal (in our work, the fourth premolar), slicing thin sections from it, and then decalcifying and staining the sections to reveal annuli (growth rings) in the cementum layer of the tooth root. The annuli occur as the result of seasonal variation in the rate of cementum deposition, and their number is related to the age of the bear. By applying this technique to teeth from known-age bears and, also, to teeth from unknown-age bears extracted over periods of two or more years, and comparing annuli, we concluded that the procedure provided accurate estimates of age (Craighead *et al.* 1970). This process is now widely used for determining age among species of wild carnivorous mammals.

RADIO-LOCATION TECHNOLOGY

Of the research technologies developed during our study in Yellowstone, the radio-transmitter collar combined with the tuned directional receiving antenna was the most innovative and proved the most valuable in documenting the more seclusive behaviors and activities of grizzly bears. The procedure was developed by Frank C. Craighead, Jr., and the senior author, with the assistance of electronic engineers. It was first used in 1961, when

Trapped bears were immobilized with a muscle-relaxing drug, administered by means of a "jab stick" or with a propelled syringe. *(Photo by Frank C. Craighead, Jr.)*

we fitted two bears with prototype transmitter collars and used converted Citizen Band transceivers to track them to their winter denning sites (Craighead and Craighead 1963, Craighead *et al.* 1963). From 1961 through 1968, 48 radio instrumentations were made, involving 23 grizzlies, to observe feeding behavior and pre-denning activities and to determine seasonal and lifetime home ranges (Craighead 1976; Craighead and Craighead 1963, 1967, 1973; Craighead *et al.* 1961, 1962, 1963, 1964, 1971). We subsequently broadened and improved the technology by using orbiting satellites to receive data transmitted by radio collars and implanted sensors (Buechner *et al.* 1971; Craighead and Craighead 1987; Craighead *et al.* 1971, 1972a, 1978a, 1978b, 1986, 1988a; Craighead 1995 [in prep.]). Today, radiotracking, both conventional and satellite-based, is used world-wide and is among the most common and valuable of methods for studying a broad range of wildlife species.

VEGETATION MAPPING

We considered that an understanding of the ecosystem-wide habitat of grizzly bears was essential to interpreting basic biological parameters of

Immobilized grizzlies could be removed from the trap for processing. *(Photo by Lance Craighead)*

the grizzly bear population, such as reproduction, social behavior, mortality, movements, and food habits. When this study began, there were no vegetation maps of Yellowstone National Park, the Yellowstone Ecosystem, or of any other wilderness ecosystems inhabited by grizzly bears. To map the vegetation of such immense areas, we adopted space-age, remote-sensing techniques, in combination with detailed, on-the-ground vegetation sampling. Specifically, we mapped vegetation using satellite multispectral imagery data from the Landsat MSS system. The methodologies we developed are described in Craighead *et al.* (1982), Craighead *et al.* (1986), and briefly, in Chapter 12 of this monograph. When satellite-mapping of wilderness vegetation becomes standardized, it will then be possible to compare quantitatively the vegetation components of different wilderness ecosystems (Craighead *et al.* 1986, Craighead and Craighead 1991). This should greatly enhance population viability assessments and improve grizzly bear recovery planning.

THE POPULATION UNDER STUDY

For purposes of consistency when comparing data accumulated over the 12-year study, the 389 marked or individually identifiable grizzlies that we observed were divided into four groupings (1-A, 1-B, and 2, below, based on how often bears visited the Trout Creek ecocenter; and 1-C, a combination of 1-A, 1-B, and other Trout Creek bears). All analyses in the first section of this monograph are based on data from 312 individual bears studied at the Trout Creek ecocenter. Data recorded on the 77 non-Trout Creek marked bears are included in some later analyses when they modify the Trout Creek results in a way that significantly affects ecosystem-wide interpretation. Our basic sample groups were defined as follows:

1. Trout Creek Bear (*n* = 312): trapped, marked, and/or censused at the Trout Creek ecocenter from 1959 to 1970.

   A. Trout Creek Flagged Bear (*n* = 274): over all years a marked individual was available for censusing, average visitation at Trout Creek ≥5% of census-hours; and if unmarked, but individually recognizable, ≥0% of census-hours.

   B. Trout Creek Nonflagged Bear (*n* = 38): over all years the marked individual was present, average visitation at Trout Creek < 5% of census-hours.

   C. Trout Creek Censused Bear (*n* = unknown, but mean = 116): included both the Trout Creek flagged and nonflagged bears, plus unmarked bears that lacked *permanent* natural markings, but could be reliably identified annually through their association with a marked female or marked littermates and/or through natural markings. These identifications, however, were not useable over a period of years.

2. Non-Trout Creek Bear (*n* = 77): trapped and marked in an area of the Yellowstone Ecosystem other than the Trout Creek ecocenter, and never observed at the ecocenter.

The term *census* is applied in a broad sense. Our observational efforts at the Trout Creek dump were censuses in the true sense of attempting complete counts of all individuals. Ecosystem-wide *censuses* were counts, including those at Trout Creek and other ecocenters, from which a population estimate could be derived.

CENSUS METHODS

Census observations were divided into two data sets for analysis: those made during the breeding season (before 15 July), and those made on or after that date. These in turn were subdivided into observations made during the long censuses of 3.5 hours and those made during the short censuses of 2.0 hours (Table 3.1). Observation effort was greatest during the breeding season, and the number of hours for that period was nearly double the number for the postbreeding season (Table 3.1). Only the long censuses were used to tally the average lifetime visitation percentages for separating the Trout Creek flagged and nonflagged populations. This arrangement enabled us to compare breeding season data with postbreeding season data at the Trout Creek ecocenter.

We attempted to keep the number of observation hours per season constant, especially during the 1 June to 15 July period. Exceptions were the initial season in 1959 and the final one in 1970. Observation data were converted to a per-hour basis when comparing one season's data with another's, when direct comparisons were made between individual animals or groups of animals, or when a discrepancy in observation hours would preclude direct comparisons. In most instances, analyses on a per-hour basis and those based on total observation hours produced comparable results.

Systematic observations and enumeration of the bear population, year after year, provided the basic information for studying population dynamics and

A gun firing a propulsive syringe dart containing drugs was used to immobilize free-ranging bears. *(Photo by Frank C. Craighead, Jr.)*

social and reproductive behavior. Our observations focused on individually color-marked or otherwise recognizable animals. During a 12-year period, 264 grizzly bears were color-marked; 125 others had permanent natural markings. A total of 380 3.5-hour censuses and 54 2-hour censuses was made at Trout Creek in Hayden Valley from 1 June to 30 August from 1959 through 1970 (Table 3.1). In addition, censuses were made at other localities throughout and adjoining the Park where bears aggregated at ecocenters to feed on garbage (Fig. 3.1).

The largest of the bear aggregations was at Trout Creek, near the geographic center of Yellowstone National Park (Fig. 3.1). Here, censuses were made and reproductive and social behavior was recorded during each observational period of 3.5 and 2.0 hours from late afternoon until dusk. The area was off limits to all visitors and park employees, except by special order of the Park Superintendent. A locked gate and barriers prevented all but authorized access. Observations were made from one or more vehicles parked at commanding sites overlooking the Trout Creek Valley and open-pit dump.

Grizzlies appeared to ignore vehicles and their occupants, but, in fact, they were acutely aware of the human presence. They adapted to this presence much like they would to a dominant bear. In the course of the study, it became evident that members of the aggregation recognized the observation site as a place to be avoided, its human observers unchallenged as long as they remained within a restricted observation area, where boundaries were defined by the bears and recognized by the humans.

Spotting scopes and binoculars were used for observing, and most observations were made within a radius of 150 meters (492 ft). Identifications of recognizable individuals were made within this distance and, occasionally, at distances up to 0.5 kilometer (0.3 mi). Normally, two observers censused and recorded behavior. Observers changed over the 12-year period of study, but at least one experienced observer was always present, and the senior author closely supervised and participated in the census throughout the entire research period. Use of standardized formsheets greatly facilitated uniformity in data-gathering. Data involving behavior, reproduction, and survivorship were routinely summarized for each adult female. These data enabled us to construct family genealogies (Fig. 3.2).

Familiarity with individual animals and family units increased with time and the accumulation of

Table 3.1. Observation hours during and after the grizzly bear breeding season at the Trout Creek ecocenter, Yellowstone National Park, 1959-70.

| | Before 15 July (breeding season) | | | | | 15 July and after (postbreeding) | | | | |
| | Number of censuses | | Number of observation hours | | | Number of censuses | | Number of observation hours | | |
Year	Long	Short	Long census	Short census	Total	Long	Short	Long census	Short census	Total
1959	22	0	77.0	0.0	77.0	18	1	63.0	2.0	65.0
1960	11	6	38.5	12.0	50.5	17	1	59.5	2.0	61.5
1961	20	5	70.0	10.0	80.0	6	0	21.0	0.0	21.0
1962	19	3	66.5	6.0	72.5	7	3	24.5	6.0	30.5
1963	24	0	84.0	0.0	84.0	1	1	3.5	2.0	5.5
1964	28	6	98.0	12.0	110.0	11	5	38.5	10.0	48.5
1965	25	1	87.5	2.0	89.5	3	1	10.5	2.0	12.5
1966	20	3	70.0	6.0	76.0	9	0	31.5	0.0	31.5
1967	31	0	108.5	0.0	108.5	4	0	14.0	0.0	14.0
1968	24	5	84.0	10.0	94.0	19	10	66.5	20.0	86.5
1969	19	0	66.5	0.0	66.5	15	0	52.5	0.0	52.5
1970	8	2	28.0	4.0	32.0	19	1	66.5	2.0	68.5
Total	251	31	878.5	62.0	940.5	129	23	451.5	46.0	497.5

Note: Long censuses = 3.5 observation hours; short censuses = 2.0 observation hours.

Team members had to work swiftly and efficiently to complete a wide variety of tasks before the immobilizing drugs wore off and the bear regained mobility. *(Photo by John J. Craighead)*

Fig. 3.2. Genealogy of grizzly bear family No. 15, Yellowstone National Park, 1959-70.

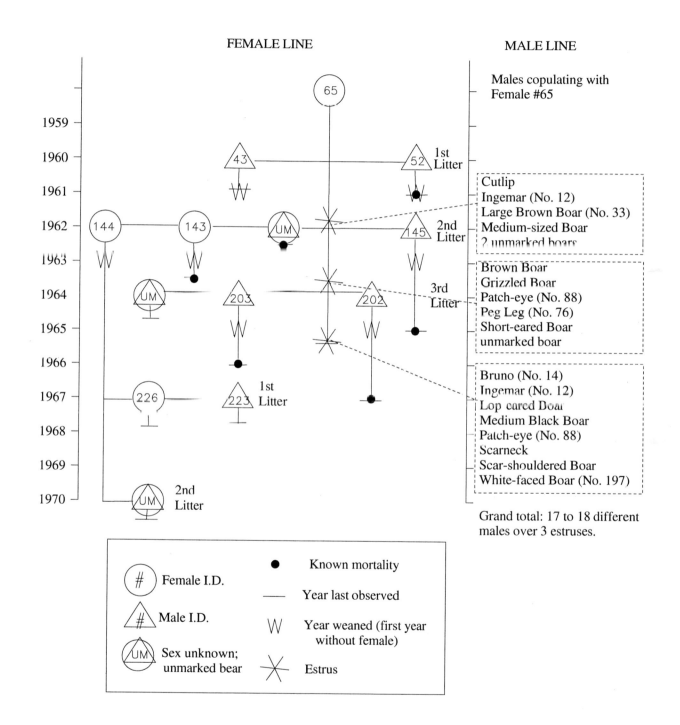

Each immobilized bear was weighed and measured to obtain data on growth rates and weight losses during the period of winter sleep.
(Photo by Frank C. Craighead)

data. This enabled us to look for specific animals and specific behaviors each year and to detect any behavioral changes. Familiarity also provided the basis for a list, each year, of the color-marked and recognizable animals that we expected to observe at Trout Creek during the course of the next field season. This information was annually incorporated into a formsheet for recording the arrival, departure, and visitation of specific members of the population.

DETERMINATION OF SOCIAL HIERARCHY

Early in the study, it became evident that a social hierarchy existed (Hornocker 1962), particularly among the bears aggregated at Trout Creek. To describe the hierarchy quantitatively, and to record changes in the hierarchical structure over a period of years, we recorded aggressive encounters between adult males, between adult males and adult females, and between adult females. The population was so large, closely aggregated, and mobile that attempts to rank all individuals linearly proved

difficult. Thus, we characterized the dominance rank of individuals in terms of a few broad categories of status, recognizing that the hierarchy was probably composed of finer divisions of individual ranks.

Major effort was made to define the hierarchical structure of adult males, primarily by observing aggressive interactions at Trout Creek during each of the 434 observation periods completed from 1959 to 1970. We focused on quantifying overt aggressive behavior such as fights, charges, lunges, and, finally, threats in which one animal of a pair distinctly dominated and the other retreated or reciprocated with clearly submissive signals. The large numbers of grizzlies frequenting the disposal area each evening, sometimes numbering close to 90 animals in a single evening, precluded quantitative documentation of more subtle behaviors such as ear movements, body posture, head alignment, and vocalizations, yet each of these behaviors alone communicates a threat between bears. Most observations of aggressive interactions involving marked and recognizable grizzlies were recorded during 1438 observation hours at Trout Creek (Table 3.1). Many other interactions were recorded at other times and at other sites when other study objectives were being pursued.

The aggressive encounters used to establish dominance relations between individual bears were of two types.

1. **Attack behavior**—encounters ending in actual physical contact, including fighting, charging, and lunging. *Fights* or battles were attack behavior in which physical contact was maintained for periods lasting from a few minutes up to 45 minutes. Prolonged contact was characterized and maintained by jaw locking and body contact with teeth and forepaws. Fights originated at a distance of a few meters to as much as 300 meters. The latter occurred when a male recognized a rival at a distance and then moved purposefully toward him until contact occurred. *Charges* were characterized by a distinct and rapid movement of a grizzly (usually the more dominant) toward another bear and terminated with physical contact that lasted less than 30 seconds. Charges usually occurred over a distance of 30 meters (100 ft) or less. *Lunges* were similar to charges, but occurred as a single, explosive movement over a much shorter distance, and no physical contact was maintained.

2. **Threat behavior**—aggressive encounters that did not end in physical contact, but where animal movement occurred. These included close-contact threats (terminating

Table 3.2. Examples of aggressive behaviors of grizzly bears recorded at the Trout Creek ecocenter, Yellowstone National Park, 1968.

Date	Designation of bear	Sex of bear	Type of encounter	Encounter behavior
6-7	Mohawk	Male	Attack	Won fight with unmarked dark brown bear.
6-15	No. 166	Male	Attack	Fought with No. 190, not clearly dominant.
	Mohawk	Male	Threat	Pursued No. 166 away from unmarked female.
6-16	No. 166	Male	Threat	Pursued female No. 167 away from dump.
6-17	No. 199	Male	Attack	Attacked No. 211, subadult male.
	Mohawk	Male	Threat	Pursued No. 166 away from female No. 109.
			Threat	Female No. 128 and yearling moved away when Mohawk entered the dump.
	Little Brown Boar	Male	Threat	Pursued Mohawk for 20 yards.
6-18	No. 166	Male	Threat	Pursued female No. 167.
	Little Brown Boar	Male	Threat	Pursued female No. 10.
			Threat	Pursued female No. 108.
			Threat	Female No. 40 moved away when he approached.
			Threat	Pursued Mohawk away from area.
			Threat	Pursued the aggregation.
6-19	No. 108	Female	Attack	Attacked dark unmarked adult male.
6-20	No. 128	Female	Attack	Attacked 7-year-old male (No. 166).
			Attack	Attacked 7-year-old male (No. 166).
6-21	No. 40	Female	Threat	Pursued young unmarked bear from dump.
6-23	No. 10	Female	Threat	Pursued sow No. 40.
			Threat	Pursued unknown aged grizzly.
			Threat	Pursued unknown aged grizzly.
6-24	No. 173	Female	Threat	Pursued 7-year-old male (No. 135).
			Threat	Pursued 3-year-old male (No. 213).

when one bear moved away or backed down), bluff charges ending in non-contact, bluff face-offs where one individual moved or stepped aside, and pursuits over longer distances. *Close-contact threats* were made when bears were within a few meters of one another: the bear, growling, ears laid back and head thrust forward, would make slow, determined movements toward an opponent. *Bluff charges* generally occurred over short distances, often terminating after only a few rapid steps. But they could be as long as 30 meters (100 ft), and were commonly accompanied by loud vocalizations. *Pursuits* were movements in which the aggressor chased the submissive animal over distances of 30 meters (100 ft) or more, but made no physical contact. *Face-offs* occurred when two bears turned and faced one another at close quarters. Face-offs usually ended when one animal conceded or both resumed their previous activity, usually feeding.

Thus, measures of dominance were based on the outcomes of fights and on displays of physical superiority and/or submissive behavior. These encounters (attacks and threats) were recorded on formsheets during each observation period and

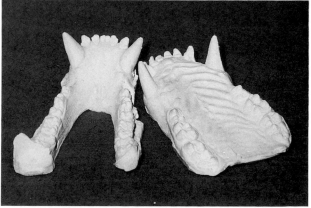

Using techniques of modern dentistry, dental impressions were taken and from these dental casts were made. *(Photos by John J. Craighead)*

Dental casts provided the investigators opportunity to study the grizzly bear's dentition characteristics at leisure in the lab. *(Photo by John J. Craighead)*

periodically summarized (Table 3.2). Additional details of the ranking methodology are presented in the chapter on population hierarchy (Chapter 6).

RECORDING AND ANALYZING REPRODUCTIVE PARAMETERS

The rate at which a wild population reproduces itself is arguably the most critical of the biological parameters that a researcher must determine. It is also the most difficult to measure. Many field biologists have experienced the challenge of maintaining consistent observations throughout the lives of individual animals that range over extensive areas, not to mention the uncertainty of projecting estimates based on often discontinuous data sets. Thus, nearly all assessments of population reproductive performance have been expressed, of necessity, in terms of the most readily obtainable data rather than the most appropriate data. For grizzly bears, the most available information is the number of females with cubs each year. That is, in the absence of other data, such as cycle length and extra-cyclic years,

essential to estimating population growth rate, only the number of females with cubs has been used to express this vital parameter.

Before describing the methods used here to derive estimates of reproduction from field data, we review methods used in the past by ourselves and others. We then explore alternative approaches to data analysis and, by critiquing their validity, arrive at the methodology we have employed.

Reproductive Cycles

The rate at which individual female bears produce offspring has been calculated most often by dividing the duration of a female's reproductive cycle (in years) into the number of cubs (litter size) she produces during that cycle (Craighead *et al.* 1969, 1974, 1976). However, the events in a female's life used to demarcate the span of a reproductive cycle have depended on the type of data obtainable by the researcher and its accuracy. Thus, results have varied from study to study, precluding the possibility of comparing this parameter among different populations of bears. Clearly a standard is needed.

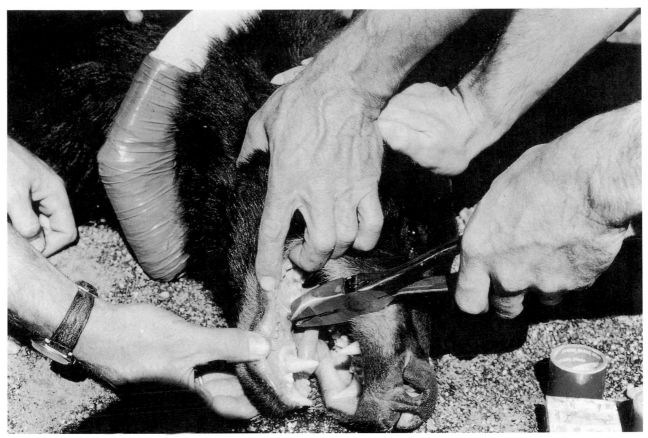

To establish the age of adult bears, one of the fourth premolars was removed, thin-sectioned, and stained to reveal annuli. Because the annuli occur as the result of seasonal variation in cementum deposition rates, each annulus represents a year in the bear's life. *(Photo by John J. Craighead)*

Cycle Length Determinants

In earlier publications of our research on the Yellowstone grizzly bear population (Craighead *et al.* 1969, 1974, 1976), we used the *pregnancy-to-pregnancy cycle*. We defined this as the period beginning with pregnancy in May to mid-July of a year and ending with weaning of offspring and subsequent pregnancy, normally during April or May two to four years later. The "pregnant" status of the female was assigned through observation of successful breeding and confirmed by observation of her litter the following spring. The duration of a female's pregnancy-to-pregnancy cycles, in sum throughout her life, was then divided by the number of reproductive cycles she completed, to generate a value for mean duration of her cycles. Combining such information from all adult females in our Trout Creek sample, we calculated a mean reproductive cycle length (total of all cycle durations, in years, divided by the total number of cycles) for the population.

Most investigations have not had, and will not normally have, the ideal observing conditions found

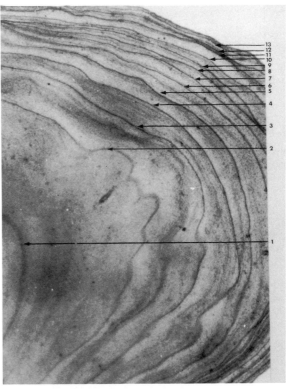

A prepared tooth section reveals annual growth rings (annuli) similar to the growth rings found in woody plants. A count of the annuli provided accurate estimates of age. *(Photo by John J. Craighead)*

Paw impressions were secured from male and female bears and used to examine age and growth characteristics. *(Photo by Frank C. Craighead, Jr.)*

in Yellowstone during the 1959-70 period. The grizzly bear populations under study in other times and places generally have not been centralized at ecocenters where individual bears could be observed continually, year after year. Furthermore, researchers have been unable to observe a large sample of females daily to determine when they become estrous. Lacking data for determining pregnancy-to-pregnancy duration, researchers have used various other events in a female's observed life history to define the cycle. Thus, no standardized, universally accepted method has emerged.

In annual reports of the Interagency Grizzly Bear Study Team, Knight *et al.* (1981-83, 1985-90) used litter-to-litter intervals to define individual cycle lengths and to calculate a mean cycle length for the population. Reynolds and his coworkers used various life history events to define reproductive intervals among Alaskan grizzly bears. The time between breeding and subsequent weaning of the litter was used to determine length of cycle early in the study of the Alaskan North Central population (Reynolds 1980). Reynolds emphasized that this method yielded a *minimum* cycle length because a female may not conceive immediately after weaning a lit-

ter, or she may lose an entire litter before it is observed. He could not always observe both estrous condition of the female (breeding) and the whelping of cubs. Subsequently, Reynolds and Hechtel (1986) redefined the reproductive cycle for the North Central females as the period between weaning of one litter and the successful rearing to weaning of the next (weaning-to-weaning).

Miller (1987), in his study of the brown bear population of Alaska's Susitna River drainage, measured the interval from production of a cub litter (observation following emergence from the den) to the successful weaning of the litter. If weaning was not observed, the cycle carried through to the next litter. Finally, in contrast to his earlier methodology Reynolds (1980), Reynolds (1984) defined the reproductive cycle for females of the Brooks Range population as the period between successive cub litters—that is, the litter-to-litter method.

We re-emphasize, at this point, that the absolute lengths (duration) of a female's cycles throughout her life, taken as a mean, constitute an individualized parameter valuable for comparison with others of her population. Moreover, the population mean cycle length (the mean of the individualized parameters) is a very important comparator among different populations, quite apart from its common role in calculations of reproductive rates.

In this monograph, we use the period from one litter to the next as the measure of a female's reproductive cycle. The birth of a litter marks the initial year of a cycle. The sequential recognitions of a marked female with a litter are reliable life events for documenting cycle length. As discussed below, however, certain assumptions based on additional observed data are necessary for a realistic estimate of the minimum mean cycle length for certain females and, cumulatively, for their population.

Minimum Mean Cycle Lengths

In general, different methods will generate the same cycle lengths when applied to a specific individual, as long as the same type of life event is used to initiate and terminate an interval (e.g., litter-to-litter, pregnancy-to-pregnancy, etc.). On the other hand, use of dissimilar life events, such as estrus as the initiator and weaning of a litter as the terminator, will frequently omit years for females that do not become pregnant in the year that a litter is weaned; the same is true of the litter-to-weaning method. All of the methods also waste useable data in instances of females that are not observed to terminate a cycle, regardless of the event preselected as terminator. In other words, when a cycle is not terminated (is left "open"), observed years since the end of the last cycle cannot be considered in determining mean cycle lengths. For some females, this "open" cycle period is quite lengthy.

In our earlier published presentation of reproductive rates among marked females of the Trout Creek subpopulation, we addressed the problem of "lost" but relevant observation data (Craighead *et al.* 1969, 1976). This was done by assuming that females having "open" cycles, as of the year in which they were last observed, would become pregnant in the next year (Fig. 3.3). This practice produces much more realistic mean cycle length values for individuals and, cumulatively, for the total female sample under study. The mean cycle length values derived in this way are minimum values, however, since it is improbable that all of the females assumed pregnant actually became pregnant in the year following their last observation. Lengthy periods of barrenness were observed among (usually) older females.

Nevertheless, the minimum mean values represent more accurately the true cycling traits of individuals and the whelping rate (litters per year of observed adult female life) for the population than do mean cycle lengths based only on "closed"

Each marked animal was given permanent lip and underarm tatoos. These permanent marks permitted identification if color and metal ear tags were lost. They proved especially valuable in collecting death statistics.

(Photos by John J. Craighead)

Fig. 3.3. Summary of life histories of 15 female grizzly bears, Trout Creek ecocenter, Yellowstone National Park, 1959-72; "age (yrs)" is the age at which female was first observed.

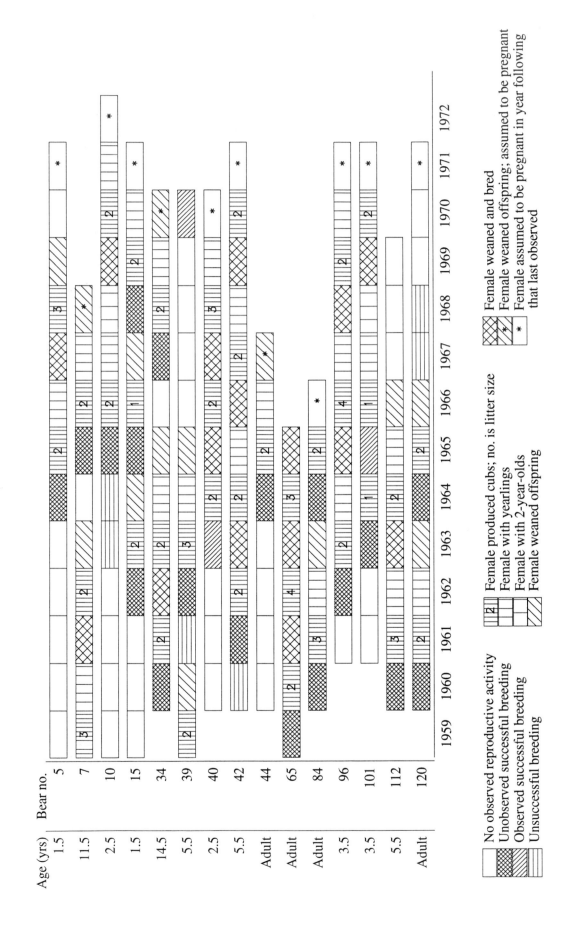

cycles—ones that have been completed. The distinction between the two approaches becomes even more relevant when reproductive rates are calculated on the basis of total cubs produced (by an individual female or by all females), divided by the total cycle-years (for an individual or for all).

Reproductive Rates Based on Cycles

Regardless of how cycle durations are derived, researchers commonly express reproductive rate as cubs-produced-per-cycle-year. Craighead *et al.* (1969, 1976) presented observed data and assumed data separately; the latter included all data relevant to calculating reproductive rate. A comparison of the data sets indicated the importance of including incomplete cycle data by assuming cycle completion at a specified time. Other researchers, however, maintaining a rigid cycle length definition, omitted incomplete cycles (Knight *et al.* 1981-82, 1985-90; Knight and Blanchard 1983; Miller 1987) Such omissions are illogical. When incomplete cycles are excluded, observed litters of cubs associated with the incomplete years of the cycle cannot be credited to an individual's productivity, her reproductive performance is misrepresented, and the reproductive rate for the population is inevitably in error.

For example, using only completed pregnancies to define reproductive periods among a sample of 19 Trout Creek females, we calculated a mean reproductive rate of 0.78 cub per cycle-year (Craighead *et al.* 1976). But when extra-cyclic years were included by assuming pregnancies in the year after some of 30 females (including the original 19) were last observed, the reproductive rate declined to 0.69 (Table 3.3).

A second extra-cyclic period of time commonly observed among adult females is the prepregnancy period (Craighead *et al.* 1969, 1976). We defined this as the years that an individual was estrous and observed to breed before the year of her first successful pregnancy. The years during which pregnancy could have occurred but did not are obviously important in any evaluation of an individual's productivity over her reproductive life. Once again, most researchers ignore these years when calculating a female's reproductive rate on the basis of cycle-years, whether they be litter-to-litter or pregnancy-to-pregnancy.

Exclusion of prepregnancy years leads to an overestimate of reproductive rate for some individuals and for the sample population as a whole. For example, when 13 years of observed prepregnancy were added to the 218 cycle-years recorded for 30 adult Trout Creek females (Craighead *et al.* 1969, 1976; see also Table 3.3), the result was a 4%

Blood serum samples were collected and serological tests made in an effort to understand the physiology of fat metabolism. The entire archive of blood samples was lost to fire before the advent of modern genetic DNA analysis. The samples would have proved invaluable in the unraveling of grizzly bear male parentage.
(Photo by John J. Craighead)

decrease in the estimated mean reproductive rate (Table 3.4). Because the nonproductive, extra-cyclic years of adult females are unquestionably relevant to a population's rate of reproduction, we concluded that the estimate incorporating prepregnancy years was the most accurate.

An important point must be emphasized here. Using reproductive cycles to calculate reproductive rates overestimates the reproductive rate for the population, regardless of adjustments for extra-cyclic years. Females must be productive to be included in the group evaluated for reproductive performance. Thus, adult females that are barren over all years of their observation are not even considered. It is true that productive females must be the source of data on important reproductive parameters such as litter sizes and cycle lengths. However, the reproductively mature, but barren females must be included in any assessment of a population's reproductive performance.

Marking methods were developed and used to identify individual grizzly bears. Over a 12-year period, a wide range of observational data were taken on 389 identifiable bears. *(Photo by John J. Craighead)*

Productivity Rates Based on Term of Observation

For evaluating the reproductive performance of all adult females throughout all observed years of reproductive life, we developed a methodology based on numbers of offspring produced during the years a female is observed as an adult. Thereby, barren-female years are included, as are the extra-cyclic years of productive females. To differentiate this way of expressing reproduction from the classic, cycle-based reproductive rate, we have termed it a "productivity rate."

For an individual female, the productivity rate expresses the number of cubs she produces per adult-year observed. By extension, the mean population productivity rate represents the number of cubs produced per year observed, per adult female in the population (total cubs produced by all adult females, divided by total adult-years observed for all females). We believe this latter rate is the most accurate composite expression of the reproductive performance for the females of a population.

For example, an individual female observed to produce eight cubs over 10 adult-years would have a productivity rate of 0.80 cub/adult-year, whereas a second female producing no cubs over five years of observation as an adult would have a rate of 0.0. Together, as a "population," their mean productivity rate would be 0.53 cub/year/female (8 cubs divided by 15 adult-years).

Use of the productivity method requires that all adult females be marked, or otherwise individually identifiable from year to year. Furthermore, the age of earliest whelping among females in the population must be known so that "prewhelping" years (years during which a female is reproductively competent prior to actual whelping) and barren years for nonwhelping females can be assigned to known- or established-age females marked as nonadults. For the 1959-70 Yellowstone population, the earliest age of observed whelping was five years.

Finally, researchers must realize that estimations of productivity rates for an individual, and therefore cumulatively for the population, are not accurate during the early years of study (depending on numbers of marked animals). Accuracy improves rapidly as more individuals are observed for progressively longer periods. In other words, no methodology can reliably correct for limitations imposed by short-term data.

Table 3.3. Reproductive rates of 30 marked female grizzly bears calculated on the basis of pregnancy-to-pregnancy cycles, including assumed pregnancy adjustments, Trout Creek ecocenter, Yellowstone National Park, 1959-70.

Bear designation	Age (yrs) when marked	Number of cycles							Reproductive period (yrs)	Number of cubs	Reproductive rate[a]
		2-yr	3-yr	4-yr	5-yr	6-yr	7-yr	Total			
No. 5	1	0	1	1	0	0	0	2	7	5	0.714
No. 7[b]	11	0	2	1	0	0	0	3	10	7	0.700
No. 10	2	0	1	1	0	0	0	2	7	4	0.571
No. 15	1	0	3	0	0	0	0	3	9	5	0.556
No. 34	14	1	1	0	1	0	0	3	10	6	0.600
No. 39	6	0	1	1	1	0	0	3	12	7	0.583
No. 40	2	2	1	0	0	0	0	3	7	7	1.000
No. 42	5	2	2	0	0	0	0	4	10	8	0.800
No. 44	Adult	0	0	0	0	0	1	1	7	2	0.286
No. 65	Adult	3	0	0	0	0	0	3	6	9	1.500
No. 84	Adult	1	0	1	0	0	0	2	6	5	0.833
No. 96	3	0	3	0	0	0	0	3	9	8	0.889
No. 101	3	2	0	1	0	0	0	3	8	4	0.500
No. 112	5	0	1	0	0	1	0	2	9	5	0.556
No. 120	Adult	0	0	1	0	0	1	2	11	4	0.364
No. 125	Adult	0	4	0	0	0	0	4	12	10	0.833
No. 128	11	2	2	0	0	0	0	4	10	13	1.300
No. 140	8	0	0	0	1	0	0	1	5	3	0.600
No. 141	1	0	1	0	0	0	0	1	3	2	0.667
No. 144	1	1	0	1	0	0	0	2	6	4	0.667
No. 150	5	0	0	1	1	0	0	2	9	5	0.556
No. 160	Adult	0	0	0	0	1	0	1	6	2	0.333
No. 163	1	1	0	0	0	0	0	1	2	2	1.000
No. 172	13	1	2	0	0	0	0	3	8	7	0.875
No. 173	2	1	1	0	0	0	0	2	5	3	0.600
No. 175	10	1	2	0	0	0	0	3	8	4	0.500
No. 175B	Adult	0	2	0	0	0	0	2	6	4	0.667
No. 180	11	0	1	0	0	0	0	1	3	3	1.000
No. 187	1	0	1	0	0	0	0	1	3	1	0.333
No. 200	3	0	0	1	0	0	0	1	4	2	0.500
Total		18	32	10	4	2	2	68	218	151	

Reproductive rate[a] = 0.693

[a] Number of cubs produced divided by reproductive period.
[b] Female No. 7 produced a total of seven cubs, not eight, as originally reported in Craighead *et al.* (1974).

Table 3.4. Average reproductive rates for 30 marked, adult female grizzly bears at the Trout Creek ecocenter, Yellowstone National Park, 1959-70 (Craighead *et al.* 1974).

Number of cycles[a]							Number of cubs	Unadjusted for prepregnancy		Adjusted for 13 prepregnancy years	
2-yr	3-yr	4-yr	5-yr	6-yr	7-yr	Total		Reproductive period (yrs)	Reproductive rate	Reproductive period (yrs)	Reproductive rate
18	32	10	4	2	2	68	151	218	0.693	231	0.654

[a] Includes cycles defined by assumption of pregnancy.

The radio collar, weighing approximately two pounds, fit snugly about the neck of the bear and seemed to be worn with indifference. No behavioral aberrations were noted and no collar was ever removed by a bear.
(Photo by John J. Craighead)

Methods of Calculating
Population Productivity Rates

We evaluated four different approaches for converting the productivity rates of individual females to a rate representative of the population. For convenience, these will be labeled "Methods I - IV." As exemplary data, we used the histories of 67 marked females observed as adults for at least one year in the Yellowstone Ecosystem from 1959 to 1970. The inclusion of adult females observed for only one year may bias a cumulative population productivity rate, unless their selection for marking in the first place was unrelated to their sex, age, or reproductive status. In our study, selection of bears for marking was a function of opportunity, rather than any individual attribute. We believe there was little trap-bias. Therefore, the ratio of females with cubs versus those without in the annual sample can be assumed to represent that of the ecosystem population in that year.

The methods take different approaches to analyzing (and, in the case of Method IV, subsampling) the single data set (67 marked, adult females). Method I entailed summing the number of adult-years the 67 females were observed and dividing that total into the number of cubs the females were

Table 3.5. Methods I and II for determining productivity rates for 67 marked, adult female grizzly bears, Yellowstone Ecosystem, 1959-70.

Bear designation	Adult-years observed	Number of cubs	Cubs per adult-year observed	Bear designation	Adult-years observed	Number of cubs	Cubs per adult-year observed
No. 5	6	5	0.83	No. 108	10	2	0.20
No. 6	2	0	0.00	No. 109	4	0	0.00
No. 7	10	7	0.70	No. 112	9	5	0.56
No. 8	6	8	1.33	No. 119	1	2	2.00
No. 10	9	4	0.44	No. 120	10	4	0.40
No. 15	7	3	0.43	No. 123	2	2	1.00
No. 16	4	3	0.75	No. 125	9	8	0.89
No. 17	3	1	0.33	No. 128	9	13	1.44
No. 18	3	1	0.33	No. 129	1	0	0.00
No. 21	6	2	0.33	No. 131	1	0	0.00
No. 23	1	0	0.00	No. 132	2	1	0.50
No. 29	2	2	1.00	No. 139	2	2	1.00
No. 31	1	0	0.00	No. 140	6	3	0.50
No. 34	11	6	0.54	No. 141	2	2	1.00
No. 39	11	3	0.27	No. 142	4	0	0.00
No. 40	7	7	1.00	No. 144	4	3	0.75
No. 42	11	8	0.73	No. 148	2	2	1.00
No. 44	7	2	0.29	No. 150	8	5	0.62
No. 45	1	0	0.00	No. 160	5	2	0.40
No. 48	4	5	1.25	No. 163	1	2	2.00
No. 58	1	0	0.00	No. 164	6	3	0.50
No. 59	1	1	1.00	No. 170	1	2	2.00
No. 64	6	0	0.00	No. 172	7	4	0.57
No. 65	6	9	1.50	No. 173	4	3	0.75
No. 72	2	1	0.50	No. 175	7	4	0.57
No. 75	4	2	0.50	No. 175B	6	4	0.67
No. 81	7	1	0.14	No. 176	1	0	0.00
No. 84	4	5	1.25	No. 180	4	3	0.75
No. 96	8	8	1.00	No. 184	2	2	1.00
No. 101	8	4	0.50	No. 187	3	1	0.33
No. 102	1	0	0.00	No. 200	4	2	0.50
No. 104	1	0	0.00	No. 204	3	0	0.00
No. 107	2	0	0.00	No. 228	3	0	0.00
				No. 259	2	3	1.50
Total					308	187	40.34

Method I

$$\frac{\text{Number of cubs}}{\text{Adult-years observed}} = \frac{187}{308} = 0.61$$

Method II

$$\frac{\text{Productivity of each individual female}}{} = \frac{\text{Cubs she produced}}{\text{Adult-years she was observed}}$$

$$\frac{\text{Sum of female productivities}}{\text{Total number of females}} = \frac{40.34}{67} = 0.60$$

Note: "Adult" is five years of age, or older.

Table 3.6. Method III (A and B) for determining productivity rates for 67 marked, adult female grizzly bears, Yellowstone Ecosystem, 1959-70.

Year	Number of females[a]	Number of cubs	Productivity rate
1959	7	3	0.43
1960	19	16	0.84
1961	22	13	0.59
1962	26	18	0.69
1963	30	21	0.70
1964	30	12	0.40
1965	29	24	0.83
1966	28	15	0.54
1967	31	20	0.65
1968	28	21	0.75
1969	29	8	0.28
1970	29	16	0.55
Total	308	187	7.24

Method III-A

$$\frac{\text{Sum of annual rates}}{\text{Number of years}} = \frac{7.24}{12} = 0.60$$

Method III-B

$$\frac{\text{Total number of cubs}}{\text{Total number of female-years}} = \frac{187}{308} = 0.61$$

Note: "Adult" is five years of age, or older.

[a] Individuals frequently were recorded as observed in more than one year; therefore, the unit for summation over all years is female-years.

observed to produce (Table 3.5). In Method II, we summed the 67 individual productivity rates (total cubs produced by a female, divided by the total years she was observed as an adult) and then derived a mean rate (Table 3.5). Method III involved either a) summing the 12 mean annual productivity values (total cubs produced in a year, divided by total marked adult females observed in that year) and taking a grand mean, or b) dividing the total number of observations of females over the 12-year period into the total number of cubs produced during the period (Table 3.6).

Finally, Method IV used a random sampling process in which productivity for each marked adult female was recorded for a single year randomly selected from the span of years over which she was observed. The value for each female's productivity in the sampling trial then ranged from zero to four (cubs). The individual values were summed and divided by the number of sampled females (67) to derive the population productivity rate. This pro-

cess was repeated to generate 13 such values; these were summed, and a grand mean was calculated (Table 3.7). For Method IV to be a valid approach, the mean of the 13 independent trials must be accepted as a fair approximation of the population productivity over the study period. The mean value of trials, up to some level of adequate subsampling, changes in a random way, depending on the number of independent trials. Therefore, what number of trials constitutes adequacy is moot.

Evaluation of Calculation Methods

As the examples show, each of the methods we have described generates similar results (Tables 3.5 and 3.6), or would if pursued far enough (Table 3.7), when the sample population is large (67 females, in this case), and when the individuals have been observed, on the average, for a long period (4.6 years, in this case). But the smaller the sample of females, and the larger the proportion having relatively short periods of observation, the more likely it is that the methods will not produce the same

Table 3.7. Method IV for determining productivity rate for 67 marked, adult female grizzly bears, Yellowstone Ecosystem, 1959-70.

Trial #	Mean productivity rate
1	0.42
2	0.45
3	0.50
4	0.50
5	0.53
6	0.53
7	0.59
8	0.61
9	0.64
10	0.64
11	0.70
12	0.73
13	0.73
Total	7.57

$$\frac{\text{Sum of subsample mean productivity rates}}{\text{Number of subsamples}} = \frac{7.57}{13} = 0.58$$

Note: "Adult" is five years of age, or older.

values for population productivity rate when applied to identical data.

Exactly what number of adult females constitutes an adequate sample of a given population, and for what number of years they must be observed to portray the population's actual productivity, are not easily specified, *a priori*. Moreover, the potential for biasing a population productivity estimate is a function of how well the sample captures the true variance in the reproductive life histories of the females and of how discontinuities in the observation data may be amplified or suppressed by the arithmetic process used to calculate the estimate. For example, adult females observed for relatively few years and producing from zero to four cubs will contribute individual productivity rates to the mean population rate that vary from zero to four (cubs) for one year of observation, from zero to two for two years, and from zero to 1.33 for three. Females average more than 1.5 cubs per year only rarely and, more commonly, average less than one cub per year when four or more years of life history data are available. Thus, the degree to which productivity rates for the females of short-term observation are weighted or discounted becomes important.

The arithmetic of Method I (Table 3.5) tends to damp the effects of short-term females. This is because the total adult-years during which each of these females was observed constitutes a relatively small percentage of the total number of adult-years that all of the females were observed (Table 3.5: 29 of 67 females were observed three or fewer years for a total of 50 years, or 16.2% of the 308 adult-years observed). On the other hand, Method II makes the contribution of the 29 short-term females equivalent to that of the 38 females observed four or more years (Table 3.5: 29 short-term females contribute 43.3% of the 67 individual productivity rates averaged to estimate population productivity). In addition, individual productivity rates of zero are poorly expressed with Method II, since females that have been cubless for several years generate the same zero productivity as those cubless for only one year of observation.

Some methods are more prone to certain types of sampling bias than others. For example, if females tend to be identified more frequently when with offspring than when alone, productivity rates for short-term females will inflate population productivity rates calculated with Methods II (Table 3.5), III-A (Table 3.6), and IV (Table 3.7) more than those calculated with Methods I (Table 3.5) or III-B (Table 3.6). Conversely, bias from a connection between high probability for long-term observation of females and extreme rates of productivity would be more strongly expressed in calculations using Methods I (Table 3.5) and III-B (Table 3.6).

Another important consideration associated with the arithmetic process of calculating population productivity rate is the need to use statistical analysis to compare rates between bear populations. Whether parametric or nonparametric, statistical testing requires a series of values (*n*) for comparison that are measured in units consistent with the series' mean value. Methods II (Table 3.5), III-A (Table 3.6), and IV (Table 3.7) generate series of values appropriate for statistical comparisons. Methods II and IV, however, share the potential for bias due to short-term reproductive histories, and III-A obviously introduces bias associated with variation in census effort and efficiency, from year to year. On the other hand, Methods I (Table 3.5) and III-B (Table 3.6), while not emphasizing term-of-observation and census efficiency biases, produce estimates of population productivity based on series of data inappropriate for between-population statistical comparisons.

Finally, a word or two about Method IV (Table 3.7). Randomly selecting a single year's reproductive performance from the observed adult life

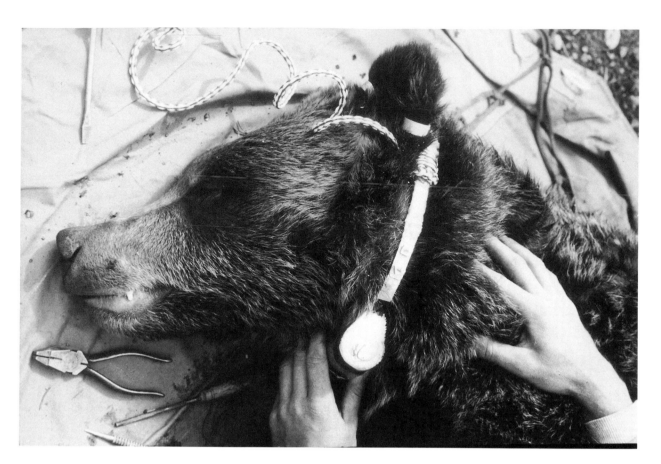

Among innovations for locating and observing this large, powerful wilderness animal, the development of a radio-location system proved the most valuable and is now a standard tool of wildlife research. *(Photos by John J. Craighead)*

The ability to radio-locate specific grizzly bears and to observe them at will proved extremely valuable in documenting pre-"hibernation" behavior, winter denning, movement, and feeding behavior. *(Photos by John J. Craighead)*

history of each adult female to provide the basis for population productivity estimates was an attempt to avoid the bias attributable to varying periods of observation. But further evaluation of this process showed that if enough random trials were performed, their mean—the population productivity by this method—would equal that calculated using Method II (Table 3.5). Thus, the number of trials needed to generate a believable rate became problematic. Would any rate different from the one generated by Method II to treat the same data be acceptable, and on what logical premise? In view of the extra time and effort necessary to using Method IV, and the fact that no advantage is obvious, we do not recommend its use.

The method used in this monograph for calculating mean population productivity rate, and recommended for use by other researchers, is Method I (Table 3.5). In instances where accumulated data must be evaluated for subperiods of the study, we apply Method III-B (Table 3.6). An important feature of both methods is that increases in period of observation translate directly into calculation of population productivity rate. A weighted version of Method II (as explained below) should be used when

comparisons between productivities of two or more populations are necessary.

Regardless of the method, however, it is axiomatic that the longer the history of observation, and the more individual females observed, the more accurate the estimate of population productivity. No method can improve on inadequate data.

Between-Population Comparisons of Productivity Rates

We have shown that Methods I and III-B best express the progressive improvement of the estimate with increased years of observation. We have also noted that statistical analysis techniques cannot be applied to data formatted for either of these methods, because the population productivity rate is not an average of individual productivity rates. To test for significant differences among productivity rates of different populations, we developed and recommend a procedure in which data are formatted for Method II, but the individual productivity rates are weighted by multiplying them by the number of years the individual was observed. Then the sum of the weighted productivities are divided by the total years all females were observed to derive the popu-

By using the radiotracking system, the investigators were able to efficiently locate individual bears in the backcountry and observe them undetected. *(Photos by John J. Craighead)*

lation productivity rate. Arithmetically, this procedure produces a result identical to that produced by applying Method I applied to the same data set, but it allows calculation of individual means and a variance, standard deviation, and 95% confidence interval. In most cases, it is also necessary to restrict the sample of females to those observed for four years or more, because productivity rate values for those observed for fewer years make the distribution of productivity non-normal. The final products of this analytical procedure are population mean productivity values with 95% confidence intervals which, when graphed, display significant difference ($\alpha \leq 0.05$) or none in terms of non-overlap or overlap, respectively, of 95% confidence intervals.

DATA MANAGEMENT AND STATISTICAL ANALYSIS

The 1959-70 field study predated the personal computer revolution, which arrived full-force in the early stages of this monograph's preparation. Thus, substantial time and effort was invested in converting existing tabulations of field data into a computerized (dBase III+) database. All subsequent tabu-

Radiotracking of bears occurred in all seasons. Tracking them in the fall eventually led the investigators to winter dens. Grizzly dens were previously unknown and undescribed. *(Photos by John J. Craighead)*

After pin-pointing the location of radio-collared bears, the information was plotted on maps to delineate home and seasonal ranges. This information was essential in delineating spatial requirements. *(Photo by Frank C. Craighead, Jr.)*

lations, as well as data provided by other researchers, were similarly computerized. The process of conversion, as well as the ease with which the new database could be searched and cross-checked led, inevitably, to the discovery of some errors in previously published data and to some revision of previous classifications of the data. Such changes may contribute, in some cases, to minor discrepancies in values reported in this versus earlier publications.

The majority of the statistical tests reported in the text used the protocols of the Statistical Package for the Social Sciences/Personal Computer (SPSS/PC+, versions 1.0, 2.0, and 3.0; Norusis 1986, 1988), Sokal and Rohlf (1981), or Conover (1980). In cases where analyses demanded statistical techniques that lay outside our comfort zone, the protocols used were those chosen by the contracted consultant (see Acknowledgements).

We made a special effort, in testing for the effect of some factor on a parameter of interest, to control for the effect of other potentially confounding factors. We did this two ways. First, when sample sizes permitted, we restricted the samples to be compared to those for which these other factors did not vary and then used univariate statistical tests. Alternatively, we applied multivariate techniques to the unrestricted sample(s). The former approach had the advantage of greater simplicity in interpretation and presentation, whereas the latter was effective in situations where some categories defined by the several factors together included a relatively small number of cases. When small samples or incomplete information precluded consideration of more than one factor, we proceeded as best we could with univariate tests. Finally, when the data were, for various reasons, not suitable for statistical treatment (e.g., much of the food habits data), we restricted ourselves to qualitative comparisons.

For standard statistical techniques, we have simply identified the test used and a reference, if not one of those mentioned above. For less commonly used techniques, or those of our own invention, we usually have provided a description in the text.

Prototype radio transmitters and loop antennas were field tested. Models actually employed in radiotracking grizzly bears were miniaturized and imbedded in a waterproof, shockproof fiberglass collar. *(Photo by John J. Craighead)*

To track individual bears through several seasons, the radio collars had to be replaced when the batteries expired or electronic malfunctions occurred. *(Photo by John J. Craighead)*

The Trout Creek waste disposal pit was centrally located in Hayden Valley, the largest open grass-shrubland complex in Yellowstone National Park. *(Photo by John J. Craighead)*

Grizzly bears inhabiting the Yellowstone Ecosystem aggregated annually at ecocenters or garbage disposal sites. These bears foraged throughout the entire Ecosystem as well as at the massive food resources deposited at refuse sites. Bears that aggregated at Trout Creek, the largest of these ecocenters, were closely studied. There was little evidence for discrete "backcountry" populations not attracted to the ecocenters. *(Photo by John J. Craighead)*

4
The Grizzly Bear Population
in the Yellowstone Ecosystem

INTRODUCTION

Rarely is it possible to study an animal population in geographic isolation. The Yellowstone Ecosystem, as we define it, has always been somewhat isolated ecologically by topography, but in recent times, its insularity has steadily increased as a consequence of human activities in surrounding areas. The grizzly bears of Yellowstone are so isolated from other bear populations by geographic barriers that, genetically and demographically, they can be considered a discrete population. Before 1970, this ecosystem population consisted of subunits that, in turn, were rather discrete, isolated not by geographic barriers, but by massive sources of food that attracted and held the bears within localized areas for much of each year.

In the following chapters we analyze data from two groups of grizzly bears: the Yellowstone Ecosystem-wide population and the Trout Creek subpopulation. The latter is a subunit of the ecosystem population that inhabited Yellowstone National Park and the adjoining National Forests before 1970. Bears that aggregated annually at the Trout Creek ecocenter (dump) are referred to as the Trout Creek bears or Trout Creek subpopulation. All or most of the remaining bears that inhabited the Ecosystem aggregated annually in one or more seasons at three other primary ecocenters—Rabbit Creek, West Yellowstone, and Gardiner—and, for a few years of the study, at four secondary ecocenters (Fig. 3.1) closed from 1960 to 1963. These subunits and the Trout Creek subunit constituted the ecosystem bear population. This population of largely ecocentered bears used the entire Ecosystem. We could distinguish no discrete "backcountry" population units that were not attracted to the dumps, or that existed independent of the aggregations.

We later discuss some movement and exchange of bears among ecocenters. Exchange was relatively minor, however, and did not alter the central fact that the subunits were discrete, held together by the massive, localized food resources and traditional movements to them. The approximate mean size of the primary ecocentered populations varied from 16 at the West Yellowstone dump to 116 at the Trout Creek dump (Fig. 3.1).

The size and structure of the Yellowstone Ecosystem population were described by Craighead *et al.* (1974), and habitat critical to the population was delineated in the late 1970s (Craighead 1980b). There is confusion among scientists and environmentalists as to what the term "ecosystem" means and how to apply it when delineating the habitat of specific wildlife populations. "Ecosystem" has been applied to natural systems varying greatly in structure and scope. It has been used to define biotic systems as small as microhabitats and as large as Yellowstone National Park and parts of the adjoining national forests.

The term "ecosystem" was first employed by Tansley (1935) to conceptualize the functional interplay of vegetation, animal species, and the abiotic resources supporting them. Two quite different views of how species are organized into interactive community assemblages had emerged earlier in the century. The "super-organism" view envisioned communities as comprised of discrete assemblages of tightly coevolved and highly interdependent species (Clements 1916). (We ignore extreme formulations of this view in which communities were held really to be organisms and focus instead on the idea of tight coevolution.) The "individualistic" view saw overlapping but often independently derived environmental needs and tolerances as the primary cause for predictable assemblages of species (Gleason 1926).

Subsequent work has validated both views to some extent. Mutualism, symbiosis, and other tightly coevolved interactions are important features of all communities (Begon *et al.* 1986). Nevertheless, the highly diffuse spatial boundaries documented for most communities, coupled with frequent temporal changes in species composition (random, cyclical, and directional), support the predominance of the individualistic model of community and ecosystem de-

Repeated censuses of color-marked bears at four primary ecocenters provided data on population size and structure. *(Photos by John J. Craighead)*

velopment (Begon *et al.* 1986). Current ecosystem concepts differ from these earlier ecological community perspectives mainly in their somewhat greater emphasis on the dynamic interplay of abiotic and biotic components. In fact, some recent authors have declined to recognize ecosystems and communities as distinct levels of biological organization (Begon *et al.* 1986), emphasizing instead the rules governing the organization of multispecies assemblages. In general, the contemporary notions of communities and ecosystems are relatively more abstract and difficult to define in a universal way. This inherent difficulty, together with the great variety of attempts at definition (O'Neill *et al.* 1986), make it important for authors to specify their meaning for these terms in particular applications.

Our definitions of ecosystem and community are influenced not only by the results of theoretical and basic research in ecology, but also by our conservation perspective and our use of remote-sensing techniques to inventory habitat over large geographic areas (Craighead *et al.* 1982, 1988a) and to assess its use by large mammals (Craighead and Craighead 1987, D. J. Craighead 1995 [in prep.]). We define the Yellowstone Ecosystem as a mosaic of plant community types that recurs more or less predictably on the Yellowstone Plateau and associated mountain ranges. We define the component plant communities as assemblages of species joined by statistical association in space and time. Thus, we speak of the Yellowstone Ecosystem as an area having similar vegetation-structure without delineating precise boundaries.

Our definitions are plant-oriented. By focusing on plants, we do not deny the existence of association patterns in plant and animal distributions. Rather, we are responding to: (1) the practical fact that the ecological and hierarchical classification of plant communities is far more advanced than that of plants and animals together (Daubenmire and Daubenmire 1968, Corliss *et al.* 1973, Mueggler and Handl 1974, Pfister *et al.* 1977); (2) the relative ease with which remote-sensing data, with computer analysis, can be used to determine plant versus animal distributions; and (3) the fact that vegetation is a large and readily identifiable component in the grizzly bear diet.

Our conservation and remote-sensing perspectives, and consequent interest in large biogeographical areas, force us to consider a range of scale not easily accommodated by a single concept. Hence, we recognize the ecosystem as a level of organization beyond that of community. Finally, our conservation interest leads us to a definition of ecosystem that identifies nonarbitrary, if diffuse, bound-

aries. Eventually, such boundaries can be delineated more definitively on the basis of vegetation ecotones or gradients separating the Yellowstone Ecosystem from adjacent ones. On occasion, we use the term grizzly bear ecosystem to refer to the biogeographic area known to have been inhabited by grizzly bears during this study; it is essentially identical in scope to the borders delineated for the Yellowstone Ecosystem, an area of approximately two million hectares (five million acres) (Craighead 1980b).

As indicated earlier, the major focus of this monograph is to describe the ecology of an ecocentered population of grizzly bears—the Trout Creek subpopulation. But to do so, we must make frequent reference to the larger ecosystem population and, when appropriate, compare biological parameters. In this chapter, we briefly describe the major biological parameters of the ecosystem population—age structure, age-specific sex ratios, age- and sex-specific survivorship, reproductive rate, and estimates of population size. This brief presentation should allow the reader to place the Trout Creek subpopulation in perspective with the larger ecosystem population. This is important, because conclusions derived from the analysis of data obtained for the Trout Creek bears will eventually be applied to achieve a better understanding of the ecosystem population and the changes it has undergone over more than 30 years.

POPULATION CHARACTERISTICS

Population Size and Spatial Structure

Using repeated census of the four primary ecocenters, we tallied the total number of *different* (color-marked or identifiable through natural markings) bears using these ecocenters annually. These counts were therefore cumulative totals of known unique bears for the year, rather than mean numbers of bears observed over all censuses. Applying a census efficiency of 77.3% (calculated by using the percentage of marked bears among those found dead outside of the Park and recorded as marked or unmarked ($n = 103$) to estimate the number of bears not counted in censuses at ecocenters [see Craighead *et al.* 1974]), we estimated an average total ecosystem population of 229 animals from 1959 to 1970. We also considered the possibility that all mortalities outside of the Park not reported as marked or unmarked ($n = 40$) were, in fact, unmarked. This would imply a census efficiency of 55.7% and an average population of 318. Using our data and somewhat different methods, Cowan *et al.* (1974) and McCullough (1981) estimated census efficiencies of 58.0% and 56.7%, respectively,

Fig. 4.1. Occupied grizzly bear habitat, 1959-70 (Craighead 1980b).

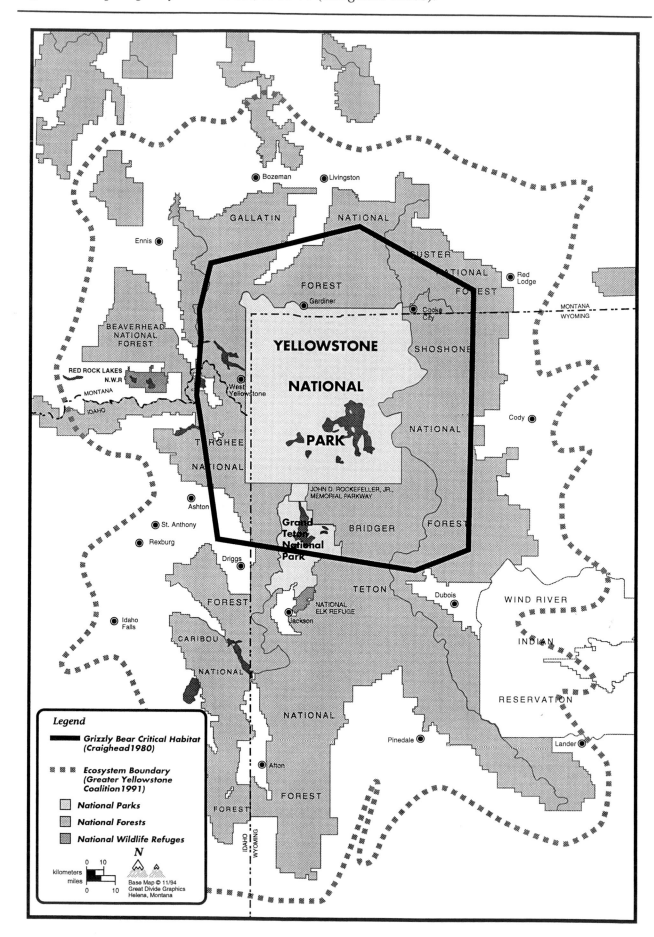

and average total populations of 309 and 312 bears. The three estimates are quite close, and we will accept the median value of this group (312 bears) as the best existing estimate of average population during the period, 1959-70.

All of these estimates were based on the observation that the percentage of marked bears in samples of deaths from outside the Park were lower than expected, based on the fraction of marked bears among samples of living bears or of dead bears within the Park. Both Cole (1973b) and McCullough (1981) took this observation, and the consequent calculation of a less than 100% census efficiency, as evidence of a substantial, discrete backcountry population—i.e., a population of "wild" bears that never or rarely visited ecocenters.

We believe that a more plausible explanation is that most uncounted bears were members of aggregations at the smaller and less regularly censused ecocenters. Recall that our census counts were cumulative totals of the number of different bears seen in a given year. These counts were therefore sensitive to censusing frequency. At the smaller and less frequently censused ecocenters, even frequent visitors may have been absent from all of a small number of censuses. For example, the likelihood that a bear with a relatively high visitation level at Rabbit Creek equal to 0.30 would be absent from all of five censuses is roughly one in six (0.70^5). Thus, census counts would surely have been significantly higher if Rabbit Creek and all other ecocenters had been studied as intensively as Trout Creek.

The reduced frequency of marked bears among mortalities outside the Park is entirely consistent with this view. Most marked bears were members of the centrally located Trout Creek aggregation. Thus, the relative absence of marked bears dying outside the Park may have reflected the spatial structuring imposed on the ecosystem population by traditional attachment to one of several ecocenters inside the Park. Many unmarked bears that died outside the Park were probably members of one of the several peripherally located aggregations with fewer marked members. Thus, we reaffirm our earlier view that backcountry populations discrete from ecocenter populations did not exist during 1959-70. Spatial structuring was the result of the tendency of most ecosystem bears to localize movements from 1 June through 30 August at and around one of several ecocenters dispersed throughout the Ecosystem (Fig. 3.1).

Population Densities

During the period 1959-70, the Yellowstone grizzly bear ecosystem was calculated to be approximately two million hectares (five million acres), based on distribution of bear mortalities and sight records, and on habitat and land-use inventories (Craighead *et al.* 1974, Craighead 1980b) (Fig. 4.1). If a mean population of 312 bears inhabited the Yellowstone Ecosystem from 1959 through 1970, then there would have been a mean density of approximately one grizzly bear per 64.8 square kilometers (25 mi²). The distribution of the bears was such that, during the summer months, high densities existed at and around the ecocenters, with extremely low densities in the backcountry and intervening habitat (Craighead 1980b). After 1968, high bear mortality and the movement of bears in search of food and secure habitats beyond the ecosystem borders (Craighead 1980b) resulted in a large, but unknown decline in the density of bears throughout a more extensive area. Movement was indicative of the expansion of home ranges (see Chapter 15; see also, Knight *et al* 1984, Blanchard and Knight 1991). This movement toward the periphery of the Ecosystem represented greater use of habitat largely unused during the preclosure period (see Chapter 18). Whether this increased use of relatively sparsely used habitat provided greater nutritional stability and security to the grizzly bear population will be explored in Chapters 15 and 16.

Age Structure—Life

The age or age class of individuals at ecocenters was recorded during each of 367 3.5-hour censuses made at three locations (the Trout Creek, Rabbit Creek, and West Yellowstone ecocenters) from 1959 through 1967. Nonadult animals captured and color-marked were of known- or established-age (cubs and yearlings) and could be recognized as they aged from year to year. Thus, changes in the age structure of the ecosystem population could also be documented from year to year (Fig. 4.2). The mean age class composition, determined annually from counts of recognizable individuals, was 18.6% cubs, 13.0% yearlings, 10.2% 2-year-olds, 14.7% 3- and 4-year-olds, and 43.7% adults. A further delineation of the adult age class shown in Fig. 4.2 was obtained by capturing and aging 52 adults (27 males and 25 females). Age was determined by extracting a fourth premolar from each captured adult, sectioning the tooth, and counting the cementum layers (Craighead *et al.* 1970 and Chapter 3). The sample of 52 aged adults was increased to 60 (32 males and 28 females) by including eight animals that were known to be adult members of the population by 1966, although they had been captured and recorded as nonadults. These adult sex-age data were used to refine the observed sex-age structure

Fig. 4.2. Observed age structure of the grizzly bear population in the Yellowstone Ecosystem, 1959-67, calculated on the basis of mean individuals observed annually (Craighead *et al.* 1974).

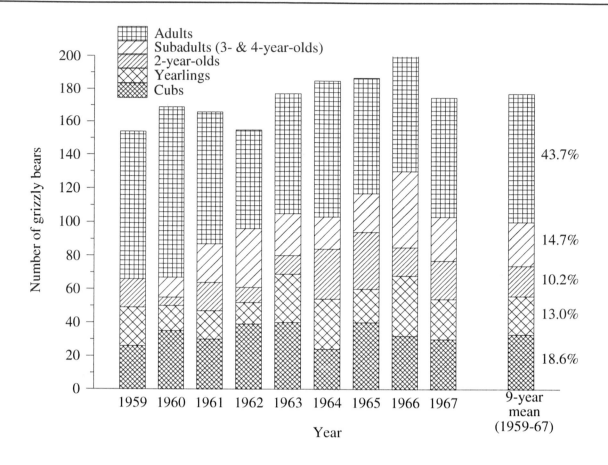

for adults. The adult sex-age structure and the sex-age structure for bears aged 0.5 to 5.5 years were then applied to the average census count of 177 animals (Fig. 4.2) to construct an age- and sex-specific life table (Table 4.1; see also Craighead *et al.* 1974). Mortality from all human-related causes was proportional to the size of the various age classes and hence, not strongly age-dependent. The data showed a relatively high juvenile mortality (cubs to four years), low and stable adult mortality to about 15 years of age, and an increasing rate of death to a maximum age of about 25 years.

Age Structure—Death

The age structure of a population can be developed from death statistics, as well as life statistics, providing deaths are not biased by differential age-specific mortality. Because deaths in the Yellowstone grizzly bear population occurred incrementally over the 12 years of field study, it was not until after 1970 that a mortality-based age structure could be compiled. We compared the age structure based on 12 years of death statistics with one based on nine years

of data from observation of the living population (Fig. 4.3). That the two ecosystem-age structures matched one another so closely is strong evidence of random, human-caused mortality among age classes. This is discussed in detail in Chapter 18.

Sex Ratios

From 1959 through 1971, nonadult grizzlies were classified as cub, yearling, 2-, 3-, or 4-year-old and by sex (Table 2 of Craighead *et al.* 1974). The sex ratios for adults were determined from observations of both marked and unmarked animals from 1964-70 (Table 3 of Craighead *et al.* 1974). Unmarked adults within binocular or spotting scope range, unlike subadults, could be sexed in the field by observing size, conformation, and reproductive behavior. Among 577 observations of adult grizzlies, most of them recognized individuals, 53.7% were females and 46.3% were males. The cumulative cub sex ratio of 0.59 males to 0.41 females and the sex ratios of all nonadult animals indicate that a differential sex-related mortality was operative among adults (Fig. 4.4). This is further substantiated by a

Fig. 4.3. Comparison of a "Living" age structure for the grizzly bear population with a "Death" age structure determined from ecosystem census data and ecosystem death statistics, Yellowstone Ecosystem, 1959-70.

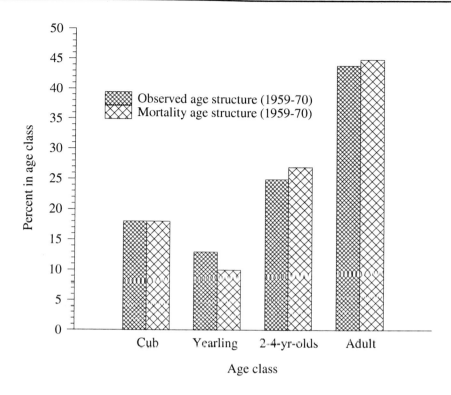

Fig. 4.4. Observed sex-age structure from cumulative observations of the grizzly bear population in the Yellowstone Ecosystem, 1959-71. Sex ratios for cubs, yearlings, 2-year-olds, and subadults (3- and 4-year-olds) are based on capture records; sex ratios for adults are based on observations of marked and unmarked animals (see Tables 2 and 3 in Craighead *et al.* 1974).

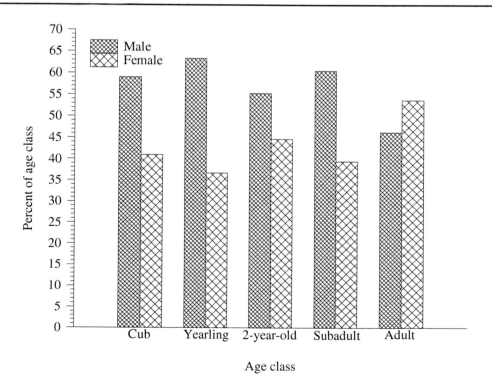

Table 4.1. Age- and sex-specific life table for the Yellowstone grizzly bear population, 1959-67 (Craighead *et al.* 1974).

				MALES		
Age (yrs)	Number in age class	Number of males	Number dying in age class D_x	Number surviving per thousand L_x	Survivorship rate P_x	Mortality rate Q_x
Cub	33.0	19.5	5.0	1000	0.7436	0.2564
1	23.0	14.5	4.6	744	0.6828	0.3172
2	18.0	9.9	1.4	508	0.8586	0.1414
3	14.0	8.5	1.5	436	0.8235	0.1765
4	12.0	7.0	3.4	359	0.5143	0.4857
5	7.7	3.6	0.2	185	0.9444	0.0556
6	7.4	3.4	0.2	174	0.9412	0.0588
7	7.0	3.2	0.1	164	0.9688	0.0313
8	6.8	3.1	0.1	159	0.9677	0.0323
9	6.6	3.0	0.1	154	0.9667	0.0333
10	6.3	2.9	0.1	149	0.9655	0.0345
11	6.1	2.8	0.1	144	0.9643	0.0357
12	5.8	2.7	0.3	138	0.8889	0.1111
13	5.2	2.4	0.3	123	0.8750	0.1250
14	4.5	2.1	0.5	108	0.7619	0.2381
15	3.5	1.6	0.4	82	0.7500	0.2500
16	2.6	1.2	0.2	62	0.8333	0.1667
17	2.2	1.0	0.2	51	0.8000	0.2000
18	1.7	0.8	0.2	41	0.7500	0.2500
19	1.4	0.6	0.1	31	0.8333	0.1667
20	1.1	0.5	0.1	26	0.8000	0.2000
21	0.8	0.4	0.1	21	0.7500	0.2500
22	0.6	0.3	0.1	15	0.6667	0.3333
23	0.4	0.2	0.1	10	0.5000	0.5000
24	0.2	0.1	0.1	5	0.5000	0.5000
25	0.1	0.1	0.1	3	0.0000	1.0000
Total	178.0	95.4	19.6			

cont'd on next page

record of 137 adult mortalities, of which 54.7% were males. That the adult female segment of the population increased with age in relation to the adult male segment had important effects on the potential growth of the population.

Reproductive Cycles and Reproductive Rates

The rate at which individuals in a population reproduce is perhaps one of the most difficult biological parameters to measure. It is best obtained by compiling individual life histories. Reproductive rates for the ecosystem population were calculated from reproductive histories of 30 adult females. Figure 3.3 (see Methods, Chapter 3) illustrates the type and sequence of information required over the

lifetime of each female. The three most important parameters are length of the reproductive cycle, litter size, and total reproductive period in years. From these, a reproductive rate can be calculated for individual females or for a population (Craighead *et al.* 1969).

The reproductive cycle for a female extends from the time of estrus in late May through June to mid-July, to the time of weaning, normally April and May, one, two, or three years later. We refer to cycles calculated in this way as "pregnancy-to-pregnancy cycles." We calculated an average reproductive cycle for the Yellowstone Ecosystem grizzly bear population by summing the cycles of 30 marked females and dividing the total number of years that each adult female was observed during these cycles (the

Table 4.1. *cont'd*

		FEMALES				
Age (yrs)	Number in age class	Number of females	Number dying in age class D_x	Number surviving per thousand L_x	Survivorship rate P_x	Mortality rate Q_x
Cub	33.0	13.5	5.0	1000	0.6296	0.3704
1	23.0	8.5	0.4	630	0.9529	0.0471
2	18.0	8.1	2.6	600	0.6790	0.3210
3	14.0	5.5	0.5	407	0.9091	0.0909
4	12.0	5.0	0.9	370	0.8200	0.1800
5	7.7	4.1	0.1	304	0.9756	0.0244
6	7.4	4.0	0.2	296	0.9500	0.0500
7	7.0	3.8	0.1	281	0.9737	0.0263
8	6.8	3.7	0.1	274	0.9730	0.0270
9	6.6	3.6	0.2	267	0.9444	0.0556
10	6.3	3.4	0.1	252	0.9706	0.0294
11	6.1	3.3	0.2	244	0.9394	0.0606
12	5.8	3.1	0.3	230	0.9032	0.0968
13	5.2	2.8	0.4	207	0.8571	0.1429
14	4.5	2.4	0.5	178	0.7917	0.2083
15	3.5	1.9	0.5	141	0.7368	0.2632
16	2.6	1.4	0.2	104	0.8571	0.1429
17	2.2	1.2	0.3	89	0.7500	0.2500
18	1.7	0.9	0.1	67	0.8889	0.1111
19	1.4	0.8	0.2	59	0.7500	0.2500
20	1.1	0.6	0.2	44	0.6667	0.3333
21	0.8	0.4	0.1	30	0.7500	0.2500
22	0.6	0.3	0.1	22	0.6667	0.3333
23	0.4	0.2	0.1	15	0.5000	0.5000
24	0.2	0.1	0.1	7	0.5000	0.5000
25	0.1	0.1	0.1	4	0.0000	1.0000
Total	178.0	82.7	13.6			

reproductive period in years) by the total number of population cycles. The reproductive rate was then calculated by dividing the total number of cubs produced by the 30 adult females by the sum of the reproductive periods for all 30 females (Craighead *et al.* 1969). This marked the first time that reproductive rates had been calculated for individual female grizzlies and for an entire population (0.654 cubs/year/female) (see Table 3.4). The implication of this low reproductive rate served a warning that the grizzly bear might require protection as a threatened or endangered species. Litter size averaged 2.29 cubs, and litters were produced at a mean interval of 3.40 years. Discussions of other less basic, but nevertheless important reproduction parameters, such as female age at first pregnancy, first

whelping, first weaning, etc., are deferred to Chapters 7 and 8 since they did not vary importantly from those obtained for the Trout Creek subpopulation.

In following chapters, we describe in detail the biology of the Trout Creek subunit. Finally, in concluding this monograph, we use the reproductive and demographic information obtained from the intensive study of the Trout Creek subunit to re-estimate the population parameters for the entire population that inhabited the Yellowstone Ecosystem from 1959 to 1970. We reaffirm, however, that the focus of this monograph is to examine the phenomenon of ecocenters. To do that, we describe the biological parameters of the Trout Creek subunit as though it were a completely discrete population.

Reproductive rates for grizzly bears were found to be low for a large terrestrial mammal. The implications of grizzlies reproducing at about one cub every two years alerted resource managers to the need for protective measures. *(Photo by John J. Craighead)*

SUMMARY

The 1959-70 Yellowstone grizzly bear population was spatially structured by the bears' traditional movements, developed over the previous century, to several human-made garbage dumps. These ecocenters developed as an increasing number of tourists visited Yellowstone National Park and their food resources attracted discrete aggregations of grizzly bears. Rather than the mutually distinct ecocenter versus "backcountry" populations suggested in other studies, we believe this structure involved subpopulations whose members moved routinely between the ecocentered aggregation to which they belonged and the spacious, broadly overlapping regions of the surrounding backcountry. We estimated an average of 312 bears in the Ecosystem for the period 1959-70, roughly a third of which were members of the Trout Creek ecocentered aggregation, a discrete subpopulation.

Age structure for the ecosystem population reflected an overall mortality schedule characteristic of many large mammals—that is, relatively high juvenile mortality (cub to four years), low and stable adult mortality to about 15 years of age, and an increasing rate of death to a maximum age of about 25 years. Mortality from all human-related causes was proportional to the size of the various age classes and hence not strongly age-dependent. The sex ratio of nonadults was biased in favor of males, whereas that of adults was female-biased, suggesting higher male mortality among adults. Litter size averaged 2.29 cubs and litters were produced at a mean interval of 3.40 years. This study resulted in the first-ever report of a reproductive rate for a population of grizzly bears—0.654 cubs/year/female. The implication of this low rate was recognized and served as a warning that the grizzly bear might require protection as a threatened or endangered species in the Yellowstone Ecosystem and perhaps elsewhere in the conterminous 48 states.

Grizzly bears that were captured and color-marked as cubs or yearlings were of known age and could be recognized from year to year. Annual censuses therefore documented changes in the age structure of the ecosystem population. *(Photo by Frank C. Craighead, Jr.)*

Fig. 5.1. Generalized aggregation pattern of grizzly bears visiting the Trout Creek ecocenter, Yellowstone National Park, 1959-70.

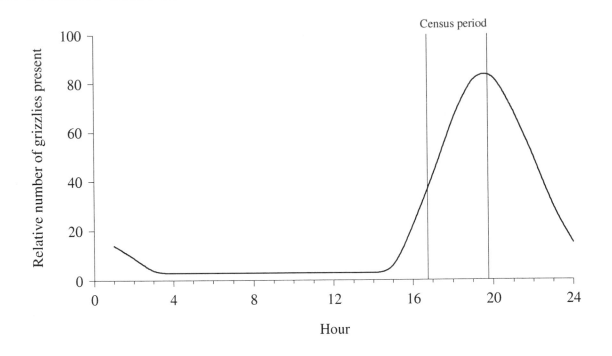

Bears aggregating at the Trout Creek dump arrived from all points of the compass, often following well-defined, time-worn paths which radiated out from the dump like spokes on a wheel hub. Some of these trails were so compacted they persisted for years. *(Photo by John J. Craighead)*

5
The Trout Creek
Grizzly Bear Aggregation

INTRODUCTION

The Yellowstone bears established traditional movements to the Trout Creek ecocenter, where they aggregated to feed and breed. It was not uncommon to find 70 or 80 bears at the Trout Creek landfill garbage dump in Hayden Valley on any given summer evening, crowded into an area of approximately 2 to 4 hectares (5 to 10 acres). Members of the aggregation moved in and out from this central core area where food was most abundant. Adult males interacted aggressively with one another, and this affected both breeding and feeding among the members of the aggregation.

The grizzly bears that aggregated at Trout Creek arrived from widely distributed locales within the approximately two-million-hectare (five-million-acre) Yellowstone Ecosystem (as defined in terms of grizzly bear distribution by Craighead 1980b; see also, Craighead and Mitchell 1982). These locales, however, could be grouped into two distinct areas—the Central Plateau of the Park and the more remote portions of the Park and ecosystem that surround it (see Fig. 10.2). Grizzlies inhabiting the Central Plateau arrived at Trout Creek from a few days to a few weeks after leaving winter dens or seasonal ranges. Those from surrounding areas were denied access to Trout Creek until late spring by formidable barriers of snow and terrain. Thus, the Central Plateau bears formed the seasonal nucleus of the aggregation and were in evidence at Trout Creek for several weeks before other bears arrived.

All bears exhibited a remarkably similar pattern of arrival year after year. Lone animals, both male and female, and some family units normally arrived in Hayden Valley in April and May. As the season progressed, they frequented the banks of the Yellowstone River and snow-free southern exposures. Increasing numbers of bears from distant portions of the Central Plateau were sighted in May, and evidence of over-snow movements indicated that some made even earlier visits to the Trout Creek dump.

The bear aggregation increased as the snow receded and when the National Park Service began dumping garbage that accumulated with the influx of park visitors. This occurred the last week in May, and normally, by 1 June, enough garbage was being deposited to attract bears regularly. It was then that we began our periodic census and observations.

Observations of bears aggregating were recorded by rangers early in the Park's history (Chapter 2). At Trout Creek, long-established trails indicated that grizzlies moved to the garbage dump from all "points of the compass." They tended to use the footprints made by earlier bears. Our observations showed that individual animals characteristically approached the feeding site from the same direction for a period of days and then changed their direction of arrival. Females with cubs were more prone to use day beds in specific locations, so their daily routes to the dump did not vary greatly. Few, if any bears established a rigid routine of approach, but most followed the well-defined, "time-worn" trails. By midsummer, the most-used trails radiated out from the dump like spokes from a wheel hub. Some of these trails were so compacted that they persisted for years. At times, bears approached the food slowly and deliberately, at other times very rapidly, with marked differences in approach behavior by individual animals. In general, all bears were cautious and suspicious of one another early in the season, becoming bolder as the season progressed and as the aggregation enlarged.

Normally, trucks emptied their loads of garbage from 1100 to 1500 hours daily. In general, grizzlies did not enter the landfill dump to feed until the garbage trucks had left the area. By 1600 hours, grizzlies began moving in from adjacent country to feed. Daily movements were as short as 3 kilometers (2 mi) and as long as 19 to 24 kilometers (12 to 15 mi). When, during a year, the aggregation attained a size of about 20 animals, the daily pattern of arrival, buildup, and departure of the bears became fairly consistent. Peak numbers occurred between 1700 and 2200 hours (Fig. 5.1). During this time in midsum-

After feeding at the dump, bears dispersed into the adjacent country, spending time in well-prepared day beds. Females with cubs were more prone to use day beds in a specific location and were seen arriving at and leaving the aggregation by predictable routes. *(Photo by John J. Craighead)*

mer, it was not unusual to see 70 or 80 individual grizzlies. Even these large numbers represented only 60-70% of the mean annual Trout Creek population, indicating that, at one time, a significant number of Trout Creek bears were using portions of the backcountry as well as the Trout Creek dump.

Animals began to leave the aggregation soon after nightfall, presumably because most of the available food had been consumed. Night censuses, employing light-enhancing spotting scopes or spotlights, showed that few bears used the dump after midnight. There were few bears present during the morning hours, and these few were normally highly habituated animals that fed almost exclusively on garbage during the summer months. The size of the aggregation increased steadily from 1500 to 2000 hours and then declined rapidly (Fig. 5.1). Between 0100 and 1600 hours, the bears were dispersed throughout the adjacent country, spending most of this time in well-prepared day beds. Individuals occasionally entered and left the dump during this time on exploratory sojourns. Normally, these animals were subadults or shy individuals that were

kept to the outskirts of the aggregation by the older and more aggressive animals; there they searched for food remnants at a time when there was little competition from other bears.

The study area was strictly zoned to exclude all personnel except our research team. When a stranger did intrude, or a researcher left the census vehicle, this disruption of the normal daily routine was immediately noted by members of the bear aggregation. The usual response was a rapid dispersal of animals to a distance of 183 to 274 meters (200 to 300 yds) and then a return to the dump when the intruder left. Each season, members of the aggregation quickly adjusted to human observers and their vehicles, exhibiting little interest or concern unless people approached too closely or behaved in some unusual way. In such instances, false bluffs, charges, and attacks could be expected and did occur.

Over a 12-year period, we observed the aggregation under a wide range of atmospheric conditions. We found no association between the daily arrival and departure of bears and precipitation,

Fig. 5.2. Cumulative numbers of individual grizzly bears visiting the Trout Creek ecocenter, Yellowstone National Park, 1964-66.

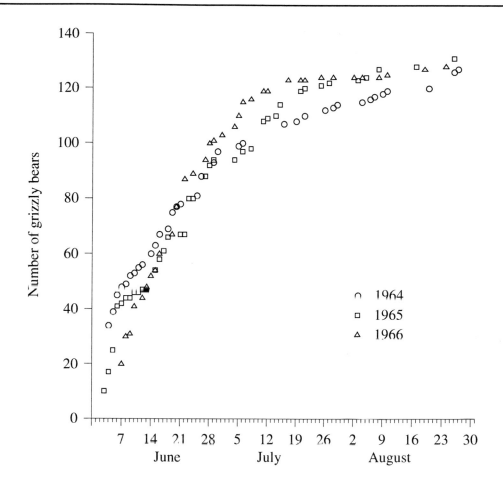

thunderstorms, clear or cloudy weather, or sudden changes in ambient temperature. The aggregation did appear to be more alert and restless on windy days, but ignored the most severe thunderstorms. We observed the Trout Creek aggregation on an evening in August 1959 approximately three to four hours prior to the Yellowstone earthquake (7.5 on the Richter Scale). At the time, we recorded no unusual behavior that would have suggested prior warning of the event nor in retrospect could we recall any. On one or two days each season, however, the entire aggregation would become hyper-alert and extremely restless. This agitation was apparent in the bears' more rapid and increased movements, more frequent and more aggressive encounters for feeding sites, and rapid, *en masse* departures from the dump, with a cautious return of individuals. This communal excitation was quite obvious to the observers, but we could never determine what caused it.

Our initial purpose in censusing grizzly bears at Trout Creek was to determine the size and the sex and age structure of the population, and to record changes in these parameters from year to year. Because the animals aggregated at Trout Creek and at other dump sites, it was possible to count and observe many individuals of a wary, secretive, and normally widely dispersed species.

In the course of the first two years of observation (1959 and 1960), it became apparent that the aggregation was a complex and interesting phenomenon in its own right. Many important facets of bear biology were manifest in the behavior of the aggregation and in the social interactions it generated. It became obvious that the size, structure, and annual continuity of the aggregation had biological causes. The large, dependable quantity of high-calorie protein food acted as an environmental magnet, attracting and holding bears for a period of three months, but factors other than food also appeared to be involved. Plainly, the phenomenon of aggregation had far-reaching effects on the dynamics of the ecosystem population and its use of available food resources.

Table 5.1. Minimum and maximum numbers of grizzly bears censused at Trout Creek ecocenter, Yellowstone National Park, 1960-68.[a]

Year	Annual minimum		Annual maximum	
	Census date	No. of grizzly bears	Census date	No. of grizzly bears
1960	6-17	35	8-17	77
1961	6-13	17	7-17	72
1962	6-11	27	7-28	73
1963	6-11	21	7-09	80
1964	6-05	22	7-21	76
1965	6-03	10	7-11	79
1966	6-09	17	8-10	88
1967	6-07	10	7-20	80
1968	6-06	8	7-08	65
Mean		19		77

[a] Data were recorded from the beginning of the population build-up in spring through the midsummer plateau to the decline in numbers in the fall; only complete censuses were included in this table; censuses taken in 1959 and 1969-70 were omitted—those for 1959 were considered incomplete, and those for 1969 and 1970 reflected management actions that, in time, affected census numbers.

THE AGGREGATION

The annual assemblage of shy "wilderness" animals to feed on the waste food of human visitors appeared, at first, to be highly artificial. Yet, the behaviors we observed in the first two seasons of study appeared to be innate and natural, not superficial adaptations to a manmade situation. Accordingly, we modified our observations and data recordings to study the behavior of the aggregation, as well as its individual members. The seasonal buildup of the aggregation naturally fell into two phases: the observed breeding season (1 June to 15 July), and a postbreeding period from 16 July to 30 August. The two phases were characterized by important changes in the behavior of the aggregation as a whole.

Seasonal Growth of Aggregation

The annual growth of the aggregation year after year was remarkably similar. The cumulative annual buildup of marked and recognizable bears at Trout Creek for the years 1964, 1965, and 1966 is shown in Fig. 5.2. The chronology was similar for the entire 1959-70 period. Each year, the aggregation grew rapidly during June, with slower growth through July and August. The maximum numbers of bears recorded in an individual annual census from 1960 through 1968 varied from 65 to 88 (Table 5.1). Mean daily census numbers peaked at 77

bears between mid-July and mid-August over the years of the study. The mean number and range of bear numbers observed weekly are shown in Fig. 5.3. Numbers began plateauing in mid-July when most of the bears had joined the Trout Creek aggregation. Throughout the season, visitation of individual bears was influenced by the availability of certain wild foods such as berries, pine seeds, rodents, and ungulates. When wild foods were available, some bears would feed on them almost exclusively for days at a time. Such foods, however, did not significantly alter the general visitation patterns of the aggregation. We concluded that wild foods were important, but that any single food resource was secondary to the food available at the Trout Creek dump. The latter, because of its quantity and dependability, appeared to act as an environmental magnet and food resource stabilizer, year after year (see Chapter 10).

Individual Visitation Patterns

Food was a major and obvious factor attracting bears and holding them in a complex social aggregation. Not obvious were the other biological factors and forces that influenced visitation to the ecocenter, permitting "solitary" animals to feed and reproduce in close proximity. The daily and seasonal coming and going of grizzlies to a centralized location was just as much a phenomenon as the structure and function of the aggregation itself. The visitation patterns of individual bears were the

During midsummer, at the peak of visitation, it was not unusual to see 70 to 80 bears in the aggregation. This large number represented only 60-70% of the mean annual Trout Creek population, indicating a significant number of bears were at any one time using backcountry resources. *(Photo by John J. Craighead)*

foundation of the visitation pattern of the aggregation as a whole.

The visitation patterns of individual bears often varied greatly across years. These patterns of visitation were displayed by both males and females and by bears of all ages (Fig. 5.4). Some individuals were not observed at all in some years, and when they did use Trout Creek, the frequency and duration of their visits varied greatly from one year to the next (Fig. 5.5). Other bears were observed at Trout Creek every year or nearly every year for which there were records, and their annual visits were consistently frequent and of long duration (Fig. 5.6). There was further substantial variation in visitation between-individuals within-years and within-individuals between-years.

We cannot cite any single factor or combination of factors to explain entirely the often great differences in the visitation patterns of individual bears. The patterns may have reflected in part the great range of individuality apparent among members of the Trout Creek aggregation, as well as ecological factors impinging on individual animals. For example, the chance establishment of home ranges in the vicinity of Trout Creek by some bears may have favored an addictive pattern of visitation. Similarly, a sporadic visitation pattern may have been adopted by animals in poor condition that could not easily cope with the established aggregation of healthy bears, or by those relatively few animals that switched between two or more open-pit dumps or periodically found rich, natural food sources.

More amenable to analysis are the potential influences of sex, age, and social/reproductive status on bear visitation. For the present, we focus on the effects of sex and age. Later (Chapter 6), we address factors affecting individual visitation by examining the influence of male dominance rank and female reproductive status on the tendency of bears to visit the Trout Creek ecocenter.

Visitation According to Sex and Age

In all our comparisons of individual visitation, we employed two types of statistical analyses. A visitation "rate" was calculated for each bear in each year of the study. This was simply the fraction of

Fig. 5.3. Number of grizzly bears visiting the Trout Creek ecocenter, Yellowstone National Park, 1959-70. Solid circles show weekly means; bars extending from the circles show weekly ranges.

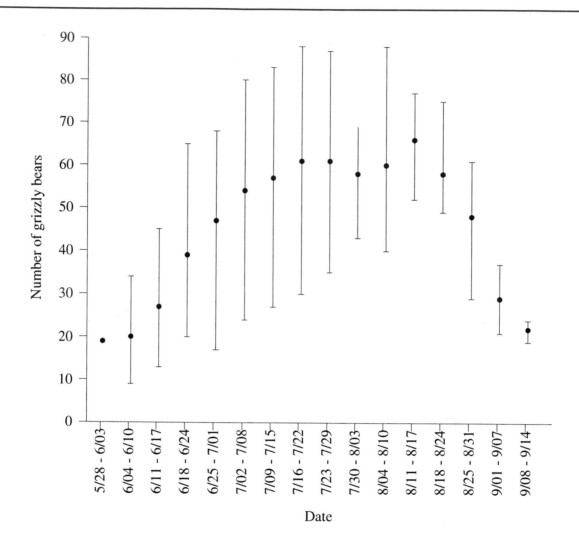

complete censuses in which an individual bear was recorded as present among the Trout Creek aggregation. In other words, if a bear was present in 14 of 50 censuses, its rate would be 0.28. These raw visitation rates were then adjusted for any yearly differences in visitation level by subtracting the mean raw visitation rate observed in the same year among all bears belonging to the two categories being compared (sex or age) from an individual's rate. The purpose of the subsampling was to remove the problem of non-independence of the data that would arise if we included individual bears two or more times in a given test. The subsampling was repeated for each comparison, regardless of whether a given age/sex category had been previously subsampled. Thus, a given subsample was never used more than once.

In the "two-sample" type of test, samples of bears (one sample for each of two categories being compared) were obtained by pooling all years of the study and then selecting bears (and their raw and adjusted visitations) at random, in such a way that a given bear appeared, at most, one time within and between samples. Again, the purpose of the subsampling was to remove the problem of nonindependence. Visitation values in the two samples and hence in the categories of interest were then compared using the Mann-Whitney two-sample test (M-W test). In analyses evaluating individual activity under different circumstances, we used a Wilcoxon matched-pairs test to compare a bear's adjusted visitation when in one category of interest to its visitation in another year when in the other category.

Fig. 5.4. An example of visitation frequencies at the Trout Creek ecocenter for grizzly bear Nos. 183 (male) and 175 (female) from 1 June to 30 August, Yellowstone National Park, 1964-70.

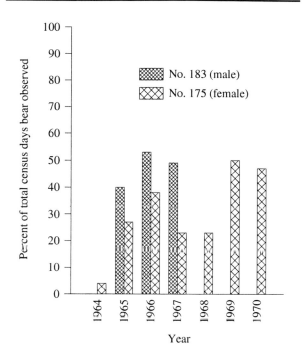

Visitation of individual bears was influenced by the availability of certain wild foods. When carrion, pine seeds, or berries were abundant, some bears would feed on them exclusively for days.

(Photo by John J. Craighead)

Fig. 5.5. Visitation frequencies at the Trout Creek ecocenter for grizzly bear Nos. 130 (male) and 5 (female), showing great variation in attendance across years from 1 June to 30 August, Yellowstone National Park, 1960-68.

Fig. 5.6. Visitation frequencies at the Trout Creek ecocenter for grizzly bear Nos. 135 (male) and 101 (female), showing consistently high attendance across years from 1 June to 30 August, Yellowstone National Park, 1962-70.

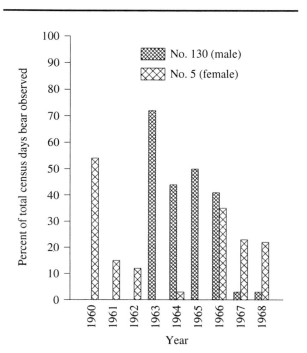

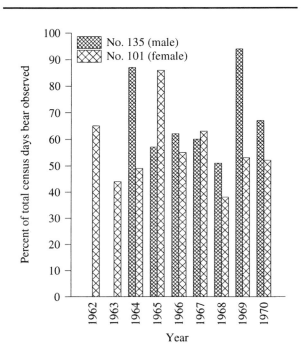

The need for subadult grizzlies to meet the energy demands of growth and yet minimize conflict with adult bears created behavioral tensions that influenced their use of the Trout Creek ecocenter. Immature bears tended to visit the ecocenter less frequently when resources were limited and competition high. They were also infrequent visitors during the breeding season when they were at risk from aggressive contending males. *(Photos by John J. Craighead)*

Visitation According to Sex

Among all adult females (irrespective of reproductive status) and all adult males (irrespective of dominance rank), there was a nonsignificant trend toward greater visitation by females than by males over the entire season, from aggregation formation in June through its dispersal in late August (M-W test, $P = 0.07$; Appendix A, Table 1). This trend was due primarily to the presence of females with offspring in the female sample, since females in this reproductive state had substantially higher season-long visitation than all males as a group (M-W test, $P = 0.01$; Appendix A, Table 2). Among nonadults aged three and four years, there was no difference between the sexes in visitation either during (M-W test, $P = 0.47$; Appendix A, Table 3) or after the breeding season (M-W test, $P = 0.73$; Appendix A, Table 4).

Visitation According to Age

Nonadults (3- and 4-year-olds of both sexes) were more frequent visitors at Trout Creek than adult males both during the breeding season and after (Appendix A, Tables 5 and 6). Because the great majority of adult males were noncontenders (not actively contending for alpha male status; see Chapter 6), this analysis amounts to a comparison of bears that were sexually inactive but growing rapidly, with adult males that were also sexually inactive (at least with regard to conventional sexual competition), but with much lower energy-for-growth requirements. When the visitation of adult females without offspring is compared to that of nonadult females, also without offspring, we find that nonadult females visited the ecocenter more frequently than adult females after the breeding season. Consistent with this result, visitation by all subadults increased significantly after the end of the breeding season (Wilcoxon matched-pairs test, $P = 0.01$; Appendix A, Table 7).

Altogether, results of visitation analysis by sex and age suggest increased use of Trout Creek by bears with special nutritional needs arising from rapid growth or maternal care. However, nonadults appeared to experience a behavioral conflict or tension. Attracted to Trout Creek to meet their energy demands of growth (apparently similar for the two sexes), they nevertheless experienced a counter urge to avoid Trout Creek during the breeding season to minimize conflict with older bears (Appendix A, Table 7). There are at least two reasons why nonadults might have joined the aggregation less frequently during the breeding season, even when nutritional needs encouraged visitation. First, immature bears might have been at greater risk of injury by the large, aggressive contending males as these bears competed for estrous females and social status. Secondly, immature bears may have "avoided" Trout Creek early in the season when food was more limited because they were at a competitive disadvantage with all adults for acceptable feeding sites. In either case, feeding away from the ecocenter early in the season likely had some advantages for the nonadults.

Visitation According to Time Since Weaning

We initially considered 3- and 4-year-olds as nonadults, because we believed they were least likely to be influenced by the visitation patterns of their mothers. A comparison of weaned 1- and 2-year-olds versus 3- and 4-year-olds, however showed identical patterns of visitation (Appendix A, Tables 7, 8, and 9). This result suggests that, for nonadults (1- through 4-year-olds), weaning represents a sharp demarcation between a visitation pattern determined entirely by the mother and a second one determined by considerations specific to the age category of the nonadult.

Size of the Aggregation

The size of the Trout Creek subpopulation, a discrete subunit of the ecosystem population, was determined from cumulative annual census counts, modified by criteria that established each bear observed as either a "resident" or a "transient" (unmarked bears that lacked permanent natural markings; see Chapter 3). Bears considered residents or members of the aggregation were further distinguished as the flagged members and the nonflagged members on the basis of frequency of visitation (≥5% or <5%, respectively; criteria are discussed in Chapter 3). Over the 12-year period of observation, the mean size of the flagged population was 78 bears, and the nonflagged population was 38 bears, for a total of 116 individual bears representing the mean of the annual cumulative counts recorded on censuses (Table 5.2). These two population segments, along with the transient bears, are used, when appropriate, to analyze biological characteristics of the Trout Creek subpopulation.

In the case of Trout Creek and other ecocenters, visitation led to formation of the aggregation, and the cumulative size of the aggregation defined the size of the population subunit. Transient bears were readily recognized as such from on-site observations, or from their identification as residents at one of the other major ecocenters where bears were marked and censuses were conducted.

The composition of the Trout Creek subpopulation varied from year to year as members died or emigrated and as bears were born. Some members were observed and identified during all or most years of the study, others had tenure of four or five years, and some individuals were observed for only a single season. Because age structure and survivorship (longevity) are factors determining variation in population size, it was essential to measure these parameters to understand the dynamics of the population.

Age and Longevity

Grizzly bears are long-lived animals. They can be aged accurately by the cementum layer technique (Craighead *et al.* 1970 and Chapter 3). The oldest

Table 5.2. Age structure of the Trout Creek grizzly bear population, Yellowstone National Park (1960-70).

| | 1960 | | 1961 | | 1962 | | 1963 | | 1964 | | 1965 | | 1966 | | 1967 | | 1968 | | Summary 1960-68 | | 1969 | | 1970 | | Summary 1969-70 | |
Class	No.	%	No.	%	No.	%	No.	%	No.	%	No.	%	No.	%	No.	%	No.	%	Mean	%	No.	%	No.	%	Mean	%
Cubs	20	20.4	18	18.4	25	25.5	27	23.3	15	11.7	30	22.4	17	12.8	19	15.0	19	16.8	21.1	18.2	12	13.8	18	12.9	15.0	13.3
Yearlings	9	9.2	9	9.2	12	12.2	21	18.1	27	21.1	13	9.7	25	18.8	17	13.4	11	9.7	16.0	13.8	13	14.9	13	9.4	13.0	11.5
2-year-olds	5	5.1	11	11.2	3	3.1	8	6.9	21	16.4	21	15.7	10	7.5	19	15.0	11	9.7	12.1	10.4	8	9.2	11	7.9	9.5	8.4
Subadults[a]	9	9.2	14	14.3	19	19.4	14	12.1	13	10.2	15	11.9	30	22.6	21	16.5	17	15.0	17.0	14.6	15	17.2	26	18.7	20.5	18.1
Adults	55	56.1	46	46.9	39	39.8	46	39.7	52	40.6	54	40.3	51	38.3	51	40.2	55	48.7	49.9	43.0	39	44.8	71	51.1	55.0	48.7
Total	98	100	98	100	98	100	116	100	128	100	134	100	133	100	127	100	113	100	116.1	100	87	100	139	100	113.0	100

[a] Subadults = 3- and 4-year-old bears.

Table 5.3. Longevity of adult female grizzly bears at the Trout Creek ecocenter, Yellowstone National Park, 1959-81 ($n = 54$[a]); [xxxx] represents nonadult age class and [——] represents adult age class.

Female number	Age first observed	Age last observed	Year (1959–1974)	Post-1974 mortality	Adult years observed	Total years observed
5	1	16	xxxx xxxx xxxx xxxx —— — — — — — — — — — —— Dead		12	16
6	0	6	xxxx xxxx xxxx xxxx xxxx —— Dead		2	7
7	11	20	—— — — — — — — —		10	10
10	2	24	xxxx xxxx xxxx —— — — — — — — — — — — —	Dead 1981	20	23
15	1	13	xxxx xxxx xxxx xxxx —— — — — — — — —		9	13
23	11	11	Dead		1	1
29	0	6	xxxx xxxx xxxx xxxx xxxx —— Dead		2	7
34	14	24	—— — — — — — — — — — —		11	11
39	6	16	—— — — — — — — — — — —		11	11
40	2	11	xxxx xxxx xxxx —— — — — — — —— Dead		7	10
42	5	16	—— — — — — — — — — — — —		12	12
45	3	6	xxxx xxxx —— —		2	4
75	15	18	—— — — Dead		4	4
81	3	12	xxxx xxxx —— — — — — — —— Dead		8	10
96	3	12	xxxx xxxx —— — — — — — —— Dead		8	10
101	3	16	xxxx xxxx —— — — — — — — — —— Dead		12	14
109	0	8	xxxx xxxx xxxx xxxx xxxx —— —— —— Dead		4	9
112	6	14	—— — — — — — — — —— Dead		9	9
128	11	19	—— — — — — — — — —		9	9
131	4[b]	5	xxxx Dead		1	2
132	0	8	xxxx xxxx xxxx xxxx xxxx —— — — —		4	9
139	0	8	xxxx xxxx xxxx xxxx xxxx —— — — —		4	9
140	8	13	—— — — — — —		6	6
141	1	7	xxxx xxxx xxxx xxxx —— —— Dead		3	7
142	1	8	xxxx xxxx xxxx xxxx —— — — Dead		4	8
144	1	8	xxxx xxxx xxxx xxxx —— — — —		4	8
148	15[b]	16	—— Dead		2	2
150	5	12	—— — — — — — —		8	8
163	1	5	xxxx xxxx xxxx xxxx Dead		1	5
172	13	19	—— — — — — — —		7	7
173	2	8	xxxx xxxx xxxx —— — — — —		4	7
175	10	17	—— — — — — — — — Dead		8	8
176	13	13	Dead		1	1
180	11	14	—— — — Dead		4	4
184	0	6	xxxx xxxx xxxx xxxx xxxx —— Dead		2	7
187	1	8	xxxx xxxx xxxx xxxx —— — — —		4	8
200	3	8	xxxx xxxx —— — — — —		4	6
204	2	7	xxxx xxxx xxxx —— — —		3	6
228	5	8	—— — — —		4	4
259	16[b]	20	—— — — — —— Dead		5	5
18	Adult	Adult	—— — Zoo		3	3
44	Adult	Adult	—— — — — — —		7	7
48	Adult	Adult	—— — — — —		6	6
64	Adult	Adult	—— — — — — —		7	7
65	Adult	Adult	—— — — — —		6	6
72	Adult	Adult	—— — —		3	3
84	Adult	Adult	—— — — — —		5	5
108	Adult	Adult	—— — — — — — — — — — —— Dead		12	12
119	Adult	Adult	——		1	1
120	Adult	Adult	—— — — — — — — — — — — —— Dead		14	14
123	Adult	Adult	—— —		2	2
125	Adult	Adult	—— — — — — — — —— Dead		10	10
129	Adult	Adult	——		1	1
160	Adult	Adult	—— — — — —		5	5

Aged bears ($n = 40$)
Mean	5.2	12.2			—	—

Total bears ($n = 54$)
Mean	—	—			5.9	7.4

[a] Includes Trout Creek flagged and nonflagged populations.
[b] Includes one or more years of observations before individual was marked.

The mean age of 12.5 years for bears of the Trout Creek aggregation was derived from known-age and established-age animals. A premolar tooth was extracted to age the older bears. Behavior was observed using radio transmitters imbedded in a collar. *(Photo by John J. Craighead)*

grizzly aged was a 25- to 26-year-old Trout Creek female (Table 5.3). Other females approached this age; two Trout Creek females and one non-Trout Creek female attained the age of 24 years. One of these females, No. 34, was still alive in the fall of 1970, and probably lived to be at least 25 years old. The oldest male, a member of the West Yellowstone ecocenter aggregation, was 23 years old. The oldest male recorded at the Trout Creek ecocenter was No. 249, a 19-year-old at last observation (Table 5.4).

Females continue to reproduce until at least age 24. Female No. 9, a non-Trout Creek female accompanied in her 24th year by one cub, was found dead. She appeared to have died of malnutrition. Two other females whelped at age 22. Female No. 34 had two cubs in 1968 at age 22 and weaned them as yearlings in 1969. But she did not have a litter the following year and was not observed after 1970. A non-Trout Creek female, No. 8, also produced a litter at age 22. Although slow to produce their first litter (minimum age five), females are potentially reproductively viable for 20 or so years. Because the average interval between litters is three years,

an otherwise "average" female, with a maximum reproductive lifetime, can conceivably produce seven litters during her life. Using an average litter size of 2.2, the potential cub production per female is 15. Females usually do not produce their first litter as early as age five, however, nor do they typically live to age 24.

The mean age of the 57 known- or cementum-aged adult grizzlies observed during the 1960 to 1968 period was 9.7 years (Table 5.5). Females averaged one year older (10.1) than the males (9.0).

We can determine the mean age that grizzlies were last observed by analyzing the fate of all Trout Creek grizzlies. Of the 146 marked and recognizable Trout Creek aggregation females, 117 (80%) were adults when first observed or attained adulthood (≥ 5 years). Among the 73 artificially marked females, 54 (74%) were marked as adults or were observed to reach maturity. We observed these 54 females for an average of 5.9 years as adults (Table 5.3). The mean age last observed for 40 of the adult females whose ages were known (known-age, cementum-aged, and established-age) was 12 years. Similar statistics for males were lower; the

Table 5.4. Longevity of adult male grizzly bears at the Trout Creek ecocenter, Yellowstone National Park, 1959-78 (*n* = 54[a]); [xxxx] represents nonadult age class and [——] represents adult age class.

Male number	Age first observed	Age last observed	Year (1959–1974)	Post-1974 mortality	Adult years observed	Total years observed
2	3	14	xxxx xxxx —— — — — — — — — — Dead		10	12
30	8	17	—— — — — — — — — Dead		10	10
41	6	17	—— — — — — — — — — — — Dead		12	12
51	1	7	xxxx xxxx xxxx xxxx —— — Dead		3	7
74	3	5	xxxx xxxx —		1	3
76	2	8	xxxx xxxx xxxx —— — — —		4	7
00	5	15	—— —— — — — — Dead		11	11
130	2	8	xxxx xxxx xxxx —— — — —		4	7
135	0	8	xxxx xxxx xxxx xxxx xxxx —— — — ——		4	9
147	1	8	xxxx xxxx xxxx xxxx —— —		4	8
158	0	6	xxxx xxxx xxxx xxxx xxxx —— —		2	7
166	3	11	xxxx xxxx —— — — — —— — Dead		7	9
171	3	9	xxxx xxxx —— — — — — Zoo		5	7
177	1	7	xxxx xxxx xxxx xxxx —— — —		3	7
178	1	5	xxxx xxxx xxxx xxxx Dead		1	5
179	13[b]	17	—— — — — Dead		5	5
181	1	7	xxxx xxxx xxxx xxxx —— — —		3	7
183	0	5	xxxx xxxx xxxx xxxx xxxx Dead		1	6
188	0	6	xxxx xxxx xxxx xxxx xxxx —— —		2	7
189	1	5	xxxx xxxx xxxx xxxx ——		1	5
190	1	11	xxxx xxxx xxxx xxxx —— — — — — —— Dead		7	11
191	1	6	xxxx xxxx xxxx xxxx —— —		2	6
194	1	6	xxxx xxxx xxxx xxxx —— —		2	6
199	2	15	xxxx xxxx xxxx —— — — — — — ——	Dead 1978	11	14
201	2	8	xxxx xxxx xxxx —— — — Dead		4	7
207	2	9	xxxx xxxx xxxx —— — — — Dead		5	8
208	1	5	xxxx xxxx xxxx xxxx Dead		1	5
210	8	11	—— — — —		4	4
211	0	7	xxxx xxxx xxxx xxxx xxxx —— — Zoo		3	8
213	1	8	xxxx xxxx xxxx xxxx —— — — Dead		4	8
217	15	16	—— Dead		2	2
232	2	5	xxxx xxxx xxxx —		1	4
249	17	19	—— — —		3	3
253	3		xxxx xxxx —		1	3
255	3	5	xxxx xxxx —		1	3
257	4	5	xxxx —		1	2
258	14[b]	15	—— —		2	2
12	Adult	Adult	—— — — — — — — —		8	8
14	Adult	Adult	—— — — — — — — —		8	8
27	Adult	Adult	—— — — —		4	4
28	Adult	Adult	—— — — —		4	4
33	Adult	Adult	—— — — ——		4	4
46	Adult	Adult	—— —		2	2
60	Adult	Adult	—— —		2	2
73	Adult[b]	Adult	—— —		3	3
85	Adult	Adult	—— — — — — — — — —— Dead		12	12
87	Adult	Adult	——		1	1
111	Adult	Adult	—— —		2	2
113	Adult	Adult	——		1	1
126	Adult	Adult	—— Dead		2	2
159	Adult	Adult	—— —		1	1
195	Adult	Adult	—— —		2	2
197	Adult	Adult	—— —		2	2
206	Adult[b]	Adult	—— — — — — — —		8	8

Aged bears (*n* = 37)
Mean   3.5   9.2   —   —

Total bears (*n* = 54)
Mean   3.9   5.8

[a] Includes Trout Creek flagged and nonflagged populations.
[b] Includes one or more years of observations before individual was marked.

Table 5.5. Age structure of the marked, adult grizzly bear population at the Trout Creek ecocenter, Yellowstone National Park, 1960-68.

		5		6		7		8		9		10		11		12		13		14		15		16		17		18		19		20		21		22		Total	
Age in years	Year	M	F	M	F	M	F	M	F	M	F	M	F	M	F	M	F	M	F	M	F	M	F	M	F	M	F	M	F	M	F	M	F	M	F	M	F
	1960	1		1	1						1																										3	5
	1961	1	3		2	1	1														1		1		1		1									4	6
	1962	1	7	1			2	1						1	1						1		1	1	1		1									5	12
	1963	1	2		2	1	1	1	2	1	2	1	3		1		1	1	3	1	1	1		1	1	1	1									5	18
	1964	1			7	1	7	1	2	1	1		3	1	3	1	3	1	1	1	2	1	1	1	1	1	1		1		1					6	21
	1965	1	5	1	2	1	1	1	1	1	1	1	1	1	2	1	3	1	1	1	1	1	1		2		1		1		1					7	20
	1966			1	1	1	1	1	7		7	1	1	1	2		2	1	2	1	2	1	2	1	2	2	2	1							1	8	19
	1967	2	5	1	1	1	1	1	1		7	1	1	1		1	3		3		2		2	1			2		3		1		2			9	24
	1968	10	2	2	5								7	1	1	1								1								1		1		20	24
Total bears		18	21	8	20	6	12	5	12	4	12	4	13	4	8	3	8	3	10	3	7	3	6	3	6	3	5	0	3	0	2	0	2	0	1	0	1	67	149

Mean age: males (n = 24) = 9.0 yrs; females (n = 33) = 10.1 yrs; all adults (n = 57) = 9.7 yrs

Note: Extreme variations in number of males and females occurring in some individual years (e.g., five-year-old males in 1968) are due to a large number of cubs and yearlings marked in years prior to the year observed; this, combined with the adults captured and cementum-aged as an adult for a given year, elevated the numbers for these years.

54 marked adult males averaged only 3.9 years of observation as adults, with an average age at last observation of 9.2 years for the 37 males that were aged animals (Table 5.4). These figures suggest that the sex difference in the age last observed (3.3 years) was due not only to a difference in the mean initial age of the male versus female sample (5.3 versus 6.6 or 1.3 years), but also to higher rates of mortality among males.

Age and Sex Structure

To obtain the best description of the sex and age structure of the Trout Creek population, we used census data from 1960 through 1968. We excluded data from 1959, because too few bears were marked and recognizable, and from the years 1969 and 1970, because closure of ecocenters by the Park Service caused an influx of bears at Trout Creek during these years.

Living Age Structure

The age structure of the population changed from year to year with variations in the annual increment of cubs and the annual mortality among all age classes (Table 5.2 and Fig. 5.7). More cubs were produced in some years than in others, and the number of yearlings and other nonadults recorded varied considerably from year to year. Some of the causes of these year-to-year variations are discussed later. It is important to note that the ratio of nonadults to adults remained relatively constant from year to year. The 3- and 4-year-old age group is composed of bears that will soon move into the sexually active adult category and contribute repro-

Table 5.6. Summary of the sex structure of the grizzly bear population at the Trout Creek ecocenter, Yellowstone National Park, 1960-68.

Age class	Number of observations of individuals			Percent[a]		
	M	F	Total	M	F	P-value[b]
Cub	46	32	78	59.0	41.0	0.14
Yearling	40	18	58	69.0	31.0	0.006
2-year-old	42	19	61	68.9	31.1	0.005
Subadult[c]	63	44	107	58.9	41.1	0.08
Adult	169	222	391	43.2	56.8	0.009

[a] Cub numbers derived from progressive summary of all individual cubs marked, 1959-70; yearling through adult numbers derived from summation of annual censuses of all marked and recognizable grizzlies, 1960-68.
[b] Binomial test.
[c] 3- and 4-year-old bears.

Fig. 5.7. Age structure (derived from census data) of the grizzly bear population at the Trout Creek ecocenter, Yellowstone National Park, 1960-68.

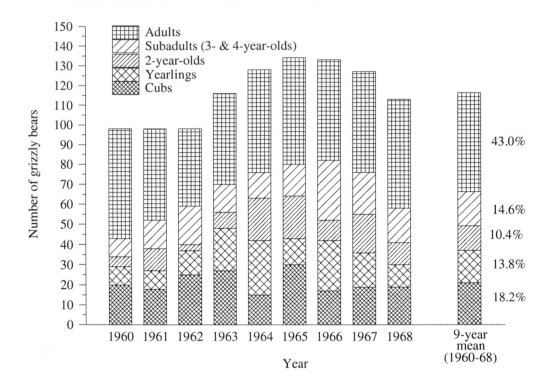

ductively to the population. The average age structure for the 9-year period was 18.2% cubs, 13.8% yearlings, 10.4% 2-year-olds, 14.6% nonadults (3- and 4-year-olds), and 43.0% adults. The aggregation was composed primarily of young bears, with 57% nonadults and 43% adults. It grew from 98 individuals in 1960 to a peak of 134 bears in 1965 (Fig. 5.7). This growth was the result of an 18.1% average annual increment, a figure that apparently exceeded the average annual losses.

Sex Ratio

The ratio of males to females was more difficult to determine than the aggregation's age composition because a positive sex determination among nonadults could be obtained only from marked animals. To develop the most representative sex structure for the Trout Creek subpopulation, we used the following data:

    a.   Sex of all cubs captured and marked.

    b.   Sex of all captured nonadults (yearlings to 4-year-olds).

    c.   Sex of both marked and unmarked adults recorded on the censuses (unmarked adults could be accurately sexed).

The sample size of sexed individuals in the aggregation increased with the number of marked

and recognizable animals in the population, and these increased steadily as the study progressed. The sex ratios were derived from a sample of 391 different marked and recognizable grizzlies. The sex structure is biased toward males in all age classes except adult (Table 5.6).

To test whether sex differentials by age class were statistically significant, we applied a two-tailed binomial test. Although apparently strongly male-biased, the sex ratio of cubs did not differ statistically from parity (Table 5.6). Although this result is consistent with a birth sex ratio of one male to one female (1:1), it is also possible that cub sex ratios were determined by some interaction of a sex-ratio bias at birth and sex-differential mortality between birth in February and our capture of cubs the following summer. A sex bias in favor of males persisted in our samples of bears from one to four years of age. For yearling and 2-year-old bears, but not 3- and 4-year-olds, this difference reflected a significant departure of the sex ratio from parity (Table 5.6). As we discuss below, the sex bias among young bears may be the result of differential mortality, but it is possible, too, that a greater susceptibility of yearling, 2-year-old, and perhaps subadult males to trapping amplified the nonparity of ratios. Among adults, a sex bias was also apparent, but

Fig. 5.8. Examples of some home ranges and seasonal movements of Trout Creek grizzly bears, illustrating the year-round attachment of some bears to the ecocenter and the wide dispersal of others, 1959-70.

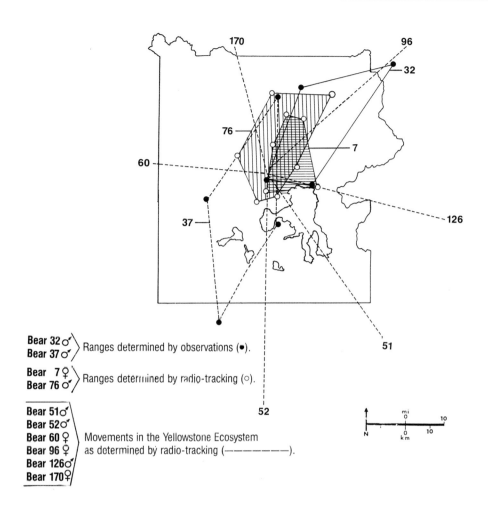

Bear 32 ♂
Bear 37 ♂ ⟩ Ranges determined by observations (●).

Bear 7 ♀
Bear 76 ♂ ⟩ Ranges determined by radio-tracking (○).

Bear 51 ♂
Bear 52 ♂
Bear 60 ♀ ⟩ Movements in the Yellowstone Ecosystem
Bear 96 ♀ ⟩ as determined by radio-tracking (−−−−−−−).
Bear 126 ♂
Bear 170 ♀

here there was a significant excess of females (Table 5.6). In Chapter 11, we evaluate the relationship of mortality to population decline, but here we specifically examine how mortality may have affected the sex ratios of age classes.

Sex-biased Mortality

Taken at face value, the results suggest a sex-differential pattern of mortality in which females are more vulnerable early in life (say birth to four years) and males more so thereafter. The female grizzly is smaller than the male, at least from about four months on (see Chapter 16), and this difference might predispose females to greater mortality in the pre-reproductive years. A sex bias in juvenile mortality has been reported in other sexually dimorphic mammals (reviewed by Trivers 1985). But in these species the larger sex has higher mortality, the difference being attributed to the greater risks of sustaining higher rates of growth. Of course, grizzly

bears, and particularly ecocentered bears, may depart from this mammalian pattern. For example, the abundant and reliable source of food available to Trout Creek bears may have allowed males to sustain high rates of growth and enjoy the benefits of large size, per se, without paying costs in survivorship otherwise attendant to rapid growth.

Male-biased mortality among adults is certainly plausible. For example, the males' larger home ranges and greater movement (Fig. 5.8; also, Craighead 1976) might have brought them into campgrounds, developed areas, hunting camps, and other sites where they were exposed to greater risk of man-caused mortality. Their violent and stressful form of reproductive competition also may have exposed them to greater natural mortality.

Human-related causes of mortality are explored in detail in Chapter 11. Suffice it to say here, however, that our analyses revealed no conclusive statis-

Fig. 5.9. Records of marked grizzly bears illustrating the natural movement patterns from the Trout Creek ecocenter ⓣ to sites within Yellowstone National Park, 1959-74.

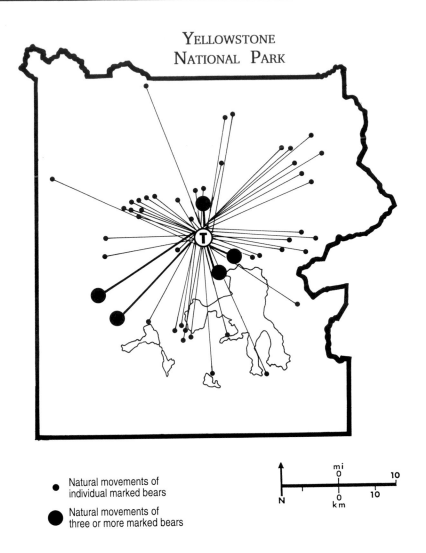

Natural movements of
individual marked bears

Natural movements of
three or more marked bears

tical evidence of male-biased mortality from all nonnatural sources of mortality such as management control, legal and illegal kills, etc. Absolute male mortality was higher than female mortality in all age categories and for most causes (Tables 11.4, 11.5, 11.9, 11.10), but these sex biases were not significantly different from those expected, given our estimates of sex ratios at those ages. This still leaves open the possibility that males are more susceptible to mortality by natural causes, or that natural and man-caused mortality combined to generate a significant sex bias among adults. Life tables based on age structure data suggest this was the case (Craighead *et al.* 1974).

Conversely, adult female movements tended to be less extensive than those of males, in part because they had smaller home ranges (Craighead

1976). Also, they used the Trout Creek ecocenter more heavily than males. By attracting and holding adult females within the geographic center of the Yellowstone Ecosystem, the Trout Creek ecocenter plausibly exerted a significant additional damping effect on female movement and hence on their susceptibility to various types of natural and human-related mortality (see Chapter 11).

Movement of Aggregation Members

As indicated earlier, grizzlies moved to ecocenters from all parts of the Yellowstone Ecosystem (Craighead 1977). It was not until many members of the Trout Creek aggregation were color-marked and radio-collared that the full extent and significance of this movement was recognized. The movements of individual grizzlies to and from Trout

Fig. 5.10. Records of marked grizzly bears illustrating the natural movement patterns from the Trout Creek ecocenter ⓣ to localities of death outside Yellowstone National Park, 1959-74.

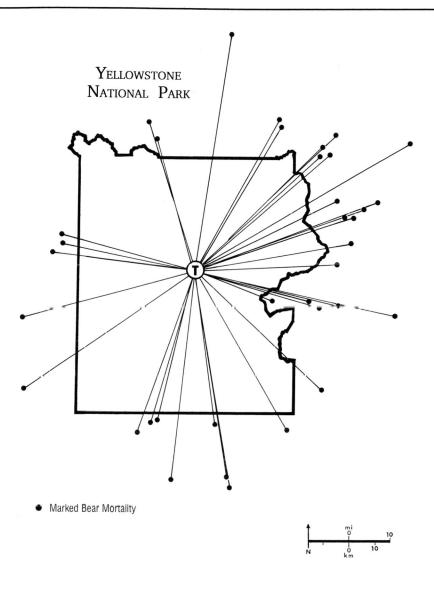

YELLOWSTONE NATIONAL PARK

● Marked Bear Mortality

Creek were divided into two groupings for analysis: movements inside Yellowstone Park (Fig. 5.9) and those outside the Park but within the two-million-hectare (five-million-acre) Ecosystem (Fig. 5.10) delineated by Craighead (1980b).

Movements between Trout Creek and other areas within the Park took any compass direction (Fig. 5.9). Movement between Trout Creek and areas outside the Park was similarly varied (Fig. 5.10). Straight-line moves in excess of 65 kilometers (40 mi) were commonly recorded (Craighead 1976, Craighead and Mitchell 1982). The aggregation at Trout Creek no doubt developed gradually over many decades until movement to the area became traditional. As both travel patterns and social behaviors became strongly established, the Trout Creek aggregation developed into a somewhat

distinct population unit—one that used the Trout Creek ecocenter heavily during June, July, and August and much of the entire Ecosystem during the remainder of the year.

Travel to and from the ecocenter was significant in several respects. It tended to establish two types of home ranges: those on the Central Plateau (Fig. 5.11) that included the ecocenter and immediate surrounding area, and those that included the ecocenter and migratory corridors to more distant fall foraging ranges and den sites (Fig. 5.8; see also Craighead 1976; Craighead and Craighead 1972a, 1972b). The corridors functioned as foraging areas for bears traveling to and from their winter dens. The intensity of foraging along the corridors varied with individual bears and with the availability of such seasonal foods as winter-killed elk, small

Fig. 5.11. Distribution of denning sites of marked Trout Creek grizzly bears on the Central Plateau and in other parts of the Yellowstone Ecosystem, 1959-70

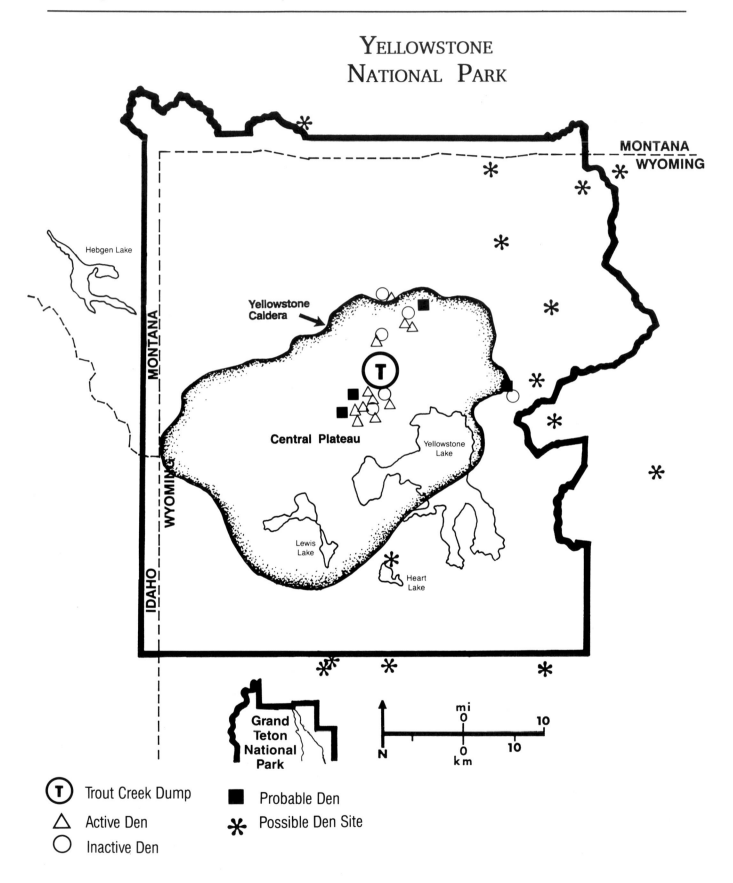

rodents, berries, and pine seeds. Most significantly, movement was curtailed by the availability of food at the ecocenter during late spring and summer (1 June to 30 August). During this period Trout Creek grizzlies seldom, if ever, left the Park and the protection it afforded; movements to campgrounds and developed areas within the Park were rare. For specifics of movements to campgrounds and developed areas, see Chapter 15 (also, Craighead and Craighead 1972a, 1972b).

Bear movements were also implied by the distribution of denning areas (Fig. 5.11). These movements were of two types—relatively short ones by bears wintering on the Central Plateau and long distant movements by those wintering beyond the Plateau. From 1961 through 1968, we radiotracked 23 individuals and located 11 dens. In addition, we located six dens that were inactive and 18 probable or possible dens (Craighead and Craighead 1972a, 1972b). Dens on the Central Plateau typically were located at elevations between 2377 and 2804 meters (7800 and 9200 ft) on northerly aspects of relatively steep (30-45 degree), timbered slopes prone to deep accumulations of winter snows. The bears moved to and entered dens during October and November, at least partially in response to an increase in inclement weather. Most movements to the dens occurred during snow storms, and many were lengthy and direct. Time of emergence from the den was, once again, closely related to weather conditions, with adult male and solitary female bears generally leaving denning areas earlier than females with cubs. The bears moved from the higher elevations, often at night when the spring snows had crusted over, to thermal areas on the Central Plateau where winter-killed ungulates accumulated. As the season progressed into late May and June, movement toward the Trout Creek dump increased and those bears that did not winter on the plateau, bounded by the Yellowstone caldera, began arriving, many from beyond the Park borders (Fig. 5.11).

The ecocenter may have influenced the choice of denning areas. The clustering of active dens within a radius of 16 to 24 kilometers (10 to 15 mi) of Trout Creek suggests such an effect, but the intensity of our field effort in locating dens was not comparable over the entire Park. Far more time and effort was expended seeking dens in the Trout Creek area of the Central Plateau than elsewhere. Thus, the distribution of known denning areas was only consistent with the movement data and the conclusion that bears of the Trout Creek aggregation used much of the Park and Ecosystem at one time of the year or another. It was clear that many members of the Trout Creek aggregation had their home ranges and denning sites confined to the Central Plateau, whereas others were essentially visitors to the ecocenter, with feeding and denning sites separated by extensive migratory corridors, some extending to the periphery of the Ecosystem. The Trout Creek and other lesser ecocenters were strong biological magnets influencing the behavior of grizzlies throughout the Greater Yellowstone area.

SUMMARY

The aggregation of grizzly bears at Trout Creek and at other ecocenters was well established by the time we started our work in 1959. The ecocentered population structure characteristic of the Trout Creek aggregation, and of the entire ecosystem-wide population, had its origins in the late 1800s when Yellowstone Park and associated open refuse dumps were first established. Virtually nothing specific is known about the nature, prevalence, or size of natural ecocenters and aggregations in the Yellowstone Ecosystem prior to settlement. In Chapter 13, we will present evidence that such ecocenters must have existed.

The Trout Creek aggregation reformed and disbanded each summer in a predictable fashion. It consisted of two segments established by movement and denning patterns. One segment was centered within the Central Plateau; the other was widely dispersed throughout the Park and Ecosystem. Those bears wintering on the Central Plateau were the first to arrive at the Trout Creek ecocenter. Those wintering elsewhere arrived much later. Bears frequented the ecocenter for about three months. Bear visitation, sporadic in the latter half of May, peaked and plateaued during mid-July to mid-August, then declined and eventually ceased in September. Seasonal changes in the size of the aggregation were clearly tied to the abundance of food at Trout Creek and to the arrival schedule of bears that wintered at the periphery of the Park and Ecosystem. For extensive periods after leaving winter dens in April and May (but before the seasonal onset of refuse dumping), and before denning in October and November (but after cessation of dumping), aggregation members foraged almost exclusively in the backcountry.

The maximum number of bears recorded in an individual annual census from 1960 through 1968 varied from 65 to 88. Although large daily aggregations of bears were not uncommon at the seasonal peak, the average number of different bears belonging to the Trout Creek aggregation over an entire year during this same period was much greater (n = 116 bears). The difference in daily versus annual

Females with offspring exhibited the greatest season-long visitation to the Trout Creek ecocenter. Their special nutritional needs outweighed the danger from aggressive competitors. *(Photo by John J. Craighead)*

size was due to substantial individual variation in bear visitation at Trout Creek. Only a small number of bears were continuously present in the aggregation. The majority of bears spent varying, but often significant, portions of the aggregation season away from Trout Creek. Radiotracking showed that they foraged in the surrounding backcountry.

Some differences in visitation frequency were unpredictable, reflecting, for example, home range placement relative to the Trout Creek ecocenter (dump), chance discoveries of rich patches of natural foods, and so on. Nevertheless, differences in visitation were predictable from an individual's sex and age. In general, bears with special nutritional needs (females with offspring or rapidly growing subadults) spent less time in the backcountry and had relatively high frequencies of visitation at the ecocenter. The average age structure of the annual aggregation was biased toward subadults, reflect-

ing high rates of recruitment and relatively high subadult survival during the study period. Sex ratios were male-biased among nonadults but female-biased among adults.

The concept of an ecosystem-wide, ecocenter-organized population structure, in which bears moved routinely between large and broadly overlapping regions of the backcountry and one of several traditional ecocenters is supported by:

1. extensive use of both the Trout Creek ecocenter and the surrounding backcountry during the aggregation season;
2. backcountry use by Trout Creek bears both before and after the aggregation season; and
3. widely dispersed backcountry locations of winter dens.

Information to develop this concept was obtained from observations of color-marked animals and from locations of radio-instrumented ones.

6

The Grizzly Bear Hierarchy

INTRODUCTION

As the bears aggregated at Trout Creek, the interactive behavior of members revealed a strong social hierarchy. It seems reasonable that social hierarchies, with their attendant behaviors, also exist among bears pursuing semi-solitary lifestyles, but under such conditions, hierarchies are less apparent and, therefore, have not been described in the literature. Only in aggregations do frequent interactions lead to strong, overt dominance relations that reveal the social order. Our observations at Trout Creek provided insights into how dominance relations and a social hierarchy develop and are maintained when grizzlies aggregate. On occasion, anthropomorphic descriptive terms may be used to convey the sense of relationships or interactions, but we have no intention to ascribe thereby any human motivations or emotions to the bears.

Social hierarchies have been described for many animal species (Brown 1975, Wilson 1975, Walters and Seyfarth 1987). Hornocker (1962) first described a social behavior and a hierarchical structure for grizzly bears at the Trout Creek garbage dump in Yellowstone National Park. Although Hornocker (1962) showed that a social hierarchy existed among the bears at Trout Creek, he did not develop a method for quantifying aggressive behaviors among the adult male members of the aggregation that could be consistently applied year after year. To decipher the many behaviors we observed among the aggregated bears, and to describe the annual structuring of the social hierarchy, we developed a rating system consistent with the quantitative recording capability of that time, 35 years ago, and accumulated a database for assigning rank. We believe the rating system expresses the relative importance of attack and threat behaviors in defining four orders of male dominance—alpha, beta, contending, and noncontending males.

From a long and intimate association with the individuals composing the bear aggregation, and a close study of their "personalities," we are confident that our four categories of dominance accurately portray the hierarchical structure that developed among adult males at Trout Creek. A social organization, similar to that described here and by Hornocker (1962), was later described for Alaskan brown bears (*Ursus arctos middendorffi*) feeding on spawning salmon (*Salmo* spp.) at McNeil Falls, Alaska (Stonorov and Stokes 1972).

Cursory observations made at Trout Creek during the month of July, 1958, revealed that a large aggregation of bears regularly used the open-pit garbage dump. At the time, there was no way to estimate their numbers accurately. Disorder rather than order appeared to govern their coming and going and the quarreling of bear with bear. It was obvious that food was a major attraction, and that some of the larger and more powerful bears dominated other bears for choice of feeding sites. One male, in particular, dominated the bears in attendance. We christened the male "Scarface" because of his conspicuous battle scars. We later learned that this Scarface was not the first grizzly bear to be so named. In 1930, a scar-faced male controlled an area at the Old Faithful bear feeding grounds and dominated the bears in the area for several years (Yellowstone Nature Notes 1930).

The new Scarface displayed all the characteristics we later found to be characteristic of alpha males in general. (The concept of an alpha male and the formation of a social structure among bears had not yet been reported in the scientific literature). Dave Condon, Chief Naturalist, Yellowstone National Park, called *our* Scarface the "boss bear" and said he had been the boss for several years. This bear played a major role in the social order in 1959 when our study began. He was no longer the "boss," but it was evident to us in retrospect that he had been an alpha male and his "reign" terminated the year our study began. Although intensive observations to record dominance behavior within the aggregation did not begin until 1959, we nevertheless recorded Scarface as the alpha male for 1958.

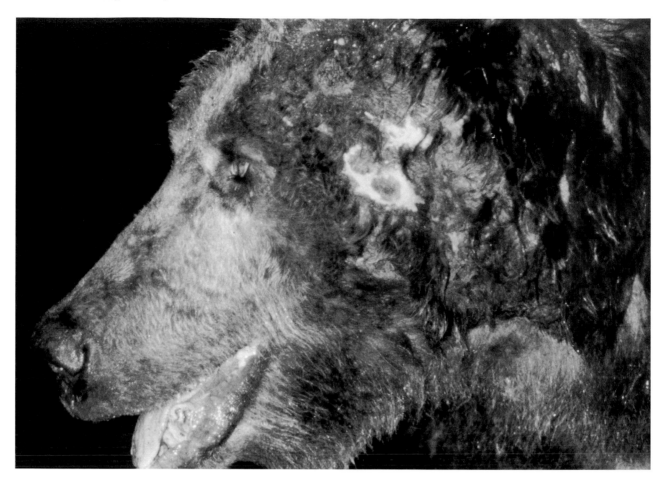

Scarface (top) was the alpha male at the Trout Creek aggregation when defeated by Ingemar (bottom) in 1959, the year our study began. *(Photos by John J. Craighead)*

Table 6.1. Hierarchy of adult male grizzly bears at Trout Creek ecocenter, Yellowstone National Park, 1959-70.

	Year					
Rank	1959	1960	1961	1962	1963	1964
Contender:						
Alpha:	Ingemar (No. 12)	Ingemar (No. 12)	Ingemar (No. 12)	Scarneck	Ingemar (No. 12)	Ingemar (No. 12)
Beta:	Cutlip	Fidel (No. 206)	Fidel (No. 206)	No. 111	Grizzled Boar	White-faced Boar (No.197)
Other-contender:	Scarface	Cutlip	Cutlip	Lrg Brn Boar (No. 33)	Short-eared Boar	Scarneck
	Old Big-headed Boar	Scarchest (No. 73)	Ivan (No. 41)	Grizzled Boar	Patch-eye (No. 88)	Patch-eye (No. 88)
			Scarneck	Patch-eye (No. 88)	Scarneck	Bruno (No. 14)
				Bruno (No. 14)		
Noncontender:	No. 27	No. 28	No. 28	No. 27	Brown Boar	No. 51
	No. 28	No. 46	No. 46	No. 28	Bruno (No. 14)	No. 195
	Bruno (No. 14)	No. 60	No. 60	No. 74	Fidel (No. 206)	Big Black Boar
	Old Blue (No. 30)	Bruno (No. 14)	No. 87	No. 85	Ivan (No. 41)	Big Brown Boar
	One Bar	Highpockets	No. 111	Brown Boar	Lrg Brn Boar (No. 33)	Crooked Fang
	Scarchest (No. 73)	Ivan (No. 41)	Bruno (No. 14)	Fidel (No. 206)	Old Blue (No. 30)	Fidel (No. 206)
	Thin Scarred Boar	Lrg Brn Boar (No. 33)	Lrg Brn Boar (No. 33)	Ingemar (No. 12)	Old Mose (No. 179)	Ivan (No. 41)
		Large Dark Boar	Medium-sized Boar	Ivan (No. 41)	Peg Leg (No. 76)	Light-muzzled Boar
		Old Blue (No. 30)	Old Blue (No. 30)	Old Blue (No. 30)	Roual	Little Brown Boar
		Old Mose (No. 179)	Old Mose (No. 179)	Old Mose (No. 179)	Square-muzzled Boar	Mean Brown Boar
		Orange Stripe	Patch-eye (No. 88)	Short-eared Boar		Medium Black Boar
		Scarface	Scarchest (No. 73)	Shorty		Old Blue (No. 30)
		Shorty	Shorty			Old Mose (No. 179)
						Peg Leg (No. 76)
						Scarcheek
						Short-eared Boar
						Tower Boar
						Zorro
Total males:	11	17	18	18	15	23

	Year					
Rank	1965	1966	1967	1968	1969	1970
Contender:						
Alpha:	Ingemar (No. 12) White-faced Boar (No. 197)	Whitefang	Patch-eye (No. 88)	Little Brown Boar	Gandolf (No. 258)	Gandolf (No. 258)
Beta:	Scarneck	Ingemar (No. 12)	Little Brown Boar	Mohawk	Black-eyed Boar	Patch-eye (No. 88)
Other-contender:	Fidel (No. 206)	Patch-eye (No. 88)	No Hump	Big Brown Boar	None	No. 166
	Little Brown Boar	Scar-shouldered Boar	No. 217	No. 166		
	Ivan (No. 41)					
Noncontender:	No. 51	No. 51	No. 130	No. 130	No. 135	No. 135
	No. 130	No. 130	No. 135	No. 135	No. 147	No. 147
	No. 195	No. 166	No. 147	No. 147	No. 158	No. 177
	No. 210	No. 210	No. 166	No. 158	No. 166	No. 181
	Big Black Boar	Big Brown Boar	No. 210	No. 177	No. 177	No. 188
	Big Brown Boar	Bruno (No. 14)	Big Brown Boar	No. 178	No. 181	No. 190
	Bruno (No. 14)	Fidel (No. 206)	Brown-faced Boar	No. 181	No. 183	No. 199
	Cassius Claw	Ivan (No. 41)	Fidel (No. 206)	No. 190	No. 188	No. 201
	Lop-eared Boar	Ivan II	Fuzz Face	No. 191	No. 190	No. 207
	Medium Black Boar	Little Brown Boar	Ivan (No. 41)	No. 194	No. 191	No. 211
	Old Blue (No. 30)	Little Grizzled Boar	Old Blue (No. 30)	No. 199	No. 194	No. 213
	Patch-eye (No. 88)	Medium Black Boar	Scar-shouldered Boar	No. 201	No. 199	No. 232
	Peg Leg (No. 76)	New Brown Boar	Scarcheek	No. 207	No. 201	No. 249
	Scar-shouldered Boar	Old Blue (No. 30)	Scarhump	No. 210	No. 207	No. 253
	Scarcheek	Peg Leg (No. 76)	Scarneck II	No. 217	No. 208	No. 257
	Scarhump	Scarcheek	Short-eared Boar	No. 249	No. 249	No. 263
	Short-eared Boar	Scarhump		Fuzz Face	Big Brown Boar	Big Brown Boar
		Short-eared Boar		Ivan (No. 41)	Fuzz Face	Big-eared Boar
				No Hump	Ivan (No. 41)	Blond Boar
				Old Blue (No. 30)	Light Brown Boar	Curl Ear
				One-eared Boar	Mohawk	Fuzz Face
				Patch-eye (No. 88)	One-eared Boar	Ivan (No. 41)
				Spaz	Patch-eye (No. 88)	Old Yellar (No. 2)
					Scarnose	One-eared Boar
						Red Boar
						Scarnose
						Wounded Leg
Total males:	23	22	20	27	26	30

Note: Adult males = 5-year-olds and older; other-contending males are listed in order of relative dominance; noncontending males are listed in numerical or alphabetical order; marked males which were recognizable before being marked are included in this table for their unmarked but recognizable years.

The male hierarchy consisted of two distinct strata of adult males. Large aggressive (contending) males (top) and large nonaggressive (noncontending) ones (bottom). They could not be distinguished by size or conformation, but behavior clearly separated the two strata. *(Photos by John J. Craighead)*

Fig. 6.1. Example showing method for annually recognizing the alpha and contenders in the grizzly bear social hierarchy at the Trout Creek ecocenter, Yellowstone National Park, using data from 1964.

Date (Aggregation size)	5 June (22)	6 June (27)	9 June (35)	10 June (30)	11 June (33)	13 June (30)	15 June (40)
Contending and noncontending adult males involved in the recorded aggressive encounters, Threat: → Attack: ▶ (Arrow points to submissive male)	Scarneck → Patch-eye (No. 88) → Ivan (No. 41) → *Unmarked* → Peg Leg (No. 76)	Ingemar (No. 12) → Scarneck → Patch-eye (No. 88) → Big Brown → Medium Black	*Unidentified* → Scarneck → *Unmarked* → Ivan (No. 41)	White-faced (No. 197) → Scarneck → Aggregation	Patch-eye (No. 88) → *Unidentified*	White-faced (No. 197) → Patch-eye (No. 88)	Ingemar (No. 12) → Scarneck
Contending adult males present, but with no aggressive encounters			Patch-eye (No. 88)	Patch-eye (No. 88)	White-faced (No. 197)		Patch-eye (No. 88), White-faced (No. 197)
Points for Beta (White-faced (No. 197))	0	0	0	4	0	2	0
Points for next closest rival (Scarneck)	6	3	0	3	0	0	1

Date (Aggregation size)	18 June (48)	19 June (55)	23 June (40)	25 June (42)	26 June (42)	30 June (56)	Outcome
Contending and noncontending adult males involved in the recorded aggressive encounters, Threat: → Attack: ▶ (Arrow points to submissive male)	Ingemar (No. 12) → Scarneck → Short-eared → Patch-eye (No. 88) → White-faced (No. 197) → *Unidentified* → Aggregation	Ingemar (No. 12) → Patch-eye (No. 88) → White-faced (No. 197) ▶ Scarneck	Ingemar (No. 12) → Patch-eye (No. 88)	Ingemar (No. 2) → Bruno (No. 14) → White-faced (No. 197) ▶ Scarneck → Aggregation	Ingemar (No. 12) → White-faced (No. 197)	Ingemar (No. 12) ▶ White-faced (No. 197) → Aggregation	Ingemar (No. 12) established as Alpha; White-faced (No. 197) recognized as Beta
Contending adult males present, but with no aggressive encounters	Bruno (No. 14)	Bruno (No. 14)	Bruno (No. 14)	Patch-eye (No. 88)	Patch-eye (No. 88)	Bruno (No. 14), Patch-eye (No. 88), Scarneck	
Points for Beta (White-faced (No. 197))	1	5	0	5	1	3	Total: 21 points
Points for next closest rival (Scarneck)	2	0	0	1	0	0	Total: 16 points

From 1958 through 1970 (13 years), seven males (including Scarface) attained alpha rank at Trout Creek. Thus, our record of alpha males began with Scarface, who was replaced by Ingemar (or Inge, No. 12) in 1959 (Table 6.1).

During each of the 12 years of the study, the number and type of dominating encounters were used to determine dominance relations and to assign levels of status among males, as well as among some females. Encounters varied so greatly in intensity, duration, and final consequences that it became apparent early in the study that simple frequencies of attack or threat alone would not provide adequate measures of male dominance. Both the chronology and types of encounters between paired antagonists throughout the breeding season were necessary to identify the alpha male and to assign status to the remaining males (Fig. 6.1).

Assigning Alpha Status

We recognized four levels of male dominance status: alpha, beta, other-contenders, and noncontenders. For convenience, males in the first three categories are sometimes referred to collectively as contending males. Typically, only one male at a time held alpha or beta rank (the single exception was a case of co-alphas in 1965). The other two categories included the rest of the adult males in the aggregation. There was a clear division of the hierarchy into males that contended for status and males that did not. Noncontenders did not initiate aggressive action, but were involved in aggressive action forced upon them by contenders. Noncontending males rarely interacted aggressively with other noncontending males. Hence, we could not determine individual rank within the noncontending stratum. Nevertheless, each year we attempted to rank-order males individually in the contending stratum.

The highest ranking or alpha male was defined as the individual observed to dominate every other male that contended overtly for status in any given season (1959 to 1970). The methodology is illustrated in Fig. 6.1, which documents the sequence of interactions that led to the identification of Ingemar (No. 12) as alpha male in 1964. A total of five males overtly contended for status that year. The male shown at the top of the column for each observation period is the male whose encounters for that 3.5-hour period indicated he was dominant among males present. Ingemar (No. 12) dominated each of the other four males at some point during the breeding season and was never observed to be dominated in turn (although he tended to avoid some individuals early in the season).

Thus, the progression and the results of paired encounters leading to the "establishment" of an alpha male is a chronology of daily interactions. In 1964, Ingemar (Inge, No. 12), the alpha male of the previous year, contended with Scarneck, the White-faced Boar (No. 197), Patch-eye (No. 88), and Bruno (No. 14). The progression of events (as diagrammed in Fig. 6.1) follows:

On the first census, 5 June 1964, an aggregation of 22 bears was present at Trout Creek, but Inge (No. 12) was not among them. Scarneck was present and dominated Patch-eye (No. 88) by threatening him three times. Patch-eye (No. 88) exhibited submissive behavior, but he in turn threatened Ivan (No. 41). An unmarked male threatened Peg Leg (No. 76).

Inge was first seen during the second observation period on 6 June. Upon arrival he threatened Patch-eye (No. 88), Scarneck, and two other adult males. He is shown as dominant for the observation period 6 June.

On 9 June, an unidentified male was the most aggressive.

On 10 June, the White-faced Boar (No. 197) arrived for the first time and clearly dominated Scarneck. Patch-eye (No. 88) was present, but did not interact with the other contending males.

On 11 June and 13 June, Patch-eye (No. 88) and the White-faced Boar (No. 197), respectively, were the most dominant males present. On 13 June, the White-faced Boar clearly dominated Patch-eye in a chase ending in submissive behavior by Patch-eye. Inge had not been present since 6 June, but appeared on 15 June. He threatened Scarneck, but did not interact with either No. 88 or No. 197, the other contenders present.

On 18 June, Inge (No. 12) encountered the White-faced Boar (No. 197) (probably for the first time) and confronted him as well as Patch-eye (No. 88). These bears fled when pursued. He treated Scarneck and an unidentified, adult male in the same manner. With less aggressive threats (growls, stance, head and ear movements, etc.), he exerted his influence over many other members of the aggregation. He did not feed during the entire 3.5-hour period, directing his energies to aggressive activity.

The following evening, 19 June, with an aggregation of 55 bears, Inge (No. 12) chased Patch-eye (No. 88), who was consorting with female No. 180, and then returned to copulate with her. On 25 June, the White-faced Boar (No. 197) attacked and fought fiercely with Scarneck. The White-faced Boar emerged dominant, but did not challenge Inge. This marked the beginning, however, of several encounters between the two that would ultimately determine the alpha male. Because of the interactions that occurred on 26 June and 30 June, Inge was rated the most

Table 6.2. Aggressive encounters involving the alpha male grizzly bears at the Trout Creek ecocenter, Yellowstone National Park, 1959-70.

Year	Bear name (no.)	Number of aggressive encounters								
		Attacks on		Attacks by	Threats toward			Threats from	Encounters per observation hour	
		Contending males	Adult non-contending males	Adult females	Bear aggregation[a]	Contending males	Adult non-contending males	Adult females	Total	
1959	Ingemar (No. 12)	2	0	1	1	3	0	0	7	2.000
1960	Ingemar (No. 12)	3	4	1	5	15	5	1	34	0.673
1961	Ingemar (No. 12)	3	3	0	2	4	20	0	32	0.400
1962	Scarneck	4	0	2	0	14	4	3	27	0.372
1963	Ingemar (No. 12)	4	0	0	4	7	3	0	18	0.214
1964	Ingemar (No. 12)	1	0	0	3	12	5	4	25	0.227
1965[b]	White-faced Boar (No. 197)	0	0	0	0	4	1	0	5	0.056
1965[b]	Ingemar (No. 12)	0	0	0	3	2	1	0	6	0.067
1966	Whitefang	3	0	0	0	1	0	0	4	0.053
1967	Patch-eye (No. 88)	1	1	0	2	2	1	0	7	0.065
1968	Little Brown Boar	0	0	0	1	2	0	3	6	0.064
1969	Gandolf (No. 258)	1	0	0	0	0	3	0	4	0.060
1970	Gandolf (No. 258)	0	0	0	1	1	3	0	5	0.156
Total		22	8	4	22	67	46	11	180	

[a] A bear aggregation occurs when numerous grizzlies concentrate in a small area, usually a feeding site.
[b] Co-alphas.

dominant male present among the assemblage of 42 bears of both sexes and all ages.

Finally, on 30 June, among an aggregation of 56 bears, Inge (No. 12) challenged his closest rival, the White-faced Boar (No. 197), forcing him to retreat following physical contact. A prolonged battle did not develop, but neither the White-faced Boar nor any of the other contending males ever seriously challenged Inge's authority again that season. The same day that Inge dominated the White-faced Boar, he scattered the entire aggregation by moving rapidly and purposefully among them, forcing bear after bear to retreat, step aside, or run. This was accomplished through overt aggressive encounters and the more subtle behaviors and signals that eventually were used to maintain the hierarchy throughout the rest of the summer. Thus, the alpha male was designated by his encounters and the specific bears he dominated (Table 6.2 and Fig. 6.1).

Assigning Beta and Contending Status

To determine rank order among the remaining contending males, we devised a point rating system based on our observations of aggressive behaviors (Table 6.3; see also Fig. 6.1). The point system is subjective, but not arbitrary. It proved helpful in evaluating the status of the contenders that did not attain alpha rank. Encounter points were assigned to these contending males as follows:

Contending male attacked by alpha male—
3 points.
Contending male attacks another contending male—
3 points.
Contending male threatens bear aggregation—
3 points.
Contending male attacks a noncontending male—
2 points.
Contending male threatens a contending male—
2 points.
Contending male receives threat from alpha male—
1 point.
Contending male threatens noncontending male—
1 point.

The weighting system was based on the observation that the intensity of aggressive interaction was at its highest for a given type of encounter whenever a contender interacted with another contender, or when a contender threatened the entire aggregation. Conversely, the intensity of interaction was lower when contenders interacted with noncontenders. Similarly, attacks represented more intense aggressive interaction than did threats. We made the assumption that intense aggressive interactions were more reliable indications of overall status than interactions of low intensity. Thus, for example, the high intensity encounters between two contenders were weighted more heavily than interactions of the same type involving a

Table 6.3. Aggressive encounters of beta and other-contending male grizzly bears, Trout Creek ecocenter, Yellowstone National Park, 1959-70.

| | | Attacks | | | | | | Toward aggregation[b] (3 points ea.) | | Threats | | | | | | | |
| | | By alpha (3 points ea.) | | On other contenders (3 points ea.) | | On adult non-contenders (2 points ea.) | | | | Toward other contenders (2 points ea.) | | From alpha (1 point ea.) | | Toward adult noncontenders (1 point ea.) | | Total number of encounters | Total encounter points |
Year	Bear name and/or no.[a]	No.	Pts.	No.	Pts.	No.	Pts.	No.	Pts.	No.	Pts.	No.	Pts.	No.	Pts.		
1959	Cutlip	0	0	1	3	2	4	0	0	4	8	2	2	0	0	9	17
	Scarface	1	3	0	0	0	0	0	0	0	0	1	1	0	0	2	4
	Old Big-headed Boar	1	3	0	0	0	0	0	0	0	0	0	0	0	0	1	3
1960	Fidel (No. 206)	3	9	0	0	0	0	0	0	0	0	9	9	0	0	12	18
	Cutlip	0	0	0	0	0	0	0	0	0	0	4	4	0	0	4	4
	Scarchest (No. 73)	0	0	0	0	0	0	0	0	0	0	2	2	0	0	2	2
1961	Fidel (No. 206)	1	3	0	0	0	0	0	0	0	0	3	3	2	2	6	8
	Cutlip	1	3	0	0	1	2	0	0	0	0	1	1	1	1	4	7
	Ivan (No. 41)	1	3	0	0	0	0	0	0	0	0	0	0	0	0	1	3
	Scarneck	0	0	0	0	0	0	0	0	0	0	0	0	2	2	2	2
1962	No. 111	0	0	1	3	0	0	0	0	1	2	6	6	0	0	8	11
	Large Brown Boar (No. 33)	2	6	0	0	0	0	0	0	0	0	0	0	0	0	2	6
	Grizzled Boar	1	3	0	0	0	0	0	0	0	0	3	3	0	0	4	6
	Patch-eye (No. 88)	0	0	0	0	0	0	0	0	0	0	4	4	0	0	4	4
	Bruno (No. 14)	1	3	0	0	0	0	0	0	0	0	1	1	0	0	2	4
1963	Grizzled Boar	1	3	0	0	0	0	0	0	1	2	3	3	1	1	6	9
	Short-eared Boar	1	3	1	3	0	0	0	0	1	2	0	0	0	0	3	8
	Patch-eye (No. 88)	1	3	0	0	0	0	0	0	0	0	2	2	0	0	3	5
	Scarneck	1	3	0	0	0	0	0	0	0	0	0	0	0	0	1	3
1964	White-faced Boar (No. 197)	1	3	2	6	0	0	0	0	5	10	2	2	0	0	10	21
	Scarneck	0	0	0	0	0	0	1	3	4	8	4	4	1	1	10	16
	Patch-eye (No. 88)	0	0	0	0	0	0	0	0	0	0	5	5	0	0	5	5
	Bruno (No. 14)	0	0	0	0	0	0	0	0	0	0	2	2	0	0	2	2
1965	Scarneck	0	0	0	0	0	0	0	0	0	0	4	4	0	0	4	4
	Fidel (No. 206)	0	0	0	0	1	2	0	0	0	0	0	0	1	1	2	3
	Little Brown Boar	0	0	1	3	0	0	0	0	0	0	0	0	0	0	1	3
	Ivan (No. 41)	0	0	0	0	0	0	0	0	0	0	1	1	1	1	2	2
1966	Ingemar (No. 12)	1	3	0	0	0	0	0	0	0	0	1	1	2	2	4	6
	Patch-eye (No. 88)	1	3	0	0	0	0	0	0	1	2	0	0	0	0	2	5
	Scar-shouldered Boar	1	3	0	0	0	0	0	0	1	2	0	0	0	0	2	5
1967	Little Brown Boar[c]	1	3	0	0	0	0	1	3	0	0	0	0	0	0	2	6
	No Hump	0	0	0	0	1	2	0	0	1	2	1	1	1	1	4	6
	No. 217	0	0	0	0	0	0	0	0	2	4	1	1	1	1	4	6
1968	Mohawk	0	0	0	0	1	2	3	9	5	10	2	2	4	4	15	27
	Big Brown Boar	0	0	0	0	0	0	1	3	0	0	0	0	2	2	3	5
	No. 166	0	0	0	0	1	2	0	0	0	0	0	0	0	0	1	2
1969	Black-eyed Boar	1	3	0	0	1	2	0	0	0	0	0	0	1	1	3	6
1970	Patch-eye (No. 88)	0	0	0	0	1	2	0	0	1	2	1	1	3	3	6	8
	No. 166	0	0	0	0	0	0	1	3	0	0	0	0	2	2	3	5
	Total	22	66	6	18	9	18	7	21	27	54	65	65	25	25	161	267

[a] Males listed in relative order of dominance determined by encounter points; see text.
[b] A bear aggregation occurs when numerous grizzlies congregate in a small area, usually a feeding site.
[c] Little Brown Boar was dominant over No. 217 and No Hump, even though tied in encounter points.

contender and noncontender. Similarly, other things being equal, attacks were weighted more heavily than threats.

Encounter points were summarized for all the males that overtly contended for status throughout the study period but did not attain alpha rank (Table 6.3; see also Fig. 6.1). The assigned encounter points for these males were converted to points per observation hour. Males with the highest score each season were designated beta, or first-contenders, to the alpha male. Those with lower scores were classified as contending males and categorized as other-contenders to distinguish them from alpha and beta males. Thus, every year, each adult male was fitted into one of four dominance status categories: alpha, beta, other-contenders, or noncontenders (Table 6.1).

The number and severity of the encounters that occurred each season depended on the relative aggressiveness of the alpha male, and the number of contending males and their relative aggressiveness. In some seasons, the alpha male was apparently able to establish his position by dominating key contenders in relatively few encounters; in other seasons, the alpha male was continually involved in aggressive encounters with each of the contenders throughout the three- to four-week period of status resolution or "period of contention."

Some adult females were extremely aggressive and may have ranked high in the social order that developed each season. Aggressive encounters involving adult females were similar to those described for adult males, and they were recorded in a similar manner (see Table 3.2).

In summary, the hierarchy was a dynamic phenomenon, reforming and changing from year to year. It developed through the medium of aggressive encounters. The male hierarchy consisted of two distinct strata of adult males: aggressive males (contenders), some of whom could be ranked through their aggressive encounters, and nonaggressive males (noncontenders) that could not be ranked because they rarely interacted aggressively with the contending males or among themselves. Individual males typically moved between the contending and noncontending levels of status during their lifetimes, often in unpredictable fashion. Position in the contending stratum was not automatically conferred on the basis of past performance and rank. Each year an alpha male emerged from the competition and held the dominant position in the hierarchy until the aggregation disbanded in late summer. The majority of the competition took place in the vast arena of Hayden Valley and persisted over a period of approximately 45 days each season. Aggression

among females and among subadults, and between members of these classes and adult males, played insignificant roles in the development of the male hierarchy. As the male hierarchy developed, however, these other segments of the population found a place in the social order of the aggregation and were affected by it in various ways.

AGGRESSIVE BEHAVIORS OF ADULT MALES AT THE TROUT CREEK ECOCENTER

Aggressive encounters among aggregation members frequently involved spectacular feats of skill, strength, and endurance. Descriptions of these from abridged field notes for the year 1963 give some sense of the atmosphere of aggressive power that created the hierarchy each season. Marked male bears are identified by a number and/or a name; unmarked but individually recognizable bears are referred to by name only.

Journal notes are as follows:

18 June 1963—Scarneck has not yet been sighted in the aggregation. Short-eared Boar is closely attending female No. 112. Inge (No. 12) is following female No. 65; she is in estrus. Inge is leaving No. 65 to challenge Short-eared Boar. He pursues him over a quarter mile [0.4 km]; Short-eared Boar turns and makes a stand. Inge closes with his opponent, parries, and jockeys for position. They move cautiously like wrestlers seeking advantage and leverage. Inge suddenly bores in, seizes Short-eared Boar by the throat, leveraging his body in a twisting turn. This maneuver throws the 750-pound male to the ground on his back. We can hear the two roaring from well over a third of a mile [0.5 km] away. Short-eared Boar regains his feet; the two stand, man-like, their jaws locked, striking fierce blows with powerful forepaws. They relinquish their holds and drop to all fours. Inge makes a bluffing charge but Short-eared Boar does not retreat. They again lock jaws and Inge forces Short-eared to back up. It is apparently not a concession, but a maneuver. Short-eared lunges for Inge's throat. Inge lowers his head and dives for the Short-eared Boar's groin. They roll and tussle. Neither can get a firm hold. Inge keeps the pressure on, driving steadily on after each lunge, backing Short-eared down the slope some 20 feet [6.1 m]. Short-eared turns and runs; he is pursued by Inge at a fast pace through the sage for half a mile [0.8 km]. Inge stands, watching his retreating contender for several minutes. He ambles slowly back to female No. 65 and together they move over the hill, the male nudging the female along until they disappear from sight.

Short-eared Boar circles back to the dump and immediately shows interest in female No. 40, sniffing her genitalia (checking). The other males have shown no interest in her this evening. Short-eared mounts No. 40 and breeds her. Number 40 is receptive and the copulation lasts 19 minutes. Number 40 moves to a feeding site; Short-eared Boar moves to the creek, enters the water, and lays down. Female No. 40 was bred earlier by No. 88 [Patch-eye] on 11 June and by Inge on 14 June. Today's breeding extends her observed estrous period to a minimum of eight days. Inge appears to be the most dominant of the large males, but he is not yet recognized as alpha. The Grizzled Boar appears to be Inge's chief contender, but he has not yet appeared today.

19 June 1963—Female No. 112 is receptive to male No. 88, but he has been unable to breed her. Fighting and chasing contenders has interrupted his advances. Female No. 65 is present and apparently still in estrus as she is attracting the large, aggressive males. Inge is alert, his eyes fastened on male No. 88, 40 yards [36.5 m] away. He appears oblivious to female No. 65 and to other bears around him who are also alert and moving outward to a fringe of nervous animals. With ears erect and head thrust forward, Inge stands motionless, watching. In prelude to battle, each male approaches the other in a slow stiff-legged walk. Inge stomps his hind feet alternately downward and briefly holds his legs rigid before taking the next step in a similar manner. Advancing in this swaggering, robot-like gait, the males move closer. Both animals salivate profusely and urinate. The approach and ultimate showdown is delayed by frequent pauses as the contenders eye one another. Each appears to look for an appeasement sign or a propitious moment to charge. Inge continues his stiff-legged walk toward No. 88 and pauses. Number 88 moves in similar manner toward Inge. Both bears are studies in concentration: both seem oblivious to their surroundings. The pace increases. They are now 50 feet [15 m] apart. Most members of the aggregation have stopped feeding and are facing the contenders. Emitting a low, guttural growl, ears laid back and head low, Inge charges. It is amazingly fast. We detect no warning signal but No. 88 apparently does, for in split seconds he is prepared to defend himself. He does not meet Inge face-on as he has other contending males. He wheels around, and as he does so, Inge's jaws close on his rump, tearing free a 3-inch [7-cm] flap of hide exposing a white slash. Without losing stride or momentum, Inge climbs over No. 88, breaking him down. Number 88 twists under Inge's weight and rolls to his back, with all four feet fending off his opponent. Inge sinks his teeth deep into No. 88's groin as he lies pinned on his back. There is a brief tussle, the roaring loud

and continuous. Inge rears back, lifting his head. As he does so, he lifts the near-700-pound [320-kg] bear clear of the ground and shakes him. The action is occurring in seconds, and I find it hard to believe my eyes. Both bears hit the ground in a tumbling roll. Number 88 has Inge by the neck as he finally regains his feet. He releases his hold and clamps on Inge's jowls, shaking his massive head. Inge rears back, fighting free at the expense of a ripped and bleeding lower jaw. The two stand erect face to face, jaws gaping, teeth bared, emitting rumbling growls. They drop to all fours, and in a synchronous lunge rise together to clash teeth, slap with forepaws, and engage and disengage their jaws as each seeks an advantage. Inge throws his weight to the right, attempting to twist No. 88 to the ground, but No. 88 counters with a lunge to the left, his teeth imbedded in Inge's neck. They break apart and silently eye one another. The pause in action is only a few seconds but seems longer. Inge is the first to move. He steps upward to the right, taking advantage of the slope. Number 88 makes a leg dive; Inge sidesteps and slaps No. 88 across the shoulder with a terrific swipe of his right paw. I hear the slap and see deep furrows ripple through the fur. Number 88 lurches backward from the blow and Inge drives forward for No. 88's throat, exposed for a fraction of a second as he seeks his balance. Rising on hind legs, the two embrace with an audible clash of teeth. Biting and swiping, they exchange roars that carry across the valley and noticeably excite the spectator bears. The roaring subsides to growling and the contestants, now only feet apart, stand motionless eyeing one another with steady stares. Neither moves in this control of wills; 88 does not retreat but slowly lowers his head to one side, and eye contact is momentarily broken. Inge's head goes up, his ears erect. Number 88 holds his submissive pose, but with eyes on Inge; still, he does not retreat. Inge deliberately turns his back on 88 and very slowly moves 60 feet [18 m] to claim female No. 65. Number 88 could attack Inge from behind but he does not. This recognition of defeat is never breached. The contest is over; for this day Inge is victorious. There will be no sneak attack, no renewal of hostilities. The other bears begin to feed as though suddenly released. I can almost feel the tension subside. Number 65 had been receiving the attention of the Large Brown Boar while Inge was fighting. This male attempts to mount the female but retreats as Inge approaches. Number 88 remains where he had fought, breathing heavily, then climbs the embankment and mixes with the other males now actively feeding. He moves on to the creek. Inge, his chest expanding and contracting like huge bellows, stands with female No. 65 at his side, eyeing the aggregation. It is a challenging pose, but no bear responds. Ten minutes elapse before Inge moves. He appears to have lost interest in the female but his presence keeps other males at a distance. Neither Inge nor No. 88 appears to be seriously injured although both show wounds from the conflict. Inge is clearly dominant over No. 88 this day, but he is not yet recognized as alpha. There are other males yet to contend with.

26 June 1963—Inge (No. 12) and the Grizzled Boar [never captured and marked] have had three encounters earlier in the season, but none have been decisive. Since then, Inge and the Grizzled Boar have avoided one another. When they have approached closely, Inge has appeared to dominate. The Grizzled Boar has not challenged Inge and Inge has been relatively tolerant of his presence. The Grizzled Boar has had successful encounters with Short-eared Boar and Scarneck. He is powerful and aggressive but keeps his distance from Inge. There are three females present that are now in estrus. Inge has been aggressive toward all bears this evening, but there have been no battles.

Inge (No. 12) is now moving among the 60 bears present. All show him deference—some by moving aside, others by lowering the head, and some by ignoring him completely. The Grizzled Boar and the Short-eared Boar are squared off, intently eyeing one another. Inge moves swiftly toward them, attacking the Grizzled Boar from behind. The Grizzled Boar is unaware of his approach until he is crushed to the ground. Inge does not carry the fight but allows the Grizzled Boar to retreat. [Inge] has not been feeding; he seldom does. He is patrolling the aggregation, exerting authority. The Grizzled Boar moves toward a choice feeding site, scattering younger bears and female No. 96 with her one cub. A growl and a rapid step is all it takes for him to clear the site. The Grizzled Boar has fresh scars on his jowls and neck that were not there a day ago; he has been in a battle we did not observe, perhaps in the backcountry. His behavior toward Inge has changed drastically. He is now relatively submissive. He probably fought Inge for female No. 65 or possibly for female No. 40. In any case, he and Inge have battled, and Inge is now the alpha male. The Grizzled Boar recognizes this, as does the entire aggregation. There was no need to have seen the showdown; evidence speaks in the actions and behaviors of every bear. All defer to the "boss," from weaned yearlings and 2-year-olds to Inge's other major contenders, Short-eared Boar, No. 88, and Scarneck. Inge has dominated all the large aggressive males and has reclaimed his former status, relinquished to Scarneck last season (1962). We might say that he has established himself as alpha in 26 days of aggressive action and constant vigilance, or that the aggregation has accepted him as the dominant male after a 26-day period of contention. He has vigi-

lantly checked the fertility of estrous females and has copulated at least three times with two different ones. He has earned the right to select females without a battle and to take choice feeding sites throughout the summer.

Encounters such as those just described occurred each year and determined the ascension and succession of alpha males from one year to the next. The organization of the remainder of the male hierarchy fell into place by the same process (Table 6.1).

RANK DETERMINATION

The male hierarchy was established by competition among the most aggressive males in the aggregation. Those males ranked as most dominant were not necessarily the same individuals from year to year, and individual males frequently achieved different levels of dominance in different years of the study (Fig. 6.1). Adult males contended for status during much of the breeding season, a period of about two months. The intensive observations (censuses) at the Trout Creek ecocenter did not begin until 1 June, so our quantitative records for each season begin with that date. The observed breeding season, then, was a period of 45 days, on the average. For purpose of the study, the postbreeding season was designated as the 45-day period after 15 July through 30 August, when bears began leaving the ecocenter and census efforts tapered off. During the first 45-day observation period, hierarchical rank was established, pair by pair, as the most aggressive males resolved dominance relations.

Each breeding season, we recognized an alpha male and his principal contenders for that status by the time the general competition was concluding. Only then could we establish with certainty the identity of the alpha male among other contending males. As the dominant male, he was recognized by all members of the aggregation, and his role became fixed and well-defined throughout the remainder of the season.

CHRONOLOGY OF DETERMINATION

In our illustration of hierarchic dynamics (Fig. 6.1), the male ascending to alpha rank each year was, in every sense, a contender and not a defender of a previous position. Thus, the males in the contending stratum each year were those that competed actively for status, and usually only one achieved the alpha position. The beta male was the bear, each season, that most seriously challenged the alpha-to-be and thus was designated "first contender." Like the alpha male, his position in the

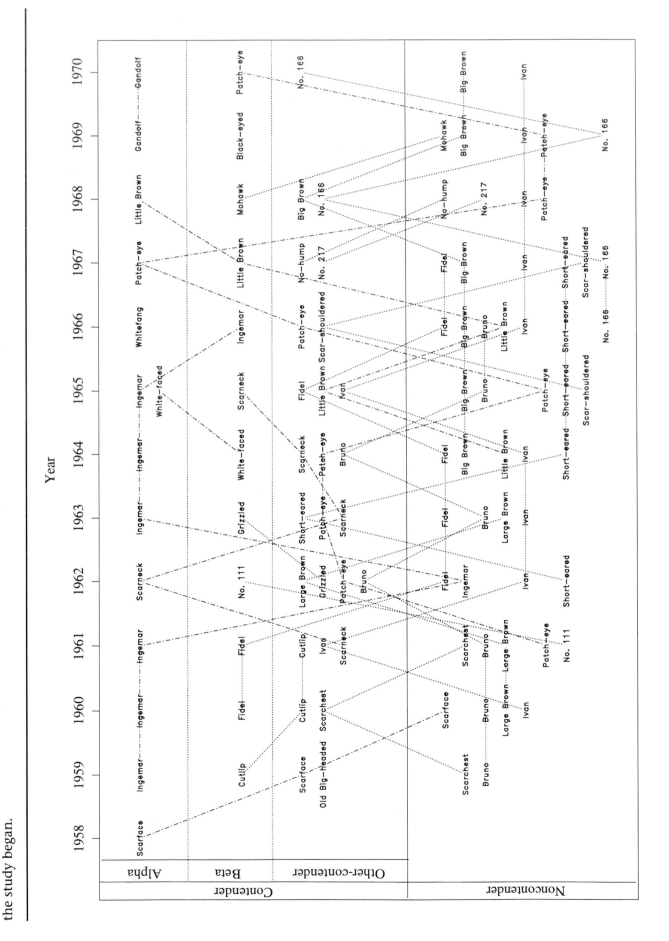

Fig. 6.2. Dynamics of the adult male grizzly bear hierarchy at the Trout Creek ecocenter, Yellowstone National Park, 1959-70. Only males that eventually reached the contending stratum are included in this figure (see Table 6.1 for complete information); Scarface was the alpha male before the study began.

Fig. 6.3. Levels of contention for seven adult male grizzly bears that reached alpha status at the Trout Creek ecocenter, Yellowstone National Park, 1959-70.

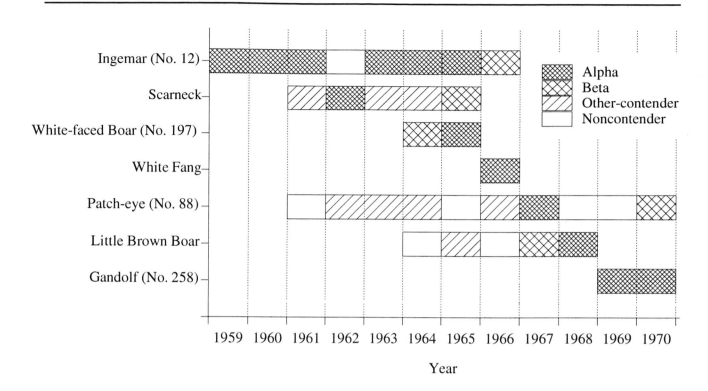

structure was determined by number and type of aggressive encounters (Table 6.3) and, ultimately, by his domineering behavior during the period of contention.

The successional events, related to rank adjustment, that occurred in the aggregation from year to year are described briefly here for ready reference and to present an overall view of dominance relationships we later describe in detail. The narratives that follow are based on data presented in Table 6.1 and Figs. 6.2 and 6.3.

1959. Cutlip, Scarface, and Old Big-headed Boar were contenders to Ingemar (or Inge, No. 12), who attained the alpha position. Inge was an 8- or 9-year-old bear of blocky proportions and commanding demeanor. Cutlip attained first-contender, or beta rank, the second most dominant position in the hierarchy. Scarface, the alpha bear in 1958, was defeated by Inge and dropped to other-contender status.

1960. Inge again attained alpha rank, with Fidel (No. 206) as his major adversary (beta). Cutlip dropped from the beta position he held in 1959 to the rank of other-contender. Scarchest (No. 73) rose from the rank of noncontender to establish himself as a contending male.

1961. Inge and Fidel re-established themselves as the alpha and beta males, respectively. Cutlip retained his position as an important other-contender, and Ivan (No. 41) emerged from the noncontending stratum to contend. Scarneck first appeared at Trout Creek en route to Rabbit Creek, where he had been the alpha male in 1960. He tarried at Trout Creek for a few days where he contended with the more aggressive males before moving on to Rabbit Creek, where he again was alpha.

1962. Scarneck arrived at Trout Creek and became the alpha male, remaining throughout the summer. He achieved alpha status through aggressive encounters with five contending males. Number 111, the beta male, had emerged from the noncontenders, as had three other contenders: Large Brown Boar (No. 33), Patch-eye (No. 88), and Bruno (No. 14). A fifth contender, the Grizzled Boar, arrived at Trout Creek for the first time. Inge was among the aggregation on 16 June, but did not challenge Scarneck. Inge returned again to Trout Creek on 19 June, severely wounded and limping badly. The following evening Scarneck chased Inge from the concentration site. Inge remained a noncontender throughout the season. Number 111 was never observed at Trout Creek again; after 1962, he joined the Rabbit Creek aggregation.

1963. Inge arrived on 14 June in excellent condition. He soon displayed his old assertive behavior, dominating four aggressive males. When the period of contention was over, about 9 July, Inge had regained the alpha role, and Scarneck, the 1962 alpha male, ranked lowest among the contenders. Patch-eye (No. 88) and Short-eared Boar were aggressive competitors, but the Grizzled Boar emerged as beta male. The Grizzled Boar was not observed at Trout Creek in later years.

1964. Inge competed with four males to win the alpha position once again. Three (Scarneck, Patch-eye, and Bruno) were contenders from previous years. White-faced Boar (No. 197), who became beta male, was new to the aggregation.

1965. Inge shared the alpha position with White-faced Boar, beta of 1964. They contended with four other males, two of which (Fidel and Ivan) had re-emerged from the noncontending stratum. Scarneck, who had consistently held a contending position since 1961 (including alpha in 1962), claimed beta status. He was never observed again in subsequent years. Little Brown Boar, first observed at Trout Creek the year before as a noncontender, rose to other-contender status.

1966. The alpha male from Rabbit Creek, Whitefang joined the Trout Creek aggregation for the first time and rose to the alpha position. He had held that rank at the Rabbit Creek ecocenter for five seasons, from 1962 through this year. Inge dropped to beta status and was not observed at Trout Creek or at Rabbit Creek in subsequent years. Patch-eye re-emerged and Scar-Shouldered Boar rose from the noncontending stratum to compete.

1967. Patch-eye became alpha male after six years in residence as a contender and noncontender. He had three contenders, two of whom (No Hump and No. 217) arrived at Trout Creek for the first time and were never serious challengers thereafter, only visiting Trout Creek for one more year. The third contender, Little Brown Boar, rose to beta rank. Whitefang, the alpha male the previous year, did not return to contend and was never observed again.

1968. Patch-eye did not regain alpha, falling to the noncontending strata. Little Brown Boar rose from beta male in 1967 to alpha male this year. His fiercest rival, Mohawk, was an aggressive male that joined the aggregation for the first time and rose to beta rank. The other contenders were Big Brown Boar and No. 166.

1969. Little Brown Boar, the alpha male in 1968, did not return to the aggregation to contend and was never observed again. Gandolf (No. 258), observed at Trout Creek for the first time, achieved the alpha position, competing with a single major contender, Black-eyed Boar, also new to the aggregation. Mohawk, Big Brown Boar, and No. 166, aggressive contenders from 1968, offered no opposition and were among the noncontending males. No bears filled the rank of other-contending males; this is the only time this occurred in the 12 years of this study.

1970. Gandolf retained the alpha role, vying with two contenders, No. 166 and Patch-eye. The latter had been alpha male three years previous. After two years as a noncontender, he had returned this year to contending status and attained beta rank. In attendance at Trout Creek since 1961, Patch-eye was in contending status for 5 of the 9 years. Number 166 was sole entry in other-contender status despite a field of 27 noncontenders—the largest number for that stratum in all the years of the study.

Understanding the behavior of the dominant animals and how they exerted power to acquire mates and food was fundamental to understanding the hierarchy and its function within the bear aggregation. Of prime interest was the process of resolving alpha status.

RESOLVING ALPHA STATUS

Attacks and threats, each breeding season, revealed that a specific bear was the alpha male. The time required for this revelation varied greatly from year to year. Decisive battles may not have been necessary to achieve alpha status, but considering the aggressiveness of the principal contenders, one or more such battles probably occurred each breeding season. We believe that we observed and recorded a high percentage of the encounters among contending bears at Trout Creek. Some unobserved encounters between males did occur, however, as evidenced by large, freshly-scarred males that were not wounded at Trout Creek, but possibly in the backcountry or at other ecocenters.

Our observations revealed that attacks among males were involved in determining the alpha position in 9 of 12 years. Threat behavior alone appeared to determine it in three, but we suspect that decisive attacks occurred that were unobserved (Table 6.4). In 1970, we observed the alpha male in a single aggressive action against the beta male. This appeared to resolve alpha status. On the other hand, while rising to alpha in 1960, Inge engaged in multiple encounters (23 observed, three of which were fierce battles). He engaged in numerous attacks

Table 6.4. Aggressive encounters of alpha male grizzly bears toward contending males and grizzly bear aggregations observed at Trout Creek ecocenter (835 observation hours), Yellowstone National Park, 1959-70.

		Aggressive encounters					
		Attacks		Threats			
Year	Alpha male	Beta	Other-contender	Beta	Other-contender	Aggregation[a]	Total encounters
1959	Ingemar (No. 12)	0	2	2	1	1	6
1960	Ingemar (No. 12)	3	0	0	15	5	23
1961	Ingemar (No. 12)	1	2	2	2	2	9
1962	Scarneck	0	4	6	8	0	18
1963	Ingemar (No. 12)	1	3	3	2	4	13
1964	Ingemar (No. 12)	1	0	2	10	3	16
1965[b]	White-faced Boar (No. 197)	0	0	2	1	0	3
	Ingemar (No. 12)	0	0	2	0	3	5
1966	Whitefang	1	2	1	0	0	4
1967	Patch-eye (No. 88)	1	0	0	2	2	5
1968	Little Brown Boar	0	0	2	0	0	2
1969	Gandolf (No. 258)	1	0	0	0	0	1
1970	Gandolf (No. 258)	0	0	1	0	1	2
	Total	9	13	23	41	21	107
	Grand total		22		85		

Note: Table does not include 215 observation hours in which no alpha male was present.
[a] A bear aggregation occurs when numerous grizzlies congregate in a small area, usually a feeding site.
[b] Co-alphas.

and threats over his 6-year reign (Table 6.4). His dominating behavior appeared to be related to innate aggressiveness rather than to responses to challenges from his competitors. He did not wait for challenges; he initiated them. Scarneck was also an extremely aggressive male and, like Inge, he seemed to seek physical contact. On the other hand, Gandolf (258) did not challenge other contenders, but responded aggressively only to the few bears that challenged him.

Attacks were the most decisive encounters, especially those against the beta male. Of the 22 attacks recorded during the 12-year period, 9 were directed against the beta or first-contenders and 13 against other-contenders. Among 85 threats, 23 were directed at beta males, 41 at other-contenders, and 21 were threats against the entire aggregation (Table 6.4). Attacks and threats directed at other-contenders were more numerous because of the large number of bears in this category. Threats directed against the entire aggregation most frequently signified which male had achieved the alpha role. We interpreted this behavior as an exercise of authority, rather than a step toward it.

The number of encounters required to "instate" the alpha male was compared on the basis of form

of encounter and rank of participants (Table 6.5). Among all contending groups, threats were employed more frequently than attacks. Total agonistic behavior decreased markedly according to social rank (Table 6.5). Statistical analysis showed that social rank had a significant effect on the mean rate of agonism (attack + threat) (median P-value of 13 Kruskal-Wallis tests < 0.00001; sample sizes were 6, 5, 7, and 27 for alpha, beta, other-contender and noncontender categories). A multiple-comparisons procedure for the Kruskal-Wallis test (Conover 1980) indicated that the overall effect was due to pairwise differences between rates for noncontenders and those for each of the contending categories ($P < 0.05$). A strong, but nonsignificant trend toward higher rates for agonistic behavior among alpha males versus among other contenders was also apparent. The small sample sizes in the three contender categories impeded any clear determination of significant differences between them specifically.

Toward the end of a breeding season, threat behaviors greatly outnumbered attacks. Once a dominant male had established himself as alpha, he normally did not have to threaten or attack his contenders to maintain rank. The erstwhile contenders typically would avoid the alpha male during the

Table 6.5. Rates of male grizzly bear aggression according to male social status, Trout Creek ecocenter, Yellowstone National Park, 1960-69.

Social rank	Mean rate[a]			Male-hours[c]	Male-years	Different males
	Attack	Threat	Agonism[b]			
Alpha	2.40	13.17	15.57	585	11[d]	7
Beta	1.19	5.71	6.90	420	9[e]	9
Other-contender	0.68	1.85	2.53	1026	22	15
Noncontender	0.03	0.18	0.21	3262	113	60

Note: based on 1050 hours of observation and 153 agonistic events from 1 June to 1 September, 1960-69; observations for 1959 were omitted because we had relatively few marked bears, and for 1970 because of disruption of hierarchical structure caused by closure of the Rabbit Creek dump.

[a] Events per male per 100 hours calculated for each different male-year, weighted by hours observed, and averaged over all male-years within each category of social rank.

[b] Sum of Attack and Threat rates.

[c] Total observation hours calculated for each different male-year and then summed over all male-years within each category of social rank.

[d] Co-alpha males in 1965.

[e] No rate available for beta male in 1965.

Scarneck was also an extremely aggressive male and, like Ingemar ("Inge"), he sought physical contact. *(Photo by John J. Craighead)*

Table 6.6. Time required each season for alpha male grizzly bears to establish dominance at the Trout Creek ecocenter, Yellowstone National Park, 1959-70.

Year	Alpha male	Date of first arrival	Date of last encounter	Period of contention (days)
1959	Ingemar (No. 12)	7/6	7/13	8
1960	Ingemar (No. 12)	6/17	7/6	20
1961	Ingemar (No. 12)	6/12	7/9	28
1962	Scarneck	5/28	6/28	32
1963	Ingemar (No. 12)	6/14	7/9	26
1964	Ingemar (No. 12)	6/6	6/30	25
1965[a]	Ingemar (No. 12)	6/5	6/27	23
	White-faced Boar (No. 197)	6/7	6/15	9
1966	Whitefang	6/10	6/22	13
1967	Patch-eye (No. 88)	6/11	7/4	24
1968	Little Brown Boar	6/15	6/18	4
1969	Gandolf (No. 258)	6/10	7/6	27
1970	Gandolf (No. 258)	6/19	6/25	7
Mean				18.9

[a] Co-alphas.

remainder of the season or, at his approach, exhibit submissive signals. As aggressiveness subsided with the resolution of male status and the decline in frequency of estrous females, subtle behavior cues such as body stance, positioning of ears and head, degree of eye contact, duration of focused attention, and vocalizations were employed more frequently and eventually replaced attacks or overt threats as a means of communicating status. These relatively nonaggressive behaviors were employed regularly by the noncontending males and other segments of the aggregation. During the postbreeding period, avoidance and deference behavior kept overt aggression at a relatively low level, such that each member of the aggregation fitted into the social structure with a minimum of persuasion. This postbreeding behavioral phase persisted from about 15 July to 30 August, when the aggregation began dispersing to fall ranges and denning sites.

TIME AND NUMBER OF MAJOR ENCOUNTERS REQUIRED TO ESTABLISH ALPHA RANK

The annual determination of rank among the male members of the aggregation no doubt began when adult males met other adult males in the backcountry after coming out of winter sleep. Rank resolution also progressed among members of the Trout Creek aggregation as soon as the bears began assembling and before many of the major male contenders had arrived. The time and effort required

for a male to achieve alpha status varied considerably from year to year. Records were kept on the interactive behavior of adult males, but we could not quantify the behavioral interactions which ultimately culminated in the alpha-to-be, until that bear arrived at the ecocenter.

As we have shown, aggressive encounters occurred throughout the breeding season. The observed period required to resolve alpha status, however, frequently was much shorter than the breeding season. Length of this period depended on when the alpha-to-be arrived at Trout Creek, where he could be observed, and the nature of his competition. We defined the period of contention for alpha status as the time period beginning with the arrival of the male that would eventually become alpha and terminating when he was recognized as alpha by his competitors and the entire aggregation. The observed periods of active contention at the ecocenter required for males to attain alpha status were as brief as 4 days and as long as 32, with a mean of 19 (Table 6.6). We recognize that competition occurred beyond our frame of observation and that such observations would have improved our data set.

In 1959, Inge (No. 12) arrived at Trout Creek on 6 July, the latest arrival recorded for any of the alpha males. He had no serious rivals, establishing his dominance over the aging Scarface and other contenders in eight days with a minimum of six encounters (Tables 6.4 and 6.6). In 1960 and 1961, Inge faced a more formidable array of aggressive

Table 6.7. Physical characteristics of adult male grizzly bears trapped at Trout Creek ecocenter, Yellowstone National Park, 1959-70.

Bear name and/or no.	Weight[a] (kg)	(lbs)	Body Length (cm)	(in)	Status
Bruno (No. 14)	235.8	520	31.6	80.25	Other-contending
Fidel (No. 206)	362.8	800	—	—	Beta
Gandolf (No. 258)	276.6	610	—	—	Alpha
Ingemar (No. 12)	281.2	620	32.7	83	Alpha, Beta
Ivan (No. 41)	249.4	550	35.8	91	Other-contending
Large Brown Boar (No. 33)	192.7	425	—	—	Other-contending
Patch-eye (No. 88)	226.8	500	30.3	77	Alpha, Other-contending
No. 13	292.5	645	—	—	Noncontending
No. 46	161.0	355	30.3	77	Noncontending
No. 85	210.9	465	29.5	75	Noncontending
No. 87	204.1	450	29.1	74	Noncontending
No. 89	288.0	635	31.7	80.5	Noncontending
No. 126	140.6	310	27.2	69	Noncontending
No. 155	299.3	660	29.9	76	Noncontending
No. 171	188.2	415	—	—	Noncontending
Old Yellar (No. 2)	244.9	540	31.3	79.5	Noncontending
Mean of contending[b] males (*n* = 7):	260.8	575	32.6	82.8	
Mean of noncontending males (*n* = 9):	225.5	497	29.9	75.9	

[a] Only June or July weights were used in this analysis; breeding season weight could not be obtained for many contenders, especially for some of the alpha males.

[b] Includes only males reaching the contending status.

males (Table 6.1 and Fig. 6.2). The time required to impose his authority over opponents was 20 and 28 days, respectively, and to accomplish this required a minimum of 23 and 9 encounters, respectively. Inge relinquished the alpha position in 1962 to Scarneck, who required 32 days and at least 18 encounters to attain alpha. In 1963, Inge regained alpha status in 26 days and had 13 observed encounters (Tables 6.4 and 6.6). He again established himself as alpha in 1964. To accomplish this required a period of 25 days with 16 observed encounters. In 1969, Gandolf (No. 258) exhibited one of the longest periods of contention (27 days), during which only one major encounter was observed. He showed no scars as evidence of encounters outside of our observation area. In the next year, the period of contention for Gandolf dropped to seven days, one-fourth the duration of the period required in 1969, and involved only two observed encounters.

There appeared to be no correlation between the observed time required to establish dominance (Table 6.6) and the number of observed encounters involving the dominating male (Table 6.4). Over his six years as alpha, Inge, an exceptionally aggressive male, averaged 22 days and a minimum of 12 encounters per season to attain that status. For all

other alphas combined, these averages were 17 days and 5 observed encounters.

One would expect highly aggressive males like Inge to resolve alpha status each year with few aggressive encounters and in a relatively short period of time. Inge, however, seemed to thrive on conflict and to encourage it. The same was true of Scarneck and Whitefang. Those males sufficiently aggressive to gain alpha rank and hold it for years may have prolonged the annual period of contention by the very nature of their aggressiveness.

SEASONS AND PHYSICAL ATTRIBUTES REQUIRED TO ATTAIN THE ALPHA POSITION AT TROUT CREEK

A male probably did not become alpha until he was fully grown and had experienced a few years of contending with other males. Large body size seemed an obvious prerequisite to dominance among bears of all ages. But so were superior fighting skills and experience, attributes that, like body size, were plausibly age-related. A comparison of weights and body measurements taken during June and July for contending and noncontending males captured at

Physical size and fighting skills were required, but to attain alpha status, aggressiveness appeared to be the overriding qualification for success. *(Photo by John J. Craighead)*

Trout Creek verified that contending males were heavier and larger, on the average, than noncontenders (Table 6.7). In this small sample, however, there was no statistical association between the male weight and his social rank category (Mann-Whitney test, $P > 0.10$). A larger and more complete sample might show an association between body size and the attainment of contending status among mature males. Nevertheless, the characteristic of all contending males, and alpha males in particular, that appeared to be most closely associated with that status was aggressiveness and a willingness to fight.

The number of years (seasons) at Trout Creek required for seven males (excluding Scarface) to attain the alpha position varied from zero to six years (Table 6.8). Inge (No. 12) rose to alpha status in the first year of our study; his prior history of contention at Trout Creek was unknown. Gandolf (No. 258) and Whitefang achieved alpha with no previous period of contention. But they had attained high rank at the Rabbit Creek ecocenter before attaining alpha among the larger aggregation at Trout Creek. This probably explains their rapid rise

to power in the first season each arrived. We also suspected that Inge had been an alpha male at Rabbit Creek before arriving at Trout Creek, but because he arrived in the first year of the study, we had no way to verify our suspicion. Thus, the smaller ecocenter at Rabbit Creek appeared to be the proving grounds for some of the high-ranking males before their arrival at Trout Creek. Our observations at Rabbit Creek were not thorough enough to determine conclusively whether all the contending bears that rose to alpha position in their first year of arrival at Trout Creek had previously held that position at Rabbit Creek, but the evidence strongly suggests it.

Scarface was the alpha male at Trout Creek in 1958, and we thus had no record of his rise to dominance. Conversations with Dave Condon, Chief Naturalist in Yellowstone National Park, however, indicate that Scarface was a "boss bear" or alpha male at Trout Creek for four or five seasons before 1959. Among the other alpha males, Patcheye was present at the ecocenter for six years before attaining alpha; White-faced Boar, for one year; the Little Brown Boar, for four years; and Scarneck, for

Young mature males moved up the hierarchical "ladder" from noncontender status to contender, with the most aggressive ones contending for the alpha position. *(Photo by John J. Craighead)*

one year (Table 6.8). This period of testing, of gaining experience, and of adjustments appeared to be a necessary prelude to attaining alpha status.

THE ALPHA MALE

Once an animal had established himself as the alpha male, this rank and the recognition it conferred exerted an intimidating effect on would-be challengers, as well as on all members of the hierarchy. Among the alpha males, Inge (No. 12) exhibited the greatest willingness to fight, and he alone patrolled the aggregation, "putting members in their place" with a lift of the head, a growl, or a false charge. He sought conflict. His daily behavior related more to detecting breeding opportunities than to feeding. Scarneck was probably the largest alpha, weighing 418 kilograms (920 lbs) in early September 1964, but we were unable to trap him for weighing during the breeding season.

RELATIVE AGGRESSIVENESS OF ALPHA MALES

Seven males held the alpha position at Trout Creek during our study, and their relative aggres-

siveness can be expressed in terms of frequency and type of encounter (Table 6.4). By this measure, Inge (No. 12) was more aggressive than any of the other alpha males at Trout Creek except Scarneck. It is difficult to compare the relative aggressiveness of Inge and Scarneck on the basis of their Trout Creek records alone, but the latter held the alpha position at Rabbit Creek for three years (1959-61) and seriously injured Inge in battle in 1962 at Trout Creek. On the basis of his combined Trout Creek and Rabbit Creek records, we considered Scarneck a highly aggressive bear. The same was true of Whitefang. These three alpha males were exceptional and markedly different from their contemporaries in regard to energy output, aggressiveness, and the will to exercise authority over the bear aggregations. All three were large, but aggressiveness rather than size appeared to be the major criterion for dominance.

With the exceptions of White-faced Boar and Little Brown Boar, all contenders had more encounters the year or years they attained alpha rank than in the years when they were beta or other-contenders, suggesting that exceptional aggressive effort by the aspiring alpha male was essential. It further

Table 6.8. Number of seasons alpha male grizzly bears were observed at Trout Creek ecocenter before attaining that status, Yellowstone National Park, 1959-70.

Alpha male	No. of years from first year observed to contender	No. of years from first year as contender to alpha	No. of years from first year observed to alpha	Remarks
Scarface	—	—	—	Scarface was alpha male at Trout Creek (1958) before the study began in 1959.
Ingemar (No. 12)	—	—	0	Dropped from alpha position in 1961 to noncontender in 1962; back to alpha position in 1963 and 1964; co-alpha with No. 197 in 1965.
Scarneck	0	1	1	Alpha male at Rabbit Creek in 1959, 1960, 1961.
White-faced Boar (No. 197)	0	1	1	Co-alpha with No. 12 in 1965.
Whitefang	0	0	0	Alpha male at Rabbit Creek in 1962, 1963, 1964, 1965, 1966; at Trout Creek as alpha male in 1966[a] only.
Patch-eye (No. 88)	1	5	6	Although an alpha male for one year, No. 88 was in the contending strata for four years prior.
Little Brown Boar	1	3	4	Contender in 1965, noncontender in 1966, beta male in 1967.
Gandolf (No. 258)	0	0	0	May have been a beta male at Rabbit Creek ecocenter.

[a] Whitefang attained alpha rank at both Trout Creek and Rabbit Creek in 1966.

suggests that when a male attained alpha position in a given year, it was not because his opposition may have been less formidable, but because he made a greater aggressive effort.

FATE OF ALPHA MALES

None of the alpha males were old animals when last observed. Their estimated mean age was between 12 and 17 years. Although they bore many battle scars, all, with the exception of Scarface, had good teeth and were strong, agile animals. Yet, with one exception, none were again observed after leaving Trout Creek (Table 6.9), despite special efforts to locate them at other ecocenters. Only one death was recorded, although the death of one of these massive, battle-scarred (in some cases tattooed) males would surely have attracted attention if shot by hunters or captured and destroyed in control operations. Scarface, Inge, and Scarneck remained as

contenders for two, one, and three seasons, respectively, after losing alpha status (Table 6.9). White-faced Boar, Whitefang, and Little Brown Boar were never observed at Trout Creek, or anywhere else, after reigning as alpha. Patch-eye was killed in a control action at the south arm of Yellowstone Lake on 5 August 1971 at 15 years of age. He had been raiding camps in the area and weighed only 181 kilograms (400 lbs) when killed; he had weighed 340 kilograms (750 lbs) on 24 August 1970 before his departure from Trout Creek. Patch-eye had contended for five years and was beta male in 1970, when the study ended (Figs. 6.2 and 6.3).

The average tenure at Trout Creek (at any level of status) of seven males following loss of alpha status was just over one year. Injuries from status conflicts did not appear to be a factor. We suspected most of them died from natural causes, perhaps related to stress and the high level of energy expendi-

Table 6.9. Number of years alpha male grizzly bears were observed following loss of alpha rank, Yellowstone National Park, 1959-70.

Alpha male	Years observed at Trout Creek after loss of alpha rank	Years observed elsewhere	Comments
Scarface	2	0	Remained a contender.
Ingemar (No. 12)	1	0	Remained a contender.
Scarneck	3	0	Remained a contender.
White-faced Boar (No. 197)	0	0	Never reobserved.
Whitefang	0	0	Never reobserved.
Patch-eye (No. 88)	3	1	Beta male when study ended; killed a year later
Little Brown Boar	0	0	Never reobserved.
Gandolf (No. 258)	—	0	Alpha male when study ended; never reobserved
Mean	1.29	0.13	
Grand mean	1.25		

Table 6.10. Tenures of alpha male grizzly bears at Trout Creek and Rabbit Creek ecocenters, Yellowstone National Park, 1959-70.

Alpha male	Period of dominance at Trout Creek	Period of dominance at Rabbit Creek
Ingemar (No. 12)	1959, 1960, 1961, 1963, 1964, 1965	Possibly 1958.
Scarneck	1962	1959, 1960, 1961.
White-faced Boar (No. 197)	1965	High in the hierarchy, probably a beta male.
Whitefang	1966	1962, 1963, 1964, 1965, 1966.
Gandolf (No. 258)	1969, 1970	High in the hierarchy, probably a beta male.

ture that characterized the alpha males. It is plausible that these males entered the winter in poor physical condition and died in dens during winter sleep.

Our observations of the disappearance of alpha males following loss of status to a more powerful contending male are not the first records of this phenomenon. In 1930, Phillip Martendale, Ranger Naturalist, reported the following:

> The night of June 25th at the bear feeding grounds at Old Faithful—while about 1000 people were seated during the lecture on wild life [*sic*]—a large grizzly lost his control of an

area which he had held for a number of years. "Scarface," a bear that I named on account of one side of his face being torn and both ears missing, and an immense old grizzly, has really been monarch of this section and all other bears gave him the right-of-way. In fact, his thousand pounds of power did knock a large black bear about ten feet [3 m] off the feed platform last summer.

On the night mentioned, three bears were feeding when "Scarface" came down out of the woods. A large dark silvertip was there, and before "Scarface" had arrived within 100 feet [30.5 m] of the feed platform, he left and started his almost human challenge. A swagger, both

Table 6.11. Annual tenure and status of alpha and beta male grizzly bears at Trout Creek ecocenter, Yellowstone National Park, 1959-70.

Bear name and/or no.	No. of seasons as:				
	Alpha	Beta	Other-contender	Non-contender	Total seasons
No. 111	0	1	0	1	2
Black-eyed Boar	0	1	0	0	1
Cutlip	0	1	2	0	3
Fidel (No. 206)	0	2	1	5	8
Gandolf (No. 258)	2	0	0	0	2
Grizzled Boar	0	1	1	0	2
Ingemar (No. 12)	6	1	0	1	8
Little Brown Boar	1	1	1	2	5
Mohawk	0	1	0	1	2
Patch-eye (No. 88)	1	1	4	4	10
Scarneck	1	1	3	0	5
White-faced Boar (No. 197)	1	1	0	0	2
Whitefang	1	0	0	0	1
Total (%)	13 (25.5)	12 (23.5)	12 (23.5)	14 (27.5)	51 (100.0)
Grand total (%)	37 (72.5)				

front and rear—moving toward his enemy of no doubt other encounters. This same swagger was used by "Scarface" but he circled away and kept his distance until finally he was chased out completely. . . . The victor came to the nearest tree, reared on his hind legs and measured his full length against the tree—a new commander of the area among the male grizzlies. The bear "Scarface" came in the next night very early—alone—secured one piece of meat—watched the woods closely, and left—not to return again with other bears. He has lost—too old (Yellowstone Nature Notes 1930).

TENURE OF HIERARCHICAL STATUS

It was evident that there was some between-year continuity in the individual males that held alpha status at all of the large ecocenters, but that status was not automatically conferred on the dominant male of the previous season. Each year the alpha male had to earn his rank. He failed to regain alpha rank if he did not return, was seriously injured, suffered defeat from a more aggressive male, or, for reasons unclear to us, simply did not compete. Considering all three major ecocenters—Trout Creek, Rabbit Creek, and West Yellowstone (Fig. 3.1)—four males exhibited long, consecutive alpha tenures that embraced most of their prime of life.

An exchange of bears occurred between Trout Creek and Rabbit Creek and, to a much lesser degree, between Trout Creek and West Yellowstone and between West Yellowstone and Rabbit Creek. This exchange included both high- and low-ranking bears. The comparatively rapid rise to power in the Trout Creek hierarchy of such bears as Scarneck and Whitefang probably occurred because they were seasoned combatants, already having earned alpha rank at Rabbit Creek before contending at Trout Creek. Whitefang was dominant at Rabbit Creek for almost as long as Inge (No. 12) was dominant at Trout Creek (6 years). Whitefang attained the alpha role at both Trout Creek and Rabbit Creek in 1966 (Table 6.10). After attaining alpha at Trout Creek (Fig. 6.2), he left the aggregation and achieved the alpha position for the fifth consecutive year at Rabbit Creek. Scarneck attained the alpha role at Rabbit Creek for the 1959, 1960, and 1961 seasons and then contended successfully at Trout Creek, rising to alpha in 1962.

Inge, Scarneck, and Whitefang, with alpha tenures (at one or more ecocenters) of six, four, and five seasons, respectively, were unequaled by other contenders. Scarface, "boss bear" before our study began, probably had an alpha reign of at least five years.

Tenure as alpha at Trout Creek among seven alpha males varied from one to six years

Table 6.12. Annual tenure and status of grizzly bears that were other-contenders at Trout Creek ecocenter, Yellowstone National Park, 1959-70.

Bear name and/or no.[a]	No. of seasons as:		
	Other-contender	Non-contender	Total seasons
No. 166	2	3	5
No. 217	1	1	2
Big Brown Boar	1	6	7
Bruno (No. 14)	2	6	8
Ivan (No. 41)	2	9	11
Large Brown Boar (No. 33)	1	3	4
No Hump	1	1	2
Old Big-headed Boar	1	0	1
Scar-shouldered Boar	1	2	3
Scarchest (No. 73)	1	2	3
Scarface	1	1	2
Short-eared Boar	1	5	6
Total (%)	15 (27.8)	39 (72.2)	54 (100.0)

[a] Only other-contenders that did not reach alpha or beta status are included in this table.

(Table 6.11). With the possible exception of Scarface, only Inge held long-term dominant status there. Observed at Trout Creek for eight seasons, he was alpha male for six of these. Patch-eye (No. 88), present for 10 seasons, was alpha for a single year. White-faced Boar (No. 197) contended for two years, attaining beta rank in 1964 and co-alpha in 1965.

The tenure for beta rank was short, generally one year; only one bear, Fidel (No. 206), held it for two years. Five of seven alpha males were known to have held beta rank in the course of their lives; three held other-contender rank and three, noncontender (Table 6.11). With the exception of two alpha males, Gandolf (No. 258) and Whitefang, all were recorded as holding lower-status positions (non-alpha contender and noncontender), either before or after attaining alpha rank (Fig. 6.2). As an average, males ultimately attaining alpha or beta status at Trout Creek were contenders for 73% of the seasons that they were present as members of the aggregation. Twenty-seven percent of the time, they were noncontenders. On the other hand, males that never exceeded other-contender status (i.e., those males that competed with the alpha and the beta contenders, but never attained those positions) were contenders for only 28% of the seasons they were present and noncontenders for 72% of those seasons (Table 6.12). Both individually and as a group, the alpha and beta males competed longer and more consistently than did those contenders that never attained the alpha or beta ranks.

CHANGES IN STATUS AND LEVELS OF CONTENTION AMONG AGGRESSIVE MEMBERS OF THE MALE HIERARCHY

The specific causes for changes in status from one season to the next were rarely evident (Fig. 6.2). A few changes in rank clearly followed injury or recovery from injury. But the majority of changes from year to year must be attributed to unknown causes that, in one way or another, attend individual motivation or aggressiveness.

Each year of the study, the aggregation consisted of more nonaggressive males than aggressive ones. The hierarchical structure of adult males consisted of an active stratum (the alpha males, beta males, and other-contenders) and a passive stratum of noncontending males. Throughout a 12-year period, only 25 males (individuals) among all those comprising the aggregation were contenders; 73 were exclusively noncontenders. Among the 98 marked and/or recognizable adult males participating in the aggregation over the 12-year period, some were members of the aggregation for only a single year, whereas others were present for as many as 8, 10, or 11 consecutive years. Of these 98 adult males, only 7% attained the alpha level during the study;

Fig. 6.4. Percentage of adult male grizzly bears achieving a given level of status as their highest level of contention (*n* = 98), Trout Creek ecocenter, Yellowstone National Park, 1959-70.

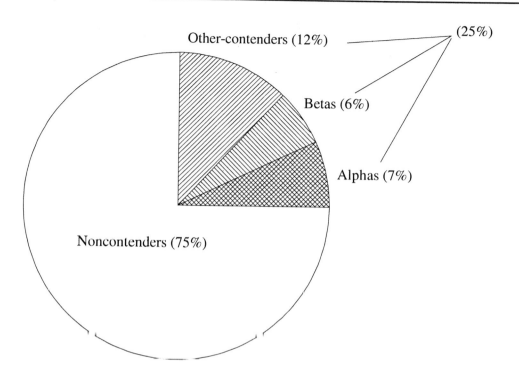

6% the beta level, rising no higher; and 12% rose to the level of other-contenders, but not beyond. Seventy-five percent never rose above the noncontender level (Fig. 6.4). These percentages probably closely approximate those that would have been obtained from a sample of males followed over their entire lives. Only a small percentage of adult males, irrespective of age, actively and overtly contended for rank.

Some bears moved through all levels of status during their lifetimes. In part, this was due to growth and maturation and consequent rise in status, but some bears dropped from higher to lower status over a period of years (Fig. 6.3). For example, Patch-eye (No. 88) was first ranked as a noncontender in 1961. In each of the next three seasons (1962 to 1964), he achieved contender status. He dropped to noncontender status in 1965, when his number of aggressive encounters decreased. In 1966, he demonstrated greater aggressive activity and in 1967 was recognized as the alpha male. The next year (1968), he became a noncontender, exhibiting little of his former aggressiveness, and continued that status through the 1969 season. In 1970, Patch-eye exhibited aggressiveness similar to earlier years and was recognized as the beta male. It is obvious that the "careers" of alpha males, as exemplified by their levels of contention within the bear aggregation from year

to year, did not conform to any fixed pattern. The pattern of movement from one level of contention to another was highly individualistic (Fig. 6.3).

CHANGES IN STATUS AND LEVELS OF CONTENTION AMONG NONAGGRESSIVE MALES

As indicated earlier, 75% of the 98 adult males were members of the passive stratum that did not overtly contend for status (Fig. 6.4). Why so many large adults did not contend at all, or contended only occasionally at the lowest level of contention (as other-contenders), is not clear. Among six less aggressive males with long records of attendance at Trout Creek, Bruno (No. 14), a large, relatively young male, contended two of the eight seasons he was observed, but was never a very active competitor (Fig. 6.5). Old Blue (No. 30), in residence for 10 years, never contended. Ivan (No. 41), a large mature male, was a known cub killer and was feared by the females, but he competed (ineffectively) in the hierarchy of males for only 2 of 11 seasons observed. Old Mose (No. 179) was an old bear in 1960 (13 years), and age probably accounted for his inactivity. Short-eared Boar and Big Brown Boar were young powerful animals in their prime, yet each contended for only one season at the lowest level of contention. Many other adult males were present

Among adult males at the Trout Creek ecocenter, 75% were members of the noncontender stratum. All male grizzlies are aggressive, but only a few seem to possess the combination of size, physical skills, aggressiveness, and willingness to fight required to attain high status in the hierarchy. *(Photo by John J. Craighead)*

Fig. 6.5. Levels of contention for six adult male grizzly bears that never attained alpha or beta rank at the Trout Creek ecocenter, Yellowstone National Park, 1959-70.

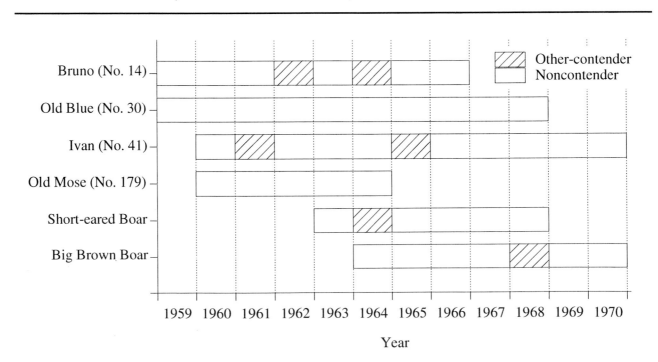

at Trout Creek for two or three seasons, yet never competed (Table 6.1).

Although all male grizzlies are aggressive, some are considerably more aggressive than others, and perhaps only a few within a population possess the combination of size, physical skills, and aggressiveness required to attain high status. In lay terms, the long-tenured alpha males, such as Inge, Scarneck, and Whitefang, appeared to be compelled to dominate; they were the Churchills and Caesars of the bear population.

RELATION OF VISITATION TO HIERARCHICAL STATUS

During the Breeding Season

Not surprisingly, visitation at the ecocenter during the breeding season varied with male hierarchical status. This is evident from a comparison of visitation by individual males before and after transitions in rank. Fourteen males made at least one transition between the other-contender and noncontender ranks in consecutive years. These transitions were of two types: (1) increase in rank from noncontending to contending status, or (2) decrease in rank from contending to noncontending status.

We selected one transition at random for each of the 14 males and calculated an adjusted visitation for the years bracketing the change in dominance status (see Chapter 5). Analysis of adjusted visitation rates for individual bears in years of contention versus years of noncontention showed that adult males visited the Trout Creek ecocenter during the breeding season significantly more often as contenders than as noncontenders (Wilcoxon matched-pairs test, $t = 2.48$, 2-tailed, $P = 0.01$). Eleven of the 14 males tested had greater adjusted visitation rates in the years when they were contenders; the mean difference in adjusted visitation between members of the two contending levels was 26%.

In a complementary analysis, we compared adjusted visitation rates for a sample of males that were contenders to those of a second sample of different, noncontending males. Thus, this is a comparison between two samples of males (Mann-Whitney test) rather than a comparison of visitation among the same males when contending and not contending (see Chapter 5 for a detailed description of these two types of analyses). As our first analysis indicated, the contender group visited more frequently, before 15 July, than did the noncontender group ($P < 0.001$; Appendix B, Table 1).

The preceding two tests compare visitation between the two main divisions of status, contenders and noncontenders. To determine whether visitation varied significantly within the contending male stratum, we performed a series of Mann-Whitney tests, in which all levels of rank within the contending stratum (including one in which alpha and beta ranks were lumped) were compared pairwise. These tests revealed that rank within the contending stratum did not measurably influence visitation during the breeding season (Appendix B, Table 2).

After the Breeding Season

For comparisons of visitation during the postbreeding season, we confined our analysis to the years 1960-61 and 1964-70. Only during these years were postbreeding censuses sufficiently numerous and evenly distributed over the period 15 July to 1 September. Unfortunately, not enough transitions between contending and noncontending status were recorded in these years to allow a comparison of visitation by individual males when contending and not contending. We therefore analyzed visitation in the postbreeding period using a Mann-Whitney test identical in concept to that described above for the breeding season. This test revealed that, as in the breeding season, contending males visited the ecocenter more frequently than noncontenders during the postbreeding season (Appendix B, Table 3).

A different question is whether males within either main stratum of the hierarchy altered visitation after the end of breeding. For example, contending males may have visited less frequently (but still more frequently than noncontenders) after 15 July because females were no longer an attractant. To answer this question, we compared visitation by a given male during the breeding season to his visitation after 15 July in the same year (Wilcoxon Matched-Pairs test). We first did so for a sample of contending males and then, in a separate test, for a sample of noncontenders. These comparisons revealed that contending males did not change visitation in the postbreeding period, whereas noncontending males increased visitation substantially, but not quite significantly by statistical criteria (Appendix B, Table 4).

Interpretation of Male Visitation

In summarizing the statistical results of the previous section, we find that (1) contenders visited Trout Creek more frequently than noncontenders in both the breeding and postbreeding season; (2) visitation by contenders did not change after the breeding season, whereas noncontenders tended to increase visitation at this time; and (3) rank within the contending stratum did not appear to influence

Table 6.13. Reproductive activities of adult male grizzly bears observed at the Trout Creek ecocenter, Yellowstone National Park, 1959–70.

Year	Alpha male (n = 12)		Beta male (n = 8)		Other-contending males (n = 21)		Noncontending males (n = 44)		All males (n = 85)	
	No. of copulations	No. of checks[a]	No. of copulations	No. of checks[a]	No. of copulations	No. of checks[a]	No. of copulations	No. of checks[a]	No. of copulations	No. of checks[a]
1959	0	1	1	0	1	1	0	0	2	2
1960	1	6	0	0	1	3	1	0	3	9
1961	3	11	3	2	4	0	11	5	21	18
1962	1	8	2	3	6	7	0	4	9	22
1963	3	7	2	0	8	4	7	3	20	14
1964	4	16	1	4	8	15	5	4	18	39
1965[b]	5,2	7,3	2	1	3	2	9	5	21	18
1966	0	0	0	0	1	0	2	0	3	0
1967	0	1	1	0	3	2	5	3	9	6
1968	0	1	2	4	3	1	3	1	8	7
1969	0	1	0	0	0	0	1	2	1	3
1970	1	0	0	0	0	0	7	3	8	3
Total (%)	20 (16.3)	62 (44.0)	14 (11.4)	14 (9.9)	38 (30.9)	35 (24.8)	51 (41.5)	30 (21.3)	123 (100.0)	141 (100.0)
Grand total	82		28		73		81		264	

Note: *n* = only those males observed exhibiting breeding behaviors.
[a] "Sniffing" of a female's urogenital region to determine her reproductive receptivity.
[b] Co-alphas.

visitation before 15 July. The broad pattern of visitation at Trout Creek is clear. The consistent availability of large volumes of food was the principal inducement for bears to make what eventually became traditional movements to the ecocenter. Because the availability of food at Trout Creek coincided with the onset of estrus among females, the bears both bred and fed, once aggregated. Without the ecocentric attraction of food, the Trout Creek population would, no doubt, have bred as semisolitary animals.

Beyond this, however, we suggest that the variation in visitation associated with male dominance status may be explained in terms of the influence that male status had on access to both resources offered by the Trout Creek ecocenter—estrous females and a concentrated source of high-energy food. Specifically, we suggest that differences in dominance status favored the use of different reproductive and feeding strategies, and that an important feature of these alternative strategies was differential visitation. It is possible that before 15 July, estrous females visited Trout Creek more frequently than non-estrous ones. But for now, we assume that state of estrus did not affect female visitation at Trout Creek. Even so, the number of estrous females at the ecocenter would be higher than in surrounding areas simply because females were more numerous there. Relatively dominant males would then obtain more opportunities to mate by concentrating their search and competitive effort near the ecocenter, thereby explaining their high rates of visitation during the breeding season.

The continued high-frequency visitation by contending males in the postbreeding season was probably a carry-over from the breeding season. These males no longer contended for estrous females, but they exploited the status they had attained during the breeding season to take and hold choice feeding sites. During the breeding season, dominant males fed less frequently and expended more energy in reproductive competition than the less dominant ones and, therefore, may have needed to spend more time feeding in preparation for winter. This may explain why visitation by contenders did not change significantly from the breeding to the postbreeding season and may also explain their greater visitation, relative to noncontenders, during the latter period.

Males of sufficiently low rank, at a competitive disadvantage in the presence of the dominant males at the ecocenter, may have realized greater mating success in peripheral areas, where estrous females were more dispersed and less numerous, but where there was also less competition with other males.

All bears (including estrous and non-estrous females) dispersed from the ecocenter around midnight (Fig. 5.1). Estrous females were not always accompanied by dominant males when they dispersed. They then became available to the less aggressive males (noncontenders; see Table 6.13).

After 15 July, breeding opportunities in the "backcountry" (a kilometer [0.5 mi] or more from the open pit dump), as well as at the dump, were infrequent. This situation reduced the searching movements of noncontenders. Gradually, food took priority over sex, and whereas their visitations at the ecocenter increased accordingly, they did not approach contender levels. Like monopoly of reproduction, the postbreeding monopoly of choice feeding sites by dominant males may have had a discouraging effect on visitation by noncontenders. "Backcountry food" may have become a greater attractant after breeding declined. Although this food source was more dispersed and less abundant than that at the ecocenter, there was less competition for it, and the numerous noncontenders may have spent relatively more time foraging for it than did the contenders.

This interpretation of male visitation patterns may also explain the puzzling phenomenon of comparatively few aggressive males (contenders) coexisting with many passive noncontenders. Females in estrus at the Trout Creek ecocenter were relatively numerous and readily apparent to all males present. Thus, small differences in status could easily have translated into large differences in mating success. If so, it clearly would have been advantageous for males with some reasonable expectation of success to compete aggressively for high rank. Males that, for reasons of age or physical condition, could expect to achieve moderate status at best would have had little to gain from aggressive competition. On the other hand, by locating and breeding estrous females at the periphery of the ecocenter or in the backcountry, low-status males attained breeding opportunities without aggressive encounters with the dominant males. The success of this dispersed, "backcountry mating" probably depended more on searching effort and perseverance than on male dominance status.

We suspect that males capable of contending for highest ranks did so. They competed intensely because there was a significant difference in reproductive opportunity (perceived as reproductive success) for those males ranked one versus two, two versus three, and so on. More subordinate males declined to take large risks by competing for rank with dominant males because, in the reproductive strategy available to them, their hierarchical status

had less influence on their opportunity to breed. The costs of gaining only a few "percentiles" of social rank by competing were not compensated by increased reproductive opportunity.

We suspect that similar reproductive patterns could characterize populations lacking a large and permanent, natural or artificial ecocenter. For example, it is possible that females in such populations "aggregate" during the breeding season on small and transitory food patches; even in the absence of such patches, they may aggregate simply to improve the likelihood of, or increase the rate of, encounters with males during their fertile period. Such "aggregations" could be relatively more dispersed than the one at Trout Creek and hence difficult to detect without intensively monitoring movement and reproductive condition.

We observed what we assumed was an aggregation on a transitory food patch in the Lincoln-Scapegoat Wilderness of Montana. An unusually large number of fecal deposits was gathered annually from a 8- to 12-hectare (20- to 30-acre) southerly exposure close to the summit of 2803-meter (9200-ft) Scapegoat Mountain (Craighead *et al.* 1982). We judged that most, if not all of the feces were deposited during late May and June, just before the beginning of our annual research activities in the area and during the breeding season. A moth belonging to the family Noctuidae, numerous in the rock slides, was available to the bears in large quantities, as were the starchy roots of springbeauty (*Claytonia megarhiza*) and emerging sedges and grasses. These food items were most frequently represented in the feces (Craighead *et al.* 1982; also, Chapter 10). The fecal scats were representative for adult males and females, as well as younger animals. We suspect that this extremely localized site may have been an area where a small number of grizzly bears aggregated to breed and, further, that preferred and easily available food was the basic attractant. Because food availability coincided with the breeding season, the result was an aggregation during the period of estrus. Thus, the catalyst for the development of this sort of regionally dispersed but locally aggregated reproductive strategy might be patches of preferred habitat, located and remembered by adult females that return to the sites seasonally, year after year (Craighead *et al.* 1982).

In any case, at any site where estrous females were sufficiently aggregated, males should adjust their search patterns accordingly. This could lead to the dichotomy in dominance behavior observed between contenders and noncontenders and to the development of a reproductive strategy similar to the one at Trout Creek in the 1960s. There, we

Table 6.14. Summary of observed breeding activity of adult male grizzly bears at four hierarchical levels over 12 breeding seasons, Trout Creek ecocenter, Yellowstone National Park, 1959-70.

Breeding activity	Mean frequency of breeding activity per season per individual				
	Alpha (*n* = 12)	Beta (*n* = 8)	Other-contenders (*n* = 21)	Non-contenders (*n* = 44)	All males (*n* = 85)
Copulations	1.67	1.75	1.81	1.16	1.45
Checks[a]	5.17	1.75	1.67	0.68	1.66

[a] "Sniffing" of a female's urogenital region to determine her reproductive receptivity.

hypothesize, an aggressive contender stratum focused its reproductive effort on aggregated females, and a passive noncontender stratum searched more widely for breeding opportunities or adopted a "waiting strategy" at the ecocenter, where the less aggressive males waited until the dominant males had attained breeding success and had temporarily lost interest or were satiated. It was frequently observed that females bred by dominant males were receptive to the less dominant ones both at the ecocenter and elsewhere, but only when a dominant male had previously bred them or was not present to disrupt the "search and waiting" strategy.

THE AGGREGATION HIERARCHICAL STRUCTURE

Competing for the highest levels of status in the aggregation was entirely an adult male endeavor. Adult females in estrus were an important catalyst for the male hierarchy. And like the males, adult females and bears in other sex-age categories also engaged in aggressive encounters and competed for rank. Although interactions among males determined the male hierarchy, interactions among all bears (adult males, adult females, nonadult females, etc.) generated what we term the aggregation hierarchy, the more inclusive social hierarchy that involved every animal present. Over the 12-year period, it developed through interactions among adult members of the aggregation's active and passive strata. The active strata consisted of 13 individual males that attained alpha or beta rank and 12 males in the other-contender category that never rose higher. The passive strata comprised 73 noncontending males and 45 adult females.

The nonaggressive males (noncontenders), most adult females, and the nonadult animals reacted defensively to attacks and threats from the aggressive members of the aggregation, but did not initiate aggressive action. Every member, by virtue of

age and aggressiveness, fitted into the hierarchical structure, the upper strata of which were defined by the adult contending males. The more aggressive adult females behaved aggressively toward members of all strata and in doing so may have attained a higher status than some noncontending males. The noncontending adult males were inactive only to the extent that they did not initiate attacks or threats toward the active stratum animals and seldom toward one another. The relatively nonaggressive females and the nonadults (1-4-year-olds) learned their levels of aggressive competence through the innumerable close approaches that occurred each day of the breeding season.

The tens of thousands of subtle aggressive and submissive signals emanating from the entire assemblage gradually brought the passive animals into the developing social order at a level that enabled each member of that stratum to interact with one another, and with members of the active stratum, with a minimum of conflict. Aggressive displays of one type or another were always evident, but following the aggregation's recognition of the alpha male and the termination of the breeding season, the entire aggregation seemed to "gel" into a relatively nonaggressive entity.

MALE REPRODUCTIVE ACTIVITY

When interpreting male visitation patterns, we assumed that dominant males had substantially greater reproductive success annually than more subordinate males; that is, they fathered a greater number of cubs during a single breeding season. This assumption was suggested by the repeated observation of dominants displacing subordinates from receptive females. Here we describe male mating patterns, and we review the evidence that access to estrous females by dominant males is a true measure of individual male reproductive success. The relationship between social status and mating success (opportunity to copulate and reproductive success) is complex and may be strongly influenced by a prolonged and complex female estrus, in which the quality of matings is more significant than the number.

An individual male often copulated with a given female numerous times during a single estrus, and individual females were often observed to copulate with more than one male during a single estrus. Collectively, the alpha and beta males obtained only 27.6% (34) of all copulations (123) observed; the far more numerous other-contenders and noncontenders accounted for the majority (Table 6.13). Moreover even when expressed on a per-bear basis, there was no clear relationship between status and copulatory performance (Table 6.14). In fact, in 5 of 12 years, alpha males were never observed to copulate, nor were the beta males in 4 of 12 (Table 6.13). There was, however, a tendency for alpha bears more actively to assay female reproductive state. Alpha males were observed to perform "checks" (that is, "sniffing" of a female's urogenital region to determine her reproductive receptivity) far more frequently than males of lower status (Table 6.14).

Other-contenders and noncontenders generally obtained copulations while more dominant males were engaged in aggressive interactions, seemingly oblivious to the opportunities to copulate. It was our impression that these subordinate males also tended to breed females late in their periods of estrus—females that would have been first identified and bred by the alpha or beta male and/or their immediate contenders. These observations suggest a straightforward resolution of the puzzling tendency for alpha and beta males to be aggressively preoccupied in the midst of sexual opportunity, resulting in a large number of copulations among lower ranking males (Table 6.14). The female estrus was often prolonged, in excess of 10 days at times, and typically was split—that is, it contained a period during which the female was neither receptive nor attractive to males (Craighead *et al.* 1969). These features suggest a complex pattern of variation in female fertility during the course of estrus. If reproductive success rests in the timing of copulations rather than the number, the data in Table 6.14 are more readily interpretable. It is conceivable that dominant males did not defend a receptive female when her fertility was low, and risk of defense (battles with other males) was less likely to be repaid in terms of reproductive success. In contrast, subordinate males, forced to accept breedings during less fertile periods of estrus, or none at all, were able to obtain relatively large numbers of these "low

quality" matings with undefended females. Such selectivity by dominant males would require close monitoring of the state of each female at the onset of her fertile period(s) and could explain the high frequency of checking by alpha males relative to other males (Table 6.14).

Under this interpretation, male reproductive success depends on the quality rather than the number of copulations. Only the most dominant and aggressive males would routinely be able to determine and monopolize periods of high female fertility. The alternative—to accept copulation frequency as a true measure of reproductive success—implies that high-ranking males risk competition for social rank without compensating reproductive benefits, an argument we regard as untenable.

Nonetheless, we cannot dismiss the fact that other-contenders and noncontenders accounted for the majority (72.4%) of all observed copulations (Table 6.13). Even if these breedings occurred during lowered fertility, as hypothesized, they must nevertheless represent some, not necessarily insignificant, degree of reproductive success. The relative reproductive success achieved by contenders versus noncontenders is not simply an academic issue. The forces that can reduce genetic diversity in grizzly populations, such as genetic drift and inbreeding, increase proportionately as dominant males monopolize lifetime reproductive success, where "monopoly" is defined as any deviation from an equal expectation of reproduction among contending and noncontending males. Monopoly of lifetime reproduction depends in turn, on (1) the magnitude of the annual monopoly of reproduction, and (2) the average tenure at high rank for individual males.

Our data demonstrate that few Trout Creek bears attained alpha or beta status at any time in their lives, that long tenures by individual bears at alpha or beta status were common, and that most bears (75%) probably did not even reach contender status in their lifetimes. Even if we grant contenders some degree of monopoly of annual reproduction, our behavioral data does not resolve what effects monopoly might have on the genetics of aggregated grizzly bear populations. Resolution would require genetic testing.

Lance Craighead used DNA "fingerprinting" techniques to examine the genetics of a dispersed population of grizzly bears in the Alaskan Arctic in an attempt to answer questions of paternity (Craighead 1994). His data demonstrated that "each cub in a litter can be sired independently and that one third of all possible litters had multiple sires." Estimates of maximum reproductive success for males indicated that no single male was responsible for more than 11-13% of total paternity. Eighty percent of 57 cubs, however, were sired by 18-24% of all potential fathers, depending on whether one uses their minimum ($n = 51$) or maximum estimate ($n = 67$) for the male population. This suggests substantial social monopoly, even in dispersed populations. The 20 litters genotyped were distributed over six years, so these data address reproductive monopoly over portions of male lifespans, rather than annual monopoly isolated as a component of lifetime success—the crucial information missing from our study. In addition, the behavioral mechanism behind multiple paternity was not investigated in this study. The author suggests takeovers of estrous females by more dominant males as one possibility. However, sneak breedings by subordinate males or changes in male consorts during the second half of split estruses (Chapter 7) are additional possibilities. Thus, substantial uncertainty still surrounds the degree of annual reproductive monopoly among male grizzlies and the social and other determinants of their annual reproductive success.

A question of related practical importance is whether the grizzly mating system, and hence genetic drift and inbreeding, differ in ecocentered versus non-ecocentered populations. The answer will require not only a paternity study, but also a comparative study of mating behavior in ecocentered and non-ecocentered situations. Nonetheless, we can speculate on which features of the mating system are most apt to change. For example, aggregations during the breeding season must increase the "availability" of estrous females to males, in the sense of reducing search time. If females do have brief periods of high fertility, it is easy to imagine an extreme monopoly on annual reproduction in ecocenters as a consequence of dominant males moving from fertility peak to fertility peak and leaving the bulk of each female's estrous period to subordinate males.

The mating pattern seen in aggregated populations would be less effective in more dispersed populations, where it might be difficult for dominant males to find estrous females and considerable time could elapse between encounters. In dispersed populations, dominant males might do better by remaining longer with each female and defending her during periods of both peak and reduced fertility. Of course, estrous females in aggregations are also more available, in this same sense, to noncontenders, to the extent that low-ranking males are able to "steal" copulations from contenders. Such activity would tend to limit reproductive monopolies in ecocentered populations.

We also speculate that synchrony of estrus among females may increase in aggregations as a result of social and pheromonal effects operating between males and females, or among females (Packer and Pusey 1983, McClintock 1981, Ims 1990). Extreme synchrony would tend to reduce the ability of dominant males to monopolize reproduction. Since harems were not a behavioral phenomenon at Trout Creek, two females in simultaneous estrus could not both be monopolized by a single male.

Thus, in this hypothetical view, increased synchrony of estrus and the increased availability of females to both contending and noncontending males are three sometimes opposing factors in determining annual reproductive monopolies at ecocenters. We cannot fully evaluate the magnitude, if any, of the individual effects, let alone their interaction. Nevertheless, we hypothesize that the formation of aggregations at ecocenters does confer genetic advantages simply by providing an opportunity for more males to breed with any given female. We suggest any future study of grizzly bear population viability should consider, separately, mating systems in populations that are ecocentered, those that are non-ecocentered, and those that might routinely switch between ecocentered and non-ecocentered status as food resources change.

STATUS OF ADULT FEMALES

Although certain females clearly were more receptive to specific males and entered and left the ecocenter with them, we never observed a female in estrus behaving aggressively to defend or to change her status as the mating partner of a particular male. A female in estrus was receptive to whichever male claimed her. Females behaved aggressively in mating relationships only to male solicitation just before or soon after their estrous period. Most female aggression was not sexually oriented, but used to defend offspring and choice feeding sites. Such aggression established the status of females in the evolving social order of the entire aggregation. As mentioned earlier, at the end of the breeding season some of the more aggressive females may have ranked as high in the social order as some of the contending males.

The "meaning" of female aggressiveness within the hierarchy is often distinct from that of the large, aggressive males because the motivation and value for aggressive acts are quite different. Males competed with males for females in estrus, as well as for food, whereas females competed with other females for food resources and protected offspring from males. The fact that an aggressive female, in

Fig. 6.6. Structure of the male grizzly bear social hierarchy and the overlapping relationship between the male and female hierarchies which combined to form the aggregation hierarchy, Trout Creek ecocenter, Yellowstone National Park, 1959-70.

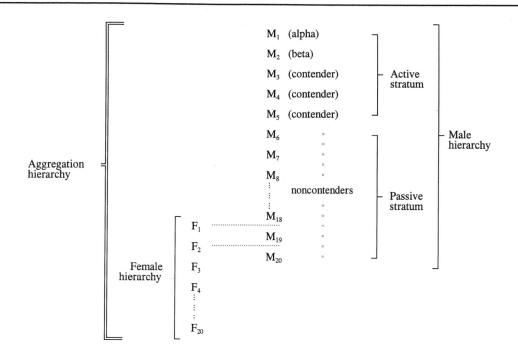

contesting a food site or defending cubs, forced an upper stratum male to concede did not necessarily indicate dominance, in the sense of inherent superior physical power. We suggest it indicated a superior motive to fight, perhaps because females had more to gain or lose in terms of fitness. In some cases, members of the female hierarchy, by virtue of size and age, were physically superior to some members of the male hierarchy, and they used this physical superiority to gain specific objectives. The behavior of young males and older noncontending males toward certain large, aggressive females suggested that these females had higher status and were dominant to the males. This was communicated to an encroaching male through discrete behavioral signals, and he generally backed down.

A female accompanied by a litter of yearlings or 2-year-olds comprised a coalition that could be physically formidable if attacked. Such aggression was never employed against upper-stratum males to gain feeding sites, but was used against low-status adult males, nonaggressive adult females, and nonadults of both sexes to gain feeding advantages. This female-family aggression, however, was used regularly against even the highest status males to defend the family unit and occasionally to keep choice feeding sites.

Like males, females in the aggregation probably could have been ranked in linear fashion relative to

other members of the same sex. We did not have the data to do this, but we suspect that a linear-ranked organization underlay the broader categories of rank that we were able to recognize. The female hierarchy developed in a less spectacular fashion than did the male hierarchy and was more difficult to detect and define. The data available did not demonstrate conclusively that females were ever dominant to males. Nevertheless, we suspect that the male and female hierarchies did in fact intertwine, forming the aggregation hierarchy. The precise nature of the interlacing can only be theorized, as illustrated by Fig. 6.6.

FEMALE AGGRESSION AND REPRODUCTIVE CONDITION

In general, the aggressiveness of females was related to their reproductive condition. Females with offspring tended to solve family problems by aggressive action against any animal that posed a threat. Documentation of female agonistic behavior as a function of reproductive status (Table 6.15) suggested that females with yearlings tended to be most aggressive, followed by females with cubs and solitary females. Statistical analysis, however, revealed no significant difference (median P-value of 0.37, based on 13 Kruskal-Wallis tests; $n = 28, 21,$ and 12 for solitary females, females with cubs, and

Table 6.15. Rates of female grizzly bear aggression according to reproductive status, Trout Creek ecocenter, Yellowstone National Park, 1960-69.

Reproductive status	Mean rate[a]			Female-hours[c]	Female-years	Different females
	Attack	Threat	Agonism[b]			
With cubs of the year	0.34	0.68	1.02	2663	70	29
With yearlings	0.41	1.31	1.72	2677	55	15
Alone	0.20	0.47	0.67	4042	127	45

Note: based on 1050 hours of observation and 100 agonistic events.

[a] Events per female per 100 hours calculated for each different female-year, weighted by hours observed, and then averaged over all female-years within each category of reproductive status.

[b] Sum of Attack and Threat rates.

[c] Total observation hours calculated for each different female-year and then summed over all female-years within each category of reproductive status.

A female accompanied by a litter of yearlings or 2-year-olds comprised a formidable coalition. Cohesive family aggression was used to gain feeding advantages over all but high status male grizzlies and to retain choice feeding sites. *(Photo by John J. Craighead)*

Table 6.16. Rates of female grizzly bear aggression according to reproductive status and season, Trout Creek ecocenter, Yellowstone National Park, 1960-69.

Reproductive status	Breeding season (before 15 July)		Post-breeding season (after 14 July)	
	Rate[a]	Female-hours	Rate[a]	Female-hours
With cubs of the year	1.29	1704	0.52	959
With yearlings	2.35	1911	0.13	766
Alone	0.89	2919	0.09	1123

Note: based on 1050 observation hours and 100 agonistic events.
[a] Rate of aggression calculated as in Table 6.15, except separately for breeding and post-breeding seasons.

females with yearlings, respectively). Female agonism was of two types—defensive or offensive—largely defined by female reproductive status and directly related to the security of the female or the family unit. Unfortunately, we did not have the foresight to record the two behavioral types separately. Thus, the mean rates of agonism shown in Table 6.15 must be interpreted accordingly. In their interaction with aggregation members, it was clear to us that cub litters were a liability to a female, whereas yearling and 2-year-old litters were an asset in her agonistic relationships.

Most, if not all of the threats and attacks by females with cubs were in defense of their cubs. We interpreted this to indicate a high degree of insecurity. Although the females would press their attacks savagely, the objective was to protect their cubs, not to gain food sites or other advantages.

On the other hand, the agonistic behavior exhibited by females with yearlings was both defensive and offensive, with the latter predominating. Defensive action was mainly against aggressive adult males and, occasionally, against females that posed a threat. Offensive action was employed to gain feeding advantages and was most frequently directed at subadults and other less aggressive members of the aggregation. The increased security with increased age of offspring was directly related to the reenforcing effect of a family unit powerful enough to take the offensive.

Our records of the agonistic behavior of females with 2-year-olds were too limited for statistical treatment. Nevertheless, females in this category appeared to be the most secure of those with offspring. We observed females with 2-year-olds displace both adult males and adult females to gain choice feeding sites. We observed no defensive behavior; even the most aggressive males did not attempt to appropriate their feeding sites.

Generally speaking, females without offspring showed lower rates of attack and threat than females in other reproductive categories. Our interpretation is that lone females, free of family responsibilities, are more secure and therefore less prone to exhibit either defensive or offensive actions. To determine if females were more aggressive during the breeding season than after, we compared rates of agonism for the two periods (Table 6.16). A Wilcoxon matched-pairs analysis of agonistic rates before and after 15 July showed that the rates did, in fact, decline during the post-breeding season ($P = 0.0004$, $n = 24$ individual females).

AGGRESSIVE ACTIONS BY ADULT FEMALES

Over the 12 years of study, it became evident that some adult females were more aggressive than others. These females were recognized for their aggressiveness by all other members of the aggregation. As in the case of the adult males, but not as clearly so, adult females gained status through their aggressive actions. A female must be continually alert, but she also must keep aggressive encounters to a minimum to conserve energy and reduce risk of injury. Nevertheless, her innate aggressiveness is essential to protecting and feeding her offspring and coping with aggressive males often twice her size.

To evaluate how adult female aggressive actions were distributed among the age- and sex-classes of recipients of those actions at Trout Creek, 1960-68, we calculated rates per individual using the mean censused populations of adults and subadults (Table 5.2), and applying the appropriate sex ratio to establish mean numbers of adult males and females (Table 5.6). Among 141 observed acts of aggression by adult females, 115 (82%) involved

Females with cubs employed threat behaviors more frequently than attacks in their interactions with other bears. *(Photo by John J. Craighead)*

other adults, male or female, at rates of 4.4 and 0.6 per individual, respectively (Table 6.17). Regardless of sex, then, nonadults were of less concern to adult females and could generally be intimidated by threats alone or even less overt signals. Threats per individual were directed toward nonadults more than three times as frequently as attacks (Table 6.17).

Because of their size and aggressiveness, adult male grizzlies posed a far greater threat to adult females and their offspring than did other females. Not surprisingly, adult females displayed attack and threat behaviors toward adult males over seven times more frequently, per individual, than toward other adult females (Table 6.17). Adult females behaved both defensively and offensively in encounters with other bears, at times even initiating attacks against adult male bears to protect family and food sites. It usually required maximum aggressive effort to route the males, and the females were in considerable jeopardy, especially when confronting males high in the dominance hierarchy. Although most adult males possessed the physical capability to inflict lethal wounds, few pressed attacks on or drew out encounters with the females, and no adult females were observed to be seriously injured by the males at ecocenters or in the backcountry. The killing of free-

ranging adult females by males is probably uncommon and has been reported only rarely in the literature (Reynolds and Hechtel 1984, Dean *et al.* 1986).

AGGRESSIVE BEHAVIOR IN DEFENSE OF OFFSPRING

Most of the overt aggression exhibited by adult females to defend offspring occurred in early June, when the aggregation was increasing and its members getting re-acquainted. Females with cubs invariably approached the dump cautiously, sometimes requiring three or four days to a week to adjust to the group. Their approach was hesitant, characterized by frequent stops and retreats before reaching the food. Some retreated a kilometer (0.6 mi) or more, returning after an hour or so. Any unusual commotion would send the females retreating to view the dump from a distance or to depart and not return that day. The more apprehensive paced back and forth at the periphery of the feeding bears, keeping their cubs close with vocal signals, moving in to feed only when much of the aggregation had finished feeding and dispersed.

The time required for the female to adjust varied with individual animals. Once adjusted, however, females with cubs moved directly to feeding sites. Some kept their cubs near them while feeding, and others placed them at a distance and then located feeding sites. Hungry from the nutritional demands of winter sleep, bearing cubs and nursing them, the more aggressive mothers were not to be denied food and frequently acted to hold choice feeding sites.

When members of the aggregation posed a threat to a mother's cubs, she normally responded quickly and decisively. A warning growl to her cubs followed by a swift bluffing charge was usually sufficient to rout the intruder. If an intruder stood his ground, as some adult males did, the bluff was followed by an attack. The male might turn and retreat or back slowly away, as repeated attacks forced him to defend himself. Very rarely did an adult male persist in the face of an aggressive attack. Those that did so became well-known to the adult females, and frequently two or more mothers would attack such males simultaneously.

Females with yearlings or two-year-olds were noticeably less nervous and more assured in their behavior toward the aggregation than females with cubs. The older family units seemed almost placid by comparison. They tended to enter the dump area with confidence and authority, taking and holding choice feeding sites, sometimes against high-ranking members of the male hierarchy. In the early phase of aggregating, females with yearlings were either in the late stages of weaning offspring or were still lactating and would retain their young for another full year. Females weaning young were especially aggressive, not only toward their own offspring, but toward yearlings of other females and other members of the aggregation. Females with large aggressive yearlings fought and fed as a cohesive family unit. This was soon noted by members of the aggregation. As a result, threats greatly exceeded attacks in defense of the yearling family (Table 6.15). Encroaching adult males would generally concede without an encounter. A serious threat to the yearlings of an aggressive female would be met by a charge of the family unit; less aggressive females, often with less aggressive yearlings, conceded food sites to adult males and females alike.

Even the more placid females became extremely nervous, excited, and aggressive when separated from their cubs. This could occur through negligence of the female, curiosity of a wayward cub, or fright dispersal at the approach of an aggressive animal, usually a large male. When a litter dispersed, the female's first concern was to find her cubs, and she would move rapidly and purposefully among the assembled bears, ignoring or snapping at those that blocked her way. Failing to find her cubs among the

Table 6.17. Summary of 141 aggressive actions by adult female grizzly bears, Trout Creek ecocenter, Yellowstone National Park, 1960-68.

Age/sex	Average no. in censused population[a]	Attacks on			Threats toward			Total attacks and threats		
		No.	%	Rate per individual	No.	%	Rate per individual	No.	%	Rate per individual
Adult males	22	40	72.7	1.8	57	66.3	2.6	97	68.8	4.4
Adult females	28	10	18.2	0.4	8	9.3	0.3	18	12.8	0.6
Adult subtotal	50	50	90.9	1.0	65	75.6	1.3	115	81.6	2.3
Nonadult males and females	66	5	9.1	<0.1	21	24.4	0.3	26	18.4	0.4
Total	116	55	100.0	0.5	86	100.0	0.7	141	100.0	1.2

[a] Age structure from Table 5.2 and adult sex ratio from Table 5.6.

Females with cubs invariably approached the ecocenter with caution, making frequent stops or retreats before reaching the food. Fights among adult males frequently caused females to retreat from the dump. *(Photos by John J. Craighead)*

group, she would course like a trained setter, nose alternately to the ground to scent, then head erect to look and listen. At times she would stand erect scenting the air. Failing to get a scent or to locate her cubs, her pace would become more rapid and erratic as she continued in ever wider arcs. Sometimes a lone nonadult would receive the brunt of her agitation, but all members of the aggregation were potential targets of her displaced aggression. The majority gave her a wide berth until she recov-

ered her cubs. Frequently, after such an exhibition, the female would leave the area with her offspring immediately.

We never observed an adult male killing cubs, but we presumed four had done so because of the behavior they elicited among females with cubs. These suspected cub killers were Nos. 41, 88, 206, and Scarface. Two were high-ranking males. These males could seldom approach the dump without attracting the attention of the adult females, and

Apprehensive females would often wait at the periphery of the aggregation of feeding bears. They kept their cubs close with vocal signals, or placed them within view, then moved in among the bears to feed. *(Photos by John J. Craighead)*

Males that had a history of attacking cubs became well known to the adult females. Frequently two or more mothers would attack such males simultaneously. *(Photos by John J. Craighead)*

Family units of yearlings or 2-year-olds, when accompanied by their mother, entered the dump area with confidence and authority, taking choice feeding sites. When necessary they defended as a cohesive family group. *(Photo by John J. Craighead)*

they were watched constantly. If one closely approached a litter of cubs, the mother would attack viciously, carrying the fight to the encroaching male, and it was not unusual for other females that had cubs to join the attack. When attacking, they lunged for the throat, and their bites drew blood. Scarface, the alpha male in 1958, and Ivan (No. 41) were attacked on separate occasions by three adult females charging simultaneously. At the time, Scarface was a noncontender, and Ivan had been observed stalking a cub that was later found dead. Aggressive females with offspring were respected by even the most dominant males.

DEFENSE OF FOOD SITE

Females with cubs or yearlings defended food sites more aggressively than females without offspring. Encounters to gain or hold a feeding site were generally less vicious than those in defense of offspring. Threats, especially from yearling families, frequently deterred an encroacher. Both attacks and threats were most frequent during the breeding period, 1 June to 15 July (Table 6.16). After that, encounters were limited to occasional contentions for choice food sites. More subtle behaviors replaced overt behaviors in signaling status and warning intruders.

DEFENSE OF PERSONAL SPACE

Grizzlies have a sense of "personal" space. The size of the space (or distance) varies with individual animals and with circumstances. It can be described as the zone within which a bear feels comfortable or unthreatened. Its transgression will elicit threat behaviors toward an encroacher. When bears aggregate, this awareness of spatial relationships is acute. In general, females with cubs seemed especially sensitive to spatial violations, and their personal space requirements were generally greater than those of other members of the aggregation. Grizzlies could feed side by side with certain members of an aggregation, but required more spatial separation when in the presence of certain other members.

When a bear's concept of personal space is violated, it becomes instantly and automatically aggressive. A seemingly placid animal (male or female) can become awesomely aggressive and attack instantaneously. Just as rapidly, it can return to "normality." This characteristic has led to the oft-stated phrase, "grizzlies are unpredictable." And indeed they are, as their response to spatial violation is virtually impossible to predict and may differ from other aggressive responses, in that no overt aggressive action by other animals appears necessary to trigger it. Personal space violations are, however, kept to a minimum by warning signals that are easily read and understood by other bears.

POSITIONS IN THE AGGREGATION HIERARCHY

The aggressive behaviors we have described probably elevated the adult females employing them most frequently and vigorously to positions relatively high in the aggregation hierarchy (Fig. 6.6). Such females were respected and permitted to take and hold choice feeding sites. Their rankings within the aggregation hierarchy became most apparent during the postbreeding season (16 July to 30 August), when contention among all members of the aggregation was for food, alone. The nonadult members of both sexes gradually found their places in the social structure. Except for minor adjustments, the aggregation hierarchy (both sexes and all ages) was complete by the end of the breeding season. The overt aggression of individuals was sharply reduced, superseded by community tolerance based on the hierarchical structure; social order was enforced through subtle signals and posturing.

PERIOD OF SOCIAL TOLERANCE

The period of social tolerance following the breeding season was a time when social rank was effectively but unobtrusively enforced. The highest ranking animals had access to the best feeding sites and feeding times. Adults fitted into the social order more rapidly than nonadults, probably because of previous experience. The choice food sites, and therefore the food resource itself, was portioned out in relation to hierarchial rank.

By 1 August, subadults were well integrated into the social structure. They could feed at the less rewarding sites on the dump and at times of the day when the more aggressive bears had left the ecocenter after feeding. At all times, they had to remain alert to the bears around them and be aware of the individual status of each. Negligence in heeding an aggressive posture or a signal from a larger and older bear could result in conflict. The learning process was continuous and repetitive, so as the summer progressed, each subadult found its

"niche" by recognizing individual animals and associating this recognition with their status.

This extraordinary feat of recognition by all members of the aggregation allowed each member to feed in relative harmony. The aggregation's behavior suggested that every adult member recognized every other adult and probably most of the subadults. Likewise, it was essential that subadults of both sexes quickly learn to recognize aggregation members and to retain what was learned. To do so reduced conflict and stress and provided the rapid learners with more undisturbed feeding time during the course of the summer. This information appeared to be retained and refined season after season and, as "stored knowledge," became an important element in the maturation of the aggregation hierarchy and in the social development of the individual.

The season for feeding is short, and the entire aggregation begins it at a low nutritional level after six months of winter sleep (Craighead and Mitchell 1982). During the winter sleep, bears neither feed nor defecate. On emergence, the immediate physiological focus of each bear is to recover from this unique winter sleep and prepare for the next. The members must regain lost weight, grow, and store fat. The winter sleep is, for some females, the climax of the previous breeding season—the birth of offspring. The postbreeding season hierarchy (aggregation hierarchy) and the learning processes associated with and implicit in it reflect individual strategies for preparing to meet the staggering nutritional requirements demanded by the phenomena of winter sleep, reproduction, and the suckling of offspring. Six months of foraging and feeding must provide six months of metabolizable fat. During the denning period, about one third of the adult females give birth to offspring and suckle them; approximately another one third are suckling yearlings; and the remaining third are without offspring, but potential candidates for breeding when spring arrives. The adult males normally leave their dens with a thick layer of unmetabolized fat. This provides the initial energy source for the aggressive behaviors of the mating season.

The abrupt changes from a semisolitary lifestyle to extreme aggregation, back to semisolitariness, and then to deep-sleep isolation, shown by the Trout Creek population and by other naturally aggregating bear populations, are unusual among mammals. Such radical changes in lifestyle in these populations may account for the initial shyness of individual bears as they begin to aggregate, their ability to learn to recognize a large number of former associates in an aggregation, and the necessity of re-establishing male dominance status and the aggregation hierarchy each year, a process that requires most of the 45-day breeding season. Thus, in contrast to many mammals with dominance relations largely settled before breeding, the bears at Trout Creek experienced a long period of social adjustment during the breeding season.

In many bear populations, members do not obviously aggregate for breeding or any other reason. Even among bears that do aggregate, food availability and the breeding season do not necessarily coincide. For example, relatively little breeding occurs at McNeil Falls, Alaska, when the bears are aggregated. The aggregation develops almost entirely as a response to food—the spawning salmon (mostly *Oncorhynchus keta*), which begin arriving at McNeil Falls in late June and reach peak numbers in mid-July. In recent years, a spawning run has developed in Mikfik Creek, a few kilometers from McNeil Falls. This run is active earlier in the season than that at McNeil River. It is attracting bears, which engage in both breeding and feeding (Larry Aumiller, pers. comm. 1987).

At Trout Creek, the timing of available food at the ecocenter and the onset of the breeding season did coincide, so the attraction was both food and procreation. When the period of male contention terminated each season, it was followed by a period of marked quiescence—a period of social tolerance characterized by few, if any, conflicts for food sites. Observations by the senior author during two seasons at McNeil Falls indicated that, as the aggregation developed, there were many confrontations, almost all related to claiming and holding choice fishing sites. As at Trout Creek, a period of social tolerance developed as the summer progressed. Status had been resolved, and food became increasingly more available. Thus, although the timing of aggregation differed at Trout Creek and McNeil River, the process of hierarchy formation appeared to be similar. For grizzly bears, assertive aggression has reproductive and survival value, and that aggressiveness is most overtly and completely expressed when food resources are concentrated by man or nature.

EVOLUTIONARY ASPECTS OF HIERARCHICAL AGGREGATIONS

The evolution of a highly efficient and retentive memory for both hierarchical structuring and foraging, and of a complex system of social signals and strategies, lead us to conclude that hierarchically structured aggregations and their attractant food sources are not recent phenomena or characteristic of only a few unusual populations. On the contrary,

the hierarchical and related behaviors we have described appear to be the result of a long evolutionary process, in which individuals of the species have routinely competed for genetic representation via aggressive actions in ecocentered aggregations and at other more ephemeral patches of food.

The ability of individuals to form stable social relationships with a large number of unrelated conspecifics ensures that they can share massive, concentrated food resources, in spite of innate aggressive behavior that would otherwise thwart communal exploitation of food or expose this large and slowly reproducing carnivore to additional mortality from serious (lethal) conflict. Whereas hierarchy-related behavior provides a mechanism that mutes the potential social costs of membership in stable aggregations, ecocenters nonetheless intensify the advantages of, and natural selection for, traits like aggressiveness, body size, memory, etc., important in competition for rank. In our view, aggregating, as a frequent, nutritional contingency in the course of bear evolution, provides a coherent explanation for the existence of a social system in which ecocentered bears compete aggressively for social and reproductive status and then use that status to settle contests for food, mates, or space with a minimum of conflict.

The hierarchy can be viewed as a population-level effect of natural selection operating on individuals to establish stable competitive relationships in aggregations of bears. Social rank replaces water, food, personal space, etc., as the direct object of social competition. The resulting hierarchy, in turn, imposes a social structure on the aggregation that influences the costs and benefits of a wide range of other individual behaviors and strategies connected with mate acquisition and foraging. As such, the hierarchy becomes a key factor to a satisfactory understanding of virtually all aspects of grizzly bear natural history.

SUMMARY

The bear aggregation at Trout Creek was not a chaotic, unstructured collection of individuals. Rather, it displayed a strong social organization as a product of individual bears competing for and acquiescing to dominance rank within a linear social hierarchy. Competition for and maintenance of social rank involved a variety of agonistic behaviors, including attacks, threats, submissive behaviors, and, among adult males, brutal fights in which the paws, claws, and teeth were used as weapons. Like the aggregation itself, the hierarchy was reformed each year, presumably because the bears

Intensity of eye contact, position of the ears and head, body posture, and vocalizations are complex social signals expressing threat and submission, without overt attack. They are socially adaptive traits that have been acquired through a long history of aggregation at natural ecocenters. *(Photos by John J. Craighead)*

could not maintain a high degree of social organization during their seven-month semi-solitary life away from the aggregation.

We recognized an aggregation hierarchy, which included all members of the aggregation regardless of age or sex, as well as component hierarchies of adult males and adult females. Most, if not all adult males outranked even the most dominant and aggressive adult females. Nonadults were low in the social order, probably ranking below all adult females. They gradually entered the hierarchical levels appropriate to their sex as they grew in size and experience.

The male hierarchy consisted of two strata of adult males: aggressive males (contenders), some of which could be ranked through their aggressive encounters, and nonaggressive males (noncontenders) that could not be ranked because they rarely interacted aggressively with the contending males or among themselves. All bears in the contender stratum could, in principle, have been assigned linear ranks, but in practice we were able to recognize an alpha and beta male each year and lump all remaining contenders into a group of coequal other-contenders. Dominance status conferred upon males a preferential access to feeding sites and to estrous females. There was, however, no relation between male dominance and increased copulatory success. In fact, alpha and beta males had lower copulation rates than other-contenders. Nonetheless, we suggest that dominant males may have achieved higher reproductive success because, by virtue of their rank, they had ready access to females during brief periods of high female fertility within a longer period of female receptivity. Thus, we suggest that timing of, rather than number of, copulations determines reproductive success. An individual male often copulated with a given female multiple times during a single estrus, and several males were commonly observed to copulate with a single female in the course of a single estrus.

Contenders had higher visitation rates at Trout Creek than noncontenders, both during and after the breeding season, although noncontenders tended to increase visitation in the postbreeding season. We suggest that this rank-reflected difference in visitation, and the curious absence of aggressive behavior by the numerous noncontenders, may have reflected differences in mating strategies between bears in the two hierarchical strata. Specifically, we suggest that the most dominant males (contenders) competed intensely to monopolize reproduction at the ecocenter. Here, estrous females were most abundant and their estrous status most readily detected. Relatively small improvements in rank at the ecocenter translated into large increases in mating opportunities and this, in turn, to reproductive success. Lower ranking males (noncontenders), with little prospect for mating opportunities at the ecocenter, searched for less frequent and more dispersed, but less contested, reproductive opportunities in the backcountry, where searching rather than aggressive effort most influenced success.

The time and effort required to establish the male hierarchy varied from year to year. In some years, this period of contention entirely overlapped the breeding season; in other years, it was complete by mid-June, before the breeding season was half over. Body size was an important determinant of success in competition for male status, but other traits, such as aggressiveness and skill in combat, seemed equally and sometimes more important.

Between years, there was both continuity and change in the social status of individual bears. Individual movements in and out of the contender stratum were sometimes explicable as a consequence of growth and maturation, senescence, or injury. In other cases, the causes of rank change were not obvious. Tenures of individual males at the alpha rank varied from one year to three to six years in the case of exceptional males. Of 98 adult males, only 7% attained alpha status during the study; 6% attained beta status, rising no higher; and 12% rose to the level of other-contenders, but not beyond. Thus, only a small percentage (25%) of adult males achieved contender status at any time during the study and, presumably, in their lifetime. The rigors of reproductive competition among contenders were apparently severe, because alpha males rarely survived more than a year after losing alpha rank.

Female agonism was of two types—defensive or offensive—largely defined by female reproductive status. In general, females with yearlings appeared to be most aggressive, followed by females with cubs and solitary females.

Among females, high rank conferred access to choice feeding sites but appeared to play no role in mating. Females and their older (1- or 2-year-old) dependent offspring attained higher status than did solitary females or some adult males. Female aggression was often unrelated to competition for either food or status, as, for example, when females with offspring reacted aggressively to male threats or to approaches by potentially infanticidal adult males. All bears, regardless of rank, reacted to violations of personal space by either attacking or ceding position. Response to encroachment on personal space differed from other aggressive actions, in that no overt aggressive action by other animals appeared necessary to trigger it.

A period of social tolerance followed the establishment of the hierarchy and the cessation of breeding. During this period, contests over feeding sites were settled by mutual recognition of the relative rank of contestants, rather than by overt aggression. The ability to recognize and remember a large number of ever-changing social relationships, and the existence of various and complex social signals expressing threat and submission, are all evidence of a long history of natural selection for socially adaptive traits among bears gathering in aggregations at natural ecocenters in pristine, presettlement environments.

7

Reproductive Biology

INTRODUCTION

It is generally recognized that black bears and brown bears have an essentially similar reproductive pattern. Studies by Rausch (1961), Wimsatt (1963), and Erickson *et al.* (1964) greatly augmented a previous scanty knowledge of the reproductive biology of the American black bear (*Ursus americanus*). More recently, Rogers (1970, 1976), Jonkel and Cowan (1971), Schwartz and Franzmann (1991), and others added significantly to that knowledge. In their review of the reproductive biology of the European brown bear (*U. arctos arctos*), Dittrich and Kronberger (1963) expanded knowledge of reproductive behavior and delayed implantation in this species. Early work on the brown/grizzly bears in North America was from Alaska. Murie (1944a, 1981) reported some breeding dates for the grizzlies of Denali National Park. Reproductive data such as mean litter sizes and cycle lengths (breeding intervals) were reported for populations at McNeil River on the Alaska Peninsula (Lentfer 1966) and on Kodiak (Troyer and Hensel 1964, 1969) and Admiralty/Baranoff/Chichagof Islands (Klein 1958).

We have described the biology of young female grizzly bears (Craighead *et al.* 1969; see also Hornocker 1962) and determined reproductive cycles and rates for the Yellowstone grizzlies (Craighead *et al.* 1974). These and other biological parameters were used to simulate population behavior (Craighead *et al.* 1974). Data from more recent studies of grizzly bear populations in Alaska (Reynolds and Hechtel 1980, 1984; Miller 1987; Schoen and Beier 1990; Sellers and Aumiller 1995 [in press]), in Canada (Pearson 1975; Russell *et al.* 1979; Miller *et al.* 1982; Nagy *et al.* 1983a, 1983b; McLellan 1984, 1989a 1989b, 1989c), on the Rocky Mountain Front, Montana (Aune and Kasworm 1989), and in Yellowstone National Park (Knight and Eberhardt 1984, 1985; Knight *et al.* 1988a) have confirmed that the relatively low reproductive rate reported initially by Craighead *et al.* (1969) is characteristic of the

species. The low reproductive rate, an evolved trait, is presumably optimally adaptive under pristine conditions, but it became a critical factor in the species' survival, as impacts of civilization, especially bear deaths caused by humans, escalated.

In this section, we document the reproductive biology of the Trout Creek grizzly bear aggregation. The discussion of female reproductive behavior is an expansion of work analyzed in Craighead *et al.* (1969 and 1976). Sample sizes used in this chapter (and in Chapter 8) vary, depending on the type of reproductive behavior analyzed. Statistical treatment required known-age females, differentiation in samples of only marked females or both marked and recognizable ones, and a record of the number of years included in the data set. For example, when we calculated litter-to-litter reproductive cycles, we used only marked adult females ($n = 30$). We were able, however, to increase the sample size to 55 females in our productivity rate analysis because we added females that produced no litters or that were without complete cycles. In some analyses, e.g., reproductive status, we added the unmarked-but-recognizable bears to the marked-bear sample, increasing it to 87 adult females. Some analyses required exact ages, considerably reducing the number of females in the sample. In our analyses of survivorship and weaning in Chapter 8, sample sizes were limited because we needed specific information for individual females during specific years. Whenever possible, we used data for the entire study period, but some data sets were not adequate for all years.

Information obtained from family histories was essential to our understanding the various facets of reproduction treated in this section (Craighead *et al.* 1974). The methodology proved so basic that we feel compelled to describe the procedure briefly before presenting the results of our research.

A total of 277 marked and unmarked but recognizable grizzlies were recorded on family history forms. From these data, genealogies for 56 families were developed (e.g., Fig. 3.2). Each family history

Table 7.1. Duration of grizzly bear mating season as defined by first and last observed copulations, Yellowstone National Park, 1961-69.

Year	Earliest date of copulation	Latest date of copulation	Period of copulation (days)
1961	9 June	9 July	31
1962	28 May	23 June	27
1963	11 June	11 July	31
1964	18 May	1 July	45
1965	26 May	27 June	33
1966	12 June	6 July	25
1967	5 June	27 June	23
1968	11 June	27 June	17
Mean			29
1969[a]	25 June	29 July	29

[a] Closure of dumps in 1969 interrupted the breeding season for this year.

was updated from year to year as pertinent data were obtained. Some data were added as late as 1981 from death records. Many of these families were followed through two or three generations. Two typical examples follow:

Data for Family 15 spanned a period of 11 years, from 1960 through 1970 (Fig. 3.2). The history began when female No. 65 whelped two male cubs in 1960. She again produced four cubs in 1962 and three cubs in 1964. Marked offspring No. 144 from the 1962 generation in turn produced litters in 1967 and 1970. Year of weaning, fate, and last year observed were recorded for each family member. The number and identity of males copulating with the female matriarch were also recorded (Fig. 3.2).

Family 14 originated from an unmarked female that whelped two cubs (No. 40 and an unmarked littermate) in 1958. Female No. 40 (second generation) produced three litters; the first in 1964, the second in 1966, and the third in 1968. One of her offspring, No. 240 (third generation), then had two cubs in 1973. These and similar histories allowed us to follow the genealogy of most marked females at the Trout Creek ecocenter.

BREEDING SEASON

In the long view, sexual reproduction perpetuates the species and generates evolutionary change. In the short view, reproduction dictates most, if not all, of the behaviors and activities of a species and a population. These behaviors and activities are cyclic, occurring in various seasonal time sequences. This section reviews some of the grizzly bear behaviors quantified over the duration of the study that permit interpretive insight. We begin with the onset of the breeding season.

Behavioral patterns associated with reproduction were noted among the Trout Creek aggregation as early as 14 May and as late as 15 July. Not including 1969, a year in which the season was delayed, mating was observed as early as 18 May to as late as 11 July, a period of 55 days (Table 7.1 and Fig. 7.1). Within any year, however, the period during which copulations were observed averaged 29 days. The latter span is shorter than the former because observations in some years began after onset of breeding, and because the actual date of the onset of breeding varied from one year to the next. Using the former extremes to define the mating season, we estimate that a period of estrous behavior probably did not persist for more than approximately 63 days in any given year. Dittrich and Kronberger (1963) reported a mating season of approximately 72 days (end of April to mid-July) for captive European brown bears. Reynolds (1979) found a mating season of approximately 44 days for grizzlies living in arctic Alaska (north slope) near the limit of their northern range.

The delay in the onset of mating for the 1969 season (Table 7.1 and Fig. 7.1) was considered abnormal and was not used to estimate the duration of mating for the Trout Creek grizzlies, even though the observational time is comparable to that of other years. We suspect that the abrupt closure of the Rabbit Creek ecocenter and the subsequent partial closure at Trout Creek in 1968-69 affected the breeding season chronology. In the spring of 1969, some members of the Rabbit Creek aggregation moved to Trout Creek, and this abrupt change in habitat, feeding regimes, social relations, and home ranges, with subsequent adjustments between individuals, may have caused the delayed breeding among these bears.

Fig. 7.1. Duration of grizzly bear mating season, defined by first and last observed copulations, Yellowstone National Park, 1961-69.

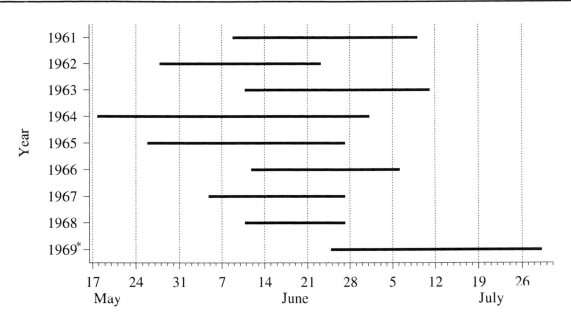

*Closure of dumps in 1969 interrupted the breeding season for this year.

FEMALE COPULATORY BEHAVIOR

When a female approached estrus, adult males frequently checked her reproductive condition by smelling her genitalia. The presence of pheromonal compounds indicate that ovulation is imminent, and that the female is receptive. If not receptive, females fended off males by growling, snarling, or attacking. If pursued, they retreated at a fast lope, frequently for a kilometer (0.6 mi) or more. Females discouraged some males even when they were receptive to others. To discourage unwanted males, females became aggressive, sat on their haunches, or retreated from the area. Preferred mates were encouraged with submissive behavior. Normally, the male initiated the mating ritual by smelling ("checking") the female's genitalia, a behavior often repeated over a period of two to three days to determine the female's reproductive receptivity. During this period of approaching estrus, the consort male, whether the alpha or his immediate contenders, fiercely defended the female or females. The alpha male explored a female's estrous condition and dominated breeding at the onset of her estrus. Normally, receptive females remained standing for the male to mount, or they swung under him in a swift, purposeful motion. The male clasped her just back of the forelegs and frequently bit or mouthed her neck. If the male was slow to initiate the copulatory act or showed no sustained interest, the female quickly swung out from under him, often growling and snapping. Likewise, the male might slap or bite her as she disengaged. Subsequently, the female might appear to lose interest, or she might initiate further action by nuzzling or biting the male's neck. If unattended by other bears, she might "display" before the male by lying on her back with legs spread or by stretching out on her belly, or she might move slowly away, turning frequently to watch or to signal the male with low growls and backward glances. These behaviors appeared to arouse the male and were often followed by the copulatory act.

In most instances, the female had control of the mating act, but at times she was forced to submit to an aggressive male. During such matings, the female frequently turned her head and snapped at the male, and the male in turn bit the scruff of her neck, holding her firmly in place and exerting his weight when necessary. After the copulatory act, the female often fought and struggled to escape from beneath the male and then departed the area on the run.

Abridged notes from 11 June 1967 describe the behavior of female No. 40 when in estrus:

> Number 40 is in estrus and has been running with the Little Brown Boar. Today she and the Little Brown Boar arrived at the dump together. Number 40 was receptive, frequently positioning herself advantageously before the male, but the boar showed little interest in mat-

Distinctive following behavior was commonly observed as males responded to a female's estrus.
(Photos by John J. Craighead)

ing. Several times No. 40 moved directly in front of the boar and quickly swung in under him. He grasped her just behind the forelegs and the two were then in the copulating position. The Little Brown Boar three times attempted to "half-heartedly" copulate, but each time released the sow. The male, not the female, initiated the disengagement. Following the third attempt by No. 40 to interest the male, she left the dump area, crossed Trout Creek, and walked several hundred yards up the stream. She entered a deep pool, bathing and frolicking for about 15 minutes. The boar joined her and also entered the water, fully immersing himself. Number 40 left the stream and walked slowly along the undercut bank, which in places was four to five feet [1.2-1.5 m] above the water level. The sow stopped twice and waited for the boar, wading downstream, to catch up. When he did so, she leaned over the bank. The Little Brown Boar then rose on his hind legs and the two mouthed one another in a playful manner, jaws wide spread, vocalizing in subdued growls. Twice the female backed away and the boar attempted to follow, but failed in each instance, thwarted by the undercut banks. The boar, standing on his hind legs and with forepaws

outstretched, tried to pull his great bulk up and over the overhang. The female moved directly to and above him and they "jawed" one another, growling softly. After this episode, she quickly retreated and sat on her haunches 20 feet [6 m] away. When the male failed to negotiate the bank, she moved slowly along with the male following in midstream. It appeared as though the female was attempting to entice the male to her. A few hundred yards downstream and 10 to 12 minutes later, the male encountered a sloping bank and climbed up to join the female. Number 40 then ran briskly in a bounding lope through the sage and grass, turning several times to glance back over her shoulder. The Little Brown Boar followed with slow, measured strides. The female sat down facing the approaching male, and as he came closer, she quickly presented her hind quarters. With equal swiftness, he pulled her under him. He did not attempt to copulate. While they were in this position, a larger light brown boar came into view. The pair quickly disengaged, and the Little Brown Boar moved several hundred feet toward the intruder (whom I recognized as No. 88). Female No. 40 followed her consort, the Little Brown Boar, at a distance of 30 or 40

A female not in estrus discourages unwanted males with aggressive behavior, by retreating from the area, or by sitting on her haunches. *(Photo by John J. Craighead)*

feet [9-12 m]. When the two males were about 100 yards [100 m] apart, both stopped and looked directly and intently at each other. The males approached one another in the characteristic stiff-legged "battle walk," heads lowered and swinging slightly, and with each hind foot planted twice in the same spot. Their courses were indirect, and they appeared evasive, although each male gave the other his full attention. They closed the distance in this manner until about 50 feet [15 m] apart and stood immobile eyeing one another. At this point No. 40 appeared extremely agitated, running back and forth well out of range of the males. She stopped frequently to rise on her hind legs to view the males. Throughout, No. 40 remained behind the Little Brown Boar. At about 50 feet [15 m] distance, the males simultaneously rushed one another, rising on their hind legs to engage. Embraced, both males twisted their bodies in attempts to throw one another. Grappling like two wrestlers, they applied weight and leverage to gain advantage. Both boars slapped with their paws, and the fur flew while they roared loudly and continuously. Number 88, the challenger, was more aggressive and slowly forced the Little Brown Boar back to the edge of the stream. Both

were striving for neck bites, and several times it appeared that one would throw the other. Finally, the Little Brown Boar dropped low and shot forward trying to come up under the challenger's forelegs. Number 88, however, stepped backwards, seized the Little Brown Boar by the back of the neck, and threw his hind legs out and back just as a wrestler does to foil a leg dive. This placed most of his 700 or 800 pounds [275-315 kg] directly on the neck of the Little Brown Boar. This position was held for 20 or 30 seconds before the smaller boar, nearly crushed to the ground, rolled sideways and freed himself. But this movement put him off balance, and the surging momentum of the challenger crushed him backwards, mouth open, teeth bared and all four feet in the air. Each strove for a neck hold, the downed male now at a distinct disadvantage. Another fast roll brought the Little Brown Boar to his feet and the two males rose together on their hind legs, the challenger (No. 88, or Patch-eye) with a firm hold on the ventral side of the defender's neck. His right foreleg rested under the left foreleg of the Little Brown Boar and against his ribs, the left paw pressed low on the left side of the defender's body. The males pushed and struggled in this

A male and an estrous female playfully mouth each other. The male, on becoming aware of an encroaching male, moves to challenge the intruder. *(Photos by John J. Craighead)*

embrace until suddenly No. 88 shoved the smaller male backwards with a sudden lurch, then leveraged his power and weight suddenly to the right. This move flipped the defending male on his back, and No. 88 stood astride him, but no longer with a grip on his neck. The Little Brown Boar immediately conceded defeat by lying back with his head to one side. The victor held his position for 30 or 40 seconds, but did not press his advantage. Slowly the Little Brown Boar regained his feet and backed away, head lowered and turned to the side. While the defeated male remained in this "humbled" attitude, the victor exposed his side to the defeated male and, only 15 feet [4.5 m] away, staged his swaggering battle dance (treading stiff-legged in place). Number 88 then turned his back on the Little Brown Boar and repeated his battle dance. The cowed boar made no attempt to resume the attack. The Little Brown Boar backed slowly away. Number 88 methodically searched the area for the female. Number 40 had run to the east over the crest of a hill. The victorious male failed to pick up her trail and he moved westward until out of sight. The defeated male walked to the stream and entered it. The victor, No. 88, was not seen again that day. However, the Little Brown Boar reunited with female

No. 40 some 30 minutes later when she returned. He mounted her, but did not breed her. The two returned to the pool where they had earlier bathed and, again, both submerged in the water. While the pair was in the pool, nearly neck deep, a large male topped a rise and moved toward them. The Little Brown Boar, alerted, then sprang ashore and moved directly toward the newcomer, prepared for an encounter. It was soon apparent that the new arrival was not a challenger. He recognized the Little Brown Boar as a superior and retreated without even a threat. This boar was later identified as Scar-shouldered Boar, one of the noncontenders during the summer of 1967.

Some females in estrus showed little preference for specific males. Others appeared to be quite selective, although all readily accepted the alpha males. Some females paired up with a male for several days, and as long as two weeks in one instance. When this occurred, the pair generally entered and left the aggregation together and bedded together. Such pairs copulated in the vicinity of the day beds, but how frequently this occurred is unknown because few of these copulations were observed. The males that paired in this manner

Occasionally, females paired with a single male for days to weeks. In these instances the pair generally joined and left the aggregation together. However, all estrous females remained receptive to other males when their consorts left them undefended. *(Photo by John J. Craighead)*

showed less sexual interest in their partners when at the ecocenter than did unpaired males, but they did attempt to discourage mating by these other males. Females paired with a specific male were, nonetheless, receptive to the advances of these other males when defenses lapsed.

Grizzly bear copulation is normally vigorous and prolonged (Craighead *et al.* 1969). The copulatory act and related overt behaviors vary considerably with the age of the female, her responsiveness, and the number and aggressiveness of the males involved. From 1961 through 1970, we observed 122 copulations during the months of June and July (Table 7.2). Copulation was frequently interrupted by challenges from other adult males. We developed a set of criteria to distinguish a completed copulation from a copulatory attempt or an aborted copulation. A completed copulatory act was characterized by:

a. Male pelvic thrusts that continued for a minimum of 10 minutes.
b. A voluntary male dismount.
c. Female receptivity to the male before and during copulation. Some females, at the end or beginning of their estrous periods, would

not allow the male to continue prolonged pelvic thrusts without trying to extricate themselves or would exhibit aggressive behavior during the copulatory act.
d. Apparent ejaculation. We considered the male had ejaculated when the female rolled out from under him after prolonged periods of pelvic thrusting.

The earliest recorded copulation occurred 18 May 1964 and the latest, 29 July 1969, the year in which the mating season appeared to have been delayed due to dump closures (Table 7.1). We were unable to determine how frequently mating occurred during the 18 May - 31 May period because spring weather conditions caused variation in the start of our observations from one year to the next. We did, however, observe eight copulatory bouts during that period, three of which were judged to be successful (Fig. 7.2). Not until June were we able to quantify the frequency of copulation. The 122 copulations recorded from 1 June to 15 July collectively define the most active part of the breeding season. Frequency of copulation increased through June, with 95% of all copulations after 30 May occurring in June, and only 5% in July (Table 7.2 and Fig. 7.2).

Table 7.2. Frequency of observed copulations among grizzly bears that occurred from 1 June to 15 July, Yellowstone National Park, 1961-70.

Date of mating	No. of individual participants		Copulations	
	Female	Male	Number	Percent
1-15 June	13	13	43	35.2
16-30 June	20	33	73	59.8
1-15 July	2	4	6	5.0
Total	25	36	122	100.0

Fig. 7.2. Observed temporal distribution of all grizzly bear copulations (130) and copulations judged to be successful (64) from 15 May to 10 July, Trout Creek ecocenter, Yellowstone National Park, 1961-70.

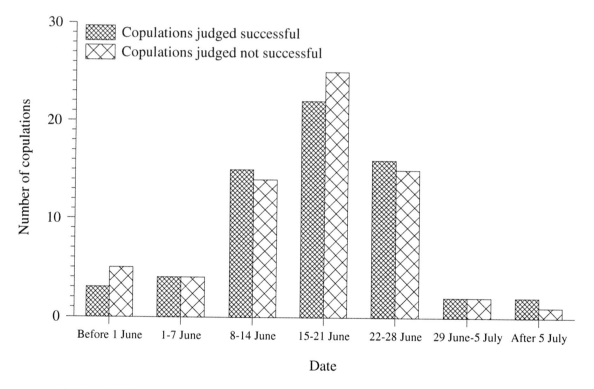

Frequency of Copulation

We observed an average of 4.6 copulations per estrous female from 1959 to 1970. These females were involved with numerous males; the mean number of males copulating per female was 3.9, with a maximum of 20 (Table 7.3). We make no estimate of the number of unobserved copulations that might have taken place after dark or away from the ecocenter. Nevertheless, we are reasonably certain that the rate per hour during our observation periods was not maintained following evening dispersal of the bears from the ecocenter. During the breeding season, bears were bedded down for most of the daylight hours.

On average, females bred every third year. But consecutive-year breeding was also observed when females failed to conceive or lost offspring early on, and alternate-year breeding characterized females weaning or losing offspring at one year of age. Females were promiscuous both within and between years. For example, female No. 29 engaged in 10 copulations in a single estrus and with eight different males. Number 39 was observed copulating eight times over a period of 12 years with six different males. Over a period of three breeding seasons, or a period of nine years, No. 65 was observed to copulate 26 times with 20 different males. In one season she copulated 11 times with eight males.

Table 7.3. Copulatory behavior of 28 female grizzly bears at the Trout Creek ecocenter, Yellowstone National Park, 1961-70.

Female number	Total yrs female observed in estrus[a]	Total copulations observed	Maximum no. copulations observed during a season	Maximum no. males copulating in one season	Total no. males copulating
6	1	3	3	2	2
7	1	4	4	4	4
8	1	1	1	1	1
10	2	2	1	1	2
15	4	2	1	1	2
29	1	10	10	8	8
39	4	8	3	2	6
40	3	11	4	4	11
65	3	26	11	8	20
81	4	4	1	1	3
84	1	3	3	3	3
96	1	1	1	1	1
101	1	5	5	3	3
109	3	8	4	4	8
112	1	6	6	4	4
120	1	3	3	3	3
125	2	5	4	4	5
131	1	1	1	1	1
140	1	2	2	2	2
142	1	1	1	1	1
176	1	1	1	1	1
180	2	5	3	3	5
187	1	1	1	1	1
200	2	7	4	3	4
204	2	2	1	1	2
Brown Sow	1	4	4	4	4
Red-backed Sow	1	1	1	1	1
Unmarked	1	3	3	2	2
Total	48	130	87	74	110
Mean per female	1.7	4.6	—	—	3.9

[a] = the number of estrous periods (estrus-interval-estrus sequences).

Number 81 was observed frequently over a 12-year period, but was observed copulating only once with each of only three different males. Because the observational time was consistent over the years, we suspect that certain females such as Nos. 29, 40, 39, and 65 (Table 7.3) copulated more frequently than other females, and with a greater number of males. Some females were observed to copulate with only a single male during the breeding season.

Despite the considerable variation in copulatory behavior, our data show that the female grizzly is typically polyandrous (Craighead *et al.* 1969). Grizzlies are probably polyandrous wherever they congregate or are sufficiently abundant to allow a female in estrus to meet more than one adult male. Pairing generally occurred for short periods of time, relative to the duration of estrus, and the maintenance of this bond depended on the ability of the male to defend his female against contenders. A female going out of estrus quickly lost the attention of her mate if another female in estrus appeared. Although pairing was observed among the grizzlies in Yellowstone, it was never a partnership that lasted throughout the mating season. Rather, it was a tenuous short-term arrangement for the convenience of mating or, in a few instances, a partnership that persisted throughout the period of an individual female's estrus. Neither the male nor female in these

Fig. 7.3. Percent distribution of durations of 64 successful grizzly bear copulations observed at the Trout Creek ecocenter, Yellowstone National Park, 1961-70.

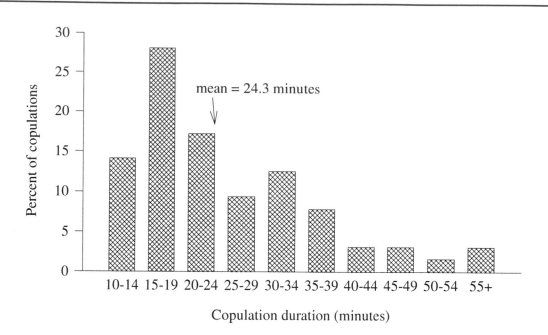

Copulation in the grizzly is often vigorous and prolonged, with an average duration of about 25 minutes. *(Photo by John J. Craighead)*

partnerships remained unresponsive to the sexual condition or sexual advances of other bears (Craighead *et al.* 1969).

Duration of Copulation

We observed a total of 64 uninterrupted, complete, and apparently successful (by our criteria) copulations (Table 7.4). The duration of copulations depended, at least in part, on the vigor of the male, the receptivity of the female, and the privacy of conditions during the copulation. If competing males were nearby, both the male and female appeared less at ease, and copulation was dilatory but uninterrupted. The mean duration of copulations considered successful was 24.3 minutes, and more than half were 24 minutes long or less (Fig. 7.3). One 60-minute breeding occurred on 17 June 1965 when No. 197, a co-alpha male, bred female No. 65 (Table 7.4). During the same evening, Scarneck (the beta male) copulated with female No. 65 for 20 minutes and then bred female No. 41 for an unrecorded time period. On 9 June 1967, an unmarked male (noncontender) copulated with female No. 40 for 60 minutes. Normally, the female rolling out from beneath the male indicated that the copulatory act was terminated. Occasionally, the male terminated or interrupted the act if he was threatened by the approach of a more dominant male.

Table 7.4. Date and duration of 64 successful copulations (those lasting less than 10 minutes excluded) involving 23 individual female grizzly bears observed at the Trout Creek ecocenter, Yellowstone National Park, 1961-70.

Female number	Date	Duration of copulation (min.)	Female number	Date	Duration of copulation (min.)
15	28 June 1961	25	7	12 June 1961	12
	27 June 1967	10	164	18 May 1964	20
39	21 June 1962	12		18 May 1964	15
200	12 June 1965	15	180	19 June 1964	42
	13 June 1965	21		21 June 1964	29
	13 June 1965	32	65	15 June 1961	42
	27 June 1966	30		26 June 1961	21
	6 July 1966	10		26 June 1961	35
29	12 June 1964	12		23 June 1963	32
	13 June 1964	53		16 June 1965	21
	15 June 1964	16		17 June 1965	60
	15 June 1964	18		17 June 1965	20
	23 June 1964	17		19 June 1965	38
40[a]	14 June 1963	21		27 June 1965	20
	18 June 1963	19		27 June 1965	13
	9 June 1967	60	84	26 June 1963	16
	22 June 1967	35	120	21 June 1967	30
101	5 June 1965	30			
	7 June 1965	15	125	21 June 1967	10
	10 June 1965	15		19 June 1970	27
	15 June 1965	38	109	12 June 1966	15
	17 June 1965	16		24 June 1967	20
187	20 June 1967	15		25 June 1967	15
				13 June 1968	27
6	6 June 1962	10		13 June 1968	24
	13 June 1962	16		17 June 1968	16
81	28 May 1962	23	142	27 June 1968	15
	27 June 1964	45			
	9 June 1967	30	Brown Sow	11 June 1968	20
				15 June 1968	13
10	9 July 1963	16		16 June 1968	28
	1 July 1964	16		23 June 1968	35
112	15 June 1963	31	Unmarked Sow	4 July 1961	25
	16 June 1963	47	Red-backed Sow	4 June 1964	31

Total number of copulations = 64

Mean duration of copulations = 24.3 minutes

[a] Female No. 40 was also observed to copulate on 26, 28, and 30 May 1965, but durations of copulation were not recorded.

Table 7.5. Records of estrus and conception for 12 marked adult female grizzly bears, Trout Creek ecocenter, Yellowstone National Park, 1961-68.

Female number	Age of female	Inclusive dates of observed estrus	Observed period of estrus (days)	Offspring produced in following year
6	3	6 June to 13 June 1962	8	None
6	5	23 June to 30 June 1964	8[a]	None
15	3	28 June 1961	1	None
15	4	11 June to 13 June 1962	3	2
15	6	26 June 1964	1[a]	None
15	9	27 June 1967	1	None
29	3	12 June to 15 June 1962	4	None
29	5	7 June to 23 June 1964	17	2
40	5	12 June to 27 June 1963	16	2
40	7	26 May to 30 May 1965	5	2
40	9	3 June to 27 June 1967	25	3
65	Adult	9 June to 28 June 1961	20	4
65	Adult	17 June to 27 June 1963	11	3
65	Adult	12 June to 27 June 1965	16	Unknown
101	7	5 June to 17 June 1965	13	1
109	6	5 June to 1 July 1967	27	None
109	7	13 June to 17 June 1968	5	None
125	Adult	11 June to 19 June 1964	9	3
131	4	25 June to 11 July 1961	17	None
131	5	15 June to 23 June 1962	9	None
176	13	18 June to 30 June 1964	13	Died[b]
180	11	19 June to 27 June 1964	9	3
180	14	18 June to 27 June 1967	10	Died[b]
200	3	10 June to 13 June 1965	4	None
200	4	22 June to 6 July 1966	15	2
Mean			10.7	

[a] Condition of female determined by capture.
[b] Female died before cubs could be produced.

DURATION OF ESTRUS

Female grizzlies in estrus were readily identified by the number and behavior of the males they attracted (Craighead *et al.* 1969). We defined estrus as the period during which females attracted males and both the male and female demonstrated sexual interest. Pre- and postestrous periods were characterized by complete lack of sexual interest by both male and female. Because adult males were always present in the observation area, no female could come in estrus without attracting one or more of these males. Therefore, any female without attendant males was considered anestrous.

There is evidence that some females may experience relatively brief heat periods (Craighead *et al.* 1969). For example, in 1962, we observed female No. 6 in estrus from 6 to 13 June. The 6th of June probably marked the beginning of her heat period, which lasted eight days. For eight consecutive days after 13 June, she attracted no males and exhibited no mating behavior. She attracted no males during four observations between 29 June and 4 July.

Number 6 was three years old and experiencing estrus for the first time. Females 15, 29, and 200, also exhibiting first estrus at the age of three years, had relatively short heat periods as well (Table 7.5). Female 40 exhibited a short estrous period in 1965 at the age of seven years, but had considerably longer periods in 1963 and 1967 (Table 7.5). On the other hand, female 15 was observed in estrus on only one day in each of three different years (Table 7.5). It is interesting that in both 1961 and 1967, her observed period of estrus was one day, copulations on those days were long and "successful" (Table 7.4), yet no offspring were recorded for her during the following years. Although observations of No. 15 were incomplete because of her erratic visitation, it would appear that her estrous periods were brief and her fertility low. Her erratic visitation record could have been related to an abnormal estrus that did not drive her to seek mating opportunities among the assembled males.

In 1966, female No. 200 exhibited estrus over a period of 15 days (Craighead *et al.* 1969). Because she was seen only six days of the mating season and mated on three of these, it is quite probable that the estrous period exceeded that observed. In contrast, female No. 29 exhibited estrus for 17 days in 1964 and, when first observed on 7 June, had attracted two large males that checked her genitalia and showed interest, but did not attempt to mate with her. Our suspicion that she was just coming in estrus was confirmed two days later when she copulated, the first of six copulations that we observed prior to 23 June. From 7 June to 3 July, we observed her 20 out of 27 days and believe her actual estrous period must have conformed very closely to the observed period (Table 7.5).

In 1963, we observed female No. 40 in heat for 16 days (Craighead *et al.* 1969). We watched her before and during estrus, as well as during the postestrous period, and believe the duration of her estrus was accurately recorded in that year. We watched the same female mate on 26, 28, and 30 May 1965 (although the durations of these copulations were not recorded). She was observed in estrus for five days and was not seen again until 10 June, after which we observed and identified her on 13 days over the remainder of the mating season. During this latter period, she showed no signs of estrus. We believe her estrous period was considerably longer than the five observed days because the female was not observed prior to 26 May and from 31 May through 9 June. In 1967, female No. 40 mated again and was in estrus for a period of 25 days. Since she was also observed to attract no males for some time before and after dates

recorded in Table 7.5, we believe that estrus must have extended over these 25 days. Thus, during three breeding seasons, there was considerable variation in the duration of the estrus period of No. 40 that cannot be attributed to our observational methods. Other females showed similar variation. For example, the longest estrous period recorded in the course of this work was 27 days for No. 109 during 1967 (Table 7.5; see also Craighead *et al.* 1969). Number 109 again came in estrus in 1968, but for a period of only five days. Observational records, which were complete for both of these estrous periods, further supported our conclusion that the duration of an estrus can vary considerably in a single individual from one year to another. Observations of the duration of estrus for female No. 65 were also complete, and we believe the 20-, 11-, and 16-day periods accurately represent the durations of her estrus during 1961, 1963, and 1965 (Table 7.5).

Observations for No. 15 were incomplete during 1961, 1962, and 1967, but do show she came in estrus during these years. In 1964, at age six years, she was attracting males on 26 June. She had not been seen before this date and was seen on four days thereafter. When she was captured on 21 July, her vulva was swollen and distinctly colored. We suspected that she had been in estrus for 26 days, although it was possible she was experiencing a second estrous cycle. Receptivity persisting so late in the breeding season might indicate that she had not become pregnant, in spite of the long or double period of estrus. She produced no cubs the following year.

Overall, the evidence is strong for wide variation in duration of estrus among individual females during a breeding season, or for an individual female from one breeding cycle to the next. Therefore, we wondered if all estruses were equal, in terms of female fertility or male interest; e.g., might short estruses reflect, in some sense, a lower quality breeding opportunity for males? There is some evidence for this theory. First, there was a not-quite-significant tendency (Mann-Whitney test, $P = 0.06$) for short copulations to be associated with "short" estruses (one day), whereas long estruses (greater than one day) appeared to be associated with the entire range of copulation durations (Figs. 7.4 and 7.5). Second, cub production was positively correlated with the number of observed copulations, which in turn was probably correlated with the actual (but usually unknown) duration of estrus (refer to data analysis under chapter subheading, "Contribution of Age Classes to Total Cub Production"). Furthermore, the fact that short copulations

Fig. 7.4. Durations of 11 grizzly bear copulations observed during 11 "short" estruses (1 day), Trout Creek ecocenter, Yellowstone National Park, 1961-70.

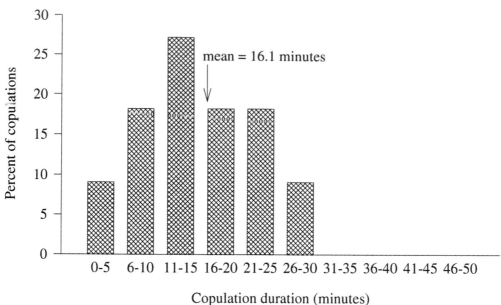

were also common in "long" estruses may reflect variation in female fertility and male interest within a single estrus. For example, short copulations may be characteristic of late and early estrus when fertility may be low, whereas copulation duration lengthens as peak fertility is approached at ovulation. This quality-of-estrus scenario seems quite plausible, considering that our *short* estrus sample was, in most cases, a terminal day in an estrous period the bulk of which took place prior to the female being observed.

The issue of within- and between-estrus variation in female fertility may be crucial to understanding male and female mating strategies and to estimating male mating success, considerations that have important genetic and population dynamic consequences. This variation deserves further study, possibly in captive groups of bears or at aggregations where animals can be individually marked.

In most cases, females exhibiting long periods of estrus whelped cubs the following year (Table 7.5). Numbers 109 and 131 were exceptions. The former produced no cubs in 1967 or 1968 following estrus periods of 27 and 5 days, respectively, whereas female No. 131, at the age of four years, exhibited an estrous period of 17 days in 1961 and nine days in 1962, but produced no offspring. Females Nos. 176 and 180 had well documented estrous periods. Offspring resulted from the first estrus recorded for No. 180, but both she and

No. 176 died before further data could be obtained on offspring following periods of estrus (Table 7.5).

No three-year-old exhibited an estrus exceeding eight days, whereas the longest for adult females was 27 days recorded for No. 109 at six years of age. Dittrich and Kronberger (1963) reported captive European brown bears in estrus for two to five weeks. Estrous periods recorded for 19 females were not included in Table 7.5 because the data were not sufficiently complete for comparative purposes. Six of the 19 females, however, were known to have produced litters the year following an estrus. Six estrous periods produced no offspring, and the results from eight were unknown.

ESTROUS AND ANESTROUS PERIODS

We recorded the estrous and the anestrous condition in young-age females from 1959 to 1970. The anestrous condition normally accompanied lactation. It was also observed to persist in some females for several breeding seasons after lactation had ceased (Fig. 7.6). In such cases, it extended the period between pregnancies.

It is evident that the length of the breeding cycle is quite variable. Three young-age females were observed to have anestrous periods of one or two years between their first potential mating season and their first established pregnancy (Fig. 7.6). Four of

Fig. 7.5. Mean durations of grizzly bear copulations per estrus period, for 52 copulations observed during 25 "long" estruses (>1 day), Trout Creek ecocenter, Yellowstone National Park, 1961-70.

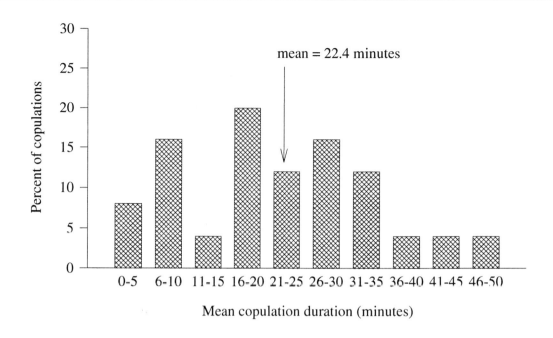

Fig. 7.6. Estrous condition of seven marked, young-age female grizzly bears, observed in successive breeding seasons, Trout Creek ecocenter, Yellowstone National Park, 1959-70.

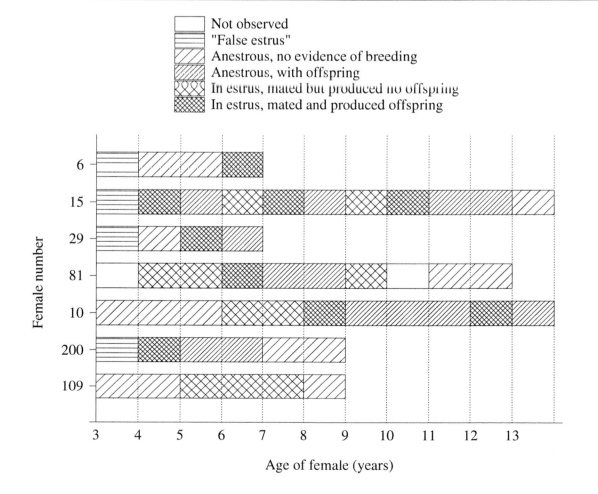

Fig. 7.7. Observed durations of mating interval and estrus for five marked, female grizzly bears, Trout Creek ecocenter, Yellowstone National Park, 1963-67.

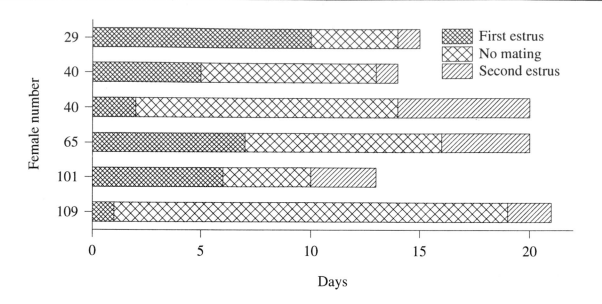

MATING INTERVAL AND ESTROUS CYCLE

seven young-age females exhibited what appeared to be pseudoestruses at three years of age in which copulations occurred, but did not result in cubs the following year. Four of the females experienced anestrus, with no observed mating at the age of four years, whereas two mated for the first time at that age and produced offspring the following year. Four females exhibited estrous periods at later ages and mated, but produced no offspring the following year: female No. 15 at six and nine years of age; No. 81 at four, five, and nine; No. 10 at six and seven; and No. 109 at five, six, and seven years (Fig. 7.6).

Hansson (1947) provided strong quantitative evidence that the female mink (*Mustela vison*) has several estrous cycles during the same mating season, with the female permitting copulation at each period of follicular maturity. The time during which a group of ovarian follicles mature and no copulations occur Hansson termed the mating "interval." Dittrich and Kronberger (1963) studied the mating behavior of the European brown bear in the Zoological Gardens of Leipzig and found that females allowed numerous copulations, but that distinct periods of copulation were separated by days without copulation. The interval between periods of mating appeared similar to those described by Hansson for mink. These pauses in sexual activity occurred in the presence of males and varied greatly in length. Prell, cited by Rausch (1961), reported an interval of variable length between a "pseudoestrus" and true estrus in both the brown bear and polar bear. Schneider, also cited by Rausch (1961), concluded that "pseudoestrus" does not occur in the polar bear, but that the period is a true estrus of long duration.

Our observations suggest that the female grizzly has two periods of estrus during a single mating season. She comes in estrus and is attractive and receptive to males. Copulations occur over a period of days, after which the female is no longer receptive and the male is not attracted to her. Then the female again becomes receptive and again attracts males.

We documented six within-year estrus-interval-estrus sequences for five females (Fig. 7.7). The females were observed during mating and throughout most of the quiescent interval. All of the females were available to large adult males during this period, but were neither attractive nor receptive. The duration of the interval, then, was the time between a terminal copulation in the first estrus and an initial copulation during the second estrus. Our estimates of the interval between mating periods ranged from 4 to 18 days but were subject to some error because we could not always maintain a continuous observational record (Table 7.6 and Fig. 7.7). Nevertheless, it was clear that there was an interval when mating did not occur, probably during follicular development following ovulation, and that it was similar, if not identical, to the "interval" in mink described by Hansson (1947).

Table 7.6. Interval between terminal copulation of first estrus and initial copulation of second estrus and length of estrous cycles observed for five female grizzly bears, Trout Creek ecocenter, Yellowstone National Park, 1963-67.

Female number	Year	Copulation dates (month of June) Terminal (first estrus)	Initial (second estrus)	Interval[a] (days)	Estrous cycle (days) First	Second
29	1964	18th	23rd	4	10	1+
40	1963	18th	27th	8	5	1
40	1967	9th	22nd	12	2	6
65	1961	15th	25th	9	7	4
101	1965	10th	15th	4	6	3
109	1967	5th	24th	18	1+	2

[a] Time, in days, between terminal copulation of first estrus and initial copulation of second estrus.

Table 7.7. Individual male grizzly bears observed to copulate with five female bears that experienced split estrous cycles, Trout Creek ecocenter, Yellowstone National Park, 1961-67.

Female number	Year	Minimum number of individual males copulating First estrus	Second estrus	Total number of different males
29	1964	8	2	8
40	1963	2	1	3
40	1967	1	2	3
65	1961	3	3	6
101	1965	2	2	3
109	1967	1	2	3

Durations of the two estrous periods varied from 1 to 10 days among the five females (Table 7.6 and Fig. 7.7), but again the duration recorded for individual bears was subject to observational error. Nevertheless, collectively, the data indicate that each period of estrus/mating probably does not exceed 10 days, and that there may be little or no difference between the duration of the first and second periods. Each cycle begins and terminates rather abruptly, probably following ovulation induced by mating. The inclusive dates of the intervals all fall within the month of June, when 95% of all copulations were recorded (Table 7.6 and Fig. 7.2).

The significance of the split estrus is the high probability that two sets of ova are produced, one set corresponding to each of the two estruses. This suggests that littermates can develop from different sets of fertilized ova and that they may often be sired by different males, because females often are receptive to multiple males within both estrous periods. For example, female No. 29 bred with eight differ-ent males during her first estrus and with two of those same males during her second (Table 7.7). Other females that displayed split periods of estrus tended to mate with different sets of males in each of their estruses. Recently, Lance Craighead has shown that members of a single cub litter can have different fathers (Craighead 1994). The genetic implications of this finding are of major importance for bear populations concentrated at ecocenters (see Chapter 6).

INDUCED OVULATION

In induced ovulation, ova are spontaneously released within a few hours after copulation; i.e., the mechanical stimulus of copulation induces the female to ovulate. The process appears to be controlled by tactile stimulation of the cervix. This elicits a neurogenic signal that releases luteinizing hormone from the pituitary, which in turn induces ovulation (Bronson 1989).

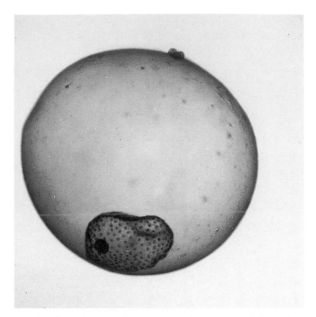

The female grizzly bear undergoes a delayed implantation. Shortly after fertilization the blastocyst ceases dividing and instead of implanting on the uterine wall, remains free within the uterus. Implantation and growth resume to time parturition with winter sleep. *(Photo by John J. Craighead)*

Rabbits (*Sylvilagus* spp.), thirteen-lined ground squirrels (*Spermophilus tridecimlineatus*), and meadow mice (*Microtus* spp.), are induced ovulators (Gunderson 1976). Induced ovulation has also been suspected or documented in a variety of carnivores, including the domestic cat (*Felis catus*), ferrets (*Mustela* spp.), mink (*Mustela vison*), weasel (*Mustela* spp.), marten (*Martes americana*), and otter (*Lutra canadensis*) (Flowerdew 1987). Wimsatt (1963) found large Graafian follicles in ovaries from 12 black bears (11 adults) in early summer and never thereafter; he found no corpora lutea. He postulated that ovulation had not occurred and that black bears may need coital stimulus to ovulate. On the other hand, Tsubota *et al.* (1987), studying five captive brown bears in Hokkaido, Japan, believed that ovulation occurs independently of coital stimulus, and the corpora lutea then function in the appropriate season. They based this conclusion on finding changes in the serum progesterone levels of a nonpregnant female completely isolated from males. The changes were similar to those observed in four pregnant females. Currently, we have no conclusive evidence that coital stimulus is necessarily the event that induces ovulation in the grizzly. It is possible that ovulation is induced by more subtle stimuli such as the sound or smell of males or more general tactile stimuli. Bronson (1989) discussed pheromones as the priming phenomena in mammals. He showed, in some species, that when males were excluded from females, the result could be prolonged and irregular estrous cycles, and, upon reintroduction of males, that cycles shortened and synchronized.

Some or all of the phenomena discussed may be operative in the reproductive biology of the grizzly. More information is needed before the complex estrous cycle of the grizzly can be fully understood. Nevertheless, the phenomena of induced ovulation could help explain certain important features of grizzly reproductive biology. For example, female promiscuity within an estrus might be explained if copulations early in estrus have little chance of fertilization, but instead function primarily to induce ovulation. It might then be advantageous for females early in estrus, and even in the face of strong mate preferences, to solicit copulation from any male, regardless of his age or social status. Similarly, the apparent "indifference" at times exhibited by dominant males in selecting mating partners might be explained by the low fertility of females at certain stages of the estrous cycle. The biological costs of mating and defending females during periods of reduced fertility may not "repay" those bears possessing the larger mating options of high hierarchical rank. Finally, induced ovulation could obviously explain the tendency of males to inseminate the same female repeatedly (Table 7.7). Female promiscuity may also play a role, such that males in sperm competition should attempt to increase the proportion of their sperm introduced into an estrous female relative to that of other males, and frequent insemination is one straightforward way of doing this.

DELAYED IMPLANTATION

Fertilized ova normally undergo mitosis to the blastocyst stage and then implant in the wall of the uterus. Some mammals, however, exhibit discontinuous embryonic development in which the blastocyst may cease dividing and remain free in the uterus for varying periods of time. This is known as "delayed implantation."

Hamlett (1935) believed there was reason to assume a delay in implantation in the black bear, European brown bear, grizzly, and polar bear. Dittrich and Kronberger (1963) subsequently provided evidence for delayed implantation in the European brown and Himalayan black bear (*Selenarctos thibetanus*), and Wimsatt (1963) provided conclusive proof in the American black bear.

Our study obtained conclusive evidence that delayed implantation occurs in the grizzly bear. Female No. 6 mated during June 1965 at the age of six and a half years. She was killed by accidental drug overdose on 27 July of the same year. Free blastocysts were flushed from the uterine horns and fixed for future histological study.

Female No. 180 mated on 18 June 1967 at the age of 14 years. She was killed 50 days later, on 7 August, and unimplanted blastocysts were recovered. The long interval between presumed time of ovulation and the recovery of the unimplanted blastocysts is evidence of developmental arrest and delayed implantation.

Delayed ovulation might also explain the protracted period between mating and the onset of gestation. Erickson *et al.* (1964), however, discounted that possibility in the black bear because of the early formation of corpora lutea following mating. Wimsatt (1963) likewise ruled out this possibility for the black bear. Large corpora were present in the ovaries of our specimens, female Nos. 6 and 81. Thus, delayed ovulation can also be ruled out for the grizzly bear.

PRODUCTION OF CUBS

Until now, we have discussed reproductive behaviors that contribute to the ultimate goal of producing offspring. Now we examine several parameters affecting the productivity of the female population of the Trout Creek ecocenter.

Because of difficulties encountered by bear researchers in maintaining consistent observations throughout a bear's life, nearly all assessments of population productivity have been derived from the simplest, most readily available data: the annual numbers of cubs that emerge with their mothers from winter sleep. This is not a true measure of fecundity or fertility as cubs may die prior to being observed and losses of entire litters may be overlooked. Furthermore, it is extremely difficult to obtain information about survival of cubs to weaning and, even more important (for an understanding of population dynamics), the survival of offspring (especially females) from a first generation to the production of a second generation. Thus, estimation of population growth rates from reproductive and mortality data for grizzly bear populations has been a difficult process, usually requiring considerable extrapolation and conjecture and almost never producing unequivocal results.

In the absence of adequate life history data for defining population growth rates, most studies of bear populations rely on cub productivity as an expression of the potential for population growth. The demographic implications of cub productivity alone are limited. If these data are to be used at all, however, it is essential that they result from long-term observation of marked or individually identifiable females that constitute a representative sample of the entire adult female population. Therefore, we use only data from observations of marked or recognizable female bears of the Trout Creek subpopulation in assessing cub production.

The rate at which an individual female produces cubs over her adult life can be determined from two basic measurements, readily observed over several years: whether a female is observed alone or observed with offspring and, if the latter, how many offspring. We look first at litter sizes documented among the Trout Creek females, then reproductive cycle length, and, finally, productivity rates over time—cubs per adult female per year.

Litter Size

We recorded production of 235 cubs in 112 litters whelped by 77 marked or otherwise recognizable females in the Trout Creek ecocentered subpopulation (Table 7.8). Numbers of litters observed annually ranged from 5 in 1959 to 14 in 1965 and averaged just over 9 litters per year for the 12-year period of study. Three 4-cub litters were recorded, but the majority of cubs resulted from 2-cub litters (56) and 3-cub litters (29). Half of all litters were 2-cub. The mean litter size over the period of study was 2.10 cubs per litter.

Reproductive Cycle Length

Litter size is obviously an important parameter of a female's productivity over her observed adult life. The frequency of her litters over the same period, however, is an equally important parameter. Frequency of whelping is expressed in terms of duration of the reproductive cycle (see Methods, Chapter 3). For an individual female, or as a mean value for the population, the greater the cycle length, the less frequent the litters; and, other things being equal, the lower the reproductive performance.

We used three approaches to determine litter-to-litter cycle lengths for Trout Creek females: observed cycles, inferred cycles, and a combination of observed and inferred cycles. Of the three approaches, use of the combination of observed and inferred cycles resulted in the least biased calculations. First, we considered only those cycles that began with and were terminated by observed litters. However, examination of our data clearly indicated that calculation of a mean cycle length for the population based on these data would produce a serious underestimate. Many adult females were observed to remain nonparous over a period of years after producing a litter; that is, these females were not observed to produce the litter terminating a particular cycle (and initiating the next) before being

Of 112 litters whelped by 77 marked or recognizable females of the Trout Creek subpopulation, only three 4-cub litters were recorded. Half of all litters consisted of two cubs. The mean litter size over the duration of the study was 2.10. *(Photo by John J. Craighead)*

Table 7.8. Litter size from 77 marked or recognizable adult female grizzly bears, Trout Creek ecocenter, Yellowstone National Park, 1959-70.

Year	No. of cubs/litter				No. of litters	Total no. of offspring
	One	Two	Three	Four		
1959	1	2	2	0	5	11
1960	3	6	2	0	11	21
1961	2	4	3	0	9	19
1962	2	6	3	1	12	27
1963	0	6	4	0	10	24
1964	1	4	2	0	7	15
1965	3	6	5	0	14	30
1966	3	5	0	1	9	17
1967	0	6	1	1	8	19
1968	1	3	4	0	8	19
1969	2	2	2	0	6	12
1970	6	6	1	0	13	21
Total	24	56	29	3	112	235
Percent	21.4%	50.0%	25.9%	2.7%		

Mean no. of cubs/litter = 2.10

lost to observation. Not considering these incomplete cycles, along with their observed litters, obviously understated mean cycle length (see Chapter 3).

In the second approach, we corrected somewhat by making the inference that when adult females were observed for at least one year after producing a litter (thereby creating the beginning of a new "open" cycle), they produced litters in the year following the year they were last observed. Thus, the minimum length for this inferred cycle was two years, on the assumption that the unobserved female weaned her litter as yearlings and bred successfully. We combined results of observed cycles and inferred cycles to calculate a minimum mean cycle length for the population. This is a minimum value because it is unlikely that all of the females assumed to whelp in the year following our last observation of them actually did so; for some the period was longer, but the precise length is unknown. Nevertheless, the mean cycle length calculated in this way is more representative than one calculated only from those cycles observed to terminate with litters.

Based on 38 observed litter-to-litter cycles for 23 females, the mean cycle length for the 12-year study period was 2.89 years (Table 7.9). Nearly half of the observed cycles were three years long. The mean length of 21 cycles terminated by inference was 4.00 years (Table 7.9). These cycles were contributed by 14 of the first group of females, plus an additional seven individuals. Inferred cycles included three that were six years long and two that were

eight. Combination of the observed and inferred data gave a total of 59 reproductive cycles for 194 years, or a minimum mean cycle length for the population of 3.29 years (Table 7.9).

Productivity

In our earlier analyses of the Yellowstone grizzly bear population (Craighead *et al.* 1974, 1976), reproductive rates were calculated for 30 marked Trout Creek females that had completed at least one reproductive cycle (see Methods, Table 3.3). This group was and continues to be important for defining cycle lengths, ages at which females first whelp, weaning ages, and the productive capacities of females. As explained in Methods (Chapter 3), however, reproductive performance of the adult female portion of the population is misrepresented when only productive females are included and barren females ignored. (It is worth noting, again, that nonproducing adult females will be underrepresented in annual counts, despite good intentions, unless they are marked or otherwise identifiable over time.)

Here, we have calculated individual productivity rates for 54 female bears observed for at least one season as an adult (at least 5 years of age, the earliest age of observed whelping) at the Trout Creek ecocenter (Table 7.10). The mean rate of productivity for the female population over the 12-year period of study, 0.61 cub per adult female per year, is derived as the quotient of total cubs divided by total adult female-years. It reflects the influence of

Table 7.9. Lengths of litter-to-litter reproductive cycles observed and/or inferred by assumption of earliest completion for 30 marked adult female grizzly bears, Trout Creek ecocenter, Yellowstone National Park, 1959-70.

Cycle length (yrs)	Observed (23 females)		Inferred[a] (21 females)		Combined (30 females)	
	No. of cycles	Total years	No. of cycles	Total years	No. of cycles	Total years
1	1	1	0	0	1	1
2	11	22	4	8	15	30
3	18	54	8	24	26	78
4	7	28	2	8	9	36
5	1	5	2	10	3	15
6	0	0	3	18	3	18
7	0	0	0	0	0	0
8	0	0	2	16	2	16
Total	38	110	21	84	59	194
Mean of reproductive cycle (yrs)	—	2.89	—	4.00	—	3.29

[a] For Trout Creek females observed for two or more years without producing a litter to complete a cycle, a litter is assumed in the year following that of last observation.

10 females barren for a total of 25 adult years (Table 7.10) and includes all adult years during which identifiable females were observed, irrespective of cycles. This mean productivity rate for the population is a considerable reduction from the reproductive rate of 0.693 cubs per year per adult female calculated earlier for 30 productive females (Craighead *et al.* 1976 and Table 3.3, Methods). We believe, however, the population productivity rate of 0.61 to be a more realistic expression of female reproductive performance, especially if such a value is to be meaningfully compared between populations.

Researchers often use data collected on individually identifiable females in a given year to compute productivity annually. The total number of adult females observed is simply divided into the total number of associated cubs observed in a year. As discussed in Methods (Chapter 3), this is a reasonable representation of the population's annual productivity, provided its limitations are appreciated. The females must be marked or individually identifiable, and the cubs, if not marked, must be distinguishable by association with the mother or siblings. Otherwise, females with litters will be over-represented and/or those without under-represented, whereas cubs may be overcounted. When a sample of reliably identifiable females is being observed from one year to the next, observational discontinuities (annual variation in sample size and quality), which

are impossible to avoid, may make it difficult to demonstrate statistical differences among years.

A grand mean of annual mean productivity rates derived over the period of study will weight means from years having few samples equally with those having many. For example, the data from each of 54 individual Trout Creek females used to calculate a mean productivity rate of 0.61 (Table 7.10), was re-expressed in terms of annual production by adult females (Table 7.11). The grand mean population productivity for the 12-year study was 0.64 cubs per year per female, when calculated by dividing number of years of study into the sum of the annual mean productivity values. Note, however, that elimination of the productivity rate for 1959, an unusually large value based on a small sample of females, gives an 11-year mean of 0.61 (6.69 divided by 11 study years). The relatively high annual productivity value of 1.0 for 1959 contributed to the grand mean of 0.64 just as much as did the 1965 value of 0.83 based on a sample of 29 females and 24 cubs (Table 7.11).

Thus, when researchers derive annual productivity rates, it is preferable that the productivity rate for the period of study be based on the sum, over all years, of females observed divided into total cubs produced. Note that these sums (Table 7.11) equal those for "Adult-years observed" and "Number of cubs," respectively, in Table 7.10.

Table 7.10. Productivity rates for 54 marked, adult female grizzly bears, Trout Creek ecocenter, Yellowstone National Park, 1959-70.

Female number	Adult-years observed	Number of cubs	Cubs per adult-year observed	Female number	Adult-years observed	Number of cubs	Cubs per adult-year observed
5	6	5	0.83	123	2	2	1.00
7	10	7	0.70	125	9	8	0.89
10	9	4	0.44	128	9	13	1.44
15	7	3	0.43	129	1	0	0.00
18	3	1	0.33	131	1	0	0.00
23	1	0	0.00	132	2	1	0.50
29	2	2	1.00	139	2	2	1.00
34	11	6	0.55	140	6	3	0.50
39	11	3	0.27	141	2	2	1.00
40	7	7	1.00	142	4	0	0.00
42	11	8	0.73	144	4	3	0.75
44	7	2	0.29	148	2	2	1.00
45	1	0	0.00	150	8	5	0.63
48	4	5	1.25	160	5	2	0.40
64	6	0	0.00	163	1	2	2.00
65	6	9	1.50	172	7	4	0.57
72	2	1	0.50	173	4	3	0.75
75	4	2	0.50	175	7	4	0.57
81	7	1	0.14	175B	6	4	0.67
84	4	5	1.25	176	1	0	0.00
96	8	8	1.00	180	4	3	0.75
101	8	4	0.50	184	2	2	1.00
108	10	2	0.20	187	3	1	0.33
109	4	0	0.00	200	4	2	0.50
112	9	5	0.56	204	3	0	0.00
119	1	2	2.00	228	3	0	0.00
120	10	4	0.40	259	2	3	1.50
				Total	275	167	

$$\frac{\text{Sum of number of cubs}}{\text{Sum of adult-years observed}} = \frac{167}{275} = 0.61$$

Note: "Adult" is five years of age, or older.

To compare productivity rates (reproductive rates) between bear populations it is imperative to standardize methodologies. We recommend the method we have described under Methods, Chapter 3. The values derived can be related to factors affecting the rates at which offspring are produced, not only for individual females, but for the population as a whole. In the next section, we present basic data for some of these reproductive parameters, evaluate their effects on individual female productivity, and, by extension, consider implications for the population.

FACTORS AFFECTING PRODUCTION OF CUBS

The productivity of a population is directly related to the productivity of each individual female,

which is a function of reproductive age, litter size, cycle length, and many interrelated factors, such as her specific genetic make-up, her life stage, the quality of her environment, and the dynamic behavior of the population of which she is a member. We now examine characteristics of the producing females of the Trout Creek subpopulation and, where appropriate, test for relationships between productivity rate and specific biological characteristics of the female and/or the dynamics of the population that have a direct bearing on reproductive performance.

Female Age at First Reproduction

Based on observations of known-age (marked as cubs) and established-age (analysis of tooth cementum) female grizzly bears (Craighead *et al.* 1970), we determined the age at which 16 individu-

Table 7.11. Annual productivity rates for 54 marked, adult female grizzly bears, Trout Creek ecocenter, Yellowstone National Park, 1959-70.

Year	Number of females	Number of cubs	Annual mean productivity
1959	3	3	1.00
1960	12	6	0.50
1961	16	13	0.81
1962	23	18	0.78
1963	28	21	0.75
1964	28	10	0.36
1965	29	24	0.83
1966	25	15	0.60
1967	30	16	0.53
1968	27	18	0.67
1969	27	8	0.30
1970	27	15	0.56
Total	275 (Female-years)	167	7.69

$$\frac{\text{Sum of annual mean productivity}}{\text{Number of study years}} = \frac{7.69}{12} = 0.64$$

$$\frac{\text{Total number of cubs}}{\text{Total number of female-years}} = \frac{167}{275} = 0.61$$

Note: "Adult" is five years of age, or older.

als first reproduced. This age, whether defined by first pregnancy or first whelping, affects a female's potential lifetime productivity and, collectively, that of the population. Most researchers have adopted the practice of reporting ages of bears in whole years. In the interest of conforming to common usage, and despite the fact that a bear's age is at a half-year mark when an individual is observed in the spring, we drop the decimal and present bear ages and events in terms of whole years (e.g., cubs = 0 years, yearlings = 1 year, first whelping = 5 years, etc.), except in cases (such as follows) where results from published research are based on half-years.

Age at First Pregnancy

We have previously reported 4.5 years as the youngest age at which females became pregnant (Craighead *et al.* 1969, 1976). Although 3.5-year-old females were observed to be estrous and to breed on occasion, no 4.5-year-olds were observed with litters.

Whereas 4.5 years was the age at first pregnancy for nine females, seven others exhibited a "prepregnancy period," that is, an observed period of nonproductive years beyond the earliest age of potential conception. Thus, in addition to the nine observed pregnancies among 4.5-year-olds, three

females first became pregnant at 5.5 years of age, three at 6.5, and one at 8.5 (Table 8 in Craighead *et al.* 1976). These data yielded a mean observed age of first pregnancy of 5.31 years for the 16 females (mean age = 4.81 years, when half-years of age are dropped). It is possible that some females with protracted prepregnancy periods did conceive earlier, but resorbed embryos or experienced miscarriages or postpartum mortalities of entire litters before our observation of them in the spring.

Age at First Whelping

We considered female bears to be reproductively mature (i.e., adult) at five years of age, the earliest they were observed to whelp litters during our study. In the literature, first whelping is uniformly the event defining adulthood.

To determine observed age of first whelping, we used only females for which we had continuous observational records for adult years before their first litter and did not include adult females observed only in the year immediately preceding their first litter. Thus, the group of 15 females selected included most but not all of those in the sample used for the earlier analysis of age of first pregnancy, plus three others for which we had complete data on whelping. Based on direct observations of first litters

Table 7.12. Age at which marked, adult female grizzly bears were observed or assumed to whelp first litters, Trout Creek ecocenter, Yellowstone National Park, 1959-70.

Age (yrs)	No. females observed first whelping	No. females assumed first whelping[a]	Total females
5	7	4	11
6	5	3	8
7	2	3	5
8	0	1	1
9	1	3	4
Ages 5 through 9	15	14	29

Mean age of observed first whelping = 5.87 years.
Mean age of assumed first whelping = 6.71 years.
Mean age of observed and assumed first whelping = 6.28 years.

Note: "Adult" is 5 years of age, the earliest age at which known-age females were observed in this study to whelp (*n* = 7).

[a] Females for which observation over the period beginning in subadulthood revealed no whelping, and whelping was assumed to occur in the year following that of the last observation.

among the group of 15, the earliest age of whelping for seven of these female bears was five years; for five females it was six years, for two females, seven years, and for one female, nine years of age (Table 7.12). These data yield a mean age of observed first whelping of 5.87 years.

Over the 12 years of study, an additional 14 female bears, aged and marked as nonadults, were subsequently observed to be barren over one or more years at the start of adulthood (5 years), but were then not observed for one or more years, during which time they may have whelped their first litter. Although exact lengths are unknown, some of these observed prepregnancy periods extended well into adulthood. Our failure to measure them reflects the unavoidable fact that the longer the actual prepregnancy interval, the more likely it is that missing life history data will exclude it from the direct observation sample. To correct partially for this sampling bias against longer periods, we calculated a minimum age at first whelping by assuming that these 14 females whelped in the year they were first missing.

Based on our expanded sample of 29 females, minimum mean age of first whelping was 6.28 years for the Trout Creek subpopulation (Table 7.12). The true mean age of first whelping was probably somewhat higher because some of the females for which whelping was assumed probably did not whelp in the year after their last observation. Although the majority of females

sampled produced first litters at five years of age, on the average, first litters were produced at least a full year after females were reproductively mature. This could indicate that the population was not fully exploiting its productive potential.

Female Age and Litter Size

We were interested in knowing whether maternal age influenced the expected production of cubs by individual females and, in a related question, whether certain adult age classes were responsible for a disproportionate share of the total production of cubs. The age-specific expected production of cubs for individuals is a function of the age-specific probability of producing a litter at all and the age-specific mean size of litters. The total production of cubs by different age classes of females is a function of these same quantities, plus the number of females in each age class (i.e., the Trout Creek female age structure). Note that we include 4-year-old females as being reproductively mature in our analyses, despite observing none to have whelped at that age. This is because several cases of whelping by 4-year-olds have been documented for the Yellowstone population over the years since completion of our study (see Chapter 17).

In earlier work, we reported a positive, but not statistically significant correlation between age of female and size of litter (Craighead *et al.* 1976). The sample of females used in the earlier calculations differed slightly from that examined here, and the

Table 7.13. Ages of 31 marked/identifiable female grizzly bears observed to have produced cubs, Trout Creek ecocenter, Yellowstone National Park, 1959-70.

Age (years)	Mean litter size	Number of litters
4	—	0
5	2.14	7
6	1.86	7
7	1.75	4
8	1.57	7
9	2.20	5
10	3.00	2
11	2.40	5
12	2.33	3
13	2.00	1
14	2.33	3
15	1.67	3
16	2.50	4
17	2.33	3
18	2.00	1
19	3.00	1
20	—	0
21	—	0
22	2.00	1
Total		57

Note: Reduced sample size of 31 females results from exclusion of many females that were not known-age or cementum-aged.

data on age and litter size were tested for significant association using the Spearman method of rank/order correlation. This test is used to assess how consistently two variables increase or decrease in value (rank/order) relative to each other, but it is not useful when one of the variables increases and decreases relative to increase or decrease in the other. As the following data suggest, the variables do exhibit the latter association; that is, one of the two variables (litter size) increases and plateaus as the other (age of female) increases. Thus, application of Spearman rank-correlation testing to those data was not appropriate, and its result was not a reliable indicator of the degree of association between female age at whelping and litter size.

In our current analysis, calculation of mean litter sizes for the specific ages at which 31 females whelped 57 litters did not reveal an obvious association between the two variables (Table 7.13). When the females were grouped into age categories based on patterns of growth and physical decline, however, and when mean litter size was calculated accordingly, it appeared that females produced larger litters in the second and third periods of their reproductive life than in the first and fourth (Table 7.14). The fourth period result was based on a single bear. But when bears of the young-aged group were compared against those of the combined mid-aged groups, the latter bears (9-20 years of age) were shown to have produced significantly larger litters than did the young-aged bears (ANOVA, Extra-Sums-of-Squares Test, $P < 0.02$; Appendix C, Table 1). Further, the analysis provided no evidence that female identity or status of ecocenters (i.e., operative or not) had any effect on litter sizes.

Table 7.14. Age class grouping of 31 marked/identifiable, female grizzly bears observed to have produced cubs, Trout Creek ecocenter, Yellowstone National Park, 1959-70.

Age class (years)	Litter size				Total cubs	Total litters	Mean litter size
	One	Two	Three	Four			
4-8	8	14	2	1	46	25	1.84
9-14	1	10	8	0	45	19	2.37
15-20	2	6	3	1	27	12	2.25
21-25[a]	0	1	0	0	2	1	2.00
Total	11	31	13	2	120	57	2.10

[a] Results based on a single bear; omitted from ANOVA for age effect on litter size (Appendix C, Table 1).

Table 7.15. Percentage of female grizzly bears with cub litters (fertility) and mean number of cubs per female (production) for females categorized by age, Trout Creek ecocenter, Yellowstone National Park, 1959-70.

Age (years)	Unrestricted sample (40 females, 171 female-years)			Restricted sample[a] (38 females, 119 female-years)		
	Female-years	% of females with cub litters	Mean no. of cubs/female	Female-years	% of females with cub litters	Mean no. of cubs/female
4	26	0.0	0.00	26	0.0	0.00
5	19	31.6	0.68	19	31.6	0.68
6	14	28.6	0.50	9	33.3	0.44
7	16	18.8	0.31	9	33.3	0.56
8	15	46.7	0.73	11	54.6	0.91
9	10	50.0	1.10	6	83.3	1.83
10	6	33.3	1.00	4	50.0	1.50
11	12	41.7	1.00	3	100.0	2.00
12	8	37.5	0.88	6	50.0	1.17
13	7	14.3	0.29	3	33.3	0.67
14	7	42.9	1.00	5	60.0	1.40
15	8	37.5	0.62	4	75.0	1.25
16	8	37.5	1.00	4	50.0	1.25
17	4	75.0	1.75	4	75.0	1.75
18	2	50.0	1.00	1	100.0	2.00
19	3	33.3	1.00	1	100.0	3.00
20	2	0.0	0.00	1	0.0	0.00
21	1	0.0	0.00	1	0.0	0.00
22	1	100.0	2.00	1	100.0	2.00
23	1	0.0	0.00	—	—	—
24	1	0.0	0.00	1	0.0	0.00

[a] Females were sampled for whelping in a year only when they were available for breeding (without dependent offspring) in the previous year.

Table 7.16. Percentage of female grizzly bears with cub litters (fertility) and mean number of cubs per female (production) for females categorized by age class, Trout Creek ecocenter, Yellowstone National Park, 1959-70.

Age class (years)	Unrestricted sample (40 females, 171 female-years)			Restricted sample[a] (38 females, 119 female-years)		
	Female-years	% of females with cub litters	Mean no. of cubs/female	Female-years	% of females with cub litters	Mean no. of cubs/female
4-8	90	22.0	0.40	74	24.3	0.43
9-14	50	38.0	0.90	27	63.0	1.44
15-20	27	40.7	0.93	15	66.7	1.47
21-25	4	25.0	0.50	3	33.3	0.67

[a] Females were sampled for whelping in a year only when they were available for breeding (without dependent offspring) in the previous year.

Female Age and Probability of Litter Production

We examined fertility by individual ages and by age categories for two samples of females (Tables 7.15 and 7.16). An "unrestricted" sample included all female-years for which a known-age female's whelping status was known, regardless of her reproductive status in the year previous (40 females followed for 171 female-years). The "restricted" sample included only those cases of the "unrestricted" sample in which the female was without dependent young in the year prior to recording whelping status (38 females followed for 119 female-years). The value of the restricted sample is that it does not compare a good thing (failure to whelp because of successful previous reproduction) with a bad thing (failure to whelp, despite availability for breeding). Both samples included females that produced no litters during the periods they were observed. As with our analysis of effect of age on litter size, no age-related pattern of fertility was obvious when females were evaluated in terms of individual ages (Table 7.15). But when ages were grouped, as was done for litter size, the percentage of females whelping litters (our estimate of the probability of whelping) was highest during the second and third periods (ages 9 to 20 years) of a female's adult life (Table 7.16). The superior performance of middle-aged females was most marked in the restricted sample.

To test for significance of differences in fertility among the age categories, we compared the data for age-class one with those for classes two and three (restricted sample, Table 7.16). Age classes two and three were lumped, and age class four deleted, to obtain sufficient sample sizes. From the restricted sample, we then drew 13 random sample sets so that each individual female appeared only once in a given set. The result of these tests showed a strong association between fertility and age in the 13 samples (χ^2-test with Yates' correction; median P-value = 0.003). We concluded that young females had significantly lower fertility than did early and late prime-aged females.

We then looked at the combined effect on cub production of age-related variation in litter size and probability of litter production by calculating the mean number of cubs produced per female for each of the four age classes (Table 7.16). The restricted sample is the more reliable source for all of these measures because it does not penalize females that do not produce litters simply because they were previously successful. Not surprisingly, this sample showed a two- to three-fold increase in cub production by age classes two and three—the early and

late prime-aged females (Table 7.16). In Section II of this monograph, we expand on the question of productivity versus female age by applying techniques of statistical regression to data from our research, combined with those from other grizzly bear research studies (Chapter 17).

Contribution of Age Classes to Total Cub Production

Prime-aged females were more likely to whelp, had larger litters, and higher mean cub production than did young-aged females. It does not follow, however, that females in the prime-aged classes were responsible for the bulk of cubs produced at Trout Creek during the study. There were progressively fewer females observed in classes of increasing age and, despite more prolific production, actual contribution to population growth declined as cohorts aged. For example, the median age of whelping for 57 litters produced by 31 known- or established-age females was 9 years (Table 7.17).

Table 7.17. Ages at which 31 marked/identifiable female grizzly bears were observed to whelp litters, Trout Creek ecocenter, Yellowstone National Park, 1959-70.

Age at whelping (years)	Number of litters	Number of cubs
4	0	0
5	7	15
6	7	13
7	4	7
8	7	11
9[a]	5	11
10[b]	2	6
11	5	12
12	3	7
13	1	2
14	3	7
15	3	5
16	4	10
17	3	7
18	1	2
19	1	3
20	0	0
21	0	0
22	1	2
Total	57	120
Median age of whelping[a,b]	9 yrs	10 yrs

[a] Fifty percent of 57 total litters were produced by females 9 years old, or younger.

[b] Fifty percent of 120 total cubs were produced by females 10 years old, or younger.

Middle-aged females (9-20 years) were more likely than young females (<8 years) to produce litters in the year following weaning of prior litters, and they had larger litters. Young females were responsible for greater cub production simply because this age class was more numerous than middle-aged females. *(Photo by John J. Craighead)*

Thus, despite the greater productivity of prime aged females as a group, 50% of all litters recorded for the 1959-70 Trout Creek subpopulation were whelped by young-aged females. Similarly, half of the total observed cub production occurred among females aged 10 years or younger (Table 7.17). This one-year (9 to 10) increase in the median age of production, litter versus cub tally, pointed once again to the generally smaller litter sizes observed among young adult females.

Effect of Estrous Duration on Cub Production

In addition to age, we wondered whether the "quality" of a female's estrus, indicated by the number of times she copulated or the number of different males she attracted, influenced cub production. This question was prompted by the great variation we observed in estrous duration and the consequent suspicion that not all of the estruses we observed were reproductively equivalent. Unfortunately, we had too few complete records to look at the effect of estrous duration directly. Frequency of observed copulation and number of copulating

males were used as surrogate measures of estrous duration.

To examine the influence of estrous "quality," while controlling for female age (above) and their visitation at Trout Creek, we used path analysis (Sokal and Rohlf 1981) to analyze cub production by 19 different females for which sufficiently complete records were available (Table 7.18). Visitation was included in the analysis to control for variation among females in our observational record of estrus and mating.

The number of times a female copulated had significant positive, direct effects on cub production the following year, independent of her visitation in a given year and her age (Fig. 7.8). In other words, when the effects of the other variables were controlled statistically, the analysis showed that more cubs were produced by females observed to copulate most frequently. Although the number of different males with which a female copulated did not directly affect cub production, this variable had a significant, positive, indirect effect on cub production, operating via the number of copulations (Fig. 7.8).

Table 7.18. Variables in the productivity of 19 marked/identifiable female grizzly bears, Trout Creek ecocenter, Yellowstone National Park, 1959-70.

Female number	Year	Age	Number of copulations	Number of different males	Mean visitation	Number of cubs produced
7	1961	13	4	4	0.5500	3
10	1963	6	1	1	0.4583	0
15	1967	9	1	1	0.1613	0
29	1964	5	10	8	0.7143	2
39	1962	8	3	2	0.6316	3
40	1967	9	4	2	0.4516	3
81	1964	6	1	1	0.6786	1
96	1962	4	1	1	0.0526	2
101	1965	7	5	3	0.8800	1
109	1967	6	3	2	0.6452	0
112	1963	8	7	4	0.7083	2
120	1967	11[a]	3	3	0.3871	0
125	1967	17[a]	1	1	0.1935	3
140	1964	10	2	2	0.7500	3
142	1968	6	1	1	0.1250	0
180	1964	11	3	3	0.3214	3
187	1967	4	1	1	0.4839	0
200	1966	4	4	3	0.3000	2
204	1968	5	1	1	0.0833	0

[a] Ages estimated from record of observations.

Because grizzly bears are probably induced ovulators, one might argue that this result could be explained if some females simply do not copulate frequently enough to induce ovulation. But this begs the question of why females have not evolved to copulate with sufficient frequency to always ovulate. A more likely explanation, in our view, is that very short estruses (which are necessarily, but incidentally, associated with low numbers of matings) are, in fact, "pseudoestruses" truncated by defects in one or more of the developing follicles or corpora lutea, or by another anomaly in the system governing the maintenance of physiological and behavioral estrus. The possibility of pseudoestrus is appealing because it simultaneously provides a cogent explanation for the otherwise curious relation between copulation frequency and cub production, as well as the seemingly inordinate variation in the duration of estrus (1-27 days; Table 7.5).

Another variable, the nutritional condition of females, was not considered in our path analysis (Fig. 7.8). The direct influence of maternal age on cub production, however, may well be due to this variable; i.e., females in good condition and able to produce more cubs tended to be of a certain age class.

Female Aggressiveness

Over the years of observation, it was plain to us that some individual females were routinely more aggressive than others. Moreover, when we examined litter sizes of females categorized in terms of their aggressiveness, aggressive females tended to produce larger litters than the nonaggressive ones. The nine Trout Creek females determined to be most aggressive (see Chapter 5) had a mean cub litter size of 2.35, compared to a mean cub litter size of only 2.06 for 30 "nonaggressive" females (Tables 7.19 and 7.20). Mean yearling litter sizes also were higher for the aggressive females (2.29) than for the nonaggressive females (2.13). The enhanced cub production of aggressive females might have been due to superior nutrition. If aggressive females were better able to obtain and hold choice feeding sites in the aggregation and, thereby, were more capable of nurturing large litters, then, as in the male (Chapter 6), aggressiveness has been an important factor in the evolution of the species. Because aggressive behavior is so characteristic of the species and yet is so detrimental to its survival under human-dominated conditions, this subject merits further study.

FEMALE VISITATION AT THE ECOCENTER

In Chapter 5, we outlined a model of the historical development of the Trout Creek aggregation in which food was the obvious primary attractant for females and males, but where the detailed pat-

Fig. 7.8. Path diagram of the relationship among female age, number of copulations, number of different males copulating with, mean visitation at Trout Creek during the breeding season, and number of cubs produced the following year, for 19 adult female grizzly bears, Trout Creek ecocenter, Yellowstone National Park, 1959-70.

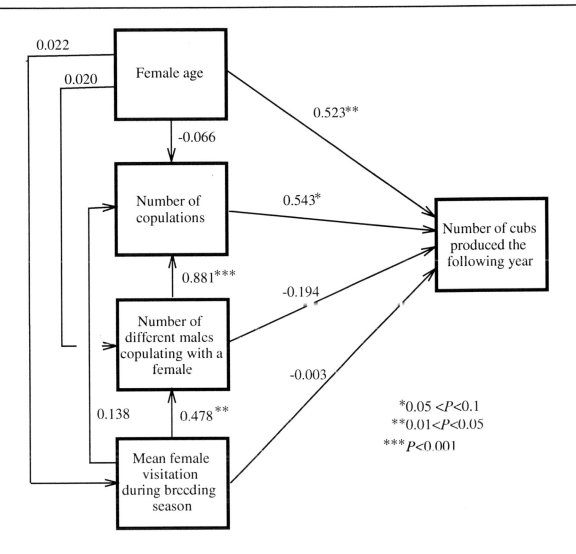

Note: Numerical values are standardized partial regression coefficients (betas) predicting:

(1) cub production from female age, number of copulations, number of different males, and mean visitation;
(2) number of copulations from age, number of different males, and mean visitation;
(3) number of different males from age and mean visitation; and
(4) mean visitation from age.

tern of visitation was also influenced by male and female nutritional needs and male social status. We now examine in more detail whether visitation among females was influenced by change in their reproductive status. Did the presence of dependent offspring affect female visitation? If so, did visitation vary with the age of these offspring? Finally, did the state of estrus influence female participation in the aggregation?

The rearing of offspring requires energy expenditure above and beyond that needed for maternal growth and maintenance, and therefore, it is reasonable to suppose that ecocenters were more important to females with offspring. We explored whether females with dependent offspring used the Trout Creek ecocenter differently than those without them. We did this by assigning females in each year of the study to one of three reproductive cat-

Table 7.19. Cub and yearling litter production by nine aggressive female grizzly bears, Trout Creek ecocenter, Yellowstone National Park, 1959-70.

Female number	Year with cubs	Cub litter size	Sex of cubs			Year with yearlings	Yearling litter size	Sex of cubs		
			M	F	U			M	F	U
7	1959	3	1	2	0	1960	3	1	2	0
	1962	2	0	0	2	1963	2	0	0	2
	1966	2	0	1	1	1967	2	0	1	1
10	1966	2	0	0	2	1967	2	0	0	2
	1970	2	0	0	2	—	-	-	-	-
40	1964	2[a]	1	0	1	—	-	-	-	-
	1966	2	1	0	1	—	-	-	-	-
	1968	3	2	1	0	1969	3	2	1	0
42	1962	2	0	1	1	—	-	-	-	-
	1964	2	1	1	0	1965	2	1	1	0
	1967	2[a]	0	1	1	—	-	-	-	-
	1970	2[a]	0	0	2	—	-	-	-	-
65	1960	2	2	0	0	—	-	-	-	-
	1962	4[a]	1	2	1	—	-	-	-	-
	1964	3[a]	2	0	1	1965	2	2	0	0
101	1964	1[a]	0	0	1	—	—	-	-	-
	1966	1	0	1	0	1967	1[a]	0	1	0
	1970	2[a]	0	0	2	—	—	-	-	-
120	1961	2	0	0	2	1962	2	0	0	2
	1965	2	0	0	2	1966	2	0	0	2
128	1962	3	0	0	3	1963	2	0	0	2
	1965	3[a]	0	0	3	—	-	-	-	-
	1967	4	1	3	0	1968	4	1	3	0
	1970	3[a]	0	0	3	—	-	-	-	-
150	1963	3	2	1	0	1964	3	2	1	0
	1967	2[b]	1	0	1	1968	2	1	0	1
Total:		26 (cub litters)					14 (yearling litters)			
Mean litter size:		2.35 cubs					2.29 yearlings			

[a] Lost one or more offspring part way through season.

[b] Adopted one or more offspring part way through season; only female's offspring are included in this table.

egories: adult females with dependent cubs, adult females with dependent yearlings, or adult females without dependent offspring. The latter group included: (1) females known to have exhibited estrus in the relevant year (mating was observed or the female was observed with offspring the following year); (2) females for which no evidence of estrus was obtained, but an undetected estrus could not be ruled out; and (3) females that weaned offspring, because offspring were almost always weaned in May or early June before any significant quantities of food were available at the ecocenter.

We employed the Mann-Whitney test to perform pairwise comparisons among the categories on the basis of raw mean visitation values and adjusted mean visitation values. We developed the latter values to control for between-year differences in the overall "attractiveness" of the ecocenter by creating a "standard" visitation value that moved "up" and "down" the visitation scale as overall visitation varied, and against which the visitation of individual bears could be compared for adjustment. To calculate adjusted visitation for a given year and make comparisons, we required that at least three bears be available in each data subgroup used to calculate the relevant annual standard visitation. Years not meeting this criterion were omitted. Thus, sample sizes for a given type of two-sample com-

Fig. 7.9. Distribution of reproductive status over 288 female-years for 87 adult female grizzly bears, Trout Creek ecocenter, Yellowstone National Park, 1959-70.

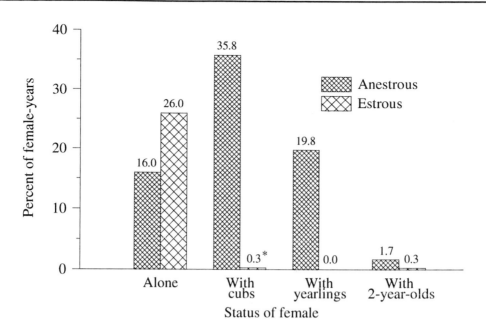

*Rarely, females were observed with cubs in two consecutive years, because the first litter was lost (or adopted) in time for the female to come into estrus and breed.

parison were sometimes smaller for analyses using adjusted values than for those using raw visitation values. Sample size differences of this kind occurred only in tests involving the postbreeding season because suitable visitation data after 15 July were available for only several, widely-spaced years (1960-61, 1964, 1968-70).

Application of the Mann-Whitney test for pairwise comparisons among these three categories (Appendix C, Tables 2, 3, 4, 5, and 6) showed that: (1) females with dependent yearlings had higher visitation than females without offspring both during and after the breeding season, (2) females with dependent cubs had higher visitation than females without offspring after but not during the breeding season, and (3) consistent with the first two results, females with dependent cubs had lower visitation than females with yearlings during the breeding season, but had similar visitation thereafter. In general, the results were the same using either the raw or adjusted visitation values; where they differed, we based interpretation on results from the latter.

Overall, these patterns of visitation support our supposition that females responded to the greater energy burden of rearing offspring by increasing

their use of the Trout Creek ecocenter. The one anomaly was the relatively low visitation of females with cubs during but not after the breeding season. It would appear that these females generally avoided Trout Creek when breeding activity was occurring and increased visitation after breeding was over. This same pattern of avoidance followed by attraction is supported statistically by a matched-pairs comparison of breeding season visitation by females with cubs with their own postbreeding visitation in the same year (Appendix C, Table 7).

Because cubs nursed more frequently and were more dependent on the mother's milk than yearlings, and because visitation increased in the postbreeding season (after 15 July), it would seem unlikely that the low breeding season visitation by females with cubs reflected a reduced maternal need for the resources offered by Trout Creek. The probable explanation is that females with cubs visited less frequently during the breeding season to reduce the risk of infanticide when offspring were young and especially vulnerable and the level of male aggression high. The extremely cautious and protective behavior of females with cubs when they did visit Trout Creek during the period of active breeding is consistent with this interpretation.

Table 7.20. Cub and yearling litter production by 30 nonaggressive female grizzly bears, Trout Creek ecocenter, Yellowstone National Park, 1959-70.

Female number	Year with cubs	Cub litter size	Sex of cubs			Year with yearlings	Yearling litter size	Sex of cubs		
			M	F	U			M	F	U
5	1965	2	0	0	2	1966	2	0	0	2
5	1968	3	1	2	0	—	-	-	-	-
15	1963	2	0	0	2	—	-	-	-	-
15	1966	1	0	0	1	—	-	-	-	-
15	1969	2	0	0	2	1970	2	0	0	2
29	1965	2[a]	0	0	2	—	-	-	-	-
34	1961	2	1	1	0	—	-	-	-	-
34	1963	2	2	0	0	1964	2	2	0	0
34	1968	2[a]	2	0	0	—	-	-	-	-
39	1959	2	2	0	0	—	-	-	-	-
39	1963	3	2	1	0	1964	3	2	1	0
44	1965	2	0	0	2	1966	2	0	0	2
48	1960	1	0	1	0	—	-	-	-	-
48	1962	1	1	0	0	—	-	-	-	-
48	1963	3[b]	3	0	0	—	-	-	-	-
75	1962	2	2	0	0	1963	2	2	0	0
81	1965	1[b]	0	0	1	1966	1	0	0	1
84	1961	3	1	1	1	1962	2	1	1	0
84	1965	2	0	0	2	—	-	-	-	-
96	1963	2	2	0	0	1964	2	2	0	0
96	1966	4	1	1	2	1967	4	1	1	2
96	1969	2	0	0	2	1970	2	0	0	2
108	1966	2	0	0	2	1967	2	0	0	2
112	1961	3	2	0	1	1962	3	2	0	1
112	1964	2	1	1	0	1965	2	1	1	0
119	1961	2	0	1	1	—	-	-	-	-
123	1962	2	2	0	0	—	-	-	-	-
125	1962	2	0	2	0	1963	2	0	2	0
125	1965	3	0	0	3	1966	2	0	0	2
125	1968	3	1	1	1	1969	3	1	1	1

cont'd on next page

THE EFFECTS OF ESTRUS ON VISITATION

Although nutritional considerations affected female visitation patterns, we suspected that other important factors were involved. Specifically, because the period of aggregation at Trout Creek extended through and beyond the breeding season, it seemed possible that the state of estrus also influenced female visitation. We hypothesized such an effect because, other things being equal, females in estrus need additional nutrition to support conception and care of offspring the following winter. Also, by frequently visiting the ecocenter, breeding females may have increased the likelihood of encountering preferred (i.e., dominant) males during estrus. They might have "preferred" dominant males as mates because they offer "good genes" or superior protection from contesting males.

The difficulty in testing for an effect of estrus lies in finding an appropriate group of females to compare with estrous females. Ideally, we would compare estrous females with anestrous ones—those that do not have dependent offspring and, for whatever reason, do not enter estrus in one or more years. Each year, beginning in 1959, we were able to identify a group of females for which we observed no evidence of estrus during the breeding season and no evidence of parturition the following spring (Fig. 7.9). Over all years (288 female-years) that 87

Table 7.20. *cont'd*

Female number	Year with cubs	Cub litter size	Sex of cubs			Year with yearlings	Yearling litter size	Sex of cubs		
			M	F	U			M	F	U
132	1970	1	0	0	1	—	-	-	-	-
140	1965	3	1	0	2	1966	3	1	0	2
144	1967	2	1	1	0	1968	1	1	0	0
144	1970	1	0	0	1	—	-	-	-	-
148	1963	2	1	0	1	—	-	-	-	-
160	1963	2	0	0	2	1964	2	0	0	2
172	1963	3	3	0	0	1964	3	3	0	0
172	1966	1	0	0	1	—	-	-	-	-
172	1968	3	0	0	3	1969	3	0	0	3
173	1967	2[a]	0	0	2	1968	1	0	0	1
173	1970	1	0	0	1	—	-	-	-	-
175	1965	1	0	0	1	1966	1	0	0	1
175	1968	2[a]	1	1	0	—	-	-	-	-
175	1970	1	0	0	1	—	-	-	-	-
175B	1960	2	1	0	1	1961	2	1	0	1
175B	1963	2	1	0	1	1964	2	1	0	1
180	1965	3	0	0	3	1966	2	0	0	2
184	1970	2	0	0	2	—	-	-	-	-
187	1969	1	0	0	1	1970	1[b]	0	0	1
200	1967	2	2	0	0	1968	2	2	0	0
259	1969	3	0	0	3	1970	3	0	0	3
Total		51 cub litters					30 yearling litters			
Mean litter size		2.06 cubs					2.13 yearlings			

Note: Larger mean yearling litter size indicates that females with smaller-sized litters were weaning (and/or losing) entire litters more frequently than were females with larger litters.

[a] Lost one or more offspring part way through season.

[b] Adopted one or more offspring part way through season; only the female's offspring are included in this table.

females were categorized for reproductive status at Trout Creek, they were judged anestrous in this latter sense 16.9% of the time, to be with dependent offspring 57.3% of the time, and to have displayed estrus 26.7% of the time (in two cases, in a year subsequent to being observed with offspring). Taken at face value, this indicates that 37.4% (i.e., 16.0%÷[16.0%+26.7%]) of the females eligible for estrus (i.e., not with offspring) did not enter estrus. There are, however, other factors involved; females may have had an undetected estrus and then (1) failed to conceive, (2) conceived, but later resorbed fertilized ova or embryos, or (3) conceived and lost their cubs during denning or after emergence, but before our censuses. Unfortunately, these factors obscure the record, so we cannot say with certainty that, among the 123 female-years in which a female was available for estrus (42.6%), she did not display estrus in 46 (37.4%) of those. It is evident,

however, that many females eligible for estrus in any given year did not enter estrus.

Under the assumption that a group of "anestrous" females was composed primarily of females that were, in fact, anestrous in the relevant year, and that assignment to this group was independent of visitation, we examined the effect of estrus on visitation at Trout Creek. First, we employed two matched-pairs comparisons. In the first, visitation by estrous females during the breeding season was compared for 12 adult females that were without offspring and in estrus in one year, and without offspring and anestrous in the next year. This analysis included any between-year transitions between states of estrus and anestrus. In the second analysis, a similar matched-pairs test was used, but comparisons were restricted to situations in which the in-estrus year preceded one or more years when the females were with offspring, and the anestrous year immediately followed loss or weaning of these offspring. The rationale for the second test was that truly anestrous years are most likely to immediately follow the energy burden of raising offspring, and such a comparison may be less likely than the first to include females and years in which an estrus actually occurred but was not detected.

Neither of the matched-pairs comparisons indicated that estrus significantly influenced visitation, although females in estrus visited more frequently in both tests (Appendix C, Table 8). In contrast, a two-sample analysis of visitation during the breeding season gave quite different results. It indicated that estrous females visited Trout Creek significantly more frequently than anestrous females without offspring (Appendix C, Table 9, column 1). A similar two-sample comparison of visitation after the breeding season also showed higher visitation by estrous than anestrous females (Appendix C, Table 9), indicating that any effect of the state of estrus on visitation persisted after the period of estrus was over. This latter result is consistent with a Wilcoxon matched-pairs analysis that indicated female visitation did not change between the breeding and postbreeding seasons as a function of their being estrous or anestrous (Appendix C, Table 10).

Thus, the two types of analyses gave contradictory results. The Wilcoxon matched-pairs comparisons of estrous versus anestrous females indicate that estrus did not significantly influence visitation. The Mann-Whitney tests indicate that estrous females generally had higher visitation over the entire season than anestrous females. One possibility is that, in the latter series of tests, we were comparing females with inherently high versus low visitation, a difference that may have determined to what extent we detected estrus. This would explain why visitation among "estrous" females remained high after the breeding season. A second possibility is that our sample sizes in the matched-pairs comparisons were simply not large enough to detect a real effect of estrus. Mean visitation among "anestrous" females was similar in samples for both types of tests.

The most troubling aspect of accepting an effect of estrus on rates of visitation lies in explaining why estrous females continued to visit at high rates after the breeding season. Visitation levels of estrous females might be governed by nutrition, and these females were preparing for the future rigor of rearing offspring. But if, as is most likely, females "delayed" reproduction for nutritional reasons, it is not clear why anestrous females did not also visit at high rates in order to regain the physical vigor necessary for reproduction. If, however, estrous females increased visitation primarily to increase encounters with males during the breeding season, it is not clear why they also visited at high rates after the breeding season. One possibility is that, as in most other mammals, hormonal changes associated with estrus induced an increased nutritional need. In short, the question of estrus effects remains unresolved. Nevertheless, our results suggest some interesting possibilities for future research on the reproductive strategy of the female grizzly bear.

SUMMARY

Before 1959, when the study began, virtually nothing had been published on the reproductive biology of the female grizzly bear. Useful information on some aspects of reproduction in two Alaskan populations was published in the late 1960s. The most extensive data of this period, however, were developed from our studies of the ecocentered Yellowstone population. In this chapter, we describe the reproductive biology of Trout Creek female bears in more detail and more comprehensively than in any of our earlier publications. Reproductive data were derived from complete or partial life histories that were recorded for 277 marked or recognizable Trout Creek bears and genealogies constructed for 56 Trout Creek families.

The breeding season at Trout Creek varied in onset, and perhaps duration, from year to year, but a typical pattern was a start of mating activity in mid- to late-May, increasing in frequency during the first half of June, and slowly declining into mid-July, when breeding ceased entirely. Although males generally initiated mating rituals, females appeared to control the frequency of copulations by reacting cooperatively when receptive and aggressively when not.

In the typical estrus, females were bred by several different males and often repeatedly by one or more of these males. Copulations that we judged successful varied greatly in length (maximum of 60 minutes, mean of 24.3 minutes) and were characterized by prolonged pelvic thrusting by the male, female cooperation, and voluntary male dismount. Most if not all of the breeding females had two distinct periods of estrus within each breeding season. Estrous periods were separated by a several-day mating interval devoid of overt sexual behavior. Durations of each estrous period and of the mating interval were highly variable, but a typical pattern was for each period to last four to seven days. During estrus, females were attended and defended by an adult male or, more often, by several different males in sequence.

We suspect, but could not demonstrate, that ovulation is induced by copulation. The need to induce ovulation via genital stimulation, however, could help explain why female grizzlies mated repeatedly and promiscuously. Implantation of the early embryo (blastocyst) and hence further development of the fetus, is delayed until fall when winter sleep begins. Thus, both birth and the great majority of prenatal development occurs in the den during winter.

Litter size of cubs at about six months of age (when our field season started) varied from one to four, with a mode of 2 and mean of 2.10. The average interval between litters was approximately 3.29 years, and the rate of cub production by Trout Creek females during 1959-70 was calculated to be 0.61 cubs per year per female. This rate was based on the total of observed female-years divided into total cubs produced.

Females produced first litters as early as 5 years of age, as late as 9 years, and on average at 6.28 years. Middle-aged females (9-20 years) were more likely than young females (≤8 years of age) to produce litters in the first year following the weaning of prior litters, and they had larger litters. Nonetheless, young females were responsible for the bulk of total cub production at Trout Creek simply because this age class was more numerous than middle-age females. In a curious and somewhat difficult-to-interpret result, we also found that females observed to copulate frequently during estrus produced more cubs than those that did not. Perhaps frequent mating helped induce ovulation. More likely, this result reflects the inclusion in our sample of females that experienced short periods of dysfunctional "pseudoestrus." Finally, there was a trend for socially dominant (more aggressive) females to have larger litters.

Both nutritional demands and offspring vulnerability governed visitation by females at the Trout Creek ecocenter. Thus, females with yearlings or with cubs of the year were present in the aggregation more frequently than females without offspring, with one exception. Females with young cubs had lower visitation during the breeding season than females without offspring, but greater visitation during the postbreeding season. Presumably, the demands of lactation usually induced females with cub and yearling offspring to visit the ecocenter frequently, but the risk of infanticide to very young cubs during the breeding season discouraged these females from visiting as frequently as they might otherwise have done.

Visitation by females at the Trout Creek ecocenter was affected by their reproductive status. Females with cubs exhibited reduced visitation during the breeding season but had visitation rates similar to females with yearlings after the breeding season. The continuous visitation by females with yearlings reflects the security of a family coalition. *(Photo by John J. Craighead)*

Table 8.1. Female grizzly bear productivity and cub mortality, Trout Creek ecocenter, Yellowstone National Park, 1960-69.

Among all females[a] (marked and unmarked cubs)

Year	Number of cub litters	Number of yearling litters	One	Two	Three	Four	Total number of offspring	Cubs missing as yearlings No.	%
1960	10		2	6	2	0	20	14	70.0
1961		3	0	3	0	0	6		
1961	8		1	4	3	0	18	6	33.3
1962		5	1	1	3	0	12		
1962	12		2	6	3	1	27	12	44.4
1963		8	2	5	1	0	15		
1963	9		0	6	3	0	21	6	28.6
1964		7	1	4	2	0	15		
1964	7		1	4	2	0	15	3	20.0
1965		6	1	4	1	0	12		
1965	14		3	6	5	0	30	9	30.0
1966		11	3	6	2	0	21		
1966	9		3	5	0	1	17	4	23.5
1967		6	1	4	0	1	13		
1967	7		0	5	1	1	17	6	35.3
1968		6	3	2	0	1	11		
1968	6		1	2	3	0	14	5	35.7
1969		3	0	0	3	0	9		
Total: Cubs	82		13	44	22	3	179	65	36.3
Yearlings		55	12	29	12	2	114		

Among returning females[b] (marked and unmarked cubs)

Year	Number of cub litters	Number of yearling litters	One	Two	Three	Four	Total number of offspring	Cubs missing as yearlings No.	%
1960	4		1	3	0	0	7	1	14.3
1961		3	0	3	0	0	6		
1961	5		0	2	3	0	13	1	7.7
1962		5	1	1	3	0	12		
1962	9		1	6	1	1	20	5	25.0
1963		8	2	5	1	0	15		
1963	7		0	5	2	0	16	2	12.5
1964		6	0	4	2	0	14		
1964	7		1	4	2	0	15	3	20.0
1965		6	1	4	1	0	12		
1965	11		2	4	5	0	25	5	20.0
1966		10	2	6	2	0	20		
1966	8		3	4	0	1	15	2	13.3
1967		6	1	4	0	1	13		
1967	6		0	5	0	1	14	3	21.4
1968		6	3	2	0	1	11		
1968	5		0	2	3	0	13	4	30.8
1969		3	0	0	3	0	9		
Total: Cubs	62		8	35	16	3	138	26	18.8
Yearlings		53	10	29	12	2	112		

[a] All females known to have had cubs in the first year of the appropriate pair of years, whether or not they were observed alive the next year or were known to have died.
[b] Only females known to have had cubs in a first year and observed alive in the second of the appropriate pair of years.

8
Survivorship and Weaning of Offspring

The reproduction of a new generation, together with mortality of juveniles, determines the rate at which adults are recruited and, ultimately, the future of the population. A delicate balance exists between factors affecting the rate at which offspring survive, and, thereby, the rate at which the population grows or declines. Having presented the data defining various parameters affecting reproduction for the Trout Creek subpopulation, we now examine the fate of offspring and factors determining their fate.

MAGNITUDE OF CUB SURVIVORSHIP

All Cubs

The majority of cub mortality escaped direct detection (e.g., observation, carcass recovery, etc.). Nevertheless, cub survivorship may be estimated by comparing the number of cubs observed with individual females (actually, cubs that survived to first observation) with the numbers of yearlings observed with these same mothers the following year. We stress once again, this method can be used only because we had consistent annual censuses involving both marked females and offspring over a 10-year period. The ability to follow marked family groups over extensive periods is essential in reducing the biases inherent in this method.

We used two data sets to estimate survivorship for the 1960-69 period (Table 8.1). First, we determined the production of cubs and those missing in the next year for all Trout Creek-flagged adult females (see Methods, Chapter 3, for definition of the flagged Trout Creek subpopulation), regardless of whether the females were observed in the following year (Table 8.1, "Among all females"). We confined the second data set to offspring of returning females only; i.e., those females known to have had cubs in one year and that were observed alive in the subsequent year, whether with offspring or not (Table 8.1, "Among returning females"). Note that unmarked as well as marked offspring constituted these data.

Among all Trout Creek females, 82 litters were produced for a total of 179 cubs and an annual mean litter size of 2.18 cubs for the 9-year period (Table 8.1). The production of cub litters ranged from 14 in 1965 to 6 in 1968, with an annual mean number of 9.1 litters. The mean size of 55 yearling litters was 2.07. For this sample, the loss of cubs was 65 of 179, for a 63.7% survivorship, from cub to yearling (Fig. 8.1A). This is a lowerbound estimate because some cubs may have dispersed with their mothers, rather than died.

When our sample was restricted to include only females observed in the year after having a cub litter, 62 cub litters and 53 yearling litters were recorded, for totals of 138 and 112 offspring, respectively (Table 8.1). Based on this more restricted sample, the mean survivorship of Trout Creek cubs to the yearling stage, over the 1960-69 period, was 81.2% (Fig. 8.1A). If we ignore the possibility of early weaning of unmarked yearlings, this is an upperbound estimate; although it removes the effect of family dispersal, it does not account for cub loss due to the death of the mother. The midpoint of these upper- and lowerbound estimates is 72.5% (Fig. 8.1A).

Marked Cubs

To evaluate the effect of early weaning of yearlings on the above estimate of survivorship, we analyzed a subset of the data (see Table 8.1) that included only marked cubs. Yearlings, marked as cubs, that were weaned early and survived would have been detected, provided they did not disperse. An "all females" data set showed a total mortality of 12 marked cubs, for an average survivorship of 76.9% (Table 8.2 and Fig. 8.1B), assuming that the cubs of missing mothers were all mortalities. Orphaned cubs, however, may survive their first year (see later in this chapter). Furthermore, some of the missing females may have dispersed with surviving offspring, although the number would be few considering the strong attachment to a home range. Thus, this is a lowerbound estimate of survival (Fig. 8.1B).

Fig. 8.1 A and B. Summary of estimates of grizzly bear cub survivorship (Tables 8.1 and 8.2); up-arrows (↑) indicate upperbound and down-arrows (↓) lowerbound estimates of cub survivorship from which midpoint estimates were derived (see text); "All Females" sample includes "Returning Females," as well as those mothers *not* observed in the year following cub production; "Returning Females" sample includes *only* those mothers observed in the year following cub production.

A.

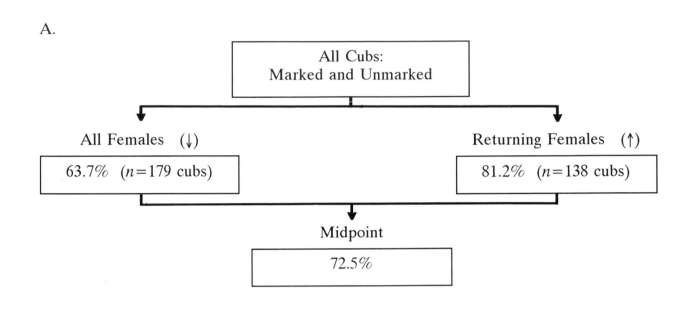

B.

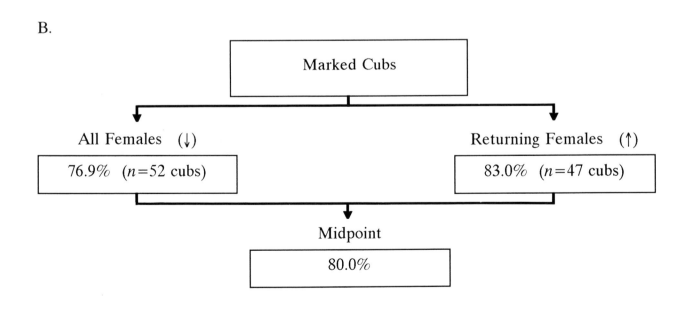

Table 8.2. Sex-differentiated mortality among marked grizzly bear cubs, Trout Creek ecocenter, Yellowstone National Park, 1960-69.

Sex of cub	Among all females[a]			Among returning females[b]		
	Marked cubs	No. missing as yearlings	% missing as yearlings	Marked cubs	No. missing as yearlings	% missing as yearlings
Male	30	6	20.0	28	5	17.9
Female	22	6	27.3	19	3	15.8
Both sexes	52	12	23.1	47	8	17.0

Note: Includes two capture-related mortalities.

[a] All females known to have had cubs in the first year of the appropriate pair of years, whether or not they were observed alive the next year or were known to have died.

[b] Only females known to have had cubs in a first year and observed alive in the second of the appropriate pair of years.

A "returning females" data set showed a total mortality of eight marked cubs, for an average survivorship of 83% (Table 8.2 and Fig. 8.1B). This however, is an upperbound estimate, because it would not account for cub mortality due to the mother's death. The true rate of cub survival for the marked sample should be between the upper- and lowerbound estimates of 76.9% and 83% (midpoint equals 80.0%), which happen to bracket the upperbound estimate (81.2%) based on the marked and unmarked cub sample (Table 8.1 and Fig. 8.1A and B).

PATTERNS OF CUB MORTALITY

As noted earlier, we have little direct information on the causes of cub mortality. What we do know mainly applies to human-related sources of death. An indirect approach is to look for patterns in the occurrence of cub loss that suggest cause or certain classes of cause. In this section, we do this by asking whether mortality varied with cub sex or litter size and, in cases where the female survived the entire pre-weaning period, whether mortality acted independently on individual cubs or uniformly on entire litters. In addition, we discuss direct evidence regarding cub death via infanticide.

Effects of Differential Cub Mortality by Sex

Analysis of the marked cub data (Table 8.2) showed that the sex of cubs had no significant effect on their rates of mortality in either the "all females" sample (G-test: $G = 0.375$, $P > 0.25$) or the "returning females" sample (G-test: $G = 0.034$, $P > 0.25$). Our sample is small, however, and any sex-differential mortality would be difficult to detect. Certainly, there are many cases of male-biased pre- and postnatal mortality in mammals (Clutton-Brock 1991), and the reduction in sex ratio (males:females) documented earlier among adult bears relative to subadults (Chapter 5) implies that, beyond one year of age, a sex-differential mortality operated at one or more points during development in the Trout Creek population.

Fate of Littermates

If mortality events act on individual offspring independently (that is, loss of one member of a litter does not affect the probability of death among littermates), then among litters with (for example) two cubs, the probability of losing both cubs is $(0.186)^2 = 0.035$, where 0.186 is the observed decimal percent of cubs lost from all 2-cub litters combined (Table 8.3). Similarly, the probabilities of retaining both cubs and of one cub dying and one surviving are $(0.814)^2 = 0.663$ and $2[(0.186)(0.814)] = 0.303$, respectively. The 2x-multiplier in this last calculation reflects the fact that there are two ways to lose one cub; cub "A" may die and cub "B" survive or the other way around. Among 35 2-cub litters, the expected number of whole litters lost (if mortality operates independently) is $(0.035)(35) = 1.2$; the number of whole litters retained is $0.663 \times 35 = 23.2$, and the number of partial litters lost is $(0.303)(35) = 10.6$. This same analytical approach was used to calculate the expected numbers of 3- and 4-cub litters that were wholly lost, retained-in-whole, or lost-in-part. Expected values were then compared with those actually observed for the appropriate variables (Table 8.4).

Note that the data contributing to Table 8.4 relate to "returning females" only, as presented in Table 8.1, so the losses of whole litters that did occur cannot be explained by the death (or dispersal) of the mother. They could be explained, however, by the undetected weaning of a few litters prior to one

Table 8.3. Grizzly bear offspring mortality as a function of cub litter size, Trout Creek ecocenter, Yellowstone National Park, 1960-69.

Litter size	Number of cub litters	Number of cubs	Number of yearlings	Percent of cubs missing as yearlings
One-cub	8	8	3	62.5
Two-cub	35	70	57	18.6
Three-cub	16	48	41	14.6
Four-cub	3	12	11	8.3
All litters	62	138	112	18.8

Table 8.4. Expected-versus-observed mortality as a function of grizzly bear cub litter size, and whether losses to individual litters were none and all or partial, Trout Creek ecocenter, Yellowstone National Park, 1960-69.

Litter size	Expected no. of litters		Observed no. of litters	
	Whole[a]	Partial[b]	Whole[a]	Partial[b]
Two-cub	24.4	10.6	28	7
Three-cub	10.0	6.0	12	4
Four-cub	2.1	0.9	2	1
Total	36.5	17.5	42	12

[a] Sum of litters entirely lost or retained in whole.
[b] Litters experiencing partial losses.

Table 8.5. Expected-versus-observed mortality of grizzly bear cubs as a function of litter size, Trout Creek ecocenter, Yellowstone National Park, 1960-69.

Litter size	Cubs missing as yearlings	
	Expected no.[a]	Observed no.
One-cub	1.5	5
Two-cub	13.2	13
Three-cub	9.0	7
Four-cub	2.3	1

[a] Expected number missing from litters of each size category was calculated by multiplying the decimal percent of cubs missing from all litters as yearlings (0.188) times the number of cubs produced in each litter size category (see Table 8.3).

year of age, possibly in late fall as cubs or as yearlings in early spring (April or May) before formation of the ecocenter aggregation. Nevertheless, the differences in expected versus observed values (Table 8.4) were not significant (G-test: $G = 2.7$, $P > 0.05$), indicating that entire litters did not tend to be retained or lost more frequently than expected, and that the majority of cub loss was due to causes that tended to operate independently on littermates.

Effects of Litter Size

If mortality is independent of cub litter size, the expected number of cubs missing as yearlings from a given litter size category would equal the observed overall percentage of cubs missing as yearlings from all litters (18.8%; see Table 8.1, "Among returning females") times the number of cubs born into that category of litter size (e.g., 70 cubs born into 2-cub litters, Table 8.3). Inspection of observed losses by litter size (Table 8.3) suggests that survival increased with litter size (Table 8.5). When litter sizes were combined into large (3- and 4-cub) and small (1- and 2-cub) categories to meet the minimum expected numbers required for statistical tests, we found that

Survival of cubs increased with litter sizes. Older, prime aged, experienced females produced most litters of more than two offspring, and this may have accounted for the greater survivability of large litters. *(Photo by John J. Craighead)*

this trend approached statistical significance (G-test: G = 1.77, 0.10 > P > 0.05).

Infanticide

We define infanticide as the killing of dependent young by other bears. We distinguish as "conspecific killing" the killing by grizzlies of other grizzly bears, regardless of the victim's age. Infanticide in mammals has been explained as social pathology, simple predation or, when practiced on another male's offspring, a strategy of males to maximize their own genetic lineage by forcing females to return to estrus more quickly (Struhsaker and Leland 1987). We first evaluate the circumstances, frequency, and potential population-level effects of infanticide among grizzly bears and then return to the issue of cause.

Infanticide was a difficult phenomenon to measure; we never directly observed older bears killing younger ones, but we found cubs that obviously had been killed by other bears. Adult males were routinely observed chasing young bears, especially cubs-of-the-year. Adult males generally pursued cubs over distances of less than 90 meters (100

yards), but occasionally the chase was longer. For example, a young adult male was observed to chase a cub for more than 1.5 kilometers (1 mi) before giving up.

An attack on a cub by male No. 88 may be typical of circumstances that characterize instances of infanticide. Female No. 180 left a cub unattended to retrieve her other two cubs feeding approximately 45 meters (50 yds) away. A large male, No. 88, grabbed the cub by the small of the back, rose to his full height with the cub in his mouth, and then dropped to all fours and started running. Adult female No. 112, accompanied by two yearlings, was feeding in the same location as the unattended cub. She immediately lunged at No. 88, forcing him to drop the cub. The cub was reunited with the parent female about 10 minutes later. Without the intervention of female No. 112, the male would have certainly killed the cub.

We observed two other attempted infanticides, one in late June 1969, the other in late July 1970. A young adult male chased one of three unattended cubs into a heavily wooded area, where the cub

Table 8.6. Known and probable grizzly bear deaths caused by other bears, Yellowstone National Park, 1959-71.

Date of death	Status of death	Age at death	Sex	Location	Remarks
9-10-62	known	2	F	Elephant Back	Found dead; wounds consistent with attack by bear.
7-15-63	known	cub	M	Trout Creek	Found dead; carcass not consumed.
6-22-65	known	cub	M	Trout Creek	Numerous punctures, broken neck and back; not consumed.
7-06-65	known	cub	M	Trout Creek	Carcass about one week old; punctures and broken bones.
Spring '67	probable	cub	unk	Trout Creek	⎤
Spring '67	probable	cub	unk	Trout Creek	3 cubs found dead together by maintenance men.
Spring '67	probable	cub	unk	Trout Creek	⎦
Spring '68	probable	cub	M	Fountain Freight Rd.	Found dead by Old Faithful maintenance men.
Spring '68	probable	cub	M	Rabbit Creek	" " " " " " "
Spring '68	probable	cub	F	Rabbit Creek	" " " " " " "
Summer '68	probable	cub	unk	Bridge Bay	Located by maintenance men.
6-19-68	known	4	F	Trout Creek	While under influence of drug, killed by large male.
6-24-70	known	cub	F	Trout Creek	Found dead; severe chest wounds, broken ribs.
5-23-71	probable	cub	M	Trout Creek	On autopsy, biologist Ken Greer felt this cub had been killed by another bear.

began bawling. The parent female ran toward her cub, and the two remaining cubs followed her into the timber. The female coursed the area for over an hour, vocalizing as she searched for her cub. We never again saw the family, and their fate was unknown. In the second attempt, an adult male attacked and picked up a cub in his jaws, and it bawled loudly. The parent (female No. 128) and another female (No. 42) immediately attacked a different male that did not appear to be involved in the cub attack. The cub was subsequently released, relatively uninjured, and returned to its mother about 10 minutes later.

Evidence of successful infanticide has been indirect. Location and examination of carcasses of bears killed by other bears is largely a matter of chance, even under the best of circumstances. Over the 1959-71 period, we recovered six bears known to have been killed by other bears (Table 8.6). Four were cubs, and the others were 2-year-old and 4-year-old bears. In addition, we documented eight cub deaths probably attributable to attacks by other bears, but conclusive evidence was lacking (Table 8.6).

After closure of the dumps and dispersal of the bear population, evidence of infanticide became even more difficult to substantiate. Mattson *et al.* (1992b) documented only one case of a grizzly cub being killed by other bears in the Yellowstone Ecosystem after 1971, two killings of older-aged grizzlies (a yearling female and an adult male), and one of a female black bear. They also reported cub remains from a grizzly bear scat collected near a trout-spawning stream in 1985.

Infanticide has been documented in Alaskan bear populations (Troyer and Hensel [1962] on Kodiak Island, Schoen and Beier [1986] on Admiralty Island, Glenn *et al.* [1976] at McNeil River Falls, Ballard *et al.* [1982] in southcentral Alaska, Dean *et al.* [1986] in Denali National Park, and Reynolds and Hechtel [1980, 1984] in the Western Brooks Range). As in Yellowstone, adult males were usually responsible, but in contrast with our observations, much of the predation was on entire family groups. Dean *et al.* (1986) reported two episodes in Denali National Park of adult males attacking females with offspring. Reynolds (1976), in the eastern Brooks Range, observed a male feeding on an adult female and two yearlings near their den. In the Western Brooks Range, Reynolds and Hechtel (1980) reported that an adult male probably killed an adult female and two 2-year-olds. They also observed mortalities of a cub, an adult female and at least one of her three cubs, three adult females that were probably accompanied by cubs, a yearling, and an adult female and cub that were likely killed and consumed by another bear (Harry Reynolds, III, pers. comm. 1992).

Infanticide has also been documented among the Canadian grizzly populations. A female reportedly lost one litter and another, two litters, at Tuktoyoktuk Peninsula/Richard Island in northern Canada (Nagy *et al.* 1983b). In both situations, it was assumed that large attending males observed with these females killed the cubs. Finally, Nagy and Russell (1978) documented predation on a young yearling male in the Swan Hills of Alberta, presumably by another bear.

Conclusions Regarding Cub Survivorship

In summary, we estimate the upper and lower bounds for cub to yearling survivorship during 1960-69 as 83.0% and 76.9%, respectively. There was no evidence for sex-differential cub survivorship or for "clustering" of cub mortality within litters. There was a fairly convincing, but nonsignificant trend for improved survivorship in larger litters. Finally, this and other studies indicate that infanticide and the conspecific killing of older bears occur in all grizzly populations. Thus, for example, infanticide has been reported in at least eight North American populations, but only five known and nine probable cub deaths attributable to other bears have been reported from the Yellowstone population over a 33-year period (1959-91), an average of less than half a cub per year.

The absence of clustering of cub mortality within litters suggests that accident, infanticide, or other such events that tend to take one cub at a time were the primary cause of cub mortality at Trout Creek. Contagious disease, abandonment of entire litters, complete maternal failure due to precipitous loss of physical condition, or other agents of mortality that might affect whole litters were apparently unimportant. Nonetheless, the tendency for cubs in large litters to survive better than those of small litters suggests that even small variations in maternal condition may affect cub survivorship. One might expect per-cub maternal investment to decline with increasing litter size, and that larger litters would suffer proportionally more mortality than smaller ones. To the extent that females in superior condition (more likely to be prime-age females or more aggressive females; see Chapters 6 and 7) produce the larger litters, however, per-cub investment and hence survivorship might actually be higher for members of some larger litters. Furthermore, the effects of reduced maternal attention among less fit females need not fall equally on all members of the litter. Mothers may favor certain

offspring, and cubs may differ in their ability to compete with littermates for limited food, so that the survival of all cubs is not jeopardized. Thus, moderate variation in maternal physical condition is, like injury and infanticide, a potential factor in cub loss that is consistent with our conclusion that mortality acted on littermates independently. Maternal condition should be less important as a cause of cub mortality in ecocentered populations, but could be paramount in populations without ecocenters, that is, with increased potential for individual variation in nutritional state.

While infanticide did not appear to be a primary cause of cub loss, it is less easy to determine whether conspecific killing may significantly limit population size. This is particularly true for killing of parent females because such losses have more serious implications for a population than infanticidal behavior alone (Knight and Eberhardt 1984). It is interesting that all reports of adult males killing adult females (*n* = 7) are from non-ecocentered populations (e.g., that of the Western Brooks Range, a population which is also at the northern extreme of the grizzly's range). It may be that males are more inclined to risk attacks on older bears, as well as cubs defended by mothers, when they are relatively food-stressed. This observation, and the indisputable fact that conspecific killing is not limited to dependent offspring, leads us to view infanticide in grizzly bears as an expression of a foraging strategy (conspecific predation) that is available in some circumstances to some bears (large, adult males), rather than as a genetically acquired mating strategy practiced by a subset of this group (adult males that are socially dominant). Unlike some other species in which infanticide has been well-documented (Struhsaker and Leland 1987), it is not at all obvious that an "infanticidal" male grizzly bear would be able to distinguish his own offspring from those of a rival or mate with victimized mothers when they return to estrus perhaps days or weeks later.

WEANING AGE

Female grizzlies at Trout Creek normally weaned their offspring as yearlings or as 2-year-olds in May or early June. To determine the distribution of ages at which cubs were weaned and to avoid confusing weaning with cub mortality in our analysis, we required that at least one cub in a litter be marked (74 of 90 cubs in these litters were marked) and, for marked females, that both the female and cubs be observed after weaning. In the case of unmarked females, it was sufficient to see only the cubs

Table 8.7. Ages at weaning for 37 litters and 81 offspring produced by 26 marked or recognizable female grizzly bears, Trout Creek ecocenter, Yellowstone National Park, 1959-70.

Litter size	Litters weaned as				Total weaned litters	Offspring weaned as				Total weaned offspring	Offspring dead or unknown
	Yearlings		2-yr-olds			Yearlings		2-yr-olds			
	No.	%	No.	%		No.	%	No.	%		
1	1	33.3	2	66.7	3	1	33.3	2	66.7	3	0
2	8	40.0	12	60.0	20	14	37.8	23	62.2	37	3
3	3	25.0	9	75.0	12	8	24.2	25	75.8	33	3
4	0	0.0	2	100.0	2	0	0.0	8	100.0	8	0
Total	12	32.4	25	67.6	37	23	28.4	58	71.6	81	6

Notes: Percentages are calculated across rows (e.g., percent of 2-cub litters weaned as yearlings plus percent weaned as 2-year-olds equals 100%). All cubs marked ($n = 75$) or members of litters of which at least one was marked ($n = 15$), but cubs ($n = 3$) dying at or near capture in first year of life not counted in litter size, and cubs ($n = 6$) dying or lost to observation before, but near weaning not counted in total offspring weaned.

following separation because our failure to resight an unmarked mother was probably due to lack of recognition rather than her death. Of 37 litters produced by 26 marked Trout Creek females, 12 (32.4%) were weaned as yearlings and 25 (67.6%) as 2-year-olds (Table 8.7). Of 90 cubs produced, nine were either known mortalities or of unknown status prior to weaning. Six of these disappeared toward the end of the period of maternal care (as judged by littermate weaning age). Thus, these offspring were counted in the determination of size for their respective litters. The remaining three cubs were never relocated after marking at one-half year of age or died at capture. Thus, we considered the effective size of these three litters to be one less than the litter size measured before capture. Of the remaining 81 offspring, 23 (28.4%) were weaned as yearlings and 58 (71.6%) as 2-year-olds (Table 8.7). We recorded one instance of a probable weaning at age three. The adult female remained with both 2-year-olds into the fall of their third year. Neither of the offspring was marked, however, and this litter does not appear in our tabulations.

We now examine female age, litter sex ratio, and the litter size as potential determinants of weaning age. Because the "decision" to carry or not carry offspring for another year dictates a substantial energetic commitment for the female, it is reasonable to assume that energy considerations will govern such decisions. For that reason, we focused on factors most likely to reflect maternal nutritional status. The sample of known-age females was too small ($n = 15$ females and 23 litters) to support any sort of multivariate analysis, so it was necessary to consider the association between weaning age and each of these three factors separately. The number

of different females was also too small to control for the effects of female identity. We assumed that weanings of different litters by the same female represent independent "decisions."

Effect of Litter Sex Ratio

Other things being equal, the duration of maternal care before weaning is a measure of maternal investment. If females invest differently in offspring according to sex, some variance in weaning age might be attributable to differences in the sexual composition of litters. Larger body weights for male cubs (Chapter 16) and precedence in the literature for male-biased maternal investment (e.g., Clutton-Brock 1991, Hogg *et al.* 1992) suggest the possibility that male cubs might be more costly, and that their presence might delay weaning. Although mothers were never observed to wean male and female littermates at different ages, weaning "decisions" might still be influenced by litter sex ratio. That is, litters with an excess of the sex requiring "greater investment" (here, males) might be weaned either later or earlier than other types of litters. When litters with sex ratios (males/females) equal to or greater than 50:50 were classified as male-biased, however, there was no association between litter sex ratio and weaning age ($n = 34$ litters, Chi-square test, $P = 0.16$). A nonsignificant result was also obtained when litters with equal sex ratios were classified as female-biased ($n = 29$ litters, Chi-square test, $P = 0.33$). In both cases, the trend was toward earlier weaning in litters classified as male-biased.

Effect of Female Age

In previous analyses of reproductive parameters, we discovered that female age was often a good predictor of reproductive performance. Age corre-

Table 8.8. Distribution of weaning age by maternal age for 23 litters born to 15 different female Trout Creek grizzly bears, Yellowstone National Park, 1959-70.

Offspring age at weaning of litter	Maternal age		
	Prime[a]	Old/Young[b]	All mothers
One year	27.3%	25.0%	26.1%
(*n*)	(3)	(3)	(6)
Two years	72.7%	75.0%	73.9%
(*n*)	(8)	(9)	(17)

[a] 9-20 years of age.

[b] Less than 9 or greater than 20 years of age.

Table 8.9. Distribution of weaning age by litter size for 37 litters born to 26 different female Trout Creek grizzly bears, Yellowstone National Park, 1959-70.

Offspring age at weaning of litter	Litter size		
	Small[a]	Large[b]	All sizes
One year	39.1%	21.4%	32.4%
(*n*)	(9)	(3)	(12)
Two years	60.9%	78.6%	67.6%
(*n*)	(14)	(11)	(25)

[a] One or two cubs.

[b] Three or four cubs.

lates with experience and may also crudely measure maternal body condition. Thus, it is reasonable to suppose that prime-aged females (experienced females in good body condition) are able to wean offspring more quickly than young females. But when adult females were categorized as prime-aged (9-20 years) or young (4-8 years), there was no association between maternal age and age of weaning (*n* = 23 litters, Chi-square test, *P* = 0.90). In fact, the distribution of weaning age was virtually identical in the two age groups (Table 8.8).

Effect of Litter Size

Litter sizes varied from one to four offspring. Increasing litter size might substantially reduce the per-cub rate of maternal expenditure and delay weaning (but see discussion earlier in this chapter regarding effects of litter size on weaning). Because only one litter each of sizes one and four had known sex ratios, litter size was classified in two categories for the following analysis—either two or less or three or more. Although there was a tendency for small

litters to be weaned earlier (Table 8.9), we found no significant association between litter size and weaning age (*n* = 37 litters, Chi-square test, *P* = 0.26).

Effect on Immediate Postweaning Survivorship

We have so far taken the perspective of the mother grizzly; that is, time of weaning was assumed to be determined or "chosen" by the female. Sample sizes were determined by the number of different litters for which the appropriate data were available. Here, we take the perspective of individual offspring and look at the relation between weaning age and offspring survival. Specifically, we ask whether weaning age, if controlled for other potentially important factors (sex of cub and litter size), influenced offspring survivorship in the first two years after weaning. If the duration of maternal care reflects the magnitude of that care, and if care has prolonged effects on the vigor of young grizzlies, we might expect to find that late weaning improved post-weaning survivorship.

Table 8.10. Age of weaning and age and cause of death for 33 grizzly bears, Trout Creek ecocenter, Yellowstone National Park, 1959-81.

Bear number	Sex	Age at weaning (years)	Age at death (years)	Cause of death[a]
6	F	2	6	Drug
29	F	2	6	Drug
37	M	1	4	Hunter
38	M	1	2	Hunter
51	M	2	7	Hunter
52	M	1	1	Hunter
53	M	1	1	NPS control
77	F	2	2	Drug
92	M	1	1	Hunter
96	F	2	12	Hunter
98	M	2	4	Rancher
100	M	2	3	Hunter
109	F	2	8	NPS control
114	M	1	1	NPS control
133	M	2	3	Hunter
141	F	2	7	NPS control
142	F	2	8	Auto
143	F	1	2	NPS control
145	M	1	3	Hunter
154	M	2	3	Hunter
157	M	2	2	Hunter
167	F	2	4	Bear
178	M	1	5	Hunter
183	M	2	5	NPS control
184	F	2	6	NPS control
185	M	2	4	WRU control
190	M	2	11	Hunter
199	M	2	15	Illegal kill
202	M	1	3	NPS control
203	M	1	2	Auto
207	M	2	9	NPS control
216	M	2	6	NPS control
245	M	2	3	Illegal kill

[a] NPS is National Park Service; WRU is Montana Cooperative Wildlife Research Unit (Craighead Study Team).

Sample sizes equalled the number of offspring for which the appropriate data were available, and littermate and nonlittermate siblings were permitted. Bears were judged to have survived to two years postweaning if they were observed any time after 15 June in their third (when weaned at one) or fourth (when weaned at two) years of life. They were judged not to have survived this period if their death was known to have occurred within a two-year period beginning 15 June in their year of weaning.

Two important methodological points must be mentioned. First, the mortality data reflected, almost exclusively, human-caused deaths (Table 8.10). Thus, the pattern of effects of litter size, sex, and

age revealed by this analysis do not necessarily mirror that which would be obtained from a similar treatment of natural mortality. Second, our method of determining individual fates was biased in favor of detecting survival. Establishing mortality required finding a carcass, whereas survival was established by observing a given bear at any point within two years post-weaning. Relatively more cases of natural mortality than survival probably went undetected. Thus, the survival probabilities reported here are not ideal absolute estimates of survivorship of bears in a given category of litter size, sex, and weaning status. If, however, most cases of human-caused deaths were detected, these prob-

Table 8.11. Logistic regression results for a model specifying survival probability for grizzly bear offspring (*n* = 51) as a function of litter size, weaning age, and sex, Trout Creek ecocenter, Yellowstone National Park, 1959-70.

	Parameter estimates			
Variable	Coefficient	Standard error	*t*	*P*-value
Intercept	1.76	0.56	3.12	0.003
Sex	0.93	0.56	1.65	0.10
Weaning age	0.72	0.39	1.84	0.07
Litter size	-0.16	0.39	-0.40	0.69

G^2 = 3.47 with 4 degrees of freedom, *P*-value = 0.48.

abilities may accurately reflect *relative* susceptibility to human-induced deaths in the first two years following weaning.

The results of a logistic regression of survivorship on sex, weaning age, and litter size did not provide statistical support that any of these parameters influenced survival for two years postweaning (Table 8.11). Nevertheless, early weaning age (i.e., weaning of yearlings) was marginally associated with increased susceptibility to human-caused deaths (*P* = 0.07). Furthermore, the data in Table 8.11 and inspection of observed survival to two years postweaning by sex, weaning age, and litter size suggested the possibility that males were more susceptible to human-caused mortality than females (Table 8.12).

Effect on Survivorship of Female Grizzly Bears from Birth to Breeding Age

Here, we again examine the possibility that the extent of maternal care, as measured by weaning age, has prolonged (both pre- and post-weaning) effects on offspring vigor. But we extend the end point of the post-weaning period from three (or four) to five years of age (the onset of adulthood) and restrict the analysis to females. Specifically, we ask whether late-weaned females are more apt to survive from birth to adulthood. To estimate and compare survivorship from birth to breeding age (five years) for females weaned at age one versus those weaned at age two, it was first necessary to adjust among individuals in the sample for differences in the age of first marking or recognition. This was done by tallying the number of "bear-months" that weaned females were observed and known to be alive, and estimating survival rates for each of the two weaning classes—those weaned as yearlings and those weaned as 2-year-olds.

The number of "bear-months" an individual female was known to be alive was calculated as the number of months she was active (not in winter sleep) from the date she was first marked (or recognized) to the date she either completed her fifth year or was known to die (see Pallack *et al.* 1990). Because all known deaths were human-caused and occurred while the bears were active, annual survival was estimated for a 7-month period of activity. It was assumed that no human-caused deaths occurred during winter sleep and that mortality rates did not differ through time.

The sample of bears used in this analysis comprised 18 marked females born from 1958 to 1965; only four were weaned as yearlings, and 14 were weaned as 2-year-olds. All were known either to survive to breeding age (five years) or to die. Many marked females of known weaning age had to be excluded because they disappeared and were never seen again. This was especially true of individuals born in the later years of the study (after 1965).

Among the sample of 18 female bears, 78% survived to breeding age. There was no evidence in

Table 8.12. Probability of grizzly bear offspring survival (*n* = 51) as a function of litter size, sex, and weaning age estimated by logistic regression (Table 8.11), Trout Creek ecocenter, Yellowstone National Park, 1959-70.

	Weaning age			
	Male		Female	
Litter size	1	2	1	2
≤2	0.57	0.96	0.90	0.97
>2	0.49	0.80	0.86	0.96

Fig. 8.2. Conditional model for variation in ages at which grizzly bear offspring are weaned; survivorship versus weaning age of offspring as a function of sow condition.

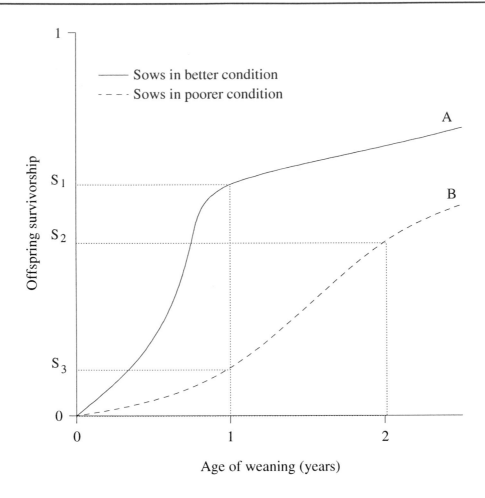

this small sample that survival to breeding age differed among females on the basis of the age at which they were weaned. The estimated probability of survival from birth to five years of age, based on monthly survival rates, was 73.3% for females weaned as yearlings versus 76.6% for females weaned as 2-year-olds. This was consistent with the accompanying analysis of weaning age and survival, which showed that survivorship from weaning to two years postweaning was high, regardless of whether offspring were weaned as yearlings or 2-year-olds (Table 8.12).

Conclusions Regarding Weaning Age

In summary, 67.6% of Trout Creek litters were weaned as two-year olds and 32.4% as yearlings. Some litters were probably weaned at age three, but this was extremely rare. This distribution of weaning age did not vary significantly with maternal age, litter sex ratio, or litter size, although small litters tended to be weaned earlier than large litters. Statistically, we found that weaning age was not a dominant influence on postweaning survivorship,

when the effects of cub sex and litter size were removed. There was, however, a trend toward reduced survivorship for offspring weaned as yearlings.

In duration of care, the litters compared differed by a factor of two; nevertheless, we could identify no clear costs associated with early weaning nor a clear "imperative" for late weaning. Similarly, we could not identify any obvious family traits (such as maternal age, litter sex ratio, or litter size) that might dictate weaning age. Why then did not all females wean litters at one year? Differences in age of weaning are unlikely to be a result of nonadaptive variation in individual behavior. The "decision" to nurture offspring for another year is clearly associated with significant biological costs and, perhaps, some heritability in timing of weaning as well. Nonadaptive behavioral variation is unlikely to persist in a species or in a population when the behavioral variants are associated with large, heritable differences in major components of fitness. Selection could be expected, in time, to eliminate the variants associated with later age of weaning if there was no advantage to this behavior.

A report of frequent weaning at age three in energy-stressed populations on the Alaskan North Slope (Reynolds and Hechtel 1984) suggests a solution to the question of how variation in weaning age might persist in populations of grizzly bears. Our explanation or "conditional" model of this behavior is based on a well-developed general theory addressing the subject of persistent behavioral polymorphisms (Maynard Smith 1982, Parker 1984). First, suppose females at any given age vary in condition (e.g., amount of fat reserves). Condition may also vary predictably with age, but it is not age-determined. Second, we know litter size is age-related such that prime-age females have larger litters than young females and possibly old females (Chapters 7 and 17). Finally, we assume the decision of when to wean litters is a function of two opposing "considerations"—the number of cubs to support and the amount of maternal reserves available for expenditure. Large litters favor late weaning, but good physical condition favors early weaning. Now suppose there is a level of investment beyond which offspring survivorship is less strongly affected by additional investment (plateaus in curves A and B, Fig. 8.2). Assume constant litter size and suppose further that females in good condition can, for a given biological cost, sustain a higher per-cub rate of investment and produce offspring with an acceptable likelihood of survival in one year (S_1 at intersection of dotted line and curve A, Fig. 8.2), whereas females in poorer condition take two years to produce offspring of comparable (not necessarily equal) quality (S_2 at intersection of dotted line and curve B, Fig. 8.2). If the expected survivorship of offspring weaned at one year by poor-condition females is sufficiently low (S_3 in Fig. 8.2), poorer-condition females may have higher lifetime success by delaying the next episode of breeding one year and investing further in existing offspring. Conversely, if females in better condition can invest maternal time and energy at a sufficiently high rate, they may, in one year, bring offspring to a level of physical condition (size, fat reserves, etc.) which additional maternal investment would not much improve, so that the best option is to wean them and prepare to breed again.

Similarly, if we fix female condition and vary litter size, curve B might apply to females with large litters, in which the same total maternal resource is divided among more cubs, and curve A, to females with smaller litters. Considering both condition and litter size, curves A and B would reflect females with large and small ratios, respectively, of total maternal reserves or number of cubs. In this explanation of the age of weaning dichotomy, choice of weaning strategy is "conditional" on a female's "estimation" of the maximum rate at which she can invest in each offspring. This model is consistent with our results. Weaning age was not related to maternal age, an indirect measure of maternal condition, because young females tended to produce smaller litters, and this tended to equalize per-cub rates of investment relative to prime females. Similarly, the effect of litter size was not pronounced because some females were in good enough condition to wean large litters early, and, conversely, some females were in poor enough condition that they required two years to wean small litters. We could not detect a marked survival cost of early weaning because, in this model, cubs from different litters were weaned only after they were brought to comparable levels of size and condition.

It is possible that a female in poor condition cannot return to estrus, which then incidentally results in retention of offspring for another year. We believe, however, that the energy required to initiate reproduction, particularly in a species with delayed implantation, must be substantially less than that required to carry offspring another year. Females that cannot support the demands of estrus should be even less able to support yearling cubs. It is true that estrus represents something of a commitment to high levels of expenditure a year or so later. But females make that commitment early in the growing season when there are few clues to their future ability to invest in reproduction. Furthermore, resorption of fetuses or abandonment of neonates are options, should poor condition persist into the winter and spring. For these reasons, we regard maternal condition as it affects the rate of offspring development (rather than ability to enter estrus) as the more likely explanation for delayed weaning.

ADOPTIONS

Grizzly bear mothers exhibited a wide range of filial concern and disciplinary control. Some females seldom left their cubs unattended, whereas others appeared indifferent to their cubs while feeding and lost track of them for short to extended periods of time. Females that were attentive to their cubs were also disciplinarians and fell into two general patterns of behavior: those that fed with their cubs in close attendance and those that "parked" their litters at some distance from the feeding site, but in view. The former constantly communicated with their cubs as they fed and were cautious when they moved among the aggregation members in response to the availability of food or pressure from other grizzlies. Disciplinarians in the latter category placed

Grizzly bear parenting varied considerably among females. Some appeared almost indifferent to their cubs whereas others were attentive. Attentive mothers either fed with their cubs in close attendance or "parked" their litters at a safe distance, issuing vocal instructions to keep them in place. *(Photos by John J. Craighead)*

their cubs beyond the feeding animals, issued vocal orders that kept them in place, and delivered more overt instructions in the form of slaps and snaps if the cubs tended to disobey by following or wandering. With the litter precisely located, the mother joined the feeding bears. She would frequently interrupt her feeding to glance at her offspring and to vocalize if the cubs became overly restless, curious, or left the area where she had deposited them. If danger threatened, she would return to her cubs, the speed of her return being in direct proportion to the nature of the threat.

Despite the generally strong maternal instinct of the mother grizzly, cubs and yearlings sometimes became "motherless" due to early weaning, abandonment, or death of the mother. Some of the motherless cubs were adopted. Adoption in brown bear populations has only rarely been reported in the literature. Barnes and Smith (1993) recorded three cases where female brown bears (*Ursus arctos middendorffi*) on Kodiak Island, Alaska, adopted cubs of the year. We recorded 9 adoptions of 11 offspring from 1959 to 1970; five were adoptions of cubs, three of yearlings, and one of 2-year-old littermates (Table 8.13). Abandoned offspring that were not adopted were not considered in this analysis. Adoptions occurred in six of the 12 years of observation, with a maximum of three in 1963. The sex ratio of adopted offspring was five females to three males, with three unknown. The youngest female to adopt a cub was five years of age. The average age of all adoptive females was 7.4, while the average age of all parent females was 11.3 years. Three adoptions occurred following deaths of the parent females, three following weaning by the mothers, and two following abandonment. The adoptive circumstances of one cub were unknown. Three females held their adoptive offspring for two years, five for one year, and one for only one week. Among the adopted offspring, two females (Nos. 5 and 139) attained ages of 16 and 8 years, experiencing 12 and 4 reproductive years, respectively. Number 5 subsequently produced one litter of two cubs, sex unknown, and a litter of three cubs consisting of two females and one male. Female No. 139 produced a litter of two cubs, sex unknown.

Most, but not all orphaned offspring were adopted. One cub of unknown sex of female No. 42 was never adopted, and one cub of female No. 148, accompanied another female, No. 48, for only a week and may not have been a bona fide adoption. The specific fate of these two cubs was unknown, but neither attained adulthood. Only one of the adopted males was known to attain reproductive age. Four of the adopted cubs were never reobserved

following the year of adoption, and two were known mortalities at an early age (Table 8.13).

The reproductive histories of females No. 40 and No. 101 illustrate the potential advantages and disadvantages of adoptive behavior. Number 40 had a mean productivity rate of one cub per year over a seven-year period of observation (Table 7.10), and a mean reproductive cycle-length of 2.33 years. In 1964, she produced a litter of two cubs. One died during the summer, and the other was weaned as a yearling early in the spring of 1965. This yearling male was not adopted, but survived through 1970. In 1966, No. 40 produced a second litter of two cubs on a reproductive cycle of two years. The two offspring were weaned in the early spring of 1967, and both yearlings were adopted by female No. 101. They wintered with No. 101, and she carried them through the summer and fall of 1968. We suspected that she wintered with them again as 2-year-olds, but had no conclusive evidence.

In 1968, No. 40 produced her third litter (three cubs), again with a two-year period between litters. All three cubs were captured and marked (Nos. 240, 241, and 256). Number 40 did not wean this third litter as yearlings, but carried them into the fall of 1969 when she was killed. She probably would have denned with them as yearlings because they were with her in October, and we never recorded an instance of weaning in late fall. Number 40, however, was shot 13 October 1969. One of her yearlings, male No. 256, was shot 8 November 1969 west of Yellowstone Park, but the others survived through the fall. Male No. 241 was killed 21 October 1970 south of Yellowstone Park, and female No. 240 survived at least through 1970.

When No. 40 weaned her offspring as yearlings in 1967, she exhibited a short reproductive cycle (two years). Her yearlings were adopted by No. 101 who was then in a long cycle (four years; Table 3.3). In this instance, adoption permitted No. 40 to produce her second litter after a two-year interval because the previous litter was attended by another adult female through the yearling stage. The reproductive advantage afforded a female that succeeds in having her offspring adopted is clear. There are, however, also clear disadvantages to females that adopt. Any maternal investment directed at adopted cubs may reduce the female's ability to invest in her own (contemporary) offspring and may also impair her future reproductive opportunities. Of course, in a species with low reproductive rate, investment in adopted cubs may directly benefit the population, if not the individual and her genetic offspring. Among species that aggregate and establish strong social structures, such as the grizzly bear (see

Table 8.13. Summary of grizzly bear adoptions, Trout Creek ecocenter, Yellowstone National Park, 1959-70.

Parent female				Adoptive female		Adopted offspring			Year of adoption	Circumstances of adoption	Length of adoption (years)	Fate of adopted offspring
Number	Age	Number of offspring	Sex of offspring	Number	Age	Number	Age	Sex				
Unk	Unk	Unk	F	UM	Ad	5	1	F	1959	Unk	2	Mortality 1974.
48	Ad	1	F	UM	Ad	49	Cub	F	1960	Female No. 48 abandoned	1	Not observed after 1961.
42	8	2	F, Unk	UM	Ad	139	1	F	1963	Adopted after weaned as yearling.	1	Last observed 1970.
148	16	2	M, Unk	150	5	UM	Cub	Unk	1963	Female No. 148 killed in 1963; one of two cubs adopted by female No. 150.	—	Last observed, September 1963; length of adoption, one week.
148	16	2	M, Unk	48	Ad	??a	Cub	M	1963	Female No. 148 killed by a truck in 1963; one of two cubs adopted by female No. 48 with a litter of three.	2	Survived at least two years.
29	6	2	Unk, Unk	81	7	UM	Cub	Unk	1965	Female No. 29 killed in August, 1965.	2	Last observed 1967.
40	9	3	M, F, Unk	101	9	216 UM	1 1	M Unk	1967	Adopted after weaned as yearlings; females 40 and 101 and their offspring had been observed together in 1966.	1	No. 216 a mortality in 1972; UM littermate last observed 1968.
42	12	2	F, Unk	150	9	225	Cub	F	1967	Female No. 42 abandoned one cub; adopted by female No. 150.	1	Sacrificed 1968.
5	12	3	2F, M	187	7	237 238	2 2	F M	1970	Adopted by female No. 187 after weaning as two-year-olds by female No. 5.	1	Last observed 1971.

Note: Ad=Adult, Unk = unknown, and UM = unmarked; offspring observed with adoptive female in late summer or fall, were assumed to have hibernated with female and weaned after hibernation.

a One cub of female No. 148 was adopted by female No. 48 (already with three cubs) before any of the offspring were marked. Then all four were marked (Nos. 181, 192, 193, 194: the Four Musketeers). It was not known which of the four marked cubs was the adopted one.

A compensating factor for the increased energy investment by females that adopted orphans may come in greater access to choice feeding locations. *(Photo by John J. Craighead)*

Chapters 5 and 6), short-term benefits to the aggregated population may, in certain instances, outweigh long-term genetic benefits to the individual. This possibility merits further investigation.

We might speculate further that a subordinate female gained compensating individual benefits from adoption because the coalition that was created improved her own ability (and that of her genetic offspring) to compete for feeding sites (see Chapter 7). Given the relative rarity of the behavior, we must also consider the possibility that we were observing "errors" on the part of females that adopt, rather than adaptation. The net effects of adoption on the Trout Creek population, which must be small in any case, remain uncertain because the reproductive gain of the "donor" females may or may not have been matched by reproductive "losses" on the part of the adopting females.

ABANDONMENT

We distinguished abandonment from early weaning by subjective assessment of the behavior of the mother, the condition, age, and behavior of her offspring, and the behavior other adult females directed toward the offspring. We classified cubs as abandoned only if separation occurred during or soon after the breeding season, and only if both the female and the abandoned offspring were seen after the separation.

We documented five abandonments by four females during the study period (Table 8.14). In addition, we recorded nine possible abandonments, three of which were probably induced by capture and transport, and others could have been simple deaths. Abandonment may result from inattentive mothers who simply lose offspring, rather than from mothers that actively and intentionally abandon them. Theoretically, a female could benefit by abandoning a small litter or an unhealthy one. This would allow her to breed at least one year earlier, and possibly produce a larger, healthier litter (Tait 1980).

Tait (1980) theorized reproductive strategies relative to the abandonment of single grizzly cubs. He varied adult survivorship rates, cub and yearling survivorship rates, and the distribution of litter sizes in six different computer runs to determine whether abandonment might affect the expected number of offspring recruited by the female. He concluded that abandonment could improve recruitment if the mother whelps two or more new offspring in the year following that of abandonment. If she did not, it gave her no advantage to abandon single cubs.

We examine this idea in relation to the reproductive history of female No. 48. She abandoned two single-cub litters from 1960 to 1962. Another adult female adopted the first abandoned cub in 1960, increasing the probability of its survival. But

Table 8.14. Summary of abandonments of grizzly bear cubs, Trout Creek ecocenter, Yellowstone National Park, 1959-70.

Female number	Age	Year	Litter size	Sex	Date family first observed	Date of abandonment	Circumstances	Fate
40	6	1964	2	M,?	4 June	9 June	Abandoned one cub which was last observed on 18 June at Trout Creek.	Unknown.
42	12	1967	2	F,?	6 June	15 June	Abandoned one cub; shared other cub with female No. 200; female No. 150 adopted abandoned cub on 22 June.	Cub survived to yearling age; died in 1968.
48	Adult	1960	1	F	26 July	26 July	Abandoned after marking on 26 July; cub adopted by Rip-nosed female 29 July; last observed with adoptive female on 25 August.	Unknown.
48	Adult	1962	1	M	23 June	28 June	Cub remained in area for three days after marking on 5 July; never reobserved.	Unknown.
65	Adult	1964	3	M,M,?	6 June	1 July	Abandoned one cub which remained at Trout Creek until 8 August.	Unknown.

female No. 48 did not breed successfully the next year, and a litter was not produced in 1961. Instead, she mated in 1961, producing another 1-cub litter in 1962. She abandoned this cub on 28 June and bred (unobserved) shortly thereafter. The abandoned cub was not adopted and was never reobserved after 5 July. In 1963, female No. 48 produced a litter of three, whelping this larger litter at least one year earlier than would have been possible had she not abandoned her single cub in 1962. This litter raised her productivity rate to 1.25, a high rate, in comparison to the population mean productivity rate of 0.61 (Table 7.10). Thus, the low probability of the abandoned cub surviving was clearly offset by the production of a larger litter in a shorter interval on one occasion (1962), but not on another (1960).

Abandonment followed by adoption may function to shorten reproductive cycles and improve the reproduction effort. Abandoning part of a litter, however, confers no such advantage because the female continues to lactate, does not wean her remaining cubs, and thus does not shorten her reproductive cycle with an early estrus.

The capture, marking, and release of females and their offspring, for either management control or scientific study, increased the rate of abandonment. Twenty-five percent of the four females known to have abandoned cubs did so immediately after handling, as did two of eight other females suspected of abandoning cubs.

FATE OF ORPHANED CUBS

Nine litters were orphaned during the course of the study. Five females were killed, leaving their litters motherless, two females were captured and shipped to zoos, and two females abandoned litters. Among the 21 orphaned cubs composing the nine litters, seven were known to have survived at least to one year of age, seven cubs died or were sent to zoos before becoming yearlings, and seven were unaccounted for (Table 8.15). Among the seven cubs surviving as yearlings, one survived beyond that age.

No cubs were orphaned in the last five of the 12 years of study, 1966 through 1970. A partial explanation of this was that no adult females with cubs died during those later years. The 21 cubs orphaned during the earlier period, however, represented 8.8% of 238 cubs produced by 33 individual adult females at the Trout Creek ecocenter from 1959 through 1970. With the exception of the single offspring of female No. 148, none of the orphaned cubs attained adulthood, and the histories of six others unaccounted for, strongly indicated

Orphaned cubs that were not adopted rarely survived into adulthood. Only one such cub was known to live beyond the yearling stage. *(Photo by John J. Craighead)*

early death (Table 8.15). This was in marked contrast to the 79% survival to adulthood of cubs attended by females for one or more years that was demonstrated earlier in the chapter. We also examined the survivorship from one to two years of age among offspring weaned at one year versus those still under maternal care at that age. Mortality of weaned offspring was determined by carcass recovery in five cases and disappearance in one case. For unweaned offspring, one carcass was recovered and two bears were never again sighted. Cubs weaned as yearlings were five times more likely to die than offspring weaned later (Table 8.16), a difference that was highly significant (Chi-square test, $P = 0.02$). It is clear that maternal investment pays dividends at all stages of care.

ADULT FEMALE BONDING

We define bonding as a strong social relationship between two or more adult females with cub litters, characterized by a free interchange of cubs for short periods of time, with close cooperative effort between the family units in traveling and feeding and in defense of resources and one another. On rare occasions, the communal care relationship includes nursing of a cub by a female other than its mother. Bonding did not occur or was undetected during the first six years of our observations. It was first observed in 1965, was recorded again in 1966 and 1967, was not observed in 1968 or 1969, and became a common event in 1970 (Table 8.17).

Female Nos. 29 and 81 established a close relationship throughout the summer of 1965, traveling and feeding together, and providing mutual protection for their cubs. One of No. 29's two cubs died early in 1965, and subsequently she was killed. Number 81 adopted the surviving cub as a yearling.

In 1967, female Nos. 42 and 200 bonded. Each had two cubs, and the cubs responded to either mother and were observed to interchange freely. Female No. 42 abandoned one of her cubs in midseason, and it was adopted by female No. 150. Female Nos. 42 and 200 remained bonded for the rest of the summer, sharing the remaining three cubs.

The most extensively documented bonding occurred in 1966 between female Nos. 40 and 101. Number 40 was first observed at the Trout Creek

Table 8.15. Fate of orphaned or abandoned grizzly bear cubs, Trout Creek ecocenter, Yellowstone National Park, 1959-70.

Year	No. of cubs (in each litter) orphaned	Circumstances of orphaning	No. of orphan cubs known surviving to yearling	No. of orphan cubs known not surviving to yearling	No. of orphan cubs of unknown status	No. of orphan cubs known surviving beyond yearling age
1959	2	Female No. 23 killed by overdose of drugs.	1	—	1	—
1960	3	Unmarked female shot near Gardiner.	3	—	—	—
	3	Female No. 31 killed by overdose of drugs.	—	1	2	—
	1	Female No. 48 abandoned her only cub.	1 (adopted)	—	—	—
1961	3	Female No. 18 shipped to zoo by NPS.	—	2[a]	1	—
1962	0	No orphaning observed.	—	—	—	—
1963	3	Old Faithful female shipped to zoo by NPS; all three cubs captured and shipped to zoo.	—	3 (zoo)	—	—
1964	2	Female No. 148 killed by truck.	1 (adopted)[b]	—	—	1
	2	Female No. 170 abandoned her two cubs after they were marked.	—	—	2	—
1965	2	Female No. 29 killed by overdose of drugs.	1	1	—	—
1966-70	0	No orphaning observed.	—	—	—	—
Total	21		7	7	6	1

Note: Not included in the table is one severely crippled cub abandoned by female No. 65 from a litter of three in 1964.
[a] One cub died and one was shipped to zoo.
[b] The single offspring of female No. 148 became a member of a juvenile band and possibly survived to adulthood.

Table 8.16. Percentage survivorship between the ages of one and two years for grizzly bears weaned as yearlings as compared with bears under maternal care during second year of life (*n* = 69), Trout Creek ecocenter, Yellowstone National Park, 1959-70.

	Weaned as yearling	Weaned as 2-year-old
Percent surviving from ages 1 to 2 years	70.0%	93.9%
(*n*)	(14)	(46)
Percent dying from ages 1 to 2 years	30.0%	6.1%
(*n*)	(6)	(3)

ecocenter with her second litter (two cubs) on 8 June. On 10 June, Nos. 40 and 101 arrived together, No. 40 with her cubs and 101 with one cub, which was her second litter. These two families were virtually inseparable during the course of the summer and fall. During 16 censuses made over the summer in which No. 40 and her litter were recorded, 101 and her litter were also observed, generally entering or leaving the feeding area together. They were not once observed apart. Among 14 censuses during which No. 40 and her family were not observed at the ecocenter, 101 and her family were absent as well. Radiotracking confirmed that they were together frequently during July and August when not recorded at the ecocenter. The two families reacted to environmental stimuli as a single family. The cubs followed either adult, irrespective of kinship, and the mothers attended to a temporarily attached cub, irrespective of kinship.

The close association of No. 40 and 101 continued through August, September, and October, terminating in mid-November. Number 40 was radio-monitored during the fall and her continued close association with No. 101 enabled us to keep track of both females and their offspring, even though No. 101 was not radio-collared. Excerpts from 1966 field notes illustrate behavior modifications the two females developed during the fall to maintain their close relationship:

19 October. The two families (No. 40 and No. 101) are still traveling together in the Trout Creek area and have been sighted frequently. Twice, the families have been seen together on elk kills. Last season the two females denned 16 miles [25.7 km] apart. Will they return to their respective denning areas of last season, or will they perhaps den in close proximity of one another?

22 October. Snowing and cold, but Nos. 40 and 101 were sighted together at Trout Creek. We have never observed bonding to persist so late in the season. This weather should move them toward winter dens. Number 40 is instrumented, No. 101 is not; we should be able to observe the final chapter in this relationship.

29 October. Number 40 and No. 101 were again observed together at the Trout Creek ecocenter. No other bears have been sighted here for weeks. Number 101 still has her one cub and No. 40 her two cubs. We trapped and marked one of No. 40's cubs and carried it back to the lab.

30 October. Returned to Trout Creek with the captured cub. Number 40 and No. 101 and families were present. Released the marked cub at a distance and it ran directly to No. 40, not to No. 101. The cubs have interchanged between females all summer, but under stress the cub sought refuge with its mother.

4 November. Number 40 and No. 101 were located on a bison carcass. All three cubs were present. All are exhibiting mild lethargy. This fact and the lateness of the season suggest that the two families will hibernate close to one another. The next storm could move them to a den or dens already prepared. No dens have yet been located.

8 November. Following the radio signal of No. 40, we located both families bedded close to a den. Their behavior suggests this is the den of No. 101. It is not large enough for two adult bears, let alone their offspring.

9 November. Located No. 40 at No. 101's den. Radiotracked her one-and-a-half miles [2.4 km] southeast. Weather forced us to return to camp.

11 November. An hour long trek through 18 inches of snow following a radio signal brought

Females with litters occasionally bonded. Bonded families fed, traveled, and reacted to stimuli as a single family unit. The cubs responded to either mother, and, on occasion, bonded females nursed their companion's offspring. *(Photos by John J. Craighead)*

Table 8.17. Bonds between adult female grizzly bears with cubs, Trout Creek ecocenter, Yellowstone National Park, 1965-70.

Year	Female number	Family number	Age of female	Number of cubs	Remarks
1965	29	118	6	2	One of No. 29's cubs died; No. 29 was killed and No. 81 adopted her surviving yearling.
	81	—	—	2	Bonded female adopted.
1966	40	14	8	2	Two yearlings weaned by No. 40 and adopted by No. 101 in 1967.
	101	42	8	1	Bonded female adopted.
1967	42	38	12	2	One cub of No. 42 adopted by No. 150.
	200	53	5	2	Bonded female did not adopt.
1968	none	—	—	—	—
1969	none	—	—	—	—
1970	101	42	12	2	No adoptions among these bonded females. One cub of No. 101 lost prior to September.
	42	38	15	2	
	128	35	19	3	
	175	50	16	1	
	173	47	8	1	

us close to No. 40. As we moved around a timbered slope, she suddenly came in view less than 40 meters [44 yds] away. She was somewhat lethargic and digging, her head partially in the den entrance. We retreated without disturbing her and observed at a distance of 100 meters [110 yds]. Both cubs were in the den. This den is not more than one airline mile [1.6 km] from No. 101's den. Apparently the prehibernation behavior of both females has been altered by their close association. Number 40 has remained close to No. 101's den during the past few weeks, she and her cub making day beds close by. She retreated to her own den (dug earlier in the fall) only when snow depth reached 1.5 to 2 feet [46-61 cm]. We suspect both females will now remain at their respective winter dens, and thus comes to a close the bonding of No. 40 and No. 101. It appears that No. 40 has made a major adjustment by abandoning her den site area along Upper Alum Creek to select a new denning site close to No. 101's. Normally, grizzlies den miles apart, isolated by rough terrain. This is our first evidence that social behavior may be a contributing factor to den site selection, and the first evidence that grizzlies will tolerate the relatively close presence of another when they prepare for the winter sleep.

13 November. Made what we thought would be a final check on No. 40 at her den. We found her bedded in front of the den and No. 101 and her cub bedded down nearby.

Sometime after 13 November, both bears confined themselves to their respective dens and soon after were denned for the winter. The families left their dens in late March or early April. Bonding did not persist into the next season. Behaviors we believed to be outcomes of bonding were extensive late fall foraging, frequent movement to and visitation at dens, change of normal denning procedures, and, perhaps, delayed lethargy among bonded bears.

Cases of bonding were most prevalent in 1970, in a complex situation involving five females and nine cubs (Table 8.17). On 19 June, No. 42 and No. 101 began appearing together and were recorded on several censuses at the ecocenter. On 1 July, they were joined by No. 128 and on 10 July by No. 175. Four days later, No. 173 and family joined the group of four females, attaching herself most closely to No. 101 and No. 42 to form a "triumvirate" within the group of five. For a period of two weeks (14 July through 26 July), Nos. 42, 101, 128, 173, and 175 arrived at the ecocenter

Numerous and complex bondings occurred in 1970 and this behavior may have been a response to the disruptive effects of closing ecocenters. Bonding behavior conferred biological benefits to the individual families during this period of stress. *(Photo by John J. Craighead)*

in various family combinations. The relationship between the females was loose and apparently in its formative stages. By 28 July, Nos. 42, 101, 173 and 175 appeared to be firmly bonded. This close relationship persisted throughout the summer. They and their cubs arrived at the ecocenter together on nine of 10 consecutive censuses. Number 128 and her three cubs were conspicuously absent after 28 July until, on 4 September, all five females were again present with eight cubs among them. The fate of No. 101's missing cub (Table 8.17) was never determined.

The five females and their cubs had formed an effective coalition where the families could feed, travel, and defend themselves as bonded pairs, or as a coalition of three or more families. We believe the numerous and complex bondings of the 1970 season may have been an adaptation to the disruptive effects of the partial closure of the ecocenter. Food abruptly became scarce, and females with litters of cubs appeared agitated and restless. They found greater security by forming a coalition of bonded females that persisted well into the fall of the year. The insecurity projected by these females was transmitted to the cubs. They tended to be skittish and overreacted to perceived threats by fleeing. When the cubs fled, they dispersed, and some were separated from mothers for as long as several days. The chance of a cub locating a group of several closely associated adult females presumably was greater than that of a reunion with the mother alone. A reduction in the length of time cubs strayed from adult protection may have served to dampen cub mortality during a period of food scarcity and social stress.

JUVENILE BANDS

Juvenile bands were composed of early weaned, orphaned, or abandoned offspring from two or more mothers. Band members may or may not have been adopted before banding together. Although we recorded formation of several juvenile bands during the course of the study, in only one band were all members marked so that its history could be thoroughly documented. This band was composed of four males that we dubbed the Four Musketeers. The history of the foursome began in 1963 when all were cubs.

Early in the summer of 1963, we observed two females, Nos. 48 and 148, running together accompanied by six cubs. This bonding continued throughout the summer. On 5 September, female No. 148 charged our observation vehicle at Trout Creek and was killed instantly on impact. Within a few days, her orphaned cubs were adopted by female No. 48, who had three cubs of her own. The female and six cubs traveled and fed together for over six weeks. On 22 September, we recaptured No. 48. Her three cubs were observed at the capture site, but only two of the three orphaned cubs of No. 148 were with No. 48 when captured. On 6 October, we observed No. 48 and her family at Trout Creek, but with only one of the orphaned cubs. She apparently wintered with all four cubs in the late fall of 1963. The fate of the two cubs from the orphaned litter was unknown.

Early in June 1964, female No. 48 weaned her three yearlings and the adopted yearling of No. 148. This was the group that became known as the Four Musketeers. All were unmarked until August of 1964, but each yearling was readily identifiable from wide, white natural neck collars. The band of four, all males, was captured and marked Nos. 181, 192, 193, and 194. One was the surviving orphan of female No. 148, but we could not specifically distinguish him from the other three. The four fed and traveled together during the summers of 1964 and 1965. They were seldom observed separately, invariably coming to the Trout Creek dump together, approaching cautiously and as a unit. Occasionally, they displaced older bears at food sites, becoming bolder and more aggressive as the summer of 1964 progressed. During the summer of 1965, as 2-year-olds, they fed frequently at the ecocenter and rated much greater respect from the aggregation than was normal for 2-year-olds. As a unit, this band appeared to have the status of young adults. Number 193 was not observed again after the autumn of 1965. The remaining three members disbanded in the spring of 1966 as 3-year-olds, but summered in the Trout Creek area. They were occasionally seen individually, but never again as a unit.

This close, three-year association of members from two families represents the strongest and longest-lived bond that we observed among family groups. The ultimate fate of the four is unknown. Number 181 was last observed as a 7-year-old, No. 192 at three years of age, and 194 at six years of age. The orphaned and then adopted yearling of female No. 148 obviously benefitted from the adoption and the later bond association, even though his survival time could not be specified. It could have been a minimum of two years or it could have exceeded seven years. There can be little doubt that this bonding was highly advantageous to weaned yearlings and, if it occurred rather regularly in a population, would have survival value.

Adoption, bonding, and juvenile bands at Trout Creek were all examples of potentially adaptive behaviors that may have improved survivorship among individuals (nonadults) or family units that were disadvantaged for one reason or another (orphaned, early weaning, social status, etc.). Although the rationale for such behavior must ultimately be found at the level of individual cost/benefit, the phenomena of abandonment, adoption, bonding, and juvenile bands may collectively have significant population-dynamic consequences for a carnivorous species characterized by low reproductive rates, omnivorous feeding habits, small population size, and complex social behavior when aggregated.

SUMMARY

Following mating, conception, and birth, the female grizzly faces the formidable task of rearing offspring to nutritional independence while protecting them from injuries, accidents, and attacks that might reduce recruitment of breeding adults into the population. Survivorship of Trout Creek cubs from 0.5 to 1.5 years was estimated to lie in the range of 76.9% to 83%. We have no data on survivorship from birth to 0.5 years. There was no detectable relationship between cub sex and survivorship to 1.5 years, and the majority of mortalities appeared to result from causes (accidents, predation, etc.) that acted independently on littermates. There was a trend toward greater survivorship of cubs born into larger litters, perhaps reflecting the ability of females

Juvenile bands represented the strongest and longest-lived bonding observed among the grizzly bears aggregated at Trout Creek. Coalitions of juveniles re-formed after winter sleep. Juvenile bands composed of yearlings or 2-year-olds rated higher in the social hierarchy than individuals of their age. *(Photos by John J. Craighead)*

in good physical condition to produce large litters of vigorous cubs and to protect them effectively.

Infanticide by adult males was documented in only *four* cases and so was too infrequent, in our view, to play a significant role as a cause of cub mortality. Our inability to document the exact magnitude and causes of cub mortality in the Yellowstone Ecosystem, however, make it impossible to determine the relative importance of infanticide as a factor in overall cub mortality. The high survival rate of cubs, in general, while exposed to numerous males suggests that infanticide is observed infrequently because it rarely occurs. Our review of infanticide in other populations of grizzly bears suggested a similar conclusion. Reports of adult male grizzlies also killing adult females and independent juvenile bears, and the nature of the grizzly mating system, lead us to view infanticide in grizzly bears as predation-related behavior rather than a mating strategy in which males force females to return more quickly to estrus.

Females usually weaned offspring in May or early June before the peak of the breeding season. Littermates were always weaned at the same time. Most litters were weaned at two years of age (67.6%), but a substantial fraction were weaned at one year (32.4%). Weaning age was not influenced by maternal age, the sex ratio of the litter, or litter size, although larger litters tended to be weaned later. Early weaning may have imposed somewhat elevated risk of immediate postweaning, human-caused mortality, but the effect was not marked. We could detect no effect of weaning age on survivorship of female offspring from weaning to adulthood, suggesting that, over the longer term, any immediate effect of weaning age on offspring survivorship is overwhelmed by other determinants of survivorship. These results are consistent with a model in which "choice" of weaning age is conditional on both maternal condition and litter size, and cubs from different litters are weaned only after reaching comparable, but not necessarily equal levels of body size and condition. From the female's perspective, slight reductions in survivorship among early-weaned offspring may be compensated by a reduction in subsequent cycle length and consequent increases in her lifetime production of weanlings.

Cubs that were abandoned, weaned early, or lost their mothers were sometimes adopted by other females. Surviving mothers of adopted cubs obviously gained from a reduction in the interval to their next litter, whereas the cost to an adopting female may have been balanced by the advantages of a larger "family" coalition in conflicts at the ecocenter or elsewhere (see Chapter 7). Adoptions were relatively uncommon, but appeared to improve survivorship. Adoptions may, however, have reflected nonadaptive anomalies in the formation of social relationships.

Like adoption, abandonment of offspring was observed, but was not common. The causes, costs, and benefits of such action are not clear, but females may occasionally terminate care of young that are doing poorly to advance production of a new litter. The crucial role of maternal care in grizzly recruitment was highlighted by the extremely poor survivorship of orphaned cubs to yearling age (33%) versus that of nonorphaned cubs (81%). No orphans were known to have survived beyond two years of age.

Adult females with offspring sometimes formed strong social bonds involving communal movement, feeding, nursing, and protection of cubs. Such bonding may have reduced cub mortality during periods of food scarcity and social stress. Following weaning, juvenile bears occasionally formed similarly stable coalitions with littermates or same-age non-kin. Such coalitions clearly were advantageous to member bears in competition for feeding sites in the aggregation and in defending against larger bears. This and female bonding were observed too infrequently, however, to reveal why some adults and juveniles bonded and others did not, or whether these behaviors might significantly affect population levels.

9
Density-Dependent Influences on Population Dynamics

It is well known that animal populations may exhibit reproductive responses to their own densities (Stubbs 1977). Such responses can be significant in maintaining or restoring a "biotic balance" when a population has declined drastically or exceeded the carrying capacity of its habitat. Responses can be subtle and difficult to detect without a thorough understanding of a species' reproductive biology and its flexibility in the face of demographic and habitat changes.

Reproductive responses are especially difficult to detect and interpret in populations of long-lived, slow-reproducing species. Although we did not address the subject in detail in earlier publications, we feel obliged to do so now because of criticisms made by Cowan et al. (1974), McCullough (1981, 1986), and Stringham (1983), which led to the use and interpretation of our data in ways which have confused rather than improved understanding of the reproductive biology of grizzly bears inhabiting the Yellowstone Ecosystem, as well as the effects of the abrupt closures of the open-pit garbage dumps (ecocenters) on the grizzly population.

Changes in population dynamics related to animal density have been termed "density-dependent" or "density-independent" (Smith 1935). Under density-dependence, as populations grow toward carrying capacity, birth rates and survivorship may decline in reaction to reduced per capita food resources, increased social stress, or some other density-dependent social or ecological factor. Conversely, under reduced densities, individuals may experience improved reproduction and survival as resources are freed or social interactions reduced. These adjustments are considered compensatory, in the sense that they tend to counter and perhaps ultimately "balance" the effects of whatever initiated changes in population numbers. Studies addressing density-dependence in species of large mammals are well-known in the literature (Cheatum 1952; Stubbs 1977; McCullough 1979, 1984, 1990; Reynolds 1993), and those pertaining to brown, black, and polar bears

have been ably reviewed by McLellan (1995 [in press]), Garshelis (1995 [in press]), and Derocher and Taylor (1995 [in press]).

The question of density-dependence in grizzly bears has special historical significance in the case of the Yellowstone population. It is also central to the theme of this monograph—the biology of grizzly bears aggregating at ecocenters and the consequences when those ecocenters are abruptly destroyed. We take special care in this section to explore in detail density-dependent compensation, if any, with regard to recruitment during the 1959-70 period.

HISTORY

When the National Academy of Sciences committee reviewed our data and preliminary population analysis in the early 1970s, it criticized our failure to include a compensatory factor or mechanism. The committee, however, offered no suggestion as to what this mechanism might be (Cowan et al. 1974). We claimed there was no evidence of compensatory behavior. Eight years later, we reviewed our data and calculations and reaffirmed our position (Craighead et al. 1982). In the meantime, other biologists using our data (see Craighead et al. 1974) sought to demonstrate compensatory mechanisms.

McCullough (1981, 1986) and Stringham (1983) claimed a density-dependent compensatory relationship between grizzly bear recruitment and adult population size, using data we had reported for the Yellowstone Ecosystem during 1959-70 (Table 1 in Craighead et al. 1974). Specifically, they used the cumulative annual counts from ecocenters throughout the Ecosystem and applied them, as density expressions, to a population of bears inhabiting more than two million hectares (five million acres) of habitat.

McCullough (1981) reported significant negative regressions between three measures of per capita recruitment (cubs/adult, yearlings/adult, and 2-year-olds/adult) and cumulative numbers of adults in the year of birth. He also found a significant regression between a measure of sex-specific per capita recruitment (cubs/adult male) and cumulative numbers of adult males, also in the year of birth. Stringham (1983) reported significant negative regressions between number of cubs and cumulative postnatal number of adult males, and between cub litter size and both pre- and postnatal cumulative number of adult males. In addition, he showed a strong negative relation between an index of reproductive rate (combining information on interval between births, litter size, and litter abundance) and the circumnatal cumulative number of adult males.

McCullough concluded that adult males were responsible for most of the suppression of cub production, but was noncommittal about whether the specific compensatory mechanism was primarily male infanticide, food resource competition, or a combination of these two processes. Stringham (1983) also emphasized the primary role of adult male density (or some close correlate) in suppressing cub production. He hypothesized that cohort sizes through 2.5 years of age were governed predominantly by the condition of young during gestation and infancy. Offspring condition, he suggested, was determined by (1) direct effects of males in reducing food availability to females and their offspring and increasing male-induced social stresses impinging on adult females, or (2) reduced abundance of natural foods that was reflected in increased ecocenter use by adult males.

These results and conclusions are important for two reasons. First, McCullough suggested that the widely reported and accepted decline in reproductive parameters reported by the Interagency Grizzly Bear Study Team (IGBST) and others in the years soon after ecocenter closures might be explained solely on the basis of an apparent increase in adult numbers at the end of our study (1968-70), with a consequent density-dependent suppression of reproduction. This suppression was, he believed, independent of changes in resource management (McCullough 1981). Second, the apparent negative effect on reproduction attributed to adult male numbers (density) has been suggested to justify trophy hunting of male grizzlies (i.e., reducing male density) as a management tool. In essence, some researchers have implied that trophy hunting, through reduction of males, could encourage reproductive compensation by escalating rates of female recruitment (Dood *et al.* 1986). Miller (1990), in a review of studies thought to support such a connection, found that none was convincing. In recent work with hunted brown bear populations in the northcentral Brooks Range, Alaska, and in southwestern Alberta, Reynolds (1993) and Wielgus and

Bunnell (1994), respectively, concluded that loss of adult males was not reproductively compensated. Subsequently, McLellan (1995 [in press]) came to the same conclusion for reproductive compensation via density-dependent mechanisms, in bear populations, in general.

Our data (Table 1 in Craighead *et al.* 1974) used by Stringham and McCullough were based on annual counts of marked and recognizable bears, not only at Trout Creek but at other ecocenters and nearby campgrounds and developed areas throughout the Yellowstone Ecosystem. Thus, the age- and sex-specific numbers of bears did not represent simple averages of census counts at ecocenters, but were cumulative totals of individual bears seen at least once in a given year. In no year did the cumulative total necessarily reflect the day-to-day density of the Trout Creek subpopulation that was the basis for the majority of the reproductive data.

When dump closures were initiated in 1968, the cumulative totals and totals for adults increased (Table 1 in Craighead *et al.* 1974) as a result of greatly increased movement among the displaced bears. Thus, our "efficiency" in enumerating bears ecosystem-wide increased, but population densities at the ecocenters did not, at least not until closure of Rabbit Creek in 1969 led to an influx of adults at Trout Creek in 1970.

In short, we were counting at the ecocenters a larger proportion of the ecosystem population after 1967. Thus, the increase in our cumulative annual ecosystem-wide census counts, a result of increased movement and mixing after ecocenter closures, did not reflect a real increase in numbers of adult males throughout the Ecosystem, as interpreted by McCullough and Stringham. Their misinterpretations were perhaps understandable, however, given the difficulty of correctly interpreting and sorting out cause and effect in data collected by others.

Our examination of the question of compensatory effects during 1959-70 is based on data not available to McCullough or Stringham. Specifically, we confine our analyses to the Trout Creek subpopulation, which was most intensively censused and studied. As our measure of density effects, we use the average number of bears observed per census for this subpopulation, rather than a cumulative annual count of individuals for the entire Yellowstone Ecosystem. This provides a more accurate estimate of the actual density effects experienced by the individuals whose reproductive performance we are measuring. For our reproductive data set, we use cub production by marked Trout Creek females and ratios of mean daily census counts of adult animals to counts of cubs, rather than ratios of cumulative annual counts of adults versus cubs. It is not our intention to repeat all of the analyses of McCullough and Stringham, but to address, with appropriate data, the basic question of whether compensatory effects on reproduction were operating during 1959-70. Our conclusion is that no such effects were detectable over the range of densities observed at the Trout Creek ecocenter during 1959-70. There is no support in our data for the conclusions that (1) reductions in reproductive parameters after ecocenter closures were a consequence of density changes unrelated to ecocenter destruction, and (2) hunting of male grizzlies would have a beneficial population effect.

ESTIMATING DENSITY AND RECRUITMENT

Density estimates are most useful when animal numbers are related to areas actually used by the species (such as specific habitat types, localized feeding or breeding areas), rather than by arbitrary geographical or standard density units. But such areas must be delimited by animal locations and may vary temporally. For example, when grizzlies aggregated at Trout Creek, local densities at peak visitation approached 25 to 30 per hectare (10 to 12 per acre). When they dispersed daily after feeding, local densities surrounding the ecocenter were more nearly 30 to 45 per 100 km² (80 to 120 per 100 mi²). When bears dispersed in the fall to denning sites, density (now ecosystem-wide) decreased dramatically to fewer than 2 bears per 100 km²

(approximately 4 per 100 mi²). Even at peak visitation, significant numbers of Trout Creek bears were using a large but unknown portion of the surrounding backcountry. Despite these obstacles to defining, let alone measuring an appropriate, absolute density for the Trout Creek population, we can use mean daily aggregation size as a relative measure of density effects in different years, provided we make one key assumption: that the majority of any density effect on individual reproduction will be due to interactions experienced while present in the aggregation, rather than to interactions occurring while relatively dispersed in the surrounding backcountry. This is certainly plausible for all likely mechanisms of density dependence (e.g., resource competition, infanticide, social stress) simply because the potential for such stressful interactions was vastly greater in the aggregation than elsewhere. Because the ecocenter was the single most important source of nutrition for Trout Creek bears, competition for food created highly stressful situations, but these were ameliorated through the hierarchical structure. Cubs of the year may be somewhat more vulnerable to infanticide when they are just out of the den than they are at aggregations. At this time, they are smaller, but the mother's defensive ability, which should not vary much seasonally, and the frequency of contact with adult males are probably more decisive factors.

Measurement of Density

The number of bears present at Trout Creek was counted in regular, daily censuses beginning in early

Table 9.1. Asymptotic mean census counts of grizzly bears for the period 4 July - 31 August, Trout Creek ecocenter, Yellowstone National Park, 1959-70; values estimated by regression are shaded (see text).

Year	Cubs	Yearlings	Two-year-olds	n^a	Adults M	Adults F	Adults Total	Mean no. all bears	n^b
1959	15.5	10.5	8.8	11	18.4	24.7	43.1	76.8	13
1960	9.6	7.5	3.5	14	17.4	23.7	41.1	67.9	15
1961	12.3	2.0	5.8	10	9.7	17.2	26.9	57.0	10
1962	—	—	—	—	10.3	16.1	26.4	58.8	5
1963	—	—	—	—	—	—	—	—	—
1964	7.8	13.0	8.1	15	7.7	17.7	25.4	57.1	15
1965	13.1	7.4	10.6	7	5.0	16.1	21.1	59.9	7
1966	11.3	12.6	4.1	14	8.2	15.4	23.6	62.6	14
1967	8.6	11.4	8.4	8	6.3	14.8	21.4	60.3	8
1968	10.1	5.2	5.3	13	5.2	10.6	16.2	42.8	13
1969	6.3	5.7	5.8	18	7.8	12.0	19.9	48.3	18
1970	9.8	6.1	5.0	17	12.5	16.1	28.6	59.8	17

[a] Number of censuses from 4 July to 31 August in which cubs, yearlings, and 2-year-olds were distinguished.

[b] Number of censuses from 4 July to 31 August in which a count of all bears present was made.

Fig. 9.1. Distribution of numbers of grizzly bears censused at Trout Creek, illustrating the rapid seasonal buildup during June and the plateauing of visitation during July and August, Yellowstone National Park, 1959-70; (O) is one individual and (②) is two individuals.

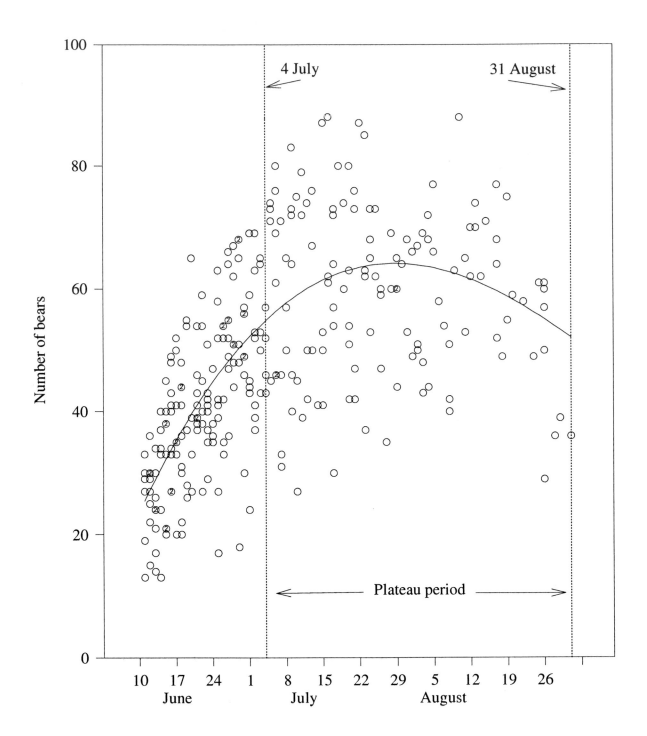

Table 9.2. Three measures of cub production by Trout Creek female grizzly bears used in tests for compensatory effects, Yellowstone National Park, 1959-70.

Year "one"[a]	Recruitment (cubs/adult female)					
	Censused females[b]		Marked females[c]		Available marked females[d]	
	Rate	n	Rate	n	Rate	n
1959[e]	0.63	11	——	——	——	——
1960	0.41	14	0.33	9	0.33	6
1961	0.72	10	0.86	14	0.82	11
1962	——	——	0.82	22	0.87	15
1963	——	——	0.70	27	0.77	13
1964	0.44	15	0.37	27	1.41	17
1965	0.81	7	0.86	28	1.15	13
1966	0.73	14	0.60	25	1.08	13
1967	0.58	8	0.50	28	1.14	14
1968	0.95	13	0.64	25	0.27	11
1969	0.53	18	0.24	25	0.82	17
1970	0.61	17	0.60	25	——	——
Mean[f]	0.62	127	0.60	255	0.92	130

[a] "One" refers to the first of a female's consecutive years under observation.

[b] Mean census count of cubs divided by asymptotic mean census count of adult females in the same year; census counts from Table 9.1; n is the number of censuses in which cubs were distinguished (see text).

[c] Cubs produced in year "one" per marked adult female alive in the same year (see text).

[d] Cubs produced in year "two" per surviving adult female "available" for breeding in year "one"; i.e., without dependent offspring in year "one" (see text).

[e] Adult females were first marked in 1959.

[f] Mean weighted by number of females (n).

June and continuing, in most years, through August (Chapter 3). These censuses revealed a rapid buildup in total ecocenter numbers through June, followed by a plateau in visitation during August (Fig. 9.1). We use, as our measure for possible density-dependent effects, the mean number of bears present during censuses in the plateau phase (4 July - 31 August) of visitation. This measure is hereafter referred to as the asymptotic mean number of all bears, or some age or sex component thereof, to indicate that it reflects aggregation size at or near the seasonal asymptote of aggregation development.

Census data adequate to measuring the asymptotic mean are available for most years and categories of sex and age (Table 9.1). Data for 1963, however, have been omitted from all subsequent analyses because only one census was made after 4 July, the beginning of the plateau phase. In 1962, age and sex were not distinguished in the 4 July to 31 August censuses, so only the mean number of all bears present is available. We therefore estimated the asymptotic mean number of adults in this year

from a regression of mean numbers of adults on the mean total census count over the remaining years (regression equation: total adults = -21.2 + [0.81 x total bears]; n = 10 years; R^2 = 0.74). Finally, the sex of adults was not distinguished in 1959, 1960, or 1962. Thus, for 1959-60, we again estimated the missing data by regressing, in this case, asymptotic mean number of adult males on mean number of adults (regression equation: adult males = -4.01 + [0.52 x total adults]; n = 8 years; R^2 = 0.71). For 1962, because we had only a mean total census count to work with, we estimated numbers of adult males and females by regressing asymptotic mean number of adult females on mean total census count (regression equation: adult females = -1.53 + [0.30 x total bears]; n = 8 years; R^2 = 0.66). The mean number of adult males in 1962 was obtained by subtracting the estimated mean number of adult females from the estimated total number of adults (above). This general approach to estimating missing population data by regression was first employed by McCullough (1981, 1986) and Stringham (1983) on our cumulative annual census figures.

Measurement of Recruitment

McCullough (1981, 1986) and Stringham (1983) estimated per capita recruitment as simple ratios of cumulative annual counts of cubs (or yearlings, etc.) versus cumulative counts for some fraction of the adult population (e.g., all adults, adult males, etc.). This procedure is subject to the same objections raised against using cumulative annual counts to estimate population density (previous section).

We use three methods for estimating recruitment that express cubs produced per adult female in our analyses for density dependence. Each estimate avoids the major pitfalls associated with cumulative counts. The first method, analogous to that of McCullough and Stringham, but using asymptotic mean census counts rather than cumulative counts, estimates annual per capita recruitment as the ratio of mean asymptotic number of cubs to mean asymptotic number of adult females (Table 9.2, column 2). The second method tallies the number of marked Trout Creek adult females alive in each year and divides this number by the number of cubs produced by these females. This method has the advantage, relative to our first method, of focusing on the reproductive performance of resident (Trout Creek-flagged) females. These females were most apt to be influenced by annual variation in aggregation size at Trout Creek (Table 9.2, column 4). Our third method tallies the number of marked adult Trout Creek females alive and without dependent offspring in year "one" and divides this total by the number of cubs produced by these females in year "two" (Table 9.2, column 6). Year "one" refers to the first of a female's consecutive years under observation. The third method has the advantage of focusing not only on resident females, but specifically on those resident females that are available for reproduction; i.e., it excludes the possibility of mistaking a year characterized by a high percentage of females with older, dependent offspring (a good year in terms of reproductive performance) for one in which reproductive success (cub production) was poor. Note that recruitment in column 6 (Table 9.2) is for year "two," whereas recruitment estimates in columns 2 and 4 are for year "one." Thus, for example, cubs per available marked female listed for 1968 (0.27) is comparable to cubs per marked female listed for 1969 (0.24).

UNIVARIATE TESTS USING MEAN ANNUAL RECRUITMENT

McCullough (1981, 1986) and Stringham (1983) both used univariate linear regression to examine the relationship between their annual estimates of population size and recruitment. This analytical method introduces additional problems regarding interpretation of results to those associated with their estimates of density and recruitment. These additional difficulties are discussed in the following section. Here, for the sake of comparing our analyses with theirs, we use the same general statistical technique of regressing annual values of recruitment on density, but use our estimates of these parameters (Tables 9.1 and 9.2). In Table 9.3, we have summarized the results of the regressions.

For each of our three estimates of recruitment, we produced eight regressions. First, we regressed each measure of recruitment on four different asymptotic mean census counts—those for all bears, all adults, adult males, and adult females. The reason for looking at four different measures of ecocenter density is that the most relevant measure will depend on the mechanism of any compensatory effect. For example, if socially dominant animals (adults) did not differentially affect resources available to subordinate nonadults (i.e., foraging was a scramble competition), the asymptotic mean number of all bears may be the most appropriate measure of density. If adults successfully monopolized resource removal, the asymptotic number of adults may be a more revealing measure of effective density. Finally, if infanticide by adult males was important, the asymptotic number of adult males or the sex-ratio of adults may be more revealing. This is the same rationale used by McCullough (1981) for using different estimates of density. In addition, for each of the four measures of density, we computed one regression in which density estimates corresponded to the year of conception for the females whose reproduction was measured (year "t + 1"; prenatal density) and another in which density corresponded to the year of parturition for such females (year "t"; postnatal density). Again, the appropriate analysis will depend on the exact mechanism of any density dependence. For example, the effects of male infanticide would most likely appear in analyses of postnatal, adult male densities, whereas resource competition could exert its effects prenatally by reducing the probability of females entering reproductive condition.

Although negative slopes were the rule among these 24 regressions, none showed significant departures from zero slope (Table 9.3). Those regressions that most closely approached significance ($P \leq 0.15$) were those in which recruitment was measured as ratios of census counts and densities were postnatal (Table 9.3, column 2). These latter results may be best explained by the tendency of females

Table 9.3. Standardized least-squares regression slopes and *P*-values (parentheses) for regressions of six measures of recruitment (columns) on various asymptotic mean Trout Creek census counts (rows) of grizzly bears, Yellowstone National Park, 1959-70.

Class censused for density estimate in year "t"[a]	Cubs/female[b] (census)		Cubs/marked female[c]		Cubs/available marked female[d]	
	Birth in year "t"	Birth in year "t+1"	Birth in year "t"	Birth in year "t+1"	Birth in year "t"	Birth in year "t+1"
All bears	-0.39 (0.27)	-0.11 (0.79)	0.03 (0.92)	0.11 (0.76)	0.25 (0.52)	0.26 (0.49)
All adults	-0.51 (0.13)	-0.29 (0.48)	-0.21 (0.57)	0.17 (0.64)	-0.33 (0.39)	-0.25 (0.51)
Adult males	-0.49 (0.15)	-0.40 (0.32)	-0.26 (0.46)	0.09 (0.80)	-0.51 (0.16)	-0.44 (0.24)
Adult females	-0.50 (0.14)	-0.15 (0.72)	-0.11 (0.76)	0.26 (0.47)	-0.04 (0.91)	-0.03 (0.95)

Note: *P*-values within shaded area were those most closely approaching statistical significance; see text.

[a] Mean (asymptotic) numbers of bears counted in censuses at Trout Creek between 4 July - 31 August in year "t" (values in Table 9.1).

[b] Computed as the asymptotic mean number of cubs counted in census divided by the asymptotic mean census count of adult females (values in Table 9.2, column 2).

[c] Computed as the number of marked, Trout Creek females alive in each year divided by the number of cubs these females produced (values in Table 9.2, column 4).

[d] Computed as the number of cubs produced per marked adult female by females without dependent offspring in the previous year (values in Table 9.2, column 6).

with cubs, or the cubs only, to avoid the Trout Creek ecocenter during the breeding season (Chapter 5) and to do so more strongly when the aggregation was large. Recruitment (now hypothetically) would then have been more severely underestimated by simple census count ratios when many bears aggregated rather than depressed by any density-dependent effects.

MULTIVARIATE ANALYSIS OF INDIVIDUAL REPRODUCTIVE PERFORMANCE

The preceding analysis has several drawbacks. First, each year contributes one data point to the regression, regardless of the number of females or censuses used to estimate annual reproduction or density. With so few years contributing data, years with small samples, and hence large errors in estimating either of these parameters, could mask the true underlying relationship. Second, because a given female is often measured for reproductive performance in two or more years, there may be significant between-year dependence (in a statistical sense) among annual values of per capita recruitment. Such dependence could similarly misrepresent the true relation between reproduction and density. Finally,

a single independent variable (a single measure of population or subpopulation density) was considered in each regression, yet several potentially intercorrelated factors, such as female age and two or more component age/sex-specific population densities, may simultaneously influence recruitment.

In this section, we present the results of an analysis that corrects these defects to some extent. We used analysis of variance in which variation in cub production by individual females available for reproduction (see above) is modeled as a function of female identity, female age, and asymptotic mean numbers of nonadults, adult females, and adult males (Extra sums of squares F-test; Draper and Smith 1981). In this analysis, we attempt to account for as much of the variation in individual cub production as possible, first by removing variation attributable to differences between individual females, then by removing variation attributable to female age, and finally, by removing variation attributable to the size of various age and sex components of the population. Sample size is too small to consider the effects on recruitment of population densities in both the year of conception and year of birth within the same test. Thus, we present the results of two analysis of variances (ANOVAs), one of which measures density as the asymptotic mean

Table 9.4. Analysis of variance in female grizzly bear fecundity in relation to female identity, female age, and densities of three age classes in the year of conception, Trout Creek ecocenter, Yellowstone National Park, 1959-70.

Source of variation	DF	SS	MS	F-ratio	*P*-value	Coeff.
Total (cor)	61	87.742				
Regression	21	53.003				
Individual female	17	50.178	2.952	4.949	0.000	
Female age	1	0.001	0.001	0.003	0.960	-0.005
No. nonadults	1	2.626	2.626	4.402	0.042	0.066
No. adult females	1	0.198	0.198	0.331	0.568	-0.083
No. adult males	1	0.000	0.000	0.001	0.979	0.059
Residual	40	22.068	0.596			

Coefficient of determination = 0.604

numbers of nonadults, adult males, and adult females in the year of conception for available females, and the other of which measures the same quantities for the year of birth. In both ANOVAs, the effect of female identity on fecundity (i.e., the statistical dependence arising from multiple measurements on individuals) was modeled using $n - 1$ indicator variables, where n is the number of different females in the data set. All other potential sources of variation (female age, nonadult density, etc.) are represented by single variables. Note that the age and sex categories explicitly considered in these tests (nonadults, adult males, and adult females) do not exactly correspond to those tested in the single variable regressions of the previous section (all bears, all adults, adult males, and adult females). This is because, when considered simultaneously in a single test, age and sex components of population density must be non-overlapping.

The results of both analyses, one using a conception year and the other a birth year density index, showed that female identity was a significant source of variation in recruitment ($P < 0.001$; Tables 9.4 and 9.5). This block of variables accounts for 60% and 73% of total variation in recruitment in the conception-year and birth-year analyses, respectively. This result justifies our concern regarding between-year statistical dependence in the univariate regressions, as discussed above, and suggests that biologically significant heterogeneity existed among Trout Creek females in terms of reproductive potential (see Kolenowsky [1990] for a similar result in black bear). But there is again no evidence of any density-dependent effect on recruitment. The only F-test that approached, let alone attained, statisti-

cal significance involved nonadult density in the year of conception, and this effect was positive; i.e., the implication is of greater recruitment in years following high densities of nonadults. This is clearly not a compensatory effect of the sort envisioned by McCullough and Stringham.

DISCUSSION

We have focused our interpretation of compensatory processes on McCullough's paper (1981) because he applied his results more specifically to management of the Yellowstone population than Stringham (1983), who was more cautious in his approach and more concerned with the role of adult males in grizzly bear population dynamics, generally, than with the status of the Yellowstone population. Both Stringham (1983) and McLellan (1995 [in press]) have effectively disputed McCullough's conclusions on statistical grounds. We have addressed the suitability of the data that he used.

McCullough's (1981) apparent discovery of strong compensatory processes in our previously published data led him to propose a rather complex model of population dynamics for the Yellowstone grizzly bear. This model featured, in addition to density-dependent recruitment, density-dependent birth sex ratios and density-dependent sexual biases in juvenile mortality, with consequent oscillatory behavior in male age structure, adult sex ratio and population size. Specifically, he proposed that recruitment varied inversely with numbers of adult males and that oscillations in numbers of adult males were driven by birth sex ratios and juvenile mortality that were more strongly male-biased at low

Table 9.5. Analysis of variance in female grizzly bear fecundity in relation to female identity, female age, and densities of three age classes in the year of birth, Trout Creek ecocenter, Yellowstone National Park, 1959-70.

Source of variation	DF	SS	MS	F-ratio	*P*-value	Coeff.
Total (cor)	72	98.000				
Regression	33	71.935				
Individual female	29	69.833	2.408	4.037	0.000	0.008
Female age	1	0.007	0.007	0.011	0.916	0.083
No. nonadults	1	1.379	1.379	2.312	0.136	-0.262
No. adult females	1	0.262	0.262	0.439	0.512	0.096
No. adult males	1	0.454	0.454	0.762	0.388	
Residual	39	22.068	0.596			

Coefficient of determination = 0.734

population density. Thus, at low population density, recruitment was high (because fewer adult males were present), more males than females were recruited to reproductive age, and numbers of adult males increased rapidly as the population grew. At high densities, recruitment declined (because many adult males were present), the sexual bias in recruitment of adults was muted (because birth sex ratios and juvenile survivorship were no longer so strongly male-biased), and the population was returned to a state of lower density, with relatively few adult males present. He proposed that the duration of a complete cycle was about 10 years (roughly the duration of our study) and that such periodicity was a consequence of the 4.5 years required to attain sexual maturity.

McCullough's model led him to two related conclusions that we feel are quite wrong. First, he concluded that "while they [the Craighead research team] attribute the lower [reproductive] rate [after 1967] to . . . closure of open pit garbage dumps, in actuality, it can be accounted for entirely on the basis of the high number of adults in the population." Second, he proposed that the Yellowstone population was in general more resilient to adversity by virtue of its ability to mount a strong compensatory response as population size declined. He went so far as to suggest that a controlled "cropping" of adults to 70 or fewer individuals within the Park could have the benefit of introducing stability in numbers, sex ratio, and age structure, but that, in the absence of such management ". . . a series of years with below average litter sizes is not sufficient cause for despair . . ." since periods of markedly high mortality or poor recruitment or both are

expected in a strongly oscillating population. He qualified this latter remark by acknowledging that other causes which are not "self-correcting" may be responsible for poor reproductive performance and, therefore, that an index of population size is desirable as a check against this possibility. Nonetheless, the comforting and potentially complacent general notion is planted that even very "bad" years, such as the tremendously increased mortality in 1968-70 followed by reduced litter sizes in 1971-73, are expected and normal features of population behavior and not necessarily cause for concern or management action.

The crucial issue here is whether strong density-dependent mechanisms were operating in 1959-70. Our analyses show clearly that, on this issue, McCullough was wrong. When appropriate measures of density and reproductive performance are used, and despite 2- to 3-fold variation in densities of bears at the Trout Creek ecocenter, there is no evidence that strong compensatory mechanisms operated on recruitment during the 1959-70 period. An increase in numbers of adult males in our cumulative census counts (a result of increased census efficiency) created the illusion of high population density following the initiation of ecocenter closures in 1968, as recruitment ecosystem-wide declined. In actuality, average daily census counts at Trout Creek in 1968 and 1969 (Table 9.1) suggest numbers of bears may have declined, not only at Trout Creek, but also ecosystem-wide. The rebound in bear density at Trout Creek in 1970 was due to bears aggregating at the only remaining ecocenter and was *not* an increase in overall numbers throughout the Ecosystem. The latter view of ecosystem population density is much more consistent with known patterns of mortality than McCullough's. More bears died in 1967-70 (144) than in the previous eight years combined (134) (Chapter 11 and Craighead *et al.* 1988b). It is unreasonable to suppose that a population of only 300 or so bears, with females maturing at age five, could sustain this level of mortality and still show significant increases in numbers of adults over a three-year period.

The same closure-related phenomena that invalidate the cumulative population counts used by McCullough and Stringham as estimates of population density and that created the illusion of strong compensation also account for many of the specific features of McCullough's oscillatory mechanism. For example, his (and Stringham's) conclusion that adult males were the main cause of suppression of recruitment is, in our view, a direct artifact of the greater movement and visibility of adult males after closure of the lesser ecocenters. The number of males appeared to rise just as the population experienced heavy mortality and was stressed nutritionally. McCullough's purported sex-ratio shift to an excess of males at high population density is explained in the same way. In brief, no feature of McCullough's oscillatory model survives close scrutiny. Evidence of strong compensatory effects due to adult males or any other population component was not apparent in 1959-70.

We do not argue that recruitment (or mortality) is fundamentally density-independent in grizzly bear populations. It seems self-evident that over some (larger) range of densities and (greater) time interval, and given a relatively constant total food resource, compensatory effects similar to those claimed by McCullough would be apparent (but not necessarily his oscillatory population dynamics). Our differences with McCullough center on his invocation of strong compensation to explain specific population phenomena during the 1959-70 period and, in particular, his combining of data from before and beyond the drastic reduction in resource availability created by the closure of the ecocenters.

Our own necessarily general view of the interaction between density, recruitment (or survivorship), and resource availability in Yellowstone during this time is summarized in Fig. 9.2. In this scheme, recruitment and survivorship decline over some broad range of density under both high and low resource availability (e.g., upper and lower curves); but at any given density, recruitment is lower and mortality higher when resources are scarce. For simplicity, we have assumed linear density effects in Fig. 9.2. But the same general point applies if, as many think more

reasonable (Taylor 1995 [in press]), density effects are nonlinear and appear only over a narrow range of higher densities. Furthermore, recruitment (or survivorship) at low bear density and low resource availability may still be less than that observed at very high bear densities when resources are abundant (compare R_x and R_y in Fig. 9.2). When the upper curve is equated with conditions of resource abundance prevailing before ecocenter closures, and the lower curve with those after closures, this figure illustrates the reasoning behind our earlier statement, quoted by McCullough (1981), that "A depressed reproductive rate probably would not be compensated by an increase in yearling survival as long as population recruitment was subjected to unusual population stresses" (Craighead *et al.* 1973b, p. 21) or our later conjecture that " . . . with a major food source removed . . . the bears may have come under increased rather than decreased stress at lowered population levels" (Craighead *et al.* 1974, p. 17). This is not a rejection of compensation, per se, but a recognition that both density *and* resource abundance changed drastically after closures, and that there is no simple answer as to how these variables interacted (or now interact) in their influence on the Yellowstone grizzly's reproduction and survival. The confounding effects of natural variation in brown bear food abundance has been invoked more generally to explain the absence of conclusive demonstrations of density effects in this species (McLellan 1995 [in press]; see also Garshelis 1995 [in press] for black bear).

Beyond the error of using the wrong data to test for compensation during 1959-70, we believe McCullough (1986) erred in implying that his results were supportive of Cole's (1974) belief that recruitment increased compensatorily after 1970 under conditions of reduced bear density. Cole's contention is equivalent to imagining that recruitment moved left along the *upper* curve (Fig. 9.2) as density declined from y to x and, therefore, was significantly higher after 1970 than before. Our view is that, if anything, recruitment after 1970 moved left along the *lower* curve and that reproduction under low density after 1970 was very plausibly less than that under the highest density before 1970. In fact, McCullough made the same logical error in his analysis of not distinguishing years with high (1959-67) versus low (1968-70) per capita resources. We avoided this problem by restricting our attention to the Trout Creek ecocenter, which was intact over this entire period.

In summary, we contend that our analysis of possible compensatory effects removes this phenomenon as a defensible explanation for the reduced

Fig. 9.2. Conceptualization of interactions between grizzly bear population density, recruitment, and resource availability.

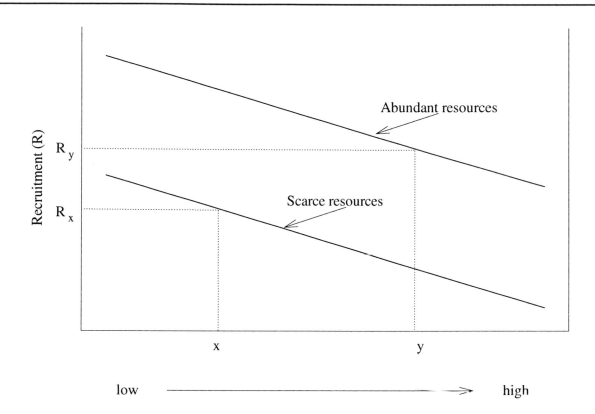

reproductive rates recorded in 1968-70 following ecocenter closures. Furthermore, neither the analyses in this chapter, nor our observations at Trout Creek, support the specific hypothesis that infanticide by adult males is sufficiently frequent in grizzly bear populations to have significant population level (regulatory) effects. With the demise of the strong compensatory effects theory, gone also is the empirical support for the notion that the Yellowstone population was and is highly resilient to reductions in animal number (density) that result from increased mortality. Given the fact that the population now subsists on a resource base much reduced from that before 1970 (see Food Habits, Chapter 16), such notions can dangerously misdirect population recovery endeavors.

SUMMARY

Animal populations may make reproductive responses to their own densities, and these responses can be important when a population has declined drastically or has exceeded the carrying capacity of its habitat. These responses are difficult to detect. We found no compensatory reproductive responses in the grizzly bear population we studied from 1959 to 1970.

McCullough (1981, 1986) and Stringham (1983), using our cumulative annual population counts, first published in 1974, claimed that strong compensatory responses to changes in bear density were operating in the ecosystem population during 1959-70. McCullough featured this result in his exculpation of ecocenter closure as the major cause of reduced reproduction during 1968-70 (and immediately after). He also used his findings to promote an optimistic view of the ability of the Yellowstone population to rebound from disturbance.

We considered the data used by McCullough and Stringham to be inappropriate for their analyses. We applied more appropriate measures of density and reproduction to reevaluate the question of density compensation among the 1959-70 Trout Creek subpopulation. We used the average number of individual bears observed per census for the Trout Creek subpopulation, rather than a cumulative annual count of individuals for the entire Yellowstone Ecosystem. This permitted us to measure for density effects each year in terms of the asymptotic mean daily census count of the entire Trout Creek aggregation or age-sex components thereof. Density measures were evaluated both in the year of conception and the year of cub production. Three measures of recruitment were examined. The strongest was the number of cubs produced per female among marked females that were without dependent offspring and available to reproduce in the previous year. Finally, the relation between all combinations of the density and recruitment estimates was evaluated in simple regressions of recruitment on density. We subjected the data to multivariate analysis in an attempt to isolate any effect of density on recruitment from the effects of other factors (like female age) that might affect reproduction and obscure a density effect in univariate regressions.

Despite analyzing appropriate measures of density and recruitment and making a fairly exhaustive search for an association between these quantities, we found no support for McCullough's claim that the reduced recruitment following ecocenter closures could be explained by changes in bear density. The data instead support our earlier and current contention that recruitment declined and mortality increased because bears were severely stressed nutritionally as a direct result of ecocenter closure. We acknowledge that compensatory behavior is likely to be an important feature of grizzly bear population dynamics when considered over a sufficiently long time span, wide range of densities, and under constant resource availability. But these effects cannot be as strong as McCullough claimed, and we must guard against undue optimism regarding the ability of the Yellowstone population to rebound from past or future disturbances.

10
Food Habits and Feeding Behavior of the Grizzly Bear Population

INTRODUCTION

We have shown that localized concentrations of food available for lengthy periods encouraged grizzly bears to aggregate. Such conditions affected bear social and reproductive behavior and population dynamics. These "ecocenters" (Craighead 1982, Craighead and Mitchell 1982) provided a predictable source for the nutritional needs of the grizzly bears using them. Nevertheless, ecocentered bears also relied on seasonal and more dispersed sources of food. In this chapter, we discuss major food categories or sources of food available to and used by the grizzly bear population inhabiting the Yellowstone Ecosystem from 1959 to 1973. We also examine the food habits for the periods 1973-74 and 1977-81 to detect changes in use of the food resources following closure of the dumps. Finally, we focus on the relative annual abundance of these resources and how the grizzly bears used them during the preclosure period.

The distribution of grizzly bears inhabiting two million hectares (five million acres) of the Yellowstone Ecosystem (Fig. 4.1) has, for years, been observed to change seasonally. The major variables dictating shifts

Ecocenters (garbage dumps) provided a predictable source for the nutritional needs of the grizzly bears. Bears also relied on seasonal and more dispersed sources of food. *(Photo by John J. Craighead)*

To understand grizzly bear feeding habits, thousands of man-hours were devoted to observing marked and radio-collared grizzly bears foraging. These observations aided the interpretation of fecal analyses.

(Photo by John J. Craighead)

in distribution are the quality and quantity of the bears' food resources. Grizzly bears are highly efficient omnivores, capable of selectively maximizing their nutrition through learned behavior and traditional migrations to proven food resources. They are also highly opportunistic and respond rapidly to new food sources. The more productive sites, remembered over the course of a long life span, become centers of activity within the life range of an individual and, in some instances, constitute traditional seasonal rendezvous for a large segment of the population (Craighead 1982).

Diversity of food resources is especially important in maintaining an adequate nutritional level in a grizzly bear population. The availability of the more concentrated and preferred plant and animal food categories normally fluctuates unpredictably from year to year. Thus, the dependability of the total food resource is subject to oscillations in both time and space. No major food category—e.g., berries, pine seeds, small rodents, ungulates—is

available in quantity year after year. Some years the bears feed well, other years poorly. Prior to 1970, the open-pit garbage dumps in and around Yellowstone Park represented an exception to the rule. Garbage was a highly dependable source of high quality food—a predictable resource not subject to the annual perturbations in availability.

SEASONAL FEEDING PATTERN

We examined the seasonal feeding habits and food preferences of grizzly bears through fecal analysis and direct observation of their foraging activity. Thousands of hours were devoted to observing color-marked and radio-collared animals. From these observations, we located sites where grizzlies excavated food plants and rodents, fed on ungulate carcasses and deposited scats at the feeding sites. Data gathered from these procedures provide a basis for understanding food preferences, seasonal feeding, and the relationship between food resources and seasonal distribution of bears.

The feeding habits of the Yellowstone grizzly bear population during the period, 1959-70, exhibited both seasonal and annual variations. Intensity of use of various food categories, as evidenced in the field and in the feces, varied according to plant phenology, long-term cyclic phenomena of plants and rodent populations, and sizes of the elk and bison herds, as well as the seasonal cyclic behavior of rodents, of elk, and of the bears themselves. Data from scat analysis for the 1968-71 period (Fig. 10.1) and data presented by Knight *et al.* (1982) for the period from 1977-81 indicate a similar pattern of seasonal use of native food categories. This pattern also closely resembles the seasonal feeding pattern described for western Montana (Craighead *et al.* 1982), suggesting that over extensive time periods, seasonal and annual variations are masked by a general feeding trend characteristic of the species. The bear's foraging begins when it leaves its winter den. From that event until it enters a den again in the fall of the year, it is basically preparing itself for half a year of winter sleep (Craighead and Mitchell 1982, Nelson *et al.* 1983). That preparation requires the bear to travel "continuously," throughout a large home range generally encompassing many vegetation types and all climatic zones of the region. Thus, the seasonal feeding habits of grizzly bears are directly related to their seasonal movements. The Trout Creek subpopulation that annually aggregated at the Trout Creek garbage dump is a case in point. As we have shown (Chapter 5), these bears traditionally moved from winter dens throughout the Park and beyond to the Trout Creek ecocenter. This

Fig. 10.1. Seasonal feeding habits determined from the percentage occurrence of food items in grizzly bear scats, Yellowstone National Park, 1968-71.

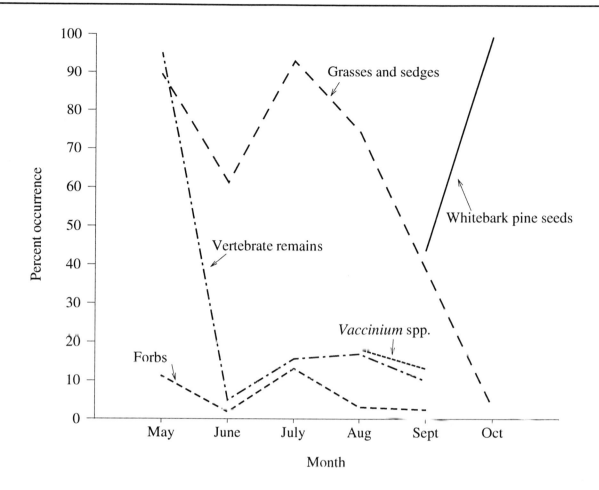

movement began as early as March, with grizzlies arriving at Trout Creek over a period of many weeks and reaching peak numbers in July. Bears denning in the vicinity of Trout Creek were the first to arrive; those denning at the periphery of the Park and beyond arrived later. The late arriving bears had to traverse mountain ranges and high forested plateaus where snow depth in early spring could range from 1.5 to 6 meters (5-20 feet). Snow depth was a formidable barrier to travel, even when the snow was crusted. There are no low elevation riparian routes to Trout Creek. Only those grizzlies that denned in the Central Plateau (Fig. 10.2) near the geographic center of the Park (Hayden Valley, Lake, and the Firehole thermal areas) had early spring access to the winter-killed ungulates of that area (Craighead and Craighead 1972a, 1972b). These bears, "hibernating" in the high central portion of the Park, left their dens as early as March, traveling when the snow was crusted, keeping to the bare south-facing ridges, or following river courses. Adult male bears and subadults were the first to move to lower elevations. Females with cubs of the year

emerged from dens in late April to late May and also moved to lower subalpine sites where snow was light and receding.

Winter-killed ungulates were an extremely important item in the diet of bears emerging from winter sleep. Grizzlies encountered this food source as they moved toward the ecocenters at Trout Creek and Rabbit Creek. Ungulate carcasses were largely localized in two distantly separated areas of the Park—the Central Plateau area and the Yellowstone-Lamar river valley area. Those grizzlies that denned in the northeastern part of the Park and Ecosystem had early access to winter-killed ungulates in the latter area, while those bears that denned in the Central Plateau had exclusive access to ungulates that died there (Fig. 10.2). Altitudinal barriers, with their great snow depths, dictated the availability of carcasses on the Central Plateau.

The areas of winter range within the Park can be broadly described as low elevational, lying within the transition climatic zone, and high elevational, confined to the subalpine zone. The small interior elk herds, and the Mary Mountain and Pelican bison

Fig. 10.2. Major thermal areas and topographic features in Yellowstone National Park.

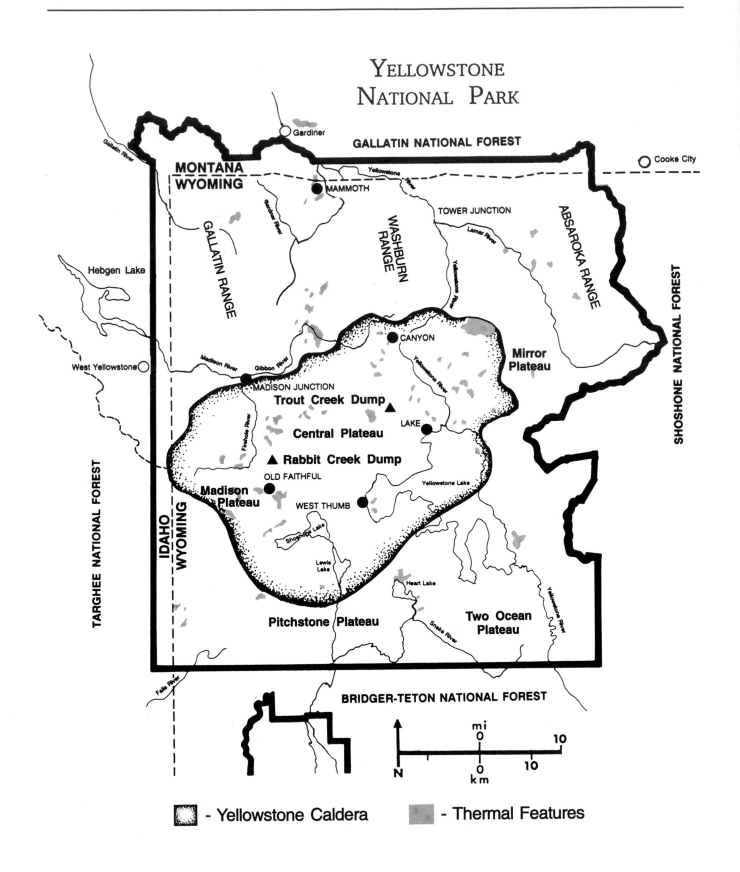

herds wintering on the high elevational Central Plateau could do so only because they occupied relatively snow-free thermal sites (Fig. 10.2). In winter and early spring, those sites were effectively isolated from the rest of the Park and Ecosystem by great snow depths that extended for many miles in all directions. For example, the snow depths recorded at the Lake meteorological station in 1983 were 72 centimeters (30 inches) in January, accumulating to 117 centimeters (46 inches) in March and 112 centimeters (44 inches) in April. Thus, the snow accumulation was at a maximum in March and April when bears were moving away from their winter dens. Some of the higher elevations annually received from 500 to 1000 centimeters (200 to 400 inches) of snow. Snow depths in these higher elevations made them virtually impassable to ungulates or bears. Grizzlies foraging in early spring (April and May) could move from peripheral areas of the Park and Ecosystem to the Central Plateau with its thermal areas and ungulate herds only by way of the Madison River drainage (Fig. 10.2). This route gave access to winter-killed elk and bison in the Upper Firehole River and Gibbon River basins. But we detected no long-distance movement of bears to the Central Plateau until most of the snow had receded.

The large number of grizzly bears we recorded feeding on carcasses on the Central Plateau were members of the Trout Creek and Rabbit Creek subpopulations, and those grizzlies encountered this early season food source as they converged on the Trout Creek and Rabbit Creek ecocenters. Thus, winter-killed ungulates of the Central Plateau of the Park were used almost exclusively by grizzly bears that denned on the Plateau or in the immediate vicinity (see Carrion-Feeding later in this chapter). Identification of marked bears at carcasses confirmed this. We recorded 178 grizzly bear sightings on 99 ungulate carcasses over a 13-year period. One grizzly returned to a carcass at least nine times over a 15-day period. Aggregations of bears occurred at some of these carcasses, and the grizzly bear social order was generally in evidence where two or more carcasses were in close proximity. Just as at the Trout Creek ecocenter, social behavior served to increase foraging efficiency by allowing numerous grizzlies to share a common food source. In the fall of 1968, we recorded up to 23 grizzly bears feeding on a single male bison (*Bison bison*) carcass. When several grizzlies aggregated in this manner, a winter-killed animal was consumed in a few days; however, some bears occasionally revisited the bones and the hide. Use of carcasses in April and May subsided rapidly

Social behavior increased the foraging efficiency of the population by allowing numerous grizzlies to share a common food source. As many as 23 grizzlies were observed feeding on a single bison carcass.
(Photo by John J. Craighead)

Meadow voles were an important item in the diet and were especially vulnerable as the snow cover receded. *(Photo by John J. Craighead)*

as garbage became available at Trout Creek and Rabbit Creek.

Another early spring food was the over-wintering rodents such as meadow voles (*Microtus* spp.), deer mice (*Peromyscus maniculatus*), and northern pocket gophers (*Thomomys talpoides*). High populations of these rodents occurred periodically, and they were especially vulnerable as the snow cover melted. A female and three yearlings were observed feeding for several weeks almost exclusively on *Microtus* spp. Bears were also frequently observed preying on pocket gophers, and the widespread and numerous scrapings and diggings were evidence of the importance of this rodent in the early spring diet. Consumption of both small and large vertebrates was greatest in early spring (April-June). During this period, grizzlies subsisted primarily on a meat diet until the snow had receded.

Grizzlies supplemented an early spring meat diet with the early emerging sedges and grasses. In May and June, as ungulate species left winter ranges and moved to higher elevations, the bears tended to follow the same pattern, feeding primarily on the emerging grasses, sedges, and forbs. Adult males, the subadults of both sexes, and females without

offspring were generally solitary foragers. Females with cubs, yearlings, or 2-year-olds foraged as family groups. A female with cubs would sometimes form a close association with a similarly aged family, which then traveled and foraged as a unit. (See Adult Female Bonding, Chapter 8).

In early June, elk began giving birth to calves in the temperate and subalpine parklands. Calving sites tend to be traditional, the elk returning to them year after year. Grizzlies whose home ranges encompassed these calving areas appeared to locate them from memory and by scent. Some grizzlies followed elk bands as they migrated. Other individuals moved to calving areas as though recalling them from past experience. The calves were highly vulnerable to grizzlies for a relatively short period of time.

Soon after calving, the cows and their offspring moved to higher elevations, their movements influenced by the recession of snow and the emergence of plants. Grizzlies followed the same general pattern so that, by July, they were feeding on the grasses, sedges, and forbs in both the subalpine and the alpine zones. Also throughout June, grizzlies were undoubtedly "fishing" for spawning cutthroat trout (*Oncorhynchus clarki*) in Pelican Creek, Beaver Dam Creek, and other small streams feeding Yellowstone Lake. Use of the fishery resource was not documented quantitatively; but while use may have been high for certain individual animals, we believed it to be minor for the population, compared with the use of the other food resources.

From late June through July, the alpine and subalpine zones were used extensively as sources of biscuit root (*Lomatium* spp.), yampa (*Perideridia gairdneri*), springbeauty (*Claytonia* spp.), and other succulent and nutritious tubers, bulbs, and greens. Insects were also important diet items during this period. The bear's selection of such insects as moths, beetles, ants, and even earthworms, is partially (but not entirely) related to their high protein and fat content. Ants (Formicidae) constituted the majority of the invertebrates in the 487 grizzly scats collected from the Yellowstone area during 1968-71.

In the Scapegoat Wilderness, a moth of the family Noctuidae occurred in 66% of the scats and averaged 65% of the mass of scats in which it occurred (Sumner and Craighead 1973). Bears fed on this moth in June through mid-July. We did not observe or record similar use of moths in Yellowstone, but it no doubt occurred, perhaps to a lesser degree than we recorded in the Scapegoat study. In 1986, scat analysis and observation showed that grizzlies were using moths extensively in the southern portion of the Yellowstone Ecosystem (Mattson *et al.* 1991b). Whether this represented a

In early June, grizzly bears preyed on elk calves for a period of several weeks. *(Photo by John J. Craighead)*

major change in feeding behavior is discussed in Chapter 16.

Because feeding, such as that on moths, tends to be aggregated and highly localized, so too is the deposition of feces. When isolated, high-elevation feeding sites are located, it follows that collection of feces for analysis becomes highly selective for the diet items that have drawn the bears, and care must be taken to avoid overemphasizing their importance in the overall diet. This is especially true of any item of diet that attracts and holds bears for a considerable period of time.

As August approached, the huckleberries (*Vaccinium* spp.) and buffaloberries (*Shepherdia canadensis*) began to ripen. Grizzlies moving within large, but undefended home ranges (Craighead 1976) traveled miles to use this energy source which, in years of peak abundance, was exploited until snow covered the subalpine country. When berries were abundant, bears tended to feed on them almost exclusively and to gain weight rapidly. In years when berry crops were poor, grasses and forbs alleviated the food shortage, but bears did not gain weight on this relatively low-energy diet. At such times, the seeds of the whitebark pine (*Pinus albicaulis*) represented a critical energy source. Grizzlies moved to the subalpine areas of their home ranges to feed on pine seeds, often using them through September, October, and, in some instances, mid-November. Following years with a good seed crop, bears foraged in the spring for cones and seeds cached by rodents and birds. We radiotracked griz-

zlies in the fall that moved more than 80 air kilometers (50 mi) to feed on the seeds of whitebark pine (Craighead 1976). Unlike the cones of most conifers, the ripe cones of whitebark pine do not open in the fall and release seeds.

Pine seeds were acquired by bears in two ways; throughout the fall (September - October), mature cones were stripped from the smaller trees and from low-hanging branches, the entire cone chewed, and the seeds, resin, bracts, and woody tissue consumed, en masse, whereas use in late fall and early spring consisted mainly of locating and excavating bird and rodent caches, which we found to be both large and numerous in Yellowstone. Caches made by Clark's nutcrackers (*Nucifragas columbiana*) and small rodents, especially golden-mantled ground squirrels (*Spermophilus lateralis*) and red squirrels (*Tamiasciurus hudsonicus*), appeared to be located by scent. The fact that many feces collected in the fall were 100% pine seed remains suggests that some seed caches located by grizzlies were quite large. For example, researchers have estimated that a single nutcracker may store between 32,000 and 98,000 seeds in a season (Hutchins and Lanner 1982, Tomback 1989). Nutcrackers open pine cones, pull out seeds, store them in a sublingual pouch (up to 150 seeds at one time), and carry them to cache sites. We found seed caches as far as 16 to 24 kilometers (10-15 mi) from the nearest seed source. Ground squirrel and red squirrel winter caches often exceeded several hundred seeds. Whitebark pine produces among the largest of conifer seeds, and

Pine seeds were preferentially sought by grizzly bears even though years of abundant crops were rare. The seeds helped to provide the high energy diet necessary for the grizzly to enter winter sleep with a heavy layer of fat. *(Photo by John J. Craighead)*

several caches of 150 or so provide a nutritious meal for a grizzly bear.

Thus, in the Yellowstone Ecosystem, the seeds of whitebark pine can provide the high energy diet necessary for the grizzly to enter winter sleep with a heavy layer of stored fat. Squirrel caches also enable the bears to regain weight rapidly in early spring following a fall season of pine seed abundance. Bumper seed crops were not frequent, however, occurring only twice throughout Yellowstone over the 12-year period of this study (see Figs. 10.7 and 10.8).

Forcella and Weaver (1977) showed a 6- to 8-year cone and seed production cycle, with great variation in cone crop abundance in any single stand, between stands, and within a geographic region. They found that seed crops varied from 0 to 6 grams per 6.5 square centimeters (0-0.2 oz per square inch) of forest canopy per year.

Generally, whitebark pine occurs only at subalpine and timberline elevations and composes a small proportion of the total mountain forests of western Montana, Idaho, and northwestern Wyo-

ming. In the Bob Marshall and Scapegoat wilderness areas, whitebark pine forest habitat types compose approximately 11% of the total vegetation (Craighead *et al.* 1982). The proportion for the Yellowstone Ecosystem is probably comparable. Since *P. albicaulis* is but one species composing the whitebark pine forest habitat types (Pfister *et al.* 1977), it obviously is not an abundant food species in grizzly bear habitat and, as a source of food for grizzlies, is both limited and undependable. Nonetheless, the fact that fecal samples, if they contained any pine seed remains at all, were almost entirely composed of them, suggests that grizzlies selectively sought them. They were the most important diet item derived from a single plant species (Craighead *et al.* 1982).

As winter approached, but before entering "hibernation," grizzlies tended to feed in the vicinity of their dens. When their foraging movements decreased, so did the variety of their diets.

We observed grizzly bears, and had reports of others, that emerged from winter dens during periods of thaw, but feeding was not observed. Nor-

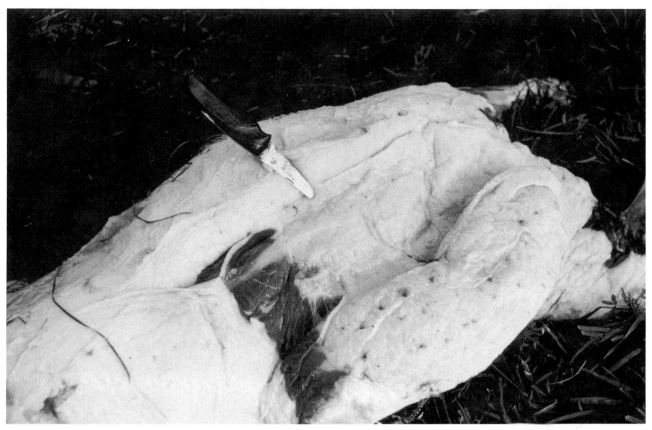

Bear carcass showing distribution of fat. To acquire layers of fat, the grizzly must concentrate foraging efforts on nutrient-rich resources. *(Photo by John J. Craighead)*

mally no food is ingested during the winter months, and we found no defecations in or around winter dens that would suggest it. With emergence from dens, the feeding pattern is reactivated, and in Yellowstone, the meat of vertebrates may be the first food consumed.

We have described the general feeding pattern. Next, we attempt to evaluate the importance of food categories in the grizzly bears' diet through fecal analysis.

FEEDING HABITS AS DETERMINED FROM FECAL ANALYSIS

The plant and animal food resources used by grizzly bears in the Yellowstone Ecosystem during the 1960s and 1970s have been evaluated by several researchers (Sumner and Craighead 1973; Mealey 1975, 1980; Knight *et al.* 1978, 1980, 1981, 1982; Craighead 1980b; Craighead *et al.* 1982). Results from fecal analyses completed in the Yellowstone Ecosystem by Mealey (1980) and Knight *et al.* (1982) for the decade immediately following dump closures permitted comparisons

with our data for the preclosure period. An overall comparison of the bears' preclosure food habits with their food habits after closure of the dumps (Mattson *et al.* 1991a) is presented in Chapter 16. Fecal analysis is the most widely used method of determining food habits, but techniques have been modified and improved to such an extent that a detailed discussion is in order before presenting the results of our analyses.

We have indicated in Chapter 3 (Methods) that we would present certain specific techniques at appropriate places throughout the text, when we believed it necessary for a clear understanding of how the information presented was derived. This is such a case. We obtained information on food habits by analyzing the contents of 487 fecal samples collected May through October from 1968 through 1971. The general analytical sequence was that described by Sumner and Craighead (1973) and Craighead *et al.* (1982). Scats were air-dried, weighed, and then rehydrated in a jar of formalin and water. They were removed to trays, dissected, and the various components identified to species, when possible. Percents of scat volume comprised by each component were estimated.

Bear feces were collected and analyzed to determine the bear's feeding habits and seasonal use of plant and animal categories.
(Photo by John J. Craighead)

Fecal items fell into 10 categories (ants, *Cirsium* spp., forbs, *Fragaria* spp., garbage, grasses and sedges, unidentified invertebrate remains, *Vaccinium* spp., vertebrate remains, whitebark pine seeds), and indices were established for evaluating the relative role of each food category in the overall diet of the bears. The indices are similar, in part, to those applied in our earlier studies (Craighead *et al.* 1982) and adopted by other researchers (Mealey 1980, Knight *et al.* 1982), but here we have redefined and extended them to reflect more realistically the relative dietary and energy contributions of the food categories. Values for Frequency of Occurrence Percent were calculated for a food category by dividing the total number of fecal samples (487) into the number of samples from which that category was identified (x 100). Average Fecal Volume Percents were determined by dividing the number of fecal samples from which a food category was identified into the sum of the percents of volume it comprised of each sample. These measures, Frequency of Occurrence Percent and Average Fecal Volume Percent, provided different information about how bears used their food resources. The former, although open to some sampling bias, indicated how routinely foods were ingested, whereas the latter provided some measure of the quantity consumed. In the absence of rigorous laboratory analysis to relate the intake volumes of foods commonly consumed by wild bears to the volumes of their remains in scats, the "importance" of foods has been consistently expressed as some function of frequency and volume of their remains in feces. We combined the two measures by multiplication to derive a Composite Fecal Index (CFI) for each food category, a reconstruction of the contribution of each to the total sampled fecal mass. Mealey (1980) applied a similar procedure to calculate his values for Percent of Diet Volume (see Table 1, Mealey 1980).

Problems with Fecal Analysis

Most researchers have been aware that fecal remains provide an inadequate representation of the relative importance of food items in the total diet of grizzly bears (Sumner and Craighead 1973, Mealey 1980, Craighead *et al.* 1982, Knight *et al.* 1987). Scats are evidence of what is excreted, not what is ingested. So, for example, the relatively low digestibility of grasses guarantees that their fibrous remains compose a larger volume of what is excreted from the bear's digestive tract than do the remains of meat ingested in an equal volume.

The bear has a "simple" gut, as compared to the "compound" gut of herbivores, yet it is not the simple gut of its carnivorous associates, the coyote and the wolf. The intestinal villi are many times larger and covered by mucosal cells that greatly increase the surface absorption area (M. Berthrong, unpublished notes). It would seem that this adaptation assures that a high percentage of what the bear digests is absorbed. Superficial observation of bear feces containing berries indicates that fruits and plant foods, in general, are incompletely digested. An omnivorous diet contributes to the bears' caloric intake, but it has obvious limitations as compared with a purely flesh diet that is "fully" digested and absorbed.

The flesh of fishes and large mammals is perhaps the most under-represented diet item because it is identified in fecal analysis only indirectly by presence of hair, hide, bone, scales, or other remains. Flesh is digested completely, and we detected none in our samples. Fur, hair, and bones are excellent analytical evidence, but they may be selectively rejected by relatively fastidious feeders. Small mammals such as rodents are generally consumed whole, and undigested remains are dependably detectable. The problem of variable digestibilities might have produced serious problems of interpretation had not direct observations in the field (regardless of ranked results of scat analysis) made it clear that certain high energy foods such as animal flesh, berries, and pine seeds were preferred food items for bears (Craighead *et al.* 1982). To improve the consistency of data from fecal analyses, however, it was obvious that a uniform, repeatable method for correcting fecal indices was needed.

The limited spatial distribution, seasonality, annual fluctuations in availability, and selectively preferred status of whitebark pine seeds, certain forb species, ungulate carcasses, rodents, and berries of *Vaccinium* spp. are another source of inadvertent bias in a comparative evaluation of food habits based on fecal analysis. A given category may have an inflated representation in a year when another category is scarce; the latter would, conversely, appear to be of minor importance. For example, Mealey's data (1980) indicated a much lower CFI value for the seeds of whitebark pine during 1973-74 than did the data of Knight *et al.* (1982) for the 1977-81 period or those of the present study for 1968-71. A reasonable explanation is that the pine nut crop was low during 1973-74 and, perhaps, by way of compensation, vertebrate foods or the ubiquitous elk thistle (*Cirsium foliosum*) were much more heavily used.

Another problem in arriving at repeatable values for relative importance of food categories is the difficulty of devising a standard field sampling procedure. Significant bias can be introduced by gathering a disproportionate number of fecal samples from a locale advantageous to a preferred food or during the time of year that a particular food is selectively used. In certain locales and/or times of year, items such as moths, pine seeds, garbage, certain forbs, and berries may comprise 75-100% of a scat. Further complications are introduced by the fact that some feces, depending on composition, degrade much more rapidly than others, and some samples are much more difficult to locate simply because of the density of undergrowth, etc., where the foods are obtained and the feces deposited. Feces containing berries of *Vaccinium* spp. exemplify both of these conditions. Nevertheless, when feces are gathered over a large area and period of years, the analytical results should provide a reasonably valid representation of the comparative importance of different food categories in the overall diet of the grizzly.

Correction Factors

An understanding of the relationship between food intake and waste output began to take shape over the last half decade because of research on captive black bears and grizzlies fed measured amounts of foods closely approximating the broad food categories eaten in the wild. Pritchard and Robbins (1990) measured food digestibility and metabolizable energy as a first step in evaluating the nutritional quality of diets of free-ranging bears. Hewitt (1989) examined similar data to derive correction factors (multipliers) that relate volume of residue in the feces to dry weight of foods representative of those ingested by grizzly bears. Ideally, correction factors could be developed for specific bear food items through repetition of the extended protocol developed by the researchers. Hewitt, in collaboration with David Mattson of the University of Idaho, is working toward that goal (Mattson, pers. comm. 1993). But until such time as a comprehensive index is available, Hewitt (1989) has recommended correction factors for application as multipliers to general categories of foods, as follows: vegetation (grasses, sedges, foliage) = 0.26; roots, bulbs, berries, fruit = 0.93; insects = 1.1; nuts and mast = 1.5; ungulates (assuming 24% of skin and hair consumed with all meat and viscera) and small mammals consumed whole = 4.0; and fishes consumed whole = 40.0.

Dietary Intake Index

To convert fecal analysis data to ingested food data, we multiplied the Composite Fecal Indices for our 10 food categories by the appropriate correction factors to derive a Dietary Intake Index. This provided an approximate expression of the ingested diet of the bear population and, as such, allowed value comparisons among food categories. The Dietary Intake Index can be used for comparisons of food habits among populations, despite the fact that application of correction factors to categories that combine multiple food plant species generalizes the results. Application of the correction factors to our specific measurements of fecal remains greatly improved our ability to assess the comparative importance of various food categories in the grizzly bear diet. As we show later, inclusion of caloric value estimates for food categories adds yet another dimension to an understanding of food habits.

RESULTS OF FECAL ANALYSIS
Frequency of Occurrence and Fecal Volume

In our study, Yellowstone grasses and sedges were by far the most frequently occurring remains in fecal samples (63.2%), followed by garbage (36.3%), whitebark pine seeds (18.3%), and vertebrate remains (15.8%) (Table 10.1). The high occurrence of grasses and sedges was probably a function of their durability on the ground and their

Table 10.1. Percentages of occurrence and volume, composite fecal indices, and dietary intake indices for food categories identified from 487 grizzly bear fecal samples, collected May through October, Yellowstone Ecosystem, 1968-71.

Food category	Frequency-of-occurrence	Frequency-of-occurrence percent	Average percent fecal volume	Composite fecal index[a]	Correction factor[b]	Dietary intake index[c]	Dietary intake rank
Garbage	177	36.3	36.3	1317.7	2.30	3030.7	1
Whitebark pine seeds	89	18.3	94.0	1720.2	1.50	2580.3	2
Vertebrate remains	77	15.8	26.7	421.9	4.00	1687.6	3
Grasses and sedges	308	63.2	69.2	4373.4	0.26	1137.1	4
Vaccinium spp.	44	9.0	46.6	419.4	0.93	390.0	5
Fragaria spp.	38	7.8	21.4	166.9	0.93	155.2	6
Forbs	33	6.8	37.7	256.4	0.53	135.9	7
Ants	20	4.1	5.2	21.3	1.10	23.4	8
Cirsium spp.	11	2.3	14.5	33.4	0.53	17.7	9
Unidentified invertebrate remains	14	2.9	0.8	2.3	1.10	2.5	10
Soil and gravel	20	4.1	41.4	170.0	—	—	—
Woody refuse	12	2.5	14.1	35.2	—	—	—

[a] Composite fecal index equals Frequency-of-occurrence percent multiplied by Average percent fecal volume.
[b] Estimate for Garbage ≈ 40% foliage + 10% rootstocks + 40% processed carbohydrates + 10% meats
≈ 0.4(0.26) + 0.1(0.93) + 0.4(1.5) + 0.1(15.0) ≈ 2.30

Estimate for Forbs ≈ 60% foliage + 40% rootstocks ≈ 0.6(0.26) + 0.4(0.93) ≈ 0.53
[c] Dietary intake index equals Composite fecal index multiplied by Correction factor.

role as a dietary staple, widely available in most areas throughout spring and summer when more preferred foods were perhaps less plentiful or absent.

Pine seeds, available in variable quantities in restricted locales during only certain seasons of the year, composed the greatest average percent volume (94%) of samples in which they occurred (Table 10.1). When available to the bears, the seeds were obviously consumed avidly and almost to the exclusion of other foods. Thus, we considered them to be a highly preferred diet item. Remains of grasses and sedges produced the next highest Average Percent Fecal Volume (69.2%), a result less likely due to preference than to their abundance and relatively poor digestibility. That is to say, any volume of grasses and sedges ingested was much less diminished by digestive processes than were comparable volumes of most other foods.

Composite Fecal Index

Not surprisingly, when measures of Frequency of Occurrence Percent and Average Percent Fecal Volume were combined by multiplication to produce the Composite Fecal Index, grasses and sedges represented by far the largest value (Table 10.1). Whitebark pine seed remains were next, followed by garbage and, much farther down, vertebrate remains and the berries of *Vaccinium* spp.

Subsequent research in the Yellowstone Ecosystem by Mealey (1975, 1980) and Knight *et al.* (1982) provided additional information on the food habits of the Yellowstone grizzly bears. We assumed that the bears' food habits following the dump closures would (with the exception of garbage) change slowly as bears adapted to the altered feeding conditions and the imposed spatial adjustments. In order to explore this assumption and make comparisons, we converted the data from the postclosure studies to our Composite Fecal Indices.

The data adapted from Mealey (1980) were based on analysis of 615 fecal samples gathered during 1973 and 1974. The remains of grasses and sedges showed the highest relative frequency of occurrence, followed by an aggregate of forb species, meat of vertebrates, species of *Cirsium*, and ants (Table 10.2). Remains associated with human garbage, when found in a fecal sample, constituted a larger proportion of the scat volume than any other category. As would be expected, however, only 0.8% of samples contained garbage. Remains of berries (*Vaccinium* spp.) represented the second highest average percent fecal volume, followed by the grasses and sedges, the seeds of whitebark pine, various species of forbs, and vertebrate remains. When Frequencies of Occurrence and Average

Fecal Volumes were combined in Composite Fecal Index values, grasses and sedges were highest, followed by the forbs category, vertebrate remains, and *Cirsium* spp. Other food categories produced significantly lower values.

Data adapted from Knight *et al.* (1982) were the result of analyses of 3028 fecal samples of both black and grizzly bears gathered from 1977 through 1981. Grass and sedge remains constituted the most frequently occurring food category, followed by the seeds of whitebark pine, forbs as a group, ants, and the meat of vertebrates (Table 10.2). Remains of whitebark pine seeds represented the highest average percent fecal volume, followed by vertebrate remains, forbs, berries of *Vaccinium* spp., and items associated with garbage. The four highest values of the Composite Fecal Index were, in descending order, whitebark pine seeds, grasses and sedges, forb species, and vertebrates. Other food remains showed relatively minor representations.

Dietary Intake Index

As discussed earlier, interpretation of Composite Fecal Index data showed a slight dietary change between the two periods. As a measure of the relative value of food categories in the grizzly's diet, however, the data can be misleading. We corrected for this by converting the data to an estimate of food quantities actually ingested. This provided a Dietary Intake Index (DII) value for each of the 10 food categories. For garbage and forbs, categories that represented combinations of foods with differing digestibility, we constructed correction factors by approximating proportions of food types that contributed to the category; that is, forbs ingested by bears during the study period were estimated to be 60% foliage (CF = 0.26) and 40% rootstocks (CF = 0.93), whereas garbage was defined to be 40% foliage, 10% rootstocks, 40% processed carbohydrates (CF = 1.5), and 10% meats (CF≈15, in the absence of hide and viscera).

Application of correction factors permitted a ranking of the food categories that is, we believe, a close approximation of their relative importance in the diet of the Yellowstone bear population. Garbage, considered by many to be an "artificial" food resource, had the highest Dietary Intake Value during the 1968-71 period (Table 10.1). It should be noted that this was true despite the fact that collection of fecal samples took place in areas removed from open-pit dumps. That is, samples were not disproportionately sought in or adjacent to the dumps.

Whitebark pine seeds ranked second in DII value, the highest among the "natural" food cat-

Table 10.2. Percentages of occurrence and volume, Composite Fecal Indices, and Dietary Intake Indices for food categories identified from 615 (Mealey 1980) and 3028 (Knight *et al.* 1982) grizzly bear fecal samples, Yellowstone National Park, 1973-74 and 1977-81, respectively.

Food category	Frequency-of-occurrence 1973-74	1977-81	Frequency-of-occurr. percent 1973-74	1977-81	Average percent fecal volume 1973-74	1977-81	Composite fecal index[a] 1973-74	1977-81	Correction factor[b]	Dietary intake index[c] 1973-74	1977-81	Dietary intake rank 1973-74	1977-81
Garbage	5	78	0.8	2.6	70.0	50.0	56.0	130.0	2.3	128.8	299.0	7	6
Whitebark pine seeds	31	1028	5.0	33.9	55.2	84.4	276.0	2861.2	1.5	414.0	4291.8	5	1
Vertebrate remains	167	585	27.2	19.3	51.0	55.2	1387.2	1065.4	4.0	5548.8	4261.6	1	2
Grasses and sedges	397	1468	64.6	48.5	55.3	45.6	3572.4	2211.6	0.26	928.8	575.0	3	4
Vaccinium spp.	26	54	4.2	1.8	63.4	52.9	266.3	95.2	0.93	247.7	88.5	6	8
Fragaria spp.	10	27	1.6	0.9	18.4	22.2	29.4	20.0	0.93	27.3	18.6	9	10
Forbs	349	968	56.7	32.0	52.7	54.6	2988.1	1747.2	0.53	1583.7	926.0	2	3
Ants	54	676	8.8	22.3	9.6	19.2	84.5	428.2	1.1	93.0	471.0	8	5
Cirsium spp.	100	212	16.3	7.0	49.4	36.6	805.2	256.2	0.53	426.8	135.8	4	7
Invertebrate remains[d]	4	46	0.7	1.5	24.1	46.7	16.9	70.0	1.1	18.6	77.0	10	9
Soil and gravel	—	353	—	11.7	—	42.2	—	493.7	—	—	—	—	—
Woody refuse	—	9	—	0.3	—	33.3	—	10.0	—	—	—	—	—

[a] Composite fecal index equals Frequency-of-occurrence percent multiplied by Average percent fecal volume.

[b] Estimate for Garbage ≈ 40% foliage + 10% rootstocks + 40% processed carbohydrates + 10% meats
    $= 0.4(0.26) + 0.1(0.93) + 0.4(1.5) + 0.1(15.0) = 2.30$
    Estimate for Forbs ≈ 60% foliage + 40% rootstocks ≈ $0.6(0.26) + 0.4(0.93) ≈ 0.53$

[c] Dietary intake index equals Composite fecal index multiplied by Correction factor.

[d] Vespidae only for 1973-74 (Mealey 1980).

Table 10.3. Rankings (1 to 10 in descending order) of Dietary Intake Indices for 10 grizzly bear food categories, Mealey (1980), Knight *et al.* (1982), and the present study, Yellowstone Ecosystem.

Food category	Mealey (1973-74)	Knight *et al.* (1977-81)	Present study (1968-71)
Garbage	7	6	1
Whitebark pine seeds	5	1	2
Vertebrate remains	1	2	3
Grasses and sedges	3	4	4
Vaccinium spp.	6	8	5
Fragaria spp.	9	10	6
Forbs	2	3	7
Ants	8	5	8
Cirsium spp.	4	7	9
Unidentified invertebrate remains	10	9	10

egories. They were followed by vertebrate remains, grasses and sedges, and *Vaccinium* spp. The remaining categories were of much less importance in the total diet.

Dietary Intake Index values derived from the data of Mealey (1980) during the period 1973-74 and Knight *et al.* (1982) for the period 1977-81 (Table 10.2) were compared with those we calculated for the period 1968 through 1971 (Table 10.1). Results support our assumption that food habits would change slowly following dump closure. The data indicate minor changes in use of specific food categories, with strong evidence that the basic food habits recorded for the 1968-71 period had not changed unpredictably from 1973 through 1981 (Table 10.3). During the period 1968 through 1971, human-derived garbage was the most important dietary source. As would be expected following phaseout of the dumps, garbage was of relatively low value in the subsequent studies. Whitebark pine seeds and the meat of vertebrates were highly valuable food sources for the grizzly bear population in all three studies, although the Mealey research (1980) did not indicate the seeds to be nearly as important as did the other studies. When converted to Dietary Intake, the data of Mealey and of Knight *et al.* (1982) indicated that the forb category was much more prominent in the diet than it was from 1968 to 1971. The grasses and sedges showed the same moderate degree of dietary importance in all three studies (Table 10.3).

The berries of *Vaccinium* apparently provided a more important dietary resource to the bear population during the 1968-71 period than during either of the later studies (Table 10.3). Variations in year-to-year use of such preferred foods as berries and pine seeds were likely a function of varia-

tions in availability of these food categories annually, seasonally, and spatially, and of their tendency to vary in abundance from one area to another. The comparatively high dietary role of *Cirsium* spp. found by Mealey indicates that, in some locales, the elk thistle was selectively and heavily used, just as *Vaccinium* spp. were in our study. The comparatively high CFI value accorded to ants by data from Knight *et al.* (1982) was explained in that study to be the result of an opportunism; i.e., use of a food in proportions approximating its availability in the habitat. Many populations of plant and animal species vary in abundance or importance from year to year, and ants are no exception.

With the exception of the grass-sedge category and some of the forbs, most seasonally important plant foods are subject to considerable fluctuations in abundance and availability from year to year. Of the 10 food categories contributing to the nutrition of the Yellowstone bears, one may be exploited much more heavily in a given year when another category is scarce. In years when some or many of the food categories are scarce, the nutritional status of bears undoubtedly suffers. Our work has shown that grizzlies do not gain weight, nor do they develop significant fat stores, when relying heavily on a diet of grasses, sedges, and forbs, particularly during the late summer to early autumn period of hyperphagia (Craighead and Mitchell 1982).

The open-pit dumps represented a highly stable and important supplementary food source, but did not appear to alter qualitatively the consumption of the wild food categories. With their closure, bears used preferred natural foods as availability permitted, but undoubtedly were often forced to shift to less preferred, less nutritious sources such as abundant, widely distributed forbs. The marked increase in use of the forb resource in the years following dump closures is unmistakable (Table 10.3).

It also appears that bears used the vertebrate food resource to a greater extent, as indicated by the relatively high values of dietary intake (Tables 10.1 and 10.2). The high likelihood of variations in sampling intensities between the three studies, however, makes questionable any direct comparisons of the absolute numeric values of the indices, as opposed to the more forgiving process of ranking categories to show relative differences. Methodologies for measuring carcass use and ungulate predation did not lend themselves to comparative quantitative analysis. This is discussed in greater detail in Chapter 16.

NUTRITIVE VALUES OF GRIZZLY BEAR FOODS

Application of correction factors, as discussed earlier, allowed us to compare the relative dietary importance of food categories with greater precision. It also provided a measure for comparing the relative importance of these categories between spatially or temporally different bear populations. But we did not have a complete picture because grizzly bear foods vary considerably in nutritive value. That is, different foods, whether plant or animal, vary in calories (kcal) per unit weight, as well as the form in which the calories occur in nature—fats, proteins, and simple (sugars) or complex (starches) carbohydrates. As a preliminary to further characterizing the bears' diet, we contracted with the Montana State University Field Chemistry Laboratory to analyze the various bear food plants for caloric values (Table 10.4). Using these, along with caloric values for vertebrate flesh and insects developed by Hewitt (1989), we were able to estimate the caloric contribution of each of the 10 food categories to the nutrition of the 1968-71 grizzly bear population.

Nutritive Value of Food Plants

Our observations of the grizzlies' food habits in Yellowstone National Park have shown that they consumed tubers, roots, and bulbs high in carbohydrates and emergent sedges and forbs high in protein. We hypothesized that food selection of various forbs, sedges, and grasses by grizzlies was largely due to availability but, in part determined by total energy content of the food. In other words, did grizzlies exhibit preferences for certain plants, plant parts, and at certain periods of the vegetation growth cycle? This posed a complex question which we could not answer with our limited observational data. Nonetheless, it was apparent that grizzlies were

at times selective and did exhibit preferences, for whatever reasons. We believed we could obtain a general perspective on selective feeding by analyzing the gross nutritive value of some of the most available plant foods, as well as some of those obviously preferred, such as berries and pine seeds.

Accordingly, energy (kcal/g) provided by proteins, fats (ether extracts) and carbohydrates (nitrogen-free extracts) was determined for some of the more important food plants (Table 10.4). Available energy of specific food plants varied from a low of 1.91 kcal/g in the roots of false hellebore (*Veratrum viride*) to 3.99 kcal/g in whitebark pine seeds (*Pinus albicaulis*).

Protein levels were highest in sedges (*Carex* spp.), the flowering heads of beargrass and cow parsnip (*Xerophyllum tenax* and *Heracleum lanatum*), emergent foliage of meadow rue (*Thalictrum occidentale*), berries of elderberry (*Sambucus racemosa*), and the seeds of whitebark pine. In general, nitrogen-free extracts, both sugars and non-sugars, were relatively low in food plants with the highest protein content (Table 10.4).

The relative composition attributable to carbohydrates was measured for 21 species of berries (Table 10.4). Highest sugar values were recorded from twinberry (*Lonicera involucrata*), grouse whortleberry (*Vaccinium scoparium*), and blue huckleberry (*V. globulare*); lowest values were from baneberry, elderberry, and red ozier dogwood (*Cornus stolonifera*). Non-sugar carbohydrates were highest in the fruit of mountain ash (*Sorbus scopulina*), hawthorne (*Crataegus* spp.), serviceberry (*Amelanchier alnifolia*), kinnikinnick (*Arctostaphylos uva-ursi*), and Oregon grape (*Berberis repens*). The nonsugar carbohydrates were also quite high in certain forbs such as woodrush (*Luzula hitchcockii*), sweet-cicely (*Osmorhiza occidentalis*), beargrass (*Xerophyllum tenax*), cow parsnip (*Heracleum lanatum*), and glacier lily (*Erythronium grandiflorum*). A high sugar content may influence selection and use of a specific food more than actual carbohydrate energy value. The bears' well-known taste for sweets seems best satisfied by berries with high sugar content, such as whortleberries and huckleberries (*Vaccinium* spp.). Although twinberry and buffaloberry (*Shepherdia canadensis*) also have high sugar values, they do not taste sweet by human standards and may not be as palatable to grizzlies as *Vaccinium* berries. A more important factor determining use than either sugar content or energy values may have been the relatively low abundance of buffaloberry and twinberry.

We recognize that the caloric values of all plants vary considerably throughout the growing season

Table 10.4. Chemical analysis and caloric values of grizzly bear food plants.

Bear food plant		Number samples	Protein %	Protein kcal/g	Ether extract %	Ether extract kcal/g	Sugars %	Non-sugars %	Total %	Total kcal/g	Total kcal/g
GRASSES AND SEDGES:											
Sedges											
Carex spp.	(leaves & base leaf)	3	11.31	0.49	2.27	0.21	4.80	33.37	38.17	1.60	2.30
Grasses											
Calamagrostis rubescens	(leaves)	1	6.00	0.26	3.15	0.30	4.70	35.38	40.08	1.68	2.24
Festuca idahoensis	(leaves)	1	3.30	0.14	1.60	0.15	—	—	41.40	1.74	2.03
Festuca idahoensis	(before heading out)	2	11.80	0.51	3.35	0.32	7.05	38.80	45.85	1.93	2.76
Melica spectabilis	(bulbets)	1	2.80	0.12	1.30	0.12	—	—	71.90	3.02	3.26
Grasses (average)		—	0.60	0.26	2.35	0.22	5.88	37.09	49.81	2.09	2.57
Grasses and Sedges (average)		—	7.04	0.30	2.33	0.22	5.52	35.85	47.48	1.99	2.52
FORBS (foliage, bulbs):											
Perideridia gairdneri	(tubers)	1	3.40	0.15	0.70	0.07	—	—	80.71	3.39	3.61
Equisetum arvense	(foliage)	2	11.58	0.50	3.97	0.37	8.40	32.39	40.70	1.71	2.58
Xerophyllum tenax	(seed pods)	1	14.76	0.63	15.64	1.47	5.40	34.21	39.61	1.66	3.76
Xerophyllum tenax	(leaf base)	2	3.86	0.17	1.25	0.12	7.20	40.03	47.23	1.98	2.27
Xerophyllum tenax	(flowering heads)	1	21.10	0.91	2.10	0.20	9.10	40.40	49.50	2.08	3.19
Erythronium grandiflorum	(bulbs)	2	3.50	0.15	0.44	0.04	—	—	81.88	3.44	3.63
Erythronium grandiflorum	(leaves and stems)	1	11.10	0.48	5.10	0.48	12.10	37.70	49.80	2.09	3.05
Erythronium grandiflorum	(seeds and seed pods)	1	15.50	0.67	2.50	0.24	6.00	38.30	44.30	1.86	2.77
Claytonia lanceolata	(corms)	2	6.80	0.29	0.66	0.06	—	—	77.05	3.59	3.94
Thalictrum occidentale	(newly emerging leaves)	1	24.40	1.05	6.00	0.56	5.90	37.00	42.90	1.80	3.41
Lomatium cous	(roots)	5	5.80	0.25	2.59	0.24	—	—	50.08	2.10	2.59
Heracleum lanatum	(flowering heads)	1	26.30	1.13	3.80	0.36	3.90	33.80	37.70	1.58	3.07
Heracleum lanatum	(foliage)	3	9.14	0.39	2.40	0.23	7.90	38.39	45.40	1.91	2.53
Hedysarum occidentale	(root)	1	9.53	0.41	1.54	0.14	3.90	30.14	34.04	1.43	1.98
Polygonum bistortoides	(root)	2	7.00	0.30	1.00	0.09	—	—	47.80	2.01	2.40
Cirsium scariosum	(stems and leaves)	1	14.90	0.64	2.20	0.21	3.60	36.80	40.40	1.70	2.55
Cirsium scariosum	(root)	1	7.20	0.31	0.90	0.08	—	—	61.30	2.57	2.96
Luzula hitchcockii	(leaves)	1	9.04	0.39	2.17	0.20	8.00	42.48	50.48	2.12	2.71
Trifolium repens	(foliage)	1	15.43	0.66	2.64	0.25	10.80	33.66	44.46	1.89	2.80
Osmorhiza occidentalis	(leaves and stems)	1	15.20	0.65	2.10	0.20	7.10	41.10	48.20	2.02	2.87
Osmorhiza occidentalis	(root)	1	9.18	0.39	4.57	0.45	11.50	34.86	46.36	1.95	2.79
Veratrum viride	(root)	1	7.93	0.34	0.94	0.09	1.40	33.81	35.21	1.48	1.91
Forbs (average)		—	11.48	0.49	2.96	0.28	7.01	34.42	49.96	2.11	2.88

cont'd on next page

and throughout the period of seed maturation. To detect significant differences for individual species would have required extensive, seasonal sampling which we were not prepared to do. Instead, we used mean caloric values for plant categories such as the grasses and sedges, forbs, pine seeds, and berries for purposes of relative comparison.

Measurement of total caloric values for individual food plants and average caloric values for food plant categories or groupings showed that pine seeds rated highest, with 3.99 kcal/g, followed by berries (3.24 kcal/g) and forbs (2.88 kcal/g); the grass and sedge category, at 2.52 kcal/g, had the lowest average (Table 10.4). The two plant food sources with the highest energy values—pine seeds and

berries—were the least available to grizzlies, both annually and over a span of years. Those with the lowest energy values—graminales and forbs—were, on the other hand, consistently available. Thus, the importance of grasses and forbs in the diet of the Yellowstone grizzlies related more to their annual abundance, seasonal availability, and wide distribution than to their average energy values.

This is substantiated by similar data from the Scapegoat Wilderness (Craighead *et al.* 1982). Considering only the food plant base, comparison of Dietary Intake Index Percents (DIIP) for the four major plant groupings in the Scapegoat Ecosystem with those measured in the Yellowstone Ecosystem (Table 10.5) showed that the DIIP for graminales

Table 10.4. *cont'd*

Bear food plant	Number samples	Protein %	Protein kcal/g	Ether extract %	Ether extract kcal/g	Sugars %	Non-sugars %	Total %	Total kcal/g	Total kcal/g
SHRUBS AND FORBS (berries):										
Vaccinium globulare	2	3.98	0.17	3.30	0.31	38.00	27.01	65.16	2.74	3.22
Vaccinium scoparium	2	6.91	0.30	4.70	0.44	40.00	26.23	66.23	2.78	3.52
Fragaria virginiana	1	8.90	0.38	3.68	0.35	25.00	28.24	53.44	2.24	2.97
Ribes cereum	2	5.21	0.22	3.40	0.32	33.60	32.13	65.73	2.76	3.30
Shepherdia canadensis	1	9.40	0.40	3.05	0.29	34.00	27.13	61.13	2.57	3.26
Rubus spp.	1	5.49	0.24	3.63	0.34	22.30	31.06	53.36	2.24	2.82
Ribes lacustre	1	5.61	0.24	3.19	0.30	33.70	23.34	57.04	2.40	2.94
Sambucus racemosa	1	14.68	0.63	19.12	1.80	6.60	21.06	27.66	1.16	3.59
Prunus virginiana	1	9.23	0.40	5.94	0.56	12.50	29.64	42.14	1.77	2.73
Rosa spp.	1	5.41	0.23	2.86	0.27	18.40	34.48	52.88	2.22	2.72
Cornus stolonifera	1	5.82	0.24	20.22	1.90	7.50	25.88	33.38	1.40	3.54
Amelanchier alnifolia	1	3.26	0.14	3.44	0.32	24.00	43.62	67.62	2.84	3.30
Sorbus scopulina	1	5.19	0.22	4.03	0.38	15.20	49.90	65.10	2.73	3.33
Symphoricarpos albus	1	4.36	0.19	5.59	0.53	30.70	28.75	59.45	2.50	3.22
Actaea rubra	1	11.25	0.48	19.32	1.82	5.40	33.11	38.51	1.62	3.92
Streptopus amplexifolius	1	8.96	0.39	3.52	0.33	29.50	20.43	49.93	2.10	2.82
Arctostaphylos uva-ursi	1	2.84	0.12	5.41	0.51	13.70	38.41	52.11	2.19	2.82
Berberis repens	1	9.20	0.40	6.10	0.57	21.00	36.40	57.40	2.41	3.38
Juniperus horizontalis	1	3.80	0.16	18.60	1.74	22.00	21.40	43.40	1.82	3.72
Crataegus spp.	1	3.00	0.13	3.50	0.33	11.90	46.40	58.30	2.45	2.91
Lonicera involucrata	1	5.94	0.26	2.40	0.23	10.40	28.17	68.57	2.88	3.37
Berries (average)	—	6.46	0.28	6.96	0.65	23.12	31.87	55.00	2.31	3.24
TREES:										
Pinus albicaulis (seeds)	1	14.20	0.61	30.45	2.86	—	—	12.45	0.52	3.99

Note: All results reported on a dry weight basis; average caloric values/gram used to obtain total kcal/gram (Schmidt-Nielsen 1975) for protein (urea free) = 4.3, and for fat (ether extract) = 9.4.

in Yellowstone was nearly triple that for Scapegoat, while the values for forbs were more than six-fold greater in Scapegoat than in Yellowstone. Graminales and forbs, however, when considered together, showed quite comparable values of 31.7% and 29.2% for the Scapegoat and Yellowstone, respectively. The DIIPs for the high-calorie groupings, berries and pine seeds (Table 10.5), were also reasonably close between the two studies. That the values would so closely match for two studies so widely separated in time and space suggests that, in general, abundant, highly dependable low-calorie food plants represent about one-third of the grizzly's vegetable calories, while the less abundant, less dependable high-calorie food plants comprise the majority. Although the proportion of the diet composed of the preferred plant resources varies from year to year with availability, the strong selectivity of the bears for high calorie plant foods is well demonstrated.

Although berries have high sugar content and were selectively sought by grizzly bears, they, like pine seeds, were not dependably abundant.
(Photo by John J. Craighead)

Plant food sources with the lowest energy values (grasses, sedges, and forbs) were widely abundant and consistently available. They formed a third of the grizzly bear's plant food caloric intake.
(Photo by John J. Craighead)

Table 10.5. Comparison of Dietary Intake Index Percents for major food plant groups determined by analysis of 282 grizzly bear scats collected in the Scapegoat Ecosystem (Craighead *et al.* 1982) and 487 scats collected in the Yellowstone Ecosystem (current study).

	Dietary Intake Index Percent[a]		
Plant food group	Scapegoat study	Yellowstone study	Average kcal/g
Graminales	8.7 ⎤ 31.7	25.7 ⎤ 29.2	2.56
Forbs	23.0 ⎦	3.5 ⎦	2.81
Berries	15.4	12.4	3.21
Whitebark pine seeds	52.9	58.4	3.99

[a] Only Dietary Intake Index values for plant food categories (Table 10.1) were used to calculate the relative percentages.

Nutritive Value of Animal Foods

An average energy value of approximately 3 kcal/g for plant foods used by grizzlies (derived from Table 10.4) is low compared to digestible energy estimates of 5.6 and 6.8 kcal/g for ungulates (Craighead *et al.* 1982 and Hewitt 1989, respectively) and 4.5 kcal/g for small mammals (Hewitt 1989). Although Hewitt has estimated available energy from invertebrates at 2.7 kcal/g, certain highly preferred insects are probably of higher caloric value. For example, noctuid moths known to be intensively used by some bear populations (Lincoln-Scapegoat Wilderness [Craighead *et al.* 1982], Mission Mountains, Montana [Klaver *et al.* 1986], and Shoshone National Forest, Wyoming [Mattson *et al.* 1991b]) may contain as much as

On average, invertebrates have low food energy values. However, some highly sought insects, such as noctuid moths, may contain high levels of abdominal fat. *(Photo by John J. Craighead)*

68% abdominal fat (Mattson *et al.* 1991b). Not surprisingly, the grizzly shows a decided preference for animal foods, but because they are unavailable for extended periods and/or are often energetically costly to obtain, the bear has evolved to be an omnivorous carnivore.

Energy Values of the Grizzly Bear Diet

To obtain a general perspective on the importance of the major food types in the overall nutrition of the bear population, we multiplied the Dietary Intake Index (DII) values for the 10 food categories by the caloric values for those categories. We concede that the results (Dietary Energy Index)

represent only a rough approximation because of sampling deficiencies for establishing caloric values for specific plants, assumptions necessary to derive some of the caloric values for the animal resource and garbage, and because the DII values are unitless while caloric values are not (Table 10.6). Even so, this exercise does provide workable relative indices characterizing the major sources of calories in the bears' diet as they relate to dietary intake.

Caloric values for food plants and food plant categories were taken from Table 10.4, whereas those for vertebrates (ungulates and small mammals at 6.0 kcal/g) and invertebrates (2.7 kcal/g) were adapted from the unpublished thesis of Hewitt (1989). We have conservatively estimated a caloric value for garbage by assuming a mix of 90% vegetation/processed carbohydrates and 10% meat and by applying caloric values of 3.0 kcal/g for vegetation (average from Table 10.4) and 6.8 kcal/g for meat (from Hewitt 1989). The caloric value for meat is higher than that applied to the general vertebrate food category because, in the absence of nondigestible hair, skin, and other proteinaceous residues, meat proteins have a true digestibility of 100% (Pritchard and Robbins 1990). We believe the estimate of caloric value for garbage to be a conservative one, considering that meat scraps, fat, and bacon grease may have constituted a proportion much greater than 10%, and that processed carbohydrates such as pastries and breads may have provided calories greater, per unit weight, than other vegetable foods.

A Dietary Energy Index for each of the major food categories and its percent representation in

Table 10.6. Caloric contributions of 10 food categories to the nutrition of grizzly bears, Yellowstone Ecosystem, 1968-71.

Food category	Dietary intake index	Caloric value (kcal/g)[a]	Dietary energy index[b]	Percent of dietary calories
Garbage	3030.7	3.4	10304.4	28.7
Whitebark pine seeds	2580.3	4.0	10321.2	28.7
Vertebrate remains	1687.6	6.0	10125.6	28.2
Grasses and sedges	1137.1	2.5	2842.8	7.9
Vaccinium spp.	390.0	3.5	1365.0	3.8
Fragaria spp.	155.2	3.0	465.6	1.3
Forbs	135.9	2.9	394.1	1.1
Ants	23.4	2.7	63.2	0.2
Cirsium spp.	17.7	2.8	49.6	0.1
Unidentified invertebrates	2.5	2.7	6.8	<0.1

[a] Caloric values for plant food categories from Table 10.4; values for vertebrate and invertebrate food categories adapted from Hewitt (1989); estimate of caloric value of garbage based on assumption of ≈ 90% vegetation + 10% meats ≈ 0.9(3.0) + 0.1(6.8) ≈ 3.4 (see text).

[b] Dietary energy index equals Dietary intake index multiplied by Caloric value.

The carcasses of winter-killed elk and bison were dependably abundant in spring when the bears' energy levels were low following winter sleep. *(Photo by John J. Craighead)*

the overall diet is presented in Table 10.6. These indicate that garbage, whitebark pine seeds, and vertebrates were approximately equivalent, representing by far the most important sources of calories (Table 10.6). This was so despite their relatively low occurrence in scats, as compared to grasses and sedges (Table 10.1). The high caloric values of pine seeds and vertebrate biomass, and their avid use by grizzlies when available, marked them as crucial food resources. Garbage was plainly a nutritional bulwark in the bears' diet. Not only did it have a high energy value, but it was readily available in large quantities each year throughout the summer months. Next in order of caloric contribution were grasses and sedges, low in calories but consistently available and consistently consumed. These were followed by the highly preferred berries of the vacciniums.

Once again, it is important to recognize the nutritional stability that garbage provided the bear population prior to dump closures. It supplemented the high-calorie, but less available, food categories. Its use was especially evident in reducing the bears' need to rely on "natural" plant foods of low caloric value during years of low seed and berry production. When the high calorie natural foods were unavailable, grizzlies increased their movements in search of food. As we shall demonstrate later, the availability of garbage greatly damped these movements by providing the necessary caloric intake within a relatively safe and circumscribed environment.

As stated earlier, our analyses provide only a rough approximation of the relative roles the various food categories played in the nutrition of the Yellowstone grizzly population from 1968 to 1971. Analyses of food habits should become more precise as work progresses on ways to correlate data from scat analysis with data on food ingestion, and as specific caloric values for specific food items are refined. The main impediments to wide application of the developing methodology are the absolute need for well-equipped analytical laboratories that can focus on the many problems inherent in fecal analysis and the determination of dietary energy values. We need to know as much about the daily caloric requirements of bears as we know for humans (1500-3500 calories per day). Such knowledge is essential to habitat evaluations and these, in turn, to population viability analyses (see Chapter 20).

USE OF THE PLANT FOOD RESOURCE

The wide assortment of plant species used as food by the grizzly is becoming increasingly evident. Mealey (1975) listed approximately 25 species for the Yellowstone area without specifically identifying grasses and sedges. We recorded more than 42 species used in the same area from 1959 to 1969 before the closure of the open-pit garbage dumps. Knight *et al.* (1982) identified approximately 99 different species and undifferentiated genera of grasses, sedges, forbs, shrubs, and trees in their analyses of over 3000 fecal samples from both black bears and grizzlies.

Craighead *et al.* (1982) listed 68 species and plant categories (genera and families) as bear food plants in the Scapegoat Wilderness between 1972 and 1978. Servheen and Lee (1979) recorded approximately 36 plant species for the Mission Mountains, Montana; Husby and McMurray (1978), 74 for northwestern Montana; and Hamer *et al.* (1980), about 41 for Banff National Park, Canada. There are undoubtedly well over 200 plant species whose seeds, fruits, foliage, flower heads, stems, cambium, roots, tubers, and rootstocks are eaten by grizzlies within their North American range.

The use of any specific food plant by a population of grizzly bears usually depends on the relative temporal and spatial abundance and availability of other food plants, as well as on the energy expended to acquire the food relative to the energy provided by it. Thus, a plant or a plant group heavily used in one area may be recorded as lightly used in another. Nevertheless, sufficient information is available to designate the most important plants or plant groups used by grizzlies in specific geographic areas.

Mealey (1975) listed the following as major food plant sources for grizzlies in Yellowstone National Park, Wyoming: grasses and sedges (Graminales); forbs—springbeauty (*Claytonia lanceolata*), elk thistle (*Cirsium scariosum*), yampa (*Perideridia gairdneri*), biscuitroot (*Lomatium cous*), common horsetail (*Equisetum arvense*); shrubs—whortleberry (*Vaccinium scoparium*); and tree—whitebark pine (*Pinus albicaulis*). Pearson (1975), working in southwestern Yukon Territory, Canada, recorded grasses (Graminae), sweetvetch (*Hedysarum alpinum*), buffaloberry (*Shepherdia canadensis*), and willow (*Salix* spp.) as important sources of food. For grizzlies in the Mission Mountains, Montana, Servheen and Lee (1979) recorded grasses, serviceberry (*Amelanchier alnifolia*), common horsetail, sweet-cicely (*Osmorhiza occidentalis*), dandelion (*Taraxacum* spp.), cow parsnip (*Heracleum lanatum*), white clover (*Trifolium repens*), cherries (*Prunus* spp., domestic), and apples (*Malus* spp., domestic) as major plant foods.

Husby and McMurray (1978), working in northwestern Montana, found the following to be important: grasses and sedges, species in the carrot family (Umbelliferae), horsetail, kinnikinnick (*Arctostaphylos uva-ursi*), buffaloberry, serviceberry, and blue huckleberry (*Vaccinium globulare*). Hamer *et al.* (1980) recorded grasses and sedges, sweetvetch, horsetail, cow parsnip, sorrel or dock (*Rumex* spp.), buffaloberry, huckleberry, and kinnikinnick as important in Banff National Park, Canada. For the barren-ground grizzly in arctic Alaska, Hechtel (1985) found boykinia (*Boykinia richardsonii*), horsetail, grasses and sedges, berries (*Arctostaphylos rubra*), and unspecified roots to be important in the vegetable diet.

It is obvious that a great number of plant species are important to the grizzly bear. The impressive range of food plants available in Yellowstone, however, does not necessarily ensure an adequate food supply for the bear population from year to year. During years of widespread failure of such preferred foods as *Vaccinium* berries or pine seeds, grizzlies generally must travel more, enlarge their home ranges, visit human-derived sources of food more frequently, and exhibit greater aggressiveness in defense of their food sources.

USE OF THE ANIMAL FOOD RESOURCE

Although the grizzly bear meets the majority of its nutritional needs herbivorously (Table 10.6), it shows a decided preference for a flesh diet when available. A wide variety of prey species are taken by the grizzly in its role as a predatory carnivore; birds, fish, and mammals, varying in size from mice to moose, are killed and eaten. Invertebrates such as ants, moths, bees, beetles, earthworms, and grubs can also represent a substantial part of the diet. Any of the vertebrate or invertebrate species may be opportunistically preyed upon at virtually any time of the year, but some species are preyed on in a more purposeful manner during certain seasons. At such times, predation tends to be concentrated and directed as it is a function of the annual biological cycles of the prey species and of the grizzly bear itself. The predatory behavior of the bear will be treated in greater detail following an evaluation of the way in which grizzly bears scavenged the carcasses of ungulates that succumbed to the rigors of the Yellowstone winter.

Fig. 10.3. Transects surveyed for ungulate carcasses, Yellowstone National Park, 1960-72 and 1981-82.

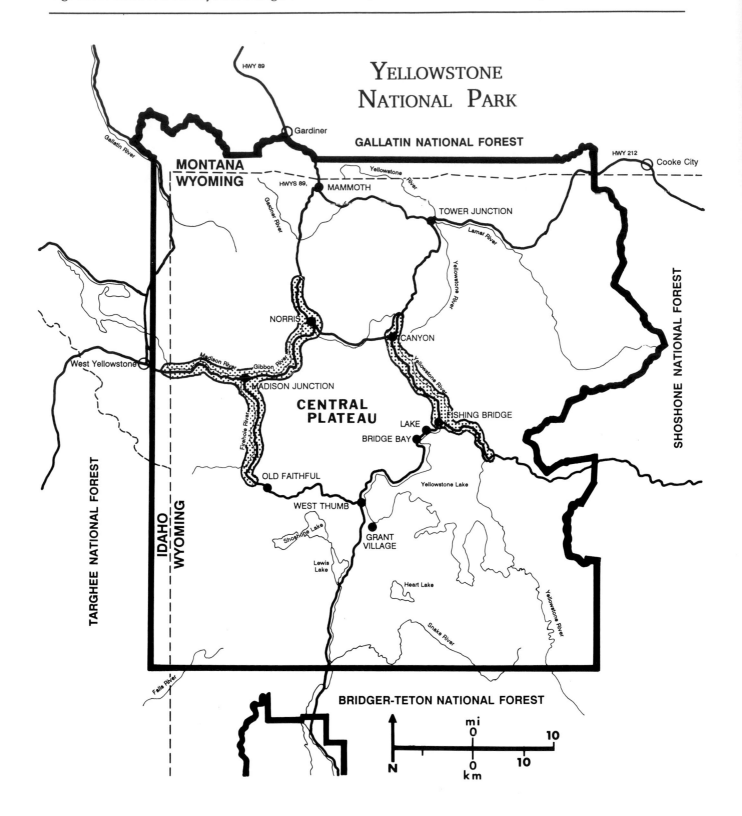

Carrion-Feeding

In the Yellowstone Ecosystem, ungulate carcasses, with an approximate energy value of 6.8 kcal/ g., provided the grizzly with an important source of calories and protein, particularly as the bears emerged from winter sleep. During this critical period, the plant food resource, with an average energy value of 3.0 kcal/g., is unavailable or poorly developed (Craighead *et al.* 1982). During the months of March, April, and May, carrion-feeding was at a peak, concomitant with the need to replenish energy reserves depleted during winter sleep, birthing, and suckling.

Ungulates such as the elk (*Cervus elaphus*), the bison (*Bison bison*), the mule deer (*Odocoileus hemionus*), and the moose (*Alces alces*) are present in the Yellowstone Ecosystem and, in localized areas, may be numerous. Populations of these species experience die-backs during extreme winters, and this is especially true of the elk and bison, which periodically exceed the carrying capacity of their winter ranges. Members of the small interior elk herds and the large Northern Elk Herd are particularly susceptible to winter-kill. The interior herds are largely confined to relatively snow-free thermal areas (Craighead *et al.* 1973a). Whereas the bison herds have been, until recent years, largely confined within the Park, portions of the great herds of elk that inhabit the Park during the summer months migrate to traditional wintering grounds throughout the Ecosystem (Craighead *et al.* 1972b). Nevertheless, substantial numbers over-winter in the Park and, along with bison, tend to congregate in areas with the least snow accumulations and the greatest associated grazing and browsing potentials.

To determine carcass use by grizzly bears, we confined our quantitative observations to the high-elevation interior of the Park—the Upper Madison River drainage and Central Plateau (Figs. 10.2 and 10.3). Located within this area were the Trout Creek and Rabbit Creek garbage dumps and many marked grizzly bears that, in spring, had access to the carcasses of ungulates dying in the area. As indicated earlier, the winter-killed animals of the Central Plateau, with few exceptions, were available almost exclusively to those grizzly bears that denned on the Plateau.

Method of Observing and Examining Carcasses

To determine carcass numbers and their use by bears, five transects were established on the Central Plateau within the Park (Fig. 10.3). Ungulate carcasses were censused during April, May, and June. Although the censuses did not provide a complete accounting of all carcasses available on the Central Plateau, they did indicate relative availability of carcasses from year to year and their sex, age, and physical condition. Location of transects along roads and rivers allowed quick and easy access and maximized the observation time at individual carcasses. Because the carcass censuses were made before the flow of visitor traffic and the recession of snow, human presence was not a significant factor in deterring bears from locating and using roadside carrion. Carcasses were located with binoculars and spotting scopes. Once a carcass was located, the general vicinity was thoroughly examined for tracks, scats, hair, and other evidence of bear activity. If bear sign was evident, the carcass was observed for extended time periods to document the number of grizzly bears frequenting the carcass, interspecific and intraspecific competition, and behavior of other carnivores at the carcass. Starlight scopes were used to observe at night.

We found that we were unable to reliably determine the species, number, sex, or age group of bears using ungulate carcasses by analyzing indirect evidence of bear use. That is, we could not impute these types of data from tracks on the terrain around many of the carcasses. Furthermore, even when the terrain did show tracks, when two or more bears were feeding on the carcass, definitive track evidence was obliterated by the bears themselves, or by other scavengers. We were able to record rigorous quantitative data only by actually observing bears using a carcass until it was consumed.

Portions of transects not easily observed from the roads were explored on foot. These included timbered islands and forest edges, meadows surrounded by timber, and terrain not visible from the roads.

Using techniques developed by Greer (1968a), we ran compression tests on femur bone marrow from all ungulate carcasses to determine the general condition of the animal at the time of death. Winter-killed animals consistently showed marrow compressions greater than 30%, whereas animals in good to excellent condition at the time of death showed compression percents less than 15%.

Characteristics of Carcasses

Of 330 carcasses observed over a span of years (1960-82), 244 (74%) were elk and 86 (26%) were bison (Table 10.7). Observations made of these carcasses included physical condition, sex, age, location, and bear use. Because not all parameters were obtained for each of the 330 carcasses observed, nor were the observations continuous each year over the span of 20 years, the sample sizes used for tabulations vary with the parameter presented.

Once a carcass was located within a transect area, it was observed through spotting and starlight scopes to record the number of bears feeding on it. *(Photo by John J. Craighead)*

Age—The very young and the old animals are among the first to die of malnutrition and exposure. The number of additional animals that will succumb increases as the severity and length of a winter increase. Of 244 elk carcasses, 71 (29.1%) were calves, 19 (7.8%) were yearlings, 97 (39.7%) were adults and 57 (23.4%) were of unknown age class. Age-class distribution of 86 bison carcasses was 2 (2.3%) calves, 11 (12.8%) yearlings, 39 (45.4%) adults and 34 (39.5%) unknown (Table 10.7).

Adult animals constituted the highest proportion of the carcasses of both species, but a much larger proportion of elk carcasses than bison carcasses was classed as calves (Table 10.7). The age classification system was partially responsible for the disparity: the term "calf" referred to elk born the year preceding the census, but did not include calves born the year of carcass census because the calving season ensued afterward; for bison, the term "calf" referred to calves born the year of census because calving began while the censuses were underway, and animals born the previous year were

considered yearlings. Thus, if the proportion of elk calf carcasses (29.1%) is compared to the proportion of calf and yearling bison carcasses combined (15.1%), the difference is not so remarkable. What difference remains, though, may indicate a greater survival potential for bison during the winter following birth because of the one- to two-month age advantage going into winter that a young bison has, on average, over a young elk.

Sex—Female elk carcasses (32.0%) were more numerous than males (9.8%), but the majority of the carcasses were of undetermined gender (58.2%). Many of the bison carcasses were also of undetermined gender (64.0%), but of those identified, male bison carcasses outnumbered the females 21 (24.4%) to 10 (11.6%) (Table 10.7).

The much higher proportion of female elk carcasses may be misleading, as regards annual carcass availability. Relatively more males tend to die in late fall or early winter as a result of the strenuous rutting season. Males, sometimes stressed already by scabies (*Psoroptes equi*, var. *cervinus*) infestations, are weakened during rutting and often

Fig. 10.4. Condition of elk (*n* = 135) and bison (*n* = 26) at time of death, Yellowstone National Park, 1960-72.

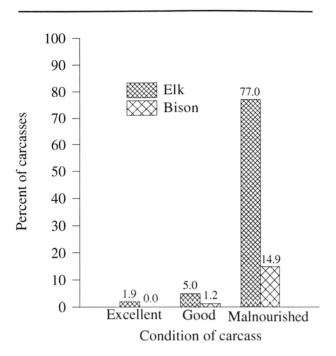

are injured in combat. The intense cold common in the Park during the autumn and winter takes its toll on the weakest males, and the remains of these animals, scavenged by coyotes and ravens, are no longer in evidence the following spring.

Physical Condition—The majority of bison and elk available as carcasses to bears during the spring died of malnutrition. Of 161 elk and bison carcasses for which cause of death was determined, only three (1.9%) were in excellent physical condition (all three deaths due to predation), 10 (6.2%) were in good condition (probable predation), and 148 (91.9%) were severely malnourished at time of death (Fig. 10.4).

Time of Death—The majority of ungulates (bison and elk) for which the month of death by apparent starvation was recorded died during the months of April (60.3%) and May (29.4%) (Fig. 10.5). Of the 100 elk for which the month of death was known, five (5.0%) died in March, 61 (61.0%) died in April, 33 (33.0%) died in May, one (1.0%) died in June. Of 26 bison for which month of death was known, 26.9% died in March, 57.7% in April, and 15.4% in May.

Many of the ungulates that die of malnutrition do so soon after the spring "green-up." The meager sustenance obtained during winter from grasses pawed from beneath the snow, from the twigs of trees and shrubs, and from aquatic vegetation (Craighead *et al.* 1973a) apparently alters the

Table 10.7. Sex and age classes of 244 elk and 86 bison carcasses recorded in transects, Yellowstone National Park, 1960-72 and 1980-82.

Sex	Elk						Bison						Grand total	Percent
	Calf	Yearling	Adult	Unknown	Total	Percent	Calf	Yearling	Adult	Unknown	Total	Percent		
Male	0	4	20	0	24	9.8	0	0	21	0	21	24.4	45	13.6
Female	2	4	68	4	78	32.0	1	1	8	0	10	11.6	88	26.7
Unknown	69	11	9	53	142	58.2	1	10	10	34	55	64.0	197	59.7
Total	71	19	97	57	244	100.0	2	11	39	34	86	100.0	330	100.0
Percent	29.1	7.8	39.7	23.4	100.0		2.3	12.8	45.4	39.5	100.0			

Note: Four moose and three deer were also recorded, but not included in table.

A "cached" carcass, covered with snow, soil, grass, or other materials, was a sure sign that grizzly bears were feeding on it. *(Photo by John J. Craighead)*

animal's digestive and metabolic processes such that ingestion of newly sprouted vegetation causes bloating, systemic shock, and death. This syndrome appears to occur only as a sequel to the predisposing effects of chronic starvation and is the reason that so many ungulates seem to starve during spring when the plant resource is rapidly improving (Thorne *et al.* 1982).

Availability of Carcasses

The length of time that individual carcasses were available to grizzlies directly affected the use of other foods and the relative value of carrion within the total food resource. Determining the temporal availability of individual carcasses, from time of death to total carcass depletion, required more intensive, long-term field observations than we could make. The larger bison carcasses, however, were generally available for a much longer period than the smaller elk carcasses. This was attributable to the difference in size and ease of dismemberment. The thick, tough hide of adult bison could not be easily breached except through the anus or when decomposition was well-advanced. Observations at 62 of the 99 carcasses (46 elk and 16 bison) showed that a given ungulate carcass could be expected to remain available for an average minimum of 9.4 days. Some remained available much longer; seven of the 62

Fig. 10.5. Month of death of elk (*n* = 100) and bison (*n* = 26), Yellowstone National Park, 1960–72.

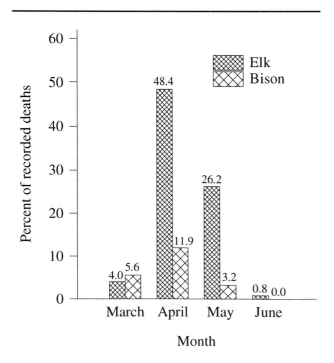

Table 10.8. Sex, age class, and condition of 130 elk and 29 bison winter-killed carcasses used by grizzly bears, Yellowstone National Park, 1960-72.

Carcass characteristics			Total number of carcasses available	Number of carcasses used by grizzlies[a]	Percent of available carcasses used by grizzlies
Sex	Elk	Male	13	8	62
		Female	43	20	47
		Unknown	74	23	31
	Bison	Male	9	9	100
		Female	5	4	80
		Unknown	15	11	73
Age class	Elk	Calf	67	17	25
		Yearling	2	1	50
		Adult	56	28	50
		Unknown	5	5	100
	Bison	Calf	2	0	0
		Yearling	1	1	100
		Adult	23	21	91
		Unknown	3	2	67
Condition (at time of death)	Elk	Excellent	0	0	—
		Good	2	2	100
		Malnourished	118	44	37
		Unknown	10	5	50
	Bison	Excellent	0	0	—
		Good	2	2	100
		Malnourished	23	19	83
		Unknown	4	3	75
Total carcasses	Elk		130	51	39
	Bison		29	24	83

[a] Based on direct (observation) and indirect evidence of carcass use by grizzlies; does not include feeding on carcasses of 23 elk and 1 bison taken by predation.

carcasses (11.3%) were used for a minimum of 20 days. For specific carcasses, however, the period of availability could be much briefer, depending on the intensity of use by bears and other scavengers. Carcasses used by as many as six to eight different grizzlies seldom lasted more than one to two days.

Use of Carcasses

We determined that a carcass had been used by grizzlies from direct observation of bears and, occasionally, by signs such as tracks, scats, and the appearance of the carcass itself. Grizzlies frequently "cached" carcasses for future use, so the remains were often covered with snow, dirt, or brush. Characteristic signs of bear feeding activity were hide and axial skeletal structure appearing as though turned inside-out, carcass progressively eaten from

hind- to fore-end, and long bones, girdles, and spine splintered or crushed.

Possible carcass-use preferences were evaluated according to species, gender, age class, and nutritional condition of the ungulate at time of death (Table 10.8). The apparent preference for bison was probably the result of sampling bias and not indicative of feeding habits. Grizzly bears did not show a significant or predictable preference based on gender or age of 159 bison and elk carcasses, except that carcasses of elk calves were used in a much smaller proportion relative to their availability than other age classes. Thirty-seven percent of malnourished elk and 83% of malnourished bison were used (Table 10.8). Of two elk and two bison carcasses judged to have been in good physical condition at

Table 10.9. Minimum number of grizzly bears observed using 89 elk and 26 bison carcasses, Yellowstone National Park, 1960-72.

Number of grizzlies per carcass	Number of elk carcasses	Number of grizzlies	Number of bison carcasses	Number of grizzlies	Total carcasses	Total grizzlies
0	15	0	1	0	16	0
1	59	59	10	10	69	69
2	5	10	3	6	8	16
3	2	6	4	12	6	18
4	6	24	4	16	10	40
5	0	0	3	15	3	15
6	2	12	0	0	2	12
7	0	0	0	0	0	0
8	0	0	1	8	1	8
Total	89	111	26	67	115	178
Carcasses used	74		25		99	
Grizzlies/carcass		1.2		2.6		1.6

Note: Data reflect only grizzlies actually observed on carcasses checked two or more times for feeding activity.

time of death, all were used by bears. This sample was too small to determine if such carcasses were preferred to carcasses of malnourished animals. Where there were numerous carcasses from which to select, however, it is reasonable to suppose that grizzlies would prefer those which were larger or less emaciated as a result of starvation.

Number of Grizzlies Observed at Carcasses

To determine the number of grizzlies using a carcass, we made direct observations of individually recognizable animals, many of them color-marked or radio-collared. We maintained repetitive observation until the carcasses were consumed. This allowed us to determine the minimum number of grizzlies using each carcass (Table 10.9). That 178 observations of grizzlies were made at 99 ungulate carcasses during a 13-year period indicates the importance of this food source to bears during the 1960-72 period. Most carcasses, whether elk or bison, attracted one grizzly, but three and four individual grizzlies were recorded at six and 10 carcasses, respectively, and one carcass attracted eight. An average of 1.6 grizzlies was recorded per ungulate carcass (Table 10.9).

Fifty-nine of 74 elk carcasses (79.7%) attracted only a single bear per carcass, and only two (2.7%) attracted as many as six. Overall, 111 observations of grizzlies were made at 89 different elk carcasses (Table 10.9), an average of 1.2 grizzlies per elk carcass.

Only one bison carcass failed to attract grizzlies; 10 carcasses (40.0%) attracted only one grizzly per carcass, three attracted as many as five per carcass, and one attracted eight. Overall, 67 grizzlies were observed at 26 bison carcasses, an average of 2.6 grizzlies per carcass (Table 10.9).

Winter-killed ungulates represented a significant early spring food that varied in abundance from year to year, but which was always annually available in sufficient quantity to be a dependable food source. Also, the fact that only 37% of the elk carcasses recorded on transects were used, suggests that this food category often exceeded the needs of the bears.

Predatory Behavior

As omnivores, grizzly bears are not highly efficient predators. Nevertheless, they possess predatory characteristics that allow them to capture and kill a wide range of prey under suitable conditions. Grizzlies are extremely fast and agile over short distances, and can maintain a fast gallop for many kilometers. In Hayden Valley, we clocked an adult bear on a 3-kilometer (2 mile) gallop that paralleled the Trout Creek road at just over 56 kilometers (35 miles) per hour. In 1930, Park Naturalist D. G. Yeager clocked the "rolling lope" of three grizzlies on a 3-kilometer (2-mile) stretch of snow-plowed road at more than 40 kilometers (25 miles) per hour (Yellowstone Nature Notes 1930). In 1937, District Ranger Ben Arnold clocked three yearlings on an oiled road at more than 48 kilometers (30 miles) per hour (Yellowstone Nature Notes 1937). Agility, combined with excellent paw-eye coordination, enables grizzlies to prey on small rodents, lagomorphs, and fish. Stamina allows them to trail and chase the large ungulates for lengthy periods

Bison carcasses were less numerous than those of elk but they attracted more bears per carcass. *(Photo by John J. Craighead)*

of time. Their olfactory sense is extremely acute, and their sense of hearing well-developed. Our observations indicate that the grizzly bear, in locating and tracking prey, whether small or large, relies primarily on its sense of smell (Craighead and Mitchell 1982).

Our research was not designed to quantify predation, but from the literature and in the course of other field work, we have assembled descriptions of sufficient predatory acts to form a general impression of grizzly bear hunting techniques and the role of predation as it relates to the diverse and, in some instances, abundant prey-base within the Yellowstone Ecosystem. Over a period of four decades (1920-58), Yellowstone personnel recorded grizzly bears in the act of killing large prey. Although anecdotal and often anthropopathic, they represent the rare instances when predation is directly observed. A spectacular battle that occurred on 20 May 1942 was described by Lester C. Abbie:

> Two combatants were fighting at close quarters, the bull elk (antlers shed) fighting for his life, the bear endeavoring to kill . . . . The bear had the elk gripped around the neck with his forepaws and was endeavoring to throw the elk . . . . The elk in turn was endeavoring to get free and by so doing was shoving the bear backward along the road. . . . The elk broke loose from the bear and started running away. He seemed partially stunned from his encounter. The bear ran right beside him and kept reaching out his paw now and then to slap the elk on the side of the head. The elk . . . appeared to be about exhausted . . . . The slapping action . . . caused the elk to stop and fight again, this time charging the bear which again reared on his hind legs. The bull's head struck the grizzly in the stomach and the bear then grabbed the elk around the neck. With all his strength the elk would butt the bear, each time raising the bear's hind feet from the ground. Shortly the elk became so tired that the bear's weight brought him to his knees.

> —(*Yellowstone Nature Notes* 1942).

Also in May 1942, Thomas Thompson noted a small herd of elk along the Gibbon River. The herd spooked, and Thompson detected a female grizzly and three yearlings moving toward the elk. When the mother and yearlings came to the spot where the elk had been bedded, they separated. Thompson described the bears as "searching on the ground first to the right and then to the left making short jumps as they searched for their prey." Thompson heard an elk calf squeal, relating that "On investigation . . . I found that they had two calves instead of one. The mother grizzly was after one and the cubs were after the other." The squealing brought back four or five cows to protect their young. Thompson disrupted the episode by blowing his car horn. The yearling stood up and ran away. The mother grizzly ran upriver after her offspring, picked up the calf elk she had killed and carried it about half a kilometer (quarter-mile), dropping it several times before disappearing with it into the timber (Yellowstone Nature Notes 1947).

In the spring of 1950, Harry V. Reynolds, Jr., reported the following in his field notes:

> The [elk] herd was pursued [in the Lamar Valley] at some distance by three mature grizzlies. Three calves in the herd were soon outdistanced by the mature elk and dropped far behind. The herd wheeled in a complete circle and picked up the calves again. This method of picking up the [straggling] calves was repeated at least four times. However, each time the pursuing bears were closer to the calves and on one such circle the calves were picked up just ahead of the [pursuing] bears. This time, one calf darted past the herd instead of running with it and was quickly downed by the three bears. I could hear the terrified bleating of the calf for some time.

Predation evidence from scat analysis and bone marrow compression tests was presented earlier. Here, we present our direct field observations—organized according to prey species, since the hunting and capture behaviors for the different prey varied considerably.

Elk and Moose

Among the ungulates, elk were by far the most important prey. This was basically a density-dependent relationship, since elk outnumbered all other ungulates combined. Elk were most vulnerable at three periods during the annual life cycle: the rutting season, the end of winter, and at calving.

During and after the rutting season, bull elk are more vulnerable to grizzly bear attacks than at other times when grizzlies may selectively hunt them. Bulls become vulnerable in fall because of their habit of bugling, their strong rutting scent, and their preoccupation with mating and holding a harem. Many also receive incapacitating wounds in combat with other contending males. Any of these situations may trigger predatory behavior in the bear, which then uses the combination of skills most appropriate to the specific opportunity.

Anecdotal accounts from field observers illustrate how grizzly bears capture elk during these periods and how they behave at their kills. In addition to having great power and speed, grizzlies are known to stalk with great skill and work cooperatively to pursue and kill large prey. As "loners," they kill and defend.

For example, in the fall of 1960, we received word from the Mt. Washburn lookout that a large grizzly had been spotted on a bull elk at the edge of a meadow. Although the lookout had not observed the act of killing, he had observed that there were no elk in the meadow where earlier there had been many, and soon afterwards he spotted a bear on a kill, with another bull elk nearby but unharmed. After feeding, the grizzly partially buried the carcass; while he was doing so, a smaller grizzly arrived, but did not challenge the larger animal. Later, the grizzly was observed defending the elk carcass against other grizzlies, ravens, and coyotes. Frank Craighead investigated the kill, and an abridgement of his notes illustrates the behavior of this grizzly at an elk kill:

> (Sept. 21, 1960) I decided to investigate the elk kill reported on Mt. Washburn. Ranger Bob Howe and I drove to the top of the mountain, but inclement weather prevented us from locating the grizzly and we decided to return in the morning.
>
> On the 22nd, the weather did not improve until mid-morning. With the dissipation of the clouds, we spotted the grizzly on his kill and observed him for a time as he alternately sprawled on and moved about the partially buried carcass. When the weather cleared, Bob and I held a compass course to the fourth meadow east of our observation point. We approached the meadow uphill and into the wind, jumping a bull elk which moved through the timber into the meadow. At the southern end of the meadow, we spotted the grizzly on his two-day-old kill. Bob and I began a cautious stalk of the bear utilizing terrain and timber to keep downwind and out of sight. We crawled to a small stand of trees in the open and took photos of the grizzly on the elk carcass. He became restless and moved to the timber a short distance from our stand, sniffing the ground and air for a scent. We were unable to see him in the timber and knew that he could approach us closely, undetected. We debated climbing a tree but decided the best course was to move into the open while the bear was in the timber. We crossed the meadow to the far side, where the grizzly could not approach us unseen and, in an emergency, we would still have trees to climb. The boar had been protecting his kill against other bears and coyotes and I was not at all certain how he might

react to human intruders. From our new position, we approached within 500 feet [154 m] of the grizzly but back-lighting and a rise in the terrain made it dangerous to approach closer. We retreated and tried a new approach from north to west. This time, noise or movement alerted him and the grizzly left the kill and moved toward us. As he entered the timber, we moved east keeping out of sight. We were again in the uncomfortable situation of having the bear within several hundred yards of us in dense lodgepole where we could neither see him, nor follow his movements. Each of us climbed a tree and waited for 20 minutes for the bear to show up. When he didn't, we descended from our precarious perches and again cautiously approached the elk kill across the open meadow. Suddenly the grizzly came out of the timber less than 250 feet [77 m] away. Bob and I retreated toward the pines a hundred yards behind us. For a moment, the grizzly hesitated and I took several steps forward, hoping an aggressive gesture might turn him. Instead, he reared up on his hind legs, lifting his forepaws as he did so, then dropped to all fours, sprinted toward us and again rose to a standing position. Within seconds, he dropped to all fours again and charged. Fortunately, the elk carcass lay between us and he stopped to claim it instead of coming on. We turned to face him as we gained the timber, recognizing we would not have time to climb the limbless lodgepole pines. As the grizzly reached the elk carcass, he got our scent, probably for the first time. In a flash, he wheeled about, ran for the far timber, and disappeared on the run. We returned to the elk carcass and determined that it had been killed by the grizzly, perhaps ambushed from the heavy timber while sparring with another bull elk. Later observations made while radiotracking other grizzlies indicated that the large harem bulls are at risk from the large grizzly boars during the elk rutting season and are especially vulnerable in dense timber or blowdowns where they go to rest or to recover from the wounds of conflict or the fatigue of battle.

As spring approached, malnourished elk were vulnerable to grizzly bear attacks during March and April, especially in areas where the snow was deep. In order to conserve energy, elk traveled the waterways, snow-free south banks of the streams and rivers, or snow-free thermal areas. We made numerous observations of grizzlies stalking or trailing both elk and moose, and we located numerous fresh kills, but observed few killings.

For example, we observed an adult grizzly with three yearlings suddenly burst out of the riverside timber in pursuit of a malnourished bull elk. The female charged the elk, while two of her yearlings raced downstream and the third turned upriver.

When the elk gained the timber on the far side of the Madison River, the female stopped at the water's edge, although she easily could have swum the distance. Two of the yearlings also stopped at the river bank, one about 0.4 kilometer, and the second about 0.5 kilometer (0.25 and 0.33 mile, respectively) from their parent. The third yearling remained in the timber several hundred yards from the open, grass-covered river bank. The action suggested a cooperative hunting effort by the family and occurred so spontaneously that we suspected the tactics had been employed before. Several days later, we found the bull elk on the bank dead and partially consumed. Tracks indicated that several bears had visited the carcass, including yearlings, but we could only confirm that the elk had been killed by bears, not how or what bears had been responsible. In the course of recording and investigating carcasses, it was not unusual to find freshly-killed elk on the river bank or partially submerged in the water. Although these acts of killing were not observed, it was evident from signs that predation had prevented a death from malnutrition.

Some grizzlies develop specific hunting strategies. Number 88 (Patch-eye), alpha male at the Trout Creek ecocenter in 1967, moved to the Firehole-Madison River drainage after the closure of Trout Creek in 1970. We observed him on three elk kills during the spring of 1971. Two of the three elk (both bulls) appeared to have been stalked and captured with a simple hunting strategy. Both carcasses were found in heavy timber less than 45 meters (50 yards) from the banks of the Madison River. We found drag-trails where the grizzly had hauled the elk from

As spring approached, winter-weakened elk traveled along the relatively snow-free south banks of rivers and streams. The disproportionate number of kills there suggested that grizzly bears developed hunting strategies to take advantage of this movement.
(Photo by John J. Craighead)

Grizzly bears are agile enough to catch small rodents, such as marmots and small voles, in the open, and they readily excavate their dens and runways.
(Photo by John J. Craighead)

the river. These observations suggested that Patch-eye, from concealment in the timber, attacked the elk wading the river, his superior strength enabling him to overpower them in the water. Bone marrow compression tests showed that neither elk was malnourished at the time of death. In the same area, we observed an adult female grizzly on a fresh elk kill in the middle of the river. We believed the female used a similar hunting strategy.

Even in thermal areas such as the geyser basins, spring snow was often 0.5 to 1 meter (2-3 ft) deep in the timber. Winter-weakened elk were able to travel more easily in the river or along its banks than in the deep, wet snow of adjacent timber. Thus, kills tended to be concentrated in riparian habitat.

Moose were scarce in comparison to elk, and they were not important in the grizzlies' diet, either as carrion or as prey. Nevertheless, on several occasions, grizzlies were observed trailing cow moose. An excerpt from the senior author's field notes documents such an occasion:

From an elevated site along Alum Creek, I observed a cow moose loping through the open grass-sage with a calf at heel. I followed them with field glasses for several miles. Suddenly, a female grizzly and her yearlings appeared two hundred yards [185 m] below me. The sow, in the lead with nose to the ground, followed precisely the trail taken by the moose; one yearling kept close to the sow, the other ranged at a distance of several hundred feet. At no time did the sow lose the scent; the family disappeared from view exactly where I had lost sight of the moose. Given the speed and staying power of the grizzly, I have little doubt that the calf could have become a victim from a concerted attack by the grizzly bear family, and that such occurrences represent normal hunting behavior.

During the elk calving season, we detected evidence of several elk calves that appeared to have been killed by grizzlies, and we made several observations of grizzly bears trailing cow elk and their calves. Only one observation was made of a confirmed kill. Robert Ruff and Dick Ellis, members of our research team, recorded the incident on 14 June 1964. They observed the following sequence of events on the Gibbon River, where it crosses the loop road in the vicinity of Wolf Creek:

> We heard a calf bawling on the south side of the river, and soon a cow elk crossed the meadow. Moments later, a large grizzly followed, carrying a bawling calf in its mouth. It crossed the road still carrying the calf, and we followed, entering the lodgepole timber in pursuit. From about 60 yards [55 m], we watched the bear, an adult female, start to feed. We later learned from another observer that the cow entered the river and crossed to the far side, but the calf was caught in midstream.

Bison

We did not observe grizzlies killing healthy bison, calves or adults, nor have we found records in the literature. But we did record several incidents where malnourished adult bison were killed. This generally occurred in the spring when the animals were immobilized by malnutrition and close to death.

Rodents

Among the most important rodents in the grizzly bear diet were the voles (*Clethrionomys gapperi* and, particularly, species of *Microtus*), available and especially vulnerable during early spring months as snow receded from subalpine meadows. When vole populations were high, individual bears and females with offspring preyed on them extensively and with great success. They were an important food item for bears soon after emergence from winter dens. Pocket gophers (*Thomomys talpoides*) and marmots (*Marmota flaviventris*) were also preyed on by bears, but to a lesser degree. These rodents were generally located by smell and, if conditions permitted, excavated from their burrows. Other rodent species indigenous to the Park were sometimes captured but were relatively unimportant food sources.

Fish

We did not observe predation on fish, either native or introduced species, nor did we find remains in feces. Bears frequented the lake shores of the Park and foraged in areas surrounding Pelican, Alum, Trout, Beaverdam, and numerous other creeks where spawning cutthroat trout were abundant. They were also frequently sighted along the Upper Madison, Firehole, Gardiner, and Gibbon Rivers, but we never saw a grizzly in the act of fishing. Bears, no doubt, used the fishery, but the level of use must have been extremely low. To obtain a general idea of food preference, we placed offerings of fresh cutthroat trout, elk meat, and raw potatoes at sites near the Trout Creek dump. The trout offerings were rejected,

as were the raw potatoes. The elk meat was immediately consumed.

Cannibalism

Grizzlies will kill and eat other grizzlies; they will also kill and eat black bears. The killing and eating of cubs is described under Infanticide (Chapter 8). Between 1959 and 1970, carcasses of two grizzly bears and several black bears were deposited by Park Rangers at open-pit dumps. All were consumed by grizzlies, although some carcasses were not eaten until they had putrefied. Meat of any kind is a high-preference food item, whether fresh or putrid. Grizzlies approached fresh black bear carcasses with caution, and the fact that some were not used immediately appeared to be a matter of cautious restraint, rather than a preference for decaying flesh. Few instances of grizzly bears killing and eating other bears have been documented in the Yellowstone Ecosystem. We recorded a 2-year-old female killed at Elephant Back in September 1962, and in June 1968, a large male grizzly attacked and killed a 4-year-old female that had been darted (drugged). He encountered her as she traversed an open meadow, still under the influence of the drug. Knight *et al.* (1981) reported that a yearling female with a foot injured by a snare was killed by bears at Nez Perce Creek in 1980 and later (Knight *et al.* 1989), that a male killed by another bear was found near the inlet of Yellowstone Lake in the fall of 1988.

ANNUAL ABUNDANCE OF FOOD RESOURCES

Determining quantitatively the annual availability of major "natural" food sources or food categories was beyond the scope of this study. It became evident early in the investigation, however, that the relative abundance of some major food categories fluctuated annually, and that some measure of these fluctuations would be essential to evaluating the nutritional role of the garbage dumps or ecocenters. Seven categories of food were recognized to be major resources: the grasses and sedges, forbs, berries, pine seeds, small rodents, carrion, and garbage. Seasonal variations in the biomass of grasses and sedges, forbs, and garbage were relatively small compared with the variation occurring in the other food categories. Because these food resources were either known to be stable (garbage) or generally widespread and abundant (grasses and the forbs, as a group), we did not measure their availability from year to year. We did, however, estimate and record the relative abundance of four major food categories that did fluctuated widely from 1960 through 1971. These categories were carrion (ungulates),

Fig. 10.6. Backcountry trek routes for survey of annual food availabilities, Yellowstone National Park, 1960-71.

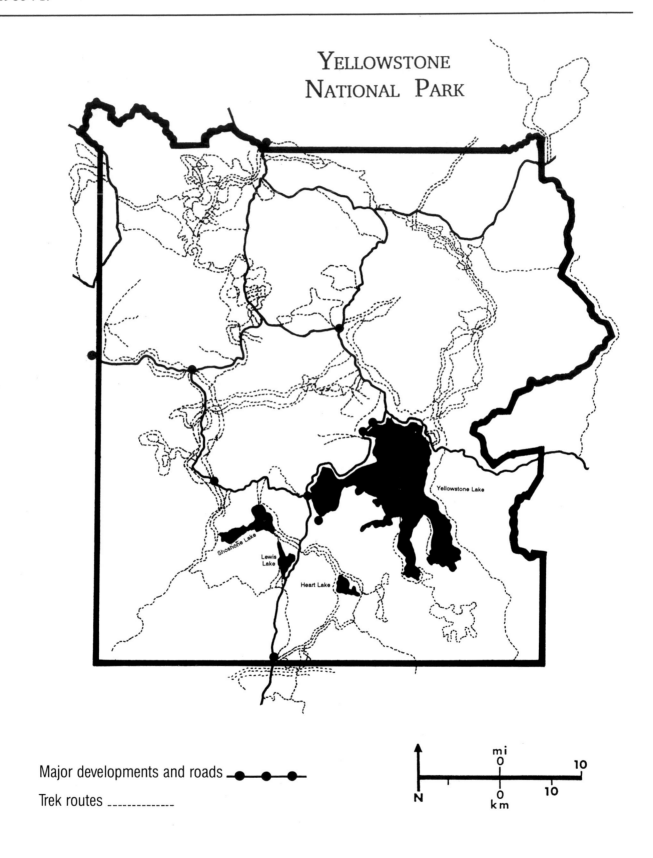

small rodents (*Microtus* spp. and *Thomomys talpoides*), berries (*Vaccinium globulare, V. scoparium*), and pine seeds (*Pinus albicaulis*).

Estimates of Relative Abundance

Relative abundance estimates of the annually variable food categories were obtained each year from transects throughout the Park (Fig. 10.6). Some transects were made specifically to estimate a particular food category; others were made in the course of related work that took us into a wide range of vegetation types. Observations of diagnostic field signs and direct inventory of the resource were made during the course of each day of travel. These were evaluated, and ratings of scarce, below average, average, above average, or abundant were then assigned to each food cat-

egory. The cumulative estimates of abundance for the major natural food categories—huckleberries (*Vaccinium* spp.), pine seeds, small rodents, and carrion—were evaluated each year and assigned a relative value from 5 to 25 (representing, by increments of five, the five subjective descriptors). The result was a record (Fig. 10.7) that depicted relative abundances of the natural food categories each year with sufficient accuracy to demonstrate relative changes in overall food abundance during the course of the field study.

Annual Composite Resource Values

The sum of the annual estimates of each of the four food categories provided a subjective composite value of total natural food availability for that year. As an example, carcasses of "winter-killed"

Fig. 10.7. Relative annual abundance of five grizzly bear food categories high in calories and/or essential amino acids, Yellowstone Ecosystem, 1960-71.

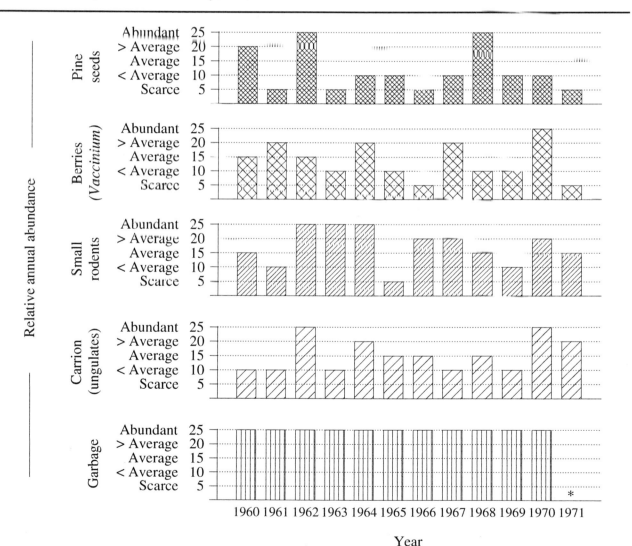

* Dumps at Trout Creek and Rabbit Creek had been phased out.

Fig. 10.8. Annual composite abundance values of four grizzly bear food categories high in calories and/or essential amino acids, Yellowstone Ecosystem, 1960-71.

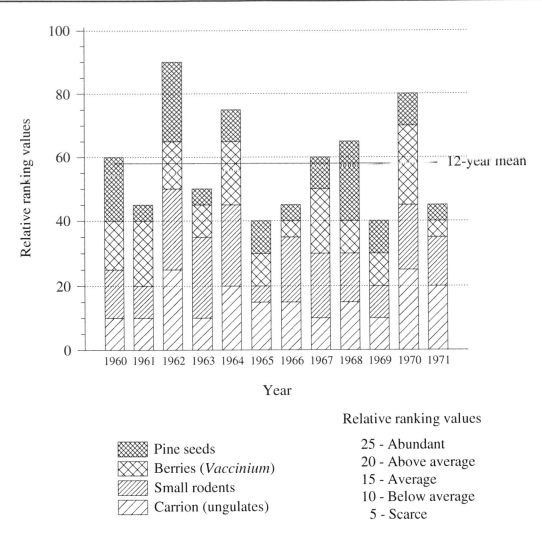

Relative ranking values

25 - Abundant
20 - Above average
15 - Average
10 - Below average
5 - Scarce

Pine seeds
Berries (*Vaccinium*)
Small rodents
Carrion (ungulates)

ungulates were numerous in 1962. Evidence of high rodent populations (*Microtus* spp. and *Thomomys talpoides*) was also observed that year as the snows receded. Concentrations of rodents attracted ravens, coyotes, gulls, various raptors, and grizzly bears, all excellent indirect indicators of abundance. Heavy and widespread girdling of woody vegetation and the distinct trail-mounding of pocket gophers further confirmed the over-wintering abundance of these prey species in specific areas. The berries of *Vaccinium* spp., their numbers estimated as the season advanced, were of average abundance in 1962, but the seeds of whitebark pine were rated abundant. Thus, the composite food resource value for natural foods in 1962 was 90 out of a possible value of 100 (four food types x maximum value per type of 25). This cumulative value was the highest recorded during the 12 years of study (Fig. 10.8). There were no years in which all four natural food categories were judged to be abundant (Fig. 10.7).

During three years (1961, 1966, and 1971), the composite values were only half that recorded in 1962, and during two years (1965 and 1969) the composite values were as low as 40, considerably below the average value of 60 for all years (Fig. 10.8).

Of the four food categories, the relative abundance of small rodents appeared to vary the least from year to year, followed by ungulate carcasses and berries of the vacciniums (*Vaccinium* spp.). The abundance of pine seeds was the least stable, being of average or greater value during only 3 of the 12 years, and scarce during 4 of the 6 years in which the composite rating was below average (Figs. 10.7 and 10.8).

Ideally, abundance of one food category will be sufficient to mitigate shortages in one or more of the other categories during any given year. But during some years, the food available to the grizzly population was low because several of the most

Fig. 10.9. Mean annual composite ratings of grizzly bear food abundance for 1960-71, Yellowstone National Park (four food resource categories plus garbage; see also Figs. 10.7 and 10.8).

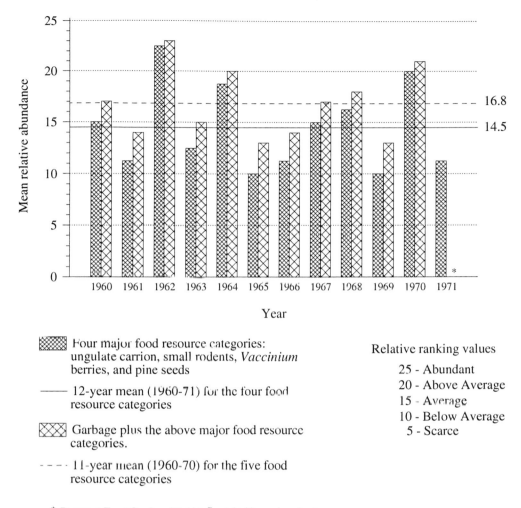

Four major food resource categories: ungulate carrion, small rodents, *Vaccinium* berries, and pine seeds

—— 12-year mean (1960-71) for the four food resource categories

Garbage plus the above major food resource categories.

- - - - 11-year mean (1960-70) for the five food resource categories

Relative ranking values

25 - Abundant
20 - Above Average
15 - Average
10 - Below Average
5 - Scarce

* Dumps at Trout Creek and Rabbit Creek had been phased out.

important food categories were subaverage (Fig. 10.8). The grasses, sedges, and forbs, highly stable energy sources but relatively low in calories, were available during the entire foraging season and served as a "survival ration" to carry the bear population through years when the four major energy sources were low.

Mean Composite and Total Resource Ratings

In order to evaluate the role of garbage, and to compare it with other food categories, we developed mean annual composite ratings for the four natural food categories (Fig. 10.9). Ratings for the four categories (Fig. 10.7) were summed for each year and divided by four to derive the annual composite mean. Then the ratings for all four categories were summed for all years and divided by 48 to estimate

the mean total resource rating of 14.5 for natural foods over the 12 years of investigation (Fig. 10.9). Thus, the mean total resource rating of the food base over the 12-year period averaged only a little over half its potential maximum value of 25. Actual realization of the maximum mean potential of 25 is clearly unrealistic because of stochastic biological processes. Nevertheless, we used this hypothetical value to compare the relative abundance of garbage with that of carrion, small rodents, berries, and pine seeds, and to evaluate its effect on the total food resource availability from year to year.

Garbage (Open-Pit Dumps)

We earlier demonstrated that garbage or human food wastes at open-pit dumps attracted and held aggregations of bears for extensive periods of time and were accordingly termed ecocenters. Although this food resource is normally considered artificial or "unnatural" compared to "natural" food resources such as carrion, rodents, berries, and pine seeds, the difference is largely one of human definition, reflecting human prejudices. By nutritional criteria and standards, both the "natural" and the "unnatural" food categories can be described and measured.

Direct Measurements

We attempted to sample garbage quantitatively as a means of estimating its nutritional value to the grizzly bear population. These attempts were not successful for the following reasons:

1. Garbage and trash were intermixed at all the dumps and could not be adequately separated for quantitative measurements.
2. Adequately sorting garbage from trash at the numerous collection sites, although attempted, far exceeded labor resources.
3. Truckloads varied daily in number and content throughout the summer season.
4. The food items were so numerous and varied (from lettuce to bacon grease) that analytical standards could not be applied with any degree of scientific accuracy as to overall nutritional quality.
5. Although caloric values of component lipids, carbohydrates, and proteins present in garbage are measurable, converting these values to practical, relative quantities available to or consumed by bears was beyond the scope of this investigation.

It became increasingly evident that assessing the nutritional value of a daily or annual supply of garbage would require some method of indirect evaluation. The most obvious indirect measurements were the number of bears attracted to the food source and the period of time the food source was available.

Indirect Measurements

Garbage was available in tremendous quantities at Trout Creek and the other disposal areas for a period of at least three months (June, July, and August) every year of our study, and was observed to attract and hold bears even when other food categories were abundant. Although unmeasurable, the quantity of food available daily at all of the open-pit dumps each year, and its consistent attraction of numerous bears, constituted grounds for assigning garbage the highest mean annual abundance rating of 25 for each of the 12 years of the study. In our judgement, this remained true during the phaseout years of 1968-70, when garbage was reduced at Trout Creek and Rabbit Creek, but was still sufficiently abundant to hold many members of

the aggregations. By incorporating this rating for the garbage resource into our appraisal of annual resource abundance (Figs. 10.7 and 10.9), we can evaluate the degree to which it damped the stochastic variation in annual, composite food resource availability and stabilized the ecosystem-wide food base.

The Role of Garbage

As we have shown, seldom were any of the natural food categories annually abundant. Carrion received a maximum rating in 2 of 12 years, small rodents in 3 of 12, berries of *Vaccinium* spp. in 1 of 12, and pine seeds in 2 of 12. Inclusion of garbage as a fifth food category with an annual abundance rating of 25 (Fig. 10.7) greatly enhanced the total resource ratings through 1970 (Fig. 10.9). Garbage was not only a massive food source, but was predictably available every year. As such, it raised the total food base within the Yellowstone Ecosystem and provided a nutritional stability unmatched by any of the other food categories. It is clear that human food wastes were important as an independent food resource. Less clear, and a subject of considerable controversy, are the effects of garbage-feeding on the behavior and health of a bear population.

Garbage Habituation and Human-Conditioning

Bears at the Trout Creek ecocenter became conditioned to feed on garbage. This habituation, however, did not preclude their feeding on a wide range of natural foods (see also Rogers 1976). Some researchers have concluded that bears also become conditioned to human scent at garbage disposal areas (Cole 1971, Herrero 1976). Our detailed observations of garbage-habituated grizzlies in Yellowstone provided no evidence to support these claims. We found that grizzlies habituated to feeding in campgrounds and developed areas in the near presence of humans lost their timidity and did become human-conditioned. Those that fed at isolated dumps such as Trout Creek did not (Craighead and Craighead 1971). Rogers (1989), in a study of black bears at garbage dumps, stated, "Human smell on garbage apparently does little or nothing to habituate bears to human presence." Moreover, Gunther (1995 [in press]), in a review of bear management problems over the 20 years following dump closures, concluded that garbage dump use by bears was not a major factor in rates of human injuries, but roadside feeding and poor campground sanitation were. Thus, current evidence suggests that bears become human-habituated in the presence of people, not from the presence of garbage alone.

Table 10.10. Human visitation-days (June through September) compared to grizzly-, bison-, and elk-days for the same time period, Yellowstone National Park, 1983 (data obtained from official Park records).

Species	Number of visitors	Mean visitor-days	Population estimate	Days	Animal-days
Human	2,375,396	1.7			4,038,173
Grizzly bear			200	120	24,000
Bison			2,000	120	240,000
Elk			30,000	120	3,600,000
Total wild animal-days			32,200	120	3,864,000

Because of this close and natural association between bears and humans, it is extremely important to the future management of bears that we determine more precisely how individual animals become human-conditioned. Human-conditioned bears are dangerous, but garbage-eating ones are not necessarily so (Craighead and Craighead 1971).

With the inevitable increase in human numbers in the future, and a corresponding increase in garbage disposal, we need to know more about the social and physiological aspects of garbage-feeding. To what extent does garbage-feeding offset the loss of bear habitat? Can dumps serve the biological needs of bears? If so, how and where should they be located? How can we erase the strong, but false public opinion that humans and their activities are not part of the natural ecology in our national parks and preserves? Some of these questions are addressed in Chapter 20.

The term "dump bear" has been used in the literature more as a disparaging term than as a class distinction from "wild," or non-garbage-feeding, bears. Barnes and Bray (1967) claimed that black bears in Yellowstone consisted of two distinct populations—backcountry bears and campground or roadside bears. Cole (1976) extrapolated this concept to grizzly bears, claiming the existence of distinct dump and non-dump populations. None of these researchers produced hard scientific evidence for their claims. On the other hand, we showed that as high as 77.3% of the ecosystem grizzly bear population visited an open-pit dump at one time or another, and that probably every grizzly in the Ecosystem visited dumps at some time in the course of a lifetime (Craighead *et al.* 1974). Adoption of Cole's unsupported classification of dump versus wild, or backcountry, bears by the Yellowstone administration led to misconceptions that found their way into the scientific literature and obstructed objective evaluation of the ecocenter phenomenon.

HUMAN IMPACT EVALUATION
Visitor-Animal Impacts

The importance of garbage as a major food resource for grizzlies can be placed in perspective by comparing "day use" of the Park by visitors versus use by elk, bison, and grizzly bears during the four-month period of most intensive resource use (June, July, August, and September).

In 1983, more than two million visitors averaged approximately 1.7 visitor-days each for a total impact of 4,038,173 visitor-days. In comparison, approximately 200 grizzly bears, 2000 bison, and 30,000 elk used the park resources during that period, for a total of 24,000, 240,000, and 3,600,000 animal-days, respectively (Table 10.10). Visitor-day use of 4,038,173 during the summer months exceeded the combined animal-day use of 3,864,000 by grizzly bears, bison, and elk. Numbers of visitors continued to increase throughout the 1980s and peaked at 3,186,590 in 1992. The visitors are concentrated along the loop roads (Fig. 10.3), which occupy much of the Park's prime riparian habitat. This habitat is used, to some extent, by elk and bison, in spite of the presence of large numbers of visitors; but much of it is avoided by grizzlies. It is evident that human visitors have been an integral part of the park biota since the Park's inception in 1872; as such, they must be considered in animal resource evaluations and management strategies.

"Natural regulation" within the Park or within the entire Ecosystem is a poorly defined term that discourages the assessment and integration of the ecological role of humans with the biological needs of the wildlife (Craighead 1991). Simply by happenstance, the waste food products of an ever-increasing human presence have, to some extent, countered the loss of habitat essential to the grizzly bear population. If humans are recognized as an integral member of the ecosystem biota, their refuse

Postmortems were performed on 13 grizzly bears to determine the effects of dump feeding and the extent of disease and parasitism in the Yellowstone grizzly bear population. *(Photo by John J. Craighead)*

No evidence was found that feeding on garbage produced health hazards for grizzly bears. Plastics, glass, styrofoam, and other foreign materials masticated and ingested during dump feeding produced no visible trauma to digestive organs. *(Photo by John J. Craighead)*

is "natural." The role of garbage (ecocenters) then becomes an ecological consideration. Aesthetic and philosophical considerations are in order, but they should not take precedence over practical and ecological issues in management actions.

Garbage-Feeding Bears and Population Health

Early on in the project, a question arose as to whether use of garbage as a major food source by bears might increase their rates of disease and parasitism. With the gracious collaboration of Dr. Morgan Berthrong, Pathologist at Penrose Hospital, Colorado Springs, field postmortems were performed on 13 grizzly and two black bears and detailed autopsies on an additional six and two, respectively (Berthrong and Craighead, unpubl. data, 1961-63). There was no organic pathology of

note observed from the black bears. Generally, this was true of the grizzlies as well. But pulmonary infarctions (vascular insufficiency resulting in tissue death in the areas affected), occasionally numerous, were noted in eight of 17 bears. Some of the infarcts may have resulted from intrapulmonary injections of immobilizing agents, but grossly identical lesions were present in uninjected bears. Microscopic examination failed to reveal hookworm or ascariform larvae, and the infarcts were presumed due to arterial emboli caused by unknown factors. Several grizzlies suffered severe bronchitis and bronchiolitis of unknown etiology which occasionally produced confluent abscesses. It is very unlikely that prevalence of such infections would be related to diet, per se, but rather would be associated with prolonged aggregation of a population infected with a pathogen transmissible by direct or aerosol routes.

Analysis of grizzly bear blood indicated high cholesterol levels in bears relative to healthy human values. However, atherosclerotic processes causing heart disease in humans were not present in the grizzly bears.

(Photo by John J. Craighead)

The skin of both black bears and grizzly bears frequently showed open wounds. Many of these were purulent, but etiology of the infections was not ascertained.

Other than frequent infection with helminths, the gastrointestinal tracts of both species were not unusual, with no evidence of inflammation, impaction, perforation, or ulceration due to metals, plastics, or other indigestible items common to unsorted refuse. Extensive analyses were performed on blood specimens drawn from 39 grizzly and three black bears (Berthrong and Craighead, unpubl. data, 1960-63). Of note was the high level (relative to the normal range for humans of ≈150-220 mg %) of cholesterol in all bears. The mean value for 26 grizzly bears was 325 mg %. Autopsy revealed no scars, fibrosis, or fresh infarction of the myocardium of any bear, however, nor were any thrombi noted. The coronary arteries of all but one bear, a 5-year-old female (No. 131), were entirely free of atherosclerotic processes. Several of that female's main coronaries showed mild, but very definite intimal fibrosis without fat deposition, and without resultant luminal stenosis or myocardial fibrosis. Thus, it appeared that the relatively high levels of serum cholesterol, whether normal for the bears or elevated as a result of a garbage diet, were not producing the atherosclerotic damage so frequently observed in humans. Additional laboratory analyses have indicated that bears carry the "excess lipids" as "good cholesterol" (high density lipoproteins), according to Berthrong (pers. comm. 1993).

The only parasites recorded from 17 grizzlies posted were tapeworms (11 bears) and roundworms (12 bears), the latter in occasionally heavy infec-

tions. None of these helminth specimens was taxonomically characterized. Neither *Trichinella spiralis*, nor *Dirofilaria* spp. were observed. Reviews of the parasites identified from North American grizzly and brown bears (Rogers and Rogers 1976, Craighead and Mitchell 1982) suggest that the tapeworms were either *Taenia* sp. or *Diphyllobothrium* sp. (most probably the latter) and that the roundworms were *Baylisascaris transfuga* and/or *Uncinaria* spp. The life cycles of all of these are such that a garbage diet would provide no advantage in transmission. That is, nothing intrinsic to the composition of garbage would increase the likelihood of parasitic infection. Nevertheless, as is the case with any parasite transmissible via feces (e.g., *Baylisascaris* spp. and *Uncinaria* spp.), the simple fact of population aggregation would enhance probability of infection.

The absence of trichinosis among the specimens is important in view of hypotheses by various researchers (King *et al.* 1960, Brown 1967, Wand and Lyman 1972a, 1972b) that feeding on garbage is directly related to prevalence of trichinosis in bears. Other researchers (Rausch 1970, Winters *et al.* 1971, Worley *et al.* 1974, Rogers 1975) have suggested that cannibalism and scavenging on carcasses of other carnivores are more realistic mechanisms for maintenance of trichinosis in bear populations than feeding on garbage. On the other hand, Keith Aune (Director of the Montana Wildlife Laboratory, pers. comm. 1993) reported that direct examination of tissues showed 23 of 69 Yellowstone grizzly bears tested since 1967 to be positive for *Trichinella*

Occasional infestations of grizzly bear by roundworms, transmissible via feces, were possibly the result of close-proximity feeding at ecocenters. *(Photo by John J. Craighead)*

Table 10.11. Summary of litter size and climate data used to calculate the relative effects of climate and dump closures on grizzly bear reproduction, Yellowstone National Park, 1959-81.

Year	Litter size Mean	(n)	Yearly climate index (Picton and Knight 1986)
1959	1.86	14	2
1960	2.06	17	-7
1961	2.31	13	3
1962	2.29	17	-1
1963	2.50	16	8
1964	1.88	8	0
1965	2.11	19	2
1966	2.13	15	-4
1967	2.50	12	7
1968	2.46	13	6
1969	2.00	14	2
1970			
1971	Litter size data not available.		
1972			
1973	1.86	14	-7
1974	1.73	15	-1
1975	1.50	4	-12
1976	1.88	17	3
1977	1.92	13	-4
1978	2.00	9	1
1979	2.23	13	-5
1980	1.92	12	1
1981	1.92	13	0
Mean 1959-81: (all years)	2.08	268	-0.30
Mean 1959-69: (Craighead study)	2.20	158	1.64
Mean 1973-81: (Knight study)	1.91	110	-2.67

spiralis (33.3%). But this rate was quite low compared to that from other regions of Montana. Aune noted a distinct gradient of prevalence and larval load (larvae/gram of tissue) in grizzlies that increased from southeast to northwest across the state. This suggests, as earlier pointed out by Worley *et al.* (1974), that bears are infected from one of their major prey species, most likely a wild rodent.

Finally, suggestions that scavenging at dumps increases rates of traumatic injuries to bears were not borne out in our many hours of direct observation. Lacerations due to broken glass or metal were not observed to pose significant health problems, nor did mastication or ingestion of plastics, styrofoam, or paper. Mouth injuries were uncommon. Although dental caries and broken teeth did

occur, we have no basis for judging whether they were more frequent than in other bear populations. Observations of bears wedging appendages in containers were rare, although in one instance a yearling grizzly wore a tin can on a forepaw with seemingly minor inconvenience for a period of about four weeks. When we removed the can, the bear was none the worse for wear. On the other hand, an adult female was suspected by the observers to have lost a forepaw as a result of entrapment in a tin can.

There was no evidence that feeding on garbage produced unique health problems for grizzly bears. Any increased potential for morbidity in the population would instead have been related to the dynamics of aggregation observed in populations of virtually any species. On the other hand, as we show later (Chapter 11), bears associated with the Trout Creek ecocenter experienced lower mortality due to human causes than did non-ecocentered bears.

REPRODUCTION AND FOOD ABUNDANCE

Numerous studies have clearly demonstrated that reproductive efficiency is directly related to nutritional well-being. Bear reproduction fluctuates from year to year in a density-independent fashion (Jonkel and Cowan 1971; Rogers 1976, 1983, 1987; Bunnell and Tait 1981; Beecham 1983; Elowe and Dodge 1989). McCullough (1981, 1986) reported a density-dependent dynamic in the 1959-70 Yellowstone grizzly bear population that keyed on the number of adult males, but careful analysis of our census data does not support his conclusions (Chapter 9). We have little doubt that annual variation in nutrition affected reproduction in the Yellowstone grizzlies during 1959-70, but this has proven difficult to demonstrate scientifically. As we have shown, major fluctuations occurred in the annual availability of the principal food categories, other than garbage (Fig. 10.7). Because of the long reproductive cycle (approximately three years), reproductive efficiency, as expressed by mean litter size, could not be consistently correlated with annual variations in wild food abundance. The annually dependable high-calorie foods and high-quality proteins provided by the open-pit dumps undoubtedly damped the effects of annual perturbations in the wild food categories (Fig. 10.9) through 1970. Thus, any detectable effects of scarcity of critical "wild" foods on reproduction or population structure from 1960 through 1970 would have been masked by the ready availability of "nonfluctuating" resources at the dumps.

The dumps were ecocenters, developing over many years as objects of migratory movements and sites of social aggregations. Their presence created food resource stability. With closure of the last open-pit dump (Trout Creek) in 1971, the rich and highly stable food resource so long available to grizzly bears throughout the Yellowstone Ecosystem was eliminated. If a relationship exists between annual food abundance and quality, and reproductive efficiency of the population, it should be evident in a comparison of mean annual litter sizes for the period before dump closure (1960-69) with that after (1973-81). In examining this, Picton and Knight (1986) concluded that climatic effects, specifically temperature and precipitation from October to May, were the factors responsible for the decline in grizzly bear litter sizes observed from 1973 through

1981. But this seems unlikely because their climatic index was derived from data at Mammoth, 1926 m (6320 ft), whereas a large proportion of the bear population lived and reproduced on the high plateaus, where elevation varies from 2134 to 3353 m (7000 to 11,000 ft) and climatic conditions are very different from those at Mammoth. Moreover, they did not attempt any analytical testing for partial effects of variables on litter size. Stringham (1986) concluded in his analysis of factors affecting litter size that the climatic index of Picton and Knight (1986) accounted for only 43% and 51% of the variations in cumulative and unduplicated litter sizes, respectively. Comparing data for the 1959-68 period with the 1969-76 period (after dump closure), he showed that closing of the dumps had about four times the impact on litter size, compared with climatic fluctuations (Stringham 1986).

We sought additional clarification of the relative effects of climate and dump closures on grizzly bear reproduction by analysis of covariance. Craighead and Redmond (1987) showed that when individual litter sizes (Table 10.11) were adjusted to account for climatic index values, the mean for the years before the dump closures (1959-69; mean of 2.20 adjusted to 2.17 cubs per litter) was significantly larger than the mean for the postclosure years (1973-81; mean of 1.91 adjusted to 1.95; $P = 0.02$). Litters born from 1970 to 1972 were omitted because they were conceived during the period of population perturbation that accompanied the major dump closures (1969-71), and because the quality of record-keeping could not be verified.

We conclude that the garbage dumps (ecocenters) were annually dependable food sources that exerted a positive effect on litter sizes. Forces other than climate and the food resource at dumps undoubtedly affected the mean numbers of cubs per litter for the two periods (see Factors Affecting Production of Cubs, Chapter 7). Picton considered the availability of carrion, a parameter certainly dependent to some extent on seasonal climatic variations, but has yet to make the data available or describe how his indices were derived. Data for other variable food types such as berries, pine seeds, and rodents were not available for the years after our study, and statistical analyses of the effects of variations in these were not attempted.

There is little doubt that food abundance plays a major role in ursine reproduction, but the subject is extremely complex and difficult to assess, partly because of the grizzly's extremely varied diet and its physiological responses to a wide range of environmental and population factors. We will discuss this subject more fully in Section II.

SUMMARY

The seasonal feeding by grizzlies on plants tended to follow plant phenology, with greatest use of emerging grasses and sedges in early spring, forbs in midsummer, berries in late summer, and pine seeds in the fall. The analysis of fecal material was used to identify specific foods consumed and the relative importance of major food resources in the grizzlies' diet. For a number of reasons, fecal analysis alone does not provide an adequate representation of the relative importance of food items in the total diet of grizzly bears; grasses and sedges tend to be overestimated and the meat from vertebrates underestimated. The fecal remains are evidence of what is excreted, not what is ingested. We, therefore, employed correction factors to develop dietary intake indices, modifying these with the inclusion of caloric values. This improved our ability to assess the comparative importance of various food categories in the grizzly bears' diet. Measurements of total caloric values for individual food plants and average caloric values for food plant categories or groupings showed that pine seeds rated highest, followed by berries and forbs, with the grass and sedge category lowest. Pine seeds and berries, having the highest energy values, were the least available to grizzlies, whereas grasses, sedges, and forbs, which had the lowest energy values, were readily and consistently available. The low-calorie food plants formed a third of the grizzlies' plant food caloric intake and the high-calorie ones constituted the remaining two-thirds. The importance of grasses, sedges, and forbs in the diet of the Yellowstone grizzlies related more to their annual abundance, seasonal availability, and wide distribution than to their average energy values. On the other hand, berries and pine seeds, in spite of great fluctuations in abundance and availability from year to year, were selectively sought, apparently for their high caloric values. In times of food shortage, the grasses, sedges, and forbs served as a "survival ration." The unreliable, high-calorie plant resources were, before 1970, amply compensated in the bear diet by garbage, which was predictably available in massive quantities each year. The garbage ecocenters provided a nutritional stability unmatched by any other food category.

Measurements of average energy values for vertebrate animal foods showed that the energy values of this food resource greatly exceed those for plants. The vertebrate food resource consisted primarily of ungulates and rodents, two high-energy foods that were more consistently available than the high-energy plant resources. Rodents were captured and consumed primarily in early spring with the recession of snow, whereas the ungulates—elk and bison—were victims of predation or consumed as carrion in spring and fall. Winter-killed elk and bison attracted numerous grizzlies at a time when other foods were scarce and the bears' energy levels were low following winter sleep. Although availability of carcasses fluctuated annually, they were always in excess of the bears' use of them. This resource represented an annually dependable food supply. Ungulate carcasses were exclusively limited to elk and bison winter range, which represented a tiny fraction of the entire Ecosystem.

Considering all food types, the most important food resources, as expressed by bear Dietary Intake Indices, were garbage, whitebark pine seeds, vertebrates, grasses and sedges, berries, forbs (as a group), and invertebrates, respectively. Garbage, grasses and sedges, vertebrates, and forbs did not fluctuate measurably in abundance. The others exhibited important measurable annual variations in abundance and were not predictably available to the bear population.

The feeding habits of the Yellowstone grizzlies changed slowly following closure of the garbage dumps. Minor changes in use of specific food categories occurred, but the basic food habits recorded for the 1968-71 preclosure period had not changed appreciably from 1973 through 1981. The most prominent changes were the expected decrease in use of garbage, an increased use of forbs and invertebrates, and a probable increase in use of calorie-rich vertebrate foods.

Bears at the ecocenters had become conditioned to feed on garbage. This habituation, however, did not preclude their feeding on a wide range of natural foods, nor did it appear to condition them to human scent. Grizzlies habituated to feeding in campgrounds and developed areas in the near presence of humans lost their natural timidity and did become human-conditioned. Those that fed at isolated ecocenters did not.

Studies of the physiology and pathobiology of grizzlies provided no evidence that feeding on garbage produced unique health problems, nor that it increased traumatic injuries. Lacerations from glass or metals and mastication or ingestion of plastics, styrofoam, or paper did not pose significant health problems.

The reproductive efficiency of grizzly bears is directly related to nutritional well-being. With the closure of the last open-pit garbage dump (ecocenter) in 1971, the rich and highly stable food resource so long available to grizzly bears throughout the Yellowstone Ecosystem was eliminated. The result was a lowered nutritional level among the bears, which was reflected in reproductive performance. Mean litter size of 2.17 cubs per litter before the dumps were closed was significantly larger than the mean of 1.95 for the postclosure period, 1973 through 1981.

The dumps as ecocenters, developing over many years, were foci of migratory movements and sites of bear social aggregations. Their presence created food resource stability that compensated for park habitat heavily compromised by human use and visitation. This stability was not reacquired, following ecocenter closures, by more intense and diverse use of "natural foods." The abundance of four important food categories (pine seeds, berries, small mammals, and ungulate carrion) oscillated over time. Never were all four food resources widely abundant in the same year. Garbage, consistently available as a fifth food category, provided an important foundation for the nutritional well-being of the bear population. For a decade following closure of the dumps, the bear population was nutritionally stressed. During this time, an expanding visitor population further compromised critical bear habitat. The full effects of dump closure over a 20-year period are discussed in Chapter 16.

Table 11.1. Observed mortality in the Yellowstone Ecosystem grizzly bear population by year, age class, and sex, 1959-70.

Year	Cubs F	Cubs M	Cubs U	Yearlings F	Yearlings M	Yearlings U	2-4-year-olds F	2-4-year-olds M	2-4-year-olds U	Nonadults[a] F	Nonadults[a] M	Nonadults[a] U	Adults F	Adults M	Adults U	Unknown F	Unknown M	Unknown U	Total[b] (known sex/age)	Total (all bears)
1959	-	1	-	2	-	-	1	-	-	3	2	-	1	6	1	-	1	2	11	16
1960	-	1	4	-	-	2	1	3	1	1	4	7	3	4	2	-	2	-	12	23
1961	1	2	4	-	3	-	4	3	-	5	8	4	1	1	1	1	1	1	15	23
1962	-	1	-	1	3	-	3	1	-	4	5	-	2	1	-	2	1	1	12	16
1963	1	2	-	-	1	-	1	3	-	2	6	-	3	2	-	1	1	1	13	16
1964	-	-	3	-	-	-	1	4	-	1	4	3	1	2	-	-	1	-	8	12
1965	-	-	3	-	-	2	1	4	-	1	4	3	3	2	-	-	1	1	10	15
1966	-	-	-	-	-	2	-	5	-	-	5	2	-	1	1	-	1	3	6	13
1967	1	1	3	-	1	-	-	5	-	1	7	3	7	5	-	7	5	3	20	38
1968	1	4	-	1	1	-	2	1	-	4	6	-	1	8	1	-	1	3	19	24
1969	-	-	2	-	3	-	2	1	-	2	4	1	8	6	-	1	3	-	20	25
1970	5	2	2	2	2	1	4	5	-	13	10	3	14	13	1	-	-	3	47	57
Total	9	14	19	6	14	5	20	35	1	37	65	26	44	51	7	12	18	18	193	278

[a] Sums, by sex, for cub, yearling, and 2- to 4-year-old age categories, plus bears recorded only as nonadults.
[b] Bears recorded only as "nonadult" excluded from total.

11
Demographic, Spatial, and Temporal Patterns of Grizzly Bear Mortality

INTRODUCTION

In early chapters, we focused on the recruitment of new individuals into the Trout Creek population during 1959-70 and on the social and physiological processes supporting reproduction. We examined the feeding habits, food base, and spatial needs that characterized the bear population in the Yellowstone Ecosystem. In short, we have described the life support system. Now we analyze the other part of the equation—death statistics for the period before dump closure. We examine the demographic, spatial, and temporal patterns of mortality in the Yellowstone Ecosystem generally and at Trout Creek specifically in an attempt to better understand the death rate among bears, and what caused them to die. If we know why bears died, we will also know how best to reduce the death rate. Death statistics, in a life system dominated by human-caused mortality, are far more amenable to change from management actions than are life statistics such as birth rate, productivity rate, or nutritive factors.

Information on grizzly bear mortality in the Yellowstone Ecosystem has been gathered by numerous agencies and individuals. Before 1959, state and federal agencies with management responsibilities within the Yellowstone Ecosystem were primarily responsible for maintaining records of grizzly bear deaths. Early data from National Park Service records and reports are presented in Table 2.5. These mortality records represent most of the deaths that occurred from control measures from 1931 to 1965. There has been no attempt to consolidate the relatively few grizzly bear deaths recorded by other agencies over this time span. With the initiation of our research study in 1959, the task of recording mortality data became the responsibility of our research team and of the Montana Department of Fish, Wildlife and Parks. The senior and second authors, along with Frank Craighead and agency associates, recorded mortality in the field and received and verified reports of mortality from the National Park Service and other cooperating agencies. Prior to 1967, we were also responsible for examining all carcasses of grizzlies killed within Yellowstone National Park. Specimens were kept at the University of Montana. From 1967 to 1987, Ken

Greer and Dan Palmisciano of the Montana Department of Fish, Wildlife and Parks, in cooperation with our research team, received the carcasses of all grizzlies killed within the Park and adjacent Montana lands. These specimens were housed at the Wildlife Laboratory on the Montana State University campus. Richard Knight and his associates on the Interagency Grizzly Bear Study Team documented the ecosystem mortality during the 1974-87 period. No single person or group was designated to document mortality during the 1971-73 period bridging our study and Knight's. The individual state agencies continued to keep records, and we helped them document bear deaths during this transition.

Following closure of the ecocenters in 1970, papers published on the status of the Yellowstone grizzly bear population used mortality statistics from a variety of different sources (Cole 1974, 1976; Knight et al. 1975, 1976). Thus, the mortality data so crucial to such evaluations varied greatly according to source and were selectively applied. Clearly, a standardized and acceptable database was needed. A cooperative effort initiated by the authors to collate information on mortality during 1959-87 into a single database culminated in 1988 with the publication of a standardized listing of 537 known, probable, or possible mortalities in Yellowstone and Grand Teton National Parks, the surrounding National Forests, and assorted private lands (Craighead et al. 1988b). This document summarizes all known mortality data for the period and is a comprehensive revision of an unpublished compendium by the senior author and Richard Knight. The database is the most complete record of grizzly mortality in the Yellowstone Ecosystem. It is the source for all analyses of mortality in this and following chapters.

The "ideal" record in the database contains information on the date of death, whether the bear was marked (and if so, its identification number), the bear's sex, age, and weight, the location and cause of death, and the record's source. In practice, there are two levels of uncertainty for many records in the database. First, there is uncertainty in some cases whether a mortality actually occurred. Thus, three levels of confidence in this regard were applied. A

Table 11.2. Age structure (number of bears censused and percent of total) of the Yellowstone Ecosystem grizzly bear population, 1959-70.

Year	Cubs		Yearlings		2-year-olds		3-4-year-olds		Adults		Total no.
	No.	% of total	No.	% of total	No.	% of total	No.	% of total	No.	% of total	
1959	26	16.9	23	14.9	17	11.0	—	—	88	57.2	154
1960	35	20.7	15	8.9	5	3.0	12	7.1	102	60.4	169
1961	30	18.1	17	10.2	17	10.2	23	13.9	79	47.6	166
1962	39	25.2	13	8.4	9	5.8	35	22.6	59	38.1	155
1963	40	22.6	29	16.4	11	6.2	25	14.1	72	40.7	177
1964	24	13.0	30	16.2	30	16.2	19	10.3	82	44.3	185
1965	40	21.4	20	10.7	34	18.2	23	12.3	70	37.4	187
1966	32	15.8	36	17.8	17	8.4	45	22.3	72	35.6	202
1967	30	17.1	24	13.7	23	13.1	26	14.9	72	41.1	175
Mean	32.9		23.0		18.1		26.0[a]		76.0[a]		176.0
Percent		18.7		13.0		10.3		14.8		43.2	
1968	32	17.7	19	10.5	15	8.3	25	13.8	90	49.7	181
1969	28	14.4	27	13.8	18	9.2	24	12.3	98	50.3	195
1970	21	11.7	18	10.1	15	8.4	31	17.3	94	52.5	179
Mean	27.0		21.3		16.0		26.7		94.0		185.0
Percent		14.6		11.5		8.6		14.4		50.8	

[a] 8-year mean since adults and subadults were not distinguished in 1959.

Table 11.3. Summary of the sex structure of the grizzly bear population in the Yellowstone Ecosystem, 1959-70.

Age class	Number of individuals			Percent	
	M	F	Total	M	F
Cub	46	32	78	59.0	41.0
Yearling	38	22	60	63.3	36.7
2-year-old	21	17	38	55.3	44.7
3-4-year-old	26	17	43	60.5	39.5
Adult[a]	38	44	82	46.3	53.7
All ages	169	132	301	56.1	43.9

[a] Adult sex ratio calculated on basis of mean of individuals observed annually, 1964-70.

"known" mortality status indicated that a carcass was recovered or that death was verified via radio telemetry. A "probable" status indicated strong evidence of death from highly reliable sources, but without verification via carcass recovery or radio telemetry. A "possible" status indicated circumstantial evidence of the death without prospect of validation (Craighead *et al.* 1988b). Second, for any level of confidence regarding occurrence of death, there may have been uncertainty about the sex or age of the bear, its location, or any of the various categories of descriptive information.

In our analyses, we considered only known and probable death records. This reduced a sample of 285 to a maximum of 278 for the 1959-70 period (Craighead *et al.* 1988b). Sample sizes for particular analyses may be smaller, differing from this total, and from each other, because of incomplete records in specific informational categories. For example, analyses of the causes of death will have larger sample sizes than those addressing sex and age at death because more mortality records lack sex or age data.

We emphasize that the database is primarily a record of human-caused mortality. Although a few cases of natural mortality were recorded, the vast majority of natural deaths, whatever the extent, went undetected. We begin each of the following sections by depicting grizzly bear mortality for the Yellowstone Ecosystem, as a whole. We then present a similar analysis of mortality for the Trout Creek subpopulation that has been the focus of preceding chapters. And finally, we examine our analyses for general trends.

SEX- AND AGE-SPECIFIC MORTALITY

Yellowstone Ecosystem

A summary tabulation of mortalities from all causes during 1959-70 (Table 11.1) was the source for all other tables presented in this section. These purely descriptive mortality data reveal relatively little about underlying biological processes leading to death or the effects these deaths had on the population. For example, we cannot tell whether the number of cubs dying in any given year, or throughout the 1959-70 period, was greater or less than expected, based simply on the proportion of cubs in the population. More generally, the raw data do not inform us whether risk of death varied as a function of any age or sex category or how the death rate in one year compared with another.

To determine whether mortality was age- or sex-dependent, we tested the null hypothesis that deaths occurred in simple proportion to the average age structure and sex ratio prevailing throughout the Ecosystem during the 1959-70 period (see Tables 11.2 and 11.3; see also Tables 1, 2, and 3 in Craighead *et al.* 1974). We know, however, that mortality was not evenly distributed over the 1959-70 period (Table 11.1). That is, a simple average of the yearly average age structures may not reflect the age structure prevailing when the majority of mortality occurred. To compensate for this, we weighted the proportion of cubs, yearlings, etc. in the population each year by the fraction of the total known-age/-sex mortality occurring in that year. For example, cubs constituted 16.9%, 20.7%, 18.1%,

Table 11.4. Comparison, by sex and age, of observed grizzly bear mortality from all causes with expected mortality, Yellowstone Ecosystem, 1959-70.

Age	Mortality estimate	Sex		Total
		Male	Female	
All nonadult[a]	OM	63	35	98
	(EM)[b]	(58.8)	(39.2)	(100.6)
Cub	OM	14	9	23
	(EM)	(13.6)	(9.4)	(32.2)
Yearling	OM	14	6	20
	(EM)	(12.6)	(7.4)	(23.1)
Other nonadult[c]	OM	35	20	55
	(EM)	(31.9)	(23.1)	(45.3)
Adult[d]	OM	51	44	95
	(EM)	(43.7)	(51.3)	(92.4)
All ages	OM	114	79	193
	(EM)[b]	(102.3)	(90.7)	

Note: OM = observed mortality; (EM) = expected mortality (see text for derivation); EM values in columns 3 and 4 equal expectation across sexes and within age class; EM values in column 5 equal expectations across age classes.

[a] Cub through four years of age.

[b] EM values do not equal sum of those for each age category because they are based on weighted average of sex ratios for all categories.

[c] Two through four years of age.

[d] Greater than four years of age.

. . . (etc.), and 11.7% of the Yellowstone Ecosystem population in 1959, 1960, 1961, . . .(etc.), and 1970 (Table 11.2). There were 193 deaths in 1959-70 for which age and sex were known (Table 11.1). Of these 11/193, 12/193, 15/193, . . . (etc.), and 47/193, respectively, occurred in 1959, 1960, 1961, . . . (etc.), and 1970. The weighted average proportion of cubs in the Yellowstone Ecosystem for the entire 1959-70 period, then, is equal to the summation (over all years) of cub deaths in a given year, divided by the total deaths for all years, times the percent representation of the cub age class in that same year, or: 11/193 (16.9) + 12/193 (20.7) + 15/193 (18.1), . . (etc.) + 47/193 (11.7) = 16.7%.

Weighted average proportions of the other age categories were calculated in similar fashion. Each proportion was multiplied by the total number of mortalities of known sex and age ($n = 193$) to give the expected distribution of mortality by age, under the null hypothesis that age did not influence the probability of mortality (expected values in parentheses, column 5, Table 11.4). Testing (G-test of goodness-of-fit) showed that the pattern of observed mortality across all four age categories (sexes combined) did not differ from these expectations ($P =$

0.15; Table 11.5). Similarly, when all nonadult age categories were combined to produce a single age class for comparison with the adult age class, we found that age had no effect on death statistics ($P = 0.50$). In other words, these data provide no evidence that age rendered the bear more or less susceptible to human-caused mortality.

To test whether sex affected the scope of mortality within each age class, and for all age classes combined, we performed a similar goodness-of-fit analysis except that the sex ratios used to calculate expected mortality within an age category were simple, rather than weighted averages of the sex ratio prevailing during the study (Table 11.3; see also Tables 2 and 3 in Craighead *et al.* 1974). Thus, for example, the cub sex ratio during 1959-70 was estimated to be 59 males : 41 females. Twenty-three known-sex cub mortalities were recorded (Table 11.1). The expected number of male cubs in this sample of 23 deaths is then (0.59 x 23) = 13.6, as contrasted with the observed number of 14 (Table 11.4). The sex ratio of observed deaths for each age category did not differ significantly from that to be expected for cubs, yearlings, other nonadults, all nonadults as a group, adults, or all age groups combined (Table 11.5). But observed deaths for males, irrespective of age, exceeded the expected values in all cases (Table 11.4), and deviation from the expected mortality toward excess death among males approached statistical significance, when all age-sex classes were combined (P = 0.09; Table 11.5).

In summary, when all causes of deaths were considered, the observed pattern of human-caused mortality throughout the Ecosystem with respect to age and sex did not differ significantly from that expected, given the prevailing age structure and sex ratios during 1959-70. That is to say, risk of death was not strongly dependent on sex or age category. The trend toward an excess of male deaths, however, was consistent with the apparent population shift in age-specific sex ratios from male- to female-biased with increasing age (Table 11.3).

Trout Creek Subpopulation

Here, we address the same issue of age- and sex-specific vulnerability to death for the Trout Creek component of the ecosystem population. Because the carcasses of the unmarked-but-recognizable bears would be recognized as Trout Creek bears only by research team members, we restricted analyses of mortality to the marked segment of the Trout Creek population. This reduced our sample size considerably (n = 39), and required that we recombine the four age classes employed in the ecosystem-wide analysis into two (adults and nonadults).

Table 11.5. Probability values associated with tests for effects of sex and age on grizzly bear mortality from all causes in the Yellowstone Ecosystem population and Trout Creek subpopulation, 1959-70.[a]

	P-values (degrees of freedom)	
	Ecosystem-wide	Trout Creek
Effect of sex (by age)		
Cub	>0.50 (1) (NS)	—
Yearling	>0.50 (1) (NS)	—
Other nonadult (2-4 years)	>0.25 (1) (NS)	—
All nonadult (cub-4 years)	>0.25 (1) (NS)	<0.01 (1) (Male excess)
Adult	0.14 (1) (NS)	>0.50 (1) (NS)
All ages	0.09 (1) (NS)	—
Effect of age:		
Four age categories[b]	0.16 (3) (NS)	—
Two age categories[c]	>0.50 (1) (NS)	0.15 (1) (NS)

[a] G-test for goodness-of-fit; the term "excess" means significantly greater mortality than expected; see text for methods of calculating expected mortality.
[b] Cub, yearling, other nonadult, and adult.
[c] Nonadult and adult.

Table 11.6. Observed mortality due to all causes in the Trout Creek grizzly bear population by year, age class, and sex, Yellowstone National Park, 1959-70.

Year	Cubs F	M	Yearlings F	M	2-4-year-olds F	M	Nonadults[a] F	M	Adults F	M	Total[b]
1959	-	1	-	-	-	-	-	1	1	-	2
1960	-	-	-	-	-	-	-	-	-	-	0
1961	-	-	-	-	-	1	-	1	-	-	1
1962	-	1	-	-	1	-	1	1	1	-	3
1963	-	-	-	-	-	1	-	1	2	-	3
1964	-	-	-	-	-	3	-	3	1	1	5
1965	-	-	-	-	-	1	-	1	2	-	3
1966	-	-	-	-	-	2	-	2	-	1	3
1967	-	-	-	-	-	1	-	1	1	-	2
1968	-	2	-	-	1	-	1	2	-	3	6
1969	-	-	-	-	-	-	-	-	3	2	5
1970	-	-	-	-	-	2	-	2	4	-	6
Total	0	4	0	0	2	11	2	15	15	7	39

[a] Sums, by sex, for cub, yearling, and 2- to 4-year-old age categories, combined.
[b] Bears recorded only as "nonadult" excluded from total.

Table 11.8. Causes of mortality in the Yellowstone Ecosystem grizzly bear population by age and sex, 1959-70.

Age category	Mgmt. control			Legal kills			Illegal kills			Natural			Accidental			Unknown			All causes		
	M	F	T	M	F	T	M	F	T	M	F	T	M	F	T	M	F	T	M	F	T
Cub	5	7	12	-	-	-	-	-	-	1	1	2	3	-	3	5	1	6	14	9	23
Yearling	9	5	14	4	-	4	1	-	1	-	-	-	-	-	-	-	1	1	14	6	20
2-year-old	1	5	6	3	1	4	2	-	2	-	1	1	1	2	3	-	1	1	7	10	17
3-year-old	5	4	9	8	1	9	1	-	1	-	-	-	1	1	2	-	2	2	15	8	23
4-year-old	5	-	5	8	1	9	-	-	-	-	1	1	-	-	-	-	-	-	13	2	15
Adult	21	16	37	24	17	41	4	1	5	-	-	-	1	10	11	1	-	1	51	44	95
Total	46	37	83	47	20	67	8	1	9	1	3	4	6	13	19	6	5	11	114	79	193

Note: See text for definition of categories of causation.

Table 11.7. Comparison, by age and sex, of observed grizzly bear mortality from all causes with expected mortality, Trout Creek ecocenter, Yellowstone National Park, 1959-70.

Age	Mortality estimate	Sex		Total
		Male	Female	
Nonadult[a]	OM	15	2	17
	(EM)	(10.0)	(7.0)	(21.6)
Adult	OM	7	15	22
	(EM)	(8.1)	(13.9)	(17.4)
All ages	OM	22	17	39
	(EM)	(18.1)	(20.9)	—

Note: OM = observed mortality; (EM) = expected mortality; EM values in columns 3 and 4 equal expectation across sexes and within age class; EM values in column 5 equal expectations across age classes.
[a] Cub through four years of age.

Death statistics for Trout Creek bears are presented in Table 11.6. The sex and age of all bears were known, so the annual totals of cubs, yearlings, and 2- to 4-year-olds shown in the nonadult category consist only of known-age and known-sex bears, in contrast to the compilation for the entire Ecosystem (Table 11.1). Values for expected age- and sex-specific mortality were calculated in the same manner described for the ecosystem-wide analysis, except that the average yearly age structure (Table 5.5) and sex ratio (Table 5.6) for the Trout Creek population (Chapter 5) served as the basis for the calculations presented in Table 11.7.

Considering mortality from all causes among all age classes, age did not affect the likelihood of death. Also, sex did not affect the likelihood of death among adults. Nevertheless, there were more deaths than expected among nonadult males (Table 11.5).

We repeated the entire analysis, excluding all research-related deaths (9 of 10 accidental deaths, or 23% of the total of 39), and obtained identical results: neither sex nor age influenced the likelihood of death ($P > 0.40$ and 0.50, respectively), but the number of nonadult male deaths was higher than expected ($P = 0.03$).

CAUSE-SPECIFIC MORTALITY
Yellowstone Ecosystem

We now extend our analyses to consider the effects of age and sex with regard to specific causes of mortality. For instance, did gender or age increase the risk of death from such specific causes as hunting, poaching, or management control? Also, did the movements of bears vary sufficiently with sex and age to be the determinant factor? We recog-

Table 11.9. Comparison, by cause-of-death, sex, and age, of observed mortality of grizzly bears with expected mortality, Yellowstone Ecosystem, 1959-70.

Age	Mortality estimate	Management control			Legal kills		
		Male	Female	Total	Male	Female	Total
Nonadult[a]	OM	25	21	46	23	3	26
	(EM)	(27.4)	(18.6)	(42.6)	(15.5)	(10.5)	(35.4)
Adult	OM	21	16	37	24	17	41
	(EM)	(17.0)	(20.0)	(40.4)	(18.9)	(22.1)	(31.6)
All ages	OM	46	37	83	47	20	67
	(EM)	(44.4)	(38.6)	—	(34.4)	(32.6)	—

Note: OM = observed mortality; (EM) = expected mortality; EM values in columns 3-4 and 6-7 equal expectations across sexes and within age class; EM values in columns 5 and 8 equal expectations across age classes.
[a] Cub through four years of age.

Table 11.10. Probability values associated with tests for effects of age and sex on grizzly bear mortality due to management control or legal kill in the Yellowstone Ecosystem population and Trout Creek subpopulation, 1959-70. G-test for goodness-of-fit; the term "excess" means significantly greater mortality than expected; see text for methods of calculating expected mortality.

	P-values (degrees of freedom)			
	Management control		Legal kill	
	Ecosystem	Trout Creek	Ecosystem	Trout Creek
Effect of sex (by age):				
All nonadult	>0.40 (1) (NS)	—[a]	0.001 (1) (Male excess)	—[a]
Adult	0.19 (1) (NS)	—[a]	0.12 (1) (NS)	—[a]
Effect of age:				
Nonadult versus adult	>0.40 (1) (NS)	0.007 (1) (Adult excess)	0.03 (1) (Adult excess)	>0.50 (1) (NS)

[a] Insufficient sample sizes for analysis by both age and sex.

nized six categories of cause: management control, legal kill, illegal kill, natural mortality, accidental death, and cause-unknown. The legal kill included mortalities from sport hunting, defense of livestock, and defense of human life. Illegal kills included grizzlies that were poached, mistaken for black bears by hunters, or shot in any other circumstance not deemed legal, as described above. The accidental category combined mortalities from research- (capture-) related activities and automobiles. With the exception of the accidental category, these six causes of death conformed to those categories presented in Craighead *et al.* (1988b).

Mortalities for the Yellowstone Ecosystem population during 1959-70 were sorted by cause, age, and sex (Table 11.8). To obtain adequate sample sizes for statistical analysis, we consider only the two major causes of death (management control and legal kill) in relation to two categories of age (nonadult and adult) in each sex (Table 11.9). Expected mortality for the four categories of age and sex were calculated as described in the previous section, except that (1) the average annual age structure was calculated by weighting yearly age structures by the magnitude of cause-specific, known sex or age mortality rather than total known sex or age mortality, and that (2) the sex ratio for nonadults was calculated as a weighted mean of the sex ratios, over the period 1959-70, of cubs, yearlings, and 2- to 4-year-olds—each component sex

Table 11.11. Causes of mortality in the Trout Creek grizzly bear population by age and sex, Yellowstone National Park, 1959-70.

Age category	Mgmt. control M	F	T	Legal kills M	F	T	Illegal kills M	F	T	Natural M	F	T	Accidental M	F	T	Unknown M	F	T	All causes M	F	T
Cub	–	–	–	–	–	–	–	–	–	–	–	–	2	–	2	2	–	2	4	0	4
Yearling	–	–	–	–	–	–	–	–	–	–	–	–	–	–	–	–	–	–	0	0	0
2-year-old	–	–	–	1	–	1	1	–	1	–	–	–	–	–	–	–	–	–	2	0	2
3-year-old	–	–	–	2	–	2	1	–	1	–	–	–	–	–	–	–	1	1	3	1	4
4-year-old	2	–	2	4	–	4	–	–	–	–	1	1	–	–	–	–	–	–	6	1	7
Adult	4	6	10	3	1	4	–	–	–	–	–	–	–	8	8	–	–	–	7	15	22
Total	6	6	12	10	1	11	2	0	2	0	1	1	2	8	10	2	1	3	22	17	39

Note: See text for definition of categories of causation.

Table 11.12. Comparison, by cause-of-death and age, of observed mortality of grizzly bears with expected mortality, Trout Creek ecocenter, Yellowstone National Park, 1959-70.

Age	Mortality estimate	Management control	Legal kills
Nonadult[a]	OM	2	7
	(EM)	(7.0)	(6.1)
Adult	OM	10	4
	(EM)	(5.3)	(4.9)
All ages	OM	12	11
	(EM)	(12.3)	(11.0)

Note: OM = observed mortality; (EM) = expected mortality.
[a] Cub through four years of age.

ratio being weighted by the average proportion of cubs, yearlings, and 2- to 4-year-olds, respectively, in the Yellowstone Ecosystem population during the period (Table 1 in Craighead et al. 1974).

Comparison of expected and observed mortality revealed no evidence that age affected a bear's susceptibility to death by control actions designed to manage the population (Table 11.10). Adult mortality, however, was shown to be greater than expected among legal kills. The sex of adult bears did not appear to influence their deaths from these two causes (Table 11.10). Among nonadult bears, death due to management control also was not affected by sex. But deaths of nonadult males were significantly greater than expected in the legal kill (Table 11.10).

Trout Creek

The Trout Creek sample of mortalities of known sex and age (Table 11.11) was adequate only to examine the effects of age on the risk of death. Two categories of age (adult and nonadult) were used (Table 11.12). In contrast to the ecosystem-wide analysis, (1) the age of Trout Creek bears did not affect their susceptibility to death by legal kill, and (2) adults were more susceptible than nonadults to mortality related to management control activities (Table 11.10).

TRENDS ASSOCIATED WITH DEMOGRAPHIC DISTRIBUTION OF MORTALITIES

Field biology is seldom amenable to controlled experiments. Gathering and interpreting mortality data are no exception. We must rely largely on quantitative input (the size of database) and the application of acceptable methods for analytical interpretation. We attempt to approach an ideal, but the ideal will always be unattainable. Thus, two general points need to be made before proceeding to detailed interpretation of our results.

First, our procedure of comparing sex- and age-specific expected and observed mortalities is most accurate for situations with no time overlap between the period for determining age and sex structure of the population and the period for determining mortality. In practice, census of the various ecocenters for age and sex composition from June through August overlapped widely with the period when mortality was recorded from April through October. Such overlap tends to create artificial differences in observed versus expected mortality.

Second, deviations from the expected mortality might occur for reasons other than assumed above, where we tested the premise that bears in certain categories of age and sex might be more susceptible to certain environmental pressures or causes of death than bears in other categories. Specifically, detecting mortalities might have been easier in certain age or sex categories than in others, and this may have created the appearance of, or masked the existence of, differential vulnerability. Detectability, in the present case, may be defined as the probability that a carcass is found, aged, and sexed. Carcasses of small, young bears were probably less apt to be found at all than those of large, old bears. Small carcasses also decay more rapidly than large ones, making age and sex determination more difficult when such carcasses are located.

Both of these general uncertainties reflect the difficulties of opportunistically recording death-related specifics throughout a large biogeographic area. With these limitations in mind, we interpreted the situations in which observed mortality differed significantly from expected mortality.

Yellowstone Ecosystem: Excess Legal Kill of Nonadult Males and Adults of Both Sexes

That mortality from legal kill fell proportionately more heavily on adults than on nonadults, and more on nonadult males than on nonadult females, can be attributed to hunter discrimination for larger (trophy) bears. Adults are, on average, larger than nonadults of all ages, and nonadult males are larger than females of an equivalent age (Chapter 16). If such discrimination occurred, it may have been

particularly effective in cases where hunters shot an animal from groups of bears they could compare directly for size; e.g., nonadult males with opposite-sex littermates. Although opportunities for direct size comparison of adult males with adult females does not often occur during the hunting season, we should still expect an excess in adult male deaths because females with cubs were not legal targets for most of the period during which legal kill occurred (see also Miller 1990).

Trout Creek: Excess Adult Mortality from Management Control

An excess of adult deaths of both sexes occurred among Trout Creek bears due to management control but did not occur ecosystem-wide. This has a straightforward explanation. Trapping of problem bears by National Park Service personnel and by our research team was pursued more intensively at the Canyon Village, Lake, and Bridge Bay developed areas than elsewhere in the Park or Ecosystem. These areas were relatively close to Trout Creek and, thus, attracted Trout Creek bears. A relatively high proportion of these adult bears that caused problems were killed.

Trout Creek: Excess Nonadult Male Mortality from All Causes

Nonadult male mortality from all causes was excessive at Trout Creek, but not ecosystem-wide. The reason for this may be reflected in the spatial distribution of bear deaths relative to the approximate center of the Yellowstone Ecosystem (i.e., the Trout Creek ecocenter [Chapter 10]). These data reveal a marked sexual difference in dispersion at death among nonadult (2- to 4-year-old) bears only. Presumably, the greater dispersion of deaths for nonadult males than for nonadult females reflects a greater tendency for nonadult males to disperse or wander (Craighead 1976, Reynolds 1995 [in press]). Sexual differences in dispersal and movement may have been a more decisive factor in differentially exposing Trout Creek nonadult males to mortality outside the Park than was the case for non-Trout Creek nonadults (males and females) that occupied home ranges located more peripherally in the Yellowstone Ecosystem. In other words, peripherally located nonadult females were not favored by the protective confines of the Park and did not have to move far to get into trouble.

THE SPATIAL DISTRIBUTION OF GRIZZLY BEAR DEATHS

Where an animal dies might appear to be largely a matter of chance. Nevertheless, it is clear that the

Table 11.13. Percent of all grizzly bear mortalities and median distance of locations of death from Trout Creek for each 2-year interval of the first (1959-64) and second (1965-70) halves of the study, Yellowstone National Park.

Years	Percent of mortalities in Park[a]	Distance from Trout Creek in kilometers (mi)	n
1959/1960	44.1	47.8 (29.7)	34
1961/1962	52.9	37.6 (23.4)	34
1963/1964	60.0	31.1 (19.3)	25
1959-1964	51.6	44.2 (27.5)	93
1965/1966	38.5	55.3 (34.4)	26
1967/1968	41.0	49.8 (30.9)	61
1969/1970	45.1	51.3 (31.9)	82
1965-1970	42.6	50.5 (31.4)	169
1959-1970	45.8	48.1 (29.9)	262

[a] Excludes research-related mortality.

Table 11.14. Causes of grizzly bear deaths, by age class, inside versus outside of Yellowstone National Park, 1959-70.

Cause	Inside the Park		Outside the Park	
	Male	Female	Male	Female
Management control				
Nonadult[a]	25	21	1	0
Adult	19	15	2	1
Legal kills				
Nonadult[a]	1	0	23	4
Adult	1	0	22	17
Illegal kills				
Nonadult[a]	0	0	4	0
Adult	0	0	4	1
Total	46	36	56	23

Note: See text for definition of categories of causation.
[a] Cub through four years of age.

landscape is, from the bear's point of view, a complex and variable mosaic of risk versus security. Some of this variation is "natural," in the sense that it arises from habitat or topographical differences, or from social relationships and the distribution of bears themselves. With the encroachment of humans, however, the dichotomy between secure and risky space becomes much more acute. Roads, towns, campgrounds, and ranches become important population "sinks" (Knight *et al.* 1988b). In this section, we document the spatial pattern of the risk of death from human-causes during 1959-70. Specifically, we examine bear movement and home range placement as factors exposing bears to higher risk of death.

All Yellowstone Ecosystem Grizzly Bears

Slightly less than half of all mortalities in the Ecosystem during 1959-70 occurred inside the boundaries of Yellowstone National Park (Table 11.13). But this fraction varied considerably during the study, ranging from a low of 38.5% in 1965-66 to a high of 60.0% in 1963-64. Some of this variation can be explained by short-term changes in the forces causing grizzly bears to die. For example, legal hunting kills outside the Park were extremely high in 1967 because Wyoming hunters, anticipating the closure of hunting in 1968, shot grizzlies in record numbers. During the second half of the study, however, there was a general tendency for mortality to shift from inside to out-

side of the Park (Chi-square test of independence, $P = 0.16$; each mortality categorized as in- versus out-of-the-Park and before- versus after-the-midpoint of the study) (Table 11.13). Furthermore, late in the study, a significant increase (Median test, $P = 0.05$) occurred in the dispersion of bear death locations from the Trout Creek ecocenter (Table 11.13).

Virtually all legal kills—hunting, defense of livestock, defense of life—and illegal kills were located outside the Park (Table 11.14). This is no surprise, considering that legal hunting and livestock grazing were prohibited inside the Park. Poaching a bear was probably much more difficult inside than outside the Park as well. More significant was the higher frequency of management control deaths within the Park versus their virtual absence outside the Park (Table 11.14). As we shall see in Chapter 18, there was a marked shift in the spatial distribution of all deaths, and of management control deaths in particular, after 1970.

There were significant differences related to age and sex in the spatial distribution of deaths during the 1959-70 period (Table 11.15). Ecosystem-wide, locations at death among cubs and yearlings (cub/yearling category) of each sex were more concentrated about Trout Creek than were adult mortalities of the same sex (Median Test, $P = 0.02$ and $P = 0.004$ for males and females, respectively; Table 11.15). Similarly, mortality locations of male cub/yearlings ecosystem-wide were less dispersed than those of nonadult males (2- to 4-year-olds) (Median Test, $P = 0.01$; Table 11.15). In contrast,

Table 11.15. Median distance from Trout Creek to location of death for all grizzly bears of known sex and age, Yellowstone National Park, 1959-70.

Age at death	Median distance in kilometers (mi)	
	Female (*n*)	Male (*n*)
Cub/yearling	26.5 (16.5) (15)	28.1 (17.5) (25)
Nonadult (2-4 years)	18.0 (11.2) (18)	49.8 (30.9) (35)
Adult (5 years and older)	42.4 (26.4) (36)	53.7 (33.4) (50)

Note: Significant pair-wise differences (Median Test) indicated by brackets, as follows: *P* = 0.004 for (a); *P* = 0.01 for (b); *P* = 0.01 for (c); *P* = 0.02 for (d); smallest *P*-value for nonsignificant differences was *P* = 0.21.

nonadult female deaths were more concentrated, though not significantly so, about Trout Creek than cub/yearling mortalities of the same sex (Median Test, *P* = 0.87). Finally, death locations of nonadult males were substantially more widely dispersed than locations for nonadult females (Median Test, *P* = 0.01; Table 11.15).

The age and sex differences in bear deaths were related to the locations where bears died, and these differences had at least two causes. First, there were age and sex differences in the distribution of home ranges about Trout Creek. The home ranges of females and dependent offspring tended to be smaller and more tightly clustered about Trout Creek than the ranges of other bears (Craighead 1976). Secondly, telemetry data showed that male grizzly bears used much larger areas than did females (Craighead and Craighead 1965, Craighead 1976, Nagy and Haroldson 1990). Trout Creek bears in age and sex classes where exploratory movement was great, and those adult bears with large home ranges or those using migratory corridors, tended to die farther from Trout Creek than did other Trout Creek bears (see also Chapter 18).

Most cubs and yearlings that died were still under maternal care at the time of death, so the relative concentration of cub/yearling death locations primarily reflected their mothers' patterns of space use. The relatively greater dispersion of adult female deaths reflects the movements of females both with and without dependent young (Table 11.15). This suggests that females without dependent young, were they distinguishable in the database, may have shown greater median dispersion from Trout Creek than adult females with offspring. If so (and accepting that differential home range placement and

movement generated sex- and age-associated variance in the dispersion of bear deaths), these data indicate that females with dependent offspring and nonadult females tended to restrict movements more tightly about Trout Creek and other ecocenters, whereas females without offspring, adult males, and nonadult males tended to roam more widely throughout the Yellowstone Ecosystem. Nonadult females commonly establish home ranges within the home ranges of their mothers (Craighead 1976, Blanchard and Knight 1991, Reynolds 1995 [in press]). This behavior may have been a factor in the trend (nonsignificant) toward fewer deaths among nonadult females than among nonadult males (Table 11.4) and may partially explain the great difference in movement between these two sex-age groups, expressed as median distance from Trout Creek to site of death (18.0 kilometers [11.2 mi] for nonadult females versus 49.8 kilometers [30.9 mi] for nonadult males, Table 11.15).

Trout Creek Grizzly Bears

There were no significant differences in the dispersion of mortality by cause and sex and age among the Trout Creek subpopulation in cases where comparisons were possible (Table 11.16; adult male versus adult female, Median Test, *P* = 1.00; adult male versus nonadult male, Median Test, *P* = 0.33). But sample sizes were too small to read any biology into these results.

The higher percentage of deaths within the Park (60%) that occurred among the Trout Creek subpopulation simply reflects the tight spatial distribution of this large, ecocentered group of bears. Therefore, we should expect that the locations of the deaths would be relatively clustered about the

Table 11.16. Median distance from Trout Creek to locations of death for marked Trout Creek grizzly bears, Yellowstone National Park, 1959-70.

	Median distance in kilometers (mi)	
Age at death	Female (*n*)	Male (*n*)
Cub/yearling	—	1.0 (0.6) (2)
Nonadult (2-4 years)	10.6 (6.6) (2)	56.6 (35.2) (11)
Adult (5 years and older)	12.6 (7.8) (18)	12.9 (8.0) (7)

Table 11.17. Summary of distribution of mortalities and median distance from Trout Creek to locations of death for marked Trout Creek grizzly bears, Yellowstone National Park, 1959-70.

Years	Percent of mortalities in Park	Distance of mortalities from Trout Creek[a] in kilometers (mi)	*n*[a]
1959-70	60.0	16.8 (10.4)	30

[a] Excludes research-related mortality.

Trout Creek ecocenter and within the Park (Table 11.17) and that median distances from place of death to the Trout Creek ecocenter would be less for Trout Creek bears than for all bears ecosystem-wide whose death locations were known. This was the case, as shown in Tables 11.15 and 11.16.

SEASONAL PATTERNS OF MORTALITY

We now explore when deaths occurred to determine if mortality was seasonally related from 1959 to 1970 and, if so, for what reasons. Mortality was recorded in each month from April through November (Fig. 11.1), but most occurred from June through October. Of course, this represents the observed, not the true temporal distribution of mortality, because natural mortality in the den (November-April) was not included. Furthermore, natural mortality outside the den was, no doubt, under-represented in spring (April-May) and fall (October-November), when the bears were dispersed and fewer observers were active.

Nonetheless, the lower recorded mortality in April, May, and November from all causes was probably also related to lower visitation of humans to Yellowstone in those months and hence, to a lower frequency of bear-human conflicts (see also Miller and Chihuly 1987). Mortality increased in June and July, concomitant with highest levels of human visitation, and began a decline in August as the number of visitors decreased (Fig. 11.2). It increased again in September and October, despite reduced human visitation, as a result of an ecosystem-wide influx of hunters and consequent increase in legal kills. Recorded mortality plunged in November with low visitor use, the closing of hunting season, and the denning of bears. Human-caused mortality resumed in spring as bears moved from den sites to ecocentered aggregations. Many of these movements originated at the periphery of the Ecosystem where bears were more exposed to lethal risks (Chapter 11).

Not surprisingly, mortality due to management control (*n* = 92) peaked in June, July, and August during the height of the tourist season (Fig. 11.3). At this time, bear-human conflicts occurred at campgrounds in and around Yellowstone. Legal kills (*n* = 97) were concentrated in September and October (Fig. 11.4), reflecting the fact that more bears were

Fig. 11.1. Temporal distribution of grizzly bear mortalities in the Yellowstone Ecosystem from all causes (*n* = 244) during 1959-70. Excludes mortalities for which date was unknown or for which only a period (summer, fall, spring) was recorded.

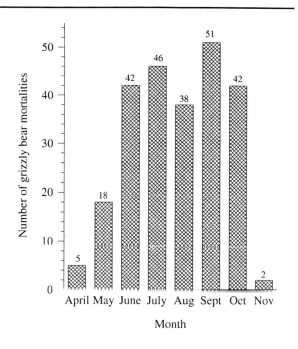

Fig. 11.2. Temporal distribution of grizzly bear mortalities relative to human visitation, Yellowstone National Park, 1959-70.

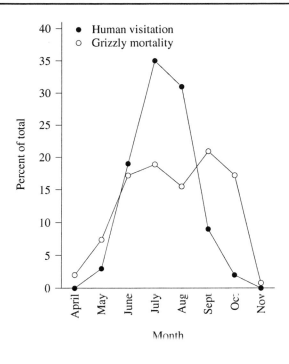

Fig. 11.3. Temporal distribution of grizzly bear mortalities in the Yellowstone Ecosystem from management control (*n* = 92) during 1959-70. Excludes mortalities for which date was unknown or for which only a period (summer, fall, spring) was recorded.

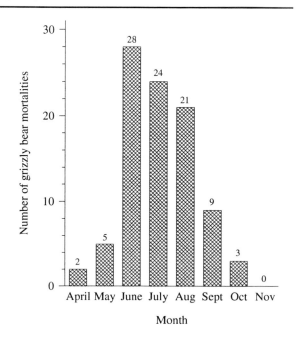

Fig. 11.4. Temporal distribution of grizzly bear mortalities in the Yellowstone Ecosystem from legal kill (*n* = 97) during 1959-70. Excludes mortalities for which date was unknown or for which only a period (summer, fall, spring) was recorded.

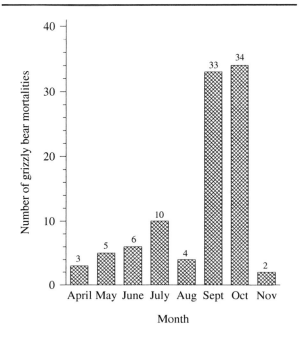

Fig. 11.5. Temporal distribution of grizzly bear mortalities in the Yellowstone Ecosystem from natural causes (*n* = 6) during 1959-70. Excludes mortalities for which date was unknown or for which only a period (summer, fall, spring) was recorded.

Fig. 11.6. Temporal distribution of grizzly bear mortalities in the Yellowstone Ecosystem from illegal kill (*n* = 12) during 1959-70. Excludes mortalities for which date was unknown or for which only a period (summer, fall, spring) was recorded.

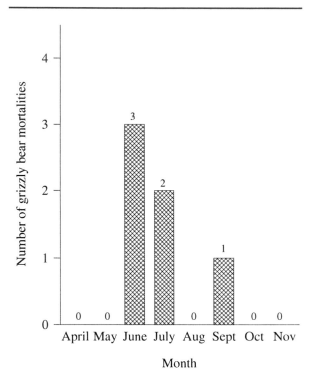

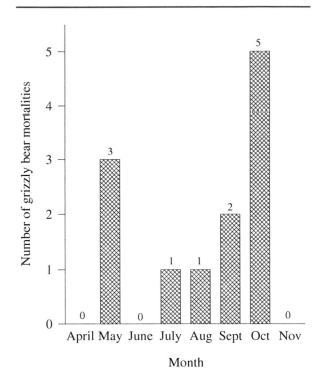

taken by fall-season (*n* = 65) than by spring-season (*n* = 8) hunters or by nonhunters acting in defense of life (*n* = 1), in defense of livestock (*n* = 14), or in the context of unofficial private control actions (*n* = 9). When it was announced in 1967 that the annual hunting season in Wyoming would end, Wyoming hunters took unusually large numbers of bears (*n* = 20, or 21% of total legal kills from 1959 to 1970). Of the 10 legal kills in July, eight (80%) were in defense of livestock, whereas eight of the 14 bears killed from April through June (57%) were killed by spring hunters (Fig. 11.4).

Natural mortality was detected too infrequently to speculate with confidence on its temporal pattern (Fig. 11.5). All recorded natural mortality in the Yellowstone Ecosystem from 1959 through 1970 (*n* = 6) was known or suspected to have been caused by other bears, and all these deaths were recorded in the Trout Creek area. Of the six bears, four were known to have been killed by large male grizzlies, and five of the six died during the grizzly bear breeding season, when large males are especially aggressive toward other bears. The six natural mortalities were either cubs (*n* = 4) or nonadults (*n* = 2).

Illegal kills were also rare, but appeared to be concentrated in the spring soon after bears emerged from dens, and during the fall hunting season (Fig. 11.6). Thus, the majority of illegal kills occurred after bears had left (fall) or before they entered (spring) the protection afforded by the ecocenters inside Yellowstone Park. Ecocenter food supplies were inadequate to hold the bears at these times. One-half of the 12 illegal kills occurred near or in the towns of West Yellowstone, Cooke City, and Gardiner where open-pit garbage dumps were operated year-round. Circumstances surrounding 10 of these kills were unknown, although poaching was suspected. The remaining two kills were known due to poachers.

Accidental mortality included road-killed grizzlies (*n* = 6) and grizzlies killed during chemical immobilization in our research (*n* = 15; Fig. 11.7). The majority of research-related deaths occurred in the early years of the study (1959-63), when intensive marking programs were conducted and drug dosages were being worked out. Most captures and capture-related deaths occurred in July and August. All road-killed grizzlies were recorded in the summer months from 1966 to 1970. Thus, the temporal pattern of accidental mortality reflected the timing of our field season (research-related accidents) and the tourist season (automobile accidents) in a straightforward way.

MORTALITY RATES

The issues of age- and sex-specific relative risk, spatial distribution, and temporal occurrence of grizzly bear deaths are important, but they do not touch on the fundamental issue of mortality rate—the absolute versus relative per capita risk of dying. Ideally, mortality rate is estimated by marking a large, random sample of bears each year, and monitoring those individuals to detect all subsequent deaths and causes of death. But in our field study, this was not possible. Our data depart from the ideal in two ways: first, the observed number of bear deaths was always fewer than the actual number; and second, the precise size of the bear population was unknown.

Nevertheless, we can use the population estimates and the mortality database of Craighead *et al.* (1974, 1988b) to estimate minimum, ecosystem-wide mortality rates. To accomplish this, we must assume that the observed number of deaths recorded each year were, on average, a constant fraction of the total number of deaths; that is, the efficiency of detecting deaths did not change between years. Ideally, estimates of the total population in each year should also equal a constant proportion of the actual population; that is, census efficiency should not vary widely from year-to-year. We believe condition one was met because our team was actively seeking mortality information each year during the entire study period, and we have no reason to suspect that the effort of other individuals contributing records of mortality varied greatly during 1959-70. The death of a grizzly bear is a noteworthy local event, attracting curious observers, the law, and the media. A record is available if the right people are contacted.

Condition two probably did not hold. Recall from Chapter 9 that there was an increased flux of non-Trout Creek bears into and through the Trout Creek aggregation following closure of other ecocenters during 1969-70. This funneling of bears through Trout Creek had the effect of increasing the "efficiency" of our census after 1968. We emphasize that, in using the term, "efficiency," we do not mean that our effort or methods changed, but rather that the visibility of bears increased as their movement increased. We are unable to adjust for this effect by calculating a separate and more representative census efficiency for the 1969-70

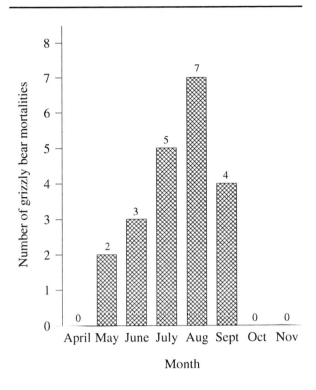

Fig. 11.7. Temporal distribution of grizzly bear mortalities in the Yellowstone Ecosystem from accidental causes (*n* = 21) during 1959-70. Excludes mortalities for which date was unknown or for which only a period (summer, fall, spring) was recorded.

period of increased movement. But we point out that the effect of applying a too-low efficiency to that period is conservative, with respect to our comparisons of mortality rate. In other words, the differences in mortality rate that we describe would be more, not less pronounced if we were able to adjust for the true efficiency of our 1969-70 annual counts. Exactly why this is true is described in more detail for each comparison of mortality rates that we present.

With these comments in mind, we proceed to compute an upperbound population estimate by multiplying our annual census counts at the various ecocenters by an inflation factor that accounts for uncensused bears. The inflation factor is a function of the estimated census efficiency—that is, the percentage of the true ecosystem-wide population accounted for in the annual censuses at the ecocenters. We originally calculated census efficiency to be on the order of 77%, but reanalysis of our data by researchers using different criteria suggested lower efficiencies (Craighead *et al.* 1974, Cowan *et al.* 1974, McCullough 1981; see also Chapter 9). To provide an upperbound population estimate, we therefore employed the smallest estimate of census

Fig. 11.8. Estimated grizzly bear mortality rate (black dots) from all causes (except research-related deaths), and 95% confidence intervals (vertical lines) plotted against year, Yellowstone Ecosystem, 1959-70.

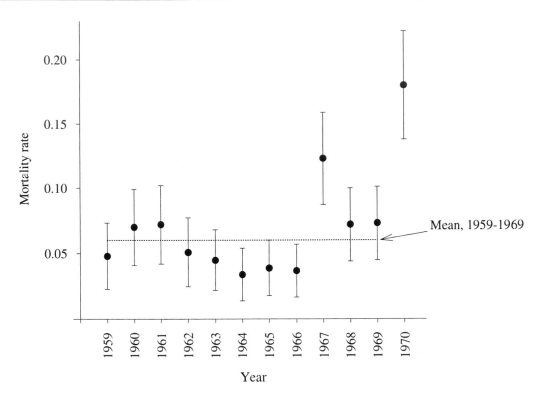

efficiency (56.73%; McCullough 1981). To calculate a common inflation factor for all years we used a multiplier equal to the reciprocal of the decimal percent census efficiency; for a census efficiency equal to 0.5673, this multiplier is 1.764. We applied a multiplier of 1.294, the reciprocal of a census efficiency of 0.773 (Craighead *et al.* 1974), to calculate lowerbound population estimates.

Annual Differences in Ecosystem-Wide Mortality Rates

We have plotted annual minimum mortality rates due to all causes, management control, legal kill, and illegal kill for the period 1959-70 in Figs. 11.8-11.11, respectively. We are, however, mainly concerned with statistically analyzing the effect of ecocenter closure on mortality rate. To do this, we compared mortality rates during the year 1970, when the last ecocenter was closed, with the appropriate average rate of mortality over the previous 11 years (1959-69). During 1969, the Trout Creek ecocenter was still operative. For all categories of death (all causes, management control, legal and illegal kill), the mortality rates were significantly higher in 1970 than before (*P* << 0.001 in all four z-tests for equality of proportions). In fact, the 1970 mortality rates were greater by factors of three to eight, depending on the cause of death (Figs. 11.8-

11.11). Examination of the annual rates of mortality from all causes (Fig. 11.8) reveals that rates during the 1959-69 period were consistently low, with the exception of 1967. The elevated overall rate of death in that year was due entirely to an unusually large legal kill (Fig. 11.10). This, as explained earlier, was due to a pending moratorium on the hunting of grizzlies in Wyoming which stimulated heavier hunting pressure.

The elevated rates of mortality in 1970 were directly attributable to disruption of the population in the wake of management actions designed to rehabilitate the "dump bears" of Yellowstone (Craighead 1980b, Knight and Eberhardt 1984). Movement of bears in search of food sources, both inside and outside the Park, increased dramatically the number of management actions that culminated in deaths of "problem bears" (Fig. 11.9; see also Chapter 15). Combined with a legal kill rate nearly equal to the high in 1967 (Fig. 11.10) and an elevated rate of poaching (Fig. 11.11), the overall mortality rate reached alarming proportions (Fig. 11.8). The principal conclusion from these comparisons is that ecocenter closure significantly elevated the ecosystem-wide mortality rate. We reemphasize that this conclusion is not affected by the probable increase in our census efficiency following ecocenter closures. This is so because, by

Fig. 11.9. Estimated grizzly bear mortality rate (black dots) from management control, and 95% confidence intervals (vertical lines) plotted against year, Yellowstone Ecosystem, 1959-70.

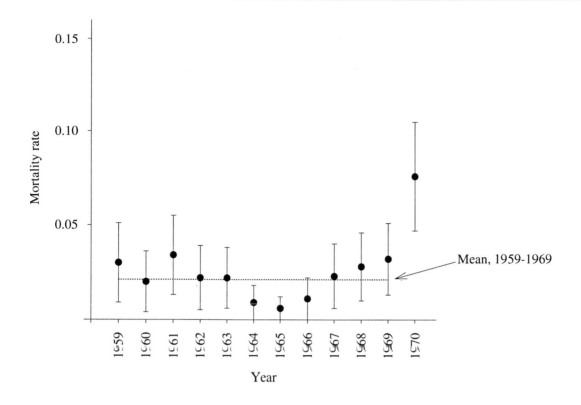

Fig. 11.10. Estimated grizzly bear mortality rate (black dots) from legal kill, and 95% confidence intervals (vertical lines) plotted against year, Yellowstone Ecosystem, 1959-70.

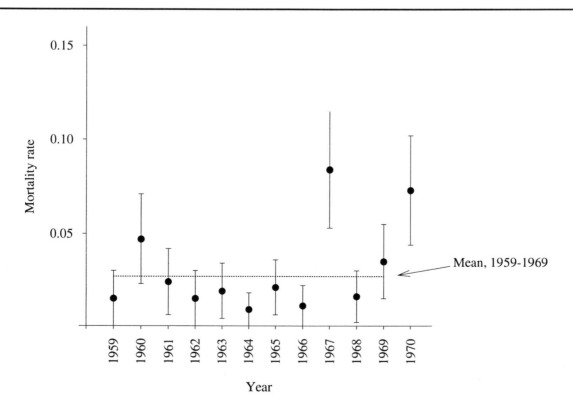

Fig. 11.11. Estimated grizzly bear mortality rate (black dots) from illegal kill, and 95% confidence intervals (vertical lines) plotted against year, Yellowstone Ecosystem, 1959-70.

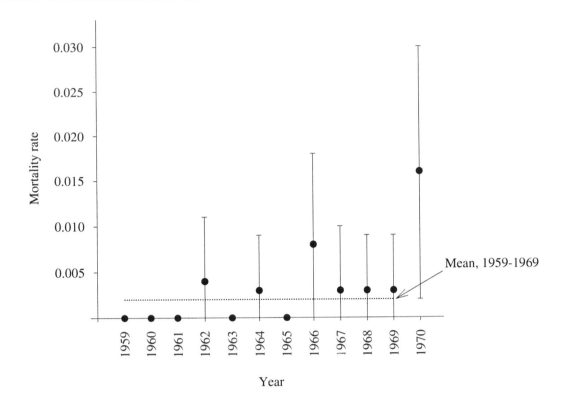

using a low efficiency estimate, we overestimate the 1970 population size and thus underestimate the associated mortality rate.

Mortality Rates of Trout Creek Versus Non-Trout Creek Grizzly Bears

There are two reasons to suspect that members of the Trout Creek subpopulation may have experienced lower rates of mortality during the preclosure period than other bears in the Yellowstone Ecosystem centered on smaller, more peripheral areas providing less protection. First, the consistent and persistent abundance of high quality food probably permitted Trout Creek bears to meet the energetic costs of growth, maintenance, and reproduction more easily, thereby reducing natural mortality during denning, as well as human-caused mortality resulting from bear-human conflicts at campgrounds and developed areas where food was an attractant. Secondly, the central location of the Trout Creek grizzly bear aggregation within the relatively protective confines of Yellowstone Park reduced the foraging movements of bears and, in turn, the frequency of bear-human conflicts that often culminated in bear deaths.

Because we knew we had not censused all bears in the Ecosystem each year (1960-70), we calculated

minimum and maximum ecosystem-wide population estimates by applying multipliers of 1.294 (minimum estimate; Craighead *et al.* 1974) and 1.764 (maximum estimate; McCullough 1981) to the annual number of bears censused at Trout Creek (Table 11.18). Next, based on Craighead *et al.* (1988b), we listed total annual ecosystem-wide deaths, not including research-related ones, and annual deaths due specifically to management control and to legal kill during 1960-70 (Table 11.18). Finally, we tallied the annual numbers of marked and unmarked but recognizable bears that, by virtue of high visitation rates, constituted the Trout Creek flagged population (see Methods) from 1960-70. Values for 1959 were omitted from the analysis because too few marked Trout Creek bears were available in the first year of the study.

We determined the total rate of mortality, ecosystem-wide, from all causes (excluding research-related deaths) and partitioned it into rates for Trout Creek and non-Trout Creek bears. First, we estimated yearly mortality rates for Trout Creek bears as the number of marked Trout Creek bears in the mortality database in each year, divided by the number of marked Trout Creek bears alive at the start of that year (Table 11.19); we then estimated the number of unmarked Trout Creek bears in the

Table 11.18. Estimated grizzly bear population size by year for the entire Yellowstone Ecosystem and for the component Trout Creek ecocenter, 1960-70.

| Year | Census[a] count | Yellowstone Ecosystem population | | | | | Trout Creek flagged population[d] | | |
| | | Adjusted population size[b] | | Mortality[c] | | | Number of bears | | |
		Minimum	Maximum	Total	Mgmt.	Legal	Marked	Unmarked	Total
1960	169	219	298	21	6	14	36	20	56
1961	166	215	293	21	10	7	56	13	69
1962	155	201	273	14	6	4	51	21	72
1963	177	229	312	14	7	6	62	14	76
1964	185	239	326	11	3	3	73	25	98
1965	187	242	330	13	2	7	76	19	95
1966	202	261	356	13	4	4	69	18	87
1967	175	226	308	38	7	26	71	15	86
1968	181	234	319	23	9	5	77	16	93
1969	195	252	344	25	11	12	65	20	85
1970	179	231	316	57	24	23	55	22	77
Total				250	89	111			
Mean	179	232	316				63	18	81

[a] Annual census counts, parkwide.

[b] The product of the annual census count, parkwide, and a minimum multiplier of 1.294 (based on census efficiency suggested by Craighead *et al.* [1974]) or a maximum of 1.764 (based on census efficiency suggested by McCullough [1981]).

[c] From Craighead *et al.* (1988b); research-related deaths excluded.

[d] Annual Trout Creek census count of marked and unmarked, but recognizable, grizzly bears; flagged bears only (see Methods).

mortality database by multiplying the yearly mortality rate determined for marked Trout Creek bears by the number of unmarked Trout Creek bears observed in the same year. Note that unmarked Trout Creek bears did not differ from marked Trout Creek bears in any way except in not being identifiable as Trout Creek bears by most observers who contributed mortality records. Thus, it is reasonable to suppose a common mortality rate for the marked and unmarked segments of the Trout Creek population.

The observed number of marked bears dying, plus the estimated number of unmarked bears dying, is the estimated total mortality of Trout Creek bears in the database in each year. Subtracting this number of deaths from the total number of deaths recorded for the Yellowstone Ecosystem (Table 11.18) gives the estimated number of non-Trout Creek bears dying in each year (Table 11.19). Finally, the rate of mortality for non-Trout Creek bears is the number of non-Trout Creek bears dying divided by the maximum and minimum estimates of the non-Trout Creek population. These latter estimates are the Yellowstone Ecosystem population estimates, minus the total marked and unmarked Trout Creek bears in each year (Table 11.18). In similar fashion, we calculated annual rates of

mortality from management control and legal kill for the Trout Creek population, and for the maximum/minimum non-Trout Creek populations (Tables 11.20 and 11.21).

Considering all causes, Trout Creek bears experienced significantly lower mortality than non-Trout Creek bears, regardless of whether the minimum or maximum estimate of the non-Trout Creek population was used (Table 11.22). It might be supposed that this difference is primarily driven by legal kill deaths, and that Trout Creek bears were protected by their greater use of the Park where hunting was prohibited. But comparisons of mortality by management control and legal kill show that, relative to Trout Creek bears, non-Trout Creek bears experienced a similar increase in mortality rate from both causes (Table 11.22). For example, using the minimum non-Trout Creek population estimate, non-Trout Creek bears showed an approximate three-fold increase in total, management control, and legal kill mortality.

Once again, this result is not compromised by our application of a common census efficiency that is known to be too low for the 1969-70 period. Recall that the Trout Creek population was estimated by direct enumeration, so was not subject to error arising from choice of census efficiency. The

Table 11.19. Annual mortality rates from all causes of marked Trout Creek versus non-Trout Creek grizzly bears, Yellowstone Ecosystem, 1959-70.

Year	Trout Creek population			Non-Trout Creek population					
	No. of marked bears		Mortality rate	Minimum population estimate			Maximum population estimate		
				Estimated no.		Mortality rate	Estimated no.		Mortality rate
	Dying	Total		Dying[a]	Total[b]		Dying[a]	Total[b]	
1960	0	36	0.000	21.0	163	0.129	21.0	242	0.087
1961	1	56	0.018	19.8	146	0.135	19.8	224	0.088
1962	1	51	0.020	12.6	129	0.098	12.6	201	0.063
1963	1	62	0.016	12.8	153	0.084	12.8	236	0.054
1964	4	73	0.055	5.6	141	0.040	5.6	228	0.025
1965	1	76	0.013	11.8	147	0.080	11.8	235	0.050
1966	3	69	0.043	9.2	174	0.053	9.2	269	0.034
1967	2	71	0.028	35.6	140	0.254	35.6	222	0.160
1968	6	77	0.078	15.8	141	0.112	15.8	226	0.070
1969	5	65	0.077	18.5	167	0.111	18.5	259	0.071
1970	6	55	0.109	48.6	154	0.316	48.6	239	0.203
Mean			0.042			0.128			0.082

[a] Calculated by multiplying the mortality rate observed for marked Trout Creek bears (column 4) by the number of unmarked Trout Creek bears (column 9, Table 11.18) in the same year, adding this number to the number of marked Trout Creek bears dying (column 2), and subtracting this sum from the total number of bears dying in that year (column 5, Table 11.18).

[b] Calculated by subtracting the total Trout Creek population size (column 10, Table 11.18) from the minimum or maximum ecosystem population size (columns 3 and 4, Table 11.18).

Table 11.20. Annual mortality rates from management control of marked Trout Creek versus non-Trout Creek grizzly bears, Yellowstone Ecosystem, 1959-70.

	Trout Creek population			Non-Trout Creek population					
				Minimum population estimate			Maximum population estimate		
	No. of marked bears		Mortality rate	Estimated no.		Mortality rate	Estimated no.		Mortality rate
Year	Dying	Total		Dying[a]	Total[b]		Dying[a]	Total[b]	
1960	0	36	0.000	6.0	163	0.037	6.0	242	0.025
1961	0	56	0.000	10.0	146	0.068	10.0	224	0.045
1962	0	51	0.000	6.0	129	0.047	6.0	201	0.030
1963	0	62	0.000	7.0	153	0.046	7.0	236	0.030
1964	1	73	0.014	1.7	141	0.012	1.7	228	0.007
1965	0	76	0.000	2.0	147	0.014	2.0	235	0.009
1966	0	69	0.000	4.0	174	0.023	4.0	269	0.015
1967	2	71	0.028	4.6	140	0.033	4.6	222	0.021
1968	2	77	0.026	6.6	141	0.047	6.6	226	0.029
1969	4	65	0.062	5.8	167	0.035	5.8	259	0.022
1970	3	55	0.055	19.8	154	0.129	19.8	239	0.083
Mean			0.017			0.045			0.029

[a] Calculated by multiplying the mortality rate observed for marked Trout Creek bears (column 4) by the number of unmarked Trout Creek bears (column 9, Table 11.18) in the same year, adding this number to the number of marked Trout Creek bears dying (column 2), and subtracting this sum from the total number of bears dying by management control in that year (column 6, Table 11.18).

[b] Calculated by subtracting the total Trout Creek population size (column 10, Table 11.18) from the minimum or maximum ecosystem population size (columns 3 and 4, Table 11.18).

Table 11.21. Annual mortality rates from legal kills of marked Trout Creek versus non-Trout Creek grizzly bears, Yellowstone Ecosystem, 1959-70.

	Trout Creek population			Non-Trout Creek population					
	No. of marked bears			Minimum population estimate			Maximum population estimate		
				Estimated no.			Estimated no.		
Year	Dying	Total	Mortality rate	Dying[a]	Total[b]	Mortality rate	Dying[a]	Total[b]	Mortality rate
1960	0	36	0.000	14.0	163	0.086	14.0	242	0.058
1961	1	56	0.018	5.8	146	0.040	5.8	224	0.026
1962	0	51	0.000	4.0	129	0.031	4.0	201	0.020
1963	1	62	0.016	4.8	153	0.031	4.8	236	0.020
1964	2	73	0.027	0.3	141	0.002	0.3	228	0.001
1965	1	76	0.013	5.8	147	0.039	5.8	235	0.025
1966	3	69	0.043	0.2	174	0.001	0.2	269	0.001
1967	0	71	0.000	26.0	140	0.186	26.0	222	0.117
1968	1	77	0.013	3.8	141	0.027	3.8	226	0.017
1969	1	65	0.015	10.7	167	0.064	10.7	259	0.041
1970	1	55	0.018	21.6	154	0.140	21.6	239	0.090
Mean			0.015			0.059			0.038

[a] Calculated by multiplying the mortality rate observed for marked Trout Creek bears (column 4) by the number of unmarked Trout Creek bears (column 9, Table 11.18) in the same year, adding this number to the number of marked Trout Creek bears dying (column 2), and subtracting this sum from the total number of bears that were legal kills in that year (column 7, Table 11.18).
[b] Calculated by subtracting the total Trout Creek population size (column 10, Table 11.18) from the minimum or maximum ecosystem population size (columns 3 and 4, Table 11.18).

size of the non-Trout Creek population, however, was obtained by subtracting Trout Creek bears from what we judged to be an upperbound estimate of total ecosystem population. Thus, mortality rates for non-Trout Creek bears are underestimates of the minimum for that segment of the population.

These data demonstrate that the Trout Creek ecocenter had a general protective effect. Of course, most non-Trout Creek bears also used other ecocenters in the Yellowstone area. But most of these ecocenters were less centrally located than Trout Creek, and all were much smaller in terms of available food resources and the number of bears using them. Nonetheless, we suspect that the protective effects of the ecocenters, in general, would have been even more pronounced during 1960-70 (Table 11.22), were it possible to compare mortality strictly among ecocentered versus non-ecocentered bears.

DISCUSSION

What are the broader messages underlying the mortality data and our analyses? First, even in the case of a grizzly bear population with a large, primarily wilderness, park at the core of its range and a relatively abundant food base in the Park and throughout the surrounding national forests, death due to humans was a significant, perhaps dominant constraint on population growth. During the period 1959-70, an annual average of at least 8-12% of the entire ecosystem population died as a direct result of human activity. This represents more than one-half of the annual population increment. For example, annual recruitment (measured at 0.5 years of age) would equal roughly 15% in a population which, like the 1959-70 Yellowstone population, was 45% adults (e.g., 45 adults/100), had an adult sex-ratio at parity (e.g., 22.5 adult females/100), an average reproductive cycle of three years (e.g., 22.5/3 = 7.5 parous females/yr), and mean litter size of two (e.g., 7.5 x 2 = 15 cubs/yr).

Compensatory natural mortality would, no doubt, have occurred eventually if these sources of human-caused mortality had been removed in preclosure times. Population size and viability, however, would also have increased. The simple, yet belatedly recognized message is that grizzly bear persistence and human activity are negatively related, even within habitat designated for relatively innocuous, nonconsumptive recreational pursuits. Management can address these problems by eliminating sport hunting and restricting economic development, livestock grazing, logging, and mining, as well as nonconsumptive recreational use in currently accessible bear habitat. Strict controls on rates

Table 11.22. Mean minimum mortality rates in the Yellowstone Ecosystem during 1960-70 for Trout Creek versus non-Trout Creek grizzly bears considering (1) all causes of mortality, (2) management-control mortality, and (3) legal-kill mortality.

	Trout Creek	Non-Trout Creek	
		Minimum population size	Maximum population size
All causes[a]	0.042	0.128[b]	0.082[b]
Management control	0.017	0.044[b]	0.029[b]
Legal kill	0.015	0.059[b]	0.038[b]

Note: 1959 is excluded because few Trout Creek bears were marked the first year.

[a] Excluding research-related deaths.

[b] Rate significantly higher than the corresponding Trout Creek rate at $P \leq 0.05$ (one-tailed Wilcoxon matched-pairs tests in which the appropriate yearly rates are paired; $n = 11$ years).

of human access to critical habitat are simple, workable management tools. Some protective measures, such as the moratorium on hunting in Wyoming and Montana, have been taken since 1970. But the situation has worsened in other respects (see also Mattson and Reid 1991). Tourism and economic development in Yellowstone National Park and surrounding forests have increased, and roads, campgrounds, and other developments have invaded de facto wilderness. Sheep allotments have been drastically curtailed, but timber sales have not been sufficiently restricted. We further examine the issues of habitat protection and bear mortality in recent years in Chapters 15, 18, and 20.

The second broad message to emerge from our analysis of death statistics is that grizzly bear populations are protected by special land-use designations (national park or wilderness) only to the extent that their habitat and other needs are adequately met throughout the population's entire range, whether specially designated or not. Sex and age classes that moved widely and spent less time in insular, protected sites experienced higher rates of mortality during our study. Similarly, regardless of sex or age, all members of the Trout Creek aggregation enjoyed lower mortality rates because the large, reliable food resource of the ecocenter made movements into less protected habitat unnecessary. Closure of park ecocenters dramatically increased bear movement into developed peripheral areas (Chapter 15; see also Mattson and Knight 1991). Because developed areas will persist and increase, managers should consider strategies to enhance potential natural ecocenters (see Chapters 13 and 20). Serious consideration should also be given to creating, on an experimental basis, tightly regulated,

carefully placed artificial ecocenters using processed human refuse, surplus grain, or animal carcasses (Chapter 20). Conversely, efforts to eliminate unauthorized refuse dumps and boneyards in developed areas of the Greater Yellowstone Ecosystem must continue, as such sites attract bears and expose them to increased risk of death. An even distribution of grizzly bears throughout the Greater Yellowstone Ecosystem is not, in itself, desirable if bears are forced to use or travel through areas with high potential for conflict with humans.

Finally, our results highlight a troubling effect of trophy hunting, as practiced during 1959-70. It induced a relatively indiscriminate harvest of bears with regard to sex. Grizzlies are sexually monomorphic, except in size. Females with offspring could be recognized as such by hunters, and this recognition protected them. But hunters could not reliably identify and avoid shooting females when they were without offspring. Consequently, our mortality data showed only a relatively weak tendency for trophy hunters to shoot males rather than females. Because males leave dens earlier, restriction of hunting to a spring season could possibly reduce female deaths in the sport-kill (Miller 1990). Nevertheless, a male:female sex ratio of 76:24, as reported in the spring kill for an Alaskan population (Miller 1990), is still disturbingly indiscriminate in view of the large per capita effect of adult female deaths on population growth rate (cf., Knight and Eberhardt 1984, 1985). In view of these considerations, and of the uncertain status of bear populations in the contiguous 48 states, we consider sport hunting of grizzly bears to be biologically indefensible under any circumstances allowing the possibility of female deaths.

SUMMARY

Demographic, spatial, and temporal patterns of mortality were examined to better understand why grizzly bears died and, using such knowledge, to determine how best to reduce the death rate. Analysis of death statistics suggested that age and sex per se were not factors in predisposing bears to human-caused mortality, when all such causes were considered collectively. Nevertheless, movement attributed to the age or sex of an animal was a determinant in the vulnerability of bears to specific categories of mortality. For example, Trout Creek nonadult males were more vulnerable, relative to nonadult females, to all causes of death, presumably because, despite their central and protected natal area, their greater movement exposed them more frequently to risk outside the Park. A similar trend was evident among males generally. Also, selection for trophies predisposed larger bears (adults versus nonadults and nonadult males versus nonadult females) to mortality by legal kill, both throughout the Ecosystem and among bears that dispersed from the Trout Creek ecocenter.

There was a trend toward increased spatial dispersion of deaths, suggesting also a dispersion of living bears, associated with closure and manipulation of park ecocenters, and a close relationship between bear mortality and human visitation in Yellowstone National Park for the entire 1959-70 period. Death from all causes was lowest in April, May, and November, and highest in June, July, and August, concomitant with high human visitation levels in the Park. Death due to management control also peaked in June, July, and August during the height of the tourist season. Legal kills were concentrated in August and September, reflecting deaths during the fall hunting season. Natural mortality was seldom detected, but when it was, it was caused by other bears.

Mortality rate, the relative per capita risk of dying, was calculated for the years 1960 through 1970. The rate during 1970 was significantly greater than the mean for the preceding 11 years. The elevated mortality rate in 1970 was directly attributable to disruption of the population in the wake of management actions designed to rehabilitate the "dump bears" of Yellowstone. Ecocenter closure elevated the ecosystem-wide mortality rate.

Annually, an average of at least 8-12% of the entire Yellowstone grizzly bear population died as a direct result of human activity, and this represented a removal equal to approximately one-half of the population increment each year. Considering all causes of death, the bears aggregated at Trout Creek experienced significantly lower mortality than did other bears throughout the Ecosystem, demonstrating a general protective effect of the Trout Creek ecocenter. Death statistics, in a system dominated by human-caused mortality, are far more amenable to change from management actions than are life statistics such as reproductive parameters or life support nutritive factors. Maintaining a low human-caused death rate is essential to the viability of the Yellowstone grizzly bear population.

12
Spatial and Vegetation Requirements

INTRODUCTION

The ecological requirements of the grizzly bear are varied and complex. The habitat must meet the biological demands of the species' omnivorous feeding habits, its complex population and social interactions, winter denning, and its aggressive intra- and interspecific behavior. Spacious habitat, with varied landforms and vegetation types providing seasonal foods in stable to periodic abundance, is essential to the grizzlies' welfare and survival. Based on our work in Yellowstone, we must question the practice of designating enclaves within an ecosystem as critical grizzly bear habitat. Information presented in preceding chapters shows that entire ecosystems are necessary to the survival of the resident populations; management units carved from the whole will not suffice. Also, adequate protection from bear-human conflict is crucial in contemporary environs. In this chapter, we attempt to describe some of the spatial and vegetation requirements and relate these to the evolving concept of an ecosystem-based approach to management.

SPATIAL REQUIREMENTS

The grizzly bear in the preclosure Yellowstone era was a mobile, wide-ranging animal that established seasonal and home ranges, but exhibited little or no territoriality. A home range is the area within which an individual grizzly annually meets all of its biological requirements (Craighead 1976). Thus, home ranges define essential habitat. In any given land area, individual home ranges critical to the survival of individual bears must also be critical to the survival of the population as a whole.

In Yellowstone, home range areas of individual bears varied greatly in size, depending on the sex and age of the animals, seasonal and annual food availability, reproductive condition of females, and other factors (Craighead 1976, Craighead and Mitchell 1982). The spatial needs of individual animals often were considerable; radiotracking confirmed that seasonal ranges of some animals were widely separated and connected by migratory corridors (Table 12.1 and Fig. 5.8). Male bear Nos. 51, 52, and 60, and female No. 96, all of which were radio-collared at Trout Creek (near the geographic center of the Park), had summer ranges in Hayden Valley and fall ranges and denning sites outside the Park. Male No. 60 and female No. 96 traveled 45 and 64.5 air kilometers (28 and 40 mi), respectively, from summer ranges to fall ranges or foraging areas (Table 12.1). Male No. 52 traveled 88 air kilometers (55 mi) in 20 days and was shot 21 kilometers (13 mi) south of Yellowstone National Park. Male Nos. 51 and 126 met similar fates 77 and 71 air kilometers (48 and 44 mi), respectively, from localities where they were last observed. Female No. 170 traveled 55 kilometers (34 mi) within a period of a few days. Male No. 32, a color-marked bear, had a home range of at least 705 square kilometers (272 mi²) that extended beyond the Park to the northeast. Male No. 37 occupied a home range of about 1217 square kilometers (470 mi²) that extended beyond the park border to the south (Table 12.1 and Fig. 5.8); he was shot 80 air kilometers (50 mi) from the site of last observation the previous fall.

Female No. 7 had a home range of 275 square kilometers (106 mi²) that was well-defined by radiotracking (Craighead 1976). Her core area, or center of activity, lay to the east of Trout Creek and the Yellowstone River. She was observed swimming across the river on six occasions. Core areas for all animals tracked by radio were small, seldom exceeding 2.6 square kilometers (1 mi²) each. Core areas, that is, areas with the highest density of animal locations (usually defined as 50%) (White and Garrott 1990), represented the most intensively used habitat within a home range. The adjacent habitat was used less frequently, but still represented important space for foraging, cover, travel, and social interaction. Male No. 76 frequently moved 11 to 16 air kilometers (7 to 10 mi) daily within a home range of 435 square kilometers (168 mi²) (Table 12.1). On

Table 12.1. Selected movements and home ranges illustrating spatial requirements of grizzly bears (see also Fig. 12.1), Yellowstone Ecosystem, 1959-70.

Bear number	Sex	Mode of detection[a]	Maximum distance moved in kilometers (mi)	Area of home range in square kilometers (mi²)[b]
51	M	R	77 (48)	—
52	M	R	88 (55)	—
60	M	R	45 (28)	—
96	F	R	64 (40)	—
126	M	R	71 (44)	—
170	F	R	55 (34)	—
7	F	R	—	275 (106)
76	M	R	—	435 (168)
32	M	C	32 (20)	705 (272)
37	M	C	80 (50)	1217 (470)

[a] R = radiotracking; C = colored ear tags.
[b] Area of home range based on 256 individual radio bearings for bear No. 7 and 129 for bear No. 76.

one occasion, this male traversed 3122-meter (10,243-ft) Mt. Washburn at about the 2743-meter (9000-ft) level and crossed the Grand Canyon of the Yellowstone River five times, traveling 93 air kilometers (58 mi) over extremely rough terrain during an eight-day period. The ground distance traversed was estimated to be three times the distance by air.

Variation in the spatial needs of an individual bear from year to year is well illustrated by the history of female No. 40. This female was radiotracked for eight consecutive years from 1961 through 1968 (Craighead 1976). She was instrumented at the age of two years and shot when 10 years old. Her life range, smaller than annual home ranges of most females, encompassed an area of about 78 square kilometers (30 mi²). Her core areas remained basically the same, year after year (none exceeded 2.6 square kilometers), but her seasonal and annual home ranges varied considerably. As a subadult during 1961 and 1962, her summer-fall range did not exceed 21 square kilometers (8 mi²). In 1963, at the age of four years, she again used the same area during the summer and fall, where she was observed breeding, and became pregnant. In 1964, she produced two cubs (one of which died) and occupied a fall range of 40 square kilometers (15 mi²). She entered her den on 10 November with her cub. In 1965, she weaned her yearling and mated; she was radiotracked for 106 days beginning 28 June and is known to have entered a den 1 November. Her home range was 52 square kilometers (20 mi²). Accompanied by two new cubs in 1966, her summer and fall range was 19 square kilometers (7 mi²). She dug a new den and wintered with her cubs. In the spring of 1967,

she weaned the cubs and bred. During the fall of 1967, she ranged within an area of 29 square kilometers (11 mi²). Her den was not located, but she emerged in 1968 with three cubs and occupied a home range of 57 square kilometers (22 mi²). She was shot by a Park Ranger in 1969.

The annual home ranges and the life range of female No. 40 no doubt reflected her frequent visitations to the Trout Creek ecocenter. This food source supplemented her "natural" food intake and that of her offspring. The result was probably a reduction in foraging area and thus in range size. Nevertheless, she made frequent and extensive seasonal movements to feed on winter-killed elk (*Cervus canadensis*) and bison (*Bison bison*) in the riparian communities and the sagebrush-bunchgrass habitat types. She also consumed huckleberries (*Vaccinium* spp.) in the subalpine fir, huckleberry, and grouse whortleberry habitat types, which were well represented within her life range. In the fall, radio locations and subsequent visual sightings showed that she traveled to the ridges for whitebark pine (*Pinus albicaulis*) seeds in the subalpine fir-whitebark pine forest types and hunted meadow voles (*Microtus* spp.) in the sagebrush-bunchgrass parklands.

Female No. 187 had a home range of 88 square kilometers (34 mi²) at the age of five. Male No. 14, a 340-kilogram (750-lb) animal, had a range of only 31 square kilometers (12 mi²) during the fall of 1964; however, data from trapping and sightings over a period of years suggested that the life range of this large adult male may have exceeded 2600 square kilometers (1000 mi²). The initial home range of a subadult may be relatively small. For example, male No. 202 had a home range of 70 square ki-

Table 12.2. Areas and percentages of landform and vegetation classes in central Yellowstone National Park, classified in Fig. 12.1 (Landsat-1 Imagery).

Vegetation class	Number of hectares (acres)		Percent of total
Grassland and Shrubland	169,429	(418,653)	22.6
Coniferous Forest	339,490	(838,868)	45.3
Bare Ground and Rock	99,362	(245,519)	13.2
Water	12,250	(30,270)	1.6
Unclassified[a]	129,532	(320,070)	17.3
Total	750,063	(1,853,380)	100.0

[a] This category is not considered a landform or vegetation class, but it does contribute to area statistics.

lometers (27 mi^2) as a weaned yearling and 324 square kilometers (125 mi^2) as a 2-year-old.

Because home ranges contained essential habitat, these areas were not readily abandoned. On the contrary, fidelity to home range exerted a strong stabilizing effect on population distribution and movement patterns. In some bears (e.g., female No. 40), biological requirements were met within relatively small areas and within a limited altitudinal range. In other cases (e.g., male Nos. 14 and 76), the home ranges were large and included an elevational range in landforms of more than 900 meters (3000 ft), supporting a diversity of habitat types. The resources within these home ranges were critical for each individual; therefore, individuals tended to maintain the home range area, despite forces compelling them to move (e.g., food stress, human activities). This was true whether the home range was a single geographic unit or two units separated by a migratory corridor. The movements and ranges of instrumented grizzlies, discussed in detail by Craighead (1976), confirm the extensive space requirements of individual bears and the strong attachment to home ranges. The welfare of the grizzly bear depends on seasonal foraging areas, travel corridors, bedding, denning, and escape areas, and activity centers within home and seasonal ranges. Even in a population where range overlap was common, core areas small, and territoriality rare, thousands of square kilometers of undisturbed habitat were necessary to support a population of several hundred animals. Some habitat types were used more than others. But we found no scientific justification for defining discrete portions of the Ecosystem as being critical, while other portions were not. The numerous biological and social needs of the grizzly bear population were all-encompassing.

VEGETATION REQUIREMENTS

To investigate more definitively the selection and use of habitat by grizzlies, the senior author's research plan called for detailed mapping and botanical description of vegetation components of the Yellowstone Ecosystem. This was to be followed with a study of how the vegetation complexes were used by the grizzly bears. Unfortunately, these plans were aborted when the Park Service abruptly terminated the research in 1971. Nevertheless, we believed that a better understanding of bear habitat, ecosystem-wide, would be imperative to eventual interpretation of our population data and perhaps even more vital to interpreting population changes following closure of the ecocenters.

Consequently, before terminating the project, we gathered vegetation data within the home ranges of several marked and radio-collared bears. This information was later analyzed with the use of satellite imagery and computer assistance. To accomplish this, we used a multispectral image of more than 0.7 million hectares (1.8 million acres) of densely occupied grizzly bear habitat. Although the vegetation mapping was broad-scale and lacked desirable botanical detail, it introduced the essential habitat factor into the Yellowstone population equation and revealed the feasibility of conducting more detailed habitat investigations elsewhere (see Craighead *et al.* 1982, 1988a).

Major Vegetation Complexes Used by Grizzlies in Yellowstone

The home ranges of five radio-collared grizzly bears (Nos. 7, 40, 76, 101, and 187) were superimposed over the satellite-mapped vegetation and landforms (Fig. 12.1). Centers of activity or core areas for bears 7, 40, and 187 were also outlined. These ranges, and known ranges of other marked

Fig. 12.1. Landsat-1 multispectral imagery map of Yellowstone National Park (centered on the Hayden Valley), showing distribution of vegetation within home ranges of five adult grizzly bears and core areas (stippled or cross-hatched) for three (Nos. 7, 40, and 187); dark areas are primarily coniferous forest, and lighter areas, primarily grass-shrubland (from Craighead 1980a).

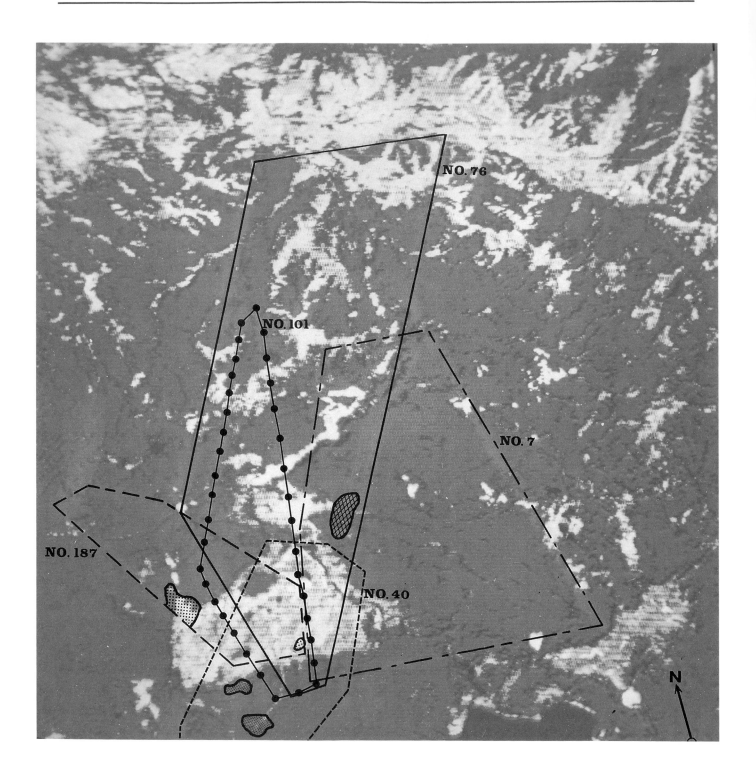

grizzlies not included in the figure, encompassed nearly the entire area displayed. The vegetation for this area was classified using the procedures described by Mueggler and Handl (1974), Pfister *et al.* (1977), and Craighead *et al.* (1982). Four vegetation and landform classifications based on the ground-truth data were developed on computer, and the acreage of each of the four classifications was compiled (Table 12.2). The two vegetation classes (Grassland/Shrubland and Coniferous Forest) were described in terms of the major habitat types and plant communities occurring in each class (see Craighead *et al.* 1982).

Most of the area depicted by the satellite map (Fig. 12.1) lies within the subalpine zone. Mountain grassland and shrubland habitat types and associated riparian communities represented about 23% of the total area. The grass and shrublands were represented primarily by two habitat types: *Artemisia tridentata/Festuca idahoensis* and *Festuca idahoensis/Agropyron spicatum*. The riparian communities were not typed, and it would have required considerable work to do so. Sedges (*Carex* spp.) predominated in most of the riparian communities and, in some situations, formed nearly pure stands over extensive areas.

Perennial herbs such as geranium (*Geranium viscosissimum*), harebell (*Campanula rotundifolia*), old man's whiskers (*Geum triflorum*), white clover (*Trifolium repens*), elk thistle (*Cirsium scariosum*), and yampa (*Perideridia gairdneri*) characterized the grass-shrubland habitat types. Instrumented grizzly bears ate the succulent greens, corms, and tubers of a wide range of species in these habitats. They also fed on small rodents—*Microtus*, *Peromyscus*, and *Thomomys* spp.—that periodically were very abundant in these habitats. Many of the elk and bison frequenting the grass-shrubland subalpine parks failed to survive the generally severe winters; some that died in the sedge marshes or in the grass-shrublands were eaten by grizzly bears from March through June (Chapter 10).

Coniferous forests comprised about 45% of the area (Table 12.2), a coverage twice that of the grass-shrubland. The distribution of coniferous forests relative to grass-forb shrublands created many miles of ecotone (Fig. 12.1). This interspersion of forest and grass-shrubland vegetation types in the subalpine parklands provided habitat conditions preferred by grizzly bears. Without exception, all core areas within defined home ranges contained both forest and grass-shrubland components and encompassed riparian plant communities. The dispersed pattern of grassland and forest enabled bears to graze sedges, grasses, and succulent forbs in season, and to feed on berries of *Vaccinium* spp. and seeds of whitebark pine, all within relatively small home ranges.

Coniferous forests were largely represented by seral stages of the subalpine fir series. Three habitat types were common: *Abies lasiocarpa/Calamagrostis rubescens*; *Abies lasiocarpa/Vaccinium globulare*; and *Abies lasiocarpa/Vaccinium scoparium*. The second and third types named commonly supported heavy ground cover of blue huckleberry and grouse whortleberry (*V. globulare* and *V. scoparium*, respectively). Although berry crops fluctuated spatially and temporally, they were a major food source (Chapter 10). The *Abies lasiocarpa - Pinus albicaulis/Vaccinium scoparium* habitat type was common on dry ridges and easterly exposures at the higher altitudes. Bears fed on whitebark pine seeds in the fall and spring. Although the seed crops also fluctuated locally and annually, pine seeds were a major dietary item (Chapter 10). This habitat type was used extensively by grizzlies because it also provided whortleberries and thereby offered a double dietary incentive.

The spruce series generally was represented in moist situations along the streams and rivers by the *Picea engelmannii/Linnaea borealis* habitat type. This habitat type was often dominated by a shrub layer of blue huckleberry, with grouse whortleberry common. The berries of these species were a preferred food of grizzlies.

The Douglas-fir series was represented by several habitat types, but the one most frequented by grizzlies was *Pseudotsuga menziesii/Calamagrostis rubescens*. This was not a preferred habitat type for feeding, although food plants such as strawberry (*Fragaria virginiana*) and glacier lily (*Erythronium grandiflorum*) were common.

Alpine meadows and spruce-fir-whitebark pine krummholz were poorly represented within the habitat study area, though small areas occurred on Mt. Washburn, the highest peak in the area. Alpine areas were more extensive in other parts of the Park, but their plant resource was only moderately used by grizzlies. In other portions of the grizzly bear's range within the contiguous 48 states—especially in the Northern Continental Divide Ecosystem of Montana—the alpine zone is heavily used in late June and July when the bears specifically seek two available food resources: the emerging alpine plants and the army cutworm moth (*Euxoa auxiliaris*) (Craighead *et al.* 1982 and Chapter 10). The most heavily used alpine plants were species of biscuit root (*Lomatium* spp.), springbeauty (*Claytonia* spp.), elk sedge (*Carex* spp.), and various graminoids.

Some grass-shrubland and forest habitat types were more important than others, seasonally and

Table 12.3. Habitat types occurring in five core areas of grizzly bears 7, 40, and 187 (see Fig. 12.1), Yellowstone National Park, 1959-70.

Habitat type	Number of core areas	Percent occurrence
Abies lasiocarpa/Vaccinium scoparium	5	100
Abies lasiocarpa/Linnaea borealis	3	60
Artemisia tridentata/Festuca idahoensis	3	60
Festuca idahoensis/Agropyron spicatum	2	40
Riparian communities	5	100

annually, as sources of food. Although some types were used, and possibly preferred, more than others, the entire complex of habitat types within the Yellowstone Ecosystem was essential and critical for the grizzly bear.

As indicated earlier, core areas or activity centers are the sites within home or seasonal ranges that are the most intensively used by individual animals. The five core areas for bear Nos. 7, 40, and 187 (Fig. 12.1) all included the *Abies lasiocarpa/Vaccinium scoparium* habitat type (Table 12.3). Riparian communities were also identified in all five core areas, and together they constituted the highest percentage of the total vegetation cover. *Abies lasiocarpa/Linnaea borealis* and *Artemisia tridentata/Festuca idahoensis* types were identified in three core areas, and *Festuca idahoensis/Agropyron spicatum* in two (Table 12.3). The presence of riparian vegetation in all core areas indicates that these types (unclassified) are of great importance to the grizzly bear (Craighead *et al.* 1982).

Denning requirements for grizzlies in the Yellowstone area were highly specific (Craighead and Craighead 1972a, 1972b), but the availability of suitable denning sites was not limiting in the study area. Of 11 active and 6 inactive dens located in four forest habitat types, 11 (nearly 65% of the total) were in the *Abies lasiocarpa/Vaccinium scoparium* type. This habitat type was in an advanced seral stage following burning. Mature lodgepole pine (*Pinus contorta*) dominated the type, with a developing understory of subalpine fir and whitebark pine. Two dens were located in *Abies lasiocarpa - Pinus albicaulis/Vaccinium scoparium*, three in *Picea engelmannii/Linnaea borealis*, and one in *Pseudotsuga menziesii/Calamagrostis rubescens*. (For detailed descriptions of forest habitat types, see Pfister *et al.* 1977.)

All dens were in secluded areas, generally on north-facing exposures, and at altitudes of 2316 to 2804 meters (7600 to 9200 ft). The dens at the highest elevations were in the *Abies lasiocarpa-*

Pinus albicaulis/Vaccinium scoparium habitat type at 2743 to 2804 meters (9000 to 9200 ft). Most of the dens were at elevations of 2377 to 2500 meters (7800 to 8200 ft). The lowest den occurred in the upper edge of the Douglas-fir forest at 2316 meters (7600 ft). The disturbance of bears by human intrusion at the time of denning could be critical (Craighead and Craighead 1972a, 1972b). Fortunately, denning occurs at a time (mid-October to mid-November) when deep snow forces both visitors and hunters out of the high country where the bears den. Thus, conflict between bear and human in this habitat was not great.

Mapping the Spatial Distribution of Vegetation

Efforts to map the spatial distribution of vegetation date back to the 16th century, when maps were beginning to be widely used. These maps were generally crude and reflected the limited knowledge of explorers and naturalists. It was not until the development of aerial photography in the 20th century that the vegetation characteristics of relatively large areas could be recorded "remotely" and studied.

In the early 1960s, scanning instruments capable of measuring a wide spectrum of the radiation reflected by vegetation and other planetary surface features were developed for aerial reconnaissance. In 1969, a prototype spectral scanning system was put into orbit around the earth aboard the Apollo 9 spacecraft. The first Earth Resources Technology Satellite, ERTS-A, later named Landsat-1, was launched in 1972. Thus, after centuries of mapping with many different criteria and classification schemes, it was now possible to use precise data—in this case, spectral reflectance values—and interpretive digital computers to classify and map the earth's surface (Colwell 1971, Varney *et al.* 1973, Hoffer *et al.* 1975).

The vegetation and landforms of several wilderness areas have been mapped and classified with the aid of aerial photographs (Dirschl *et al.* 1974,

Racine 1974, Cowardin *et al.* 1979) and/or satellite imagery (Acevedo *et al.* 1982, Adams and Connery 1983, Meyer and Spencer 1983, Talbot *et al.* 1984 Winterberger 1984). On the ground, distribution and relative abundance of plant species and communities have been determined by intensive vegetation surveys (Cadbury *et al.* 1971, Dirschl *et al.* 1974, Racine 1974). Combining these two approaches—that is, classifying and mapping a large area of wilderness with satellite imagery and, simultaneously, correlating a detailed quantitative sampling of the vegetation—was the next logical step (Craighead *et al.* 1982). Meaningful ecological comparisons between ecosystems in different geographic areas were then possible (Craighead *et al.* 1986).

The Landsat satellites gather a continuous series of digital images of the earth's surface from a polar orbit at an altitude of about 901 to 917 kilometers (560 to 570 mi). Scanning systems that record radiant energy over a wide spectrum of wavelengths collect and digitize the imagery data for later transmission—"down linking" to ground receivers. The basic spectral unit digitized by the scanning system is the picture element, or pixel. A pixel represents the brightness level of a rectangular area on the earth's surface; its size varies according to the scanning system. The pixel recorded by the Landsat Multispectral Scanning System (MSS) is currently about 0.6 hectare (1.6 acres). The newer Landsat Thematic Mapper can define an area about one-fifth that size.

Vegetation complexes are first identified in the field. In the laboratory, a digital image-analyzer measures the range of a sample of pixel values for each complex and builds a unique spectral signature. The signature is then extrapolated to the rest of the digital map, thus transforming the digitized spectral data to color-coded maps, pixel by pixel. When merged with topographic models of the area, a digital-topographic-vegetation map and data base are generated (Craighead *et al.* 1982).

The digital map portrays the study area in terms of spectral values that represent vegetation complexes. The areas of similar spectral reflectance have similar vegetation. The botanical composition within each complex then requires extensive qualitative and quantitative on-the-ground sampling and analysis. This is termed "ground-truthing." The degree to which this is accomplished determines the quality and usefulness of the database that must accompany each spectrally defined vegetation complex. A complex described to the species or to the plant community level is obviously more useful than one described to the level of series or habitat type.

Vegetation Complexes Used by Grizzlies in Other Ecosystems

Our evaluation of habitat in the Yellowstone Ecosystem answered some questions, but left others unanswered. We needed more information on the vegetation components of wilderness, how these were distributed, and how the grizzly bear used them.

It was clear that the welfare of grizzly bear populations could best be evaluated, and the carrying capacity of their habitat estimated, in terms of vegetation characteristics of entire ecosystems. When we were not permitted to complete the vegetation analysis of grizzly bear habitat in the Yellowstone Ecosystem, we (the Craighead Scapegoat Team) initiated a project to map and describe the vegetation components of the Scapegoat and Bob Marshall wilderness areas of Montana (Craighead *et al.* 1982). Subsequently, we (the Craighead Alaska Team) established a second mapping project for the Kobuk River valley of Alaska (Craighead *et al.* 1988a). In both projects, study areas were mapped using satellite multispectral imagery and botanically described (Figs. 12.2 and 12.3, pages 354 and 356). To establish a direct relationship between the spectral data (vegetation complexes) and the specific vegetation they represented, we established "on-the-ground" vegetation sample plots and transects. The cumulative botanical data from these described each vegetation complex in terms of percent cover and frequencies of individual species' occurrence. The botanical data were also used to define habitat-type groupings within the vegetation complexes, such as forest habitat-types, ecological land units, or plant series. These groupings were then subdivided further in terms of their constituent plant communities and plant species (Craighead *et al.* 1982, 1986; Craighead and Craighead 1991).

The spectral values for each vegetation complex, now botanically defined in detail, were computer-extrapolated to generate mosaic maps of 4593 square kilometers (1,773.5 mi²) of the Bob Marshall-Scapegoat wilderness complex (Fig. 12.4, page 357), and 33,768 square kilometers (13,038 mi²) of wilderness country in the Kobuk River valley of Alaska. These studies illustrate the type of habitat data needed to formulate management policies, to determine habitat use, and to integrate bear management with management of other wilderness resources. Because we believe that knowledge of the vegetation base is essential to interpreting grizzly bear population data and for estimating population viability in the various recovery zones, we have treated this subject in some detail,

Table 12.4. Area statistics for 4593 square kilometers (1773.5 mi^2) of the Bob Marshall-Scapegoat Wilderness areas, Montana.

Vegetation complex[a]	Pixels	Hectares	(acres)	Area (%)
Alpine Meadow	23,136	10,487	(25,912)	2.28
Vegetated Rock	15,185	6,883	(17,007)	1.50
Bare Rock I	28,698	13,008	(32,142)	2.83
Bare Rock II	39,631	17,963	(44,387)	3.91
Xeric Pinus Albicaulis Forest	28,093	12,734	(31,464)	2.77
Mesic Abies Lasiocarpa/				
Pinus Albicaulis Forest	87,275	39,559	(97,748)	8.61
Subalpine Parkland	56,899	25,790	(63,727)	5.61
Xeric Abies Lasiocarpa and Xeric				
Pseudotsuga Menziesii Forest	109,456	49,612	(122,591)	10.80
Mixed Coniferous Temperate Forest	438,877	198,927	(491,542)	43.31
Temperate Parkland	154,244	69,913	(172,753)	15.22
Carex-Salix Marsh	1,300	589	(1,456)	0.13
Equisetum Seepage	1,676	760	(1,877)	0.17
Scree	25,152	11,400	(28,170)	2.48
Unclassified[b]	3,809	1,726	(4,266)	0.38
Total	1,013,431	459,352	(1,135,043)	100.00

[a] For botanical details of the vegetation complexes, see Craighead *et al.* 1982.

[b] This category is not considered a vegetation complex, but it does contribute to area statistics.

even though the research was not conducted within the Yellowstone Ecosystem. The methodology is directly applicable, and the results permit between-ecosystem comparisons.

Nine vegetation complexes in the Scapegoat-Bob Marshall wildernesses and 15 in the Kobuk River valley were spectrally classified and mapped, and area statistics were calculated (Tables 12.4 and 12.5). Each vegetation complex was botanically described to the species level; the complexes, in sum, quantitatively described each wilderness ecosystem. Here, we briefly abstract the extensive habitat data from these studies in terms of their application to grizzly bear management generally, referring the reader to the original publications for procedures and informational details (Craighead *et al.* 1982, 1988a).

Bear food plants were quantified at the species level for each complex and compared. For example, abundance values for specific bear food plant species were compared individually and cumulatively within and among the nine vegetation complexes that were derived from computer classification of the spectral reflectance values representative of the Scapegoat-Bob Marshall wilderness areas (Table 12.6). Within the Alpine Meadow Complex, 21 bear-food plants comprised 50.3% of the total vegetation cover, 13 comprised 46.9% within the Vegetated Rock Complex, and so on for each color-coded vegetation complex (Fig. 12.5). Abundance of bear food plants evaluated in the nine vegetation

complexes of the Scapegoat-Bob Marshall was greatest in the Xeric Pinus Albicaulis Forest, Xeric Abies Lasiocarpa, and the Xeric Pseudotsuga Menziesii complexes, respectively.

The abundance of potential bear food plants within complexes dropped sharply when only those most preferred by the bears were considered (Fig. 12.6). Nevertheless, the relative plant food resource values for complexes remained consistent, with two exceptions: the value for the Xeric Pseudotsuga Menziesii Forest Complex decreased, whereas the value for the Mixed Coniferous Forest Complex increased. The relative importance to the bears of each of the nine complexes in the Scapegoat-Bob Marshall is indicated by the total relative abundance of bear-food plants in each (Table 12.6). The distribution maps and botanical composition summaries of these vegetation complexes can be extrapolated and evaluated holistically for entire ecosystems (Fig. 12.4, page 357) or for any part of them (Figs. 12.2 and 12.3, pages 354 and 356). This enables the forest or land manager to make biologically sound decisions about management actions that might adversely affect the most critical habitat of the grizzly bear or any other threatened species (Craighead *et al.* 1982).

Applications of the Mapping System

Landsat's multispectral scanning system is well adapted to recording large vegetation units com-

Table 12.5. Area statistics for 33,768 square kilometers (13,038 mi²) of the Kobuk River Ecosystem, Alaska.

Vegetation complex[a]	Pixels	Hectares	(acres)	Area (%)
Alpine Tundra	295,317	133,483	(329,836)	3.96
Vegetated Rock	246,093	111,234	(274,859)	3.30
Alpine Shrubland	184,396	83,347	(205,950)	2.47
Upland White Spruce	168,165	76,011	(187,823)	2.26
Upland Mixed Spruce	385,532	174,260	(430,596)	5.17
Riparian Spruce	87,306	39,462	(97,511)	1.17
Tussock Tundra	1,607,445	726,565	(1,795,342)	21.52
Shrub Tundra	856,367	387,078	(956,470)	11.47
Tall Willow	85,372	38,588	(95,351)	1.15
Riparian Mosaic	285,503	129,047	(318,875)	3.83
Lichen Tundra	162,120	73,278	(181,070)	2.18
Tidal Flats	20,532	9,280	(22,931)	0.28
Tidal Marsh	2,242	1,013	(2,503)	0.04
Sedge/Grass Marsh	62,624	28,306	(69,944)	0.84
Semi-Vegetated	64,580	29,190	(72,128)	0.87
Sand Dunes[b]	13,058	5,902	(14,584)	0.18
Clouds[b]	401,625	181,535	(448,573)	5.38
Gravel Bar and Open Beach[b]	3,346	1,512	(3,736)	0.05
Bare Rock[b]	20,690	9,355	(23,116)	0.28
Water[b]	2,170,632	981,126	(2,424,360)	29.06
Burns[b]	310,937	140,544	(347,284)	4.17
Unclassified[b]	36,861	16,662	(41,172)	0.49
Total	7,470,749	3,376,778	(8,344,018)	100.00

[a] For botanical details of the vegetation complexes, see Craighead *et al.* 1988a.
[b] These categories are not considered vegetation complexes, but do contribute to area statistics.

Fig. 12.5. Percent abundance of total grizzly bear food plants, as they related to the ecospectral classification of the Scapegoat Study Area, Montana (from Craighead *et al.* 1982).

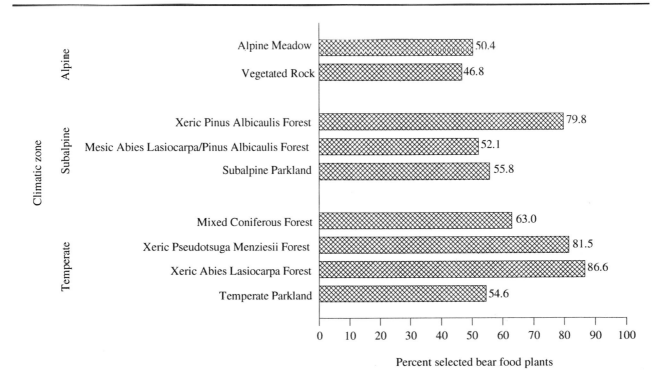

Table 12.6. Relative percents of bear food plants abundance in the vegetation complexes of the Scapegoat Study Area, Montana, 1972-76.

Bear food plant	Percent cover								
	Alpine Meadow I	Vegetated Rock II	Xeric Pinus Albicaulis Forest V	Mesic Abies Lasiocarpa/Pinus Albicaulis Forest VI	Subalpine Parkland VII	Xeric Abies Lasiocarpa Forest[a] IXA	Xeric Pseudotsuga Menziesii Forest[a] IXB	Mixed Coniferous Forest X	Temperate Parkland XI
Graminales									
Carex spp.	19.5	19.2			13.1				5.9
Festuca idahoensis	15.2	10.5	T		7.7		0.4		5.2
Gramineae	2.8	2.8			6.0				8.2
Calamagrostis rubescens	0.2		T		1.9	3.1	57.7	0.7	0.2
Poa spp.	0.1	0.2							0.2
Carex geyeri			5.4	7.6		0.8	16.4	1.1	3.2
Melica spectabilis					0.1				
Agropyron spicatum							2.4		3.0
Festuca scabrella							0.4		10.9
Phleum pratense									2.2
Bromus sp.									0.7
Deschampsia cespitosa									0.7
Poa pratensis									0.4
Danthonia unispicata									0.2
Calamagrostis canadensis					2.2				
Phleum alpinum									0.1
Subtotal	37.8	32.7	5.4	7.6	31.0	3.9	77.3	1.8	41.1
Shrubs									
Arctostaphylos uva-ursi	5.1	7.4			0.1	T		0.3	1.0
Vaccinium scoparium	1.4		43.6	28.3	3.2	30.9		32.5	0.2
Ribes spp.	0.1								
Ribes lacustre		T	T	0.1	T	T	T	T	1.0
Vaccinium globulare			0.3	0.1	0.6	18.7		6.4	
Shepherdia canadensis			0.3		0.5	0.8		4.8	0.6
Vaccinium myrtillus					0.1				
Amelanchier alnifolia					0.1		0.4		0.9
Lonicera involucrata					0.1	T			0.6
Lonicera utahensis						T			0.2
Symphoricarpos albus							1.0	0.3	0.7
Rosa sp.							0.4		0.2
Berberis repens					T		T	0.1	0.1
Prunus virginiana									0.2
Subtotal	6.6	7.4	44.2	28.5	4.7	50.4	1.8	44.4	5.7

Cont'd on next page

prising entire ecosystems, yet researchers can readily use available computer software to map and quantify smaller sites of special interest. Geographic Information System (GIS) computer software provides a powerful tool for integrating spatial data such as elevation, slope, and aspect. These attributes can be expressed digitally, displayed, and analyzed. The many parameters of grizzly bear habitat that cannot be recorded remotely will require *in situ* delineating and sampling. Vegetation sites thus defined in the field can be integrated with digital maps and delineated as polygons. These can be as discrete as a single, large, plant community, a combination of several communities, or the core area of an animal's home range, and can be delineated within spectrally-defined vegetation complexes.

The resulting map can be used to interpret similar small, discrete, land areas located throughout the ecosystem, even though the data cannot be computer-extrapolated to other areas. Both the small polygon-defined sites and the large, spectrally-defined vegetation complexes can be used in total-area computer computations and in statistical analyses of vegetation interspersion, juxtaposition, and habitat diversity.

Table 12.6. *cont'd*

Bear food plant	Percent cover								
	Alpine Meadow I	Vegetated Rock II	Xeric Pinus Albicaulis Forest V	Mesic Abies Lasiocarpa/Pinus Albicaulis Forest VI	Subalpine Parkland VII	Xeric Abies Lasiocarpa Forest[a] IXA	Xeric Pseudotsuga Menziesii Forest[a] IXB	Mixed Coniferous Forest X	Temperate Parkland XI
Forbs									
Hedysarum spp.	1.6	0.8							
Erythronium grandiflorum	1.1	0.2	T	T	1.0		T		
Polygonum spp.	1.0								
Juncus parryi	0.7	3.3			0.2				
Lomatium spp.	0.4				0.6				
Hedysarum sulphurescens	0.3	0.3			0.1				
Polygonum bistortoides	0.2				0.3				
Polygonum viviparum	0.1								
Cirsium scariosum	0.1		T		0.1	T		T	
Claytonia lanceolata	0.1	0.2	T		0.8				
Fragaria virginiana	0.1	0.2	T	0.2	1.5	T	1.7	0.1	2.8
Lomatium cous	0.1	0.8							
Hieracium spp.	0.1				T				
Claytonia megarhiza		1.0							
Agoseris spp		T			T				
Heracleum lanatum		T	T	0.3	1.1				
Xerophyllum tenax			30.1	15.2	12.1	32.4		16.5	2.7
Lomatium dissectum			T		0.1				
Hedysarum occidentale				0.1	0.2				
Equisetum arvense					1.2				
Osmorhiza occidentalis				0.9			T		
Rubus parviflorus					0.1	T	0.7		
Angelica dawsonii					T				
Perideridia gairdneri					T				1.1
Trifolium sp.									0.1
Hieracium gracile									0.2
Juncus spp.									0.9
Subtotal	5.9	6.8	30.1	15.8	20.4	32.4	2.4	16.6	7.8
TOTAL BEAR FOOD PLANTS	50.3	46.9	79.7	51.9	56.1	86.7	81.5	62.8	54.6
OTHER SPECIES	49.7	53.1	20.3	48.1	43.9	13.3	18.5	37.2	45.4
TOTAL	100.0	100.0	100.0	100.0	100.0	100.0	100.0	100.0	100.0

Note: T = trace
[a] Subcomplexes of the Pseudotsuga Menziesii Forest Complex.

To illustrate the application more specifically, assume that we are mapping a 26-square-kilometer (10-mi²) study area of heavily used grizzly bear habitat, but want to evaluate a relatively small area of habitat within the larger area. We first delineate the vegetation complexes in terms of their unique spectral signatures, extrapolate these to the larger study area, and then generate area statistical readouts (Tables 12.4 and 12.5). The habitat sites of special interest are then identified in the field, located on the thematic map, and delineated as signature polygons within the spectrally-defined vegetation complexes. These are color-coded and incorporated into the digital database. Readouts of area statistics for both the large, 26-square-kilometer (10-mi²) area and the small units of special interest can then be obtained from the map database. For details of these procedures, see Craighead *et al.* (1982, 1988a).

Vegetation Components of Home Ranges and Core Areas

As mentioned earlier, our project objectives in Yellowstone included plans, first, to satellite-map grizzly bear habitat and, second, to examine in detail how the bears made use of the habitat. We pursued the first objective in the Lincoln-Scapegoat/Bob Marshall wilderness complex, developing and perfecting the satellite-mapping techniques discussed above. Pursuing the second objective was more

Fig. 12.6. Percent abundance of selected grizzly bear food plants, as they related to the ecospectral classification of the Scapegoat Study Area, Montana (from Craighead *et al.* 1982).

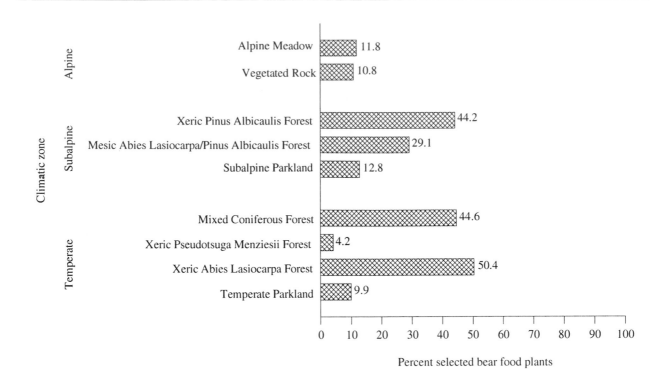

difficult because the grizzly bear was declared a threatened species in the lower 48 states during the course of our work. The opportunity for our research group, a non-government entity, to trap and handle bears was greatly diminished. Therefore, we initiated a project, in Alaska's Kobuk River valley, where grizzlies were plentiful and available to us for radiotracking studies.

From 1983 to 1986, we used Landsat MSS imagery data and the techniques developed on the Scapegoat project to classify and map the vegetation/landforms of a 33,768-square-kilometer (13,038-mi²) area in the Kobuk River valley northeast of Kotzebue, Alaska. The 16 vegetation complexes and land classes, delineated and ground-truthed, provided a database of plant species, their frequencies of occurrence, and their relative percentages of total ground cover (Craighead *et al.* 1988a).

On completion of the satellite vegetation map and database, our next objective was to place transmitter collars on grizzly bears resident to a representative portion of the satellite-mapped study area and track them to determine how they used the habitat. Previous experience adapting transmitters (platforms) for the Nimbus satellite to track elk (Buechner *et al.* 1971, Craighead [F.C., Jr.] *et al.* 1972a, Craighead [F.C., Jr.] *et al.* 1978a, Craighead

[J.J.] *et al.* 1978b, Craighead and Craighead 1979, Fuller *et al.* 1984) and to monitor a black bear during winter sleep (Craighead *et al.* 1971) convinced us that satellite-tracking had a future. The size and remoteness of the study area, the ruggedness of the terrain, and the great expense of conventional VHF radiotracking from fixed-wing aircraft or helicopters reinforced this decision. Accordingly, we chose to collaborate with Telonics, Inc. (Mesa, AZ), a specialty electronics firm, in developing satellite transmitters for our tracking work. In 1984, Telonics produced a first-generation transmitter and provided us with prototypes for field-testing.

Before investing heavily in these transmitter collars for grizzly bears, we evaluated the satellite-tracking system for accuracy of locations, field dependability, ease of use, and cost (compared to conventional VHF systems). We accomplished this by instrumenting a large mammal that was more readily available and more easily handled than a bear. In 1984, we collared two pregnant female caribou with Telonics prototype satellite transmitters. The females were members of the Western Arctic Caribou Herd that was grouping up in the vicinity of Kotzebue, Alaska, preparatory for their northward migration to calving grounds. Our objectives in the tracking study were to determine the herd's migratory routes and monitor their daily

Fig. 12.7. Habitat use as evidence of habitat selection by grizzly bear No. 3, using Method No. 1 (within study area), Kobuk River study area, Alaska, 1986-88.

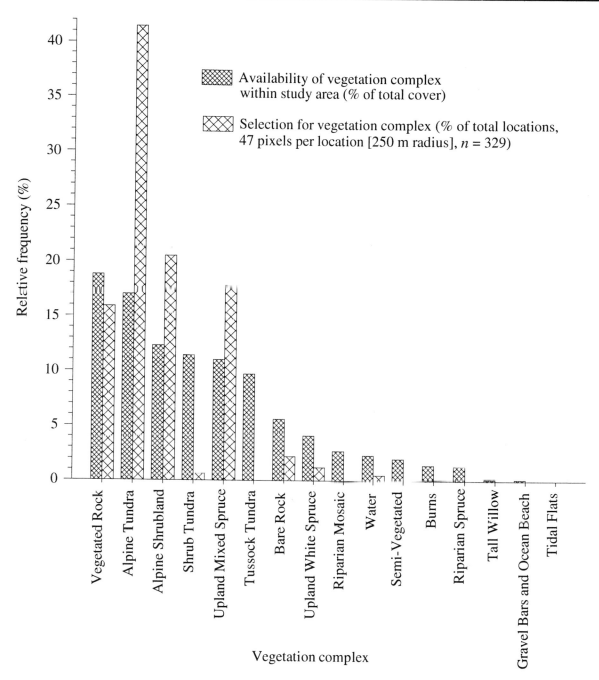

activity patterns (via motion sensors in the collars) while they moved from their wintering grounds to the calving areas and back and, at the same time, to test the performance of the tracking system. Results of that study showed that the satellite transmitter collars generated animal locations with an acceptably high accuracy, but with much greater ease (an armchair at the computer terminal) than that of VHF-tracking equipment, and at 7% of the cost/ day/location. Moreover, the transmitters provided valuable information on the migratory behavior of

the caribou herd that had been almost unattainable using conventional tracking techniques (Craighead and Craighead 1987).

Subsequently, from 1986 to 1988, a total of six adult grizzly bears were captured and collared with satellite transmitters in a 5931-square-kilometer (2290-mi²) portion of the satellite-mapped study area. The bears were located a total of 1624 times over the three-year period via the NOAA/Tiros satellites (D. Craighead 1995 [in prep.]). In order to study how bears utilized the habitat within their

Fig. 12.8. Habitat use as evidence of habitat selection by grizzly bear No. 3, using Method No. 2 (within home range), Kobuk River study area, Alaska, 1986-88.

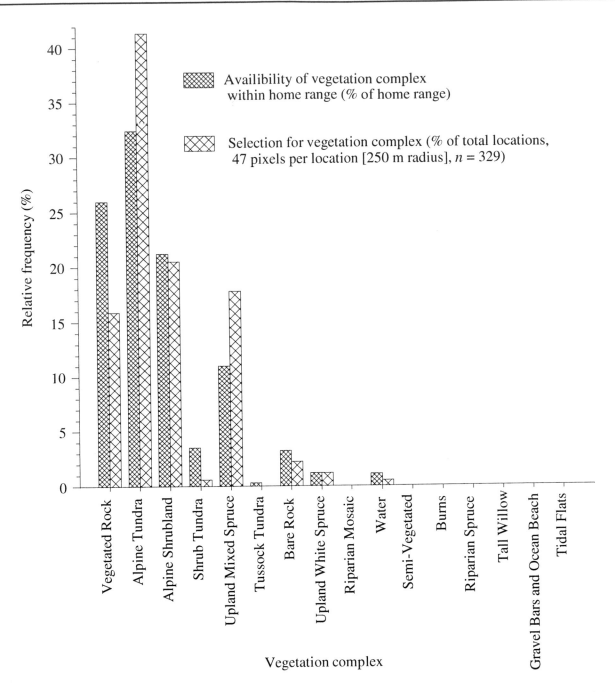

home ranges, a series of algorithms were created to integrate the digitized location data from the NOAA/ Tiros system with the digitized vegetation data from the Landsat MSS system. The algorithms allowed computerized superimposition of bear locations and home ranges on the vegetation map, thereby providing a mechanism for statistically quantifying habitat selection.

Location data were analyzed in three ways. First, the percent cover of each vegetation complex in the entire study area was computed and compared with

the relative percent vegetation cover around individual bear locations occurring within those complexes (Method One, Fig. 12.7). Secondly, the percent cover of the vegetation complexes within each bear's home range was compared with the relative percent vegetation cover around each bear's locations (Method Two, Fig. 12.8). And thirdly, the percent coverage of each vegetation complex for the entire study area was compared with their percent coverage in each bear's home range (Method Three, Fig. 12.9).

Fig. 12.9. Habitat use as evidence of habitat selection by grizzly bear No. 3, using Method No. 3 (comparison of availability between study area and home range), Kobuk River study area, Alaska, 1986-88.

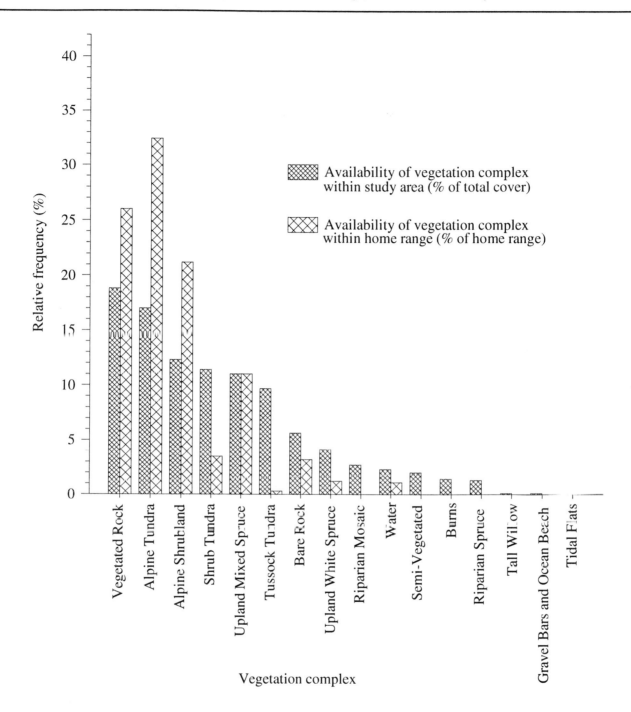

These measures were analyzed to determine whether apparent disproportionate use of certain vegetation complexes suggested by bear locations or home range compositions was a function of habitat preference or avoidance, or simply an artifact of the distribution of vegetation complexes throughout the study area. It is important to note that our methods were designed to show habitat use as it relates to the vegetation base. The vegetation base supports not only bear food plants, but a wide variety of animal life also dependent on the vegetation and used by the bear.

Because grizzly bears are omnivorous, it may be assumed that some of their feeding efforts were directed at specific prey species. Hechtel (1985) reported that several bears in his study spent 24% of their feeding and foraging time in the pursuit of arctic ground squirrels (*Spermophilus parryii*), and tundra voles (*Microtus oeconomus*). Both of these species are niche-specific and can be linked to specific vegetation classes, and so should reflect habitat selection by bears in the same way as would food plant species. Banfield (1958) described a similarity of distribution for bears and squirrels. Hechtel (1985) suggested this was probably a reflection of similar ecological requirements for food and substrate (dens/burrows), rather than the bear's dependence on squirrels for food. Further, he stated that squirrels competed with bears for the roots of *Hedysarum alpinum* and grasses and sedges, but were, themselves, an important part of the bears' diet. Squirrels also played an indirect role in the bears' feeding behavior; the lush vegetation surrounding ground squirrel mounds was often grazed by bears (Hechtel 1985). The complexity of bear feeding habits is apparent, yet the plant base is, in itself, the best classification standard for describing grizzly bear habitat.

Thus, selection or avoidance of vegetation complexes is a response by the bear to the total environment, not to the plant environment, alone. The use of vegetation nomenclature is simply a convenient way of classifying bear habitat, recognizing that both plant and animal components are involved. The plant base can be readily quantified; the animal base cannot.

Habitat preference was examined for six bears. As an example, bear No. 3, an adult female radio-collared on 23 May, was tracked for 71 days. During that period, she was located 329 times, with an average time interval between locations of 5.2 hours. Using a harmonic moment estimator, her home range was calculated to be 39,637 hectares (97,944 acres).

Using method one, bear No. 3's use of habitat, as determined by individual satellite locations, is shown in Fig. 12.7. The Alpine Tundra, Alpine Shrubland, and Upland Mixed Spruce complexes stood out as having been used proportionally more than their availability would suggest. Vegetated Rock was used in proportion to its availability, but all other vegetation complexes were under-selected (Fig. 12.7).

Employing method two, the percent cover of the vegetation complexes within bear No. 3's home range was computed and compared with the relative frequencies of its locations occurring within those complexes (Fig. 12.8). Only Upland Mixed Spruce and Alpine Tundra were over-selected. Vegetated Rock and Shrub Tundra were under-selected (Fig. 12.8).

It is reasonable to assume that the vegetation characteristics (vegetation complexes) within a home range are the result of selection, determined as the bear uses the habitat in establishing its home range. Thus, method three compared the vegetation within

bear No. 3's home range area with what was theoretically available to her throughout the study area. Three vegetation complexes emerged as being disproportionally present or selected: Alpine Tundra, Alpine Shrubland, and Vegetated Rock. With the exception of Upland Mixed Spruce, all other complexes were disproportionally low (Fig. 12.9).

The results from all three methods (1, 2, and 3) indicate that there was some degree of selection and avoidance occurring on the part of bear No. 3, not only in establishing her home range, but in her use of it as well. This selectivity can be only partially explained by available bear food plants (plant statistics derived from on-the-ground sampling). When location data from other bears were similarly analyzed, again, it was apparent that each individual selected strongly for specific vegetation complexes and the habitat types and plant communities characterizing them. Use of habitat was quite different for each instrumented bear, however, suggesting a rather high order of habitat selectivity by each individual, but with few common denominators. The population as a whole (six bears) was not vegetation-specific in its habitat use, but was highly cosmopolitan, using all vegetation complexes to a greater or lesser degree (Craighead 1995 [in prep.]). A larger sample might have revealed a stronger selection for specific vegetation complexes, but would not have masked the strong tendency to use all complexes within the study area. The main conclusion for management consideration is that habitat needs of this grizzly bear population were very broad, and its area requirements, expansive. Specific enclaves of habitat selected on the basis of "designated habitat" criteria for delineating preferred or critical habitat would have been meaningless; the entire ecosystem was essential.

The grizzly bear is omnivorous, opportunistic, mobile, and a generalist in its niche selection. As we have shown in Chapter 6, it is capable of assimilating information and retaining it long-term. Furthermore, it has a long period of parental attachment, during which time the offspring learn specific foraging strategies and feeding locations from their mothers. Such filial learning could explain the strong habitat selection exhibited by individuals. The individuals would continue to use the preferred sites, the most productive ones in terms of fulfilling immediate requirements of food, water, and shelter, until they were no longer productive, at which time alternative foraging areas would be sought. Thus, analysis of habitat use for a single bear over a long period, or for several bears over a season, will tend to obscure selection for specific vegetation types. This is to be expected, given the grizzly bear's learned foraging behavior, as well as its opportunistic foraging strategy. It also confirms the dependence of bear populations on large, diverse areas of intact wilderness for meeting all biological requirements. The current practice of identifying "critical habitat" and classifying it into management situation categories (USFWS 1993) is an approach that may help a few individual bears over the short term, but will, over the long term, surely vitiate the collective resources and space necessary for population viability (Craighead 1995 [in prep.]).

SUMMARY

The spatial and vegetation requirements of the grizzly bear must be well understood in order to accomplish recovery of the species from threatened status in the lower 48 states. Spacious habitat, with varied landforms and vegetation types providing security and adequate food resources, is essential to its welfare and survival.

During our study, a significant proportion of the Yellowstone grizzly bear population trekked annually to and from one or more of the open pit dumps. The dumps and immediate environs occupied only a small part of the home ranges of most bears. As the area in which an individual bear must meet all of its biological requirements, the home range defined the habitat essential to its survival.

Home ranges of individual bears varied greatly in size, depending on the sex and age of the animal, the reproductive status of females, and the annual fluctuations in availability of food resources. Radiotracking confirmed that seasonal ranges of some animals were widely separated and connected by migratory corridors that extended from the geographic center of the Park to well outside its borders. Movements from summer feeding areas to fall ranges and denning sites were also extensive. Home ranges of males exceeded those of females. Range core areas—centers of activity—were small for all bears tracked by radio, seldom exceeding 2.6 square kilometers (1 mi²) each. One male frequently moved 11 to 16 kilometers (7 to 10 mi) daily within a home range of 435 square kilometers (168 mi²). Variation in the spatial needs of individual female bears from year to year was clearly evident in the delineation of a "life range" using radio location data. The life range of one large adult male may have exceeded 2500 square kilometers (1000 mi²).

Fidelity to home range exerted a strong stabilizing effect on population distribution and movement patterns. Because home ranges contained essential habitat, these areas were not readily abandoned. The composite of home ranges of individuals encompassed the entire area of the Park, and

beyond, and delineated the expanse of habitat critical to the survival of the population. Some habitat types were used more than others within the total range, but we found no scientific justification for singling out discrete portions of the habitat as critical and, by omission, suggesting that other portions were not also essential. The numerous biological and social needs of the grizzly bear population were, spatially, all-encompassing.

Our study in Yellowstone was terminated by the National Park Service before plans for detailed vegetation analysis and satellite-mapping of grizzly bear habitat could be fully implemented. Before terminating the project, however, we gathered vegetation data within the home ranges of several marked and radio-collared bears. Later, we analyzed the data, classified them ecologically, and transposed them to a multispectral satellite image of more than 0.7 million hectares (1.8 million acres) in the center of Yellowstone. Four relatively general vegetation and landform spectral classifications, based on botanical data gathered in the field, were developed on computer, and the acreages of each compiled. Coniferous forests comprised an area approximately twice that of the grass-shrublands, and their relative distributions created many miles of ecotone. This interspersion of the classes, particularly in the subalpine parklands, provided habitat conditions highly preferred by grizzly bears.

Coniferous forests in the Yellowstone satellite scene were largely represented by seral stages of the subalpine fir series, with Douglas-fir and spruce series at lower elevations, the spruce series particularly in moist environments along rivers and streams. Several of the forest habitat types in the subalpine fir series were important foraging areas because of heavy ground cover of the blue huckleberry and grouse whortleberry and the availability of whitebark pine seeds. Presence of berries and several food plants associated with riparian conditions attracted foraging bears to habitat types in the spruce series.

The grass-shrublands were particularly valuable feeding areas during the spring and early summer as a source of emergent grasses and sedges, and forb species such as yampa, clover, and dandelion. The alpine meadows and meadow krummholz, high elevation habitats poorly represented within the satellite scene we analyzed, provided food plants such as elk sedge, springbeauty, and biscuit root.

Denning habitat was principally in the subalpine fir series, generally on north-facing exposures. Dens were in remote areas at elevations from 2316 to 2804 meters (7600 to 9200 ft). The lowest den occurred at the upper edge of the Douglas-fir series at 2316 meters (7600 ft).

The limited evaluation of habitat that we were able to complete in the Yellowstone Ecosystem before termination of the study answered some questions, but left others unanswered. More information was needed on the vegetation components of wilderness, how these were dispersed, and how the grizzly bear used them. To answer these questions, we initiated a project to satellite-map and describe botanically the vegetation components of the Scapegoat and Bob Marshall Wilderness areas in Montana and, a decade later, of wilderness in the Kobuk River Valley, Alaska. Both studies entailed intensive, on-the-ground vegetation sampling to describe the computer-defined vegetation complexes in terms of percents cover and frequencies of occurrence of individual plant species.

The nine vegetation complexes derived for the 4593-square-kilometer 1773.5-mi^2) Scapegoat-Bob Marshall Wilderness study area, and the 16 for the 34,109-square-kilometer (13,170 mi^2) Kobuk River study area, quantitatively described each wilderness ecosystem to the species level. Analysis of bear scats from the Scapegoat-Bob Marshall study area, and long-term field observation and satellite tracking of individual bears in the Alaskan wilderness area, revealed which plants were used as food by bears and, of those, which were most highly preferred. The food plants were then quantified for abundance within each complex, and the complexes ranked for plant food value.

In the Scapegoat-Bob Marshall study, the importance to bears of the eleven vegetation complexes was expressed as a function of the presence of food plants. For example, the Alpine Meadow Complex supported 14 bearfood plants that composed 50.4% of total vegetation cover, 46.8% in the Vegetated Rock Complex, and so on, for the remaining seven complexes. The Xeric Pinus Albicaulis Forest, Xeric Abies Lasiocarpa, and Xeric Pseudotsuga Menziesii Complexes were most important, in terms of availability of all food plants. When only preferred food plants were considered, however, the importance of the Xeric Pseudotsuga Menziesii Complex declined sharply, and that of the Mixed Coniferous Forest Complex increased.

Although distribution of the plant food resource unquestionably affected how grizzly bears used their habitat, it, alone, was an inadequate definition of what habitat was "critical" to the population's survival. An answer to that question required that individual bears be radio-tracked to determine how they used the habitat, day-by-day, throughout the year.

From 1986 to 1988, six adult grizzly bears (four females and two males) were collared with transmitters for the NOAA/Tiros Satellite System and tracked, on a daily basis, within the Kobuk River wilderness study area. A series of algorithms were created to integrate the digitized location data with the digitized vegetation data from the habitat analysis of the study area performed earlier via the Landsat Satellite System. The algorithms allowed computerized superimposition of bear locations and home ranges on the vegetation map, thereby providing a mechanism for statistically quantifying habitat use and assessing whether bears selected for (or avoided) specific vegetation complexes.

Location data clearly showed that each of the individual bears selected for specific vegetation complexes and against others, and that individual selectivity could be only partially explained by relative availability of food plants. Despite the rather high order of selectivity demonstrated by each individual bear, there were few common denominators between individuals. In other words, the bear population, as a whole, was **not** vegetation-specific in its habitat use, but was quite cosmopolitan. This supported a similar conclusion derived from our earlier research in Yellowstone. The most important implication of these conclusions for management is that habitat needs of these grizzly bear populations were very broad, and their area requirements, expansive.

Thus, it is pointless to specify that enclaves of *"agency-designated habitat"* be reserved, based on management criteria for what is judged to be preferred or critical habitat; entire ecosystems, with their diversity of flora and fauna, are essential to the survival of grizzly bear populations. Since ecosystems vary in their capacities to support grizzly bear populations, it is essential that digitized resource data, derived from satellite vegetation-mapping, be incorporated in grizzly bear population viability analyses of the future.

13
Ecocentered Relationships of Other Grizzly Bear Populations

INTRODUCTION

The discrete grizzly bear aggregation at Trout Creek represented a subunit of a larger population within the Yellowstone Ecosystem. Members of the Trout Creek aggregation were drawn annually to a relatively circumscribed area of food abundance (ecocenter). Members of this aggregation, as well as those at other ecocenters, were studied as identifiable individuals during the 12 years of our research in the Park and adjacent areas.

The "magnets" that so reliably drew the bears to these circumscribed areas were refuse dumps. The major dump at Trout Creek and a smaller one at Rabbit Creek, both deep in the interior of Yellowstone National Park, were considered eyesores—incursions of civilization incongruous with, and unwelcome in, the unspoiled wilderness. But this condition had been tolerated for years by park administrations. Grizzlies had thrived despite a growing eco-aesthetic sensitivity that bore no relation to the realities of animal ecology. Even the third important ecocenter, the dump at West Yellowstone (Fig. 3.1), was considered highly unaesthetic and therefore undesirable, despite being outside park boundaries. To the bears, these dumps represented dependable sources of large quantities of high-quality food, which partly compensated for habitat encroachment by an ever-expanding human visitor population.

Grizzlies are very successful as omnivorous, opportunistic foragers that exploit any rich food source within their range. If a particular source is plentiful and predictably available, a population of bears from surrounding areas will aggregate in rough proportion to the magnitude and nutritional quality of the food source. The result is a traditional movement and a traditional period of aggregation keyed on the period of food availability.

In Yellowstone, the period of food availability, and thus the aggregation, coincided with the bears' breeding season. The ecocenter affected not only feeding behavior, but virtually every other measurable aspect of population biology.

To portray the pervasive ecological influence of the Park's dumps on the grizzly bear population, we termed them *ecocenters*. Although these particular ecocenters were of human origin, the term embodies a much broader biological concept. We believe ecocenters, whether termed "artificial" or "natural," have been basic in the evolution of the grizzly bear. Consequently, the grizzly, with its extensive broadly overlapping home ranges occupied by solitary individuals or family units, has developed behaviors for social interaction, which any study of its biology must take into account. This is not to say that all pristine bear populations are ecocentered or that all individuals within ecocentered populations participate in aggregations. Rather, ecocentricity, throughout the paleontological past, to the present, appears to have been a strategy which greatly influenced the biology of grizzly bear populations. The Trout Creek ecocenter served as a buffer against periodic failures in one or more "natural" components of the overall food base and thereby stabilized reproduction and insulated the subpopulation against catastrophic environmental events.

We further examine the ecocenter concept by (1) refining our definition of the term and clarifying our usage of it; (2) examining the paleontological and historical evidence for ecocenters; and (3) comparing "artificial" ecocenters with "natural" ones (e.g., salmon runs, bison jumps, or whale carcasses).

THE ECOCENTER CONCEPT

We have defined an ecocentered population as one that congregates at a high-quality food source in a relatively confined, predictable portion of the entire ecosystem during an extensive, annually predictable time period (one to three months). The general feeding pattern of grizzlies includes movements in synchrony with the phenology of local flora and fauna, and the return to discrete feeding sites learned and remembered from earlier experiences.

This type of feeding behavior is characteristic of all members of a bear population and is not to be confused with the ecocenter phenomenon, which exhibits definite characteristics readily recognized in larger aggregations.

The definitive characteristics of an ecocenter are:

1. Highly nutritional food in large quantities acquired with relative ease, and annually available for a period long enough to justify, energetically, long-distance travel by individuals;
2. A relatively circumscribed area of food availability that encourages group-feeding and the formation of a social hierarchy among assembled individuals; and
3. Presence of food during approximately the same time period each year, allowing individuals in the population to establish a food-place-time association and thus "learn" to return to the site.

We use the term "ecocenter" to refer to the place where food is concentrated, as well as to describe the population of bears routinely attracted to it . . . i.e., an ecocentered population. Because both the size and discreteness of ecocenters vary considerably, we defined three ecocenter categories. This rough classification establishes some general distinctions useful in a discussion of this phenomenon. These categories are:

1. *Primary ecocenter*—a relatively large food biomass in a spatially discrete communal feeding area attracting a large aggregation of bears.
2. *Secondary ecocenter*—a smaller food biomass or less spatially discrete communal feeding area, or both, attracting a relatively smaller aggregation.
3. *Non-ecocenter*—food resource distribution favoring solitary animals or small, dispersed groups with no specific association with a communal feeding area.

Generally speaking, secondary ecocenters, like primary ones, have predictable locations and periods of food availability, but they provide food in smaller quantities per unit area, can be subject to minor periodic variation in availability, or are of reduced nutritional quality. These sites do have positive effects on the health and stability of the entire population, but benefits are less pervasive than those conferred by a primary ecocenter. Examples within the Yellowstone Ecosystem include disposal areas for carcasses of cattle or wild ungulates, insect aggregation sites, and possibly, the spawning beds of native fish.

We consider food sources without predictable time or place (e.g., berry or pine seed crops), irrespective of their quality or quantity, to be *non-ecocentered*. These sources are fortuitous or remembered finds and are a result of foraging excursions by individual bears. Despite attracting groups of bears at times, such sources are routine nutritional finds that do not induce traditional annual movements or individually exert a definable long-term effect on the dynamics of the overall population.

Plainly, the degree to which any food source will influence the biology of a bear population is a continuum, and the importance of any ecocenter is best defined by its effect on the entire population. This definition raises a thorny conceptual issue that is often settled arbitrarily: what constitutes a "population" of bears? In circumstances of isolated populations such as that of the Yellowstone Ecosystem, the occupied habitat of the species is usually considered definitive. Other areas—for example, those in Alaska such as the Noatak and the Kobuk River valleys, Kodiak Island, or the north slope of the Brooks Range—do not impose obvious geographic limitations, and bears may be found widely distributed over large areas. Thus, analytic methods which do not require spatial closure of the "population" under study are necessary (Miller *et al.* 1987). A seasonal aggregation of 5 to 10 bears that would, in some circumstances, be a very minor subset of a population, would in others, be a significant percentage of the population that might reasonably have access to a specified ecocenter. Thus, classification of, as well as assignment of biological values to ecocenters, must be approached with caution. Researchers must have a strong sense of the size and distribution of the population with which they are working and the overall structure and characteristics of the habitat. This said, we suggest that the distinction defining a primary ecocenter is the effect it has on the entire population of a given area. It must annually influence the biology of a significant proportion of the individuals.

A PERSPECTIVE ON ECOCENTERS

The bears' use of the Yellowstone refuse dumps was not an aberrant result of their conditioning to the presence of humans, but rather a basic biological phenomenon characteristic of the species. Both paleontological and historic evidence support this conclusion.

Prehistoric Evidence

The brown bear lineage (*Ursus arctos*) diverged throughout the Pleistocene from forest-dwelling Etruscan bear stock (*U. etruscus*) of Asia and underwent significant adaptive radiation during the Ice Ages (Thenius 1959 and Kurtén 1968 *in* Herrero

The term ecocenter was coined to describe areas of concentrated resources, as well as aggregations of bears using these resources. The high-quality food available at an ecocenter and its attraction to foraging bears is illustrated in this "natural" ecocenter at the McNeil River Falls, Alaska. *(Photo by John J. Craighead)*

1972). The bear line lost the typical carnivore dentition early in its evolution, developing omnivorous habits and a dentition better adapted for both meat and vegetable consumption. This adaptation from strictly carnivorous to omnivorous feeding was likely a response to food stresses. The same competitive pressures encouraged the development of minor anatomical adaptations in the digestive tract, as well as metabolic mechanisms that mediate storage of fat and its hydrolization during a period of winter sleep.

Physical adaptations were no doubt accompanied by behavioral adaptations to meet the same environmental stresses. As the great glaciers of the Ice Age repeatedly advanced and receded, vast areas of the northern hemisphere were deforested and left as tundra or open plains. The brown bear group evolved to utilize an open-ground type of habitat, often as forest-to-nonforest ecotones. The bears' food resources must have included herd mammals, particularly winter-killed carcasses of species referred to as "megaherbivores" (e.g.,

wooly mammoth, rhinoceros, and the prehistoric bison), carcasses of which could feed many bears for days, even weeks.

A minimum of nine families and 20 genera of megaherbivores were evolving in the same time and space as the Etruscan predecessor of the present-day grizzly bear (Owen-Smith 1989). The most numerous of the megaherbivore genera belonged to Order Proboscidea. These included three species of gomphotheres, the wooly mammoth (*Mammuthus primigenius*), plus two species of mastodonts (*Mammut* spp.), and three other species of the elephant family. The rhinoceros family was also well-represented in at least five species. Other megaherbivores extant at this time were the giant ground sloth (*Eremotherium* sp.) and two giraffe species. Within any given regional fauna, Owen-Smith estimates that two to six species of megaherbivores were present. Although too small to be classified as true megaherbivores, two species of bison (*Bison antiquus* and *B. latifrons*), now extinct, existed during the period. The presence of

The social behaviors permitting grizzlies to exploit spatially and temporally stable resources at ecocenters also reduced aggression whenever group feeding occurred. *(Photo by John J. Craighead)*

large numbers of bison and megaherbivores offered bears potential food sites at which to gather, perhaps in large numbers, to feed on carcasses. These sites probably played the role of primary or secondary ecocenters for the grizzly bear progenitors. There is also reason to believe that salmon (*Salmo* spp., *Oncorhynchus* spp.), trout (*Salvelinus* spp.), and other anadromous species present during the Pleistocene were abundant, and "pooled up" at stream barriers and spawning sites as they do today. These sites too would have been centers for bear aggregations.

The strong response of grizzly bear populations to ecocenters was probably well-established in North America during prehistoric times. Grizzlies were ecologically associated with the tremendous bison herds that roamed the Great Plains, and the bears used bison carcasses seasonally abundant at specific sites. The Great Falls of the Missouri in Montana and Head Smashed In Jump in Alberta, Canada, are examples of such sites. Furthermore, "artificial" ecocenters had long been available where Paleo-

Indians, subsisting on the bison herds, applied techniques of communal mass-harvesting. Frison (1978) has presented evidence for many sites throughout the northwestern plains where bison pounds (corrals), surrounds, and jumps may have provided dozens to hundreds of dead and injured animals in a single harvesting event. Many prehistoric mass killing sites have been identified in the region adjacent to what is now Yellowstone National Park. Frison (1978) has located some of the major prehistoric sites in Wyoming, while Dr. Dee Taylor, University of Montana, Missoula, documented some of those in Montana (pers. comm. 1989). Bison kill sites dated to more recent times were commonplace in Montana. One such was described on 29 May 1805 by members of the Lewis and Clark expedition after examining a buffalo jump on the banks of the Missouri in the vicinity of its confluence with the Musselshell River, Montana (Biddle 1962).

On the north we passed a precipice about one hundred and twenty feet [37 m] high, under

The aggregations of grizzlies at Trout Creek in Yellowstone National Park did not differ fundamentally from the brown bear aggregations at McNeil River, Alaska, except for the food source—garbage at the former, salmon at the latter. *(Photo by John J. Craighead)*

which lay scattered the fragments of at least one hundred carcasses of buffaloes, although the water which had washed away the lower part of the hill must have carried off many of the dead.

The Indians then select as much meat as they wish, and the rest is abandoned to the wolves, and creates a most dreadful stench. The wolves who had been feasting on these carcasses were very fat, and so gentle that one of them was killed with a spontoon.

Bison jumps were especially concentrated in Park, Madison, Gallatin, and Beaverhead Counties of Montana, whereas pounds and surrounds were most common in Beaverhead County (Table 13.1).

Kill sites were numerous in areas of Wyoming surrounding Yellowstone. At the larger, more traditional sites (Table 13.2), the natives of the Paleo-Indian period used natural topographic features to trap or slow the bison so they could be killed. For example, animals might be driven into arroyos, where they were trapped and killed, whereas others were harried into natural traps such as sand dunes or bogs. In the more recent Late Plains Archaic and

Late Prehistoric Periods, more sophisticated corrals and buffalo jumps were used.

One of the earliest-known Late Prehistoric pounds, Wardell, was near Big Piney in the Green River Basin of Wyoming (Frison 1978). Juniper posts were used to build a corral into which bison could be hazed while moving between their grazing areas and water. Frison estimated that, over a period of about 500 years, more than a thousand bison were killed and rendered at the Wardell site alone.

Some of the traditional Wyoming bison jumps used by ancestors of the Crow Indians were east of the Bighorn Mountains along the drainages of the Little Big Horn River, Rosebud Creek, Tongue River, Big Goose Creek, and Piney Creek (Table 13.2). The early Crow stampeded the bison several kilometers along one or more drive lines to jump-offs from 15-meter (50-ft) cliffs. Such vertical jumps appeared to have been used perhaps for centuries by the Native Americans. It is probable that grizzly bears in the vicinity of jumps, pounds, or surrounds were attracted to and used

Grizzly bears have been responding to man-made ecocenters for thousands of years. Paleo-Indians in the Americas practiced communal mass harvesting of bison at traditional Piskun sites. Offal from the slaughter attracted aggregations of grizzly bears. *(Painting by William R. Leigh, Courtesy of Buffalo Bill Historical Center)*

Table 13.1. Location and numbers of historic bison kill areas in six Montana counties in proximity to Yellowstone National Park.[a]

County	Bison jumps	Bison pounds/surrounds	Total bison kill areas
Gallatin	7	2	9
Sweetgrass	4	0	4
Jefferson	3	0	3
Yellowstone	3	0	3
Park	18	2	20
Madison	10	1	11
Beaverhead	6	4	10

[a] Dr. Dee Taylor, Univ. of Montana, pers. comm. 1989.

them as ecocenters. Furthermore, the high numbers of coyote (*Canis latrans*), bobcat (*Lynx rufus*), wolf (*Canis lupus*), skunk (*Mephitis* spp.), and badger (*Taxidea taxis*) bones, as well as those of bears, found at the various jump sites verify these sites as attractive to a wide range of carnivores and imply that bison remains may have been deliberately utilized as bait to attract them. Carnivore bones had been scored and cracked in a way similar to those of the herbivores, suggesting the carnivores also served as food items (Frison 1978).

Historical Evidence

There are specific early records of grizzly bears gathering at ecocenters. Many of these pertain to the California grizzly (*Ursus arctos californicus*), a subspecies extremely numerous before the large European migrations to California in the nineteenth century, but now extinct. For example, G. C. Yount commented in 1831 that "they [the grizzly bears] were on the plains, in the valleys, and on the mountain" and "it was not uncommon to see fifty or sixty within the twenty-four hours" (Storer and Tevis 1955). The coastal abundance of marine mammal carcasses, especially whales, concentrated grizzlies for extensive periods of time. The first

The coastal abundance of marine mammal carcasses, especially whales, concentrated grizzlies for extensive periods of time. This is an artist's depiction of Sebastian Vizcainos' 1602 description of an aggregation of California grizzlies on a beached whale. *(Painting by Ron Jenkins)*

Table 13.2. Traditional Wyoming bison kill sites in close proximity to the Yellowstone Ecosystem.[a]

Name	County	Cultural period	Method of kill	Bison killed
Wardell	Sublette	Late Prehistoric	Corral	1000
Glenrock	Converse	Late Prehistoric	Buffalo jump	Unknown
Big Goose	Sheridan	Late Prehistoric	Buffalo jump	Unknown
Piney Creek	Johnson	Late Prehistoric	Buffalo jump	200
Vore	Crook	Late Prehistoric	Buffalo jump	20000
Hawken	Crook	Early Plains Archaic	Arroyo trap	Unknown
Ruby	Campbell	Late Plains Archaic	Corral	Unknown
Muddy Creek	Carbon	Late Plains Archaic	Corral	Unknown
Agage Basin	Niobrara	Paleo-Indian	Arroyo trap	Unknown
Casper	Natrona	Paleo-Indian	Sand dune trap	Unknown
Horner	Park	Paleo-Indian	Unknown	Unknown
Finley	Sublette	Paleo-Indian	Sand dune trap	200

[a] Table adapted from Frison, 1978.

As early as 1916, waste generated by Yellowstone Park visitors was sufficient to attract large numbers of grizzly bears. This natural response of the grizzly to a plentiful resource persisted until park managers decided to cease the "unnatural" feeding at open pit garbage dumps in 1969.
(Photo by J. E. Haynes 1916, Montana Historical Society)

specific record of this phenomenon was a description by S. Vizcaino in 1602 of an aggregation on a beached whale carcass (Storer and Tevis 1955). In 1849, J. W. Revere reported "countless troups" of grizzlies feeding on whale carcasses and, in what may be the first written record of grizzlies feeding at an "artificial" ecocenter, on the remains of commercially rendered whales along the beaches of Monterey (Storer and Tevis 1955). The California grizzly also congregated in large numbers at human cultivations of plant foods such as clover patches, corn fields, and plum tree orchards. Native oak stands also attracted grizzlies in numbers during good mast years (Storer and Tevis 1955).

In summary, the ecocentric behavior we have described for grizzly bear aggregations at Trout Creek has a clear evolutionary rationale and is not simply an aberrant response to modern "artificial" ecocenters. The grizzly has most probably displayed this tendency to aggregate at sites providing large, stable quantities of food, and to develop traditional

migrations to them, for many thousands of years. As the great bison herds (and the bear-attracting Native American practices associated with them) declined, sparse, free-roaming bison groups remained in and around the future site of Yellowstone, and the bear population probably continued to be dependent on winter-killed carcasses as a food source. When hotels were first opened in the new Yellowstone National Park in the late 1800s and garbage dumps appeared, they immediately attracted grizzlies, evolutionarily and culturally conditioned to concentrate at ecocenters, regardless of the origin of the food source. The central role that the major dumps came to play in the biology of the bears, then, is not surprising.

SOME CURRENT BROWN/GRIZZLY BEAR ECOCENTERED POPULATIONS

The Trout Creek ecocenter was the largest and most central of the eight aggregation sites within or

Table 13.3. Some examples of recent grizzly/brown bear aggregations in the United States.

REGION / Site	Bear numbers	Food source	Time period
ALASKAN PENINSULA			
McNeil River	20-120	Salmon	July/August
Mikfik Creek	5-10	Salmon	Early June
Brooks River	10-20	Salmon	July
Nak Nak Dump	2-8	Garbage	Summer
Hallow Bay	30-40	Salmon	June/July
SOUTHEAST ALASKA			
Admiralty Island	(multiple sites) 5-10	Salmon	July
Kadashan River	10-15	Salmon	July
Greens Creek	5-14	Salmon	July
INTERIOR ALASKA			
Prairie Creek	50-60	Salmon	July/August
KODIAK ISLAND, ALASKA			
Red Lake	60-80	Salmon	July/August
O'Malley Lake	40-55	Salmon	July/August
Dog Salmon Creek	50-60	Salmon	July/August
Sturgeon River	40-50	Salmon	July/August
ARCTIC COAST ALASKA	10	Marine mammal carcasses	Summer
LOWER 48			
Cooke City, MT[a]	15+	Garbage	June–August
Rocky Mtn. East Front, MT[b]	Unknown	Livestock boneyards	Spring/Summer
Brent Creek, WY	9+	Livestock boneyards	Spring/Summer

[a] Cooke City dump reportedly was closed and bear-proofed in 1979.

[b] Carcasses of cattle that died on ranches along the Front are transported to various boneyards in the backcountry by personnel of Montana Dept. of Fish, Wildlife and Parks.

peripheral to Yellowstone National Park during the period of our study (Fig. 3.1). Other probable examples of bear aggregations at either primary or secondary ecocenters are found in Alaska and, sporadically, in the lower 48 states (Table 13.3). Specific bear visitation data based on systematized, direct observation are not available for the majority of these, nor are data on site size or on population parameters such as sex/age structure and reproduction. Nevertheless, cursory studies and reliable anecdotal information have suggested that ecocenters function at some level for all the sites listed in Table 13.3.

Primary Ecocenters

We have selected three sites that appear to function as primary ecocenters for bear populations which have been the subject of good research stud-ies—the Alaskan brown bear populations that sea-sonally congregate at McNeil River Falls and at Karluk Lake on Kodiak Island, and those on Ad-miralty and Chichagof Islands, Alaska. Visitation data and reproductive parameters for the McNeil River Falls and Admiralty/Chichagof Island popu-lations are available for our comparisons. The published data for Karluk Lake were derivative, however, and the raw data were not available for our analysis.

McNeil River Falls—Alaska

McNeil River Falls is on the Alaska Peninsula near the southwestern tip of Kamishak Bay about 186 miles (300 km) southwest of Anchorage, Alaska. Several good studies have been done on the behav-ior, aggregation dynamics, and reproductive biol-ogy of the brown bear population, which annually

Fig. 13.1. Mean daily observed bear visitation at McNeil River Falls, Alaska, for the period 1980-87 (based on unpublished data; L. Aumiller, R. Sellers, and C. Smith, Alaska Department of Fish and Game, 1992).

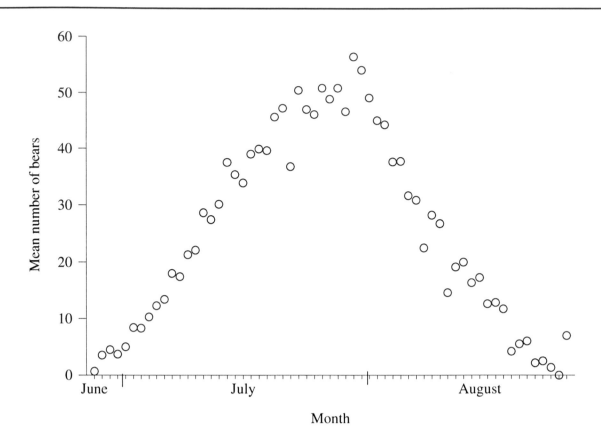

gathers over a two- to three-month period to feed on migrating chum salmon (*Oncorhynchus keta*) (Stonorov and Stokes 1972, Egbert and Stokes 1976, Glenn *et al.* 1976, Luque and Stokes 1976, Egbert 1978). In addition, the largely unpublished data of Larry Aumiller, Chris Smith, and Richard Sellers, Alaska Department of Fish and Game, extend the period of continuous study to more than 20 years. They have generously provided some data to us, through 1990.

The brown bear aggregation at McNeil River Falls is an impressive one, comparable to that observed among the grizzly bears at Trout Creek. Within an area about 25 meters by 165 meters (27 yds by 180 yds) in the vicinity of the falls, more than 60 to 70 and as many as 120 brown bears routinely gathered during July and August to fish for the chum salmon. The aggregation enlarged slowly in late June, with a mean number of 34 brown bears feeding daily on the early salmon runs (Fig. 13.1). The numbers increased steadily in July, reaching their mean daily peak of 56.5 around the end of July, then gradually declined to only a few

bears by the end of August. Current population counts (Larry Aumiller, pers. comm. 1992) indicate the aggregation has grown even larger in recent years. Like that at Trout Creek, the McNeil population is protected by a park, the 5504-hectare (13,600-acre) McNeil River Game Sanctuary, and is essentially unhunted.

Karluk Lake System, Kodiak Island—Alaska

Numerous ecocenters with a food base of salmon are used by the brown bears of Kodiak Island (Table 13.3). The magnitude and duration of the bear aggregations suggest that many of the sites are primary ecocenters frequented by a majority of the endemic population. It is probable, given the absence of obvious geographic barriers, that the ranges of the "populations" actually overlap.

Troyer and Hensel (1964) studied the Kodiak bear population within a 250-square-kilometer (96.5-mi^2) study area encompassing Karluk, Thumb, and O'Malley Lakes, and their tributaries. Population density during the red salmon (*Oncorhynchus nerka*) spawning run approximated one bear per square kilometer (3 bears/mi^2). Greatest bear den-

The senior author had the opportunity to observe the McNeil Falls brown bear aggregation on several occasions. Observing bear behavior at McNeil River, Alaska, was like a nostalgic return to the Trout Creek ecocenter of Yellowstone. In all important respects, the behavior of bears at McNeil mirrored the ecocentered behavior of grizzly bears at Trout Creek.

A female with cubs cautiously enters the salmon-infested river after placing her young on shore. Wary of potential danger and sensing the approach of another bear, she retreats from her fishing spot to defend her cubs from the intruder. As in Yellowstone, females with cubs were cautious in their approach and use of the ecocenter. *(Photos by John J. Craighead)*

sities occurred around tiny O'Malley Lake where, for a period of several weeks in July and August during the peak of the salmon run, 40-55 bears gathered daily to fish and to feed.

Admiralty and Chichagof Islands—Alaska

The third primary ecocenter, Admiralty and Chichagof Islands, is located in the Alexander Archipelago of southeastern Alaska. The Admiralty brown bears began aggregating on the salmon spawning streams around the middle of July, and remained on the streams for about one month (Schoen and Beier 1986, 1989). The population density in the 344-square-kilometer (133-mi²) study area was estimated at 0.4 bear per square kilometer (1 bear/mi²). At peak aggregation, approximately 136 brown bears congregated at several sites in this study area.

Secondary Ecocenters

Many of the seasonal brown/grizzly bear aggregations (Table 13.3) represent, at the least, secondary ecocenters. Because most of these Alaskan sites and their bear aggregations, visitation phenomena, population structure, or reproduction have not been studied systematically, their true dynamics are unknown. One Alaskan site that may represent a secondary ecocenter has been well-studied in the

Females with cubs were effective disciplinarians, maintaining order with facial expressions and vocalizations. *(Photo by John J. Craighead)*

A large male uses expressive body postures and vocalizations to pass through another male's fishing site without conflict. Social signals understood by all bears of a population are adaptive traits allowing efficient use of a common food resource. *(Photos by John J. Craighead)*

Susitna River drainage (Miller 1983). In the lower 48, the Cooke City dump may have functioned as a secondary ecocenter in the Yellowstone Ecosystem before 1979, but its effects on the ecosystem-wide population were probably minor. We briefly describe available data regarding bear activity at Susitna and at the Cooke City dump. In general, the grizzly bears inhabiting the Yellowstone Ecosystem after 1970 are considered non-ecocentered.

Susitna Hydroelectric Project—Alaska

An environmental impact study of the area that would be affected by the proposed Susitna Hydroelectric Project included extensive data on movements, reproduction, and structure of the brown bear population. Miller and McAllister (1982) and Miller (1983) reported that Prairie Creek, a tributary of the Talkeetna River (which, in turn, flows into the Susitna) attracted at least 10 bears to feed on spawning king salmon (*Oncorhynchus tschawytscha*) during July and August. Additional radio-tracking data accumu-

lated in later years showed that the annual bear aggregation actually averaged 50-60 bears, which moved to the ecocenter from throughout the 12,127-square-kilometer (4682-mi²) impact study area and perhaps beyond (Miller 1987). Because this number represented only 15% to 18% of the estimated population of 327, the likely influence of the ecocenter on the total population probably meets our criteria for a secondary ecocenter. Moreover, other secondary ecocenters, perhaps of smaller impact, also appeared operative in the area. Less detailed data for Gold Creek, a tributary of the Susitna below Devil's Canyon, suggest an annual brown bear aggregation during spawning runs of pink and chum salmon (*Oncorhynchus gorbuscha* and *O. keta*) (Miller 1987).

The Cooke City Dump—Montana

During 1959-70, a few bears aggregated at the Cooke City dump (Fig. 3.1), but we did not consider it a bona fide aggregation. In an interesting development in the history of Yellowstone grizzly

As in Yellowstone, females often placed their young at a safe distance from the ecocenter, returning to them with fish and maintaining discipline with persuasive vocalizations. *(Photo by John J. Craighead)*

bear research and management, this site probably functioned as a secondary ecocenter, and also as a research site, throughout the 1970s until its closure in 1979. This was reportedly a period when garbage dumps had been eliminated and when only "wild, free-roaming" grizzlies inhabited the "Yellowstone backcountry." Although no quantitative data on bear aggregations were obtained by the Interagency Study Team (IGBST), grizzlies were trapped and radio-located there, and two adult females were marked and their reproductive histories recorded. Not until 1985 did the IGBST begin publishing reproductive data for specifically identified individual females (Knight *et al.* 1985). In that report, females No. 8 and No. 10 were noted to be routinely nourished by "supplemental food from the Cooke City dump," and to have produced, between them, seven litters for a remarkably high, combined productivity rate of 0.800. But these circumstances were stated to be "abnormal," so the records of the two females were deleted by the IGBST from the Yellowstone population statistics. With this deletion, a reproductive rate of 0.482 was reported, based on 16 other marked females that presumably had not been feeding on garbage (Knight *et al.* 1985).

In 1989, the IGBST finally included the reproductive histories of females No. 8 and No. 10 with those of other marked females, apparently acknowledging that these "dump bears" were not aberrations (Knight *et al.* 1989).

Developing Ecocenters

Functional primary ecocenters are, by definition, relatively long-lived. They develop gradually through a process that depends on the nature, quantity, and periodicity of the food source, and on the aggregation dynamics created as foraging bears encounter the site, move in to exploit it for the first time, and become habituated to return to it in subsequent years.

Several ecocenters are now in the process of development. One is on the southern boundary of Glacier National Park, Montana, where a series of railway accidents dumped tons of grain corn. Two others are disposal sites for livestock carcasses ("boneyards") in Montana and Wyoming. Finally, a series of sites in Wyoming east of Yellowstone National Park, where since 1986 growing numbers of bears have been observed aggregating to feed on insects, may well develop into secondary ecocenters. Other invertebrates, as well as spawning fish and

A female with yearlings feels less threatened at the ecocenter than females with cubs. This female waits a turn for her family to fish at a favored site. *(Photo by John J. Craighead)*

winter-killed ungulate carcasses, are food resources capable of attracting and holding bears in the Yellowstone Ecosystem, but we currently do not have evidence to show that they function as ecocenters.

Grain Spills

Although grain spills are generally accidental and temporary food sources, they can attract a substantial number of bears within a relatively short period of time. Three derailments of Burlington Northern railroad cars loaded with corn south of Glacier National Park in 1989-90 present a case in point. The result was spillage of about 8528 metric tons (9400 short tons) of corn from 19 December to 22 February. Burlington Northern attempted to clean up their tracks and the immediate area, but large quantities of corn were available in the locality for several years, despite repeated efforts to remove it and to bury any that remained. Grizzly bears moved in from the surrounding forests to feed, the aggregation gradually enlarging as more and more bears found the corn. The spills occurred in

an area where the railroad bed was confined by steep slopes, and within a relatively short distance of human habitation. This created a problem for both bears and people.

Forest Service biologists estimated that 14 to 16 individual grizzlies visited the spills during the summer of 1989. In addition, eight grizzlies were struck and killed by Burlington Northern trains, as were unrecorded numbers of black bears, also attracted to the grain. These deaths occurred from 12 June 1989 to 19 October 1990, a period of approximately 16 months (two summer seasons). The number of documented deaths was a minimum as other grizzlies may well have been injured on the tracks and died at a distance, undiscovered. The situation probably attracted poachers who may have killed additional bears. A controversy and, ultimately, a suit resulted from this incident (see Amended Complaint, National Wildlife Federation and Great Bear Foundation versus Burlington Northern Railroad). Properly placed and managed, grain spills could have proven advantageous to the grizzly population, but because of failure to recognize the "drawing power" of the grain and the hazards to the bears attracted to that particular site, it had a highly detrimental effect.

The Burlington Northern corn spill incident shows that ecocenters can develop very rapidly. Approximately 25 to 30 grizzly bears, and at least that many black bears, were attracted to the spill sites within a period of about 16 months. When and if additional grain spills occur in this or similar areas, they would be handled most effectively by a rapid, thorough cleanup, cooperatively accomplished by U.S. Forest Service and Burlington Northern Railroad personnel. The grain could either be hauled away to landfills (as in the first incident) or, at additional expense, helicoptered to prime bear habitat, isolated from humans, where grizzlies could use the grain in relative safety.

Boneyards

When livestock die of disease or injury, western ranchers often haul their carcasses to dump sites on remote parts of the ranch. The number of carcasses annually deposited at these sites, often termed "boneyards," is roughly proportional to the size of the rancher's livestock herd. Where bears are indigenous, this practice attracts them and, at times, leads to frequent bear-human conflicts.

East Front of the Continental Divide—Montana. Research on the grizzly bear population along the east slope of the Rocky Mountains in northwestern Montana was begun in 1977 as part of the Border Grizzly Project, University of Montana, Missoula

At the height of the salmon run, most bears readily satisfied their appetite: the senior author watched an experienced female catch 23 salmon in three hours. At first she consumed entire salmon, but as she became satiated, she ate only the skin and roe. *(Photo by John J. Craighead)*

(Schallenberger and Jonkel 1980). Research maintained from 1979 through 1987 by Aune and various co-workers (Aune 1985; Aune and Brannon 1986; Aune and Stivers 1981, 1982, 1983; Aune *et al.* 1984) resulted in a cumulative final report (Aune and Kasworm 1989).

The area of study was inclusive along the east face of the Continental Divide from State Highway 2 on the north boundary to State Highway 200 about 177 kilometers (110 mi) to the south. Although the eastern boundary was fixed only approximately, the area of study was estimated by Aune and Kasworm (1989) to be 4662 square kilometers (1800 mi²).

Over the 11-year period of study, 61 grizzly bears were captured in 105 capture actions, pertinent biological data were recorded, and the bears were marked or radio-collared for future observation (Aune and Kasworm 1989). Twenty-one of the 61 bears were captured in management (rather than research) actions, however, and 20 of these were then transported out of the study area (one bear died at capture).

Initially, we classified the East Front population as a non-ecocentered population because no significant bear aggregations were reported by Aune and Brannon (1986). But a reinterpretation of their research suggests otherwise. Of 341 actual observations of grizzlies from 1980 through 1987, 182

were of a single bear, 76 were of two bears, 60 were of three, 19 were of four, 3 were of five, and 1 was of seven. Other data from scat analyses and frequency of use of specific habitat components suggest that many bears were habituated to feed in relatively circumscribed areas (boneyards), perhaps even to a degree commensurate with that observed elsewhere at secondary ecocenters.

Aune and Kasworm (1989) reported on analyses of 1094 grizzly bear scats collected from 1979 through 1987. The scats were dissected, specific food items were identified, where possible, and their frequencies of occurrence and relative volumes were recorded. Calculated for each item were an Importance Value (product of percent frequency times percent volume divided by 100) and an Importance Value Percent (Importance Value of a food item multiplied by 100 and divided by the sum of Importance Values for all food items), after Craighead *et al.* (1982). In order of their Importance Value Percents, the five top-ranked food items were unidentified grass/sedge, ants, dandelion (*Taraxacum* spp.), cattle (*Bos* sp.), and chokecherry (*Prunus virginiana*). Within the food category "mammals," meat of cattle was, by far, the most frequently encountered mammal remains (42% of total frequency), and comprised the largest total volume (45%). The next highest ranking food item identi-

With neck outstretched and ears flat, signaling aggressive intent, a female with cubs then vigorously defends her fishing site against an intruding male. *(Photos by John J. Craighead)*

fied in the "mammals" category, deer meat, achieved an Importance Value Percent only 18% that of cattle (0.67 versus 3.79). Clearly, grizzlies were feeding frequently on cattle carcasses and consuming relatively large quantities.

Vegetation and landforms comprising the area of study were defined, mapped, and grouped into larger descriptive units termed habitat components (Aune and Stivers 1983, Aune *et al.* 1984). Based on this description of the bear's habitat, the relative frequency of use of 14 habitat components was interpreted using 2633 radio locations accumulated over a 10-year period (Aune and Kasworm 1989). Further, grizzly bear locations were categorized as to a bear's observed behavioral activities. The four habitat components in which feeding activity centered on "carrion" were the closed timber, *Populus* stand, riparian shrub, and prairie grassland components. Feeding on carrion was the second most commonly observed feeding activity in the closed timber component (after digging for and feeding on whitebark pine seeds [*Pinus albicaulis*] which became available in late fall). But carrion-feeding was the single most commonly observed feeding activity in the other three habitat components.

Interestingly, 67% of all radio locations of grizzly bears were ascribed to the four habitat components in which carrion-feeding was a major activity. Unfortunately, Aune and Kasworm (1989) routinely refrained from specifying what kinds of mammal carcasses constituted the carrion on which feeding was observed, so the degree of association of bears with cattle carcasses could not be determined conclusively. Furthermore, livestock carcass disposal sites (boneyards) were mentioned only rarely in the text and were absent entirely from the numerous maps showing cumulative grizzly observations. The authors mentioned casually that "domestic cattle are available from livestock boneyards along the east front." Based on our knowledge of "carrion centers," it seems quite possible (if not probable) that East Front boneyards functioned as secondary ecocenters, given the frequency of feeding on "carcasses" and the food value of cattle indicated by scat analysis.

Boneyarding has been practiced since homestead settlement of the land along the East Front of Montana's Bob Marshall Wilderness. Bears were often killed at these feeding sites, particularly when they turned from carrion-feeding to livestock-kill-

Like two sumo wrestlers, these males fight for a prime fishing site. After throwing his opponent and establishing dominance, the victor claimed the fishing location without further conflicts. *(Photos by John J. Craighead)*

ing. Keith Aune (pers. comm. 1991) stated that his observations suggest that cattle carcasses, whether widely distributed or concentrated at boneyard depositories, are highly important food sources for many, but not all grizzlies. Although bear visitation to boneyards has been recognized by biologists for some time, not until recently has the Montana Department of Fish, Wildlife and Parks implicitly recognized that East Front boneyards are in fact secondary ecocenters that convey both advantages and disadvantages to the local grizzly bear population.

To retain the nutritional advantage, yet reduce the risk of bear deaths and bear-human conflicts, the Department, with ranchers' consent, phased out historic boneyard sites and initiated a program to redistribute livestock carcasses to more isolated sites in prime grizzly bear habitat. During 1988-90, the Department phased out 17 boneyards along the East Front (Madel 1991). Twelve small, private refuse dumps in or near choice bear habitat were also phased out. A total of 498 livestock carcasses were collected and transported to three isolated areas within grizzly bear habitat over a three-year redistribution period. The program was successful in reducing bear-human conflicts along the East Front by 84% (from 19 to 3).

In effect, the Department used the redistribution program to create secondary ecocenters at selected sites remote from human activity. Madel (1991) showed that the new sites attracted and held bears for extended periods of time. The frequency of use during the spring and summer months increased dramatically as foraging bears discovered the newly placed carcasses. The actual number of grizzlies using the redistributed carcasses was not reported.

In retrospect, the grizzly bear population along the East Front has historically been influenced by secondary ecocenters of two types—boneyards and private refuse dumps. Both could be considered "artificial" insofar as they are human-derived. At the same time, agency biologists have routinely described the indigenous bear population of that region as a "genuine wild and free-roaming population" unaffected by "artificial" foods. In reality, the population has been, and continues to be, significantly affected by ecocenters. This fact should be publicly acknowledged in the scientific literature. The Montana Department of Fish, Wildlife and Parks has done a very effective job of applying the ecocenter concept as a management tool to increase the foodbase and hold the grizzlies in prime habitat away from human population centers where encounters are likely.

Brent Creek Area—Dubois, Wyoming. Developing ecocenters generally induce a redistribution of an existing grizzly bear population. IGBST capture-recapture and mortality records, plus our interpretation of bear movements (Chapter 15), indicate the grizzly bear population in the Yellowstone Ecosystem has moved away from the geographic center of the Park to the peripheries of the Ecosystem. Grizzlies foraging in outlying areas predictably concentrated at any consistently available, high-quality food sources. A case in point is the resident population and seasonal aggregation of bears at boneyards in the vicinity of Brent Creek, Wyoming, south of the currently defined boundaries of the Yellowstone Grizzly Bear Recovery Zone. The area, in general, also provides a variety of excellent food plants for bears.

Jon Robinett, ranch manager for the Diamond G Ranch, 14.5 kilometers (9 mi) north of Dubois, documented the increases in grizzly bear activity during the 1990-92 period. He reported that 22 different marked and unmarked grizzlies routinely used a relatively discrete area between 1990 and 1991. The number frequenting the area during 1992 increased to 27 grizzlies, mainly due to cubs born to resident females. The grizzlies concentrated in this area because of cattle boneyards, but predation on cattle was also significant in some years; 22 head of cattle were killed at the Diamond G Ranch in 1991. Ranch manager Robinett quickly learned and applied the ecocenter concept. By judicious movement and placement of boneyards, he manipulated grizzly bear movements and quite effectively protected his cattle herds (pers. comm. 1994).

Grizzly bear capture records in 1991 (Knight *et al.* 1992) showed that at least seven of the 22 resident Brent Creek grizzlies had previously been captured and radio-collared by the IGBST and the Wyoming Game and Fish Department (Table 13.4). They collared three additional grizzlies during the spring of 1991 at nearby DuNoir Reservoir. Four marked grizzlies had extensive histories that showed residence in and movement from other areas, often quite distant. The first of these, No. 34, a 19-year-old adult male, was initially captured at Togwotee, Wyoming, in 1978. He moved north into Yellowstone Park and denned about 16 kilometers (10 mi) south of the South Arm of Yellowstone Lake in both 1978 and 1979 (Knight *et al.* 1980). In 1980, he remained in a 228-square-kilometer (88-mi²) home range northwest of Togwotee in the southern portion of Yellowstone Park (Knight *et al.* 1981). His home range was not determined after 1980 because of radio failure in April 1981. In 1994, No. 34 was killed by a hunter near Moran, Wyoming.

Another grizzly resident at Brent Creek was No. 128, a 6-year-old female. She was first captured in August 1986 at West Yellowstone and transported to Badger Creek. The IGBST radio-monitored her movements during 1987-89, although her home ranges were not reported. She cast her collar in October 1990, and remained in the Brent Creek area through 1992, producing cubs in that year.

Number 168, a male, was first captured in 1989 as a 3-year-old at Pacific Creek in the Bridger Teton National Forest. He subsequently moved south to the Brent Creek area where he was observed from 1990-92.

The fourth "Brent Creek grizzly," No. 181, a 2-year-old male, was captured as a yearling in April 1990, near the NPS Lake Ranger Station and released at Slough Creek. He was again captured in August 1990 at Yancey's Hole and released at Nez Perce Creek. Since that time, he has resided outside the Park in the Brent Creek area, approximately 114 air kilometers (71 mi) from his initial capture site.

Although not recorded specifically at Brent Creek, two males were captured in 1978 at Togwotee, Wyoming, 40 kilometers (25 mi) northwest of the Brent Creek area. That both had extensive home ranges encompassing large portions of Yellowstone Park was evidence that some members of the Yellowstone population had moved south and west out of the Park beyond boundaries of the Recovery Zone (USFWS 1993). Number 35, a 3-year-old male when first captured in 1978 at Togwotee, was tracked by the IGBST for four consecutive years. He ranged from Montana near the extreme northwest corner of Yellowstone Park south to the Jackson Hole area. He denned about 19 kilometers (12 mi) northwest of Jackson Hole in both 1978 and 1979 (Knight *et al.* 1980). His life range of 5374 square kilometers (2075 mi²) was the largest recorded for any of the grizzlies monitored by the IGBST study team (Blanchard and Knight 1991). His collar was removed in 1981, but he was sighted in the West Yellowstone area again in 1983, approximately 118 air kilometers (73 mi) from his denning area northwest of Jackson Hole.

Number 36, a 10-year-old male when first captured at Togwotee in 1978, was radio-located 96 times during the next three years (Knight *et al.* 1982). His life range of 1256 square kilometers (485 mi²) extended north to the east shore of the Southeast Arm of Yellowstone Lake, and south to an area east of Moran, Wyoming. He subsequently moved into the Brent Creek area for at least portions of the year.

Grizzlies were attracted into the Brent Creek area not just because it is excellent grizzly bear habitat, but because the cattle ranches and their boneyards provide dependable, high quality food during critical periods when natural foods are in short supply. We have shown that significant numbers of bears are populating the area, some from the extreme northern portions of the Yellowstone Ecosystem. The consistent availability of carrion at the boneyards will most likely continue to attract and hold increasing numbers of grizzly bears, as new litters are raised in the Brent Creek area, and the aggregation could justifiably be classified as ecocentered. Movement of some of these bears into the adjoining Wind River landscape could lead to eventual recolonization of this large ecosystem. Current evidence suggests this may already be happening. Grizzlies have been reported in the Wind River region, but none of the sightings have been confirmed (Joe Nemick, Wildlife Management Coordinator, Wind River District, Wyoming Game and Fish, pers. comm. 1993).

In general, indirect evidence implies that bear densities are declining within the Park and increasing outside its boundaries (Chapter 15). It is extremely important that land managers understand this apparent increase in bear numbers as a redistribution from other parts of the Ecosystem and adopt conservative management guidelines. If they make the error of assuming that range expansions are evidence of surplus bears and relax management guidelines, extractive industry on public lands and private development elsewhere will increasingly encroach on critical grizzly bear habitat. Also, as bears move into national forests and developed areas where management guidelines are oriented, not to their needs, but to economic mandates, risk of death is increased. This is particularly true of the Wind River area, which supports extensive grazing allotments under federal jurisdiction. Recovery Zone boundaries must be reevaluated annually and adjusted to include areas where grizzlies have redistributed.

The Brent Creek area deserves immediate special attention. The 27 grizzlies that used the area in 1992 included many that moved south from Yellowstone Park to the Brent Creek site. In 1994, Jon Robinett (pers. comm. 1994) reported that he observed 23 of these resident bears, and that numerous others, mostly adults, had moved through over the summer, apparently dispersing farther southward. Adjustment of Recovery Zone boundaries should be strongly considered to protect these bears and their habitat, and timber sales and other potentially conflicting uses must be critically evaluated and approached cautiously. This and similar management problems are discussed more fully in Chapter 20.

Table 13.4. Marked grizzly bears frequenting the Brent Creek, Wyoming area, Yellowstone Ecosystem, 1991.

Bear no.	Sex	Age in 1991	Date of first capture	Area of first capture	Date of capture (1991)	Area of capture (1991)
34	M	19	7-3-78	Togwotee, WY	5-17	Dunoir R.
128	F	6	8-8-86	West Yellowstone, MT	7-13	Horse Cr.
152	M	18	7-16-88	Five Mile Cr., WY	7-30	W. Fork Long Cr.
164	M	7	5-16-89	Horse Cr., WY	5-5	Dunoir R.
168	M	5	8-13-89	Pacific Cr., WY	8-31	Dunoir R.
174	M	5	6-1-90	W. Fork Long Cr., WY	5-13	Dunoir R.
181	M	2	4-15-90	Lake, YNP	Sighted at Brent Creek	
184	M	10	4-28-91	Dunoir R., WY	4-28	Dunoir R.
185	M	5	4-30-91	Dunoir R., WY	4-30	Dunoir R.
189	F	10	5-27-91	Dunoir R., WY	5-27	Dunoir R.

Insect Aggregations

At certain times and places, insect swarms provide enough high fat and protein food to attract grizzly bears and hold them for considerable periods of time. Such sites with their aggregations of bears may represent valid secondary ecocenters, but, so far, evidence is inconclusive.

Moths: Scapegoat Mountain—Montana. We first observed the phenomenon of bears aggregating to feed on insects in the Lincoln-Scapegoat Wilderness (Sumner and Craighead 1973). The army cutworm moth (*Euxoa* sp.) migrates to high elevations in July and feeds on nectar until fall (Pruess 1967). On Scapegoat Mountain, the moths were available for over two months. Moth migrations to the high country are highly cyclic, however, and the periods between migrations vary greatly (Metcalf and Flint 1951), suggesting an unpredictable food source from year to year and, thus, a food source that may not fully qualify as an ecocenter. The moths congregated during the day in a talus slope composed of large rocks on the southwestern exposure of Scapegoat Mountain. They emerged from under the rocks to feed at night. The moths did not fly in daylight, even when disturbed.

From 24 July to 24 August 1972, a female and three cubs fed almost exclusively on this insect. In feeding, the bears simply overturned the rocks that sheltered the moths. On one steep talus slope, the bear family dislodged almost every large rock within a 4-hectare (10-acre) area. We counted over 200 moths concentrated under a single boulder, covering an area of 0.2 square meter (2 ft²). Although it fed selectively on the cutworm moth, the family also consumed a less abundant, but unidentified purple moth. The bears also foraged on two plants: springbeauty (*Claytonia megarhiza*) and sedge (*Carex* spp.). In 1993, the range in the size of scats

collected and their distribution at the site indicated that a family, as well as several large adults, had fed on the moths.

Bedding areas, where the bears rested between feeding bouts, lay directly below the feeding site. Analysis of 95 scats collected there during two summers showed cutworm moth remains in 66% and averaging 65% of the total volume. The purple cast of some of the scats indicated that the purple-colored moth was also eaten.

Moths: Absaroka/Clark's Fork—Wyoming. The IGBST first recorded use of alpine insect aggregation sites in the Absaroka-Clark's Fork area of the Yellowstone Ecosystem in 1986 (Mattson *et al.* 1991b). Later, in 1987-89, they visited these sites and observed grizzlies eating the army cutworm moth (*Euxoa auxiliaris*). Analysis of 116 bear scats from 6 known and 12 suspected moth aggregation sites showed heavy feeding on this insect. French *et al.* (1995 [in press]) reported seeing 22 bears feeding on moths at one site. Frequency of moth remains in scats was as follows:

Year	% Occurrence	% Volume
1987	64	42
1988	92	77
1989	74	57

Use of moths in the Yellowstone Ecosystem is addressed again in Chapter 16.

Other Insects. Grizzly bears often have been reported to feed on beetles and ants to such an extent that the insects' indigestible remains are well represented in their feces. It is unlikely that bears aggregate to feed on ants, but there is evidence to suggest that this may occur at times in the case of beetles.

Chapman and others (1932) documented large concentrations of ladybird beetles (*Hippodamia caseyi*) on McDonald Peak in the Mission Mountain Wilderness of Montana. Nevertheless, nine of 15 grizzly bear scats collected there in the 1950s were largely or entirely composed of cutworm moth remains (*Euxoa auxiliaris*), and beetle remains were not in evidence (Chapman *et al.* 1955). More recently, Klaver *et al.* (1986) observed grizzlies on McDonald Peak feeding on both cutworm moths and ladybird beetles, but it appeared that ladybird beetles were not used as extensively as the moths. They documented an average of 10 bears at the site during the period of insect aggregation.

Thus, while grizzly bear populations have been shown to concentrate at McDonald Peak in the Mission Mountains when the beetles are abundant, scat analysis has not consistently shown significant use of the beetle. Servheen (1983) did not record any beetles in scats collected in the Missions in 1979, but did record a high relative percent composition (>80%) for moths.

Grizzly bear use of beetles has been reported only rarely in the Yellowstone Ecosystem. Analysis of 243 scats by Murie (1944b) showed a total insect diet volume of 3.3% (Mattson *et al.* 1991a). No specific species were identified. In our analyses, we found virtually no use of ladybird beetles, and only 11 of 487 scats (2.3%) collected from 1968-71 contained any trace of beetles. Ant remains were found more consistently, occurring in 4.1% of the scats and constituting an average 5.2% by volume (Table 10.1). Mealey (1980) found similar results from analysis of 615 scats collected in 1973 and 1974. Ants occurred in 8.8% of the scats, and the average fecal volume was 9.6% (Table 10.2). Knight *et al.* (1982) found considerably greater use. In an analysis of 3028 scats collected in 1977-81, ants occurred in 22.3% and, in the scats in which ants occurred, they showed an average fecal volume of 19.2% (Table 10.2). Beetles, on the other hand, occurred in only 0.4% of the scats (Knight *et al.* 1982).

To evaluate insect aggregations as a food source, it is necessary to consider their dependability and availability. Although ants have been shown to play a consistent role in the grizzly diet, the distribution of ant biomass is not conducive to attracting aggregations of bears. Only the cutworm moth has been shown, in recent years, to attract aggregations of bears. We observed concentrations of moths in only one of four years of study in the Scapegoat area, although the intensity of field effort was similar in all years. In the Mission Mountains, Servheen (1983) found moths in scats during 1979, but reported ants as the only invertebrate in scats collected in 1978 (Servheen and Lee 1979). Most of the scats he collected during 1978 were from the subalpine zone

because grizzly activity in the alpine zone was minimal, possibly because moth concentrations were not high in that year.

As mentioned earlier, feeding at moth aggregation sites in Yellowstone was not documented for Yellowstone bears until 1986 (Mattson *et al.* 1991b), although intensive field work had been conducted since 1959. These sites, if they were active before dump closures in 1970, may not have been as attractive to grizzlies because the large quantities of food available at the ecocenters tended to hold the bears in the lower subalpine areas. With little incentive to move, they may not have located moths aggregated in the alpine zone. Mattson *et al.* (1991b) questioned whether a sampling bias may have accounted for the lack of direct fecal evidence for moth-feeding before 1986. None of the local sheepherders or outfitters interviewed, however, recalled seeing aggregations of bears in the region of the currently active feeding sites before 1981, with one exception: in the mid-1970s, a hunter reported observing bears feeding at what was later identified as a moth aggregation site. Hence, it is equally plausible that the grizzly population, slowly redistributing toward the periphery of the Ecosystem (Chapter 15) following closure of the ecocenters, may have discovered the various moth aggregations.

After 1986, four radio-collared grizzlies were annually radio-located at moth aggregation sites in the Absaroka Mountains between Clark's Fork of the Yellowstone River to the north and the Wind River to the south (Mattson *et al.* 1991b). One grizzly was located at the sites for six consecutive years. It is not known whether the moth aggregations were present during each of the six years since the bears were radio-located in some years, but not observed feeding at the sites. Nonetheless, it is evident that alpine moth aggregations can periodically serve as a highly nutritious food for grizzlies, attracting and holding them in an area for periods of several months. How important these are as major nutrient sources, or whether they qualify as secondary ecocenters, has yet to be demonstrated. We discuss noctuid moths as a component of the grizzly bear's diet in Chapter 16.

Fish Aggregations

The bear aggregations at McNeil River Falls, the Karluk Lake System, and on Admiralty and Chichagof Islands, responding to an abundance of salmon (*Salmo* spp.), represent primary ecocenters. We have found no scientific evidence, present or past, to suggest similar fish-oriented aggregations in the lower 48 states, although they may have occurred in Pre-Columbian times at sites along the Columbia River drainage, where spawning runs were temporarily blocked at falls and rapids. No bears were observed to aggregate to feed on fish within the Yellowstone Ecosystem during the period 1959-70, and fish were not a major food source (Chapter 10).

Cutthroat trout (*Oncorhynchus clarki*) and long-nose suckers (*Catostomus catostomus*) spawn in tributaries of Yellowstone Lake, and grayling (*Thymallus arcticus*) spawn in the upper reaches of the Gibbon River. When these species gather on their spawning grounds, they have the potential to attract aggregations of bears. Data gathered since dump closures in 1970 indicate that grizzlies do make use of spawning fish (Reinhart and Mattson 1986, 1987; Reinhart 1988, 1990b; see also Chapter 16). But direct observations supporting such use are rare. Numbers and species of individual bears active in the study areas, both temporally and spatially, are inferred from conformation and measurements of forepaw tracks. Based on such track analyses, Reinhart (1988) estimated that 28 to 44 different grizzlies in 1985 and 25 to 40 in 1987 frequented the widely dispersed tributaries of Yellowstone Lake to feed on migrating fish. The extent of individual bear movements, site discreteness, frequency and duration of on-site fishing activities of individual bears, and the degree of aggregation were not reported. Moreover, the relatively large area in which the surveys were performed (although size was not specifically reported) suggests dispersed feeding rather than aggregation phenomena. Fish, as a food source, are discussed in greater detail in Chapter 16.

Winter-Killed Ungulates

Winter-killed ungulates are relatively widely dispersed throughout the riparian habitat and thermal areas of the Park. Although a major food resource, they do not normally cause grizzlies to aggregate annually for extended periods of time at localized sites. During the 1959-70 period, carcass sites did not qualify as ecocenters. Throughout the postclosure period, data on carcass-feeding have been minimal.

For reasons similar to those discussed for documenting use of fish, survey data on use of carcasses of winter-killed ungulates during the postclosure period are not amenable to analysis regarding possible aggregation behavior. Direct observation of identifiable grizzlies was rare, and track analysis was the primary indicator of bear activity on elk and bison carcasses. Carcass availability, their locations, and relative amounts consumed by various species (as indicated by tracks) constituted the main

focus of April and May surveys on a study area of unspecified size along the Firehole River from 1985 through 1988 (Mattson and Henry 1987, Henry and Mattson 1988, Green 1989), and on another of unspecified size on the Northern (elk) Winter Range from 1987 through 1988 (Green and Mattson 1988, Green 1989). No estimates of numbers of grizzlies actually using carcasses were reported for 1985 and 1986. In 1987, 11 grizzlies, and in 1988, seven grizzlies were thought to have visited carcasses along the Firehole River. Carcasses on the Northern Winter Range study area attracted an estimated minimum of six grizzlies in 1987, and 10 in 1988. Neither study described the food resource in terms of spatially focused sites, and actual quantities of carcasses consumed by grizzlies, black bears, and other species were not differentiable, nor were indicators of bear aggregations noted. From the evidence presented, we judge that no ecocentric feeding behavior was involved. Carcasses as a source of food are discussed more fully in Chapter 16.

SOME CURRENT NON-ECOCENTERED GRIZZLY POPULATIONS

Several thorough, long-term studies provide sex, age, and reproductive data on grizzly bear populations having no reported aggregation foci within their ranges. Here, we describe the areas in which the studies took place, and the general methodologies of the studies themselves.

Western Brooks Range—Alaska

Research on the grizzly bear population inhabiting Alaska's North Slope was begun in 1973 on a 5200-square-kilometer (2008-mi^2) study area in the mountains and foothills of the western Brooks Range (Reynolds 1974). Population size and structure were reported from an intensive capture and marking program, after which the population was followed principally by radio-tracking (Reynolds 1980, Reynolds and Hechtel 1984). A second intensive research effort was conducted from 1986 through 1988 (Reynolds 1989).

Over the period of this research, 171 bears were captured and marked or radio-collared. Radio contact was maintained on 25 bears from capture until their deaths, and on an additional 21 bears from 1977-78 through 1988 (Reynolds 1989). No evidence of ecocentered feeding was reported.

Northcentral Range—Alaska

The grizzly bear population inhabiting the northcentral Alaska Range is subject to much greater hunting pressure than that in the western Brooks Range. In an effort to determine the effects of hunting on bear population dynamics, Reynolds and his colleagues initiated investigations of the northcentral population in 1981 on a study area of 3900 square kilometers (1500 mi²) (Reynolds 1982). Subsequently, they reported extensively on population density, sex and age structure, movement, and reproduction, based on 101 grizzly bears captured and marked or radio-collared from 1981 through 1989 (Reynolds 1982, 1989, 1990; Reynolds and Hechtel 1983, 1984, 1986, 1988; Reynolds *et al.* 1987). Again, no evidence of ecocentered feeding was reported.

Greater Yellowstone Ecosystem, 1975-89

Although we have classified the Greater Yellowstone Ecosystem (1975-89) as an area with a non-ecocentered grizzly population, there was a secondary ecocenter operating at the Cooke City dump during part of the period. The honeyards of ranches and insect aggregation sites appear to be developing secondary ecocenters, but data are, as yet, insufficient to reach firm conclusions.

SUMMARY

We have earlier demonstrated that the aggregating of bears at Trout Creek is not an isolated response of a single population of grizzly bears to novel circumstances, but is instead an *evolved* phenomenon. In this chapter, we elaborated on our definition of an ecocenter, looking back in time for evidence that ecocenters have long been a general feature of grizzly bear natural history.

An ecocenter is any food source that is sufficiently nutritious, abundant, localized, persistent, and predictable to regularly attract a significant number of bears to feed communally, and thereby to markedly affect basic population parameters. Bear food resources lie on a continuum of abundance, localization, and persistence, but, for convenience, we divided this continuum into three food resource categories: primary ecocenter, secondary ecocenter, and non-ecocenter. These categories reflect a progressive decline in ecocenter size with decreasing influence on the population.

The grizzly lineage has a long history of coevolution, both in Asia and North America, with humans (who may *create* ecocenters) and a variety of large-bodied or strongly aggregating vertebrate prey species (which may *be* ecocenters). For example, until 10,000 years ago, grizzlies coexisted regionally with at least two to six species of terrestrial megaherbivores (e.g., mammoths, sloths, etc.) weighing over 1000 kilograms (2205 lbs) and numerous species of smaller, herd-dwelling ungulates (e.g., bison). A few winter-killed carcasses of

a megaherbivore or the localized grouping of carcasses of the smaller herd-dwelling ungulates, predictably occurring at specific locales, would easily have qualified as ecocenters, and there can be little doubt that bears fed on such carcasses and returned periodically (annually) to such sites. Coastal populations of bear were contiguously distributed with several species of whales. There are indisputable historical records of large numbers of bears aggregating on carcasses of beached whales. Spawning runs of one or more species of anadromous fish must have occurred in many streams and rivers within the geographic range of the grizzly, and bears would have aggregated, as they do today, at locales where geologic barriers or spawning beds concentrated fish.

For at least the past 10,000 years in North America, aboriginal people created ecocenters by capturing and killing large numbers of bison in corrals or driving them off cliffs. In short, ecocenters as large or larger than Trout Creek probably have been features of the grizzly bear's environment long enough that we should not be surprised to find that natural selection has molded bear life history, behavior, and physiology to exploit foci of massive food abundance.

The frequency of ecocenters in contemporary bear populations is almost certainly much reduced. The megaherbivores are extinct, as are several species of herd-living Pleistocene ungulates. Overharvesting, pollutants, and dams have destroyed or reduced many spawning runs of the anadromous fish. Aboriginal hunter-gatherer cultures have been replaced by European agronomists in North America, and surviving species of native social ungulates are much reduced in numbers and range.

Nonetheless, it is clear that contemporary examples of ecocenters, other than those we have described in Yellowstone during 1959-70, are widespread. "Natural" examples of ecocenters include spawning runs of anadromous salmonids at McNeil River Falls, Prairie Creek, and at several sites on the islands of Kodiak, Admiralty, and Chichagof, all in Alaska. Stranded whales along North Pacific and Arctic coasts often attract aggregations of bears. Concentrations of insects may function as ecocenters in alpine areas (e.g., army cutworm moths on Scapegoat Mountain, Montana). The precise role of these natural food resources in the biology of grizzly bears should be investigated from a perspective of their ability to attract, hold, and nourish sizeable aggregations.

Remaining examples of existing or developing ecocenters often involve human-made concentrations of foods. These include refuse dumps (used by polar bears, as well as brown bears), cattle ranch boneyards, and accidental grain spills. We would not expect ecocenters to occur in all grizzly populations, even in pristine times and environments; in fact, we have identified three contemporary populations that appear to be non-ecocentered (Brooks Range and Northcentral Range in Alaska and the Yellowstone Ecosystem after 1970).

We may decry the aesthetic (and other) losses associated with the replacement of mammoths and salmonid fishes by ecocenters of human origin. But we must not be blind to the potential practical value of ecocenters, whatever their origins, to the persistence of contemporary grizzly bear populations. This is particularly true for bear populations in fragmented environments where ecocenters could represent an effective management tool for reducing the size of areas required for population persistence and for minimizing bear-human conflict and the oft-resulting bear deaths (Chapter 20).

COLOR GALLERY

COLOR GALLERY

Large male grizzly at culvert trap. *(Photo by John J. Craighead)*

Immobilized bear. *(Photo by John J. Craighead)*

Weighing grizzly bear. *(Photo by Frank C. Craighead)*

Attaching color ear marker. *(Photo by Frank C. Craighead)*

Color marked bear. *(Photo by John J. Craighead)*

Night visitors. *(Photo by J. & F. Craighead)*

Taking paw impressions. *(Photo by Frank C. Craighead)*

Releasing cub. *(Photo by Frank C. Craighead)*

Radio collaring large male grizzly. *(Photo by J. & F. Craighead, courtesy of National Geographic)*

Radio tracking #14. *(Photo by John J. Craighead)*

Camping near den site. *(Photo by John J. Craighead)*

Radio tracking. *(Photo by John J. Craighead)*

Tracking bear #40. *(Photo by John J. Craighead)*

Bear #40 with radio collar. *(Photo by J. & F. Craighead)*

Den of bear #40. *(Photo by John J. Craighead)*

Alpha Male. *(Photo by John J. Craighead)*

Typical breeding season wounds. *(Photo by John J. Craighead)*

Defeated male. *(Photo by John J. Craighead)*

Contending male. *(Photo by John J. Craighead)*

Female in Hayden Valley. *(Photo by J. & F. Craighead)*

Alert female.
(Photo by J. & F. Craighead, courtesy of National Geographic*)*

Trout Creek bear.   *(Photo by J. & F. Craighead, courtesy of* National Geographic*)*

Juvenile band. *(Photo by John J. Craighead)*

Bonded pair. *(Photo by John J. Craighead)*

Foraging bear. *(Photo by Tom Mangelson, courtesy of* Images of Nature*)*

Covering carcass of bull elk. *(Photo by J. & F. Craighead, courtesy of* National Geographic*)*

Polygonum bistortoides

Lomatium montanum

Pinus albicaulis

Vaccinium scoparium

Claytonia lanceolata

Erythronium grandiflorum

Claytonia megarhiza

Cirsium scariosum

Fragaria virginiana

Perideridia gairdneri

Common bear food plants. *(Photos by John J. Craighead)*

Fig. 12.2. Digital vegetation map of the 200-square-kilometer (77-mi²) Scapegoat Study Area, Lincoln-Scapegoat Wilderness, Montana (from Craighead *et al.* 1982); north at top of map; 61-meter (200-ft) contour intervals, U.S. Geological Survey 1:24,000 quadrangle.

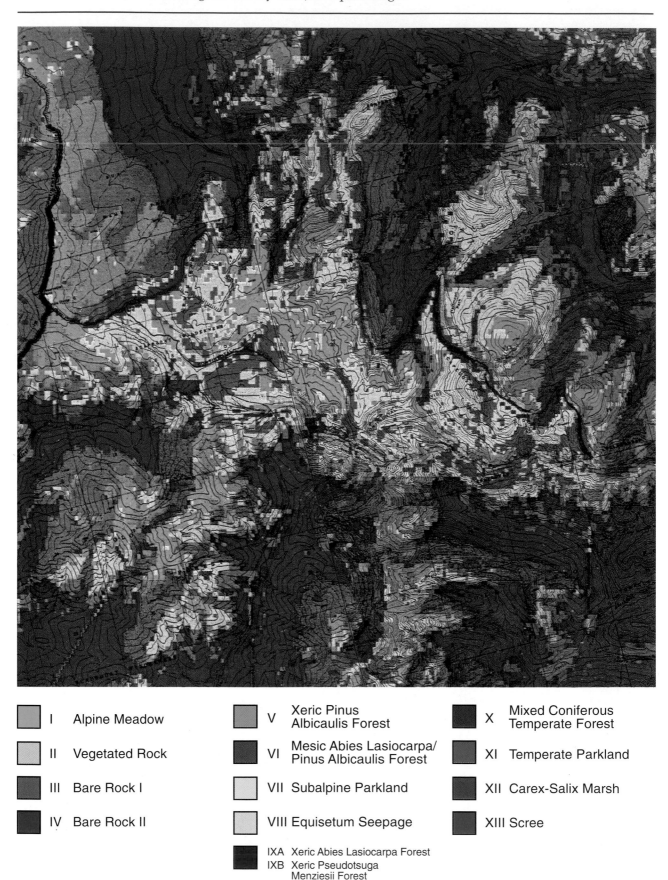

I   Alpine Meadow

II   Vegetated Rock

III   Bare Rock I

IV   Bare Rock II

V   Xeric Pinus Albicaulis Forest

VI   Mesic Abies Lasiocarpa/ Pinus Albicaulis Forest

VII   Subalpine Parkland

VIII   Equisetum Seepage

IXA   Xeric Abies Lasiocarpa Forest
IXB   Xeric Pseudotsuga Menziesii Forest

X   Mixed Coniferous Temperate Forest

XI   Temperate Parkland

XII   Carex-Salix Marsh

XIII   Scree

Hiking to Scapegoat Wilderness study area.
(Photo by Karen Craighead Haynam)

A ground truthing campsite
(Photo by John J. Craighead)

Checking accuracy of satellite imagery.
(Photo by Karen Craighead Haynam)

Field camp below Scapegoat Mountain.
(Photo by John J. Craighead)

Observing grizzly bears feeding.
(Photo by Karen Craighead Haynam)

Fig. 12.3. Digital vegetation map of the 4945-square-kilometer (1909-mi²) Kobuk River Study Area, Alaska (from Craighead *et al.* 1988a); north at top of map; 61-meter (200-ft) contour intervals, U.S. Defense Mapping Agency 1:250,000 quadrangle.

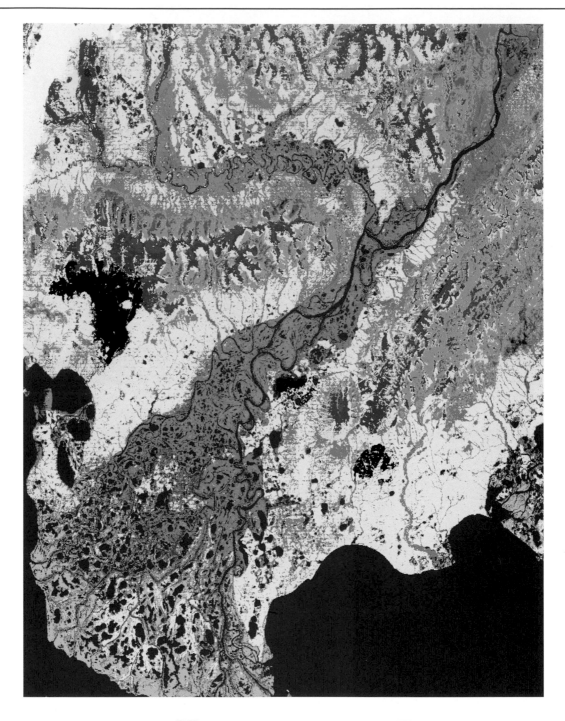

■ Alpine Tundra Complex	■ Shrub Tundra Complex	■ Burns
■ Vegetated Rock Complex	■ Tall Willow Complex	□ Tidal Flats
□ Alpine Shrubland Complex	■ Riparian Mosaic Complex	■ Tidal Marsh Complex
■ Upland White Spruce Complex	□ Gravel Bars and Ocean Beach	■ Sedge/Grass Marsh Complex
■ Upland Mixed Spruce Complex	□ Bare Rock	■ Semi-Vegetated Complex
■ Riparian Spruce Complex	■ Lichen Tundra Complex	□ Sand Dunes
□ Tussock Tundra Complex	■ Water	□ Clouds

Fig. 12.4. Computer-generated map of the vegetation complexes in the 4593-square-kilometer (1773.5-mi^2) area of extrapolation in the Bob Marshall-Scapegoat wilderness complex, Montana (from Craighead *et al.* 1982); A, B, and C denote the Scapegoat (primary), and the Danaher and Slategoat (secondary) study areas; D indicates area to the east of dashed line that would require development of new spectral signatures for accurate mapping; 61-meter (200-ft) contour intervals, 35 U.S. Geological Survey 1:24,000 quadrangles.

	I	Alpine Meadow
	II	Vegetated Rock
	III	Bare Rock I
	IV	Bare Rock II
	V	Xeric Pinus Albicaulis Forest
	VI	Mesic Abies Lasiocarpa/ Pinus Albicaulis Forest
	VII	Subalpine Parkland
	VIII	Equisetum Seepage
	IXA	Xeric Abies Lasiocarpa Forest
	IXB	Xeric Pseudotsuga Menziesii Forest
	X	Mixed Coniferous Temperate Forest
	XI	Temperate Parkland
	XII	Carex-Salix Marsh
	XIII	Scree

Fig. 15.9. Map showing extent and severity of the 1988 Yellowstone Ecosystem fire.

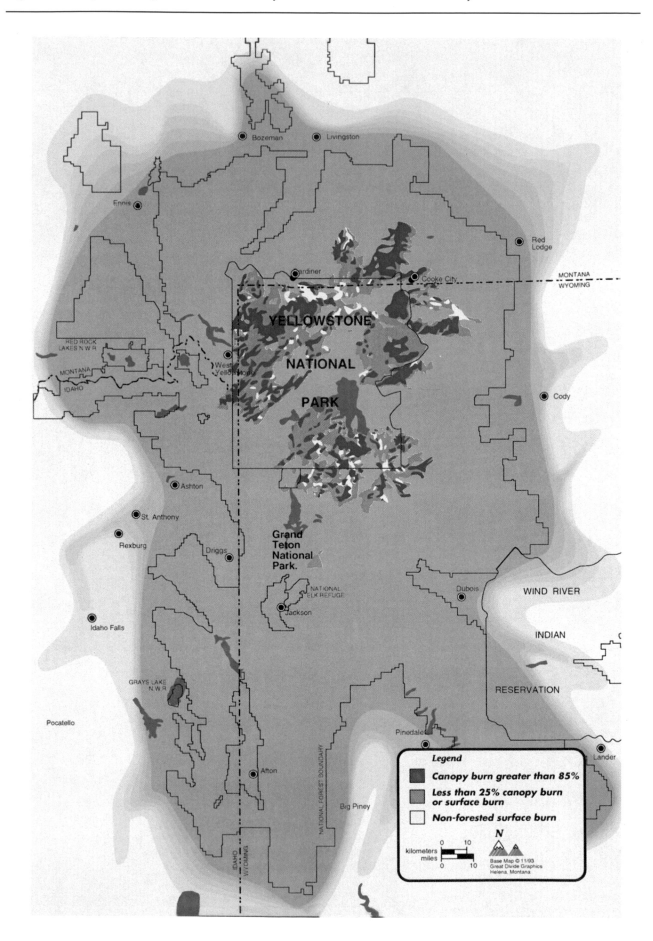

Fig. 20.4. Map showing proposed protection of public land under the Northern Rockies Environmental Protection Act. (*Courtesy of Alliance for the Wild Rockies*)

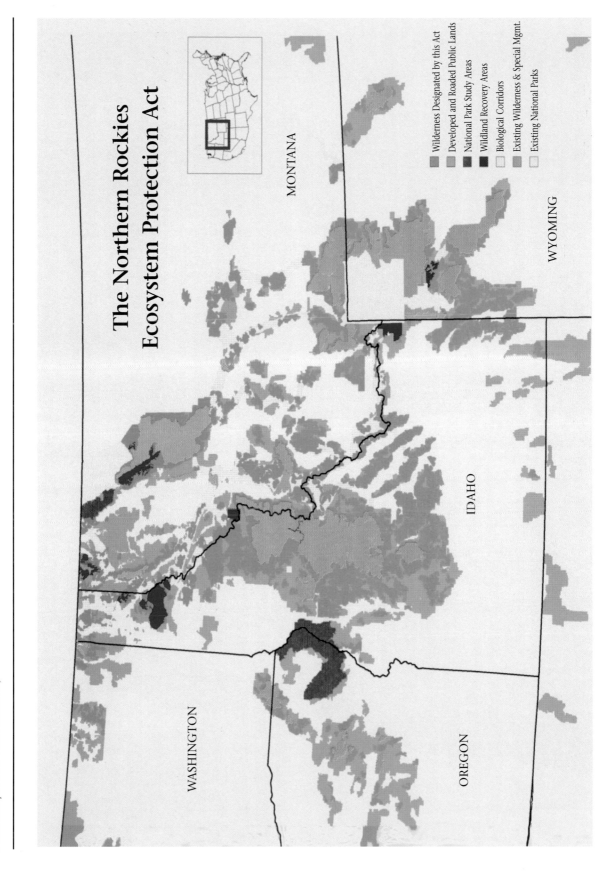

(Photo by John J. Craighead)

SECTION II: POPULATION COMPARISONS

14
1974–1992:
The Era of Decreed Research

INTRODUCTION

We have demonstrated that certain biological parameters of ecocentered bear populations are significantly different than those of non-ecocentered ones. Breeding, social, and feeding strategies are modified by centrally located access to high protein food and by the cohesiveness (density) of the aggregation attracted to the food. A highly structured society is evident and well defined at ecocenters. The behaviors evident there are also present, although less observable, at a much lower level of interaction among populations where members are dispersed. The social hierarchy governs both breeding and feeding behavior of aggregation members. Feeding at the ecocenter for extended periods (2 to 3 months) tends to decrease the size of bears' home and life ranges, and produces a seasonal movement pattern among aggregation members characterized by spring and fall seasonal ranges connected by migrational corridors. Nutritional levels are higher due to the annual availability and stability of the food resource and this, in turn, affects bear growth and reproductive dynamics. In general, ecocenters confer distinct biological advantages to bear populations that aggregate to feed and breed.

We now examine the period following closure of the major dumps in and adjacent to Yellowstone National Park and the effect that destruction of these ecocenters had on the grizzly bear population throughout the entire Ecosystem. To accomplish this, we compare preclosure biological parameters with postclosure ones, wherever this is feasible. Basically, this means comparing data from our 1959-70 study with the government's 1974-92 Interagency Grizzly Bear Study (IGBS). Before attempting to make statistical comparisons of the data, we briefly review research methods employed in each of the two studies. Where techniques are shown to differ, we have established rules for data comparison. The results of comparison enable us to detect certain changes in basic population parameters, population behavior, and habitat quality. Finally, we draw conclusions concerning the status of the

postclosure grizzly bear population and the biological role of ecocenters.

Our basic research method for determining population parameters of the preclosure population consisted of making numerous annual censuses of individually marked or recognizable bears aggregated at defined and predictable locations. All population data were obtained and recorded by the principal investigator or by other members of the research team. Where this technique could not be applied, data were accumulated by tracking and observing radio-instrumented animals, a technique developed during the course of the study (Craighead and Craighead 1965; see also Chapter 3). Through its use, the location of specific animals or family units could be predicted and their activities and behaviors observed. Data from trapping and retrapping marked animals supplemented the data recorded by direct observation. Habitat parameters and utilization of the foodbase were also quantified and analyzed. Replication of observations of individual animals and of family units and their feeding behaviors provided a strong quantitative database.

The Interagency Grizzly Bear Study Team (IGBST) ushered in the new era of bear research, federally decreed and almost entirely government-financed. The Team, headed by Richard Knight, relied principally on trapping and retrapping of marked animals, radio locations, aerial censuses, and observations by team members, nonteam professionals, and lay persons. The IGBST research personnel changed from year to year, limiting the annual number of quality observations of unduplicated female bears. Also, wide dispersal of the postclosure population and a National Park Service prohibition on the marking of bears from 1971-74 made repeated observation of individually identifiable animals much more difficult than during the 1959-70 period. This resulted in a weak database for understanding certain population parameters. Food habits data and feeding behavior were well-recorded by the IGBST, providing a strong database for comparative analysis.

In our comparison of biological parameters, we have used the published data of Knight and his coworkers, whenever possible. When this was not feasible because of basic differences in methodology, continuity of research effort, or when published data suitable for statistical comparison were lacking, we have used raw data supplied by R. Knight or, when appropriate, routinely recorded data made available by administrative officials. Because of methodological differences in the two studies, certain desirable comparisons were not feasible and, therefore, were not attempted. In such cases, we examine why these were not feasible. Later in discussion, we analyze the effects of these omissions on our ability to compare the preclosure and postclosure populations. We also discuss the effect these omissions have on drawing a credible conclusion concerning the biological status of the present population. As a matter of perspective, it is first necessary to briefly review the management history and rationale leading to dump closures.

PRECLOSURE MANAGEMENT HISTORY (1959–70)

Prior to 1963, the Yellowstone Park administration had three official guidelines for managing black bears and grizzly bears, but no clear management objective other than preserving bears as a major tourist attraction, and controlling them when they threatened lives or property. The management guidelines may be summarized as follows:

1. To sustain populations of grizzly bears and black bears under natural conditions as part of the native fauna;
2. To minimize conflicts and unpleasant or dangerous incidents with bears through control actions; and
3. To encourage bears to lead their natural lives with minimum interference from humans.

Recall that from its inception, the National Park Service was without any internal research apparatus for achieving a scientific understanding of the resources it was charged with managing. As a result, management policy was reactive, rather than proactive. The great fluctuation in annual bear control kills between 1930 and 1965 shows that no management guidelines existed for predetermining rational levels of the kill from year to year (Fig. 2.2). Incidence of personal injury and property damage (by both black and grizzly) was the determinant of bear control measures. Thus, annual variation in numbers of kills was directly proportional to ad-

ministrative apprehension following visitor complaints. Obviously, this type of control, based on direct response to public pressure, could not constitute a rational management program. Bear control had to be based on systematically gathered biological facts, as well as annual damage and injury reports. Moreover, black bear management proceeded on the widely-held assumption that two discrete black bear populations existed—roadside or campground and backcountry—and that the nuisance black bears killed in control measures would not affect the population of backcountry bears. This assumption was later supposedly verified by the work of Barnes and Bray (1967) and, unfortunately, was misconstrued by Cole (1970, 1971, 1972a, 1972b, 1973a, 1973b) to include grizzly bears, although substantiating data were not presented.

Killing programs aimed at reducing the Northern Elk Herds, overpopulated in Yellowstone as a result of predator control and various other protective actions, so inflamed public opinion in the early 1960s that the Secretary of the Interior was compelled to convene independent advisory committees to evaluate wildlife management in the Park, and to recommend future policy. These committees, later dubbed the "Leopold" and the "Robbins" committees after their chairmen, presented their findings and recommendations in March and August, 1963, respectively.

The Leopold and the Robbins Reports both recommended that park resource management be based on scientifically sound data and, where necessary, on regulation of the human uses for which the parks were intended. As a primary goal, the Leopold Report recommended "that the biotic associations within each park be maintained or, when necessary, recreated as nearly as possible in the condition that prevailed when the area was first visited by the white man." Unfortunately, this goal, and the policies and methods recommended for attaining it, were ignored or misinterpreted. Instead, a "hands off, let nature take its course" policy prevailed. This policy of "naturalism" rested on a misinterpretation of the Committee's advice to maintain or recreate "the primitive conditions" of prepark times and the erroneous notion that biotic communities can regulate themselves within artificial boundaries where humans are highly intrusive.

The "hands off" management policy adopted by the National Park Service, but not recommended in the Leopold Report, could not be applied effectively to black bears and grizzly bears because of their aggressive tendencies and destructive habits. A more specific program of action was needed to control bears while assuring healthy populations.

To this end, another advisory committee report was prepared in the wake of meetings and discussions with senior members of our research team. Entitled *A Bear Management Policy and Program for Yellowstone National Park* (1969), it was coauthored by A. Starker Leopold, Stanley A. Cain, and Charles E. Olmsted. It addressed several problems related to bear management. Unfortunately, the report was never previewed by or made available to us. Its authors unwisely accepted the preset agenda of Park Service administrators that all open-pit dumps were to be summarily closed, an action opposed by the senior author. Consequently, the report by Leopold *et al.* (1969) was both inadequate and confusing, and eventually it prompted controversy between scientists and administrators and their advisors. For example, an excerpt of the report states:

> The ultimate objective agreed upon by all participants, is the incineration or sanitary disposal of trash and garbage in a manner that will completely isolate this food source from bears. Ideally all refuse should be hauled out of Yellowstone National Park for incineration and burial. Future plans for such a program of disposal are being considered now.
>
> But in the meantime, there is disagreement as to the sequence of steps leading to the elimination of garbage from availability to grizzlies. One view is to cut off all garbage quickly, forcing the bears to turn immediately to natural foods. The opposite contention is to phase out garbage feeding over a period of time, "weaning" the bears gradually. The issue hinges on which of these procedures will result in the least number of bears going into campgrounds. Unfortunately, prognosticating how garbage-fed grizzlies will respond to deprivation depends on judgement rather than data. Much is known about the habits of garbage bears and of wild-living bears, but little is known about the process of transition. In any event, the transition must be made, and everything possible done to protect both park visitors and the population of bears during the process.

The Committee, in effect, ignored recommendations from "Management of Bears in Yellowstone National Park," a report documenting bear research in the Park over the previous eight years (Craighead and Craighead 1967). Further, it was apparently in ignorance of information having a direct bearing on the issue, that of excessive campground incursions and grizzly bear mortalities which followed the earlier abrupt closures of the Canyon Village and Old Faithful feeding stations in 1941 (Craighead and Craighead 1971). Finally, by failing to make a firm recommendation as to how, much less whether the dumps should be closed, the Committee endorsed the projected schedule for abrupt closure that the Park Service and its advisory scientist, A. Starker Leopold, were advocating.

Subsequently, in reports and in conference, we continued to advocate a gradual 10-year phaseout of the open-pit dumps, accompanied by an intensive program for continued monitoring of 66 marked grizzlies and additional unmarked, but recognizable, ones to determine how the garbage-fed bears would respond to food withdrawal. Neither of these recommendations was accepted by the Committee or park administrators. Instead, the abrupt closure of the dump at Rabbit Creek in late 1969 was followed by closure of Trout Creek in late 1970 and the West Yellowstone dump the following year. There was no adequate scientific follow-up during the enormously disruptive 1970-73 period, and marked bears were either killed in control measures or were captured and their identifying tags removed.

POSTCLOSURE MANAGEMENT HISTORY (1971–75)

From 1959 to 1968, there were relatively few problems with grizzly bears throughout the entire Yellowstone Ecosystem; however, following the rapid phaseout of the open-pit dumps, bear-human encounters and bear visits to campgrounds and developed areas increased dramatically (Craighead and Craighead 1971). The new bear management strategy, programmed to phase out the open-pit dumps quickly without first adequately sanitizing the campgrounds, forced grizzlies into congested visitor areas and greatly increased the probability of bear-human conflicts. Troublesome campground bears were literally "created" (Chapter 15). As a consequence of these conditions and a greatly accelerated grizzly bear mortality rate, a controversy developed between those who had advocated abrupt closure of open-pit dumps and those who had recommended gradual closure with careful monitoring.

On September 19, 1972, the senior author briefed Park Service Administrators, the Fish and Game directors of Wyoming, Idaho, Montana, and Colorado, Undersecretary of the Interior Nathaniel Reed, his biological consultant, A. Starker Leopold, and several University of Montana faculty members on the grizzly bear situation in Yellowstone (Craighead 1972). The technical presentation of data was introduced as follows:

> As we all know, this problem of managing grizzly bears is very complex, and there is bound to be difference of opinion. That is why we are here. We must find answers.

Before presenting our data, I would like to present a background of information that I believe is essential to an understanding of why my brother, Dr. Frank Craighead, my colleagues, and I are concerned about the Yellowstone grizzlies. We find it necessary to reiterate why our position is at variance with that of the National Park Service.

The scientific evidence gathered by our research group does not support the Park Service's bear management policies. However, the very essence of the scientific method and of scientific progress is the freedom to disagree. Only with constructive criticism can we find the solution to managing and preserving the grizzly in the Yellowstone environment. We have come by our differences honestly. Our loyalty is to the resource. I will take time to define the issues and attempt to put basic differences in perspective.

We began researching grizzlies in 1959 and, in the course of the study, developed numerous techniques for studying these large carnivores. At the request of the National Park Service, we prepared a management report in 1967. In that report [*Management of Bears in Yellowstone National Park*], we advocated management policies and a management program based on our research findings and on conclusions supported by nine years of intensive research.

The National Park Service and its advisory board, however, established management policies and put into action a grizzly bear management program based on hypotheses supported by virtually no research of the local problem by its biologists. I regret that I must take issue with Dr. Leopold's remark this morning that there was a consensus of opinion on the Committee report [*A Bear Management Policy and Program for Yellowstone National Park*, Leopold *et al.* 1969]. With all due respect to Dr. Leopold, on this point we do not agree.

The hypotheses or assumptions on which present National Park Service management is based were essentially:

1. That open-pit garbage dumps in and outside of Yellowstone National Park could be rapidly phased out with relatively little effect on the grizzly bear population;

2. That most of the grizzlies denied these long-established feeding areas would establish themselves in the Yellowstone backcountry and cease to frequent their former haunts; and

3. That those grizzlies that did not adjust as anticipated would be dispatched, but it was assumed that the numbers of these would not be great.

This National Park Service grizzly bear management program has now been in effect for four years.

Our contention four years ago and today is that management of a resource such as the grizzly bear should not be based upon unproven assumptions. It is true that no one could predict exactly what would occur when the new management program went into effect. But much data existed to show that the basic assumptions of the Park Service bear management policy were highly questionable.

We made our scientific evidence available to the National Park Service in the form of publications, reports, and briefings. We provided the Service with two detailed reports documenting the grizzly bear problem in Yellowstone and, in these reports, we made specific recommendations for managing grizzly bears. These recommendations were made on the basis of years of research and at a time when the National Park Service had little scientific expertise in the ecology and management of grizzlies. We recommended that a completely natural population of grizzlies be established by closing the open-pit garbage dumps over an 8- to 10-year period or longer. We emphasized the importance of establishing a population of grizzlies that could feed entirely on natural foods.

But, we especially warned against a rapid phase-out of the open-pit garbage dumps, particularly the Trout Creek dump located in the geographic center of the Park, and we advocated continued research by our research team to document what occurred during the phaseout. We believed that only by comparing the dump phase-out period with conditions that existed before the phase-out could the success of the management program be accurately evaluated. We still believe this to be so.

In our reports and at numerous meetings, we stated that if the open-pit dumps were phased out rapidly as the National Park Service indicated it planned to do—and I repeat—this action was to be taken on the basis of the unsupported hypotheses:

1. That only a relatively few grizzlies were addicted to feeding in campgrounds, and that there existed a relatively large nucleus of backcountry grizzlies that had never developed either campground or garbage dump feeding habits. (Our backcountry observations based on thousands of man-hours showed little evidence for a large backcountry population of grizzlies); and

2. That with dumps closed, most of the grizzly bear population would revert to living in the backcountry, their traditional migration to these dump sites would cease, and the few remaining in developed areas and campgrounds could be disposed of.

Now—if this rationale was accepted as the basis for managing grizzlies, we said in our reports that the following could be expected to occur:

First—There would be a heavy influx of grizzlies into developed areas and campsites both inside and outside the Park where they would cause trouble;

Second —That this would lead to increased mortality of grizzlies and, since our data showed that most of the Yellowstone grizzlies had learned over extended periods of time to use the dumps and had established strong habit patterns and traditional migrational movements to these sources of food, these ties and habits developed by a basic reward relationship could not be altered quickly;

Third —The grizzly population would not be able to adjust by occupying the backcountry, but instead, most of these animals would still continue to seek food during the summer period in the areas they had so long frequented. I think Mr. Cole's maps support this;

Fourth—The opportunity for bear-human encounters would increase. This could be partially, but not entirely offset by sanitizing tourist areas, closing campgrounds at critical times, increasing ranger patrols, and alerting the public to the dangers inherent in the management program. We emphasized the necessity of completely sanitizing campgrounds before closing the dumps;

Fifth —Grizzlies that had frequented the open-pit dumps (dumps that had been isolated from the public) would, we predicted, enter campgrounds, the offspring of females would learn this habit from their mothers through the basic reward relationship mentioned this morning, and generations of grizzlies would become conditioned to campgrounds just as they had become conditioned to the open-pit dumps in the past; these animals would be more dangerous than those that had frequented the open-pit dumps because they would become man-conditioned, and they would not be isolated;

Sixth—The disruption caused by the sudden depletion of an important and historic food source would place heavy stress on the population causing:

1. lowered reproduction; and
2. increased movement, resulting in increased deaths

Both factors would adversely affect the population size as well as grizzly bear-human relationships. The probability of serious bear-human encounters would increase.

Mr. Chairman, we contend that this is the situation that prevails today. Grizzlies frequenting developed areas and campgrounds continue to be a serious problem, the risk to visitors has not decreased appreciably, and the grizzly population is declining rapidly. It is this latter problem that I will discuss in detail.

The technical briefing and data analysis that followed served only to alarm the agencies and inflame the controversy (see also Craighead *et al.* 1974). To resolve it, the National Academy of Sciences (NAS) was asked by the Secretary of the Interior to select a committee to investigate the situation and make recommendations. The NAS committee, although composed largely of former students of Starker Leopold, well aware of his position, strongly supported our research and warned that park managers were not dealing with separate populations of backcountry and "dump" bears. But the NAS Committee offered no continuing role to the Craigheads, nor did Starker Leopold or Park Superintendent Jack Anderson, in spite of a written request to Starker Leopold from the senior author. Their attitudes are blatantly revealed in their personal correspondence, to none of which were the Craigheads privy (Starker Leopold, Letter to Jack Anderson, June 1970; Jack Anderson, Letter to Starker Leopold, June 1970). See also, correspondence between John Craighead and Harry Woodward (John Craighead, Letter to Harry Woodward, May 1972; Harry Woodward, Letter to John Craighead, May 1972); and Nathaniel Reed's letter to Senator Dingell (October 1972). However, a program of research and management was proposed, and reorganization of the Interagency Grizzly Bear Study Team (IGBST) to remove it from control of the National Park Service was strongly recommended (Cowan *et al.* 1974).

Created in 1973, the IGBST was composed of research biologists from the National Park Service, the U.S. Fish and Wildlife Service, and the U.S. Forest Service, but was clearly under the aegis of the first, as it was the "payroll" agency. The intent of the reorganization recommended by the NAS was that these federal members be joined by representatives of the Montana, Idaho, and Wyoming wildlife departments and, importantly, by researchers from the private sector. These fundamental recommendations of the NAS Committee were only partially met. Researchers were recruited from the three states, and the leader, Richard Knight, was housed on the campus of Montana State University. But no private sector scientists were included. Knight continued as a National Park Service employee, and Yellowstone Park maintained financial control of the operation. For a time, IGBST research was confined within the park boundaries, and collaboration at the field level with agencies responsible for the remainder of the larger Ecosystem was nonexistent. Our previous work was, by administrative decree, ignored and its import was misrepresented by high-level civil servants. Such activities were never rectified by members of the NAS Committee or by federal agencies responsible for wildlife research on public lands.

Although annual reports of Cole (1971, 1972b, 1973a, 1975) and National Park Service newsletters routinely reassured the public as to the health and size of the largely "backcountry" population of grizzlies, there was great apprehension on the part of certain members of the research community as to the true status of the population. This concern initiated long overdue bureaucratic action. In 1975, the grizzly bear in the lower 48 states was declared a threatened species under the Endangered Species Act of 1973. In accordance with tenets of the Act to perpetuate threatened and endangered species and, where possible, to extend their populations, all federal agencies conducting land management programs were required to prevent destruction or adverse modification of critical grizzly bear habitat. Thus, a critical habitat determination was required, which involved delineating an area essential for the survival and recovery of the species. This area had to meet all specific physical, seasonal, and behavioral requirements of the bears, as well as assure sites for breeding, reproduction, and shelter. Unavoidably, delineating critical grizzly bear habitat entailed "invading agency turf," both state and federal, and thus, enlarged and intensified the grizzly bear controversy. During the interim, the Yellowstone population had continued to decline (Craighead *et al.* 1974).

A proposed federal rule-making, delineating critical grizzly bear habitat in the contiguous 48 states, was published in the Federal Register on 4 November 1976 (U.S. Congress 1977). The extensive land areas proposed by the U.S. Fish and Wildlife Service as habitat necessary for the grizzly bear's survival involved four regions totalling about 13 million acres. Public hearings that followed revealed widespread public and agency opposition. The opposition was due, in part, to alleged economic, social, and jurisdictional conflicts, and to a lack of sound scientific data corroborating the delineation.

Concern over the biological status of the Yellowstone grizzlies and the jurisdictional and policy considerations of implementing the directives of the Endangered Species Act prompted a Senate Hearing in December, 1976, in Cody, Wyoming. Of paramount concern was whether the delineation of critical habitat would extend beyond the perimeters of Yellowstone National Park and, if so, how this would affect multiple-use management in the surrounding National Forests. Other issues of biological significance were clearly of secondary importance

(Senate Hearings, 94th Congress, Second Session, 1977). Thus, although scientific information for a federal delineation of critical habitat in the Yellowstone area was available from our research, it was virtually ignored by the federal agencies and the IGBST. We had delineated an area of critical habitat using:

1. Observations of marked grizzly bears;
2. Movement and distribution data derived from the place of death of marked and unmarked animals;
3. Analyses of the spatial and habitat needs of the species; and
4. Basic demographic and biological parameters of the grizzly bear population.

The delineation based on these data encompassed an area of approximately 5 million acres (2 million hectares), with minor exclusions for towns, campgrounds, and developed areas, and represented a discrete ecosystem for grizzly bears (Craighead 1980b).

We originally had planned to map the vegetation of the entire Ecosystem using remote-sensing techniques to achieve a better understanding of bear habitat and spatial requirements. With our expulsion from the Park in 1971 and the allocation of research responsibilities to others, ultimately the IGBST, the plan was dropped, and we (the Scapegoat team) conducted the habitat research elsewhere (Craighead *et al.* 1982, Craighead *et al.* 1988a; see also Chapter 12).

REORGANIZATION OF THE INTERAGENCY GRIZZLY BEAR STUDY TEAM

The reorganized Interagency Grizzly Bear Study Team, headed by Richard Knight, began its follow-up study of the Yellowstone grizzlies in 1974. Unfortunately, the intensive study of the population during the postclosure period (1970-73), so strongly recommended by virtually all principals, had not occurred. The status of the population to be studied by the IGBST was unknown and remained a point of much concern and speculation. The IGBST, itself, developed its research objectives based on the recommendations of the NAS Committee on the Yellowstone Grizzlies (Cowan *et al.* 1974). These objectives were stated in the 1974 IGBST Annual Report (Knight *et al.* 1975) as follows:

1. To determine the status and trend of the grizzly bear population; and
2. To determine the use of habitats by bears, and the relationship of land management activities to the welfare of the bear population.

For each objective, several specific avenues of study were identified and assigned relative priorities. These are too detailed and extensive for listing here. In retrospect, however, it is apparent that the research plan was crippled in several important respects.

1. Field-level coordination between the concerned federal and state agencies was inadequate.
2. The three-year gap in research (1971-73) broke continuity in recording biological parameters of the grizzly bear population following closure of the dumps.
3. For many years, our wealth of baseline data for the preclosure grizzly population was virtually ignored (ignored, some say, by administrative directive).
4. The research methodology was not designed so that postclosure data eventually could be compared directly with preclosure data.
5. No researchers from universities or private sector resource groups were invited to participate in planning the research, analyzing the data, or peer-reviewing results. All aspects of the research were strictly in-house.

For these and other reasons, many of the data concerned with objective #1 and some of those for objective #2 proved difficult or impossible to acquire. Moreover, many of the postclosure data sets did not lend themselves to comparisons with data from the preclosure populations. Consequently, it has been necessary in the chapters that follow to define criteria for what and how comparisons would be made and, all too frequently, to substitute criticism of methodology for forthright data analysis.

The general thrust of the IGBST research conducted during the 1970s was critically examined by Roland Wauer, Chairman of the Interagency Grizzly Bear Steering Committee, and a surprising memorandum was issued. In August, 1982, Wauer predicted the bear's disappearance from Yellowstone unless the official "all is well" line of reporting was abandoned and major action was taken to improve the research design and objectively analyze the biological data being collected (Frome 1984). This was a necessary and critical step, but it was too late to establish a methodology for comparing pre- and postclosure data. Although IGBST research improved, as did the reporting, criteria for comparatively analyzing data between the two periods were absent.

To further complicate the management of grizzlies in the Yellowstone Ecosystem, the information flow to key decision-makers was biased and inadequate, and enforcement of the Endangered Species Act, ostensibly prohibitive, was transformed into a discretionary permit system (Mattson and Craighead 1994). In the chapters that follow, we attempt to analyze and evaluate the information needed for creating a recovery plan fulfilling the intent of the Endangered Species Act.

Table 15.1. Location, frequency of capture, and number of individual grizzly bears captured in campgrounds and developed areas, Yellowstone National Park, 1959-67 and 1968-70.

Campgrounds and developed areas[a]	1959-67		1968-70[b]	
	Frequency of capture	Individual bears captured	Frequency of capture	Individual bears captured
Canyon Village	48	33	42	24
Lake	40	26	73	48
Old Faithful	9	9	23	17
West Thumb	17	11	7	6
Grant Village	—	—	30	25
Norris	1	1	2	2
Lewis Lake	0	0	1	1
Indian Creek	2	2	0	0
Slough Creek	0	0	3	3
Tower Fall	0	0	3	3
Mammoth	0	0	1	1
Madison Junction	0	0	5	3
Total	117	82	190	133
Corrected total[c]		75		124
Total individuals involved for period		57		70
Mean per year	13	6	63	23

[a] Canyon Village includes Canyon Village and Otter Creek Campgrounds. Lake includes Lake, Fishing Bridge, Bridge Bay, and Pelican Creek Campgrounds. Grant Village opened in 1967 when the National Park Service began phasing out open-pit dumps.

[b] In addition to marked animals, a minimum number of unmarked individuals, determined by criteria of sex, age, and time of capture, were recorded for the 1968-70 period.

[c] Corrected for seven individuals trapped in two different areas in a year during the 1959-67 period, and nine during the 1968-70 period.

15

Grizzly Bear Distribution and Movements Following Catastrophic Events

In 1967, prior to the NPS decision to begin phasing out the open-pit garbage dumps, the senior author and F.C. Craighead, Jr., submitted a report to the National Park Service, *Management of Bears in Yellowstone National Park* (Craighead and Craighead 1967). In that report, they recommended a gradual phasing out of the open-pit garbage dumps, cautioning that:

> Because phasing out of refuse dumps will disperse the grizzlies by destroying an attractive, if not essential, food source, the transition from pits to incinerators must proceed gradually, enabling the grizzlies to develop new feeding habits as well as altered social behavior and movement patterns. If the transition is slow and follows a recommended procedure, it is possible that no severe changes in population level, distribution, or behavior will result. If, on the contrary, the phasing out operation is abrupt, and a carefully planned procedure is not followed, the result most certainly will be increased grizzly incidents in campgrounds, accelerated dispersal of bears to areas outside the Park, and greater concentrations of grizzlies at the public dumps in Gardiner and West Yellowstone, where food will be available but where adequate protection will not. The net result could be tragic personal injury, costly damages and a drastic reduction in the number of grizzlies.

This advice was ignored and, instead, the phaseout of the Trout Creek dump began in the summer of 1968. The predicted movement of bears into park campgrounds and an escalating death rate caused the Park administration to slow the phaseout process during 1969-70, with abrupt closure of Trout Creek occurring during spring of 1971.

The questions we seek to answer here are how the closure of the ecocenters affected the postclosure distribution and movement patterns of bears, and of what significance, if any, was this? Also, of what significance was the complete disregard of 12 years of our documented data by the NPS and its scientific advisors? We examine these questions over four time periods: first, the dump phaseout period (1968-70); second, the immediate postclosure period (1971-74); third, a period of population adjustment (1975-87); and, finally, the cataclysmic fire and postfire period (1988-92). Wherever possible and appropriate, we compare data from our study with those reported by the Interagency Grizzly Bear Study Team (IGBST).

In Section I, we described the major grizzly bear aggregations within the Yellowstone Ecosystem and showed how the members of these aggregations dispersed in fall and reconvened in spring. We also demonstrated that the Trout Creek aggregation was a discrete subunit of the ecosystem population, as were also the Rabbit Creek and West Yellowstone aggregations. Movements between these ecocenters (earth-filled garbage dumps) were of a transient nature, and exchange was minimal, confined largely to adult males. Census data revealed that a majority of the ecosystem population of approximately 300 animals visited the ecocenters during the summer months of June, July, and August in any given year, and further, that probably every member of the ecosystem population fed at one or more of them at some time during the course of their lives. Some visited the ecocenters frequently, others infrequently. The population used the entire five-million-acre (two-million-hectare) ecosystem, while making seasonal movements to and from the ecocenters. These annual movements, terminating in the formation of specific aggregations (Fig. 3.1), created a period during the summer when bear densities were locally high at the ecocenters and very low ecosystem-wide. In Chapter 18, we will discuss the mortality induced by dump closures. Here, we examine the movements of bears induced by cataclysmic events, and which were, in turn, responsible for most of the recorded bear deaths.

Fig. 15.1. Annual number of captures and number of individual grizzly bears captured in campgrounds, Yellowstone National Park, 1959-70.

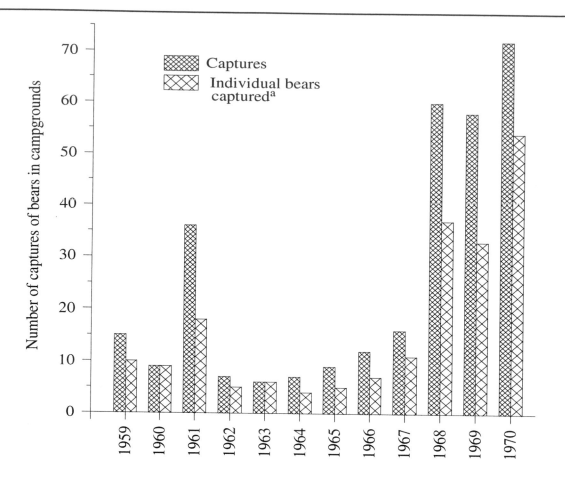

[a] Some bears were captured in more than one year; also, annual totals were corrected to account for seven individuals trapped in two different areas in a year during the 1959-67 period, and nine during the 1968-70 period.

GRIZZLY BEAR MOVEMENTS FOLLOWING PHASEOUT OF THE EARTH-FILLED GARBAGE DUMPS (1968-70)

During the summers of 1959 through 1967, approximately 1000 125-liter (33-gal) cans of unsorted edibles and trash were deposited daily at Trout Creek. The refuse was lightly covered with soil, usually on the day of deposit. The great volume of food attracted and held grizzlies in the area throughout the summer months.

In 1968, the volume of garbage taken to Trout Creek was drastically reduced. Edibles were partially separated from trash and dumped, but not buried. The sorted food consisted of approximately 50% inedible trash. Our records, made when the refuse was dumped, showed that a maximum of eight cans of garbage were deposited each day from 3 June

through 14 June; between 15 June and 15 July, the number gradually increased to 40 per day. Most of the refuse not dumped at Trout Creek was handled by the newly installed incinerator at Bridge Bay which had an operating capacity of 272.7 kilograms (6000 lb) per hour. Thus, over the remainder of the summer, the number of cans of garbage deposited at Trout Creek did not increase significantly, except when the incinerator broke down.

This significant reduction in garbage induced a dispersal of Trout Creek bears to adjacent campgrounds and developed areas in search of food. These bears continued to visit Trout Creek as in the past (see 1968 census, Chapter 5), but enlarged their foraging ranges. Our documentation of marked animals showed that many found their way into campgrounds, traveling among Trout Creek and the Canyon Village and Lake Campgrounds (Table

15.1). In addition to the recognizable animals, many other grizzlies also entered the campgrounds. It appeared that the movement of practically all Trout Creek grizzlies, marked and unmarked, was greatly expanded. A measure of this movement and disruption of long-established habits was reflected in the frequency of capture and the numbers of individual grizzlies captured in campgrounds each year during the 1959-67 period, as compared to similar data for 1968 through 1970. The number of individual grizzlies captured in 1968 was approximately double the previous high recorded in 1961, whereas frequency of capture was the highest yet recorded (Fig. 15.1).

The next year, 1969, we continued on-site measurements of the volume of garbage dumped at Trout Creek. Between 130 and 170 cans were deposited daily from 15 June through August, only a fraction of the amount deposited seasonally between 1959-67. Most of this was in plastic bags, and only about half the contents was edible. As in 1968, the garbage and trash were not buried, so grizzlies that arrived at the dump first were able to consume most of the food by evening, and little remained to hold animals that arrived later. From 1959 through 1967, the general procedure had been to deposit trash with garbage, and to cover it partially with soil each day. This provided a long feeding period, allowed numerous animals to share the food, and kept the grizzly bears concentrated in the area.

The dispersal of grizzlies that began in 1968 continued in 1969 with closure of the dump at Rabbit Creek and a relatively poor year for natural foods (Fig. 10.8). The slight increase in the garbage dumped during 1969, designed to rectify the situation at Trout Creek, was ineffective. The frequency of capture and the number of grizzlies captured in campgrounds and developed areas, remained high (Fig. 15.1). In 1970, the Rabbit Creek Dump was completely closed, and frequency of capture of grizzlies in campgrounds climbed still higher to a total of 72, twice the preclosure high experienced in 1961. The number of individual grizzlies captured was 54, the greatest number ever recorded (Fig. 15.1).

The frequency of capture of bears in campgrounds and developed areas over the nine-year period from 1959-67 was 117, an average of 13.0 captures per year (Table 15.1). Over the next three years, 1968-70, it was 190, an average of 63 per year. The mean numbers of individual grizzlies captured for the same periods were 6 and 23, respectively (Table 15.1). Thus, from 1959 through 1967, an average of six grizzlies were campground-ori-ented each year, but during the period of open-pit dump closures (1968-70), an average of 23 grizzlies were campground-oriented each year. Grant Village, first opened to the public in 1967, presumably had no problem bears that year, as none were captured (Table 15.1), but the following year, it was visited by grizzlies. Twenty-five individuals were captured at Grant Village during the three-year period of dump phaseouts. Similarly, no grizzlies were recorded as captured at Slough Creek, Tower Fall, Mammoth, or Madison Junction from 1959 to 1967, but a total of 12 captures were recorded at these campgrounds from 1968 to 1970 (Table 15.1). Thus, data on the number of individual grizzlies captured in campgrounds during the three years of revised management clearly showed that the new practices created problem bears that then had to be dealt with by the park administration.

Dispersal of Individual Female Grizzlies in Response to Reduced Food at the Trout Creek Dump During 1968-69

Five female grizzlies frequenting the Trout Creek dump were color-marked from 1960 to 1962 (Table 15.2). They were captured in campgrounds or developed areas for the first time following the reduction of food at Trout Creek in 1968. The long interval between marking and first capture in a campground in 1968-69 can be explained if these bears suddenly were forced to increase their daily movements and extend their home ranges in search of food.

The home ranges of Nos. 39 and 40 had been established earlier by radiotracking (Craighead 1976). Female No. 39 extended her home range in 1968 to include Lake Campground and the Bridge Bay developed area. This more than doubled the size of her previously established annual home range. During 1968-69, she was captured five times at Bridge Bay, and twice in adjacent developed areas. These were her first campground captures within an eight-year period (Table 15.2).

We radiotracked No. 40 for eight consecutive years, establishing for the first time, a life range perspective. While this female was radio-monitored, she was never tracked into a campground or developed area. But in 1969, she entered the Lake developed area and was shot (Craighead 1979). First-time captures in campgrounds showed that female Nos. 34, 109, and 128 also extended their ranges, and that campground entries by these and other bears were probably induced by food stress (Table 15.2). All of the bears recorded in Table 15.2 at least doubled their home areas within a two-year period following dump closures.

Table 15.2. Campground captures of five Trout Creek female grizzly bears during the phaseout of open-pit dumps, Yellowstone National Park, 1968-69.

Bear number	Year marked	Year of first capture in campground	Interval (in years) between marking & first capture in campground	Age at first capture in campground	Number of capture areas	Frequency of capture by years	Number of offspring produced between marking & first recapture	Status of female 1969
34	1960	1969	9	19	1	1969-1	6	Alive
39	1960	1968	8	13	2	1968-2 1969-5	5	Alive
40	1960	1969	9	11	1	1969-1	7	Control kill
109	1961	1969	8	8	1	1969-2	0	Alive
128	1962	1969	7	18	1	1969-1	10[a]	Alive

[a] No. 128 produced three more cubs in 1970.

Because many grizzlies of all ages first entered campgrounds in 1968 and 1969, it is unlikely that the advanced age of three of the five female grizzlies listed in Table 15.2 was a factor in that behavior. They were in excellent condition when last captured, and all had good reproductive records. Number 128 had produced 10 cubs before her capture in 1969. She bore three more in 1970 at an age of at least 19 years, for a total of 13 cubs during the nine years she was marked. From 1960 to 1970, the five females, between them, bore a total of 28 offspring (Table 15.2). Number 40 was shot in 1969 following an initial capture in a campground, and three of the other females were never identified again after 1970, nor did they show up as documented mortalities. Number 128, with only two yearlings from her three-cub litter the previous year, was captured at Pelican Campground in 1971 and never reobserved (see Chapter 18). It became evident that management practices "forcing" productive females into developed areas and campgrounds where they were subject to a death penalty could rapidly alter the population level. This concern was conveyed to park administrators (Craighead and Craighead 1967).

Availability of the staple natural foods was not a probable cause of movement in 1968, because availability was not importantly different from that in earlier years. Our data on the use of natural foods by grizzlies, and on the relative abundance of these foods over a 12-year period (Chapter 10), fully support this contention. By contrast, availability of natural foods was much reduced in 1969, and movement patterns developed in 1968 as a result of garbage reductions at Trout Creek were expanded. We conclude, therefore, that the five females recorded in Table 15.2, as well as an additional 11 marked grizzlies of both sexes, were captured for the first time in campgrounds in 1968 and 1969 primarily because of the acute food shortage at Trout Creek and reduced natural foods throughout the Ecosystem in 1969. During this period, a minimum of 35 unmarked individuals (based on sex, age, and natural markings) were also captured in campgrounds or developed areas for the first time.

Records of the movements of 34 marked Trout Creek grizzlies and two recognizable cubs were obtained by capture or by observing their individualized color markings (Table 15.3, Fig. 15.2). Sixteen of these individuals visited the Lake developed area, eight entered Canyon Village, and 10 moved to the Rabbit Creek dump. Two moved as far as Cooke City; three, to Tower; three, to West Yellowstone; one, to Norris; and one, to Grant Village. Eight of the marked bears visited two camp-

Table 15.3. Dispersal of 34 marked grizzly bears and 2 recognizable cubs from Trout Creek following management changes at the dump, Yellowstone National Park, 1968-69.

Bear number	Campgrounds and developed areas visited by marked grizzlies							
	Lake	Canyon Village	Grant Village	Norris	Cooke City Dump	Tower Dump	West Yellowstone Dump	Rabbit Creek Dump
7								X
10 + 2[a]						XXX[a]		
30	X							
34					X			
39	X							
40	X							
88								X
109	X							
112		X						
128	X							
130			X					X
141	X	X						
144		X						
147								X
166							X	X
183	X							
188		X						
189								X
190							X	
194								X
201								X
207	X	X						
208	X	X						
210	X							
211	X			X				
213							X	X
217	X							
218								X
228		X						
236	X	X						
240	X							
241	X							
242					X			
256	X							
Total	16	8	1	1	2	3	3	10

[a] The cubs of bear No. 10 dispersed with their mother and could be recognized.

grounds or developed areas (Table 15.3), thus accounting for the total of 34 bears (Fig. 15.2). Movement of grizzlies to campgrounds and developed areas following reduction of food (garbage) at Trout Creek is, thus, strongly documented, and this movement should have alerted National Park Service administrators and their advisors to the serious implications that closing the long-established

dumps would have for the grizzly bear population. It did not.

Dispersal of Individual Grizzlies in 1969-70 Following the Closure of Rabbit Creek

The closing of the Rabbit Creek dump proceeded on schedule. Grizzlies using this ecocenter

Fig. 15.2. Dispersal of 34 marked grizzly bears and two recognizable cubs from Trout Creek following management changes at the dump, Yellowstone National Park, 1968-69. Numbers on the map indicate numbers of bears dispersing; eight of the bears visited two campgrounds or developed areas, thus accounting for the figure total of 44.

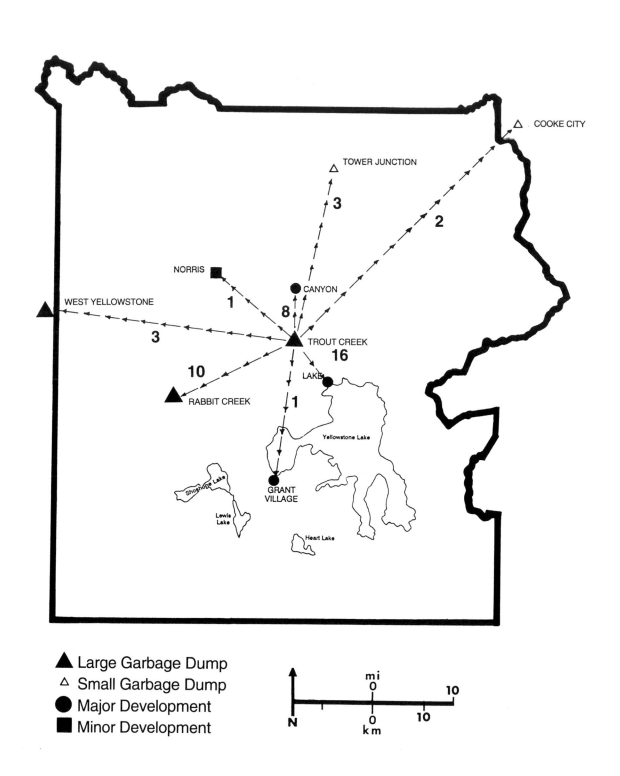

Table 15.4. Movement of grizzly bears following closure of the open-pit dump at Rabbit Creek in 1969, Yellowstone National Park, 1970.

Bear number	Sex	Number of years observed at Rabbit Creek Dump	New area visited[a]	Air miles from Rabbit Creek Dump to new area (km)	
2	M	11	Old Faithful	5	(8.0)
			Grant Village	14	(22.5)
8	F	11	Grant Village	14	(22.5)
57	M	10	West Yellowstone Dump	18	(29.0)
139	F	7	West Yellowstone Dump	18	(29.0)
			Trout Creek Dump	19	(30.6)
147	M	7	Trout Creek Dump	19	(30.6)
164	F	7	Old Faithful	5	(8.0)
171	M	6	Old Faithful	5	(8.0)
213	M	5	Trout Creek Dump	19	(30.6)
217A	M	3	West Yellowstone Dump	18	(29.0)
			Trout Creek Dump	19	(30.6)
219A	M	3	West Yellowstone Dump	18	(29.0)
224A	M	3	West Yellowstone Dump	18	(29.0)
226A	M	3	West Yellowstone Dump	18	(29.0)
—[b]	M	4	West Yellowstone Dump	18	(29.0)
—[c]	M	10	Old Faithful	5	(8.0)

[a] New area visited was determined by observation or capture of marked or recognizable animals.

[b] Unmarked male identified by size, conspicuous wound over left eye, and patches of white hair on shoulders.

[c] Large unmarked male identified by wound damage exposing the upper left canine.

during 1969 dispersed widely following its complete shut-down in 1970. We documented the movements of 12 color-marked and 2 recognizable grizzlies that had been observed and recorded at Rabbit Creek for between 3 and 11 years prior to its closure (Table 15.4); therefore, all were considered resident members of this population unit.

Seven of the 14 grizzlies moved 29 air kilometers (18 mi) to the West Yellowstone dump, which was still operative outside of the Park (Fig. 15.3). Two joined the grizzlies at Trout Creek, 31 air kilometers (19 mi) away; these also visited the West Yellowstone dump. Grizzly Nos. 2 and 8 moved 23 air kilometers (14 mi) to the Grant Village Campground and developed area. Three grizzlies, including No. 2, moved to the Old Faithful area only 8 air kilometers (5 mi) distant. Comparison of the 1970 census data taken at Trout Creek and West Yellowstone with other years suggests that many of the unmarked grizzlies from Rabbit Creek also moved to these other open-pit dumps and intervening campgrounds and developed areas.

In 1970, five grizzlies from Rabbit Creek were captured in campgrounds or developed areas, and four were dispatched (Table 15.5). Two of these were adult females. None of the five had previous campground records. The long intervals between marking and first capture in a campground or developed area show conclusively that closure of the Rabbit Creek dump initiated movements to campgrounds and developed areas similar to the movements that occurred in the early 1968-69 period of garbage reduction at Trout Creek. The official management solution was to eliminate the offending animals. There is circumstantial evidence that No. 8 severely mauled a park visitor in Grant Village on 3 September 1970. This "resident" of Rabbit Creek, color-marked in 1959, had not been captured or observed in a campground during an 11-year period prior to her death. Another old, but unmarked adult female member of the Rabbit Creek subpopulation that moved into the Old Faithful developed area was captured in October of 1970. She was ear-tagged and released at Gibbon Meadows, approximately 29 kilometers (18 mi) from Old Faithful. Two years later in June, 1972, the female mauled and killed a 25-year-old camper near the Old Faithful developed area.

The rapid phaseout of the long-established open-pit dumps initially increased the local movement among bears comprising discrete subpopula-

Fig. 15.3. Movement of 14 marked or recognizable grizzly bears following closure of the open-pit dump at Rabbit Creek in 1969, Yellowstone National Park, 1970. Numbers on the map indicate numbers of bears dispersing; three of the bears visited two or more dumps or developed areas, thus accounting for the figure total of 17.

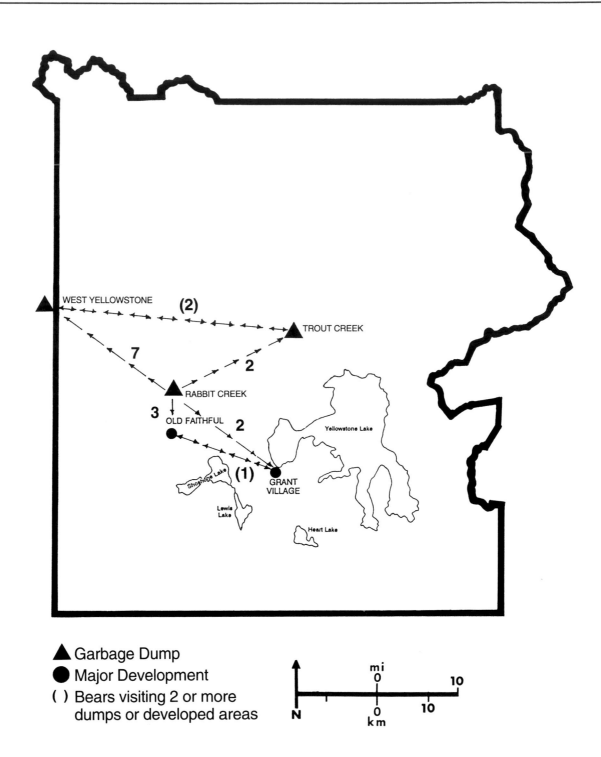

Table 15.5. Campground captures of five grizzly bears following closure of the open-pit dump at Rabbit Creek in 1969, Yellowstone National Park, 1970.

Bear number	Sex	Year marked	Interval (in years) between marking & first capture in campground	Age at first capture in campground	Capture areas	Frequency of capture	Status in 1970
2[a]	M	1959	11	15-16	Old Faithful Grant Village	2	Control kill
8[a]	F	1959	11	19	West Thumb Grant Village	2	Control kill
164	F	1963	7	12	Old Faithful	1	Alive—Yellowstone
171	F	1964	6	9	Old Faithful	3	Alive—zoo
—[b]	M	1960	10	22-25	Old Faithful	1	Control kill

[a] Exact age unknown.

[b] Large unmarked male identified by wound damage exposing the upper left canine.

tions at Trout Creek, Rabbit Creek, and West Yellowstone. Most of this movement was to and from the dump sites and the adjacent campgrounds and developed areas. The movements were relatively short in time and distance and did not substantially alter the size or composition of the respective aggregations. Grizzlies were forced into areas of high visitor use, and this increased the probability of bear-human conflicts. As the phaseout progressed through 1969-70, nonlocal movements among Trout Creek, Rabbit Creek, and West Yellowstone bears intensified, as did movement to areas outside of park boundaries. These longer-range movements between ecocenters and beyond gradually changed the size and structure of the ecocentered populations, a change best illustrated by changes recorded at Trout Creek for the years 1969 and 1970, particularly the great influx of adult male bears. With the rapid closure of the dump at West Yellowstone in 1971-72, the stage was set for extensive changes in the distribution of bears throughout the entire Ecosystem. This was, in turn, accompanied by significant increases in bear-human conflicts and in grizzly bear deaths. The administration's declared policy, formulated to deal with this situation, was to kill or ship to zoos all three-time offenders, but was later amended to apply to two-time offenders.

GRIZZLY BEAR MOVEMENTS DURING THE IMMEDIATE POSTCLOSURE PERIOD (1971-74)

The Park Service completed the closure of the Trout Creek dump during the spring and summer of 1971. At the same time, the West Yellowstone dump was moved and fenced. This action created more acute bear problems in campgrounds and developed areas in and near the Park than had existed in the previous 100 years of park history. Practically all of the Yellowstone bears were "on the move."

The excessive mortality within the Ecosystem recorded for 1970 continued throughout 1971 and 1972, finally tapering off in 1973 and 1974. It is obvious that overall grizzly bear mortality was shifting from inside to outside the Park (Table 15.6), as were specifically the deaths due to management control (32 of 38 [84%] inside the Park during the 1968-70 period versus 9 of 30 [30%] during the 1971-74 period). The decline from the earlier period to the later one in percentage that management control contributed to total mortality (35.8% versus 27.5%; Table 15.6) seems anomalous, given the substantial increase in total control actions, and is discussed more thoroughly in Chapter 18.

Thus, the grizzly bear population was under immense stress following closure of the dumps, and the immediate reaction of the bears was to move widely in search of food. The specific questions we seek to answer are how far did they move from the long-established ecocenters, what was the nature of these movements, how permanent was the dispersal, and how rapidly did a new distribution pattern evolve? These questions cannot be answered from data on the movements of individual bears, except when they appeared in the death statistics (Table 15.6). They can be addressed with reasonable precision, however, from campground capture records and by comparing bear density-distribution patterns that we mapped for 1968 against patterns mapped by Knight *et al.* (1976) and Mealey (1978) for the years 1973 and 1974.

Table 15.6. Grizzly bear deaths inside and outside Yellowstone National Park during and immediately following dump phaseouts, 1968-70 and 1971-74.

Year	Inside Yellowstone National Park			Outside Yellowstone National Park			Inside and outside Yellowstone National Park		
	Total deaths	Control deaths	Percent control	Total deaths	Control deaths	Percent control	Total deaths	Control deaths	Percent control
1968	15	4	26.7	9	4	44.4	24	8	33.3
1969	12	10	83.3	13	0	0.0	25	10	40.0
1970	25	18	72.0	32	2	6.2	57	20	35.1
Total	52	32	61.5	54	6	11.1	106	38	35.8
1971	8	4	50.0	39	11	28.2	47	15	31.9
1972	10	4	40.0	19	6	31.6	29	10	34.5
1973	1	0	0.0	16	3	18.8	17	3	17.6
1974	2	1	50.0	14	1	7.1	16	2	12.5
Total	21	9	42.9	88	21	23.9	109	30	27.5

Movement Inferred from Campground Captures

We have shown that, during 1968-70, the movement of bears induced by the closure of the Rabbit Creek and Trout Creek garbage dumps was primarily to local campgrounds and developed areas, with long range movements increasing with time. Most of the local movement occurred in the vicinity of Trout Creek where, in 1970, the bear aggregation was still intact.

During the period 1971-73, this localized movement continued. In 1971, 32 captures (not including recaptures) were made (Table 15.7). Over 56% of these were at sites along the northern and northwestern shores of Yellowstone Lake with 19% at Canyon Village. In 1972, 40% of 20 captures were along the northern and northwestern lake shore, and another 40% were at Canyon. The total number of captures decreased annually, with only 7 in 1973 and 10 in 1974, but 71.5% and 60% of these, respectively, were still highly localized at the northern and northwestern end of Yellowstone Lake, with another 14% and 30%, respectively, at the Canyon Village and campground (Table 15.7). We conclude that grizzlies deprived of food at Trout Creek were still in the near vicinity of that ecocenter in 1973-74, foraging at unsanitized campgrounds and developed areas.

Movement Inferred from Bear Density Distribution

Density distribution patterns during the preclosure period were clearly defined from the annual census records (Chapter 4). We consider the density-distribution pattern for 1968 to be representative of the major summer concentrations of grizzlies from 1959 through 1968. Bear numbers from census and movement data showed, as would be expected, high densities at and adjacent to the major ecocenters, with minor concentrations at preferred, natural habitat sites (Fig. 15.4).

The IGBST (Knight *et al.* 1975) and Mealey (1975) obtained bear distribution patterns from a wide range of observers, but there was no accurate way to convert the mass of sightings to number of individual animals. Nevertheless, we believe their "suspected" high density areas can be used to indicate the broad dispersal pattern of grizzlies during the 1971-73 period when our research team had left the Park, all dumps within the Park had been closed, and closure of the West Yellowstone dump was in progress.

A comparison of the data of Knight *et al.* (1975) and Mealey (1975) with our data for 1968 shows no important change in the locations of high den-

Table 15.7. Location and number of marked and unmarked grizzly bears[a] captured (recaptures excluded) in Yellowstone National Park campgrounds and developed areas, 1971-74 (data from National Park Service log books).

Year	Place of capture	Number of captures	Percent of captures	
1971	Fishing Bridge	11	34.4	} 56.3%
	Pelican Campground	7	21.9	
	Canyon Village	6	18.8	
	Grant Village	5	15.6	
	Other sites	3	9.4	
	Total	32	100.1	
1972	Fishing Bridge	8	40.0	
	Canyon Village	8	40.0	
	West Entrance	3	15.0	
	Other sites	1	5.0	
	Total	20	100.0	
1973	Bridge Bay	3	42.9	} 71.5%
	Lake Developed Area	2	28.6	
	Canyon Campground	1	14.3	
	Other sites	1	14.3	
	Total	7	100.1	
1974	Lake Developed Area	3	30.0	} 60.0%
	Bridge Bay	2	20.0	
	Fishing Bridge	1	10.0	
	Canyon Village	2	20.0	
	Canyon Campground	1	10.0	
	Other sites	1	10.0	
	Total	10	100.0	
1971-74	Fishing Bridge - Bridge Bay - Lake Devel.	30	43.5	} 53.6%
	Pelican Campground	7	10.1	
	Canyon Village/Campground	18	26.1	
	Grant Village	5	7.2	
	West Entrance	3	4.4	
	Other sites	6	8.7	
	Total	69	100.0	

[a] Unmarked animals differentiated by sex, age, and time of capture.

sity concentrations of bears within the Park, as of 1974 (Fig. 15.4). Summer concentrations were still in the Hayden Valley and Canyon Village area and along the northwestern shore of Yellowstone Lake between Bridge Bay and the Pelican Valley, northeast of Fishing Bridge; in the Lamar Valley and Mirror Plateau area east of Tower Junction; in the Swan Pass and Mt. Holmes area to the west of Mammoth; and in the area of the West Entrance to Yellowstone Park. An apparent decline along the Firehole River between Madison Junction and Old Faithful may have been due to lack of investigative effort in 1973-74 (Knight, pers. comm. 1976), rather than to an actual change in the distribution patterns (Fig. 15.4). A suspected increase in bear numbers along the eastern shore and at the southern end of Yellowstone Lake (Knight *et al.* 1975) may have been a real response to natural foods in the area, especially fish. Bear sign was observed in this area in earlier years, but few bears were seen, and the area was not considered to be a concentration site. The Cooke City, Montana, dump was considered

Fig. 15.4. Comparative grizzly bear concentrations in Yellowstone National Park, 1968 and 1973-74.

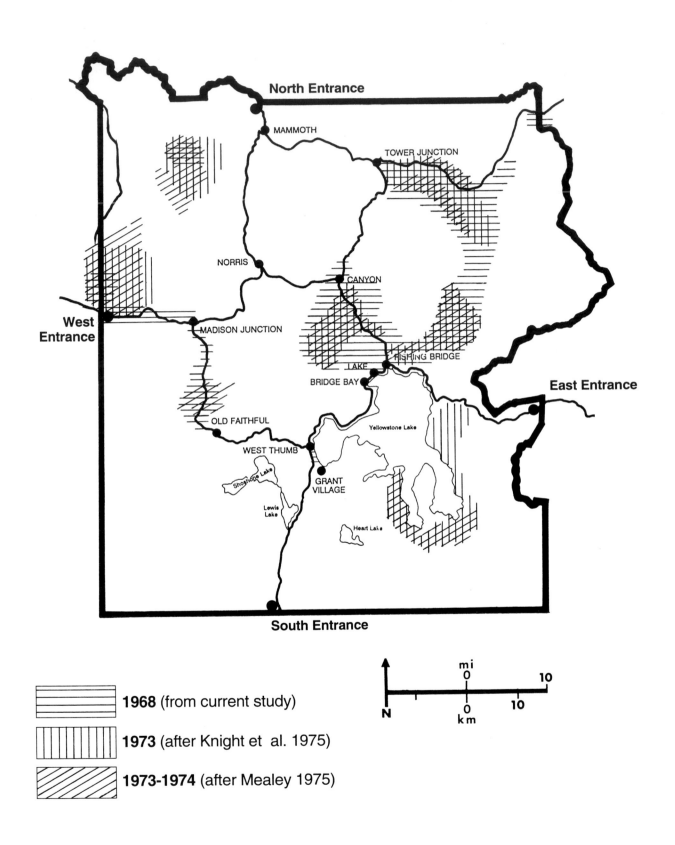

a minor concentration site from 1959 to 1970, but this dump was never closed. It was not considered a concentration site by Knight, although Greer trapped and instrumented five grizzlies there (Knight *et al.* 1976).

Thus, the pre- and postclosure distribution patterns were quite similar, indicating no major change in the grizzly bear concentrations within the Park from 1968 through 1974. This conclusion is further supported by the trapping data presented earlier in Table 15.7. Therefore, park management actions to disperse bears from the established concentration sites (ecocenters) and nearby campgrounds and developed areas within the Park were apparently unsuccessful. Habitat occupancy was not altered importantly, although more intensive use of the habitat for foraging no doubt occurred. Distribution data presented by Knight *et al.* (1976) indicated further that the situation had changed little by 1975. Local changes probably occurred from year to year in response to food availability, and there may have been population shifts between the long-established high bear-density areas. Even so, because those areas originally provided suitable breeding sites, natural food sites, and abundant food from the open-pit dumps, such stability of distribution was predictable. Although the dumps had been closed, the other attractions—including fidelity to established home ranges—still existed. Thus, competition between bear and human for the same space was still acute in many of the campgrounds and developed areas of the Park. The distribution of these areas of high grizzly bear density (Fig. 15.4) indicates that in spite of the extensive movement of bears during the years 1968-74, the broad pattern of distribution had not altered significantly. Grizzlies were still centered in the vicinity of the defunct open-pit dumps.

GRADUAL DISPERSAL OF THE YELLOWSTONE GRIZZLY BEAR POPULATION, 1975-87

We have shown that the grizzly bear population of Yellowstone Park was dangerously destabilized between 1969 and 1973-74 as a result of manipulation and closure of the historic feeding sites at open-pit dumps within and adjacent to the Park. Control actions greatly reduced bear numbers, while age and sex structure of the population and traditional ecocenter-oriented behaviors of remaining individuals were disrupted (Craighead and Craighead 1971, Craighead 1980b). Ecosystem-wide, the population was also seriously destabilized.

Not until 1974, following the inception of the Interagency Grizzly Bear Study Team, did serious research begin. The IGBST was permitted to radio-collar bears beginning in 1975 (Knight *et al.* 1976) and, as importantly, to track them from aircraft. The IGBST applied this methodology effectively, collaring and then radiolocating 97 individual bears 6299 times through 1987 (Blanchard and Knight 1991). This enabled them to define home ranges in great detail. The process of bears establishing home ranges is basically one of finding and utilizing a habitat area that meets the animals' biological requirements, of which food is paramount. Establishment of secure home ranges is an indicator of population stability.

We now examine the process of establishing new home ranges by those individuals remaining in the population and expected to accommodate to the new conditions. To examine this process, we will compare how mean home ranges of the population developed over the years following the highly disruptive 1969-74 period. We compare IGBST data gathered since 1975 (Knight *et al.* 1976, 1977, 1978, 1980, 1981, 1985; Blanchard and Knight 1991) with comparable data from the period, 1963-68 (Craighead and Craighead 1971, Craighead 1976, Craighead 1980b). Progressive change in mean home range and life range parameters was revealed by looking at those parameters for the period 1975-80, and determining how they changed as data for subsequent years were included. Thus, the mean areas of home and life ranges for the population from 1975-80, 1975-83, and 1975-87 were compared with those parameters for the 1963-68 preclosure bear population. Although the three postclosure periods over which we evaluate the range parameters are somewhat arbitrary, their selection is largely a function of when data were made available in agency reports. Home and life range information for the 1975-80 period was taken from IGBST Annual Reports (Knight *et al.* 1976, 1977, 1978, 1980), whereas progressively cumulative data were provided from Knight *et al.* (1984) and Blanchard and Knight (1991) for the periods 1975-83 and 1975-87, respectively. Although sample sizes for numbers of radiolocations and numbers of individuals representing age and sex cohorts varied, and adults were considered to be four years of age or older in Blanchard and Knight (1991), the comparison between periods successfully illustrates dramatic and progressive increases in life and home range areas for most sex and age cohorts of the Yellowstone population.

Change in Area of Mean Life Ranges

Over the four periods, mean life ranges for adult bears increased substantially (Fig. 15.5). We had

Fig. 15.5. Mean life ranges of adult grizzly bears (*n* = number of individuals), Yellowstone Ecosystem, 1963-68, 1975-80, 1975-83, and 1975-87.

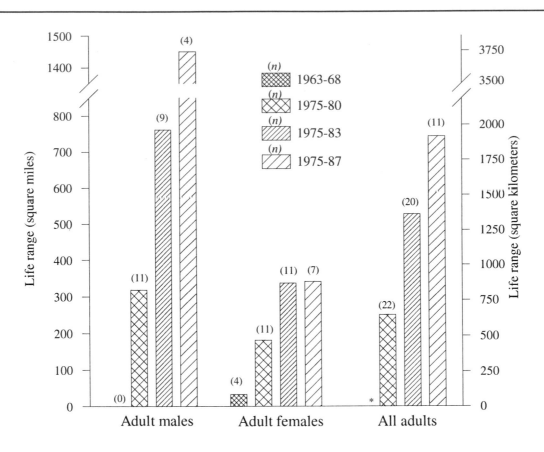

* Not shown, as it would reflect results for females only.

insufficient data to calculate life ranges for adult males for the 1963-68 period, but it is plain that males were expanding their life ranges substantially between 1980 and 1987, as the mean area for 1975-87 was more than quadruple that for the 1975-80 period. Mean life range for adult females also increased progressively, although not so graphically as that for the males. Although mean life range area for males was nearly doubled when data for 1984 through 1987 were included with those for 1975-83, mean life range for females increased by only 1.2%. The mean life range area for females during the 1975-87 period, however, was tenfold greater than that recorded for the 1963-68 period (Fig. 15.5).

The substantial increases in life ranges exhibited for both sexes over the time period documented were undoubtedly related to the bears' dependence wholly upon natural foods that varied in abundance and quality from season to season, and from year to year. Males continued to assimilate substantial new territory into their life ranges throughout the entire 1975-87 period, but life ranges of females appeared to stabilize somewhat in the later years. This decline in rate of life range expansion by females reflects a real decline in the size of mean annual ranges recorded for adult females, as discussed next.

Change in Area of Mean Annual Ranges

Mean annual ranges calculated for the Yellowstone grizzly bear population showed substantial increases in area for each of three time periods through 1983, but declined somewhat when fourth period data, 1984-87, were incorporated (Fig. 15.6). Nevertheless, as a consequence of the continuous expansion of annual ranges of both sexes and all age classes, mean home range areas by 1987 were nearly five times larger for all bears than those calculated for bears during the 1963-68 period. The decline in mean range for females (Fig. 15.6) and, specifically, for adult females (Fig. 15.7), with inclusion of 1984-87 data, suggests that their ranges had begun to stabilize by the mid-1980s, approximately 15 years after closure of the major ecocenters. Although mean range areas for adult males

Fig. 15.6. Mean annual ranges of adult and nonadult grizzly bears (*n* = number of individuals), Yellowstone Ecosystem, 1963-68, 1975-80, 1975-83, and 1975-87.

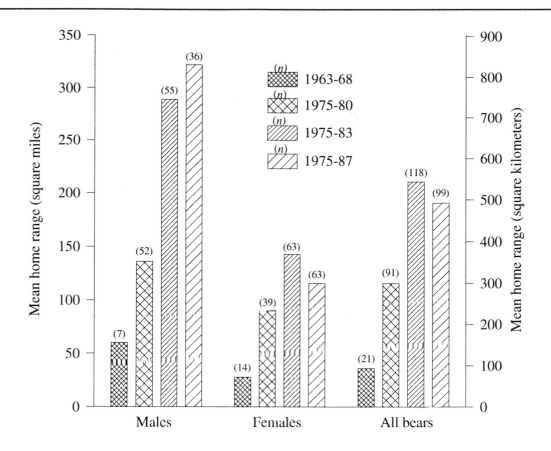

(Fig. 15.7), and for nonadult females and nonadult males (Fig. 15.8), increased over the same period, impending stability is implied by an apparent decline in the rates of range extension for the first two groups. On the other hand, nonadult males continued to expand their ranges substantially in search of food and security.

Male bears exhibited considerably larger mean annual ranges than did females (Fig. 15.7), except in the comparison of nonadults for the 1963-68 period (Fig. 15.8). But the nonadult comparison for that period was probably not valid because the annual range area for females was based on but a single individual.

Mean annual range area for adult males exceeded that for nonadult males over all periods (Figs. 15.7 and 15.8). Adult females exhibited larger mean range areas than did nonadult females for the 1975-80 and 1975-83 observation periods, but a smaller mean area for the overall 1975-87 period (Figs. 15.7 and 15.8). For the same reason of inadequate sample mentioned above, the comparison of female age classes for the 1963-68 period was not meaningful.

Thus, adult female home ranges appear to have begun stabilizing by the mid-1980s, some 15 years postclosure. Additionally, Blanchard and Knight (1991) reported a strong tendency for nonadult females to share substantial portions of the mother's home ranges into adulthood, a phenomenon earlier observed by F. C. Craighead, Jr. (1976) for the preclosure population and interpreted as indicative of stability. But because inadequacy of the food resource was, in the first place, the driving force behind home range expansions, and noting that adult males and nonadults of both sexes continued to expand their ranges as though food availability were still a limiting factor, we must question whether reduction in mean range sizes for adult females actually relates to stability at all. Plausibly, the females have been competitively inhibited from further range expansion and have settled for smaller home ranges where the food resource is but minimally adequate to meet biological requirements, and security for offspring is greater. This interpretation is certainly consistent with observations reported by Blanchard and Knight (1991) and Mattson *et al.* (1991a) that females with cubs tended to be com-

Fig. 15.7. Mean annual ranges of adult grizzly bears (*n* = number of individuals), Yellowstone Ecosystem, 1963-68, 1975-80, 1975-83, and 1975-87.

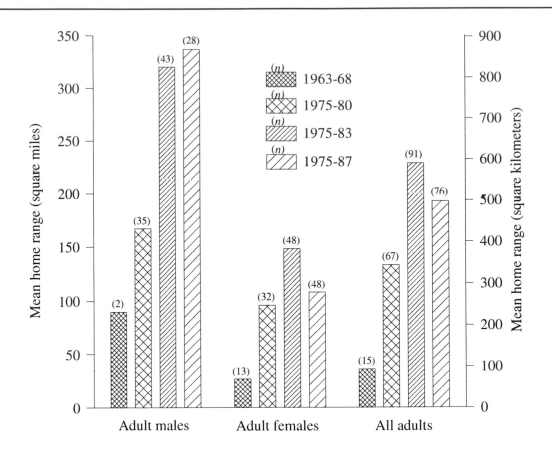

petitively excluded from crucial, prime seasonal foods such as berries (*Vaccinium* spp.) and pine seeds (*Pinus albicaulis*) during low-crop years.

Given the great fluxes, seasonally and annually, of food resources in the Yellowstone Ecosystem (Chapter 10; see also Mattson *et al.* 1991a), it is probable that adult female bears will experience periods of potentially severe nutritional stress that will drive them into conflict with human interests. Given this, and the fact that the ranges of other bears continued to expand, it is our opinion that the postclosure Yellowstone population had not stabilized relative to the food resource by the mid-1980s, a problem seen to be intensified by the fires of 1988.

Despite the fact that the size of mean annual home ranges of adult females began declining toward the end of the 1975-87 periods, we reemphasize that the mean area was four times that recorded for adult females during the 1963-68 period (Fig. 15.7). Quality and quantity of nutrition was not a concern for females with cubs in the earlier period because their home ranges consistently included one of the open-pit dumps within or adjacent to the Park. Indeed, Blanchard and Knight

(1991) found that adult females known to use the open-pit dump at Cooke City, Montana, for at least one month a year after 1975 tended to have smaller mean home ranges (214 square kilometers [83 mi²]) than did other adult females (425 square kilometers [164 mi²]). They also averaged significantly shorter distances traveled per day, exceeded average weight standards, and were highly productive. Mean daily movements were also significantly shorter for adult males that used the dump than for those that did not. Thus, the positive, stabilizing effect of an ecocenter in the relationship between the food resource and the health and ranging behavior of grizzlies is as clearly demonstrated for the postclosure period (1971-87) as for the preclosure period (1959-70).

POST-FIRE DISPERSAL OF THE YELLOWSTONE GRIZZLY BEAR POPULATION, 1988-91

After at least a decade of continuous increase in the mean of annual home ranges for all ages/ classes and both sexes, the Yellowstone grizzly

Fig. 15.8. Mean annual ranges of nonadult grizzly bears (*n* = number of individuals), Yellowstone Ecosystem, 1963-68, 1975-80, 1975-83, and 1975-87.

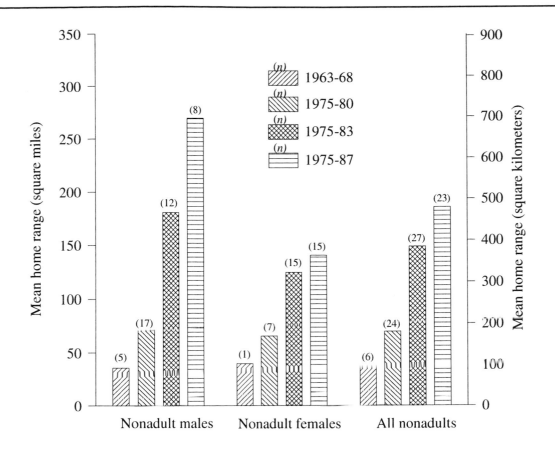

bear population showed signs of achieving, on the average, home ranges that provided many individuals in the population with minimal to adequate food resources and security. This process of stabilization, not yet completed by 1987, had, as noted earlier, required more than 15 years following the population disruption induced by closure of the open-pit dumps in and around the Park. There is evidence that population size was, and still is, considerably smaller than during the years prior to closure of the dumps, but particulars are lacking (Chapter 19). Whether stability eventually would have been achieved, under circumstances of habitat availability and the dynamics of the population existing by 1987, will never be known. The Yellowstone fires of 1988 altered basic characteristics of grizzly bear habitat within the Ecosystem, as well as responses of the bear population to its environment. The effects of this stochastic event will be felt for many years to come.

In 1988, wildfires swept over more than 400,000 hectares (985,000 acres) of Yellowstone Park, and an additional 170,000 hectares (416,000 acres) of the surrounding Ecosystem (Fig. 15.9, page 358). Forest canopy fires accounted for more than

half the total acreage burned (Division of Research and GIS Laboratory, Yellowstone Park 1989). There is no question that some food resources within the burns were immediately decimated, while others, such as ungulate carcasses and small rodents, were immediately available to the grizzly bears. Many of the grass and deep-rooted forb species recovered the following year where fire damage was light, but important food plants of the forest floor and canopy such as fruiting shrubs and whitebark pine (*Pinus albicaulis*) would be lost to the bears for several to many years. It is unlikely that the extent of damage to the whitebark pine seed resource and the impact on the grizzly bear population of its impairment will ever be entirely known. An IGBST survey in 1989 of 151 canopied feeding sites known to be frequented by grizzly bears showed that 54% of whitebark pine cover had been destroyed (Knight *et al.* 1990), and the researchers concluded the fires had severely reduced future opportunities for bear use of pine seeds at the sampled sites. The same canopy fires also destroyed alpine fir, successively setting the stage for high post-fire reseeding of whitebark pine (Despain *et al.* 1989). Unfortunately for the bears, the new forest would not produce cones for at least 40 years. Fewer pine seeds will be available to the grizzly bear population for many years to come.

With long-term reduction in availability of pine seeds, an unreliable, but very important food (Chapters 10 and 16), and the unpredictable, but extensive, shorter-term alteration wreaked by the fires on other food resources—rodents, berries, and some tubers—it was predictable that some bears, at least, would expand their home ranges in an attempt to secure an adequate, dependable food supply. Because reestablishment of traditional annual ranges could be expected to take several years, we would expect the IGBST to have followed through time those bears directly affected by loss to fire of some known portions of their traditional home ranges. Changes in their pre-fire range parameters could then be compared, contrasted, and finally, considered in context with post-fire parameters. Also, bears whose home ranges were not fire invaded or were affected by the fires in more indirect ways could be evaluated. Of particular interest would be any changes in size and conformation of annual ranges induced by this catastrophic event. Equally important would be any redistributions of ranges in the Yellowstone Ecosystem that might suggest how bears were meeting their needs outside areas of the most intense burns.

Unfortunately, the data available for our evaluation provide little opportunity for relating fire damage to home range size. For the period 1988-91, IGBST annual reports (Knight *et al.* 1989, 1990, 1991, 1992) and one other relevant publication (Blanchard and Knight 1990) provide no information on intensity and extent of burns on pre-fire-documented home ranges of individual bears, nor do they identify which bears were directly affected by the burns with follow-up on home range statistics for these bears from one year to the next. Secondly, home range statistics reported each year were for only a few bears selected according to unknown criteria from all those radiotracked, and only rarely were the same individuals identified in more than one of the four post-fire years. Furthermore, conclusions as to variation in home range parameters in a given year usually were expressed in terms of derivative data; e.g., significant difference (or none) from mean annual range values for sex and age cohorts from 1975-87. Lastly, locations, perimeter changes, or relocations of home ranges within the Ecosystem were at no time displayed or discussed in any of the publications.

Considering that annual variations in home ranges are expected "normal" responses to food availability and family composition (Craighead

1976), we can learn very little about fire-induced changes in range sizes or distributions without information that compares burned and unburned home ranges. Thus, in the absence of directly applicable data, we must determine whether changes in home range characteristics occurred in the years following the fires on the basis of ancillary information presented in the various IGBST reports.

In 1988, the year of the fires, Blanchard and Knight (1990) reported that 21 of 38 radio-monitored grizzly bears inhabited home ranges invaded by one or more of the fires. None of these bears was individually identified, nor were locations of, or extent of damage to, their home ranges revealed. Thirteen of the 21 grizzlies used the burned areas soon after passage of the fire fronts, feeding on carcasses of ungulates killed in the fires. Later, they fed on newly-emerged grass and sedge species, ants and various other insects, and root stocks of forbs. Home range information provided for only 6 of the 38 monitored bears, 5 females and a male, for the year 1988 showed range areas larger than the means of their pre-fire ranges, but because the differences were not statistically significant, Blanchard and Knight (1990) concluded that the fires had no apparent effect on annual range sizes. Considering that means of annual ranges tend to change gradually following a cataclysmic event, this conclusion seemed premature. The initial changes in range size reported for only a few bears in the year of the fires would hardly be revealing, particularly when there is no information provided on the extent to which the fires modified the vegetation within their home ranges.

That some bears were displaced and, perhaps, food-stressed in 1988 is indicated by the 18 captures or recaptures of 14 grizzlies in management actions between the first of August and the first of October of that year, compared to an average of 13 captures or recaptures per year for that same period from 1980 to 1987 (Blanchard and Knight 1990, Knight *et al.* 1988c). The drought and a poor pine seed crop in 1988, in combination with the fires, had unquestionably put the bear population in jeopardy. Many bears had already begun seeking foods associated with humans by July, before the fires compounded the drought-induced food shortage. At least 11 grizzlies were observed by residents near home sites and campgrounds east of the Park in Sunlight Basin, Wyoming (F. Hammond, WY Game and Fish Department, Cody: personal communication in Blanchard and Knight 1990). Thus, it is certainly puzzling that the researchers reported no indications of significant alterations in home range parameters.

In 1989, 26 of 40 radio-monitored bears were located on burned sites within their home ranges at some time during the year (Knight *et al.* 1990). Significantly, on-site investigation found sign of feeding activity at only 29% of the radiolocation sites in burned habitat compared with 54% at unburned sites. By fall, the majority of tracked bears were located in unburned whitebark pine habitat foraging for pine seeds. Once again, inexplicably, bear identities, locations of ranges, and proportions of them burned were not reported. Home range areas were reported for only 5 of the 40 bears monitored in 1989, a male and four females, and no mention was made of how their ranges related to the fires of the previous year. In all five cases, the range areas were found to be smaller than areas recorded for the bears in 1987. This is unexpected if the home ranges were invaded by fire, yet the authors offer no explanations.

If fire damage was slight within the five home ranges reported for 1989, then we might not expect major changes in home range size and perhaps would see some reduction of them compared to prior years if burned areas offered some initial feeding advantages. On the other hand, the whitebark pine cone crop of 1989 was, by far, the largest recorded since IGBST surveys began in 1980, whereas that for 1987, the year of comparison used above, was far below average. Noting that Blanchard (1990) has concluded that availability of pine seeds is a very powerful determinant of bear movement patterns (as also have we [Craighead *et al.* 1982]), it is logical to assume that reduction in range sizes, 1987 versus 1989, for these five bears was purely a reflection of the greater ease of securing pine seeds. Whatever influences the fires of 1988 may have had are obscured, except insofar as we can be confident that bears congregating in the whitebark pine habitat to take advantage of the bumper crop were not doing so in the burns and, very likely, their 1988 home ranges were not severely fire-invaded.

The IGBST annual reports for 1990 and 1991 did not distinguish radiolocations of bears made in burned versus unburned sites. Why comparative data for the three years 1988, 1989, and 1990 are not presented is puzzling. In fact, no mention was made of the fires or their lingering effects (Knight *et al.* 1991, 1992). Ranges reported for six grizzly bears tracked in 1990 did not differ significantly from cohort means for the 1975-87 period, and range areas for three were significantly smaller. As in previous reports, the reasons for selecting a small number of bears for evaluation from some larger number (unstated in Knight *et al.* 1991) of radiotracked individuals were unclear. Moreover,

1990 was an extraordinarily poor year for pine seeds, but there was no indication in the IGBST report (Knight *et al.* 1991) that this translated into bears foraging farther afield and, consequently, defining larger home ranges as Blanchard (1990) has suggested occurs during years of poor seed crops. The seven captures in management actions after 1 August in 1990, however, certainly are in keeping with findings in the same paper (Blanchard 1990) of significant inverse correlation between mean number of cones produced per survey transect and numbers of grizzly bears trapped in management actions after 1 August for 1980-86. Apparently, then, many grizzly bears were moving widely enough during late summer into fall of 1990 to encroach on human activities, but not enough, at least in the case of the six grizzlies selected for range analysis, to alter home range sizes appreciably.

The IGBST report for 1991 once again supported the perception of no consistent trend in home range parameters. Annual range areas recorded for 12 grizzly bears did not differ significantly from 1975-87 cohort means (Knight *et al.* 1992). The ranges for four other bears were significantly larger, but three of the four had been transported long distances from their traditional home ranges and were thought to have ranged widely in attempting to return. The fourth, an adult male, exhibited an annual range of 2140 square kilometers (826 mi²), the largest ever recorded for that cohort by the IGBST.

Overall, we find it impossible to conclude much about range sizes and patterns from IGBST home range data available for the 1988-91 post-fire period. But other kinds of data may shed some light on the movement dynamics of the population. For example, during 1989 when sites of feeding activity were categorized as to whether they had burned or not, unburned ones were preferred by the bears, particularly during the autumn when whitebark pine seeds were the crucial food (Knight *et al.* 1990). In their "white paper" regarding habitat preservation, Mattson and Servheen (1992) reported that bears used unburned habitat out of proportion to its availability and that the 1988 fires may have produced important long-term habitat losses.

We have assumed (and demonstrated earlier in this chapter) that the number of management captures is indirectly indicative of degree to which grizzly bears move into campgrounds and developed areas in search of food. There were 21 management captures and recaptures logged in "fire-year" 1988, 3 each in 1989 and 1991, and 13 in 1990, a year of very poor pine seeds production (Knight *et al.* 1989, 1990, 1991, 1992). It is perplexing that no

Table 15.8. Grizzly bear deaths inside and outside Yellowstone National Park during and following the 1988 fires, 1988-93.

Year	Number of deaths[a]	Number of deaths outside YNP				
		MT[b]	ID[b]	WY[b]	Total	% of all deaths
1988	14	2	1	8	11	78.6
1989	3	1	0	1	2	66.7
1990	9	1	0	6	7	77.8
1991	0	0	0	0	0	—
1992	8	3	0	3	6	75.0
1993	4	2	0	2	4	100.0
Total	38	9	1	20	30	78.9

[a] Includes only "probable" and "known" mortalities, and one removal to a zoo in 1990; not included are one "possible" mortality in 1988 and two in 1991.

[b] The states of Montana (MT), Idaho (ID), and Wyoming (WY).

significant changes in mean home range were reported in conjunction with the increased movement implied by captures in 1988 and 1991. Was it that only bears with little or no fire damage within their home ranges have been reported by the IGBST or, alternatively, that data were not obtained for those bears whose pre-fire home ranges were critically fire-invaded? If the former, then we must question the objectivity of the researchers and their reporting procedures or, if the latter, their lack of insight as to what was needed to explore the reactions of instrumented bears to post-fire conditions.

A major objective of IGBST research was to determine if and when the bear population stabilized following dump closures, and presumably, following catastrophic fire damage. Because establishment of secure home ranges is a strong indicator of population stability, we are amazed at the seeming failure of IGBST personnel to focus on this highly revealing biological parameter.

A last indirect indicator of how the bear population responded to food resources following the fires may be found in the annual death records for the 1988-93 post-fire period (Dood and Pac, Annual Grizzly Bear Mortality Reports 1988-93). Of 38 known or probable deaths, 30 (79%) were outside of the Park (Table 15.8). For the pre-fire period of 1980-87, 66% of 85 known or probable deaths were located outside the Park (Craighead *et al.* 1988b). This indicates that the demographic shift out of the Park observed during the 1980s accelerated as a result of the fires. For this to occur, bears of the post-fire period had to enlarge their home ranges.

Because information crucial to interpreting the response of the bear population to the 1988 fires—namely, the size and distribution of home ranges in the Yellowstone Ecosystem, the degree to which each suffered fire damage, and the extent of adjustment to such damage—was missing from the annual reports and publications of the IGBST, we have found it necessary to make inferences from indirect sources. It is logical to expect that the rapid expansion of annual ranges among all population cohorts during the pre-fire decade (described earlier in this chapter) continued, as bears annexed yet more of the National Forests beyond the Park's confines. For example, two reasonable responses by bears to partial fire-loss within their home ranges could have been 1) consolidation of their range outside of the burn (e. g., decline in annual range areas reported by the IGBST for some adult females) or 2) further extension of annual home ranges to encompass unburned portions of the Park and adjoining National Forests to the east and south (e. g., the moderate to large increase in annual range areas reported for some males). The implication of these scenarios is a distributional shift of bears to the ecosystem peripheries, a process anecdotally suggested by many observers to be occurring throughout the 1980s.

To explore the question of a shift in population distribution to the periphery of the Ecosystem, Blanchard *et al.* (1992) performed a detailed analysis that compared bear distribution during the 1973-79 period with that during the 1980-89 period. The Ecosystem was partitioned into a core area (the Park) and peripheral areas along all park boundaries. The entire Ecosystem was then gridded into 510 cells of 100 km² each, and data were recorded for each cell. Cautioning that non-uniformity of sampling throughout the study periods may have biased results somewhat, the authors concluded that they had insufficient evidence to determine whether overall bear distribution had shifted to the ecosystem peripheries. Nevertheless, evidence did seem sufficient to show that significantly fewer grid cells than expected were occupied in the Park during both time periods. Also, more cells than expected that were inhabited within the Park during the 1970s were left unoccupied during the 1980s. Occupancy was lower in all peripheral areas, compared to that within the Park, but higher to the east and south of the Park than to the north and west. No comparison was made between bear distribution at ecosystem peripheries pre- and post-fire, very important information, if the data were available. On the other hand, significantly more cells east of the Park were occupied by females with young during the 1980s (26%) than during the 1970s (16%).

The work of Blanchard *et al.* (1992) suggests a weak trend in which home ranges have shifted somewhat to eastern and southern portions of the Ecosystem, but does not show this conclusively, except for the shift of females with young into the region east of the Park. Even so, such a redistribution of the population would be consistent with the home range expansions we have described for the mid-1980s, with recently reported concentrations of bears such as those at moth aggregations and boneyards east and south of the Park (see Chapter 13), and with death locations for that period (see Table 15.8).

Protection for members of the grizzly population that have extended their ranges into areas at the periphery of the Ecosystem will be secondary to many other resource-use mandates. In terms of effective strategies for the recovery of this threatened species, it is essential that management policies consider mechanisms for restricting movement of bears beyond those areas where their habitat and safety are secured by law and management priority.

SUMMARY

It is important to keep in mind that within a span of less than 20 years, two catastrophic events in the Yellowstone Ecosystem—closure of the open-pit garbage dumps and extensive wildfire—had direct and widespread effects on the grizzly bear population. These events provided an unprecedented opportunity to employ well-designed scientific studies to measure and compare the responses of the bear population to the two very different cataclys-

mic events. Unfortunately, the Interagency researchers did not take advantage of these opportunities, at least insofar as the published record would indicate. Critical parameters amenable to investigation were the changes in distribution of bears in time and space and the movements associated with these changes. In spite of numerous scientific advisory committees (NAS, Leopold Committee, IGBST Steering Committee, etc.), no comprehensive research plans were developed to measure and integrate data on movement, distribution, changes in home range size, or home range stabilization for the years following either the ecocenter closures or the 1988 fires. As a result, our attempt to reconstruct movement, spatial distribution, and the response of grizzlies to drastic changes in proximate and ultimate food sources has been based largely on some solid, but "piecemeal" data and much indirect evidence. In spite of these difficulties, we constructed a generalized scenario of the spatial redistribution following each of the catastrophic events that we believe is reasonably close to reality.

The closures of the ecocenters deprived most of the ecosystem grizzly bear population of massive, localized food resources, but left the broad "natural" food base intact. The immediate response by bears to the dump closures was significant movement into campgrounds and developed areas adjacent to the ecocenters. During 1968, following the first step to close Rabbit Creek dump, movement was localized, with funneling of grizzlies into the nearby campgrounds and developed areas. After the closure of Rabbit Creek dump in 1969, movements to more distant ecocenters and to the periphery of the Ecosystem increased significantly. This was a period of extremely high mortality. Over a period of years, the surviving bears moved outward from these high density centers. The population showed some indications of stabilizing at reduced density levels as new home ranges evolved, or old ones were enlarged. This required a period in excess of 15 years, during which grizzlies made adjustments to exploit more fully the "natural" food base. The population, much reduced from that prior to ecocenter closures, was nutritionally challenged, pressing the ecosystem limits, and exposed to greater risk of human-caused death.

During the period 1975 through 1987, progressive increases occurred in the life and home range areas for most sex and age cohorts of the Yellowstone bear population. By 1987, mean home range areas were nearly five times larger for all bears than those calculated for bears during the 1963-68 period. Adult female home ranges began stabilizing by the mid-1980s, some 15 years postclosure. Some nonadult females shared home ranges with their mothers. Mean annual range area for adult males exceeded that for nonadult males over all periods.

In the case of the 1988 fires, the catastrophic habitat alterations deprived bears whose established home ranges were fire-invaded of major food resources and, at least temporarily, damaged the entire nutritional base of the Ecosystem. Most critical was fire damage to whitebark pine and fruit-bearing shrubs. Unfortunately, data were not available to relate fire damage to changes in home range size. The immediate response to the fires was movement of bears into fire-damaged areas to take advantage of displaced rodents, ungulate carcasses, and tubers, corms, and bulbs exposed where ground fires were light. This was followed by movements of bears to unburned areas, both within and beyond their previously established home ranges. Finally, dispersion continued into unburned peripheral portions of the Ecosystem, where bears attempted to reestablish home ranges that could supply basic biological needs. Mortality did not increase substantially, as it had done following closure of the ecocenters because protective legislation and management practices were already in place. How long it will take to complete spatial adjustments, and at what levels of population density, cannot be answered at this time. There is little doubt that stress on the bear population, severe for over two decades, will persist for another decade, or more, and perhaps even intensify.

16
Comparative Food Habits and Feeding
Behavior Following Ecocenter Closures

INTRODUCTION

Many of the measurable parameters that characterize the behavior and demographics of grizzly bear populations are influenced by the quality, quantity, and distribution of the food base, and its fluctuations seasonally and annually. Our analysis of food habits and feeding behavior of the bear population inhabiting the Yellowstone Ecosystem from 1960 to 1971 (Chapter 10) showed that the abundance of four important food categories (pine seeds, berries, small mammals, and carrion) varied over time, that never were all four food types observed to be widely abundant in the same year, and that garbage, commonly available as a fifth category, provided an extremely important foundation to overall nutrition. Garbage, in fact, was so important that the dump sites functioned as ecocenters affecting the behavior, movements, home ranges, and reproduction of the population. Predictably, the precipitous dump closures were enormously disruptive.

The key questions now, and the focus of this chapter, are: to what extent has the food base changed since 1970; how is it used by the bear population; and if a difference in use has developed, to what extent has it compensated for the loss of ecocenters? In brief, have fundamental changes occurred in the food base and feeding behaviors, and what are their potential impacts on the grizzly bear population? In the discussion to follow, we compare food resources and feeding behavior during the 1959-70 preclosure period with those during the 1973-91 postclosure period. The periods compared are constrained by periods of IGBST research effort, with occasional overlaps.

Comparison of the population's food habits from 1968-71 with food habits during 1973-74 (Mealey 1980) and 1977-81 (Knight et al. 1982) showed that, other than near elimination of garbage as a food resource, the Dietary Intake Index rankings of the major food categories did not change appreciably during the first decade of the postclosure period (Chapter 10 and Table 10.3). At the same

time, however, grizzly bear use of the forb food category appeared to have increased substantially in the 1973-74 and 1977-81 periods over that recorded during our study (Chapter 10, Tables 10.1 and 10.2). This relative increase suggests that the population's response to loss of the garbage resource was, at least in part, more intensive use of forbs. Our focus, then, was to determine whether this trend continued and whether other feeding trends developed.

Mattson et al. (1991a) presented cumulative food habits data for the period 1977-87 based on analysis of 3631 fecal samples (D. Mattson, pers. comm. 1992; Mattson et al. 1991a reports 3423 scats). Although some percentage of these were from black bears, rather than grizzly bears, the authors believed that the percentage of black bear scats was low. Another complication was that a substantial, but unknown, number of the scats gathered and analyzed for the 1977-81 period (Knight et al. 1982) was included among results representative of the total 1977-87 period. Also, raw data were not presented by Mattson et al. (1991a), and derivative data were arranged by month (April through October) as averages across the 11 years of study. To standardize sample size, they rejected a given month's fecal samples for inclusion in the across-years averaging if the samples numbered fewer than 20 in any given year. Their analyses provided across-years means for each of the seven months. We were unable to rearrange the food habits data to focus specifically on the years 1982 to 1987. Consequently, in using the cumulative values for the entire 1977-87 period, we ran some risk of redundancy with comparisons we made in Chapter 10.

In comparing food habits between the pre- and postclosure periods, we recognized a sampling disparity with regard to study length and sample sizes. For this reason, we confined our comparisons to seven major food categories to conform generally with those in Table 2 of Mattson et al. (1991a). This change from the 10 categories used in Chapter 10 (Tables 10.1, 10.2, 10.3, 10.6) was necessary because many of the data presented for the 1977-87 postclosure period were not amenable to reformulation. Moreover, data presented in Mattson *et al.* (1991a) and unpublished data provided by D. Mattson (pers. comm. 1992) showed that their Relative Volume Percent parameter was equivalent to our Composite Fecal Index parameter, divided by 100. Therefore, we restructured our food habits data (Table 10.1) to accommodate their seven-category format. The composite fecal indices were recalculated, excluding nonfood items, and divided by 100. Some of the food groupings in Table 2 of Mattson *et al.* (1991a) were combined, and some minor components were specifically placed in separate categories. We then used the monthly volume values (percent) from Table 1 of Mattson *et al.* (1991a) to allocate estimated values to the "Garbage" and "Forbs/Horsetail" food categories of 1.0% and 2.2%, respectively. The unpublished data from Mattson indicated that only an extremely small portion of the 2.2% value represented fecal volume of mushroom/puffball remains and, in view of the large Composite Fecal Index Value characterizing the Forbs/Horsetail category, we considered the volumetric contribution of these fungi to be substantively insignificant.

Despite absence of raw data, the study by Mattson *et al.* (1991a) is both comprehensive and useful, especially with regard to seasonal feeding patterns. These values allowed us to make certain between-study comparisons not otherwise possible.

SEASONAL FEEDING

Fecal analyses and observation showed that the bears' use of the seven food categories was related to plant phenology and seasonal cycles of ungulates, rodents, and insects. Use of individual species was additionally related to their abundance in any given year. For example, available biomass of certain invertebrate species, and of plant species such as yampa (*Perideridia gairdneri*), elk thistle (*Cirsium scariosum*), biscuit root (*Lomatium* spp.), spring-beauty (*Claytonia lanceolata*), and lupine (*Lupinus* spp.), varied from year to year. In years when a species was very prolific, bears tended to use it more extensively. Such fluctuations in annual resource availability were reflected in annual amounts of individual species identified in fecal remains.

In general, the emergent grasses and sedges showed heavy use in early spring and continued to be a relatively important food category into late summer in both the pre- and postclosure periods, 1959-70 (Fig. 10.1) and 1977-87 (Mattson *et al.* 1991a). The foliage and roots of various forb spe-

cies were consistently consumed from late spring throughout the summer months during both study periods, but a late season peak in use of roots was noted during the 1977-87 period. Certain forbs such as *Perideridia gairdneri* and *Lomatium* spp. were heavily exploited during both study periods. The specifically heavy use of others, such as *Claytonia* spp. during the 1968-71 period and *Taraxacum* spp. during the 1977-87 period, illustrates the bears' ability to concentrate feeding activity on a species periodically abundant, or to substitute one food plant for another, depending on availability.

The seasonal use of whitebark pine seeds (*Pinus albicaulis*) during our study (1968-71, Fig. 10.1) was quite similar to that documented during 1977-87 (Mattson *et al.* 1991a, 1994). The seeds were used most heavily in late summer and throughout the autumn, although Mattson *et al.* (1991a, 1994) also documented minor use from bears raiding red squirrel (*Tamiasciurus hudsonicus*) caches during June. Peak consumption of berries such as *Vaccinium* spp. and *Shepherdia canadensis* occurred in late summer months in both studies. Use of winter-killed or weakened ungulates was highest in April and May during both study periods, whereas heaviest predation on rodents such as voles (*Microtus* spp.), pocket gophers (*Thomomys talpoides*), and deer mice (*Peromyscus maniculatus*) tended to coincide with the recession of snow and the emergence of vegetation during May and June.

Although predation on elk (*Cervus elaphus*), and on spawning cutthroat trout (*Oncorhynchus clarki*) and long-nosed suckers (*Catostomus catostomus*), was not documented quantitatively during our study, such behavior was known to occur. We recorded predation on malnourished ungulates during April, May, and June, concomitant with use of winter-killed animals. Predation on elk calves, heaviest during June, dropped off rapidly in July as the calves gained in strength and endurance and were able to elude bears. In the autumn months, grizzlies killed and fed on mature bull elk weakened or injured by the rigors of the rut. These various periods of directed predatory activity coincided with those reported by Mattson *et al.* (1991a), and by workers specifically studying predatory behavior of grizzly bears at elk calving areas (French and French 1990) and more generally (Gunther and Renkin 1990) in Yellowstone Park.

During the period of our study, grizzlies were incidently observed in the vicinity of cutthroat trout spawning runs (e.g., those at West Thumb and Fishing Bridge) in June and July, but no direct predation was documented. Manpower constraints precluded gathering direct observation data or specifically seeking out fecal remains along the major rivers and the tributaries of Yellowstone Lake. Thus, data on intensity of use of fish during the 1959-71 period is unavailable for comparison with those presented by Mattson *et al.* (1991a) and other researchers (Reinhart 1988, 1990b, 1991; Reinhart and Mattson 1986, 1987, 1990), which suggest, based on indirect evidence, an increase in use of the fishery in the decade of the 1980s.

Grizzly bears tended to feed on insects most heavily in June and July during the 1968-71 period. Their use appeared to be opportunistic, concomitant with foraging for various preferred forbs in the alpine and subalpine zones. Remains most commonly identified from fecal samples were of ants (Formicidae), moths (Noctuidae), and ladybugs (Coccinellidae). Mattson *et al.* (1991a) recorded use of insects to be highest in July and August during 1977-87, with ants being especially prevalent in fecal samples in most years, and army cutworm moths (*Euxoa auxiliaris*: Noctuidae) making a first appearance in 1987.

The loss of garbage as a stable, high-quality dietary resource undoubtedly intensified the bears' use of the remaining food resources. Mattson *et al.* (1991a) reported that the grizzly population exhibited large-scale "diet-switching" from year to year, as some foods were unusually plentiful or others, extremely scarce. Such opportunistic behavior is certainly characteristic of the grizzly bear, but "diet-switching," as a term, implies a directed volition on the part of the bears that does not well describe the process we observed (Craighead *et al.* 1982).

In Chapter 10, we showed the relative abundance of major food resources from year to year (Figs. 10.7 and 10.9). We did not include forbs, as they were generally abundant and widely available during all years of the study. Certain forbs used by grizzlies, however, did vary greatly in abundance over periods of time, and these variations, as well as the response of the bear population to them, were noted. For example, yampa was extremely abundant during some years, as was elk thistle, and two species of biscuit root. Bear use was evident in their numerous "diggings," each of which might exceed several acres in extent. This type of feeding behavior prompted us to interpret the differences in forb use (Mealey 1980, Knight *et al.* 1982, and the present study) to be opportunistic feeding by the bears. If a choice forb was not plentiful, the bears simply turned to one that was.

Similarly, we observed changes in the grizzly's diet associated with the annual variations in abundance of berries and pine seeds (cf., Fig. 10.8), the most nutritious components of the plant food base.

Population-wide changes in diet, however, were not so well-defined during the preclosure period, perhaps as a result of the stabilizing influence of food resources available at the ecocenters.

With the exception of use of garbage, seasonal foraging by the Yellowstone grizzly bear population during 1977-87 (Mattson *et al.* 1991a) was quite similar to what we documented for the population during the period of the current study (Chapter 10) and during 1973-74 (Mealey 1980), as well as for the Scapegoat and Bob Marshall Wilderness Areas during 1972-76 (Craighead *et al.* 1982). We next compare the relative importance of the major food categories in the bears' diet as revealed by analysis of fecal remains.

COMPARATIVE USE OF FOOD RESOURCES

To identify changes in the roles of any of seven food categories in the overall diet of the grizzly bear population following closure of the open-pit dumps, we compared results of fecal analyses performed during 1977-87 with those from 1968-71 (Chapter 10). The fecal sampling effort during the preclosure period (487 samples over four years) was not nearly so extensive as that for the postclosure period (2568 samples over 11 years). Thus, we will focus on differences in use of the food categories, rather than individual species (except for whitebark pine seeds).

Garbage

For the 1968-71 period, the Garbage category of foods exhibited the highest Dietary Intake Index (DII) value among seven food categories (Table 16.1). We ask, then, how did use of food resources by the grizzly bear population change following the 1969-70 closures of the open-pit garbage dumps? Not surprisingly, comparison of DII values derived from fecal analysis data reported by Mattson *et al.* (1991a) for the 1977-87 period showed the Garbage category to be the least important of seven food categories (Table 16.1), and a corresponding shift toward more intensive use of one or more of the remaining food groups.

Whitebark Pine Seeds

Despite fluctuations in their availability seasonally and from year to year, pine seeds ranked highest among all food categories in terms of the DII for the 1977-87 period and were second only to the Garbage category during 1968-71 (Table 16.1). The relatively high values for average percentages of fecal volume indicate that pine seeds were preferentially consumed in large quantities during both study

Table 16.1. Composite Fecal Indices and Dietary Intake Indices for food categories identified from 487 and 2568 grizzly bear fecal samples collected in the Yellowstone Ecosystem, 1968-71 (Craighead Study) and 1977-87 (Mattson *et al.* 1991a), respectively.

Food category	Composite Fecal Index[a]		Correction factor[b]	Dietary Intake Index[c]		Dietary Intake Ranking	
	1968-71	1977-87		1968-71	1977-87	1968-71	1977-87
Garbage[d]	13.2	~1.0	2.30	30.4	2.3	1	7
Whitebark pine seeds	17.2	25.7	1.50	25.8	38.6	2	1
Vertebrate remains	4.2	9.3	4.00	16.8	37.2	3	2
Grasses and sedges	43.7	23.8	0.26	11.4	6.2	4	5
Fleshy fruits	5.9	4.3	0.93	5.5	4.0	5	6
Forbs and Horsetails[e]	2.9	28.0/~2.2	0.53	1.5	16.0	6	3
Invertebrate remains	0.2	5.8	1.10	0.2	6.4	7	4

[a] Composite Fecal Index values from a recombination of those values for food categories shown in Table 10.1 (divided by 100) for the 1968-71 study and from Table 2 (Relative Volume Percents) of Mattson *et al.* (1991a) for the 1977-87 study.

[b] Estimate for Garbage ≈ 40% foliage + 10% rootstocks + 40% processed carbohydrates + 10% meats
$$≈ 0.4 (0.26) + 0.1(0.93) + 0.4(1.5) + 0.1(15.0) ≈ 2.30.$$
Estimate for Forbs and Horsetails ≈ 60% foliage + 40% rootstocks ≈ 0.6(0.26) + 0.4(0.93) ≈ 0.53.

[c] Dietary Intake Index equals Composite Fecal Index multiplied by Correction factor.

[d] Estimated for 1977-87 period; see text.

[e] Forbs and Horsetails category combines use of roots and foliage and, for 1977-87, an estimated value for horsetail, including a minor contribution from mushrooms/puffballs; see text.

periods. When available, they were used to the near exclusion of other foods. These results are consistent with the findings of Knight *et al.* (1982) for the 1977-81 period, but they are not consistent with the findings of Mealey for 1973-74 (1973 was a year of poor whitebark pine seed production [Mealey 1980]).

Vertebrates

As prey or as carrion, meat ranked third in the Dietary Intake Index for our 1968-71 study, and second for the 1977-87 postclosure study, when the deceptively low representation of meat as fecal remains was adjusted to approximate actual ingestion (Table 16.1). In our evaluations of the fecal analysis data of Mealey (1980) and Knight *et al.* (1982) for the 1973-74 and 1977-81 periods, respectively (Chapter 10), the meat of vertebrates scored, by far, the highest DII value for the former period and was second highest for the latter (Table 10.2). Even when considered cautiously, the much larger DII value for the vertebrate food resource for the 1977-87 study (Table 16.1; 37.2 versus 16.8 for the 1968-71 study) would suggest that the disparity between their rankings (second vs. third, respectively) was a large one. However, the observed heavy use of winter-killed ungulates described earlier for the preclosure period (see Chapter 10) suggests that our fecal sample data may well have understated the level of use of this important component of the vertebrate food category.

Mattson *et al.* (1991a) showed substantial between-years variation in early-spring fecal volumes of ungulate remains during the 1977-87 period of study. This may imply that in some years, bears emerging from their dens found few winter kills. During the preclosure period, however, bears found spring carrion abundant. This difference, and possible causes for it, is discussed later in this chapter.

Forbs

During the 1977-87 period of study, the Forb and Horsetail food category was ranked third and appeared to be a much more important dietary resource for the bear population than it was during the 1968-71 period, when it ranked sixth out of seven food categories (Table 16.1). This is consistent with the trend toward increased use of the forb resource during the 1973-74 (Mealey 1980) and 1977-81 (Knight *et al.* 1982) periods (Compare Tables 10.1 and 10.2). The magnitude, timing, and progressive nature of the change indicate that the grizzly bear population responded to the loss of the garbage resource, in part, by increased use of forbs, both root stocks and foliage. The forbs, as a group, are of lower nutritional quality and digestibility than are preferred food types such as meats, pine seeds, or berries. Because of their seasonality and fluctuation in annual availability, we suggest that the high-quality foods were not available or attainable in sufficient quantity to offset the loss of garbage and that bears were then forced to greater use of forbs.

Invertebrates

The Invertebrate category ranked fourth in DII for the 1977-87 period, although its value is so close to that of the Grass and Sedge category that the rankings are essentially equivalent (Table 16.1). Nevertheless, comparative rankings between the two periods of study show that invertebrates were much more important in the diet of the bear population during the 1977-87 period than they were during 1968-71 (ranked last). This was due to a relatively high frequency of occurrence of ants in fecal samples, particularly during the summer of 1977 (Mattson *et al.* 1991a). Knight *et al.* (1982) suggested that the high frequency of ants reflected their relative availability in the habitat and opportunistic use by bears. Mattson *et al.* (1991a) concluded, however, that bears turned to ants when more preferred foods were scarce. Either explanation is certainly plausible, and they are not mutually exclusive. In any case, it seems unlikely that ants are a major nutritional source for grizzlies, but rather, may be sought for their acidic flavor (formic acid).

Grasses and Sedges

The Grass and Sedge food category ranked fourth in the DII during 1968-71 and fifth during 1977-87 (Table 16.1). Mealey (1980) considered grasses and sedges to be the most important food items consumed by the Yellowstone bear population during 1973-74 because relatively low protein digestibility (42.8%) was offset by high intake. The magnitude of use, however, would seem to reflect what was available, rather than what was preferred. When more nutritious food sources are readily available, it would be advantageous for grizzlies to make use of them, simply because of higher nutrient return for equal foraging effort. In early spring, as a matter of necessity, grizzlies seek the emerging grasses and sedges, which at that time are relatively high in protein, while other food plants have not yet emerged. Nevertheless, we suggest that in the total diet, grasses and sedges constitute a "survival" ration used until more nutritious, preferred foods of a seasonal nature become available. Before closure of the dumps, bears were observed to intensify their foraging on grasses and sedges, as well as on forbs, when pine seeds and berries were scarce.

Fruits

Fleshy fruits ranked fifth and sixth in dietary intake for the 1968-71 and 1977-87 periods, respectively (Table 16.1). They were represented primarily by the berries of *Vaccinium* spp. The reduced relative importance of fleshy fruit in the bears' diet during the postclosure years may indicate a failure in annual crops over the 1977-87 period (cf., Fig. 2, Mattson *et al.* 1991a) cumulatively greater than that during the 1968-71 period. Perhaps even more significant, if it occurred, would be an overall decline in berry production between the pre- and postclosure periods. There is evidence that such a decline may, indeed, have occurred in the northern half of the Park where, during the preclosure period, the Northern Elk Herd was relatively small, ranging in size from 5000 to 9000 animals (Houston 1982). In the early postclosure period, the herd averaged about 12,000 animals, but increased to about 16,500 for the period 1981-88 (Singer 1991; see Fig. 16.1). Kay (1995) and Kay and Wagner (1994) have demonstrated that ungulate browsing in recent times has virtually eliminated shrub seed production in areas frequented by wintering elk. Deciduous shrubs such as serviceberry (*Amelanchier alnifolia*), chokecherry (*Prunus virginiana*), and buffaloberry (*Shepherdia canadensis*), all found on the winter range of the Northern Elk Herd, have been especially hard hit. Shrubs measured inside exclosures throughout the Yellowstone Ecosystem produced up to 20,000 times more berries than did unprotected plants. Kay asserts that these circumstances are not unique to the Yellowstone. We suggest that the Interior Elk Herd has had similar impact on the growth and production of *Vaccinium* spp. (huckleberry and whortleberry) throughout the Central Plateau of Yellowstone.

Mattson *et al.* (1991a) stated that the Scapegoat grizzly bear population was one of only three North American brown bear populations that "ate as few fleshy fruits" as the 1977-87 Yellowstone population. Craighead *et al.* (1982), however, showed fleshy fruits to be considerably more important in the diet of the Scapegoat population than in that of the Yellowstone bears during the preclosure period (1959-70). Moreover, the data of Aune and Kasworm (1989) indicated that fleshy fruits (their shrub category) played a larger role in the diet of the bear population of the Bob Marshall-Scapegoat Wilderness complex (Montana) than they did for the 1977-87 Yellowstone population. For whatever reasons, the use of fleshy fruits by the Yellowstone grizzly bear population was unusually low during the 1977-87 period.

CHANGE IN USE OF FOOD RESOURCES

Annual variations in the nutritionally superior plant resources—berry species and the seeds of whitebark pine—characterized both study periods. The substantial disparity in rankings of the Forb and Horsetail category (Table 16.1) for the two study periods and, to some extent, the increased variety of forb species identified from fecal specimens suggest that bears relied much more heavily on this resource during 1977-87 than during the earlier study. At least in part, then, the grizzly bear population compensated for loss of garbage (estimated at 3.4 kcal/g; Table 10.6), by more fully exploiting the forb resource (estimated at 2.9 kcal/g; Tables 10.4 and 10.6). In this connection, it is important to point out that the nutritive disadvantage becomes even more pronounced in late summer and early fall when many of the forbs mature and often dry out, becoming less palatable and nutritious. Unfortunately, this is also the period when forbs serve as a critical substitute for scarcities of berries or whitebark pine seeds.

Fecal analyses indicated that animal-derived foods—the vertebrate and invertebrate categories combined and individually—composed a larger part of the grizzly bear population's diet during the 1977-87 period than during 1968-71. For the vertebrate category, this may have reflected increased use of any of the components—ungulate carcasses or prey such as ungulate calves and spawning fish—but, as we discuss later in this chapter, which one(s) remains unclear. Greater use of invertebrates during the 1977-87 study largely reflected a high use of ants.

In summary, it appears that the grizzly bears were able to compensate partially for loss of the garbage resource by intensifying their use of forbs and of certain vertebrates and invertebrates, as prey or carrion. But in the absence of garbage as a consistent nutritional foundation, the bears were obliged to range widely for adequate food and, in the process, expended more energy for less energy return. The energy intake-vs.-output balance would be especially critical during the hyperphagic period of late summer through early fall in years when berries or pine seeds were in short supply. It is likely that the substantial increases in annual range sizes recorded for the bear population during the '70s and '80s were a direct result of this heightened foraging activity (see Chapter 15; see also Blanchard and Knight 1991). The need to range widely in seeking adequate food directly threatens the security of the Yellowstone bear population, because grizzlies are forced to move from the Park where protection is high to multiple-use areas of the National Forests where it is not. Bears that range more widely experience higher rates of mortality (Chapters 15 and 18; see also Knight *et al.* 1988b).

STATUS OF FOOD PLANT RESOURCES

Grasses and Sedges

In Chapter 10, we showed the annual instability of some of the major grizzly bear food resources, so of particular interest were Mattson's findings of similar food resource instability during the postclosure period (Mattson *et al.* 1991a). The grasses and sedges, a relatively stable food source, however, were widely available for use in relatively large quantities during both periods. Mealey (1980) concluded that grasses and sedges were the most important food item in the diet of grizzlies during 1973-74. Use is normally constrained by quality that varies seasonally and annually. Dramatic declines in volume of graminoids in fecal samples have been documented during extremely dry years (e.g., 1977; Blanchard and Knight 1991). Given their relatively low digestibility and decline in moisture content and nutritional value as they mature, we have suggested that the grasses and sedges serve more as a "survival ration," a hedge against seasonal and annual scarcities of more preferred foods such as ungulate carcasses, berries, and pine seeds.

Abundance of the grasses and sedges, and their availability to grizzlies, may have varied between the two periods of study, but we do not have adequate data for comparison. Merrill *et al.* (1993) suggested that green phytomass is positively related to winter precipitation, and Frank and McNaughton (1992) presented evidence that ungulate grazing stimulates forage production. We observed variations in the abundance of specific components of the plant biomass, but are reasonably certain that available biomass of grasses and sedges exceeded demand in both periods. Great increases in ungulate numbers, however, no doubt heightened competition for this resource in certain areas of the Park (see Chapter 20). The high likelihood of substantial increases in ecosystem-wide biomass of the grasses and sedges as a result of the 1988 fires suggests that this greater abundance may offset the heavy ungulate grazing pressure. Even so, if, as we have hypothesized, the grasses and sedges serve mainly as a dietary substitute, use of this resource by bears will relate less to its abundance than to the availability of more preferred foods.

Forbs

As was the case with the grasses and sedges, we do not have the detailed data necessary for comparing the abundance of the forbs, as a group or as individual species, between the two periods of study. Abundance and edible biomass of individual species do vary over time in response to factors such as annual temperature and moisture regimes, ungulate grazing pressure, and fire. For example, we noted, as did Dave Mattson (pers. comm. 1994), periodic dramatic increases in yampa abundance. This species and others, such as biscuit root, springbeauty, elk thistle, and bear grass, had periodic increases in abundance and were preferentially consumed during those years. Taken over time, variations in the abundance of the constituent species evened-out such that the overall forb resource was a reasonably stable one.

As discussed earlier, comparison of the grizzly population's diet during 1968-71 with that during 1977-87 suggests that, with closure of the garbage dumps, the bears relied more heavily on the forbs as a food resource and were able to accommodate periodic scarcities of some species by exploiting more abundant ones. It is likely that the forbs, like the grasses and sedges, will increase in total biomass as a result of the 1988 fires and continue to serve this crucial function. Nevertheless, during the late summer-early autumn of years when berries or pine seeds are in short supply, many forb species are of reduced moisture content and nutritional quality and may meet only marginally the metabolic needs of grizzlies.

Berries and Fruits

Berry crops were small during 1977-87, and fleshy fruits were represented poorly in the diet of grizzly bears (Mattson *et al.* 1991a). In contrast, our fecal analyses for 1968-71 showed berries, and particularly *Vaccinium* spp., to be important in the overall diet of the bears, even though above average crops occurred in only 4 of 12 years from 1960 through 1971 (Tables 10.1 and 10.3). The extended period of poor crops, implied by the relative scarcity of fleshy fruits in 1977-87 fecal samples (except 1982; Mattson *et al.* 1991a) suggests a more-or-less gradual, ecosystem-wide decline in major berry-producers like the vacciniums. This may have been related to the near-climax condition of most of the ecosystem forests before the 1988 fires. The vacciniums produce more heavily in open-canopied forest habitat types. Mattson *et al.* (1991a) postulated that the availability of fleshy fruits was the major factor limiting bear numbers before the 1988 fires. If so, post-fire vegetation surveys should be done to determine the occurrence and distribution of *Vaccinium* spp., *Shepherdia canadensis*, and *Fragaria* spp. throughout the Ecosystem. Except where the burns were extremely hot, the vacciniums will recover and fruit biomass, improve over time.

Whitebark Pine Seeds

The seeds of whitebark pine have been shown, in all food habits analyses, to be a highly preferred and nutritious staple in the diet of the Yellowstone grizzlies. The extent to which the seeds are used is probably limited only by their availability, year to year. Unfortunately, nearly a quarter-million hectares (more than a half-million acres) of the 1988 burn in the Yellowstone Ecosystem consisted of forest canopy fires. The extent to which whitebark pine stands were eliminated and the future production of pine seeds impaired has serious implications for this food-plant resource.

Although the degree of fire damage to the pine seed resource may never be entirely known, a 1989 survey in the Park of 151 known grizzly bear feeding sites under forest canopy revealed that 54% of whitebark pine canopy cover had been eliminated (Knight *et al.* 1991). The researchers concluded the fires had severely reduced future opportunities for grizzly bears to use pine seeds in burned areas where whitebark pine had been a component of the forest. A post-fire bumper seed crop, higher than any of the nine previous years, served to mitigate this loss in 1989-90. Nevertheless, after surveying the burns, we concluded that both the short- and long-term role of whitebark pine seeds as a nutritional mainstay in the diet of the ecosystem bear population has been severely damaged.

Also of great concern is the decline of whitebark pine throughout the Northern Rockies as a result of white pine blister rust (*Cronartium ribicola*), mountain pine beetle (*Dendroctonus ponderosae*), wildfire, and successional replacement by more shade-tolerant conifers (Arno 1986, Arno and Hoff 1989, Keane and Arno 1993). The blister rust is a serious pathogen of whitebark pine in many regions of North America, but has been less so in the Yellowstone Ecosystem, where Carlson (1978) reported infection rates of zero to six percent. This may well have been due to the blister rust eradication programs pursued in the Park during the 1960s. In a more recent study, Keane and Arno (1993) concluded, "Based upon past history, blister rust biology, and evidence from [their] study, . . . there is a strong possibility of further southward and eastward advancement of heavy rust mortality in whitebark pine." In addition, the mountain pine beetle continues to pose a significant threat. Generally, the beetle first infests lodgepole pine (*Pinus contorta*) and moves into the whitebark pine, as the infestation becomes epidemic throughout the forests. Bartos and Gibson (1990) found 13,760 hectares (34,000 acres) of infected whitebark pine forest in Yellowstone in 1983, and although heavy outbreaks have not been reported recently, the threat remains clear.

Fig. 16.1. Summer and winter ranges for the Northern Elk Herd, Yellowstone Ecosystem (after Houston 1982), which averaged approximately 16,500 animals from 1981 through 1988 (Singer 1991).

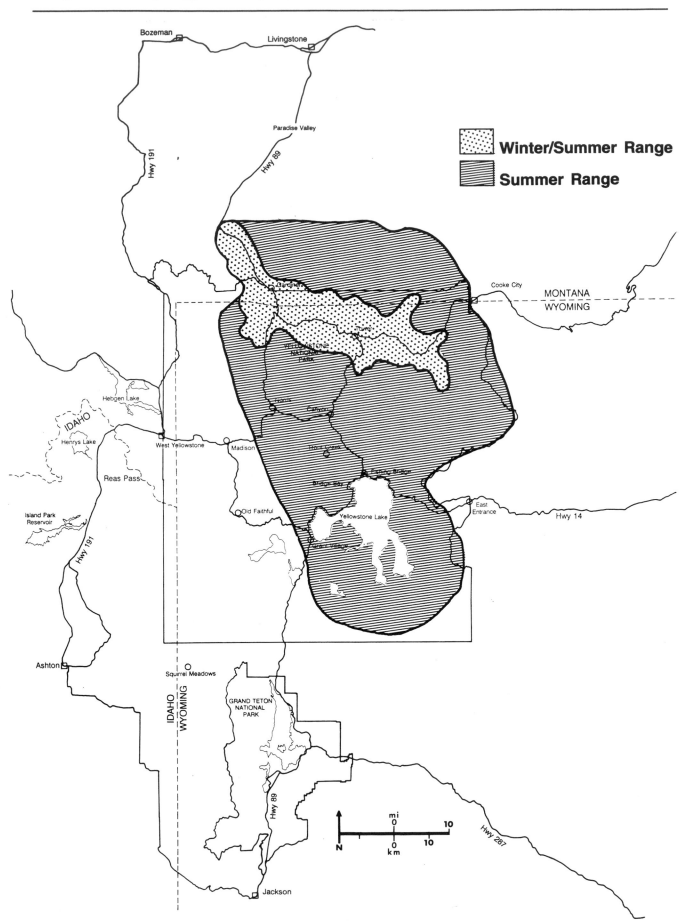

Arno (1986) contends that periodic wildfires are necessary for the regeneration of whitebark pine, where it is not the seral dominant (e.g., the subalpine fir forest habitat type). Also, fire is necessary to eliminate over-mature whitebark pine, which is most susceptible to the pine beetle, in stands where the pine is dominant (e.g., the whitebark pine forest habitat type). Keane *et al.* (1990) used computer modeling to show that suppression of fire for a 500-year period would result in loss of the shade-intolerant whitebark pine from the subalpine fir forest habitat type. If fire was brought into the scenario every 150 years, however, the pine regenerated quickly to pre-burn condition within 70-80 years. Conversely, Romme and Turner (1991) pointed out that, under circumstances of global warming, wildfires may hasten the inevitable loss of whitebark pine from lower elevation forest stands.

We know that fire is important in the ecology of whitebark pine, particularly where the pine is not a climax dominant. But therein lies the paradox relative to the role of pine seeds in the grizzly bears' diet. Whitebark pine is comparatively slow-growing in poorly developed soils, at cool temperatures, and for short growing seasons (Arno and Hoff 1989), and may require a century or more to mature enough to produce cones (Mattson and Jonkel 1990). Therefore, fire, pine beetles, or any other agents causing widespread loss of mature, cone-bearing trees, represent a serious threat to the nutritional stability of the Yellowstone grizzly bear population and, thereby, to its security.

Several recent IGBST studies, as well as our own, have provided a clear picture of just how critical the whitebark pine resource is to the security of the population (Knight *et al.* 1988b, Blanchard 1990, Blanchard and Knight 1991, Mattson *et al.* 1992a). Blanchard (1990) showed a statistically significant inverse relationship between annual cone production and numbers of marauding grizzlies subsequently trapped in management actions. Also, more bears, and particularly subadult males and females with cubs, were found to aggregate in close proximity to human facilities during years of poor pine seed crops and, as a result, mortality-at-large and for adult females was two times greater, and for subadult males, three times greater, during those years (Mattson *et al.* 1992a). Such a close association between grizzly bear behavior and the availability of pine seeds suggests a population under nutritional stress. Should the availability of the ecosystem-wide whitebark pine resource decline further, the effect on the bear population could be almost as catastrophic, given current conditions, as was the destruction of the ecocenters.

STATUS OF ANIMAL FOOD RESOURCES

As a prime source of protein and fat calories, vertebrates were quantitatively crucial in the diet of the Yellowstone grizzly bear population, both before and after closures of the garbage dumps. As we mentioned earlier, however, fecal analyses for the 1968-71 period may not have been as good an indicator of the use of ungulate carcasses, a substantial contributor to the vertebrate component of the food animal resource, as was direct observation of carcass use by bears over the 1960-72 period. To assess changes in feeding behavior between the periods, we next examine the comparative availability and use of each of the animal food resources—the scavenging of winter-killed ungulates, and the predation on ungulates, rodents, elk, fishes, and invertebrates.

Availability and Use of Ungulate Carcasses

The National Park Service regulated the growth of elk and bison populations for decades, keeping the numbers in check during the first half of the 20th century (Houston 1982). Thus, during this period, the herds did not seriously overpopulate their ranges, and the winter-kill of ungulates was relatively small.

Historically and within recent times, the elk herds inhabiting the Yellowstone Ecosystem have been the primary vertebrate foodbase for the ecosystem population of grizzly bears. Thus, the size and distribution of these herds are relevant to understanding the population dynamics, distribution, and food resource base of grizzly bears, as well as for other carnivores dependent on this food resource. Only two of the elk herds summering within Yellowstone National Park winter largely within its boundaries. Figure 16.1 shows the winter range of the Northern Elk Herd, where the carcasses of winter-killed elk are annually available to the ecosystem-wide population of grizzly bears, and Fig. 16.2, the winter range of the Interior Elk Herd where carcasses were available to grizzly bears denning on the Central Plateau—the Trout Creek and Rabbit Creek subpopulations (see Chapter 10). It is evident from Figs. 16.1 and 16.2 that elk constitute a potentially massive food resource for bears; however, they are largely available only in early spring as carrion and in early summer as prey (calves).

We now examine the increases occurring in the elk herds over a period of more than two decades and the parallel growth of the bison herds over the same time period. In the late 1960s, Yellowstone Park administrators adopted a policy of natural regulation (Cole 1969), and following the moratorium

Fig. 16.2. Summer and winter ranges for major elk herds in the Yellowstone Ecosystem (adapted from Singer 1991); numbers represent herd sizes. The Northern Elk Herd ranges are not included (see Fig. 16.1).

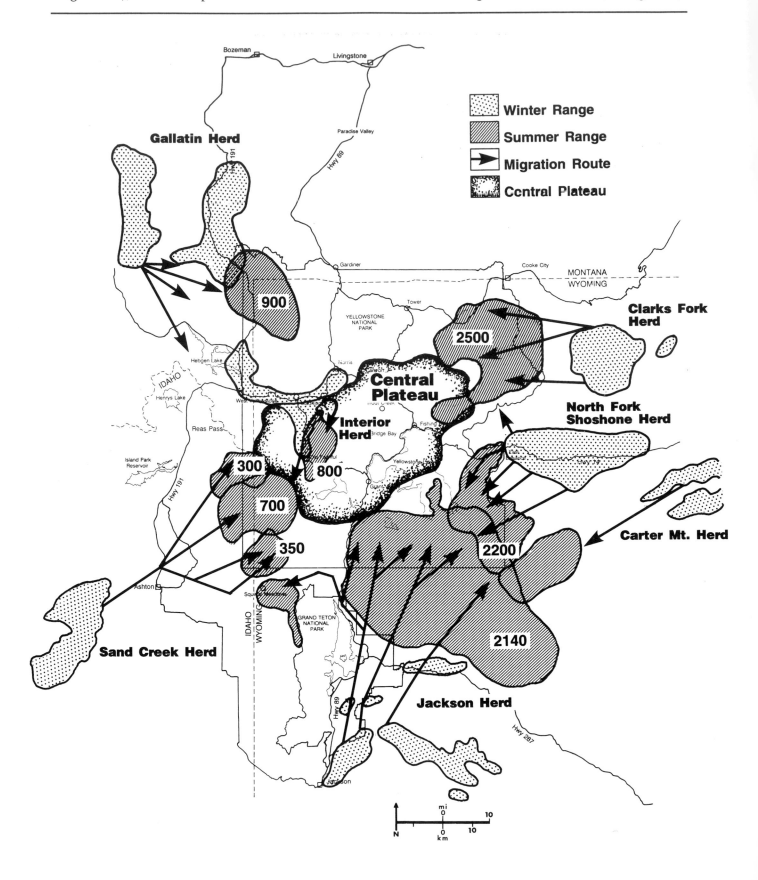

on elk removals in 1969, the elk herds grew steadily (Houston 1982). The Northern Elk Herd increased from an average of fewer than 5000 animals prior to 1968 to averages of about 9000 for the 1970-74 period, about 12,000 for the 1975-80 period (Houston 1982), and finally, about 16,500 for the 1981-88 period (Singer 1991). Singer (1991) estimated that the Interior Elk Herd (Madison-Firehole Herd) averaged about 800 animals on the summer and winter range during the 1965-88 period, consistent with an estimate of 475 animals made by Craighead *et al.* (1973a) for the wintering herd only. The Interior Herd, wintering in the snow-free thermal areas of the Central Plateau (Fig. 16.2)—Hayden Valley, Firehole, and Madison drainages—was isolated within that region by the deep snowpack at the higher elevations enclosing the plateau. As discussed in Chapter 10, the Trout Creek and Rabbit Creek subpopulations of grizzlies used elk carcasses nearly exclusively from the Interior Herd. Although many elk herds in the Yellowstone Ecosystem increased enormously in size from 1960 to 1988, the size of the Interior Elk Herd appears to have remained relatively constant throughout the period.

Over the years, bison populations have been much smaller than elk populations, and their distribution has been confined to the Park. Nevertheless, they have constituted a large foodbase for grizzly bears and, like elk, bison have been primarily available as winter kills and to a lesser degree as prey (calves).

Over the same period that the ecosystem-wide elk herds increased, the bison herds more than quadrupled from a mean population of 463 animals for the 1961-68 period (Meagher 1973) to a mean of approximately 2300 animals in 1988-89 for the combined Lamar, Mary Mountain, and Pelican herds (Thorne *et al.* 1991). Thus, more bison carcasses were available to Trout Creek and Rabbit Creek grizzlies in the Madison-Firehole area as a result of the growth of the Hayden Valley and Firehole bison herds, subunits of the Mary Mountain herd (Meagher 1973).

Our study to document the availability and use of carcasses by grizzly bears was confined to the Hayden Valley-Lake area and the Firehole, Rabbit Creek-Madison Junction areas (Fig. 10.3). The frequencies of occurrence and volumes of ungulate remains (expressed as percentages) were generally highest in fecal samples during April/May (Fig. 10.1), as was, also, the number of observations of carcass-feeding. This was also true for the postclosure period (Mattson *et al.* 1991a; D. Mattson, unpubl. data 1992). For bears emerg-

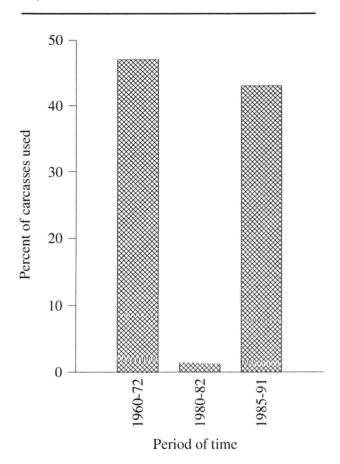

Fig. 16.3. Use of ungulate carcasses by grizzly bears, Yellowstone National Park, 1960-72, 1980-82, and 1985-91.

ing from winter sleep, carrion is especially important in metabolic recovery. This was true even when the dumps existed, because daily garbage availability, increasing as tourist numbers increased, was relatively meager during early spring. Before dump closure, grizzlies of the Trout Creek and Rabbit Creek subpopulations moved from winter dens to feed on ungulate carcasses, then to the dumps, which, by early June, were receiving garbage.

During April and May of 1971 and 1972, immediately following elimination of the ecocenters, we continued our carcass surveys through areas where carcass numbers and their observed use by bears had been high during the preclosure years (Fig. 10.3). Carcass use, although somewhat reduced from the preclosure period, continued to be reasonably high during 1971-72. A decade later, during April and May of 1980, 1981, and 1982, we conducted carcass surveys along the same transects (Fig. 10.3). Although 91 elk and 56 bison carcasses were examined, indirect evidence showed only two bison carcasses to have been used by grizzlies. No bears were sighted. This radical decline in carrion use rate (Fig. 16.3) would suggest a radical shift in grizzly bear distribution that precluded early spring carcass-feeding in historically heavily frequented areas or a reduction in the size of the Trout Creek and Rabbit Creek subpopulations beyond anything indicated by the recorded mortality (Craighead *et al.* 1988b) or both. The extreme mortality that occurred among the marked Trout Creek and Rabbit Creek bears (Chapter 18) and the movement and reestablishment of home ranges of surviving bears toward the periphery of the Park and Ecosystem (Chapter 15) would support such a scenario.

In the years following dump closures, overall availability of ungulate carcasses and the degree to which they were used by grizzlies were not specifically documented by IGBST until the 1980 and 1981 field seasons (Knight *et al.* 1981, 1982). The studies were not confined to any particular area of the Park, nor was methodology explained. Therefore, the data were not useful for our current comparison of carcass use which, as we have indicated earlier, was confined to thermal areas and drainages of the Central Plateau.

Studies that were comparable to our surveys were performed in the Firehole River drainage from 1985 through 1990 (Mattson and Henry [in Knight *et al.* 1987], Henry and Mattson [in Knight *et al.* 1988c], Green [in Knight *et al.* 1989], Mattson and Knight 1992). The surveys were intensive from mid-February through mid-May each year. For purposes of analyses, the authors presented data for 1989, reflecting the massive ungulate die-off following the 1988 fires, separately from those for 1985-88 and 1990-91. Carcass use by bears was rarely documented through direct observation, despite the extensive manhours devoted to off-road searches. Instead, use was inferred, as well as ranked for intensity (incidental use or moderate to heavy use), from bear scats and tracks and the nature and degree of carcass manipulation (e.g., major bones and skeletal structures disarticulated or cracked, everted hide, etc.). Mattson and Knight (1992), for purposes of their analyses, did not include carcasses incidentally used, but counted only those showing moderate to heavy use as being scavenged. Because we have no way of reconciling our direct observations of bears on carcasses with their indirect evidence and scaled distinctions, we have interpreted their data to show carcass use, or none, regardless of intensity.

A total of 646 carcasses was located over the 1985-91 period, of which 388 (60.1%) were elk and 258 (39.9%) were bison (Table 1 in Mattson and Knight 1992). By comparison, of 330 carcasses located on our transect surveys (Fig. 10.3) during 1960-72 and 1980-82, 244 (73.9%) were elk and 86 (26.1%) were bison (Table 10.7). Most of the disparity in carcass totals between our surveys and those of the IGBST is attributable to survey methodology and to the massive post-fire die-off of the herds during 1988 and 1989 (416 carcasses). The relatively larger proportions of bison carcasses recorded over the period of the Firehole study, but especially during the 1985-88 and 1990-91 periods, no doubt relate to the rapid growth of the bison herds during the 1980s (Fig. 16.1) and correspondingly greater competitive stress during winter months.

Mattson and Knight (1992) showed that frequency of carcass use, regardless of carcass species or year of survey, was significantly greater in the "backcountry" (>5 km from the Old Faithful developed area or >400 m from the Norris Junction-Old Faithful road) than in areas of higher human activity. During the 1985-88 and 1990-91 periods, frequency of carcass use by scavenging bears, regardless of carcass location, was similar for elk (52.0%) and bison (48.2%). This was much greater than that observed during the post-fire year of 1989 (16.5% of elk and 12.0% of bison). Undoubtedly, this large reduction in relative use in 1989 was primarily the result of the great number of carcasses available following the 1988 fire, rather than a decline in foraging bears, although the fire displaced many bears from their historic ranges. During the preclosure 1960-72 period, frequency of use of bison carcasses was much greater (82.8%) than for those of elk (39.2%). This disparity with the findings of Mattson and Knight (1992) is probably due less to species preference than to the relative size and number of bison carcasses compared to those of elk. In our surveys, all bison carcasses drew and held bears for extended periods, whereas some of the smaller and more numerous elk carcasses were quickly consumed and others, apparently in excess of the needs of the bears, were not even visited.

When frequency of carcass use was considered irrespective of species or location of carcasses, 50.4% of those available were used during the 1985-88 and 1990-91 periods, but use was only 14.5% in the post-fire year of 1989. Both of these values are vast improvements on the 1.4% frequency of use recorded for the 1980-82 period, and the overall use for the period (43%) approaches the rate for 1960-72 (Fig. 16.3). As discussed earlier, these results suggest that the subpopulations of grizzly bears inhabiting the Central Plateau were, for a time, much smaller than those recorded before 1970, or that bear distribution had shifted away from historically important areas of carcass availability. We have presented evidence earlier that both factors were involved. In brief, it would appear that the use of carcasses in the Central Plateau decreased drastically in the early 1980s, but gradually increased to near preclosure levels in the mid to late 1980s. We interpret this as an indication that the grizzly bear subpopulations inhabiting the Central Plateau are recovering from catastrophic declines following closure of the open-pit dumps. Carcass use should increase in the future, as the grizzlies of the Central Plateau increase in number.

With the exception of the Lamar and the lower Yellowstone Valleys, most winter-kills occurred on elk winter range in the relatively snow-free thermal areas (Craighead *et al.* 1972b). During the preclosure period, many bears wintered near the thermal areas and had ready access in the spring to carcasses that were not available to bears wintering elsewhere (Chapter 10). The bear population became smaller and more widely distributed following the closure of the dumps, and both denning sites and home ranges redistributed toward the park peripheries (Judd *et al.* 1986; see also Chapter 15). Thus, it is not surprising that, during the 1980s, bears, impeded by spring snow-depths of 1.2 to 2.4 meters (4-8 ft), made less use of thermal areas in the Central Plateau where winter-killed carcasses had been traditionally available to the Trout Creek and Rabbit Creek grizzlies, year after year.

The relatively high dietary intake value for vertebrate flesh (Table 16.1) suggests that this food resource played a greater role in the diet of the ecosystem-wide grizzly bear population during the 1980s, and it is plausible that a substantial part of this increase was from ungulate carcasses in areas other than the Central Plateau region; e.g., the Lamar and Yellowstone Valleys. We are concerned, however, that this important stable food source may be denied to many bears, as the "outward" redistribution of the population continues to and beyond the Ecosystem peripheries and away from ungulate wintering areas within the Park. Moreover, Singer (1991) estimated that the Northern and the Interior Elk Herds declined by about 40% and 50%, respectively, as a result of the 1988 fires and the unusually severe winter in 1989. As in the case of the dwindling pine seed resource, the unavailability of winter kill carcasses would have serious future nutritional implications for the bear population.

Predation

In Chapter 10, we described specific acts of predation as representative of what occurred during the preclosure period. For reasons stated, we were unable to quantify predation by grizzlies on any of their major prey species. During the postclosure period, IGBST members and other researchers have accumulated a respectable quantitative base for some prey species, especially elk, fish, and certain invertebrates. We next examine these data and attempt to determine whether they indicate major diet changes and a shift in resource use that could compensate for the loss of the preclosure ecocenters.

Rodents

Mattson *et al.* (1991a) reported from scat analysis that rodents constituted only a minor dietary component. As a percentage of diet volume, this is true. But his general conclusion is not consistent with our observations of rodent use during the 1968-71 period. We considered rodents a major food resource because of relative stability from year to year. The ready availability of rodents in spring, as snows recede, assures their seasonal importance in the grizzly bears' diet. No mention was made of rodent use in IGBST reports for the 1989 and 1990 field seasons (Knight *et al.* 1990, 1991). Use, following the 1988 fires, should have been extremely low, as predicted by Craighead (1991). Mattson (pers. comm. 1991) has noted that the IGBST recorded the greatest use of pocket gophers and voles in the unburned areas of Hayden and Pelican Valleys and very little use elsewhere in the Ecosystem. Continued development of grass-forb meadows on some areas covered originally by forest canopy before the 1988 fires will probably be conducive to increased populations and wider distributions of several rodent species in the future. But, at this time, we find little evidence to suggest an ecosystem-wide increase in this prey base during the postclosure period through the 1988 fires, or greater predation on them by bears.

Ungulates

Reports by Gunther and Renkin (1990) and French and French (1990) documented grizzlies successfully taking elk calves in 26 of 70 (37.1%) and 29 of 60 (48.3%) predatory episodes, respectively, in Yellowstone Park. Because their presentation does not show annual use and represents an intensive effort to record predatory acts, we can conclude only that elk calves were a seasonally important food, as in the past, not that they were more- or less-used during the two study periods. Predation on calves is limited to the period during and shortly after the calving season from late May through the end of June. By mid-July, bears are rarely successful in catching calves.

During our studies of the bear population before 1970, occurrence of predation on the calving elk herds was well-known, but we were unable to direct quantitative effort to its study. Thus, we have no quantifiable data as to frequency of predation or rates of success, nor have other workers published such data for that period. Nevertheless, whereas they acknowledged the absence of specific data for the pre-1970 period, French and French (1990) concluded; ". . . the lack of comprehensive documentation in the past leads [them] to believe that Yellowstone grizzlies are currently [1986-88] preying on elk calves more." We contend that this conclusion, as quoted, is scientifically insupportable.

Observations of successful predation on rut-weakened bull elk during autumn have not been quantified, but analyses of bear feces indicated that it occurred at some level (Mattson *et al.* 1991a). This food source probably constitutes a quantitatively minor component of the bears' diet, given the unpredictability of its abundance and spatial and temporal availability.

Because programs for reducing the elk herds ended in 1968, the herds have increased (Figs. 16.1 and 16.2), and the prey base for the grizzly has been strengthened. Indeed, in a density-dependent situation, bears may now be preying on elk calves relatively more frequently. But considering the seasonally-restricted availability of the resource and the relatively high cost-to-benefit ratio of the hunting effort, it is unlikely that increased predation on elk calves has appreciably bolstered the nutritional level of the Yellowstone grizzly population, or will in the future. Also, permitting population increases of elk, at or in excess of carrying-capacity in portions of the Ecosystem (Tyers 1981; Kay 1985, 1990), could cause unacceptable range damage and, in the long run, lower the density of elk and their calves as a prey base. This applies to bison, as well.

Fish

Studies showing grizzly bear use of spawning cutthroat trout and, to a lesser degree, long-nose suckers in Yellowstone Park have appeared intermittently since 1975. During the 1974 field season, six spawning streams were stated to support bear fishing activity, as evidenced by tracks, fish parts, and scats (Knight *et al.* 1975). A total of 17 grizzlies were observed in close proximity to Yellowstone Lake and its tributaries, but the annual report speculated that further study of tributary streams with good spawning runs was not likely to reveal additional use by grizzly bears. At the same time, W. P. Hoskins, in a study of Yellowstone Lake tributaries (IGBST unpubl. report 1975), indicated that 59 of 124 streams (48%) supported cutthroat trout spawning runs.

A decade later, studies of cutthroat trout and long-nose sucker spawning runs in the Yellowstone

Lake tributaries (Knight *et al.* 1986, 1987, 1988c; Reinhart and Mattson 1990) indicated that the distribution of the runs was unchanged from that reported by Hoskins. Reinhart and Mattson (1990) developed rather obscurely derived formulas to incorporate measures of scat content, volumes of fish carcass remains, and intensity of bear trail use with results of bear track analyses. From the resulting indices, they concluded that cutthroat trout were an important, high quality food to a substantial number of bears from May through July, noting that 55 of 59 (93%) of the spawning streams had evidence of bear activity, and 36 (61%) had conclusive evidence of bear fishing. In comparison, Hoskins (1975) indicated that only 17 (29%) and 11 (19%) of the streams in 1974 and 1975, respectively, showed activity or fishing. Reinhart and Mattson (1990) further concluded that a minimum of 44 individual bears were active on spawning streams during 1987, in spite of human activity at nearby visitor facilities. Thus, spawning fish do appear to have become a more important, more highly exploited food source for the grizzly population over the 11-year period from 1980-90, and use has apparently increased progressively throughout the postclosure period. Fish represented only one percent of the total volume of scats collected during 1977-87 (Table 2 in Mattson *et al.* 1991a) and comprised the smallest relative volume of nine major food type categories. But application of a correction multiplier of 40 (Hewitt 1989), to adjust fecal remains to approximate volume ingested, indicates that fish were a substantial component in the diet of some bears during the spawning season.

On the other hand, graphs depicting annual frequencies of occurrence of fish in fecal samples collected over the period of study (1977-87) indicate fish remains only in 1979, 1983, and 1985-87, with the largest representations, by far, in the latter three-year period (Fig. 2 in Mattson *et al.* 1991a). Considering that methodical procedures for detecting bear use of spawning fish were not well developed in the 1974-75 studies, and that fecal sample data in Mattson *et al.* (1991a) did not suggest annually increasing volumes of fish remains, it is conceivable that the apparent increase in use of the spawning trout resource reported by Reinhart and Mattson (1990) is entirely or partially a function of increased intensity of research effort. In other words, we did not specifically study bear use of the fishery before 1970, although we knew bears used it at some level. Likewise, data for the years 1970-73 and 1976-84 were either nonexistent or the result of less intensive study than that for 1985-87. Thus, as noted earlier regarding the extent of predation on elk calves, lack of baseline data for the preclosure period has prevented between-period comparisons. The data do indicate that fish are an important part of the vertebrate food resource available to bears, and a seasonally important item in their diet, but not that they were more important in the postclosure period than in the preclosure one.

Despite encouraging reports of improvement in the cutthroat trout fishery during the '70s and '80s (Gresswell 1988), we believe it unlikely that the dietary resource represented by spawning trout and suckers will be sufficient to offset the instability of crucial food categories such as berries and whitebark pine seeds. The future dietary role of fish is uncertain; for example, the effects of the 1988 fires on the trout fishery may not be well understood for many years and are certainly not predictable. Reinhart reported that 34 of 58 (59%) Yellowstone Lake tributaries documented as supporting trout spawning runs before 1988 were partially or wholly within burns (Reinhart 1990a). He showed that volumetric densities of spawners on burned streams in 1989 were significantly reduced ($P-0.05$, Wilcoxon) from densities on unburned streams, but that bear use of *or presence at* (our italics) all cutthroat trout spawning streams in 1989 did not differ significantly from use documented in 1985-87. This latter result was somewhat ambiguous because it was unclear from the report what values were compared in the analyses. But it is not unexpected that bears would return to sites of previously successful fishing, despite reduced opportunities in 1989. In other words, the bears were present, but the densities of fish were not.

The IGBST report for the 1990 field season (Knight *et al.* 1991) included no data regarding the fish resource, so post-fire trends in spawning runs will continue to be a source of concern. It is essential that the spawning runs and their use by bears be closely monitored, especially because continued expansion of visitor accommodations and ever-greater off-road visitation are likely to further impede the bears' access to the fishery (visited by humans as well as bears), whatever its condition.

Invertebrates

Although invertebrates, and particularly the insects, have been reported as components of the grizzly bears' diet in virtually all studies of their food habits, what species were used and their relative importance in the overall diet have been variable among studies, and from year to year within studies. The Dietary Intake Index Values we determined for the 1968-71 period were only 23.4 and 2.5 for Ants and Unidentified Invertebrates, respectively (Table 10.1), as compared with 93.0 and 18.6 during

1973-74 and 471.0 and 77.0 during 1977-81 (Table 10.2). These data imply that ants and other invertebrates were progressively more important in the diet of the Yellowstone population as years passed following closure of the dumps. Based on unpublished data provided by Mattson (pers. comm. 1992), relative percentage frequency of occurrence of ants in 2569 fecal samples collected from 1977 to 1987 was greater than that for vertebrate remains, an indication of greater availability or of preferred use throughout the feeding season. Because the majority of the volume of invertebrate remains over those years was from ants, it would appear that their importance as a diet item has remained high.

One possible explanation for the apparent increased importance of the ants is that bears were preferentially seeking ant colonies in partial compensation for reductions in the total biomass available to them after dump closure. Other explanations include bias in distributions of scat collection sites for any of the studies or increase in the abundance and availability of ant colonies and concomitant simple opportunism on the part of the bears. Assuming that fecal sampling was representative of the feeding habits of the grizzly population during all periods of study, we are inclined to believe that use of ants was not entirely opportunistic, but was preferential, particularly in years such as 1977 when ants comprised extraordinarily high percent volumes of fecal samples (Mattson *et al.* 1991a).

The strong tendency of grizzly bears to concentrate their feeding on locally or seasonally available insect aggregations is well-known in the food habits literature. Use of alpine aggregations of beetles such as the ladybird (a.k.a. ladybug; Coccinellidae) has been reported by Chapman *et al.* (1955) and Klaver *et al.* (1986); nests of colonizing wasps (Vespidae) reported by Mealey (1980) and Servheen (1983); and noctuids such as the army cutworm moth (*Euxoa auxiliaris*) concentrated in alpine areas of talus and scree reported by Chapman *et al.* (1955), Craighead *et al.* (1982), Servheen (1983), and Klaver *et al.* (1986).

More recently, studies in the Yellowstone Ecosystem have indicated that substantial numbers of grizzly bears were feeding on large aggregations of the army cutworm moth (Mattson *et al.* 1991b). Twelve suspected and six known sites of bear feeding activity were identified on talus and scree shoulders of mountain massifs, typically above 3000 meters (9850 ft) in elevation. The sites were in the Shoshone National Forest to the southeast and outside of park boundaries, but Mattson *et al.* (1991b) provided no maps indicating specific locations. Although all but one of the sites were discovered after 1986, the authors speculated that bear use had increased progressively from 1981 through 1989. Aerial surveys were used to determine numbers of bears frequenting the sites. The numbers located at an unspecified number of the feeding sites in any given year were 2, 1, 0, 8, 3, 10, 25, 41, and 51 for the years 1981 through 1989 (adapted from Table 2, Mattson *et al.* 1991b), but family groups were counted as observation of a single bear. That these were different bears within or between years was not indicated, nor was the number of feeding sites contributing to total observations in any of the years. Thus, the data may suggest a progressive increase in use of the moths by bears, but we cannot determine from this data set the number of individual bears involved, nor the extent to which bears may have aggregated at any one site, if at all. Nevertheless, based on a sample of 13 bear-years from four radio-collared grizzlies, Mattson *et al.* (1991b) did show that bears exhibited feeding site fidelity. None of the individuals shifted among sites within years, and only one bear shifted sites between years.

French *et al.* (1995 [in press]) present data that also suggest increased bear use of the moth aggregations in the Shoshone National Forest. During 1988-90, they located 10 moth aggregation areas, reported 22 grizzly bears feeding on moths for periods of one to eight days, and estimated a minimum of 93 different bears on the sites over a 10-day period.

If, as these studies suggest, so many bears are gathering in such restricted locales over two- to three-month periods, then the moth aggregations may portray many of the characteristics of secondary ecocenters (Chapter 13). Several aspects remain unclear, however. First, what proportion of the total bear population is involved and how many individual bears are active at each site? Second, how widely are sites distributed relative to each other, and is ingestion of moths at these sites quantitatively significant in a bear's annual diet? And finally, how annually dependable are the moth aggregations at particular sites?

As in past years, the role of invertebrates in the diet of the Yellowstone grizzlies is likely to continue to vary in importance according to fluctuations in invertebrate populations. For example, there was virtually no use of moths by bears in 1993 (presumably because there were few moths at the traditional feeding sites during the unusually cold, wet summer; David Mattson, pers. comm. 1994). Although the army cutworm moths, sometimes containing as much as 64% abdominal fat (Mattson *et al.* 1991b), are extremely nutritious and, in aggregations, represent efficiently obtainable food biomass, we think

it unlikely that their aggregations will significantly improve the nutritional opportunities of the grizzly population, as a whole. Richard Knight, leader of the IGBST, has reported army cutworm moth aggregations on peaks within the Park that were, at the time of the communication, unexploited by grizzlies (R. Knight, pers. comm. 1992). But situation of sites known to be used by bears in Shoshone National Forest outside park boundaries (Mattson *et al.* 1991b; French *et al.* 1995 [in press]) and, in some cases, outside federally-defined grizzly bear recovery boundaries (D. Mattson, pers. comm. 1992), raises, first, the issue of bear security. If bears in significant numbers are drawn out of those insular areas of the Yellowstone Ecosystem where their survival is a primary management concern, history suggests their survival is threatened. And second, does the intensive feeding on moths represent increased exploitation of an important food source, under-used in the preclosure period, or does it indicate a nutritionally stressed population resorting to an available, but quantitatively insignificant food resource?

NUTRITIONAL CONSIDERATIONS

The literature on growth and development in bears strongly suggests that a demonstrable decline in growth rates, and in weights and measures by sex and age class, can be taken as evidence for nutritional stress. To determine if such evidence existed for the Yellowstone bears, we compared weight parameters for grizzly bears of the preclosure population with those of postclosure bears.

Body weight and other morphometric measurements of grizzly-brown bears have been reported for numerous North American populations. Rausch (1953, 1963), in a re-evaluation of the species' systematics, examined the skulls of bears, both within and among populations, and concluded that the North American grizzly-brown bear was of a single species (*Ursus arctos*) showing considerable mensural variation, in some cases identifiable as a geographic cline. Strong correlations between weight and girth have been reported for grizzly bears (Russell *et al.* 1979, Glenn 1980, Nagy 1984, Blanchard 1987), as well as between weight and skull length (Stringham 1990b). Growth from birth is rapid in both sexes (Glenn 1980, Blanchard 1987). No sex-related variation in weights was demonstrated for spring cubs, but males were found to be heavier, on the average, than females by the age of one year among brown bears of the Alaskan Peninsula (Glenn 1980), by two years of age among Yellowstone grizzlies (Craighead and Mitchell 1982, Blanchard 1987), and not until four years of age among the bears of Kodiak Island (Troyer and Hensel 1969).

Whether cubs show sex-related weight differences at birth has not been documented, and sample size of older cubs and of yearlings is, for most studies, so small that significant mean differences have not been detected. In general, for age classes beyond yearling, male grizzlies are larger and do show higher growth rates than females of the same age class (Troyer and Hensel 1969, Glenn 1980, Craighead and Mitchell 1982, Kingsley *et al.* 1983, Blanchard 1987). Upper limits on growth and the age at which these are attained are species- and sex-specific, but do vary among populations and among individuals within populations (Troyer and Hensel 1969, Kingsley *et al.* 1983, Blanchard 1987).

The most available data on growth and development of bears in different populations are weights, either estimated or obtained by use of scales. It is obvious that weights will be influenced, independent of genetic determinants discussed above, by the quantity and quality of available food resources. For example, the weights of bears vary seasonally, as losses in body mass during winter sleep and after emergence are rapidly replaced over the summer and fall months. Males tend to replace their weights more rapidly than females. Most individuals attain maximum annual weights shortly before denning (Troyer and Hensel 1969, Craighead and Mitchell 1982, Kingsley *et al.* 1983, Blanchard 1987). The reproductive status of females appears to influence their growth and weight gains to the extent that reproductive needs compete for energy, resulting in reduced rates of growth and fat storage.

Several studies have explained differences in weights within populations of grizzly bears (Russell *et al.* 1979, Knight and Eberhardt 1985) or black bears (Rogers *et al.* 1976) on the basis of superior nutrition afforded by garbage. In 1983, at a conference on the pros and cons of supplementary feeding, the senior author presented a comparison of the growth and development of the pre- versus postclosure Yellowstone Park grizzly populations, concluding that weights for all age classes of both sexes had declined following dump closure (in Knight *et al.* 1983). Blanchard (1987) reported similar conclusions based on comparison of pre-1971 weights reported by Craighead and Mitchell (1982) with weights for Yellowstone grizzly bears recorded by the IGBST from 1975 to 1985. Subsequently, Stringham (1989, 1990b) analyzed these data, along with those from other studies, and showed that ecocenters such as dumps are correlated with not only increased rates of growth, but with higher reproductive rates, as well.

Indeed, if growth rate and body weight predict and strongly constrain reproductive performance and survivorship, as suggested by Stringham (1990b), then growth-related parameters may be particularly useful guides to population trends over time, provided the effects of age, sex, and other confounding variables can be controlled. In particular, growth data suitable to statistical analysis may provide better insight into the demographic consequences of ecocenter closure than do many difficult-to-estimate demographic measures such as litter size, productivity rate, age at first reproduction, etc. (see Chapter 17). Here, we statistically compare body weights of bears from the pre- and postclosure populations. Multivariate techniques are used to control for the effects of age, sex, and date of weighing. Post-1970 weights (1975–1985) were kindly supplied by R. Knight. Preclosure weights were obtained from 1959 to 1970 and represent the larger pool from which the weights reported by Craighead and Mitchell (1982) were taken.

A total of 518 weights (342 and 176 before and after closure, respectively) comprised the data set. All preclosure weights used for analysis were scale weights, whereas postclosure weights included both scale and estimated weights. Blanchard (1987) has documented the reliability of estimated weights in the postclosure period. Multiple regression was employed to test for pre- and postclosure difference in weight. We regressed weight on 10 independent variables (sex, the natural logarithm of age in years, date of weighing, study period, and all six first order interaction terms [sex by age, sex by weigh date, etc.]). Because body weight is strongly dependent on bear identity and multiple measures of weight were made for many bears, one randomly sampled weight per bear was used per regression. To make better use of the entire data set, however, we performed the random sampling process 20 times and applied the multiple regression procedure to each of these samples. We then selected from the 20 regressions the regression model that was most "typical" in terms of all 10 regression coefficients. For each term in the model, we (1) ranked the 20 coefficient values from small to large, (2) calculated the squared deviation of each rank from the median rank (equaled 10.5), and (3) averaged this deviation across all 10 regression terms for each of the 20 models. The model with the smallest value of this average was most typical in the sense that it contained fewest extreme coefficient values.

In Table 16.2, we have summarized the results of this most typical regression. All of the main variables (age, sex, weigh date, and study period) were involved in significant interactions. That is,

Table 16.2. Summary of "most typical" regression model relating weights of grizzly bears to their sex, age, weigh date, and study period (1959-70 versus 1975-85), Yellowstone National Park.

Sum of squares:[a]

Term	In	Out	F	P-value
Age*Study period	4,435,000	4,407,405	$(-2.769)^2$	0.0061
Sex	4,407,405	4,378,845	$(-2.779)^2$	0.0059
Study	4,378,845	4,368,483	$(1.652)^2$	0.0999
Weigh date*Study period	4,368,483	4,280,790	$(-4.788)^2$	0.0000
Weigh date	4,280,790	4,186,435	$(4.762)^2$	0.0000
Sex*Age	4,186,435	3,455,019	$(12.718)^2$	0.0000
Age	3,455,019	0	$(21.581)^2$	0.0000

Source of variation:

Term	D.F.	Sum of squares[a]	Mean square	F	P-value
SST	252	5,316,975	—		—
Regression	7	4,435,000	—		—
Age	1	3,455,019		465.7	<0.0001
Sex*Age	1	731,416		161.7	<0.0001
Weigh date	1	94,355		22.7	<0.0001
Weigh date*Study period	1	87,693		22.9	<0.0001
Study period	1	10,362		2.7	0.0999
Sex	1	28,560		7.7	0.0059
Age*Study period	1	27,595		7.7	0.0061
Error	245	881,975	3,599.9		

[a] SS attributable to term X; i.e., SS_x in minus SS_x out; terms removed in descending order of T values observed in model trial #4.

Regression equation: $R^2 = 0.834$

Weight = -41.1 + 0.63ln(AGE) + 0.55(SEX*AGE) + 0.23(DATE) - 0.30(DATE*STUDY) + 0.79(STUDY) - 0.13(SEX) - 0.16(AGE*STUDY).

bear age, sex, weigh date, and study period all influenced body weights, but did not do so independently of each other. Thus, we have restricted our interpretation to the interaction terms. Specifically, we found that (1) males weighed more than females, but this effect of sex on weight increased as bears aged (sex by age interaction); and (2) preclosure weights were greater than postclosure weights, but this effect on weight increased with weigh date (i.e., throughout the nonhibernating season; study by weigh date interaction) and with bear age (study by age interaction). All of these statistical effects may be interpreted as consequences of higher growth rates for males versus females during both studies, and higher growth rates, in general, for bears before versus after final ecocenter closures in 1970. Thus, as bear age increased, the discrepancy in weight between males and females also increased (sex-age interaction), as did the dis-

crepancy in weight between ecocenter and non-ecocenter bears (age-study interaction). Similarly, within a growing season (April - October), the weight advantage of ecocentered bears increased (weigh date-study interaction). Interestingly, the effect of sex on weight did not depend upon weigh date. This suggests that, within a season, males and females were growing at the same rate, and that the greater lifetime growth rates of males (implied by the significant interactive effect of sex and age; above) are not the result of differences in seasonal energy gain, but result from differences in male and female energy expenditure. For example, females may expend a relatively greater proportion of their seasonal energy gain in reproduction as opposed to growth, and vice versa for males.

In any case, these analyses clearly show that growth rates declined after destruction of the Yellowstone ecocenters. In Table 16.3, we have

Table 16.3. Average weights of grizzly bears (one weight per age category per bear) by age category, weigh date category, and sex for the 1959-70 and 1973-87 periods, Yellowstone National Park.

1959-70 population (*n* = 233)

Period	Sex	Cub	Yearling	Subadult (2-4 years)	Adult
< 15 Aug.	Male (*n*)	60.0 (24)	143.3 (21)	254.0 (34)	532.2 (9)
	Female (*n*)	56.2 (11)	121.9 (8)	202.9 (30)	301.0 (19)
≥ 15 Aug.	Male (*n*)	100.4 (8)	169.5 (15)	333.1 (18)	739.3 (4)
	Female (*n*)	90.5 (4)	146.2 (6)	259.2 (13)	377.2 (9)

1973-87 population (*n* = 110)

Period	Sex	Cub	Yearling	Subadult (2-4 years)	Adult
< 15 Aug.	Male (*n*)	43.3 (4)	134.0 (5)	246.8 (16)	426.5 (17)
	Female (*n*)	47.7 (3)	113.3 (3)	203.3 (9)	294.7 (7)
≥ 15 Aug.	Male (*n*)	83.4 (5)	168.3 (4)	310.9 (14)	385.0 (1)
	Female (*n*)	—	163.8 (4)	269.2 (6)	289.8 (12)

averaged weights into crude categories of age and weigh date to give some idea of the magnitude of the differences. Note that sample sizes are small for some categories, and that individuals placed together within categories must actually often differ substantially in age or weigh date. Nonetheless, for 12 of 15 categories of age, sex, and weigh date, preclosure weights were greater than postclosure weights.

The weight analysis provides no evidence that, by 1985, the per capita food resource in the Yellowstone Ecosystem was comparable to pre-1971 levels. Although total population size must have declined substantially after 1970, these results imply that less net energy was available per bear during this period than was the case before ecocenter closure, that switching from garbage to greater use of other foods was not nutritionally compensatory, and that growth rates suffered accordingly.

THE NUTRITIONAL IMPERATIVE

Grizzly bears that inhabit the Yellowstone Ecosystem are jeopardized to the extent that food resources, and particularly, certain foods that impart the greatest nutritive value in exchange for the least foraging effort, are subject to unpredictable variation in quantity and quality. Garbage dumps

(ecocenters) greatly mitigated this jeopardy before 1970. Indeed, Mattson *et al.* (1991a) hypothesized that

. . . density of grizzly bears in the Yellowstone area is limited in large measure by the lack of fleshy fruits, though this was mitigated by the availability of edible human refuse during the three decades prior to 1970. Irregularity of pine seed production is probably the next most limiting habitat factor, primarily because during years when pine seed production is low, grizzly bears come into greater conflict with humans (Knight *et al.* 1988b) and more bears are killed (Knight *et al.* 1988a).

In comparing the preclosure to the postclosure periods, we have shown substantial expansion of home ranges among all sex and age classes (Chapter 15 and Blanchard and Knight 1991) and reduced growth rates in all sex and age classes (above, in this chapter, and Blanchard 1987). These changes indicate a postclosure population frequently stressed by fluctuations in food availability and open to serious destabilization through stochastic incidents. Assertions of increased predation on elk calves, fishing on spawning runs, and harvesting of army cutworm moths are encouraging indicators, though not conclusive evidence, that the population is compensating for loss of the garbage resource. To offset the nutritional loss of the ecocenters, each of these food sources must be annually dependable, and each must make substantial quantitative contributions to the bears' food intake. There is no rigorous evidence of this, nor that, even in aggregate, they constitute a foodbase comparable to that once provided by the ecocenters. Each of these nutritional resources is prone to fluctuation in annual availability or is of unknown long-term stability.

It appears likely that the bear population, in ranging more widely for food, is also decentralizing from protection afforded by the Park. If so, management strategies should be developed that provide greater protection for the grizzly bear in the outer reaches of the Ecosystem where multiple-use activities currently have priority. Unfortunately, given the aggressive nature of the bear and the necessity of pursuing an extractive-use economy, adequate protective measures may be difficult to devise and enforce.

In short, we see no convincing evidence that the postclosure grizzly bear population has bridged the nutritional gap created when the ecocenters were destroyed. The bears have changed some of their feeding habits, used some resources more effectively, and altered their distribution and use of space throughout the Ecosystem. But, in spite of this, the present population emerges as a nutritionally-challenged population, rather than a nutritionally-stabilized one. At best, the population has only partially recovered. Analysis clearly showed that weights and growth rates declined after destruction of the Yellowstone ecocenters and have remained depressed, and also that bears required larger per capita foraging areas (home ranges). These results imply that the decline in per capita resources between the two study periods has not been compensated by the use of "new" food resources or altered feeding habits. Whether there has been sufficient compensation to assure a viable population sustaining itself at a lowered nutritional level remains to be seen.

SUMMARY

The focus of this chapter was to determine if fundamental changes occurred in the foodbase and feeding habits of grizzly bears following closure of the garbage dumps, and if a difference in resource use had compensated for the loss of these. To accomplish this, we compared bear food habits documented during the 1959-70 preclosure period with those documented over various timespans of the 1973-91 postclosure period.

Seasonal foraging behavior in the late postclosure period (1977-87) was quite similar to that documented for the population in the years immediately following closure (1973-74; see Chapter 10), as well as that in the earlier preclosure period (1959-70). It was related to plant phenology and seasonal cycles of ungulates, rodents, and insects. Although the seasonality of grizzly bear feeding behavior showed little change, the loss of the garbage resource (ecocenters) tended to intensify the use of the remaining resources. This was most evident in the bears' use of specific forbs abundant from one year to the next. It was less evident in the use of calorie-rich preferred foods such as berries, pine seeds, small mammals, and large vertebrates.

Changes in the relative use of seven major food categories were revealed by fecal analysis modified to reflect food intake and expressed as a Dietary Intake Index (DII). The DII values reasonably approximated quantitative use of the major food resources and, we believe, provided a good basis for the comparison of food habits. As expected, garbage was the least important of the seven food categories during 1977-87. Whitebark pine seeds scored the highest DII values, followed by vertebrates, forbs, invertebrates, grasses and sedges, and fleshy fruits. The major detectable differences in intensity of use between periods were an increased use of the forb, vertebrate, and invertebrate food resources.

When the six nongarbage-derived food categories were combined in terms of being plant-derived or animal-derived, the food plant resource was found to have contributed more than half of the grizzly bears' diet in both the 1968-71 and the 1977-87 studies (72.2% and 59.8%, respectively). In terms of both individual and combined DII values, however, the vertebrate and invertebrate foods appeared to have increased in dietary importance, compared to the 1968-71 study. To assess this change, we examined the comparative availability and use of the animal resource components—ungulate carrion and rodents, elk, fishes, and invertebrates acquired by predation. The tremendous growth of elk and bison populations following closure of the ecocenters greatly augmented the potential ungulate foodbase for grizzly bears. Annual availability of this food resource, however, was essentially limited to short seasonal periods of carrion-feeding and predation on calves. Carcass surveys completed during 1981-82 indicated a drastic decline in use-rate of ungulate carrion, compared to the use-rates a decade earlier and during the preclosure period. Evidence suggests that the decline in carcass use in the Central Plateau during the 1970s and early 1980s was the result of a shift in grizzly bear distribution away from this region and an ecosystem-wide decline in bear numbers from preclosure levels, but particularly a drastic decline in the ecocentered bear subpopulation of the Central Plateau. Carrion use began to increase from 1985 to 1991, approaching the usage rates recorded along our transects for the period 1960-72. This suggests a gradual recovery of the grizzly bear subpopulation of the Central Plateau so devastated following ecocenter closures (see Chapter 18).

There is no indication that predation on rodents changed significantly from the pre- to the postclosure periods. Nevertheless, with an increase in suitable grass-forb habitat likely, following the 1988 fires, rodents may increase in the next decade.

Lack of baseline data on use of fish for the preclosure period precluded between-period comparisons. Fish are an important part of the vertebrate food resource and, as such, are a quantifiable item in the grizzlies' diet, but evidence does not confirm whether there was an important difference in use between the pre- and postclosure periods. Recent surveys have indicated that restrictive regulations on angling, over the last decade, have greatly improved numbers and age structure of the Yellowstone cutthroat trout population. But it is premature to conclude that the dietary resource represented by spawning fish will expand enough or be sufficiently used by grizzlies to offset the instability of crucial, high-calorie food categories such as berries and whitebark pine seeds.

Invertebrates, such as ants, beetles, wasps, and noctuid moths, appeared to become progressively more important in the grizzlies' diet as years passed following closure of the dumps. These foods may be preferentially sought for taste and ease of capture, as well as for what is sometimes, in the case of moths, substantial protein and fat content. The concentrations of noctuid moths at high elevations have drawn bear aggregations and thus represent efficiently obtainable food biomass. It remains to be seen, however, whether moth aggregations, of unknown annual dependability, will significantly improve the bear population's nutritional level or adequately compensate for loss of the ecocenters.

To offset the nutritional loss of the ecocenters, foods such as elk calves, spawning trout, ants, and army cutworm moths must provide annually and seasonally dependable food, and each must make substantial contributions to the bears' food intake. As yet, there is no convincing evidence of this, nor that in aggregate, they constitute a food base comparable to that formerly available at the ecocenters.

Weight analyses of grizzly bear sex and age classes showed that growth rates declined after destruction of the ecocenters. There was no evidence that by the mid-1980s a new equilibrium had been established between bear numbers and food resources such that the per capita food resources available to the bear population were comparable to the pre-1970 levels. Rather, the results implied that, although total population size must have dropped substantially after 1970, there was still less nutritional energy available per bear during the postclosure period than was the case before ecocenter closure, and that growth rates had suffered accordingly.

In conclusion, we found no evidence that the postclosure bear population had found and exploited nutritional resources comparable to those lost when the ecocenters were destroyed. Bears changed some of their feeding habits, used some resources more effectively, and altered their distribution and use of space throughout the Ecosystem. But in spite of this, the population emerged as a nutritionally-pressured population, rather than a nutritionally stable one. Its encroachment into multiple-use areas in search of food resources on the periphery of the Ecosystem and beyond greatly increases the potential for bear-human encounters and, thereby, for increased bear deaths. This is of great concern because annual human-caused mortality is the one factor most amenable to management actions, and its control is a straightforward way of preserving demographic viability.

Table 17.1. Mean ages at which marked, female grizzly bears were observed or inferred to whelp first litters, Yellowstone Ecosystem, 1959-70 and 1975-89.

Period of observation	Observed first whelping		Inferred first whelping[a]		Observed/Inferred first whelping	
	No. of females	Mean age in yrs.	No. of females	Mean age in yrs.	No. of females	Mean age in yrs.
1959-70	15	5.87	17	6.41	32	6.16
1975-81	6	5.67	1	5.00	7	5.57
1982-89	7	5.71	13	5.77	20	5.75
1975-89	13	5.69	14	5.71	27	5.70

[a] "Interred whelping" applies to females for which observation over the period extending from subadulthood through adulthood revealed no whelping; for these females, whelping was assumed to occur in the year following that of the last observation.

Table 17.2. Hierarchical log-linear analysis of the association between fertility, maternal age, and study period among female grizzly bears, Yellowstone Ecosystem, 1959-70 and 1975-89.

Source of variation	DF	Chi-square change	P-value
Fertility × age	3	19.3	0.002
Fertility × study period	1	2.7	0.10
Age × study period	3	4.2	0.24
Fertility × age × study period	3	2.2	0.53

Table 17.3. Mean fertility of female grizzly bears available for breeding[a] according to female age and study period, Yellowstone Ecosystem, 1959-70 and 1975-89.

Study period	Female fertility[b] by age category (yrs) (No. of cubs)				Mean fertility
	Young age	Prime age		Old age	All ages
	4-8	9-14	15-20	21-25	
1959-70	0.23	0.64	0.69	0.50	0.39
(n)	(79)	(28)	(16)	(4)	(127)
1975-81	0.40	0.40	—	0.33	0.39
(n)	(15)	(5)	—	(3)	(23)
1982-89	0.31	0.67	1.00	—	0.40
(n)	(32)	(9)	(1)	—	(42)
1975-89	0.34	0.57	1.00	0.33	0.40
(n)	(47)	(14)	(1)	(3)	(65)

[a] At least four years of age and without dependent offspring.
[b] Proportion of adult females without dependent offspring producing a litter the following year.

Table 17.6. Mean sizes of 119 litters (±SE) produced by 72 known-age female grizzly bears tabulated according to study period and age, Yellowstone Ecosystem, 1959-70 and 1975-89.

Study period	Mean litter size by maternal age category (No. of litters)				
	Young age	Prime age		Old age	
	4-8	9-14	15-20	21-25	All ages
1959-70 (*n*)	1.84±0.15 (25)	2.43±0.13 (21)	2.38±0.27 (13)	1.50±0.50 (2)	2.15±0.10 (61)
1975-81 (*n*)	1.91±0.16 (11)	2.33±0.33 (6)	2.00±0.00 (2)	1.33±0.33 (3)	1.95±0.14 (22)
1982-89 (*n*)	2.05±0.12 (19)	2.50±0.15 (12)	2.00±0.32 (5)	— —	2.19±0.10 (36)
1975-89 (*n*)	2.00±0.10 (30)	2.44±0.15 (18)	2.00±0.22 (7)	1.33±0.33 (3)	2.10±0.08 (58)

Table 17.7. Analysis of variance in litter size relative to ecocenter status (present or absent) and age of female grizzly bear, Yellowstone National Park, 1959-70 and 1975-89.

Source of variation	DF	SS	MS	F ratio	P-value
Ecocenter status	1	0.41	0.41	0.13	0.72
Female age	3	5.48	1.83	5.89	0.001
Interaction	3	1.77	0.59	1.91	0.14
Residual	64	19.84	0.31	—	—

Note: Based on 119 litters born to 72 different females; table shows the results of the ANOVA giving the median *P*-value for effect of study among 13 such tests based on 13 random samples of one litter per female; effect of age was significant in all 13 tests.

For our analysis of weaning age in the pre- and postclosure periods, we restricted the sample to cases meeting the following criteria: (1) mothers marked or recognizable, (2) litters with at least one marked cub surviving to weaning, and (3) families for which both the female and marked offspring were observed pre- and postweaning. The latter criterion means that weaning was never inferred from observation of only the mother without offspring or only the offspring without mother. Such cases could reflect death of the offspring or mother, respectively, rather than weaning.

Tabulation of 44 grizzly bear litters showed that 31.8% were weaned as yearlings versus 68.2% as 2-year-olds. The proportion of litters weaned at two years versus one was virtually identical, pre- and postclosure (Table 17.8). Nevertheless, because female age was such a strong determinant of fertility and litter size, we proceeded to test for effects

Table 17.8. Percent of grizzly bear litters weaned as yearlings versus 2-year-olds, Yellowstone Ecosystem, 1959-70 and 1975-89.

Study period	Offspring age at weaning (No. of litters)	
	Yearling	2-year-old
1959-70 (*n*)	32.3% (10)	67.7% (21)
1975-89 (*n*)	30.8% (4)	69.2% (9)
Both studies (*n*)	31.8% (14)	68.2% (30)

Table 17.9. Length of litter-to-litter reproductive cycles exhibited by adult (not necessarily known-age) female grizzly bears, Yellowstone Ecosystem, 1959-70 and 1975-89.

Cycle length in years	Percent (no.) of cycles by study period			
	1959-70	1975-81	1982-89	1975-89
2	29.0% (11)	33.3% (2)	22.2% (2)	26.7% (4)
3	50.0% (19)	66.7% (4)	77.8% (7)	73.3% (11)
4	18.4% (7)	—	—	—
5	2.6% (1)	—	—	—
≥6	—	—	—	—
Total	100.0% (38)	100.0% (6)	100.0% (9)	100.0% (15)
Mean length	2.87	2.67	2.78	2.73

between studies in a subset of these data for which female age was known. We found that weaning age was independent of both female age and study (*P* = 0.76 and 0.77 for weaning age versus maternal age and study, respectively; *n* = 36 litters, hierarchical log-linear analysis). Sample sizes were too small to control for individual female effects by subsampling for one measurement of weaning age per female. Thus, the analysis treated different litters produced by the same female as independent "experiments" in the determination of weaning. Nonetheless, the data suggest very similar patterns of weaning for the two periods.

COMPOSITE INDICES OF REPRODUCTION

Age of female at maturity, fertility, fecundity, and weaning age are fundamental measures of reproductive performance in the sense that they combine in individual life histories to determine the contribution of reproduction (as opposed to juvenile mortality) to population recruitment. In practice, however, we could not always estimate true reproductive performance because we could not isolate, for example, true fertility and fecundity from the effects of neonatal mortality. Nonetheless, it was important to ask whether interactions among these components, in conjunction with mortality, altered or reinforced our conclusions regarding differences in reproductive performance between study periods that were based on the analysis of each parameter individually. In this and the following section, we evaluate two composite indices of reproductive performance—cycle length and production of cubs (female productivity).

Length of Reproductive Cycle

We have defined reproductive cycle length as the litter-to-litter time interval (Chapter 3). As such,

cycle length is a composite of fertility (probability of producing a new litter, once offspring are weaned), weaning age, and cub mortality. The first two components affect cycle measurement in a consistent manner. For example, increased fertility will always shorten cycles and increased time to weaning will lengthen them. On the other hand, offspring mortality can shorten or lengthen estimates of cycle length, depending on when litters are lost. Mortality of dependent offspring can shorten perceived cycles if deaths occur after first observation of the litter, because mothers are freed prematurely to begin a new cycle. Cub mortality may increase estimates of cycle length when it occurs before observations are made because cycles measured subsequently then become combinations of two or more litter-to-litter intervals. We have not included one-year intervals in our analyses, because these can only result from early litter loss or abandonment.

Based on 38 litter-to-litter cycles for 24 females, ecosystem-wide, the mean cycle length for the 1959-70 preclosure period was 2.87 years. For 15 cycles measured on 12 females during the 1975-89 postclosure period, the mean was 2.73 years. Division of the postclosure data in terms of early (1975-81) and late (1982-89) subperiods revealed little difference in cycle durations (Table 17.9).

To test for study period differences attributable to ecocenter status, however, it was again necessary to account for the ages at which females whelped. This reduced the 1959-70 sample by eight cycles, but the period means remained very similar (Table 17.10). Regression analysis comparing the pre- and postclosure periods showed that study (ecocenter status), but not female age, had a significant effect on litter-to-litter cycle length (Table 17.11). Cycles longer than four years were observed only during the preclosure period. Cycle length, however, is a "waiting-time" type of variable—

Table 17.10. Mean lengths (±SE) of 45 litter-to-litter reproductive cycles for 36 known-age female grizzly bears, Yellowstone Ecosystem, 1959-70 and 1975-89.

| Study period | Mean length of reproductive cycle (yrs) by female age[a] (No. of cycles) | | | | |
| | Young age | Prime age | | Old age | |
	4-8	9-14	15-20	21-25	All ages
1959-70 (*n*)	2.91±0.21 (11)	3.09±0.16 (11)	2.88±0.40 (8)	—	2.97±0.14 (30)
1975-89 (*n*)	2.80±0.20 (5)	2.75±0.16 (8)	2.00 (1)	3.00 (1)	2.73±0.12 (15)

[a] Female age determined at the mid-point of cycle interval.

Table 17.11. Analysis of variance (weighted regression) in reproductive cycle length relative to ecocenter status (present or absent) and age of female grizzly bear, Yellowstone National Park, 1959-70 and 1975-89.

Source of variation	DF	SS	MS	*F*-ratio	*P*-value
Ecocenter status	1	0.78	0.78	4.27	0.05
Female age	1	0.16	0.16	0.89	0.35
Residual	27	4.91	0.18	—	—

Note: The analysis uses mean cycle length and mean female age (measured at the midpoint of each cycle) for each of 30 females on which 45 cycles were measured (age categories one and four combined, and two and three combined).

researchers observe a litter and then "wait" for the next. The longer the researcher is able to wait, the more likely it is that relatively long cycles will be detected. Thus, it is important to note that females composing the 1959-70 sample were under observation, on the average, for substantially longer periods (9.8±1.8 years) than those in the 1975-89 sample (6.2±1.8 years). Consequently, mean cycle length in the postclosure period may have been underestimated.

Female Productivity

As described in Methods (Chapter 3), rates of cub production (female productivity) can be calculated in several ways (Table 17.12). In all cases, however, the measurement is based on a ratio of the number of adult years a female or group of females is observed and the number of cubs produced in that time by these females. The rate at which a female or group of females produces cubs is a composite measure of reproduction, manifesting the interaction of age at maturity, fertility, fecundity, weaning age, and cub mortality and is expressed as "cubs produced per adult female per year." Despite being sensitive to the effects of cub mortality, productivity rates are the closest thing we have to a direct field measurement of the contribution of reproduction (as opposed to juvenile mortality) to population recruitment. Note that measurement of productivity rates is not affected by the "waiting-time" bias mentioned in the treatment of cycle length, because years that are not bounded by litter production still contribute to the rate calculation (see Chapter 3).

In this section, we calculate productivity rates for marked females only, and we assume adulthood to begin at four years of age, the earliest age that females in any population have been reported to whelp. Annual and individual ecosystem-wide productivity rates for 1959-70 are presented in Tables 17.13 and 17.14. Because the raw data on individual and annual productivity rates for the 1975-89 population are, as yet, unpublished, we do not present them in tabulated form, but rather, show only the results of our analyses.

Ecocenters and Productivity Rates

We examined whether production of cubs is affected by the presence of ecocenters by comparing the productivity of five populations (Fig. 17.1).

Table 17.12. Mean population productivity rates of female grizzly bears, varying as a function of method of calculation, age at which females were considered adult, and study period, Yellowstone Ecosystem, 1959-70 and 1975-89.

Method of calculation	Productivity rates, expressed as cubs per female			
	1959-70	1975-81	1982-89	1975-89
Direct population mean rates:[a]				
Adult at 4	0.55 (337)	0.62 (69)	0.65 (123)	0.64 (192)
Adult at 5	0.61 (308)	0.65 (62)	0.72 (109)	0.69 (171)
Unweighted individual mean rates ± SE, 4-year filter:[b]				
Adult at 4	0.58±0.06 (40)	—	—	0.66±0.08 (20)
Adult at 5	0.63±0.06 (38)	—	—	0.66±0.08 (18)
Weighted individual mean rates ± 95% CI, 4-year filter:[c]				
Adult at 4	0.58±0.12 (40)	—	—	0.65±0.16 (20)
Adult at 5	0.62±0.12 (38)	—	—	0.65±0.17 (18)
Unweighted annual mean rates ± SE:[d]				
Adult at 4	0.55±0.05 (12)	0.58±0.12 (7)	0.64±0.09 (8)	0.61±0.07 (15)
Adult at 5	0.60±0.05 (12)	0.58±0.13 (7)	0.71±0.10 (8)	0.65±0.08 (15)

[a] Total number of cubs produced by all adult females was divided by the total number of years the females were observed as adults (female-years); sample sizes (in parentheses) are total adult female-years.

[b] Productivity rates calculated for individual females (cubs produced divided by adult-years observed) that were observed as adults for four or more years were summed and divided by the number of females; sample sizes (in parentheses) are the number of females observed four or more years as adults.

[c] Same sample as above[b] except that the individual rates were weighted by the number of adult-years each female was observed.

[d] Annual productivity rates (total cubs produced divided by number of adult females observed) were summed and divided by the number of years of the study; sample sizes (in parentheses) are the number of years of the study.

Of the five populations, two (Susitna, AK, and Yellowstone, 1959-70) were considered to be ecocentered and three (Northcentral, AK, Brooks Range, AK, and Yellowstone, 1975-89), non-ecocentered (see Chapter 13). All are interior, as opposed to coastal populations of brown bears. For this analysis, production of cubs was measured as the mean productivities of individual females weighted by the number of years they were observed as adults. The effect of female age on productivity was controlled by including the mean age of each female during the period for which her production was known. The fitted curve for age-specific production of cubs (productivity rates according to female age) was highest for the postclosure Yellowstone population; that for the Brooks Range, lowest; and that for the preclosure Yellowstone population, intermediate (Fig. 17.1). There were significant differences between the five populations in the production of cubs, when the effects of age were removed (Table 17.15; $P = 0.013$). However, this result only identified an overall effect of study area, not whether any particular pairwise comparison of productivities was statistically significant. Pairwise tests (using the 1959-70 Yellowstone population as reference) revealed that the difference in pre- versus postclosure Yellowstone productivities approached, but did not exceed, conventional statistical significance (Table 17.16; $P = 0.08$). We will, nevertheless, examine the implications of this result in more detail below, because the trend is strong and the higher rate of productivity exhibited by the non-ecocentered, 1975-89 grizzly bear population is exactly opposite of expectation. Certainly, most of the other measures of female reproductive performance—age at first whelping, fertility, fecundity, and weaning—raised no suspicion that productivity by postclosure females might exceed that of preclosure ones.

Comparison of the populations in terms of mean productivity and mean body weights of adult females and males (as a rough measure of nutritional status) raised additional questions. With the exception of the postclosure Yellowstone population, production

Table 17.14. Annual productivity rates for 67 marked, adult female grizzly bears, Yellowstone Ecosystem, 1959-70.

Year	Adult at four years			Adult at five years		
	No. of females	No. of cubs	Rate	No. of females	No. of cubs	Rate
1959	7	3	0.43	7	3	0.43
1960	19	16	0.84	19	16	0.84
1961	29	13	0.45	22	13	0.59
1962	31	18	0.58	26	18	0.69
1963	32	21	0.66	30	21	0.70
1964	30	12	0.40	30	12	0.40
1965	30	24	0.80	29	24	0.83
1966	36	15	0.42	28	15	0.54
1967	34	20	0.59	31	20	0.64
1968	30	21	0.70	28	21	0.75
1969	29	8	0.28	29	8	0.28
1970	30	16	0.53	29	16	0.55
Total	337	187		308	187	

$$\frac{\text{Total no. of cubs}}{\text{Total no. of females}} = \frac{187}{337} = 0.55 \qquad = \frac{187}{308} = 0.61$$

Fig. 17.1. Productivity rates (cubs per year per adult female) relative to female age and population, as predicted by regression analysis, for five interior grizzly/brown bear populations (females observed for at least one year; weighted for number of years observed); filled and open symbols indicate ecocentered and non-ecocentered populations, respectively; based on life history data provided by R. Knight and B. Blanchard (Yellowstone 1975-89), S. Miller (Susitna), H. Reynolds, III (Brooks Range and Northcentral), and our records for Yelllowstone, 1959-70.

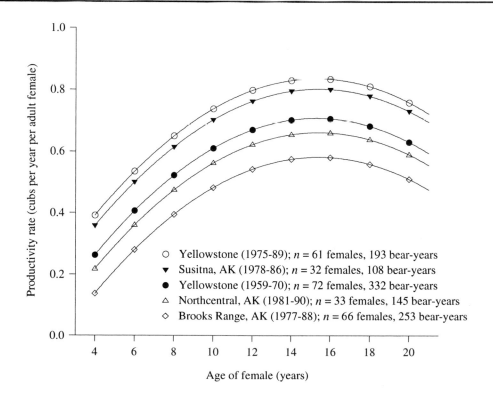

○ Yellowstone (1975-89); n = 61 females, 193 bear-years
▼ Susitna, AK (1978-86); n = 32 females, 108 bear-years
● Yellowstone (1959-70); n = 72 females, 332 bear-years
△ Northcentral, AK (1981-90); n = 33 females, 145 bear-years
◇ Brooks Range, AK (1977-88); n = 66 females, 253 bear-years

Table 17.13. Individual observed productivities, for 67 marked, adult female grizzly bears, used to compute two distinct productivity rates, Yellowstone Ecosystem, 1959-70.

Female ID number	Age (yrs) at start of observation period[a]	Adult at four years			Adult at five years		
		Adult-yrs observed	No. of cubs	Cubs/adult-yr observ.	Adult-yrs observed	No. of cubs	Cubs/adult-yr observ.
5	1	7	5	0.71	6	5	0.83
6	cub	3	0	0.00	2	0	0.00
7	11	10	7	0.70	10	7	0.70
8	8	6	8	1.33	6	8	1.33
10	2	10	4	0.40	9	4	0.44
15	1	8	3	0.38	7	3	0.43
16	A	4	3	0.75	4	3	0.75
17	A	3	1	0.33	3	1	0.33
18	A	3	1	0.33	3	1	0.33
21	A	6	2	0.33	6	2	0.33
23	11	1	0	0.00	1	0	0.00
29	cub	3	2	0.67	2	2	1.00
31	5	1	0	0.00	1	0	0.00
34	14	11	6	0.55	11	6	0.54
39	6	11	3	0.27	11	3	0.27
40	2	8	7	0.88	7	7	1.00
42	5	11	8	0.73	11	8	0.73
44	A	7	2	0.29	7	2	0.29
45	3	2	0	0.00	1	0	0.00
48	A	4	5	1.25	4	5	1.25
54	3	1	0	0.00	—	—	—
55	3	1	0	0.00	—	—	—
58	A	1	0	0.00	1	0	0.00
59	A	1	1	1.00	1	1	1.00
62	3	1	0	0.00	—	—	—
64	A	6	0	0.00	6	0	0.00
65	A	6	9	1.50	6	9	1.50
72	A	2	1	0.50	2	1	0.50
75	15	4	2	0.50	4	2	0.50
81	3	8	1	0.13	7	1	0.14
84	A	4	5	1.25	4	5	1.25
96	3	9	8	0.89	8	8	1.00
101	3	9	4	0.44	8	4	0.50
102	A	1	0	0.00	1	0	0.00
104	A	1	0	0.00	1	0	0.00
107	A	2	0	0.00	2	0	0.00
108	A	10	2	0.20	10	2	0.20
109	cub	5	0	0.00	4	0	0.00
112	5	9	5	0.56	9	5	0.56
119	A	1	2	2.00	1	2	2.00
120	A	10	4	0.40	10	4	0.40

cont'd on next page

of cubs, mean adult male weight, and mean adult female weight (Table 17.17; columns 2, 3, and 4, respectively) were all higher in ecocentered versus non-ecocentered populations. In fact, age-specific productivities were perfectly correlated across populations with estimates of mean adult weight (Table 17.17; column 5), again excepting that for postclosure Yellowstone.

A Postclosure Rebound in Productivity?

As we have shown, differences in pre- versus postclosure Yellowstone production of cubs approached statistical significance. The question arises, did this occur gradually throughout the 1975-89 study period, or was there a period of depressed production followed by a clearly defined period when the population rebounded?

Table 17.13. *cont'd*

Female ID number	Age (yrs) at start of observation period[a]	Adult at four years			Adult at five years		
		Adult-yrs observed	No. of cubs	Cubs/adult-yr observ.	Adult-yrs observed	No. of cubs	Cubs/adult-yr observ.
123	A	2	2	1.00	2	2	1.00
125	A	9	8	0.89	9	8	0.89
128	11	9	13	1.44	9	13	1.44
129	A	1	0	0.00	1	0	0.00
131	4	2	0	0.00	1	0	0.00
132	cub	3	1	0.33	2	1	0.50
139	cub	3	2	0.67	2	2	1.00
140	8	6	3	0.50	6	3	0.50
141	1	3	2	0.67	2	2	1.00
142	1	5	0	0.00	4	0	0.00
144	1	5	3	0.60	4	3	0.75
148	15	2	2	1.00	2	2	1.00
150	5	8	5	0.63	8	5	0.62
160	A	5	2	0.40	5	2	0.40
163	1	2	2	1.00	1	2	2.00
164	6	6	3	0.50	6	3	0.50
165	1	1	0	0.00	——	——	——
167	cub	1	0	0.00	——	——	——
170	A	1	2	2.00	1	2	2.00
172	13	7	4	0.57	7	4	0.57
173	2	5	3	0.60	4	3	0.75
175	10	7	4	0.57	7	4	0.57
175B	A	6	4	0.67	6	4	0.67
176	13	1	0	0.00	1	0	0.00
180	11	4	3	0.75	4	3	0.75
184	cub	3	2	0.67	2	2	1.00
187	1	4	1	0.25	3	1	0.33
200	3	5	2	0.40	4	2	0.50
204	2	4	0	0.00	3	0	0.00
227	1	1	0	0.00	——	——	——
228	5	3	0	0.00	3	0	0.00
259	16	2	3	1.50	2	3	1.50
Total		337	187		308	187	

$$\frac{\text{Sum of no. of cubs}}{\text{Sum of adult-yrs observed}} = \frac{187}{337} = 0.55 \qquad \frac{187}{308} = 0.61$$

[a] A = adult of unknown age.

It has been frequently suggested, most recently by Knight *et al.* (1990), as evidence of improving status of the Yellowstone population, that the productivity of the adult female population was greatly depressed for some period of years following the elevated mortality of the 1970s that resulted from closure of the Trout Creek dump, but that rates subsequently rebounded. If true, this would suggest an even more marked and anomalous increase in productivity among recent Yellowstone grizzlies (relative to the preclosure population) than was apparent from our analysis in the previous section comparing productivities for the entire postclosure period (Table 17.15). In order to examine this possibility with respect to the production of cubs, we first calculated annual productivity rates and

Table 17.15. Analysis of variance in production of cubs relative to age of female and population studied ("Study") for five grizzly/brown bear populations (females observed for at least one year; weighted for number of years observed; see Fig. 17.1).

Source of variation	DF	SSE	MSE	*F*-ratio	*P*-value
Total	235	146.09	—	—	—
Regression	14	19.84	1.42	2.48	0.003
Age X Study interaction	8	3.63	0.45	0.79	0.61
Study	4	7.36	1.84	3.22	0.013
Female age	2	8.85	4.42	7.75	< 0.001
Error	221	126.25	0.57	—	—

Table 17.16. Estimated regression coefficients for regression analysis of cub production relative to female age in five grizzly/brown bear populations, using the preclosure Yellowstone population (1959-70) as the reference.

Variable	Coefficient	SE of coefficient	*T*-statistic	*P*-value[a]
Intercept	-0.0635	0.1478	-0.4294	0.67
Age	0.0989	0.0247	3.9990	0.0001
Age (squared)	-0.0032	0.0009	-3.3840	0.0007
Yellowstone (preclosure)	0	0	—	—
Yellowstone (postclosure)	0.1267	0.0736	1.7215	0.08
Northcentral	-0.0413	0.0811	-0.5091	0.61
Brooks Range	-0.1210	0.0701	-1.7265	0.08
Susitna	0.0966	0.0885	1.0921	0.28

[a] Values reported for study areas are for pairwise comparisons with the reference population, preclosure Yellowstone.

then, unweighted mean annual rates for the subperiods 1975-81 and 1982-89. Assuming female adulthood to begin at four years of age, the mean annual productivity rates were 0.58 and 0.64 cub per female per year for the 1975-81 and 1982-89 subperiods, respectively (Table 17.12; row 7, columns 3 and 4). However, when annual rates from the two subperiods were compared statistically, they were not significantly different (Mann-Whitney Test, $P = 0.86$, $n = 15$ years). Note that the difference in productivity between the two subperiods is much reduced when annual means are weighted by the number of females sampled (Table 17.12; row 1, columns 3 and 4). This shift reflects the undue effect on unweighted subperiod means of years in which only a small number of females were observed— e.g., 1975 when four females had a collective

productivity of zero. Thus, despite superficial appearances to the contrary, no significant rebound between the two subperiods was detectable.

DISCUSSION

In this chapter, we have attempted to compare reproductive processes, isolated from the issue of mortality, in the pre- and postclosure periods. In practice, our estimates of age at reproductive maturity, fertility, and fecundity, and hence of cycle length and cub productivity, are all potentially influenced by mortality of cubs in the first six months or so, post-parturition. Recognition of this complication must temper any attempt at interpretation of the comparisons (below). The principal results of our analyses of the pre- and postclosure Yellowstone populations are as follows:

Table 17.17. Productivity rates and mean adult weights for the five interior, grizzly/brown bear populations depicted in Fig. 17.1.

Population	Productivity rate[a] (No. females)	Adjusted body weight in kg (lbs)[b] (No. of bears)		
		Males	Females	Both sexes[c]
Yellowstone (1975-89)	0.823 (58)	193 (426) (65)	134 (295) (63)	164 (362) (128)
Susitna, Alaska	0.793 (31)	269 (593) (12)	144 (318) (21)	207 (456) (33)
Yellowstone (1959-70)	0.697 (49)	245 (540) (33)	152 (335) (72)	199 (439) (105)
Northcentral Alaska Range	0.655 (33)	224 (494) (24)	135 (298) (32)	180 (397) (56)
Brooks Range, Alaska	0.576 (65)	182 (401) (26)	117 (258) (35)	150 (331) (61)

[a] From Fig. 17.1; adult females observed one or more years.
[b] Data compiled by McLellan (1995 [in press]); adults were bears five or more years of age and weights were adjusted to a common date (sample sizes in parentheses).
[c] Unweighted average of means from columns 2 and 3.

1. There were no statistical differences between the pre- and postclosure periods in age at first reproduction, fertility, litter size, weaning age, or cub productivity. However, in the postclosure period, cycle lengths were shorter, and there was a trend toward a higher rate of cub production.
2. During both periods, there were strong age-related patterns in fertility, litter size, and production of cubs. Young females (4-8 years) and old females (21-25 years) had lower fertility, smaller litters, and reduced productivity.
3. The cause of previously reported reductions in the size of litters born during 1975-81 versus 1959-70 was probably due to a younger female age structure caused by adult mortality following dump closures.

Thus, contrary to our initial expectation and some previous publications, there was no statistical evidence of systematic reductions in individual reproductive performance following dump closures. In fact, the strongest trends were in the direction of improved postclosure reproduction. What are we to make of these results?

First, we cannot ignore the potential for sampling biases peculiar to one or the other study. For example, we have already mentioned that preclosure bears were followed for more years and that this observational bias may have contributed to the recording of longer cycles and later age at first reproduction prior to 1970. We have also noted that

mortality occurring before the weaning of offspring can artificially increase the frequencies of litter production and, in turn, the observed rate of production of cubs (*sensu* this monograph and the brown bear literature, generally). Such productivity measures, therefore, carry the risk of confusing a bad thing (frequent cub production due to high cub mortality) with a good thing (high production due to rapid weaning). Unfortunately, we have no direct way, at present, of deciding whether study differences in detection of cub mortality have compromised our comparisons of productivity in the pre- and postclosure periods. We need to step back from the details of individual analyses and consider these results from the broader perspectives of (1) emerging models of bear reproductive ecology (Garshelis 1995 [in press], McLellan 1995 [in press]) on the one hand and (2) first principles of demography on the other.

Bear Reproductive Ecology

Much thought and effort have been devoted to identifying what factor(s) regulate bear populations (recently reviewed by McLellan 1995 [in press], Garshelis 1995 [in press], and Derocher and Taylor 1995 [in press] for grizzly/brown bears, black bears, and polar bears, respectively; see also Chapter 9). The failure to find consistent and clear evidence of density-dependent regulatory mechanisms involving reproduction, and a growing appreciation for the effects of large annual variations in the availability of bear food resources (e.g., Chapter 9; see also Jonkel and Cowan 1971, Rogers 1987, Pelton 1989, Mattson *et al.* 1991a), have led to the view that, whereas density effects probably exist, reproduction in bears generally varies in density-independent fashion with fluctuations in food quality and quantity. This view is perhaps best documented in black bears (Garshelis 1995 [in press]), but similar views have been suggested for both grizzly/brown bears and polar bears (McLellan 1995 [in press] and Derocher and Taylor 1995 [in press], respectively).

The underlying reproductive physiology would seem to be that, as per capita food resources vary, more or less energy and nutrients are available for individual growth and reproduction, after maintenance needs have been met (Stringham 1990b). Variation in per capita resources might occur through changes in bear density, total food production, or both. Thus, McLellan (1995 [in press]) found a positive relation between average adult female weight and mean litter size across 16 grizzly/brown bear populations. Garshelis (1995 [in press]) reported a significant negative correlation between mean yearling weight and female age at first reproduction in an analysis of 20 black bear populations. A similar result was obtained in a within-population analysis of age at first reproduction in black bears (Noyce and Garshelis 1995 [in press]). Stringham (1990a) compared seven black bear populations and found that litter size and the production of cubs increased with increased adult weights (male and female), whereas cycle length and age at first reproduction declined. Identical results were obtained in a companion study of 11 grizzly/brown bear populations (Stringham 1990b).

Blanchard (1987), using a subset of the data available to Stringham (1990b), reported a significant positive association between adult female weight and cub litter size among eight populations of grizzly/brown bears. Blanchard (1987) also reported data indicating that (1) postclosure Yellowstone bears with continued access to a refuse dump (Cooke City) weighed more than postclosure bears using only natural foods; (2) the weight difference was comparable to that observed in the pre-versus postclosure weight comparison we presented in Chapter 16; and (3) the production of cubs for two of the dump-using females was almost twice that of nine females not known to frequent the mini-ecocenter at Cooke City (0.800 versus 0.469, respectively). Note that Stringham (1990b), in the only grizzly/brown bear study to examine statistically the effect of weight of adult females on the production

of cubs, did not include data from the postclosure Yellowstone population, presumably for lack of an adequate data set.

In the Hudson Bay region, survivorship of dependent polar bear cubs (spring to fall) was higher for mothers weighing more in spring (Ramsay and Stirling 1988). Reductions in adult weight, the proportion of females with cubs, and estimates of food abundance were all found to be associated with cub survivorship among polar bears in the region of the eastern Beaufort Sea, Canada (Stirling *et al.* 1976, 1977). Finally, Derocher and Stirling (1994) speculated that age-related declines in the reproductive performance of female polar bears were due to undetermined processes that impair accumulation of fat stores—in other words, an inability to gain sufficient weight.

Not all of these studies succeeded in using weight to predict performance in all aspects of reproduction (age at first reproduction, litter size, cycle length, and production of cubs). None attempted to control for the effects on reproduction of maternal age, and all of the brown bear analyses lumped coastal-versus-interior populations, despite the heightened possibility of genetical differences in life history strategies between these populations and important differences in their food resource base. However, each of the reproductive parameters was related to weight in at least one analysis across all bear species, and in no case did the data suggest a life history in which reproductive success increased with reduced weight.

Thus, in examining pre- and postclosure Yellowstone populations, we arrive at a dilemma arising from three seemingly solid results. First, grizzly bear food resources in the Greater Yellowstone Ecosystem declined significantly after 1970. Before this time, bears used both natural foods and human-derived ecocenters (see Chapter 10). After 1970, only natural foods were available. Second, Yellowstone bears weighed less in the postclosure period (Chapter 16). Third, the literature strongly suggests that body weight in bears is a reliable measure of per capita food resources and hence, a measure of the capacity for reproduction. The paradox is that there was a strong trend toward increased productivity in the postclosure period when Yellowstone bears weighed less and hence, operated on a smaller per capita food resource base. In fact, production of cubs by the postclosure Yellowstone population was the highest of age-adjusted productivities measured for five interior grizzly/brown bear populations (Fig. 17.1), which otherwise showed perfect correlation between productivity and mean adult weights (Table 17.17).

The implications of this curious result for the postclosure Yellowstone population must be examined more fully.

We reject the suggestion that preclosure bears were obese and consequently suffered reproductive impairment (S. French, pers. comm. 1990). Weight differences within age, sex, and weigh-date categories were on the order of 20% or less between pre- and postclosure populations (see Table 16.3). Even if we assume that this difference was entirely due to increased fat storage, significant reproductive problems such as ovulatory infertility generally do not appear in humans until body weights approach 30% above ideal (Grodstein *et al.* 1994). Furthermore, this idea carries the absurd implication that bears faced with a relative surplus of natural foods misallocate all of this surplus energy into fat storage, rather than an appropriate mix of growth, reproduction, and storage. Finally, bears have evolved to accumulate enormous stores of fat, highly metabolically active brown fat in particular, in preparation for winter sleep. Hence, we might reasonably expect them to have a reproductive system resistant to the adverse effects of extreme fat deposition that may appear in other species (Stringham 1990b).

We also reject the possibility that aggregation-induced social stress, operating, for example, on reproductive endocrine pathways, inhibited preclosure reproduction. First, aggregated bears did grow larger. Thus, there was no evidence of physiological dysfunction attributable to excessive rates of agonistic conflict over feeding sites, increased vigilance, or some other sort of environmental stressor. The ecocenters were not ecological islands, and bears were free to regulate social stress by foraging away from the aggregation. Second, aggregated grizzly bears display complex social behaviors, which clearly are evolved traits (Chapter 13). It is reasonable, therefore, to expect that bears have also evolved an endocrine system tolerant of sociality. Finally, the suggestion that social stress at ecocenters inhibits reproduction carries with it the improbable implication that bears aggregate at natural concentrations of food (salmon runs, etc.) at a net reproductive cost.

Of course, one way out of the dilemma is to accept the statistical conclusion (based upon a significance level of $\alpha=0.05$) of no difference in rate of producing cubs between pre- and postclosure populations and, in addition, to assume that a reduction in postclosure reproduction actually existed, but was not large enough to be detected, given the weight differences involved and the life history samples available to us. However, this would be special pleading. The suggestion of a positive trend was strong enough, and the samples large enough, that the apparent anomaly of equal or higher, postclosure reproduction is deserving of serious scrutiny. Perhaps the anomaly reflects an adaptive, but as yet unrecognized plasticity in the grizzly bear life history—a plasticity that is governed, for example, by unusual combinations of, or drastic changes in, bear density, spatial or temporal aspects of the food resource, or mortality schedules. Alternately, there may, in fact, have been study biases that favored the capturing and marking of females with cubs or overlooked the extent to which preweaning mortality contaminated this particular measurement of reproduction. In either of the latter cases, the comparison of the production of cubs would be potentially misleading.

It is important to note that the relatively large difference in productivity rate between the two periods was not foreseeable on the basis of results from comparisons of fundamental reproductive parameters. Postclosure values for these parameters were either only slight improvements (age at first reproduction and fertility) or inferior (litter size and weaning age) to preclosure values. Note that only early cub mortality was likely to compromise estimates of the fundamental reproductive parameters, whereas the production of cubs (productivity rate) was subject to the more extended effects of cub loss right up to time of weaning. For example, to classify incorrectly a female as infertile in the year following conception, her entire litter must be lost before first observation, which was usually sometime in summer (i.e., before cubs were four to six months old). Similarly, to misclassify a four-cub litter as a litter of three, one cub must be lost prior to first observation, again at about one-half year of age. By contrast, premature return of a female to producing cubs can develop as a consequence of whole litter loss, gradual or abrupt, occurring anytime throughout the period of maternal care up to 2.5 years of age. Not only would this elevate her (apparent) productivity rate, but it would cause a decrease in the other composite reproductive parameter we have examined, mean length of reproductive cycle. And, in fact, our earlier analysis showed the mean reproductive cycle length for postclosure females to be significantly shorter than that for preclosure females.

These results and considerations suggest that differential mortality may be at the root of the productivity anomaly, and that the difference should be sought among older (> 0.5 year) dependent offspring. Although cub-to-yearling survivorship in the preclosure period appears to have been similar

(0.812; Chapter 8) to that reported for the postclosure period (0.830; Knight *et al.* 1992), many authors attempting to model the postclosure population (see Chapter 19) have expressed concern about unrecorded mortality and the accuracy of survivorship estimates (Knight and Eberhardt 1984, 1985; Eberhardt *et al.* 1986, 1994; Knight *et al.* 1988b).

Demographic Considerations

Even if we could remove every last vestige of sampling bias and any occult mortality from our measures of reproduction for the pre- and postclosure populations, reproductive parameters could offer only partial insight into population status. Mortality rates and density estimates are other essential parameters to be considered. For example, taking at face value our statistical result of no difference in the rates of production of cubs, and ignoring the discrepancy in pre- versus postclosure body weights, imagine a population (such as the postclosure Yellowstone population) that experienced, first, a food-rich pre-catastrophe period, then a catastrophic decline in its food resource concomitant with an exceptionally heavy mortality (or reduced reproduction or both), and finally, a longer period of relative stability in which fewer bears enjoyed the same per capita resource as did a larger number of individuals in the food-rich, pre-catastrophe period. In comparing the first and third time periods, we might observe identical patterns of reproduction (cycle length, litter size, etc.) and mortality, and hence, equivalent population growth rates (probably near zero), yet miss the critical point that the latter population is much smaller and, consequently, less viable.

In fact, it is easy to imagine a sequence of abrupt, step-like population declines, each characterized by short transitory periods of increased mortality and/or reproductive failure, and each associated with habitat losses or changes that reduce the food base. These sequences go undetected simply because they are masked by the more extensive data samples (on reproduction or mortality) compiled over the relatively long periods of apparent stability when per capita levels of food intake have equilibrated at similar levels, but at different population densities.

In short, reproductive parameters might be useful in measuring short-term fluctuations in per capita food resources (as is body weight), but they do not necessarily reflect population status or trend. If we recognize the limits of comparing populations using reproductive parameters, then we can appropriately place greater emphasis on estimates of population density as an indicator of status and trend in any grizzly bear population, including that of the Greater Yellowstone Ecosystem. We argue that reproductive parameters in the pre- and postclosure populations were not, in themselves, reliable indicators of population status or trend. Rather, such parameters are indicators of per capita food intake, which may be an indication of population stability, but not necessarily of population viability.

Habitat quality (expressed as total food resources), interacting with bear density, is the critical factor determining grizzly bear productivity. Ecocenters which improve habitat quality must, among other effects, improve the capacity for reproduction. It follows that we cannot analyze for either long- or short-term population viability without incorporating into our models quantitative measures of habitat quality. This implies that we cannot set population recovery goals, determine recovery status, or predict persistence times until habitat parameters become an integral component of population viability analyses. We will discuss this further in Chapter 20.

SUMMARY

We compared four fundamental and two composite measures of reproduction in the pre- and postclosure periods. There were no study differences in age at first reproduction, fertility, litter size, or cub weaning age (fundamental parameters), but cycle length was shorter and production of cubs was higher (composite parameters) in the postclosure period. There was no evidence that productivity changed across the 1975-89 period. Rather, the apparent improvement in productivity was a feature of the entire postclosure study. Average body weight was shown to be a good indicator of productivity.

The production of cubs in postclosure Yellowstone was higher than in any of three Alaskan populations, two of which we consider to be ecocentered. Curiously, the high postclosure productivity of the Yellowstone population was not associated with larger mean body weights, although mean weight was a good predictor of productivity among the other four populations that we examined, including preclosure Yellowstone. A review of the black, brown, and polar bear literature revealed that this positive association between mean body weight and various measures of reproductive performance was a consitent pattern. Thus, the productivity difference between the pre- and postclosure Yellowstone populations represents a fundamental dilemma: postclosure bears were smaller but more fecund. We reject resolutions of the dilemma that suggest ecocentered bears had reduced productivity because they were obese or suffered from aggregation-induced, endocrine-mediated social stress. Our analyses suggested that the increase in postclosure productivity was apparent rather than real and was likely due to failures to detect preweaning cub mortality in the IGBST study. A composite measure of reproduction, like productivity, is especially sensitive to such a bias.

For this reason, we caution against placing too much emphasis on reproductive parameters, especially the production of cubs parameter, in evaluating population status or trend. A population that is declining as a result of a series of abrupt reductions in the food base may quickly equilibrate at comparable levels of per capita resource, and hence, comparable per capita reproductive performance. Reproductive parameters, therefore, reflect the current relationship between the size of a population and its total resource base, but they are not necessarily reliable indicators of the size or the long-term trend of the population. Where ecocenters are present, habitat quality is high, and carrying capacity is increased. However, per capita productivity may be elevated only in the relatively brief period of population growth following the initial improvement in habitat quality.

Table 18.1. Observed mortality in the Yellowstone Ecosystem grizzly bear population by year, age, and sex, 1971-87.[a]

Year	Cubs F	Cubs M	Cubs U	Yearlings F	Yearlings M	Yearlings U	2-4-yr-olds F	2-4-yr-olds M	2-4-yr-olds U	Adults F	Adults M	Adults U	Nonadults[b] F	Nonadults[b] M	Nonadults[b] U	Unknown F	Unknown M	Unknown U	Total
1971	-	1	-	1	-	-	3	2	-	15	16	-	1	-	1	-	2	5	47
1972	1	2	2	-	-	-	-	-	-	7	12	-	2	-	-	1	-	2	29
1973	-	-	-	-	-	-	1	1	-	4	3	5	1	-	-	-	-	2	17
1974	-	-	1	-	1	-	2	3	-	4	3	-	-	-	-	-	-	2	16
1975	1	-	-	-	-	-	-	-	-	1	1	-	-	-	-	-	-	-	3
1976	-	-	-	-	-	-	-	-	-	5	1	-	-	-	-	-	-	1	7
1977	-	3	-	-	3	-	1	4	-	2	4	-	-	-	-	-	-	-	17
1978	-	-	-	-	-	-	-	2	-	3	2	-	-	-	-	-	-	-	7
1979	-	-	-	-	-	-	-	-	-	1	2	-	-	-	-	-	-	5	8
1980	-	-	1	2	2	-	-	3	-	1	1	-	-	-	-	-	-	-	10
1981	-	-	2	-	-	-	-	3	-	5	2	-	-	-	-	-	-	1	13
1982	1	1	1	-	-	1	-	6	-	4	3	-	-	-	-	-	-	-	17
1983	1	-	-	-	2	-	-	1	-	1	2	-	-	-	-	-	-	-	7
1984	1	-	-	-	1	-	3	2	-	2	2	-	-	-	-	-	-	-	11
1985	-	-	3	-	-	4	2	-	-	2	2	-	-	-	-	-	-	-	13
1986	-	2	1	-	1	-	2	-	-	1	2	1	-	-	-	-	-	1	11
1987	-	-	-	-	-	-	-	-	-	2	1	-	-	-	-	-	-	-	3
Total	5	9	11	3	10	5	14	27	0	60	59	6	4	0	1	1	2	19	236

[a] See Table 11.1 for 1959-70 data.
[b] Nonadults are bears 1-4 years old whose exact age was not determined.

Table 18.3. Yellowstone Ecosystem grizzly bear mortality by age, cause, and sex, 1971-87.[a]

Age[b]	Management control Male	Management control Female	Management control Total	Legal kill Male	Legal kill Female	Legal kill Total	Illegal kill Male	Illegal kill Female	Illegal kill Total	Natural Male	Natural Female	Natural Total	Accidental Male	Accidental Female	Accidental Total	Cause unknown Male	Cause unknown Female	Cause unknown Total
Cub	3	1	4	2	-	2	1	-	1	1	3	4	1	1	2	1	-	1
Yearling	7	1	9	-	-	0	-	-	0	2	1	3	-	1	1	1	-	2
2-year-old	4	2	6	1	1	2	3	2	5	-	-	0	1	-	1	-	-	0
3-year-old	-	1	1	1	3	4	6	2	8	-	-	0	-	-	0	-	-	0
4-year-old	4	1	5	-	2	2	3	-	3	1	-	1	1	-	1	2	-	2
Adult	28	19	47	14	17	32	15	15	33	1	1	2	-	6	6	1	2	3
Total	46	25	72	18	23	42	28	19	50	5	5	10	3	8	11	5	2	8

[a] See Table 11.6 for 1959-70 data.
[b] Totals for age classes may exceed the sum of male and female mortalities by inclusion of mortalities for animals of unknown sex.

18
Grizzly Bear Mortality
Following the Closure of Ecocenters

INTRODUCTION

With the destruction of the Yellowstone ecocenters, it was obvious that the large, discrete aggregations of grizzly bears characteristic of the pre-1970 period would disappear and that there would be at least a local redistribution in the areas surrounding the former dumps. In Chapter 15, we presented evidence that this redistribution was, in fact, extensive. Over a period of years, local movements from ecocenters into nearby campgrounds and other park developments gave way to major shifts and expansions of home ranges toward the periphery of the Park and Yellowstone Ecosystem. Here, we focus on the consequences of how these changes in home range and movement affected grizzly bear mortality. First, we examine pre- versus postclosure differences in the demographic and spatial distributions of deaths. We close by attempting to quantify mortality rates in the immediate postclosure period.

THE DEMOGRAPHIC DISTRIBUTION OF MORTALITY

In this section, we first compare observed frequencies of death in the various categories of age and sex, relative to those expected if death occurred in simple proportion to the prevailing age structure and sex ratio during the postclosure period. We then compare these results with those of a similar analysis for the preclosure period (Chapter 11). By making this comparison, we hope to illuminate the nature of the stresses imposed by ecocenter closure and the bears' responses to the new circumstances.

Table 18.2. Yellowstone Ecosystem grizzly bear mortality by year and cause, 1971-87.[a]

Year	Management control	Legal kills	Illegal kills	Natural mortality	Accidental	Unknown
1971	18	20	7	-	1	1
1972	16	8	3	-	1	1
1973	3	8	3	-	-	3
1974	3	9	3	1	-	-
1975	-	-	3	-	-	-
1976	1	-	4	1	1	-
1977	4	3	6	-	2	2
1978	2	-	4	-	1	-
1979	1	0	6	-	1	-
1980	1	-	3	4	2	-
1981	-	1	9	2	-	1
1982	6	1	6	2	1	1
1983	6	-	-	-	-	1
1984	5	1	3	1	-	1
1985	2	-	3	7	1	-
1986	4	3	2	1	-	1
1987	2	-	1	-	-	-
Total	74	54	66	19	11	12

[a] See Craighead *et al.* (1988b) for 1959-70 data.

Table 18.4. Age structure and sex ratio of grizzly bears captured during 1975-89 (based on trapping data provided by R. Knight). See Fig. 4.4 for 1959-71 data.

Sex	Number of captures by age class[a]				
	Cub	Yearling	Nonadult	Adult	All ages
Female	19	17	45	86	167
Male	14	19	49	81	163
Total	33	36	94	167	330
Sex ratio (M:F)	42:58	53:47	52:48	49:51	49:51
Percent of total by age class	10.0	10.9	28.5	50.6	100.0

[a] Individual bears were counted only once in a year, but could be counted in more than one year.

We have tabulated mortality data for the 1971-87 period in the same way as those for the 1959-70 period (Chapter 11). This enables us to compare deaths by year, age, and sex (Table 18.1), by year and cause (Table 18.2), and by age, cause, and sex (Table 18.3). These tables are the source for all subsequent analyses.

All Causes of Mortality

Using methods similar to those described in Chapter 11, we have calculated the expected distribution of mortality among age and sex categories to test the null hypothesis that mortality occurred in simple proportion to the age structure and sex ratio documented during 1971-87. Note that methods employed by the Interagency Grizzly Bear Study Team (IGBST) to obtain age structure and sex ratio data from 1975-87 were different from those used during the 1959-70 period. In the earlier period, data from trapping and from capture-by-tranquilization were used to estimate sex ratios, whereas age structure was determined from repeated census counts of marked, unmarked-but-recognizable, and unmarked bears. For the 1971-87 period, however, trapping data provided by Richard Knight, Interagency Study Team Leader, were used to estimate both age structure and sex ratio (Table 18.4). Because sizes of the yearly samples of marked bears were much smaller, it was necessary to compute an age structure based on captures for the entire period, rather than computing structures on an annual basis as was done for the 1959-70 period. Thus, the sex-age structure is an average for the years 1975-89 (Table 18.4). Individual bears were counted only once in a year, but could be counted in more than one year.

In contrast to a similar analysis performed for the 1959-70 period (Tables 11.4 and 11.5), deaths from all causes among adults were more frequent during 1971-87, and among all nonadult classes, less frequent than expected (Tables 18.5 and 18.6; $P=0.006$, d.f.=1). For all age classes combined, death among males was also in excess of that expected (Tables 18.5 and 18.6; $P=0.05$, d.f.=1), a circumstance consistent with a statistically nonsignificant trend toward greater male mortality at each age level, and overall, that also was present in the 1959-70 period.

We emphasize that these results, as well as those reported for the 1959-70 period, have nothing to say about whether the mortality rate of a given age or sex class differed between periods. These are tests for the relative vulnerability to, or detectability of, death across demographic classes of two populations, each characterized by a specific age and sex structure. We also caution that the apparent trend toward greater vulnerability (or detectability) of adults versus nonadults (Tables 18.5 and 18.6) might be due to age-differential trappability (i.e., to error in the estimated age structure). For example, the expected number of adults will be an underestimate (and the deviation between observed and expected values an overestimate) if adults were less trappable than nonadults. This potential complication is less likely for the 1959-70 period when age structure was determined by repeated census of a large percentage of the free-ranging population. This concern aside, the apparent excess of adult deaths during 1971-87, but not 1959-70, is consistent with our hypothesis of reduced food resources causing increased adult movements, which in turn exposed adult bears more frequently to specific causes of death (next section).

Table 18.5. Comparison, by sex and age of grizzly bears, of postclosure Observed Mortality (OM) from all causes with Expected Mortality (EM), Yellowstone Ecosystem, 1971-87. EM values in columns 3 and 4 equal expectation across sexes and within age class; EM values in column 5 equal expectation across age classes.

| Age | Mortality estimate | Sex | | Total | Percent total |
		Female	Male		
Cub	OM	5	9	14	7.5
	(EM)	(8.1)	(5.9)	(18.7)	
Yearling	OM	3	10	13	7.0
	(EM)	(6.1)	(6.9)	(20.4)	68
Other nonadult	OM	14	27	41	21.9
	(EM)	(19.7)	(21.3)	(53.3)	(92.4)
Adult	OM	60	59	119	63.6
	(EM)	(60.7)	(58.3)	(94.6)	
All ages	OM	82	105	187	100.0
	(EM)[a]	(95.4)	(91.6)		

[a] EM values do not equal sum of those for each age category because they are based on a weighted average of sex ratios for all categories.

Table 18.6. Probability values associated with tests for effects of sex and age of grizzly bears on mortality from all causes in the Yellowstone Ecosystem, 1959-70 and 1971-87. G-test for goodness-of-fit; the term "excess" means significantly greater mortality than expected; see text for methods of calculating expected mortality.

| | *P*-values (degrees of freedom) | |
	1959-70[a]	1971-87
Effect of sex (by age)		
Cub	>0.50 (1)	0.10 (1)
	(NS)	(NS)
Yearling	>0.50 (1)	0.08 (1)
	(NS)	(NS)
Other nonadult	>0.25 (1)	0.08 (1)
	(NS)	(NS)
Adult	0.14 (1)	>0.60 (1)
	(NS)	(NS)
All ages	0.09 (1)	0.05 (1)
	(NS)	(Male excess)
Effect of age		
Four age categories[b]	0.15 (3)	0.006 (3)
	(NS)	(Adult excess)

[a] Based on analyses in Chapter 11; see Tables 11.4 and 11.5.
[b] Cub, yearling, other nonadult, and adult.

Table 18.7. Comparison, by cause-of-death, sex, and age of grizzly bears, of postclosure Observed Mortality (OM) with Expected Mortality (EM), Yellowstone Ecosystem, 1971-87. EM values in columns 3-4 and 6-7 equal expectations across sexes and within age class; EM values in columns 5 and 8 equal expectations across age classes.

Age category	Mortality estimate	Management control			Legal kills		
		Male	Female	Total	Male	Female	Total
All nonadult	OM	18	6	24	4	6	10
	(EM)	(12.0)	(12.0)	(35.1)	(5.0)	(5.0)	(20.3)
Adult	OM	28	19	47	14	17	31
	(EM)	(23.0)	(24.0)	(35.9)	(15.2)	(15.8)	(20.8)
All ages	OM	46	25	71	18	23	41
	(EM)[a]	(34.8)	(36.2)	—	(20.1)	(20.9)	—

[a] EM values do not equal sums of those for "all nonadult" and "adult" categories because they are based on a weighted average of sex ratios for both categories.

Table 18.8. Probability values associated with tests for effects of age and sex of grizzly bears on mortality due to management control or legal kill in the Yellowstone Ecosystem, 1959-70 and 1971-87. G-test for goodness-of-fit; the term "excess" means significantly greater mortality than expected; see text for methods of calculating expected mortality.

	P-values (degrees of freedom)			
	Management control		Legal kill	
	1959-70[a]	1971-87	1959-70[a]	1971-87
Effect of sex (by age):				
All nonadult	>0.40 (1) (NS)	0.02 (1) (Male excess)	0.001 (1) (Male excess)	>0.50 (1) (NS)
Adult	>0.19 (1) (NS)	0.15 (1) (NS)	0.12 (1) (NS)	>0.60 (1) (NS)
Effect of age:				
Nonadult versus adult	>0.40 (1) (NS)	0.01 (1) (Adult excess)	<0.03 (1) (Adult excess)	0.001 (1) (Adult excess)

[a] Based on analyses in Chapter 11; see Tables 11.9 and 11.10.

Cause-specific Mortality

Repeating the above analyses for specific causes of death, we found that observed adult mortality from legal kills exceeded expectation during the 1971-87 period (P<0.001; Tables 18.7 and 18.8). This also was true during the 1959-70 preclosure period (Table 18.8). But during 1971-87, there was no evidence of sex-differential vulnerability to legal kill for nonadults or adults (P>0.25 and 0.50, respectively). In contrast, nonadult males were more vulnerable than nonadult females during 1959-70 (Table 18.8). In addition, among deaths attributable to management control actions during the 1971-87 period, those of adults were greater than expected, and those of nonadults less than expected (P<0.01; Tables 18.7 and 18.8). During the postclosure period, management-related mortality also varied from that expected based on sex ratio. That is, nonadult male deaths were greater than expected, whereas those of females were fewer (P<0.025; Table 18.8). In contrast, no effects of age or sex were found in the analyses of the 1959-70 management-related mortality data (Table 18.5).

Influences of Age and Sex on Mortality

Interpretation of these results demands caution, in view of the different methods employed for estimating sex ratio and, especially, age structure in the two periods and the great diversity of factors potentially influencing mortality. Nevertheless, these differences in patterns of cause-specific mortality between the two periods do have a coherent explanation that is consistent with the spatial distribution theme of Chapter 15: that ecocenter destruction, in reducing both the amount and spatial concentration of food, caused grizzlies to range more widely and take more risks in pursuit of food. The vulnerability of adults to all causes of mortality and, specifically, to control actions was plausibly a consequence of the increased movement of adults in search of food, which exposed them to greater risk from hunters and to bear-human conflict. Bears with higher potential rates of growth or larger absolute energy requirements (e.g., adults of both sexes, but particularly males) would have been more aggressive in seeking food in developed areas or at hunter camps when natural foods or foods at protected ecocenters were relatively unavailable.

THE SPATIAL DISTRIBUTION OF MORTALITY

In Chapter 11, we showed that the median distance from sites of 262 deaths to the approximate center of the Yellowstone Ecosystem at Trout Creek was 48.1 kilometers (29.9 mi) for the 1959-70 period, and that 54.2% of the death locations were within park boundaries (see also Table 18.9). Furthermore, we showed that there was a trend for more deaths to occur outside the Park in the second half of the 1959-70 period and a corresponding, significant increase in the dispersion of deaths from the Trout Creek ecocenter. We attributed these latter differences to increased bear movement in response to phasing out the ecocenters and related manipulations beginning in 1968 (see Chapter 15). If this view is correct, we should expect to see an even more marked increase in the dispersion of deaths in the 1971-87 period, when ecocenters were entirely absent. This was the case (Table 18.9). The median distance from Trout Creek of 234 mortalities during 1971-87 was 58.7 kilometers (36.5 mi), a highly significant increase over the median for the earlier period (Mann-Whitney U Test, $P \ll 0.001$). Fully

Table 18.9. Percentage of mortalities of grizzly bears outside versus inside Yellowstone National Park, and median distance of mortalities from the Trout Creek Ecocenter, by 2-year increments,[a] 1959-70 and 1971-87.

Year	% out of YNP	Median distance from Trout Creek in kilometers (mi)		n
59-60	55.9	47.8	(29.7)	34
61-62	47.1	37.7	(23.4)	34
63-64	40.0	31.1	(19.3)	25
65-66	61.5	55.4	(34.4)	26
67-68	59.0	49.7	(30.9)	61
69-70	54.9	51.3	(31.9)	82
1959-70 period	54.2	48.1	(29.9)	262
71-72	75.7	57.0	(35.4)	74[b]
73-74	90.9	61.0	(37.9)	33
75-76	80.0	70.0	(43.5)	10
77-78	70.8	60.7	(37.7)	24
79-80	61.1	57.0	(35.4)	18
81-82	76.7	65.0	(40.4)	30
83-84	72.2	50.1	(31.1)	18
85-86	58.3	59.4	(36.9)	24
1987	66.7	47.2	(29.3)	3
1971-87 period	74.4	58.7	(36.5)	234

[a] Excepting final entry, which is for the single year, 1987.
[b] Total omits two bears included in total for 1971 in Table 18.1 because the locations of death were unknown.

Table 18.10. Location of deaths of grizzly bears resulting from management control actions, 1959-70 and 1971-87.

Management control subcategory	Period			
	1959-70		1971-87	
	No.	Percent	No.	Percent
In the Park:				
Intentional	68	70.1	20	27.0
Accidental	11	11.3	9	12.2
Unknown	6	6.2	1	1.4
Subtotal	85	87.6	30	40.5
Out of the Park:				
Intentional	8	8.2	39	52.7
Accidental	0	0.0	3	4.1
Unknown	4	4.1	2	2.7
Subtotal	12	12.4	44	59.5
Total	97		74	

74.4% of the deaths were outside park boundaries, and the between-period difference also was highly significant statistically (Chi-square test of independence, $P<0.001$). The magnitude of the redistribution (20.2%) suggests that the difference was highly significant biologically, as well. Deaths from management control actions (versus all causes) shifted even more dramatically from predominantly inside the Park in 1959-70 (88% of 97 cases) to predominately outside the Park in 1971-87 (60% of 74 cases; G-test, $P<<0.001$; Table 18.10).

These data show that bears of the postclosure period were not only using more peripheral areas of the Ecosystem (see Chapter 15), they were also dying there, particularly in management actions. Again, however, these data tell us nothing about the postclosure mortality rate. Thus, the spatial shift in management activity might have been the result of a constant per capita risk of management actions that was simply redistributed over all possible locations of bear-human conflict. But arguing by analogy from the elevated mortality rates observed for non-Trout Creek versus Trout Creek bears during 1959-70 (Chapter 11), it is more likely that the per capita risk of death associated with management control increased for some years after 1970. This increase was disproportionately expressed in developed areas outside the Park because bears used these more frequently after losing the security conveyed by ecocenters emplaced within the Park. In the following section, we look more directly at mortality rates in the postclosure years.

MORTALITY RATES
Absolute Versus Per Capita Mortality

In Chapter 11, we showed that annual ecosystem-wide mortality rose sharply in 1970 (57 deaths) after closure of the largest ecocenter at Trout Creek. Absolute numbers of deaths remained relatively high in 1971 (47), but then declined rapidly to pre-1969 levels (Table 18.1). A minimum of 158 bear deaths was recorded over the period 1969-72, during and immediately after ecocenter closures (Craighead *et al.* 1988b). This is a number equivalent to approximately 68% of the best-estimate, ecosystem-wide population in 1968 (231), and 51% of the most optimistic, upperbound estimate (309) for that year (Craighead *et al.* 1974). Deaths by management control, legal kill, and illegal kill continued to account for the majority of human-caused mortality after 1970 (Table 18.11). Although we cannot calculate the per capita mortality rate for the early postclosure years, these considerations make it clear that it must have been excessively high in 1971 and remained high for some time thereafter.

The decline in absolute mortality recorded for 1975 does not imply a comparable decline in mortality rates. If absolute mortality is a valid indicator, the postclosure population-at-risk was clearly much reduced for at least several years. Unfortunately, in the absence of rigorous estimates of population size for the postclosure period, we cannot know from these data if or how quickly mortality rates stabilized, or at what level. Because

Table 18.11. Distribution and causes of mortality of grizzly bears within the Yellowstone Ecosystem, 1959-70 and 1971-87.

| | Period | | | |
| | 1959-70 | | 1971-87 | |
Cause	No.	Percent	No.	Percent
Management control	97	35.0	74	32.7
Legal kills	114	41.2	54	23.9
Illegal kills	13	4.7	66	29.2
Natural mortality	6	2.2	9	4.0
Accidental[a]	24	8.7	11	4.9
Unknown	23	8.3	12	5.3
Total	277	100.0	226	100.0

[a] Roadkills and research-related mortalities.

this was one of the major research objectives recommended by the NAS Oversight Committee (Cowan *et al.* 1974) and subsequently adopted by the Interagency Grizzly Bear Study Team (IGBST) (Knight *et al.* 1975), we are surprised and concerned that the objective was not pursued more vigorously by these controlling bodies. Clearly, annual death rates during the postclosure period of study (IGBST) would have been key parameters in determining population trend. We believe it important to arrive at an understanding of mortality rate and its significance during the postclosure period, although the nature of the data greatly complicates our task.

Fate of Marked Bears in the Early Postclosure Period

Translocation of Marked Bears

From 1959 to 1989, marked grizzly bears, captured in management control actions within the Yellowstone Ecosystem, were translocated to other sites in the Ecosystem on more than 300 occasions (Table 18.12). Although the way in which these translocations relate to bear deaths may not be immediately apparent, they do relate directly to the issue of mortality sinks peculiar to the 1971-73 period. Thus, before discussing our reconstruction of postclosure mortality for this period, we will summarize translocation outcomes for three periods; 1959-70, 1971-73, and 1979-89. Our own translocation records, gathered in cooperation with the managing agencies at the time, are the source of the 1959-70 data. Those for the 1971-73 period were provided by the National Park Service. Translocations for the third period have been published (beginning consistently in 1979) in IGBST annual reports. Case-specific records for the 1974-78 pe-

riod were not consistently available, so this period was omitted from our evaluation. Translocation outcomes were determined from our life histories (1959-70), NPS logbooks (1971-73), Knight life histories (1981-89) and the Craighead *et al.* (1988b) mortality database (all periods).

For purposes of comparison, we define a successful transplant as one in which the bear was known to survive at least one year after translocation and was not retrapped in a management action during that time. Many such bears were trapped for management reasons two or more years later and hence were really failed translocations in the longer term. Our goal, however, is not to define some unambiguous standard of transplant success, but rather to use definitions that can be applied both to the brief 1971-73 period and to the longer intervals of study before and after. Translocation failure, in our treatment, is defined by a management control recapture within one year of the translocation in question. We did not distinguish between recapture following return to the original capture site versus recapture in the area to which the bear was relocated. Thus, "failure" was a problem bear returning soon after translocation to cause trouble and attract management attention. Translocations in which the bear was never again recorded as observed after transplant were classified as unknown in outcome. We excluded translocations in which the bear was known to die from non-management control causes in the same year as transplant. These were, in effect, "aborted" experiments. Finally, we also excluded translocations to areas outside the Yellowstone Ecosystem.

Our tabulation of outcomes by time period shows substantial similarity between the 1959-70

Table 18.12. Percentage and number of translocations of marked grizzly bears that were successful, unsuccessful, or of unknown outcome during 1959-70, 1971-73, and 1979-89, Yellowstone Ecosystem; excludes translocations out of Yellowstone Ecosystem, and cases in which nonmangement-control mortality occurred in year of translocation.

Period	No. of different bears translocated	Translocations			Total no. of translocations
		% Success[a] (*n*)	% Failure[b] (*n*)	% Unknown[c] (*n*)	
1959-70	76	17.6 (26)	74.3 (110)	8.1 (12)	148
1971-73	47	20.8 (11)	17.0 (9)	62.2 (33)	53
1979-89	72	27.7 (36)	59.2 (77)	13.1 (17)	130

[a] Bears known to survive one year post-translocation without retrap for management action.
[b] Bears retrapped for management action within one year of translocation.
[c] Bears never again recorded as observed after translocation.

and 1979-89 periods (Table 18.12). Most translocations, 74% and 59%, respectively, were immediate failures, and the great majority of relocated bears, 92% and 87%, respectively, were observed in one way or another after translocation. Thus, only 8% and 13%, respectively, disappeared entirely. Conversely, translocation outcomes in 1971-73 stand out in marked contrast with the other two study periods. A similar percentage of transplants were "successful" (roughly one-fifth), but there was a drastic redistribution of cases from the "failure" to the "unknown" category (Table 18.12). In short, 1971-73 problem bears did not return to cause trouble after translocation, as they did consistently during both 1959-70 and 1979-89. Rather, they simply disappeared. If we use the average percentage of transplants during the 1959-70 and 1979-89 periods that were never again recorded as observed (10.4%) to estimate an expected value for that parameter during 1971-73, we find that 28 more bears disappeared following translocations during 1971-73 than should have been the case (33 - [47 × 0.104]).

Craighead Study Bears

Following closure of the major ecocenters in 1969 and 1970, mortality rate from all (primarily human) causes increased dramatically (Fig. 11.8) and remained high in 1971, the year after the Trout Creek closure. We would most like to know whether, and for how long, these high rates of mortality persisted after 1971.

We have pointed out that a serious flaw in the postclosure study was a three-year data gap from 1971 through 1973. This gap and the absence of reliable population size estimates after 1970 have confused analyses of all postclosure biological parameters, including that of mortality rate. The large number of marked bears at the end of our study, in combination with mortality records for 1971-87 (Craighead *et al.* 1988b) and IGBST trapping records after 1974, provided an opportunity to address the crucial question of postclosure per capita mortality.

In the final year of our study (1970), 66 marked bears were known to be alive (36 males and 30 females, mostly Trout Creek bears). Known-ages of these bears ranged from 2-19 years among males, and 2-24 years among females. For purposes of the following analyses, three adult bears of unknown age were assigned the median age (7 and 12 years for males and females, respectively) observed among known-age bears.

Subsequent to 1970, there were two systematic ways in which these marked bears might come to the attention of researchers. First, they might have appeared in the mortality records (Craighead *et al.* 1988b) and second, beginning in 1975, they might have been live-trapped by the IGBST. Provided that mortality and live-trap samples were random annual samples from the population, the proportion of marked bears in these samples is a maximum likelihood estimate of the proportion of marked Craighead bears in the population in any given year. From 1975 to 1981, the proportions of our marked bears in the mortality and live-trap samples, and in these samples combined, were 2.9%, 4.6%, and 3.4%, respectively (Table 18.13).

These values suggest either a relatively small number of 66 marked bears present in 1970 survived to 1975 and beyond, or that recruitment in the intervening period was substantial enough to dilute significantly these marked bears within the population. The latter possibility can be discounted. All indicators suggest that, far from increasing, the Yellowstone Ecosystem population declined, at least through 1981 (Knight and Eberhardt 1985). At this point, we are left with the qualitative, and now widely accepted conclusion that ecocenter closure increased absolute mortality and reduced population size. At the same time, however, an obvious question is whether, given the age structure of marked bears in 1970, the fraction of marked bears in the mortality and live-trap samples after 1975 was consistent with "normal" rates of mortality that might have been expected in the intervening period (1970 to the year in question). The marked bears known alive in 1970 and the uncontroversial assumption that the postclosure population was not greater than 300 through 1975 allow us to make quantitative statements about postclosure mortality rates relative to a preclosure standard.

For our analysis, we project, beginning in 1970, the number of marked bears that could be expected to survive to each year from 1975 through 1981 under different levels of survivorship. Then we ask which projected level of survivorship is most consistent with the number of marked animals actually observed in the live-trap and mortality samples.

The "base" level of survivorship used in these projections was calculated from sex- and age-specific survivorship values reported for the 1960-67 period (before any ecocenter closure) by Craighead et al. (1974). For example, a 2-year-old male in 1970 must have survived 5 years to age seven to have been available for observations in 1975. The expected number of marked 2-year-old males alive in 1970 surviving to 1975 (based on 1960-67 levels of survivorship) is the product of the yearly survivorship probabilities for the ages 2-7 (in this case $0.86 \times 0.82 \times 0.51 \times 0.94 \times 0.94 = 0.32$) times the number ($n = 2$) of 2-year-old marked males alive in 1970 ($2 \times 0.32 = 0.64$). Expected numbers for other age and sex categories represented in the 1970 marked population were calculated similarly, and then summed to project the total number of marked bears expected to be alive in 1975. Similarly, survival values were projected for each year through 1981 by extending the product of survivorship probabilities (as described above) to include the appropriate probabilities for each additional year. We plotted expected male survivorship to 1975 versus male age in 1970 (for example, Fig. 18.1),

Table 18.13. Number and percentage of 66 marked grizzly bears alive in 1970 subsequently appearing in Interagency Grizzly Bear Study Team trap records or the Yellowstone Ecosystem mortality database (Craighead et al. 1988b).

Year	Marked bears in trap records			Marked bears in mortality database			Marked bears in trap & mortality records[a]		
	Total captured	Number marked	Percent marked	Total mortalities	Number marked	Percent marked	Total captured and/or dead	Number marked	Percent marked
1975	11	0	0.0	3	0	0.0	14	0	0.0
1976	14	1	7.1	7	0	0.0	21	1	4.8
1977	26	1	3.8	17	0	0.0	43	1	2.3
1978	9	0	0.0	7	1	14.3	16	1	6.3
1979	17	1	5.9	8	0	0.0	25	1	4.0
1980	25	0	0.0	10	0	0.0	35	0	0.0
1981	36	1	2.8	13	2	15.4	49	3	6.1
Total	138	4	2.9	65	3	4.6	203	7	3.4

[a] In a total of 203 captures and mortalities, three bears marked by the Craighead team were captured or recovered as mortalities a total of seven times.

Fig. 18.1. Example of the use of life table data and regression to estimate the expected survivorship of marked grizzly bears alive in 1970 to some future year if mortality rates characteristic of the 1960-67 period prevailed after 1970. Survival of male bears to 1975 illustrated here.

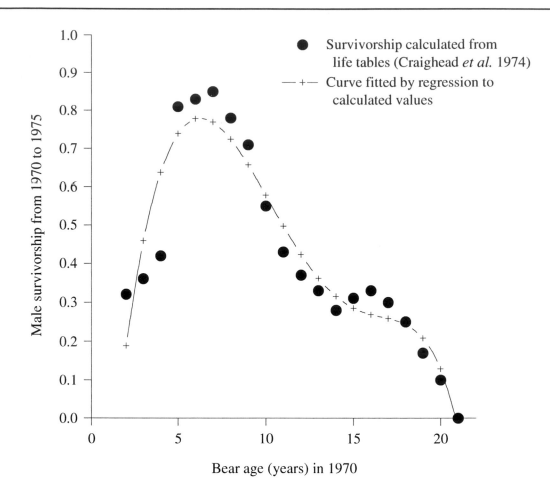

marked bears alive in these years would ideally be calculated as the ***observed proportion*** of marked bears in the mortality and live-trap samples times total population size in that year. Because population size estimates were not available for the years after 1970, we calculated a "maximum" and "minimum" estimate of this number by assuming the population could not have exceeded 300, nor been less than 100 (Table 18.14). Finally, we calculated an index of goodness-of-fit between observed and expected numbers for each model by calculating the average squared deviation between estimated and projected values over the seven years from 1975 to 1981.

and made similar plots relating expected survivorship versus age-in-1970 for both sexes and for all years, 1975 through 1981. We then fitted a least-squares regression curve to each of these 14 scatterplots (2 sexes times 7 years) to yield a set of 14 equations giving a "base" probability of survivorship to any year for either sex of any age. The range of R^2 values for these 14 equations was 0.90 - 0.97 for males and 0.97 - 0.99 for females.

Next, we created nine additional "levels" of survivorship by multiplying each base equation by 0.9, 0.8, 0.7, . . . and 0.1. That is, lower levels of survivorship were created by assuming that survivorship to a point in the future was reduced for all ages by a constant fraction of 1960-67 values.

We then used these 10×14 = 140 equations to project the total number of 1970 marked bears ***expected*** to be alive in each of the seven years from 1975 to 1981, given each of the 10 levels of survivorship (the 1960-67 base level and 9 lower levels) (Table 18.14). The ***observed*** number of

Assuming a population size of 300 over the period, 1975-81, the survivorship model that best projects numbers of surviving marked 1970 bears (according to our index-of-fit) was one in which survivorship was 40% of that observed for the 1960-67 period. Similarly, the best model, assuming a population size of 100 during 1975-81, placed

Table 18.14. Projected numbers of marked grizzly bears alive in 1970 surviving to 1975 through 1981 under 10 different levels of survivorship in relation to the number of these bears estimated to have survived based on Interagency Grizzly Bear Study Team trap and Craighead *et al.* (1988b) mortality records, Yellowstone National Park.

Year	Projected numbers under survivorship level[a]										Estimated numbers for population size[b]	
	1.0	0.9	0.8	0.7	0.6	0.5	0.4	0.3	0.2	0.1	100	300
1975	34.8	31.3	27.8	24.3	20.9	17.3	13.9	10.4	7.0	3.5	2.9	8.6
1976	30.5	27.4	24.4	21.2	18.2	15.2	12.1	9.1	6.1	3.0	3.1	9.4
1977	26.1	23.4	20.9	18.3	15.6	13.1	10.5	7.8	5.2	2.6	3.4	10.2
1978	22.0	19.8	17.7	15.4	13.2	11.0	8.8	6.6	4.4	2.2	4.9	14.6
1979	18.2	16.4	14.6	12.7	10.9	9.1	7.2	5.4	3.6	1.8	1.7	5.0
1980	14.6	13.2	11.8	10.3	8.8	7.4	5.9	4.4	3.0	1.5	3.6	10.7
1981	11.7	10.5	9.3	8.2	7.0	5.9	4.7	3.6	2.3	1.2	3.1	9.3
Goodness-of-fit index:[c]												
N=100	435	341	260	188	128	81	44	19	5	2		
N=300	234	169	116	74	44	25	17	21	36	63		

[a] Expressed as a fraction (1.0, 0.9, 0.8, etc.) of 1960-67 levels.

[b] Based on proportion of marked 1970 bears in IGBST trap records and Yellowstone Ecosystem mortality database relative to hypothetical minimum (100) and maximum (300) population levels.

[c] Equals squared difference between projected and estimated numbers within each year, averaged over all 7 years, and rounded to nearest whole integer; smallest index value indicates the closest fit.

survivorship at only 10% of the base level. Either model represents a substantial reduction relative to the pre-1968 survivorship standard. These analyses and records of known mortality lead us to conclude that mortality rate for at least a segment of the population remained unacceptably high for several years after 1970.

Interpretation of Translocation and Mortality Data

What are we to make of the anomalous bear translocation record and inferred rate of marked bear survival for the immediate (1971-73) postclosure period? First, in contrast with results for the translocation programs during the 1959-70 and 1979-89 periods, failed translocations were relatively rare during 1971-73, and the frequency of unknown outcomes (bears never again recorded as observed after being translocated) was greatly elevated. The unmistakable implication is that bear survival following translocation during 1971-73 was very poor. Two possible explanations are: (1) bears died as a result of translocation in ways or at sites that precluded discovery for the record, or (2) bears were known to have died, for whatever reason, in the process of translocation, and records were not kept. In the first case, we must be concerned about what kind of methodology could have produced such poor results. In the second, we have that same concern and more. Could bears have died, whether

accidently or purposefully, and no records have been kept? As improbable as such a thing might seem, the possibility must be entertained, especially because it also offers one explanation for the poor postclosure survival of bears marked during the 1959-70 study and still alive in 1970.

With respect to the poor postclosure survival of marked bears, alive in 1970, there are two extreme possibilities. First, these survivorship probabilities may have been representative of all bears in the ecosystem population during the postclosure period. If so, severely elevated rates of mortality that were comparable to or, more likely, in excess of the high 1969-70 levels (Chapter 11) must have persisted for some time after 1970, and the Yellowstone Ecosystem grizzly population must have experienced a collapse beyond anything documented or yet imagined. Alternatively, survivorship of marked 1970 bears may not have been representative of the entire population; i.e., marked bears may have somehow been at greater risk than unmarked bears for mortality after 1970. Most of the marked bears were Trout Creek ecocenter bears. Thus, it might be argued that these bears were habituated to garbage and, when the dumps were closed, were chronically involved in conflicts with humans, some fatal, at campgrounds and at other developed areas. If this were a major factor, and if, as should be expected, all management control actions were dutifully recorded, we would expect to find most of the 1970 marked bears in the mortality database (Craighead *et al.* 1988b). But only 21 of the 66 marked 1970 bears were recorded as mortalities before 1975 (1971-74), and only 13 of these deaths were due to management control actions. This suggests that (1) unrecorded natural mortality peculiar to marked bears accounted for the difference; (2) not all 1970 marked bears were recognized as such during the postclosure trapping; or (3) not all management control fatalities were recorded. The first suggestion is logically unlikely. The second suggestion is unlikely in view of the fact that, in 1970, all postclosure marked bears bore vinyl ear tags, numbered metal ear tags, and a lip and axillary tatoo. Where tags were lost, punctures and ear scars clearly identified a bear as having been marked. Trapping personnel had been thoroughly informed on the importance of, and indications for, recognizing marked animals. (R. Knight, pers. comm., 1976). Thus, the chances of recognizing all recaptured marked bears were excellent. The third suggestion implies either an undetected "breakdown" in the mortality recording system (highly unlikely, also) or a decision not to record the deaths of a significant number of grizzlies killed in "control" measures during 1971-73.

As disturbing as this latter scenario might be, it is consistent with the foregoing information describing the absence of marked bears translocated during 1971-73 from all subsequent records, and with a population discovered in the early years of the IGBST study (1975-80) to be markedly depauperate of bears in the older age classes (Knight *et al.* 1981). A program for covertly eliminating problem grizzly bears is not as incredible as it might seem, when one recalls that the park administration professed to believe that there were two distinct populations of grizzlies: garbage bears and backcountry bears. By park administration definition, bears marked at the Trout Creek ecocenter could be considered "garbage addicted" and therefore "problem bears." Elimination of the marked, "man-conditioned" bears would allow the hypothetically "wild, free-roaming" unmarked population of the backcountry, purported to existed at that time, to regenerate itself, leaving only "unspoiled" bears to inhabit the Park and the Ecosystem. To understand fully the present, we must know what happened in the recent past (1971-74). We urge the IGBST to examine this period more closely.

SUMMARY

Our analysis of the vulnerability of the different demographic classes to mortality following closure of the dumps (ecocenters) indicated that deaths of adults and nonadult males were more frequent than expected, given their representation in the population, when all causes of mortality and management control actions were considered. This contrasted with the 1959-70 period when similar comparisons revealed no conclusive evidence of differential vulnerability according to sex and age. Our analysis of the spatial distribution of death in the two periods indicated an increased dispersion of deaths over the Yellowstone Ecosystem after 1970. Together, these results are consistent with the view, supported by other lines of evidence (Chapters 15 and 16), that postclosure bears searched more widely over the Greater Yellowstone Ecosystem for food, and that this movement increased the frequency of fatal bear-human conflicts, particularly for bears with large absolute energy and nutrient requirements, such as adults or nonadult males experiencing rapid growth.

We identified a curious difference in the outcomes of programs for translocating problem bears during 1959-70 and 1979-89 versus 1971-73. During 1971-73, documented translocation failures were relatively rare, and there was a sharp increase in the frequency of unknown outcomes (bears never again seen). To our knowledge, this pattern has never been acknowledged, let alone explained, yet it is deserving of thorough review and evaluation. It is unlikely that the excess of "unknown" outcomes reflects such successful translocations that the bears involved completely avoided subsequent contact with humans. If so, however, we need to know what methodologies were responsible so they can be applied routinely. Alternatively, this excess may reflect increased bear deaths, unknown, or known-but-unrecorded, as a result of transplant. If so, we need to know what methodologies or circumstances were responsible so that we can ensure they are never again a part of management policy.

Our analysis of the fate of 66 marked bears alive in 1970 indicates that they survived at rates that were 10-40% of preclosure (1960-67) survivorship levels. Only 21 of the 66 were recorded as mortalities during the 1971-74 period, and, of the 45 unaccounted for by 1975, only 3 individual bears subsequently showed up in the trapping or mortality records of the IGBST. A seemingly straightforward explanation of the disappearance of 42 marked bears from all records is that marked bears trapped by the IGBST were not recognized as such. The bears were clearly marked, however, and the trapping procedure called for identifying all previously marked bears. Another possibility is that deaths of all marked bears had been duly recorded, and these deaths were not peculiar to marked status; that is, survivorship of marked bears, relative to the 1960-67 bear mortality standard, was representative of the entire ecosystem population. This would imply an unsuspected population decline of huge proportions operating naturally (e.g., disease, starvation in dens) or unnaturally (e.g., large-scale poaching). A third possibility is that grizzly bears were killed in "control" measures during 1971-73, and records were lost or possibly not kept. This latter explanation is consistent with other anomalous demographic data, such as the unusual numbers of marked bears never again observed after their translocation during the 1971-73 period, and a population age structure for the 1975-80 period markedly low in adults of both sexes (Knight *et al.* 1981).

Regardless of whether a clear explanation is ever uncovered, the low survival rates of 66 marked bears alive in 1970 suggest a demographic event having significant impact on the structure of the population that remained by 1975. The size and structure of that residual population, and the dynamics of it into the future, can only be satisfactorily explained when we have more information on translocations, death rates, and causes of death during 1971-74. Death events during those years, in conjunction with a persistent food stress, could have significantly affected reproduction by initiating compensatory reproductive mechanisms not operative during the preclosure period. It is important to remember that reproductive assessment was a primary research objective of the IGBST. We will examine this subject in the next chapter.

19
Assessment of Size and Trend of the Yellowstone Grizzly Bear Population

A major objective of the Interagency Grizzly Bear Study Team (IGBST) field study, as recommended by the National Academy of Science Committee (NASC), was to determine the size and trend of the grizzly bear population inhabiting the Yellowstone Ecosystem following the closure of the open-pit garbage dumps (see Chapter 14). Here, we examine whether this primary research objective was accomplished by describing and comparing the censusing techniques employed and, where feasible, the results obtained. We then evaluate and discuss the estimates of population size and trend that have been calculated for the Yellowstone grizzly bear population based on various kinds of census data.

Measuring the size and trend of an animal population through time and space is one of the more difficult tasks of field biology. When it is accomplished with reasonable scientific accuracy, the resulting data provide a solid basis for population analysis and interpretive insights not possible from other biological parameters. Measuring these characteristics of a grizzly bear population distributed throughout an entire ecosystem presents an especially unique challenge. Our approach to this task between 1959 and 1970 was quite different from that employed by the Interagency Grizzly Bear Study Team (IGBST) from 1975 to date, both in technique and result.

CENSUSING THE POPULATION

The most direct method of determining the status of a bear population over time is to estimate population size periodically. We arrived at annual estimates of population size during the 1959-70 period of study by means of census counts and estimates of census efficiency (Craighead *et al.* 1974). For several years after closure of the last major dump in 1970, no data appropriate for measuring population status are available because, amidst the politically charged atmosphere surrounding the bear management program, no research plan to methodically monitor the bear population had been devel-

oped. The IGBST settled on a research approach beginning in 1974 (Knight *et al.* 1975). Captured bears, whether taken in management actions or a research trapping program, were radio-collared so they could be followed remotely, particularly from aircraft. As data accumulated from the marked bears, population characteristics such as age structure, sex ratios, and reproductive parameters could be estimated. At the same time, however, the numbers of "unduplicated" adult females with cubs observed annually were recorded for use as an indication of population size and trend. Although both ground and aerial observations of marked and unmarked bears contributed to the annual data sets, no systematic methodology for avoiding duplication in the "unduplicated" sample was presented.

Concerned about indications of a population decline, the Interagency Research Steering Committee (IRSC) later recommended an intensive, saturation capture-mark-and-release program to derive a population estimate (Kilgore *et al.* 1981). The senior author, in personal communication with IGBST leader Richard Knight (1978-79), had previously objected to such an approach as being too intrusive on a population already intensively studied and had pointed out that the capture-recapture data would be difficult to obtain, with the population so widely dispersed. Other bear biologists also had foreseen difficulties and predicted that such a census would not achieve the required accuracy (Harry Reynolds, III, Charles Jonkel, and Stephen Stringham, pers. comms. 1979). The IRSC plan for saturation trapping was never implemented, but a less ambitious program to capture and radio-collar bears was continued.

In 1983, the Ad Hoc Committee for Population Analysis (of the IRSC) reviewed the IGBST population data (Knight *et al.* 1983). They concluded that "the minimum number of breeding females is the most reliable index available to estimate population size" and recommended basing the index on three-year sums of annual counts of unduplicated females with cubs. This report essentially established by fiat an index to population size of absolutely unproven validity. Moreover, it provided no protocol for distinguishing *unduplicated* observations. Annual censuses measuring only the adult female segment of a bear population have the advantage of focusing on a single component of the population and thereby, conserving time, energy, and money. But they are an appropriate index of population health only if proven to be so, and only if the counts, themselves, are accurate and unbiased.

Nevertheless, the Ad Hoc Committee proceeded to calculate the size of the ecosystem grizzly bear population (Knight *et al.* 1983) by back-summing counts of adult females with cubs for 1982, 1981, and 1980 to arrive at a total census of 32 breeding females alive in 1980. On the assumption of a 50:50 adult sex ratio, 64 adults were hypothesized to be present, plus an additional six adult females to account for variation in age-of-first reproduction. The 20 cubs produced by the 32 females over the 1980-82 period were added, as were 49 nonadults (ages one to five years). The number of nonadults was calculated by applying age-specific survivorship estimates (based on survivorship of marked bears, 1974-82) to nonadult age cohorts descendent from an initial cohort of 20 cubs, the mean number of cubs per female-year produced by 32 females from 1975 to 1983. Because little scientific rationale or documentation was provided in support of this exercise, we doubt the credibility of the minimum population estimate of 139 bears that was the result. Moreover, we note that the IGBST Annual Report for 1985 (Knight *et al.* 1986), and all subsequent, indicated a three-year back-sum of 36 breeding females for 1980, instead of 32. Apparently, additional observations of three females with cubs were discovered, well after the fact, for 1980, plus an additional one for 1982. This oversight relates to the question of reliability, as well as applicability, of *unduplicated* sightings, a question we explore in detail later in the text.

We note, also, that the use of this method to sum the annual counts of adult females with cubs—that is, back-summing the counts over a three-year period—unavoidably would result in an over-estimate of the number of adult females in the first year and inflate the "minimum" population size calculated. For example, some unknown number of *adult* females calculated by back-summing to be present in Year (T) were very likely subadults: five-year-olds in Year (T+1) and five- or six-year olds in year (T+2). The documentation of females producing first litters at four years of age (Eberhardt *et al.* 1994) further confuses the issue of *adults* in Year (T). This

method of summing annual observations was used in nearly all studies employing the sums as an index.

With the publication of the Ad Hoc Committee report (Knight *et al.* 1983), use of the annual counts of unduplicated adult females with cubs as an index of population size and trend seemingly was accepted by the Interagency Grizzly Bear Committee. As we show below, however, that its value as an index remained unproven, and its adoption, subsequently, by the U. S. Fish and Wildlife Service as the fundamental measure of progress toward population recovery (USFWS 1993) raised serious concerns. If the index was not valid, and its measure not reliable, then projections of population status based on it were potentially in error. The population could be declared "recovered," when in fact, its status was unknown. During the mid to late 1980s, several studies explored the validity of the index as a predictor of population size and trend. In the following section, we examine these to determine if any proved it valid.

Validity of the Adult-Females-with-Cubs Index

Knight and Eberhardt (1984) created a stochastic model of the adult female grizzly bear population of Yellowstone Park and, using repeated computer simulations, projected population sizes from 1970 through 2000. Their model projected a continuous decline in population over the 30-year period. They noted that 3-year sums of adult females with cubs (their "observed" index of abundance) generally followed the declining trend of the population projections for the 1970-80 period, but they made no formal attempt to evaluate the relationship.

In a later, quite rigorous analysis of the dynamics of the Yellowstone population, Knight and Eberhardt (1985) stated that a reliable index of population trend was needed and suggested that a running sum of annual numbers of unduplicated females with cubs *may* serve that purpose. Once again, however, they provided no evidence of reliable correlation between the counts and population trend.

Subsequently, Eberhardt *et al.* (1986) addressed the population index issue more directly, using a prospective model of the dynamics of the adult female population over the 1959-89 period. They adjusted the model to "fit" the annual observed numbers of adult females with cubs by manipulating the number of adult females used as the starting population for the model and the value of Q_s, a single value in the recruitment equation that was a compression of three essential, hard-to-measure parameters—namely, survival of juveniles from age six months to five years, proportion of female cubs in litters, and mean reproductive cycle length. In none of the adjustments attempted did the index reflect the sharp decline in population trajectory associated with the high mortality recorded from 1970 to 1974. And the one population trajectory modeled using actual observed values for basic population parameters (bottom, Fig. 2 of Eberhardt *et al.* 1986) indicated a progressive decline in population from 1970 to 1985, a trend entirely unpredicted by the index. In fact, the χ^2 value for fit between the index and expected values generated from the projection model indicated that the fit was poor. What conclusions Eberhardt *et al.* (1986) reached regarding the validity of the index were somewhat obscure, but they appeared to place little confidence in its ability to predict trend and, certainly, they failed to demonstrate that ability. In our opinion, the suitability of annual numbers of unduplicated adult females with cubs for use as an index of population trend remained unproven.

Subsequently, no additional work by the IGBST was published that sought to prove a reliable relationship between the numbers of adult females with cubs observed annually and population size or trend. Yet, it came to be accepted as the principal index of population status. For example, in their report, "Equivalent Population Size for 45 Adult Females," the Yellowstone Grizzly Bear Population Task Force (Knight *et al.* 1988a) assumed an annual count of 15 adult females with cubs per year as the basis for calculating an equivalent population size for the grizzly bear population in the Yellowstone Ecosystem. The purpose of this exercise was to project, using field data, the size of a grizzly bear population that would correspond to 45 adult females, the three-year total for annual counts of 15 unduplicated adult females with cubs. Data for estimating proportions of female cubs, (estimated from both birth and death ratios), reproductive cycle lengths, and age-specific survivorship rates were based on bears marked or radio-tracked from 1975 to 1987. Average annual litter size was determined from the *unduplicated* observations of females with cubs over the same period. Using these estimates, Knight *et al.* (1988a) projected a stable sex and age structure and calculated rates of population growth of +0.075 and +1.5% for scenarios in which 49% of cubs were female (birth ratio) or 42% of deaths were female (death ratio), respectively. Summing of adult female age classes from the projected sex and age structure yielded proportions of 0.2634 (assuming sex ratios at birth) and 0.2473 (assuming sex ratios at death) for the adult female cohort. Proportions of adult females in the population were divided into 45, the presupposed number of adult females observed with cubs over a three-year period. This exercise provided population estimates of 171 and 184.

The estimates of +0.07% and +1.5% for annual growth rate were encouraging, but the Task Force cautioned that the accuracy and precision of the estimates were low (Knight *et al.* 1988a). Nevertheless, the assumptions and methodologies set forth in the 1988 Task Force report were incorporated, in their entirety, into the revised Grizzly Bear Recovery Plan (Appendix C of USFWS 1993) as the basis for estimating minimum population size, not only for the Yellowstone grizzly bears, but for all bear populations targeted for recovery. It is important to recognize that the annual count of *unduplicated* adult females with cubs is *the* most crucial parameter among the four established in the Recovery Plan as criteria for population recovery. It is the basis for projection of population size, and the other three parameters—distribution of family units and allowable known mortality among adult females and among the population in total—derive from the counts, directly or indirectly.

Recent work by Mattson (1995, unpubl.) addresses the possibility of bias in the counts of unduplicated females with cubs in an attempt to evaluate their validity as an index of population size and trend. He makes the assumption that the counts are accurate (i.e. all females were in fact unduplicated) and then asks whether search effort and observability, rather than changes in population size, could account for the observed variation in the counts. His analyses show that, indeed, annual variation in counts of unduplicated females with cubs is explainable by variation in search effort and observability, and he concludes that the raw counts vary without any known relationship to population size. He notes further, that the trend in unduplicated counts, when the influences of effort and observability are statistically controlled, is consistent with a declining, rather than increasing, population. Mattson's analyses also indicated that use of the unduplicated-females-with-cubs index to predict population size and subsequently, permissible mortality, allowed a >10% risk of over-estimating allowable deaths (mortality >6%), when population and mortality ratios varied over a possible spectrum of values. He then suggested a different treatment of the index that would reduce risk of allowing excessive mortality to 10% or less, but cautioned that this procedure would not remedy errors inherent to the methodologies for estimating unduplicated females, proportions of adult females, or lengths of reproductive cycles.

It is our contention that annual number of unduplicated adult females with cubs is unproven as an index of trend. But even if it had been proven, its use in the Recovery Plan as the basis for projecting population size raises the issue, once again, of what constitutes "unduplicated." The research protocol for achieving the annual counts has yet to be published. Are the counts reliably free from error and identifiable biases? We have reason to believe that they are not, and we discuss the basis for our concerns in the sections that follow.

Accuracy of the Annual Counts of Adult Females with Cubs

Early censuses of females with cubs relied on ground observations from a wide variety of sources. In later years, the IGBST census consisted of counts of both marked and unmarked females with cubs from combined ground and aerial identifications. This process may have been designed independently or evolved in response to McCullough's suggestion of "developing a standardized set of routes and flying times to equalize the effort between years of counts of unduplicated bears" (McCullough 1986). Regardless, neither McCullough nor members of the Interagency Research Review Committee defined *unduplicated count*. Also, they did not adopt the multiple-recognition standard that we used, a criterion essential to precise, *unduplicated* counts of unmarked animals. Except for identifications of radio-collared animals, application of the term, *unduplicated*, was a subjective determination.

Conditions for censusing the population differed greatly between the two studies, and these, to a large extent, dictated the specific techniques employed. During our study, bears aggregated at ecocenters, where they were readily captured, marked, and reobserved. But motorized access to backcountry roads for observation purposes was greatly restricted, and aerial observations, prohibited. Our research team developed the techniques for radio-locating bears (Craighead and Craighead 1963, 1965) and used them to obtain a wide range of behavioral data. But relatively few bears were radio-collared, and we could not rely on radiotracking as a census technique.

During the IGBST study, bears did not aggregate and were more widely distributed during the summer months (see Chapter 14). Use of backcountry roads and aircraft was permitted. The IGBST team also used radio location as a census technique, but employed it in conjunction with other less accurate techniques for annually enumerating adult females with cubs.

Regardless of these differences, the objective of both research teams was to identify individual animals, and to use observations of recognizable bears as a database for estimating the adult female segment of the population. Potential techniques for individual identification were:

a. Identification of individuals from year to year by means of artificial markers, including ear tags, colored markers, and radio signals;

b. Identification of individuals from year to year by recognition of permanent, unique natural marks or distinctive conformation;

c. Identification of unmarked animals as individuals within a single year by means of nonpermanent natural markings, such as pelage color or patterns; and

d. Identification of unmarked animals as individuals by isolation of observations in time and space.

Identifications by the IGBST were made by direct observation from the ground, from the air, from recaptures, and from radio signals. Our identifications were made by direct observation from the ground.

With close-range multiple observations, techniques (*a*) and (*b*) are highly reliable and provide accurate identification of individual animals. Technique (*c*) can be effective for individual identification over the course of a single field season, if based on multiple sightings, but is highly unreliable for year-to-year recognition. Technique (*d*) requires numerous assumptions and is subject to a wide range of possible errors. Any of the four identification techniques, with the exception of radio-location, are considerably less accurate when applied from the air. Before comparing census data obtained by the two research teams, we consider which of the various techniques (*a,b,c,d*) they employed, as well as their observational procedures, and how thoroughly they verified individual identifications through repeated observation in a given year.

We explained our census methodology used to describe the Trout Creek population unit from 1959-70 in Chapter 3. Only techniques *a* and *b* were used; that is, the census of adult females (with and without offspring), and of other age and sex classes as well, was based on multiple observations of marked or permanently recognizable animals. All of our identifications were made from the ground at close range by our research team. Because of the marking system employed and the replicated sightings of recognizable individual bears by trained personnel, the accuracy of identification was extremely high.

The IGBST employed all four census techniques (*a,b,c,d*) to characterize the ecosystem-wide population from 1976 to 1990. Identifications made using radio signals were highly reliable. Identification of

Table 19.1. Results from Census Technique (a) during the Craighead and IGBST grizzly bear studies, Trout Creek ecocenter, 1960-70, and Yellowstone Ecosystem, 1976-90, respectively.

	Craighead Team				IGBST[b]		
Year	Females identified with cubs[a]	Number of observations of marked females	Mean observations per female identified	Year	Females identified with cubs[c]	Number of observations of marked females	Mean observations per female identified
1960	2	8	4.0	1976	2	56	28.0
1961	5	29	5.8	1977	5	110	22.0
1962	9	114	12.7	1978	2	122	61.0
1963	8	54	6.8	1979	2	80	40.0
1964	5	98	19.6	1980	3	81	27.0
1965	11	101	9.2	1981	4	70	17.5
1966	8	104	13.0	1982	4	101	25.3
1967	6	68	11.3	1983[d]	4	56	14.0
1968	6	72	12.0	1984	2	41	20.5
1969	3	39	13.0	1985[d]	2	12	6.0
1970	9	99	11.0	1986	9	84	9.3
				1987	3	59	19.7
				Subtotal	42	872	20.8
				Mean	3.5	72.7	
				1988	6	—	—
				1989	6	—	—
				1990	5	—	—
Total	72	786	10.9	Total	59		
Mean	6.5	71.4		Mean	3.9		

[a] 39 individual females were color-marked.

[b] Sources are IGBST Annual Reports (Knight *et al.* 1977-78, 1980-83, 1985-91) and IGBST list of unduplicated females (unpubl. 1990).
[c] 41 individual females were radio-collared.
[d] One non-radioed bear included in each of these years.

those bears not radio-collared, however, was often compromised by combining aerial observations with those from close range on the ground, lack of multiple identifications, inconsistent census techniques, great variation in the training or reliability of observers contributing identifications, and questionable criteria for identification based on isolation in space and time.

We next attempt to analyze the results obtained by the two research teams as a function of the census techniques employed. Obviously, population numbers cannot be directly compared, but the techniques employed to obtain data can be. This can be accomplished, even though the population units examined are different—ecosystem-wide for the IGBST and the Trout Creek subpopulation for our team.

Applying census technique (*a*), we censused 39 artificially marked, adult females that, over an 11-year period, were identified as having cub litters 72 times (Table 19.1). Likewise, using technique (*a*), the IGBST followed 41 radio-collared adult females that had cub litters 59 times from 1976 to 1990 (Table 19.1) The IGBST obtained multiple observa-

tions on these radioed females in a given year, with a mean observation per female of 20.8 per year of the study. In making direct observations at the Trout Creek ecocenter, we averaged 10.9 observations per marked female with cubs. Thus, multiple observations of individual marked females with cubs contributed much accuracy to the samples for both teams. The number of females identified with cubs using technique (*a*) is virtually unchallengeable.

Females with cubs identified using census technique (*a*) constituted only a portion of the total census data in both studies. Both included additional data on females with cubs identified on the basis of permanent natural characteristics (Table 19.2). We made multiple observations of 30 such individuals to conclude that they had litters 31 times over the period from 1959 to 1970. Annual observations per individual naturally marked female were a minimum of three and averaged 6.4 per individual per year, over the period of the study. On the basis of multiple identifications, the precision of these observations can be considered equivalent to those made of females bearing artificial markers. Thus,

Table 19.2. Results from Census Techniques (b), (c), and (d) during the Craighead and IGBST grizzly bear studies, Trout Creek ecocenter, 1960-70, and Yellowstone Ecosystem, 1976-90, respectively.

	Craighead Team (census technique b)				IGBST[b] (census technique b, c, and d)		
Year	Females identified with cubs[a]	Number of observations	Mean observations per female	Year	Females identified with cubs[c]	Number of observations	Mean observations per female
1960	8	63	7.9	1976	14	23	1.6
1961	3	12	4.0	1977	8	11	1.4
1962	4	24	6.0	1978	7	9	1.3
1963	2	19	9.5	1979	11	12	1.1
1964	2	10	5.0	1980	9	13	1.4
1965	3	11	3.7	1981	9	17	1.9
1966	1	6	6.0	1982	7	12	1.7
1967	1	7	7.0	1983	9	11	1.2
1968	1	3	3.0	1984	15	36	2.4
1969	2	24	12.0	1985	7	15	2.1
1970	4	19	4.8	1986	16	40	2.5
				1987	10	19	1.9
				1988	13	21	1.6
				1989	10	21	2.1
				1990	19	41	2.2
Total	31	198	6.4	Total	164	301	1.8
Mean	2.8	18.0		Mean	10.9	20.1	

[a] 30 individual females bore permanent natural markings.

[b] Sources are IGBST Annual Report (Knight *et al.* 1977-78, 1980-83, 1985-91) and IGBST list of unduplicated females (unpubl. 1990).

[c] The number of individual females is unknown, as is the number of observations based on permanent natural markings, as opposed to seasonally variable markings or time-space patterns.

Table 19.3. Frequencies of observation of unmarked female grizzly bears with cubs recorded in the IGBST annual censuses as "identified—unduplicated," Yellowstone Ecosystem, 1976-90 (Knight *et al.* 1977-78, 1980-83, 1985-91).

	Females identified in a year						
	Multiple observations		Single observations		Total females	Total observations	Mean observations per female
Year	No.	%	No.	%			
1976	5	36	9	64	14	23	1.6
1977	1	12	7	88	8	11	1.4
1978	2	29	5	71	7	9	1.3
1979	1	9	10	91	11	12	1.1
1980	3	33	6	67	9	13	1.4
1981	3	33	6	67	9	17	1.9
1982	3	43	4	57	7	12	1.7
1983	2	22	7	78	9	11	1.2
1984	8	53	7	47	15	36	2.4
1985	3	43	4	57	7	15	2.1
1986	10	62	6	38	16	40	2.5
1987	6	60	4	40	10	19	1.9
1988	5	38	8	62	13	21	1.6
1989	4	40	6	60	10	21	2.1
1990	13	68	6	32	19	41	2.2
Total	69	42	95	58	164	301	
Mean							1.8

Table 19.4. Artificially marked and unmarked grizzly bear females with cubs in the IGBST sample of unduplicated females, Yellowstone Ecosystem, 1976-90 (IGBST Annual Reports [Knight *et al.* 1977-78, 1980-83, 1985-91]).

Year	Marked females		Unmarked females		Total females
	No.	%	No.	%	
1976	2	13	14	88	16
1977	5	38	8	62	13
1978	2	22	7	78	9
1979	2	15	11	85	13
1980	3	25	9	75	12
1981	4	31	9	69	13
1982	4	36	7	64	11
1983	4	31	9	69	13
1984	2	12	15	88	17
1985	2	22	7	78	9
1986	9	36	16	64	25
1987	3	23	10	77	13
1988	6	32	13	68	19
1989	6	38	10	62	16
1990	5	21	19	79	24
Total	59	26	164	74	223

31 reliable observations of unduplicated females with cubs (Table 19.2) could be combined with the 72 observations of color-marked females (Table 19.1) for a total of 103 annually individualized identifications over the 11-year period of study. This year-to-year identification of adult females with cubs at Trout Creek employing techniques (*a*) and (*b*) resulted in a thorough record of the reproductive condition of 69 (30 + 39) individual females whose identifications were certain.

The IGBST, using techniques (*b*), (*c*), and (*d*), made 164 identifications of an unknown number of females (Table 19.2). Unfortunately, the IGBST database did not lend itself to separating these identifications in terms of specific technique applied. Some identifications were made using technique (*b*), but the majority were made using techniques (*c*) and (*d*) (R. Knight and B. Blanchard, pers. comm. 1989). The mean observation rate for all females was 1.8, and many females were identified only once (Tables 19.2 and 19.3). The precision of these identifications is not equivalent to those made of the 41 individual females bearing radio collars (Table 19.1), and thus, the two sets of data should not be combined for analysis. Nevertheless, they were considered equivalent by the IGBST in their censuses, in spite of the wide variation in reliability. Of the combined total of 223 annually individualized identifications of *unduplicated* females with cubs, only 26% were marked animals (59 radio-collared; Table 19.4).

Based on the Team's annual reports (Knight *et al.* 1977-78, 1980-83, 1985-91), we now examine the IGBST census techniques more critically. Comparisons of numbers and percents of artificially marked to unmarked females identified annually in the *unduplicated* sample of females with cubs show that only 13% and 12% of the total females identified bore artificial markers (radio collars) in 1976 and 1984, respectively, whereas identifications by artificial markers were at a maximum of 38% in 1977 and 1989 (Table 19.4). The 164 females without artificial markers, reported as identified and counted in the annual censuses of adult females with cubs, represented 74% of the *unduplicated* sample over the 15-year period. Furthermore, analyses of the records of observation show that 58% of the 164 unmarked females recorded in the censuses were identified (recognized) with but a single observation (Table 19.3). Single identifications of unmarked or of naturally marked individuals are weak criteria on which to base a census. When over half of all the unmarked bears censused by the IGBST are so identified, then the accuracy of the census can justifiably be questioned.

As we have shown, 76% of all *unduplicated* identifications of adult females with cubs by the IGBST were of unmarked females. Had all these observations been made on the ground at close range by research personnel only, the IGBST census results would still have equivocal validity, because

Table 19.5. Census methods employed by the IGBST for identifying unmarked female grizzly bears with cubs in the Yellowstone Ecosystem, 1976-90 (IGBST Annual Reports [Knight *et al.* 1977-78, 1980-83, 1985-91] and IGBST list of unduplicated females [unpubl. 1990]).

Year	Number of females observed						Total females
	Ground only		Air only		Ground and air		
	No.	%	No.	%	No.	%	
1976	8	57	5	36	1	7	14
1977	8	100	0	0	0	0	8
1978	6	86	0	0	1	14	7
1979	5	45	5	45	1	9	11
1980	3	33	5	56	1	11	9
1981	2	22	6	67	1	11	9
1982	4	57	2	29	1	14	7
1983	3	33	5	56	1	11	9
1984	7	47	4	27	4	27	15
1985	4	57	2	29	1	14	7
1986	6	38	5	31	5	31	16
1987	5	50	3	30	2	20	10
1988	4	31	6	46	3	23	13
1989	3	30	4	40	3	30	10
1990	8	42	4	21	7	37	19
Total	76	46	56	34	32	20	164

many identifications were judged to be unduplicated based on undocumented, and hence, questionable space-time criteria. Further compromising the credibility of their census accuracy, however, less than half (46%) of the 164 IGBST identifications of unmarked females were made from the ground (Table 19.5). The margin for error in identifying unmarked bears from the air in the Yellowstone Ecosystem is substantial. Furthermore, census procedures were not standardized; neither the aerial nor the ground censuses were conducted along predetermined routes, according to a grid pattern, or along transects, and the aerial censuses were not conducted regularly, nor in any specific temporal sequence. It seems reasonable that aerial identifications could be presented as *unduplicated* on occasions of simultaneous observations in space and time. But the uncertainties of time and space distinction are great, and as a rule, aerial identifications that rely on a single observation should not be represented as *verified*. When such identifications are included, they detract from, rather than add to the credibility of census precision.

It is axiomatic that data in any scientific study are obtained and verified by the research personnel. This is particularly true of a wild animal census, which requires specific experience, skills, and training. Nevertheless, the IGBST relied on a wide range of observers, from study team members to *unknown*

sources (Table 19.6). Only 50% of the identifications of unmarked females with cubs presumed to be *unduplicated* were attributed to research team members and bear biologists. The remainder were identifications made by less well-trained observers, and in more than half of those cases, the identity of observer was not reported. These multiple inconsistencies, alone, are ample reason for us to suggest a high probability of error in the annual counts.

In summary of the previous two sections, we conclude that IGBST census methods were insufficiently rigorous to provide a reliable index of population size or trend. The likelihood of errors in the "unduplicated" counts is too great, as is the probability that methodological bias has rendered the counts useless as an index of population size or trend, even if all counted females were, in fact, unduplicated.

ESTIMATES OF POPULATION SIZE AND TREND, 1959-74

Throughout the 1959-74 period, we made estimates of the size and trend of the grizzly bear population in the Yellowstone Ecosystem, based on data from annual censuses (1959-70) of color-marked and otherwise individually identifiable bears (Craighead *et al.* 1974). We applied the ratio of marked and unmarked bears recorded as dying outside the Park to establish values for census efficiency of 77.3% and 55.7%, depending on what proportion of 40 bears recorded as mortalities, but of unknown mark status, was considered to be marked. As discussed in Chapter 3, we believed at the time that efficiency was most likely 77.3% and that the ecosystem-wide population averaged 229 animals. Reassessment, however, has indicated census efficiency to have approximated the lower value. As a result, we have accepted the estimate of 56.7% suggested by McCullough (1981) for a value of census efficiency, and based on that, we consider 312 grizzly bears to be the best estimate of mean population size in the Yellowstone Ecosystem from 1959 to 1970 (see Chapter 3).

Based on annual cumulative totals of individual bears observed each year from 1959 to 1970, reproductive rate and sex and age structure were determined for the Yellowstone population, and used in combination with records of annual mortality to construct a life table (see Table 4.1) describing the population over the 1959-67 period. Population growth rate was estimated to be an average of 2.4% per year for the period (Craighead *et al.* 1974). Then, a deterministic model of the population was developed on the basis of these data to project a population trend over the 1970-74 period. The results suggested, at the least, a 30% decline in the population by 1974. Certainly, such a decline was not unexpected, in view of high grizzly bear mortality recorded from 1970-72, but it did much to crystalize widespread concern for the long-term survival of the bears. On 28 July 1975, the Yellowstone grizzly bear population, and others in the contiguous U. S., were officially listed as threatened, and planning for population recovery was begun. Not until seven years later, however, was the formal recovery plan actually in place (USFWS 1982).

ESTIMATES OF POPULATION SIZE AND TREND, 1975-94

The first estimate of population we found in the literature was not based on annual counts of unduplicated adult females with cubs. In the 1974 IGBST Annual Report, Knight *et al.* (1975) offered a "best estimate" of from 237 to 540 bears in the Ecosystem. Neither rationale nor basic data were provided in support of this estimate and, considering the high bear mortality of the early 1970s, we doubt the remaining population could have been even as large as their lowerbound estimate of 237.

With the exception of that of Eberhardt *et al.* (1994), all subsequent estimates of size or projections of trend of the Yellowstone Ecosystem grizzly

Table 19.6. Numbers and types of observers used by the IGBST as sources for reports of unmarked grizzly bear females with cubs in the Yellowstone Ecosystem, 1976-90 (IGBST list of unduplicated females [unpubl. 1990]).

Year	Study team	Bear biologists	NPS	USFWS	Fish & Game	Out-fitters	Ranchers	Attack victims	School fieldtrip	Locals	Photo-graphers	Film crews	Unknown	Total no. of observers	Annual no. of observer categories
1976	4	1	11	3	1	2	—	—	—	—	—	—	—	21	5
1977	—	—	6	2	1	2	—	—	—	—	—	—	—	11	4
1978	2	1	1	1	2	2	1	—	—	—	—	—	—	9	6
1979	6	—	2	—	—	—	—	—	—	—	—	—	3	12	4
1980	8	1	—	—	—	—	—	—	—	—	—	—	4	12	2
1981	14	2	1	2	—	—	—	1	—	—	—	—	—	17	4
1982	4	—	—	—	—	—	—	—	—	—	—	—	4	12	4
1983	6	3	1	—	—	—	—	—	—	—	—	—	4	11	3
1984	19	—	—	—	—	—	—	—	1	—	—	—	13	36	4
1985	11	—	—	—	—	—	—	—	—	1	1	—	1	14	4
1986	26	—	1	—	—	—	1	—	—	—	4	—	9	40	4
1987	7	—	—	—	—	—	—	—	—	—	3	—	8	19	4
1988	10	—	1	—	—	—	—	—	—	—	2	—	9	21	3
1989	9	2	1	—	—	—	—	—	—	—	—	1	6	18	4
1990	9	2	—	—	—	—	—	—	—	—	6	—	19	37	5
Total	135	12	24	8	3	6	2	1	1	1	16	1	80	290	

Table 19.7. Estimates of size and trend of the grizzly bear population based on annual number of unduplicated females with cubs, Yellowstone Ecosystem, 1980-93.

Source	Population size	Year estimated	Population trend	Period
IGBST Annual Report, 1980 (Knight *et al.* 1981)[a]	247	1980		
IGBST Annual Report, 1981 (Knight *et al.* 1982)[a]	197	1981		
Ad Hoc Committee for Population Analysis (Knight *et al.* 1983)[b]	183-207	1980	Declining	1959-80
Knight and Eberhardt (1984)[c]	——	——	Declining	1959-80
Knight and Eberhardt (1985)[d]	——	——	Declining ($\approx$1.7%/yr)	1959-80
Eberhardt *et al.* (1986)[e]	——	——	Declining	1970-85
Yellowstone Grizzly Bear Population Task Force (Knight *et al.* 1988a)[f]	170-180	——	Increasing ($\approx$0.7-1.5%/yr)	1975-87
IGBST Annual Report, 1989 (Knight *et al.* 1990)[g]	——	——	Increasing	1975-89
IGBST Annual Report, 1990 (Knight *et al.* 1991)[g]	——	——	Increasing	1975-90
IGBST Annual Report, 1991 (Knight *et al.* 1992)[g]	——	——	Increasing	1975-91
Grizzly Bear Recovery Plan (USFWS 1993)[h]	236	1992	——	——
Mattson *et al.* (in press, 1995)[i]	197 95% CI: 142-252	1993	Increasing ($\approx$1.0%/yr)	1975-93

[a] Population size estimated by back-summing numbers of unduplicated adult females with cubs observed over 3-year periods, applying adult sex ratio to determine number of adult males, and age structure to determine number of subadults.

[b] Interagency Grizzly Bear Steering Committee; a minimum population size of 139 animals calculated by back-summing the numbers of unduplicated adult females with cubs observed from 1982-80; range then projected based on an unexplained "estimated maximum [census] efficiency" range of 67-76%.

[c] Trend estimate based on an index of abundance (3-year back-sums of unduplicated adult females with cubs observed, 1959-80) matched with projected population sizes from a stochastic model of the adult female grizzly bear population.

[d] Trend estimate based on regression of logarithms of 3-year back-sums of unduplicated adult females with cubs observed, 1959-80.

[e] Trend estimate based on a prospective model (a difference equation) of the dynamics of the adult female population, in combination with observed mortality and litter size, and matched with 3-year back-sums of unduplicated adult females with cubs observed, 1959-83.

[f] Population size estimated by dividing proportions of adult females in population under two conditions of sex ratio (49% females in litters and 42% females in the mortality record) into 45, the number of adult females corresponding to a tally of 15 unduplicated adult females with cubs over a 3-year period; trend estimates based on solution of standard Lotka equations.

[g] Trend estimates based on increases in running 6-year averages of annual numbers of unduplicated females with cubs observed.

[h] Population size estimated by dividing proportion of adult females in population (calculated using method of Knight *et al.* 1988 and assuming a M:F ratio of 51:49) into the 3-year sum of unduplicated adult females with cubs observed in 1990, 1991, and 1992, less known mortalities of adult females over the same period.

[i] Population size estimated as above[h] except that value for 3-year sum of unduplicated females with cubs observed was the lower bound of a 95% C. I. on the 3-year sum for 1991-93; trend estimate based on solution of a standard Lotka equation.

bear population have been based, directly or indirectly, on the annual counts of adult females with cubs (Table 19.7). As discussed earlier, many of the papers that advanced these estimates presented well-executed statistical exercises in an attempt to derive meaningful information on the status of the population, insofar as limited data on basic parameters, particularly mortality rates, would permit. None of them, however, succeeded in validating annual counts of unduplicated adult females with cubs as an index appropriate for use in calculating population size or tracking trends, nor did they validate the census methods used to obtain the data. As we have shown, there is good reason for concern about the accuracy of the counts and biases associated with observability and degree of census effort. Thus, we have been disinclined, over the years, to place much credence in population estimates and trends based on the counts, and we are disturbed that the counts continue to serve as the basis, along with other indices of distribution and mortality derived from them, for judging whether a grizzly bear population has recovered enough to be delisted from protected status.

Recently, Eberhardt *et al.* (1994) made an estimate of trend in the Yellowstone grizzly bear population based on reproductive rates calculated from 22 marked (radioed) individual females and survival rates from 400 female bear-years. Using a model approximating Lotka's equation (Eberhardt

1985) and bootstrapping techniques for estimating confidence limits, they claimed the population to be increasing 4.6% per year (95% C.I. = 0% to 9%). This approach, to its credit, made no use of the unduplicated counts of adult females, but examination of the data used to generate the model, and how these were structured to create some of the rate parameters, nevertheless, caused us to question their results. For example, reproductive rates were calculated based only on productivity of females observed to complete at least one reproductive cycle (at least two litters), thereby ignoring one-litter females and barren females. As we showed earlier (Chapter 3), this approach inevitably over-estimates population productivity. Also, Pease and Mattson (1995 [in prep.]) have demonstrated a behavioral structuring in the Yellowstone population—human-habituated and -nonhabituated grizzly bears—and they estimate that annual mortality rates of habituated bears are roughly double those of wary ones. They contend that, to avoid risk of serious error, estimates of population trend must consider the habituation process in calculating mortality rates, since habituation of previously wary bears represents a significant, continuous mortality sink. Furthermore, they point out that use of the formula, "marked female deaths divided by bear-years observed (radioed)," as the basis for calculating survival would lead to its overestimation if uncollared intervals (periods when collars were lost and later replaced) were included in the total bear-years observed (pers. comm. 1995). Eberhardt *et al.* (1994) expressed different misgivings about this measure of survivorship, noting that only 19 of the 58 individuals used to estimate adult survival were followed until death, but whether they made the error of over-estimating bear-years, and hence survivorship, described by Pease and Mattson, cannot be determined from their paper.

CURRENT STATUS OF THE YELLOWSTONE GRIZZLY BEAR POPULATION

On the basis of our foregoing critique of the quality of the data applied to estimating population size and trend, and of the methods used to obtain them, we caution that none of the size and trend estimates offered for the grizzly population in the Yellowstone Ecosystem since 1970 are reliable. The analytic approach used by Eberhardt *et al.* (1994) was certainly on the right track, but it was compromised by the quality of the underlying data on reproduction and survivorship. Unfortunately, achieving an accurate measure of age- and sex-specific survivorship will probably continue to be a crucial weak-point in modeling the population, but, for management purposes, calculating maximum allowable observed mortality using methods suggested by Mattson (1995 [in prep.]) would reduce the probability of over-estimating allowable mortality in the population model.

Unquestionably, our most serious criticism of determining population status relates to use of annual counts of unduplicated females with cubs as an index of trend and as the basis for calculating "minimum" population size for purposes of measuring progress toward recovery. We have shown that the index has never been proven to be a valid indicator of population status and, in fact, that work by Mattson (1995 [in prep.]) shows that it is not. Furthermore, the index is derived using methods open to a wide range of error and bias. Yet, it continues to serve as the fundamental metric for judging population recovery. We recognize that determining the size and trend of the Yellowstone grizzly bear population has been technically challenging. Continuing to apply an inappropriate measure to determine progress toward recovery, however, is inexcusable and endangers the population's prospects for long-term survival.

We urge that no actions be taken to abrogate the population's threatened designation until appropriate measures of its status show it to be healthy in terms of long-term viability. In this monograph, our comparative analyses of the grizzly bear populations inhabiting the Yellowstone Ecosystem prior to and following closure of the open-pit dumps have revealed information that will be important in redesigning a plan for recovery and improving prospects that the population will remain viable. As we suggest in Chapter 20, the size and the sex- and age-structure of the preclosure population should serve as recovery standards. Also, the ecocenter concept must be seriously considered in any future planning for the Yellowstone grizzly bear population.

SUMMARY

In 1983, the Ad Hoc Committee for Population Analysis (of the IRSC) reviewed the IGBST grizzly bear population data and concluded that the "minimum number of breeding females is the most reliable index available to estimate population size" and recommended basing the index on three-year sums of annual counts of unduplicated females with cubs. This report essentially established by fiat an index to population size of absolutely unproven validity. Moreover it provided no protocol for distinguishing unduplicated observations. There were serious concerns that the index was not valid and the measure of it not reliable.

The IGBST employed four census techniques to characterize the ecosystem-wide population from 1976 to 1990. Identifications made using radio signals were highly reliable. Identifications of those bears not radio-collared, however, were often compromised by combining aerial observations with those made on the ground, lack of multiple identifications, inconsistent census techniques, great variation in the training or reliability of observers contributing identifications, and questionable criteria for identification based on isolation in space and time. Seventy-six percent of all unduplicated identifications of adult females with cubs by the IGBST were of unmarked females. Fifty-eight percent of those unmarked females were identified (recognized) with but a single observation. We conclude that IGBST census methods were insufficiently rigorous to provide a reliable index of population size or trend.

The unduplicated females with cubs index continues to serve as the fundamental metric for judging population recovery. It is the basis for projection of population size. The other three parameters—distribution of family units and allowable known mortality among adult females and among the population in total—derive from the counts directly or indirectly. Estimates of the size and trend of the postclosure population have been both confusing and unreliable. Our best appraisal of the situation is that a mean preclosure ecosystem population of 312 grizzly bears was reduced by at least half that number by the early 1980s. The current size is unknown and the trend questionable. We urge that no action be taken to abrogate the population's threatened designation until appropriate measures of its status show it to be healthy in terms of long-term viability.

20
Population Recovery and
Long-Term Management

INTRODUCTION

We feel that we would be remiss if we did not offer suggestions for managing the Yellowstone grizzly bear population and other population units south of the Canadian border. We do this fully recognizing that the agencies responsible for grizzly bear management have not been and are not now likely to be receptive to suggestions from the private sector. We believe, however, that the prospects of long-term grizzly bear survival, under present land management concepts, philosophy, and actions are not good. An entirely new perspective is needed; a holistic approach that relates management not only to the specifics of bear biology, but also to the socioeconomic conditions of our times. Social realities, past and present, have had a direct bearing on our wildland ethic. These realities are major determinants of any program to preserve the grizzly bear and its habitat in perpetuity.

When Lewis and Clark encountered the Northern Rockies in 1804, the area was wilderness, its flora and fauna supporting small, widely dispersed tribes of Native Americans. This condition extended to the Pacific Ocean, encompassing, overall, a pristine area of millions of square hectares. The grizzly bear was found throughout the entire area, but populations were largely concentrated in localized sites of high animal (prey) density. As a fierce, powerful co-dominant with the Native Americans, the grizzly bear was worshiped, feared, and always respected. The two species coexisted. The grizzly population was very roughly estimated to be about 100,000 throughout the West.

Explorers and fur trappers followed in the wake of the Lewis and Clark Expedition (1804-6), and they dominated the scene until the fur resources were depleted. Then followed a sequence of change and development familiar to all Americans. The hunters and trappers were succeeded by the homesteaders and developers. This movement coincided with the "demise" of the North American bison, the food base for many Native American tribes and an important item in the diets of all the larger predators, including the grizzly bear. Gradually, native grasslands were occupied by domestic livestock or were plowed and sowed to domestic plants. Miners crisscrossed the western lands in search of gold and silver. Mining towns of thousands of prospectors sprang up overnight, and most disappeared almost as quickly. The more accessible forests were cut to supply lumber to the growing towns and settlements, and huge blocks of land were granted to the railroads to encourage the development of transcontinental transportation. Rangeland and water rights were acquired at the expense of the native tribes and defended at gunpoint. The Army maintained a constant and shameful campaign of extermination against the Native Americans who sought to hold their homelands and lifestyles. The encroaching Caucasians finally displaced the Native Americans and, in their ruthless land acquisition, also took a tremendous toll on wolves, lions, bears, and other predators that interfered with occupancy of the land.

By the turn of the 20th century, the pristine wilderness of the Northern Rockies was no longer pristine except in isolated blocks of mountain terrain. The population of invading whites dominated the landscape, their numbers measured in the millions and their "handiwork," a preview to the future. By 1923, the California grizzly, estimated to have numbered about 10,000, was extinct (Storer and Tevis 1955), and grizzly bears in the Northern Rockies, harassed with guns, traps, and poison, had been greatly reduced. Even at that time, the major threat to the pristine landscape and its flora and fauna was not the wasteful destruction and use of the natural resources, but the rapidly growing Caucasian population with a greed ethic that sanctioned genocide to all life forms that threatened European man's manifest destiny. The raw, ugly disdain for life exhibited in the "annihilation" of Native Americans, the bison, and the large predators appears to be innate in the human species. Though veiled in modern America, this proclivity

458   *The Grizzly Bears of Yellowstone*

for destruction survives and must be reckoned with in any plan to conserve the all-too-exploitable wilderness of today and its dominant occupant, the grizzly bear.

The population and environmental changes that followed the settling of the West continued at an accelerated pace to create the conditions we find in the Northern Rockies bioregion today. Awareness of the rapidity of human-caused changes in the landscape, and of what this portended for the future of the native flora and fauna and that of the remaining pristine wildlands was eloquently voiced by Henry David Thoreau, John Muir, and others. Muir believed that nature was sacred and should not be encroached on; his contemporary, Gifford Pinchot, advocated sustainable resource management. Their views clashed early in the conservation movement. Muir formed the Sierra Club to champion his views, and Pinchot was instrumental in creating the United States Forest Service. Governmental reaction to the grassroots movement to preserve land resources of the West was expressed in the creation of Yosemite State Park in 1864, followed by Yellowstone National Park in 1872. Nearly a half century elapsed before the creation of the National Park Service (NPS). Preservationists' concepts and sustainable resource management were championed by Gifford Pinchot, Aldo Leopold, and others. Leopold, like Muir, believed that wilderness had inherent value. He advocated preserving and restoring it where possible. Gradually, a conservation ethic evolved that tempered the philosophy of unrestricted resource exploitation—protecting the pristine became a national movement, a force that was to grow in size and intensity with the growth of population and the economics that supported it—a national economic concept that envisioned no limit to natural resource capital (Rees and Wackernagel 1992).

The realization that natural resources were limited and possessed non-extractive as well as extractive values began to take hold during the last half of the 20th century in a society geared to an unlimited economic expansionist concept. However, there were few philosophical, social, or governmental constraints to temper the burgeoning economic growth and development.

The Wilderness Society was founded by Bob Marshall in 1935, and the concept of wildlands preservation was ably advanced by Olaus Murie, Howard Zahnheiser, and colleagues. In 1964, the first national forest wilderness area was established and named after its chief advocate, Bob Marshall. The pros and cons of "locking up" large areas of pristine landscape were fiercely argued in scientific, bureaucratic, and political circles.

On 3 September 1964, Congress passed the Wilderness Act, listing potential wilderness areas, and establishing preservation policies for implementing and managing them as a National Wilderness System under the stewardship of the Departments of Agriculture and Interior. It was clearly an act reflecting the will of the people. Through their representatives in the U. S. Congress, the American people expressed their desire to preserve some semblance of natural conditions. Congress defined wilderness as ". . . an area where the earth and its community of life was untrammeled by man, where man himself is a visitor who does not remain." Wilderness was further defined by four major conditions that specifically described it and the permissible uses that would leave wilderness unimpaired. The prohibition of certain uses and the designation of special use provisions, unfortunately, diffused the grass-roots vision of wilderness as inviolate natural sanctuaries. Regulations developed by the agencies following passage of the Act further compromised this vision. But, clearly, a historic step had been taken in recognizing pristine conditions as innately valuable and wilderness per se as an important component of our natural resource capital. The Wilderness Act gave National Wilderness System protection to some 3,650,000 hectares (9 million acres) of National Forest land that had been designated previously as Wilderness Areas and Wild Areas. Forests administratively classified as Primitive Areas were not placed in the National Wilderness System by the Act. They were subject to future review with a ten-year deadline.

The Wilderness Act and subsequent development of our wilderness system assured federal protection for a significant, but relatively small part of the vast habitat the grizzly bears required to survive, habitat that thereafter only congressional action could alter. The grizzly quickly became the symbol of wilderness and its habitat, a place for spiritual, aesthetic reflection and rejuvenation. Even more important, wilderness was increasingly recognized as a national resource protected for future generations, a gift that could be passed on unscathed by the expansionist economic forces of the times. But even as the Wilderness Act came into being, powerful social and economic forces were impinging upon the wilderness sanctuaries and the as yet unclassified de facto wilderness. The declared pristine landscape was being or was soon to be altered by forces largely unrecognized and unmeasured: radiation fallout; acid rain; ozone depletion; introduction of exotic plants; gas, oil, and mineral exploration; decline in biodiversity; and a steady increase in recreational impact. Outside of designated wilderness, but on adjoining federal land, logging, mining, livestock grazing, outdoor recreation, and the development of campgrounds and recreational sites were proliferating. Many of these activities required the roading of pristine habitat unprotected by wilderness designation. The mountain West's intermountain valleys and limited riparian habitat were rapidly changing from farm and ranch land to clustered suburban areas that brachiated up the waterways to the wilderness borders.

When the Wilderness Act was enacted in 1964, the grizzly was being legally hunted in Montana and Wyoming with no restrictions on the annual kill in Montana and no research studies to determine the effect of the kill or whether a harvest could be biologically justified. Each year, new roads penetrated the mountains, often paralleling the descending streams and rivers and leading hunters, fishermen, and other recreationists into the backcountry of the alpine and subalpine zones. People and grizzlies then had to share more of the common ground. Some of these people were prepared to shoot a grizzly on sight, for coexistence was not then a concept or an option. Since humans may well have a more limited capacity to adjust to bears than bears have to adjust to humans, the security of the grizzly was not assured with the passage of the Wilderness Act, although it was a major step forward.

The grizzlies' insecurity became strikingly evident with the abrupt closure of the Yellowstone ecocenters. In 1974, just ten years after the Wilderness Act provided the highest level of governmental protection to grizzly bear habitat, grizzly bear deaths had skyrocketed within the Yellowstone Ecosystem, and in 1975 the great bear was hastily listed as threatened under provisions of the Endangered Species Act (ESA) (U. S. Congress 1973). This action became necessary largely because of management decisions left to misguided bureaucrats. It became all too clear that agencies responsible for the preservation of the grizzly had jeopardized the species more critically than all the other impinging forces collectively. Inability to accept and make changes not on the official agenda is inherent within the bureaucratic system and will cause similar catastrophic events in the future. What then is the prospect for the grizzlies' long-term survival? What, in fact, does population recovery really mean and is it feasible under current vision and planning? Will the grizzly be continually forced to adjust to humans or can humans learn to adjust to bears?

The growth or decline of grizzly bear numbers is directly tied to demographic dynamics of the human population. The bear is confronted with the predicaments of an exponentially increasing human presence and a steadily increasing human appropriation of federally managed natural resources. This presents an eventual "no win" situation for both bear and human, but the climax arrives more quickly for the bear. This type of dilemma was discussed by Wiesner and York (1964) in relation to the nuclear arms race, and by Hardin (1968) as it applies to human population growth in a finite world. These authors concluded that, for these problems, "there are no technical solutions."

Perpetuating the grizzly bear, likewise, is an objective not entirely amenable to technical solution. According to Hardin (1968), a technical solution may be defined as one ". . . that requires a change only in the techniques of the natural sciences, demanding little or nothing in the way of change in human values or ideas of morality . . ." or, we might add, inherent human behavior. In our judgment, technical solutions must be pursued but viewed and interpreted in the broad perspective of a "finite world," where space and resources are finite for both grizzly bears and humans. In this context then, can the grizzly bear maintain genetically and demographically viable populations in the face of its shrinking habitat? Can human values and behavior change sufficiently to permit the possibility of technological solutions—if only in the short term (100 years)? The answers to these questions carry enormous implications for both bear and human. Clearly, we must find means to alter our overexploitative use of natural resources, our acquisitive behavior, and our uncontrolled population growth if we are to have any chance of preserving the grizzly bear in perpetuity by applying management strategies based solely on technical knowledge gained from biological studies.

In the sections to follow, we review the currently approved version of the federal grizzly bear recovery plan (USFWS 1993), evaluate current availability and future need for grizzly bear habitat, and offer suggestions for a formal analysis of grizzly bear population viability that incorporates an ecocentric perspective. We end the chapter, and this monograph, with a strategy for grizzly bear population recovery in the Yellowstone Ecosystem and elsewhere, and some thoughts on translocating grizzly bears to the large recovery zones that are currently uninhabited.

THE FEDERAL PLAN FOR RECOVERY OF GRIZZLY BEAR POPULATIONS

In Chapter 19, we discussed the origins of the primary parameter for recovery adopted by the U. S.

Fish and Wildlife Service—the annual number of unduplicated adult females with cubs—and found its value as an index of population status to be unproven, and its accuracy as a metric to be doubtful. Here, we scrutinize the origins of the recovery plan, and the plan itself, in more detail. The grizzly bear recovery plan was issued in 1982, and draft revisions were submitted for review in 1990 and 1992. Both revisions were withdrawn following extensive criticism.

The first-draft revision (USFWS 1990) states: "The population objective for the Yellowstone Grizzly Bear Ecosystem was established through consultations with scientists and managers . . . and outside persons familiar with conditions in the area. . . Important input came from the scientists of the Interagency Grizzly Bear Study Team (IGBST) as to the number of bears that could be expected to be supported in this ecosystem." It is difficult to determine whether individual members of the IGBST, or members of the Interagency Research Review Committee (IRRC), or members of the Interagency Grizzly Bear Committee (IGBC), or all three committees, were responsible for the determination of recovery goals (number of bears, etc.) and for the field research on which the recovery plan is based. Neither the second-draft revision, nor the final plan approved 10 September 1993 (USFWS 1992, 1993, respectively) clarified this question of ultimate responsibility.

It seems plausible that the IGBST was specifically charged with obtaining data to meet certain objectives of the planning committee, and the IRRC, in the role of a special technical review committee, was responsible for reviewing the data and findings of the research team. We suppose that research results were next integrated into new planning for the Ecosystem, perhaps via the IGBC, especially as results relate to the task of population assessment. The assessments made by these several committees and, possibly, selected reviewers, were then, we would imagine, integrated into dynamic management strategy for recovery of the grizzly bear population. In Yellowstone, the thrust of the IGBST research has been to determine the status of the population, but our analysis suggests that the data they have developed are not adequate to the task (Chapter 19). It is important that lines of responsibility and authority be clarified, and that the creation of a plan for recovery be an open, peer-reviewed process, rather than an insular, "in-house" agency one. Otherwise, politics will continue to supplant biological rationale as the basis for management.

The final recovery plan prepared by the USFWS (USFWS 1993) has apparently incorporated many of the findings and, perhaps, recommendations of the IGBST, yet the current draft plan remains weak. The objective for each of four grizzly bear populations listed as threatened, including that in the Yellowstone Ecosystem, is their delisting as recovery targets are achieved. Recovery criteria are based on, but not limited to, annual identification of some minimum number of females with cubs, uniform distribution of family groups throughout the recovery zone, and holding human-caused mortality to some specified annual limit, criteria we believe to be inadequate expressions of population status. The estimated cost of recovery for at least six populations is $26,000,000, but specific dates or time frames for accomplishing the recoveries are not provided. Our objective here is to evaluate progress toward meeting the specific goals for recovery of the population inhabiting the Yellowstone Ecosystem, with reference to the other populations, when appropriate. Since the broad strategy for recovery is inherent in the Endangered Species Act of 1973, a brief review of the congressional mandate is in order.

The Endangered Species Act (ESA) of 1973

The Endangered Species Act (P. L. 94325, as amended; U. S. Congress 1973) is truly visionary legislation intended to provide protection and consideration for "endangered" species—plants and animals in danger of extinction throughout all or a significant part of their range—and for "threatened" species—those likely to become endangered. That precipitous or long-term decline in species may be symptomatic of broader ecosystem dysfunction is acknowledged in Section 2(b) of the ESA through an explicit requirement that ways and means be provided for conservation of ecosystems on which these species depend. In essence, management agencies are specifically directed to maintain the functional integrity of ecosystems that include endangered or threatened species. This has not been accomplished largely because the prime directive has been amended through direct efforts of the management agencies. Not only is the directive clear in the law, it was specifically emphasized by the ESA's author, Representative John Dingell, when he stated:

> Once this bill is enacted, the appropriate Secretary, whether Interior, Agriculture or whatever, will have to take action. . . The purposes of the bill include the conservation of the species and the ecosystems upon which they depend, and every agency of the government is committed to see that those purposes are carried out . . . [T]he agencies of government can no

longer plead that they can do nothing about it. They can and they must. The law is clear. (119 Cong. Rec. 42913)

But this clarity of the law was altered.

Section 4 of the ESA requires federal agencies charged with management of wildlife to design and implement plans for the recovery and eventual delisting of any species officially recognized ("listed") as endangered or threatened. The Grizzly Bear Recovery Plan, completed in 1982 (USFWS 1982) and revised in 1993 (USFWS 1993), is a case in point. In response to Section 7 of the ESA, the plan provides a methodology for preserving habitat necessary to the survival of the bear, but it is unacceptable principally because of how the intent of Section 7 has been thwarted and reinterpreted by the management agencies, and Section 4, frequently ignored.

Recast Interpretation of the Endangered Species Act

Section 7 of the ESA dictates that federal agencies responsible for land management (the U. S. Forest Service [USFS]; the U. S. Department of the Interior, Bureau of Land Management [BLM]; and the National Park Service [NPS]) may not undertake activities that jeopardize the continued existence of listed species, nor adversely modify habitat critical to those species. Furthermore, it specifies that the agencies must consult with the appropriate wildlife management agency (U. S. Fish and Wildlife Service [USFWS], in the case of the grizzly bear) whenever a proposed course of action may affect a listed species and must act in accordance with that agency's written biological opinion.

Rohlf (1991) has pointed out that new and recast interpretations adopted by the agencies in 1986 further weakened the intent of Section 7 of the ESA in several ways. For example, resource agencies were allowed to conclude that resource extractive plans and actions could proceed ("no adverse affects" designation) without formal consultation or supportive biological rationale. Also, damage to critical habitat became permissible when such action might diminish a species' chances for population recovery, but did not threaten its ultimate survival; whatever foggy distinction that implies, it obviously thwarts the intent of the ESA.

Thus, on the face of it, provisions of the ESA would seem strongly protective of the grizzly bear's habitat needs which, as we have shown earlier, are a mosaic of diverse plant communities recurring over an entire ecosystem, not enclaves within them (Chapter 12). But latitude for interpretation of Congress' intent has permitted actions since the bear's listing in 1975 that have destroyed and fragmented substantial amounts of its habitat.

When the grizzly bear was listed in 1975 as a threatened species throughout the contiguous 48 states, the response of federal management agencies to the requirements of Section 7 was first to delineate critical habitat as entire ecosystems (U. S. Congress 1977, Craighead 1980b; also, see discussion of Cody, Wyoming Hearing, Dec. 1976, in Chapter 14). Later, realizing what this entailed in loss of resource development options and influenced by political pressures from private sector resource users (outfitters and guides, cattlemen, loggers, and oil and gas developers), the U. S. Forest Service (USFS) redefined habitat under their jurisdiction. The Service's response reflected the ESA Amendments of 1978 which defined the concept of "critical habitat" (U. S. Congress 1978, Section 2), but stated that "critical habitat *may* be established for species now listed. . ." (Sec. 2[2], emphasis ours), and later, that the requirement of specifying critical habitat did not apply to the grizzly bear (listed in 1975) or to any other species listed prior to 1978 (Sec. 11[1]). Thus, designation of critical habitat for the grizzly bear was badly compromised and legally opened to redefinition and interpretation by the agencies.

The USFS's redefinition was expressed in terms of Management Situation (MS) categories. A MS category was defined for parcels of habitat as a function of (1) presence or foreseeable presence of grizzly bears and (2) quality of the habitat for meeting the needs of grizzly bears. The USFS, BLM, and NPS, in collaboration with the USFWS, adopted this habitat-parcelling as an appropriate recast delineation of critical grizzly bear habitat. The result was a method of habitat inventory (Management Situation Criteria) much more reflective of the designs of extractive industries than of the needs of the grizzly bear. As of 1995, the congressionally mandated delineation of critical grizzly bear habitat had not been achieved, based on biological rationale, very likely as a result of the ESA Amendments in 1978 that lifted the requirement of specifying critical habitat for all species listed prior to 1978.

Of primary importance in the evaluation of habitat is our ability to consider objectively the grizzly's needs for cover, security, and especially, food. The five Management Situation categories now in use specify management recommendations for enclaves of habitat under different conditions of bear population numbers and habitat quality. Thus, protection of entire ecosystems as critical habitat has been rejected outright, although this is clearly in violation of the original intent of Congress.

Management Situation 1 applies to areas containing grizzly bear population centers and critical habitat, and it imposes the greatest degree of restraint on permissible activities. Applied to areas characterized by lower levels of perceived habitation or quality of habitat for grizzly bears, MS-2, -3, and -5 designations prescribe progressively less consideration for bears and their needs. The MS-4 designation is reserved for areas such as the Selway-Bitterroot and North Cascades Ecosystems where bears do not currently exist, but habitat is suitable and, we presume, for areas beyond recovery zone boundaries that bears may invade as population recovery progresses. It is important to note that only in areas labeled MS-1 are the needs of the grizzly bear accorded primacy over other land-use values. Under the MS-2, -3, and -5 classifications, consideration for the bears and their habitat is progressively diminished to none.

The process of assigning MS designations is highly subjective and, in some cases, arbitrary. Past history has shown the classification system to provide a ready opportunity for political manipulation and many incursions into recovery zone habitat by extractive industry. We conclude that the security of grizzly bears inhabiting de facto wilderness is in jeopardy so long as land management agencies can proceed with resource-extractive plans, while indicating "no adverse affects" (to the grizzly bear or its habitat), yet providing no credible scientific documentation to support such decisions.

A Critique of Federal Recovery Plan Guidelines

Our critique focuses on the recovery plan as it relates to the Yellowstone Ecosystem, but the relevance of our comments often will be much broader. By delineating recovery zones, the recovery plan acknowledges the need for space, but nowhere establishes "objective measurable criteria" to describe and quantify the habitat that the zones encompass or the human activities that degrade it as habitat for grizzlies. Emphasis, however, is directed almost exclusively toward establishing measurable recovery criteria for the population. But as we showed in Chapter 19, the rationale for these recovery goals are either weak or completely lacking.

The major thrust of the recovery plan for the Yellowstone Ecosystem population (USFWS 1993, p. 30) is achievement of goals set for specific demographic parameters, as follows:

1. The presence of 15 females with cubs over a 6-year running average, both inside the recovery zone and within a 16-kilometer (ten mile) perimeter immediately surrounding it;

2. The occupancy of 16 of 18 bear management units (BMU's) by females with young from a running 6-year sum of verified sightings; no two adjacent BMU's unoccupied; initiation in 1993 of a study to determine the potential and present habitat capability of the Plateau and Henry's Lake BMU's to support females with cubs; and

3. Known, human-caused annual mortality not in excess of 4% of the population estimate, based on the most recent three-year sum of females with cubs, with no more than 30% of that mortality being females.

When these goals are achieved, the population will be considered to have recovered, and delisting can occur.

We demonstrated in Chapter 19 that the number of females with cubs recorded by the IGBST, whether as an annual accounting or on a 6-year running average, is an unreliable parameter that has not been shown to be a valid indicator of population status. The annual census data were too often subjective, and their interpretation, fraught with questionable assumptions. Of equal concern is how and why a goal of 15 females with cubs was determined to be an indicator of recovery. Our assessments show that, before recovery was a necessity, the mean number of females with cubs each year over a 9-year period (1960-68) was approximately 25 for the ecosystem population (see later in this chapter).

We question the rationale for uniform distribution of females with young. It is stated to be ". . . designed to demonstrate adequate dispersion of the reproductive cohort within the recovery zone . . ." (USFWS 1993, p. 20). We can interpret this to mean only that, unless the family units are evenly distributed throughout the recovery zone, recovery has not occurred. We have shown throughout this text that grizzlies aggregate where food is abundant, and this results in a nonuniform (clustered) distribution. "Adequate dispersion," as determined by habitation of human-defined management units, is not a defensible recovery target, nor is it "evidence of adequate habitat management" as stated in the recovery plan (USFWS 1993, p. 20). Bear distribution is governed by a wide range of ecological and biological factors, and bears are not bound to select habitat on the basis of human-delineated, ecologically arbitrary management units. The Yellowstone grizzlies are not territorial, and they will tend to aggregate or establish home ranges where food is most abundant, whether this is at an ecocenter or an area where berries, pine seeds, or carrion are readily available.

The management plan sums up the subject of distribution of family units with the surprising statement that ". . . adequate distribution of family groups indicates future occupancy of these areas because grizzly bear offspring, especially female offspring, tend to occupy habitat within or near the home range of their mother after weaning" (USFWS 1993, p. 20). It is true that this tendency has been demonstrated (Craighead 1976, Aune and Kasworm 1989, Blanchard and Knight 1991), but the data do not support the generalization accorded them in the recovery plan. The inference of that statement is that management units occupied by a family group will continue to be occupied into the future by female offspring simply by habit or tradition, rather than by the ability of the habitat to support one or more bears. This is unwarranted supposition. For example, the catastrophic fire that occurred in 1988 surely altered the distribution of family groups, as did food scarcity in certain previous years (Chapters 15 and 16). Conversely, the superabundance of carrion following a severe winter or the development of ecocenters (garbage disposal, boneyards, etc.; see Chapter 13) do so, as well.

The third major target—controlling human-caused mortality—is of special concern because a theoretical model on sustainable mortality (Harris 1984) may have been generalized too broadly. Moreover, allowable levels of mortality are inappropriately linked with the first target, the numbers of adult females with cubs. The recovery plan states: "As the grizzly population increases, the number of sustainable, known human-induced mortalities also increases. The known number of females with cubs is used to calculate what is believed to be a minimum population estimate. . ." (USFWS 1993, p. 48). We interpret this to indicate that, if unduplicated counts of females with cubs increase as an average over 6-year periods, an acceptable target for human-induced mortality can be elevated. The reverse is true, if the counts decline. Such modification of targeted goals amounts to a subjective juggling of data for mortality and productivity parameters that, in themselves, were shown in Chapter 19 to be of dubious relevance and reliability. Mattson (1995 [in prep.]) argues that this approach lends itself to setting goals for sustainable mortality that may be excessive.

The recovery plan goes on to state (USFWS 1993, p. 62) that ". . . of all females with cubs in the NCDE [Northern Continental Divide Ecosystem] in a given year, on average 60% will be detected/seen and reported. . . (Aune and Kasworm

1989)." Although this metric was not used to adjust the number of adult females with cubs observed in Yellowstone, it was used in computations for this parameter in the Northern Continental Divide zone (USFWS 1993, p. 62). In personal communication (1991), Aune objected to the use of the 60% reporting efficiency of females with cubs as a misrepresentation of his conclusions.

By applying the criteria that relate to each of the four major recovery goals, the USFWS describes the present grizzly bear population in the Greater Yellowstone Ecosystem as possessing the following population characteristics, computed as running 6-year averages from 1987 to 1992 (except goal No. 2). These data for goal fulfillment (USFWS 1993, pp. 43-46), ascribed to IGBST annual reports (Knight *et al.* 1986-92), are:

1. Annual average unduplicated females with cubs = 19.8 (Recovery target ≥15).
2. Number of BMUs with family groups (1986-91 running sum) = 16 of 18 (Recovery target ≥16 of 18 BMUs occupied).
3. Annual average known human-caused bear deaths, females only = 1.8 (Recovery target ≤3).
4. Annual average known human-caused bear deaths = 4.0 (Recovery target ≤9).

Parameter No. 1 shows 19.8 unduplicated females with cubs. This, according to the plan, "is the prime indicator of population level." We have shown earlier (Chapter 19) that it very likely is not. According to the tenets of the recovery plan, however, goal No. 1—evidence of an average 15 females with cubs—has been exceeded, and the recovery goal is met. Parameter No. 2, the habitation of family groups in 16 of 18 BMUs, is of questionable value, as we have demonstrated. Nevertheless, it, too, meets the recovery goal established for the population. Parameters No. 3 and No. 4 show the average annual deaths for adult females and for all bears to be 1.8 and 4.0, respectively. To determine whether these mortality values met the target criteria, a "minimum" ecosystem population was estimated using the process of analysis devised by Knight *et al.* (1988a; see USFWS 1993, Appendix C), and the annual number of unduplicated adult females with cubs averaged over the 3-year period, 1990-92. Based on percentages of that estimate, then, the allowable mortality was calculated. An average of 23.7 females with cubs recorded for the 1990-92 period and 71 adult females, less 4 known mortalities, presumed alive in 1993 (USFWS 1993, p. 43), provided a total "minimum" population estimate for 1993 of 236 bears. Application, then, of the allowable death rate of 4% to 236 bears yielded a value of 9 human-caused bear deaths and, at a rate of 30% of those 9, 3 adult female deaths. In other words, if we assume there were 236 bears in 1993, then the 6-year averages of recorded bear mortalities (1987-92) equaling 4 bears and 1.8 for females only, were less than the calculated targets of 9 and 3, respectively, and once again, recovery goals were met.

The minimum estimated population of 236 bears computed for 1992 is 54 to 65 bears in excess of the minimum population of 171 to 182 bears calculated by the method of Knight *et al.* (1988a) four years earlier, based on a 3-year total of 45 females with cubs (i.e., the 15 females with cubs target). This suggests that the population has recovered. We question, however, whether the goal calculated by Knight *et al.* (1988a) of 171-182 bears or the 1992 estimate of 236 represent recovered populations. We believe they do not and wonder why the 9-year mean of 312 bears calculated for the preclosure population (see Chapter 19) is not a more plausible goal for recovery. By what criteria are 171-182 bears, or any number other than 312, a scientifically justifiable recovery goal? The recovery plan does not explain.

If we accept the IGBST statistic of 19.8 adult females with cubs, as a 6-year mean, and the recovery coordinator's interpretation of these data, as well as the values reported for the other recovery parameters, then the population inhabiting the Yellowstone Ecosystem has recovered. One wonders why the USFWS felt constrained to state that the goals ". . . are based on the best available information on the population necessary to be viable and as self-sustaining as possible in this ecosystem. These goals [apparently already met] will be revised as necessary or as new information becomes available" (USFWS 1993, p. 41). Why not delete the numerous pages devoted to detailing minor recommendations for bringing about recovery in the next decade and simply state that recovery has already occurred based on fulfillment of the stated recovery targets? Is it possible that the recovery coordinator is as dubious as we of the data on which population trend and status are based? If so, pouring huge sums of federal money into the existing research program is not the solution. Also, pursuing a seriously faulted recovery plan for another decade will only place the Yellowstone grizzly bear population in greater jeopardy. We believe a better plan can be devised and, in the sections that follow, we develop one. We recommend that it be seriously studied by the responsible agencies, in spite of its origination in a private sector which has too long been excluded from making contributions at the planning level.

GOALS FOR GRIZZLY BEAR POPULATION RECOVERY

To preserve the grizzly south of the Canadian border, we must first effect recovery of the isolated populations now classified as either threatened or endangered. We have reviewed the government's recovery plan and found it inadequate to the task. We agree on the broad goals for recovery, but disagree on how to attain them.

We believe the following goals must be attained to achieve population recovery in Yellowstone and elsewhere in the conterminous United States:

1. Specific protection of adequate and appropriate habitat that meets the grizzly bear's need for space, high quality food, and security from human intrusion (minimize effects of environmental stochasticity);
2. A population sufficiently large and a mechanism for demographic (genetic) enhancement sufficiently dependable that persistence of the population for at least 100 years is highly probable (minimize effects of demographic and genetic stochasticity);
3. Little or no human-caused mortality;
4. The development and application of new management and monitoring techniques that are more precise and less intrusive on the populations than current ones; and
5. Negotiation of an agreement between U. S. and Canadian authorities for cooperative protection of the grizzly bear and its habitat where populations are confluent between the two countries.

In the sections that follow, we describe how these goals may be achieved. However, as indicated earlier, many of the issues associated with grizzly bear population recovery are sociopolitical as well as biological, and this duality must be accounted for in any biological strategy for population recovery. We address first the problem of protecting and retaining adequate habitat, and second, of determining population viability and persistence times. Third, based on our knowledge of ecocentric behavior in grizzly bear populations, we suggest a management strategy that should enhance the likelihood of population recovery and persistence. Fourth, we reanalyze the structure of the preclosure grizzly bear population to establish recovery goals for the current population that must serve until rigorous viability analyses show otherwise. Fifth, we suggest specific techniques and strategies for effecting population recovery in the Yellowstone Ecosystem and the other recovery areas within the conterminous 48 states. Finally, we discuss some positive responses from society that appear necessary, if grizzly bears are to survive under circumstances of a growing human population and its unrelenting use of natural resources.

PRESENT AND FUTURE HABITAT NEEDS

The inexorable growth of human population and attendant resource exploitation has rendered the vast majority of the bears' historic range inaccessible to them. Relict populations of bears have persisted in the lower 48 states only in areas where protected cores exist in the form of national parks or designated wilderness enhanced by contiguity with substantial, unroaded expanses of de facto wilderness in the national forests. Where unroaded areas of national forests are occupied by grizzly bears, or where they provide opportunity for range expansion or potential corridors for genetic exchange between occupied areas, their preservation will be crucial to achieving long-term population viability.

Over the last decade, the timber industry has seriously depleted the resources of their privately held lands and of the more accessible, lower-elevation areas of the national forests in the Northern Rockies bioregion. Timber and other extractive industries, particularly mining, have mounted increasingly strong political and economic pressure for access to the largely unroaded de facto wilderness that is critical grizzly bear habitat. De facto wilderness has survived not because of protection by congressional mandate or for its intrinsic value to bears and other wildlife, but simply because it has previously been considered too remote, rugged, and expensive to develop. The harvesting of resources and the intrusions of habitat envisioned for these lands by the timber, mining, petroleum, and motorized recreation industries are inimical to the needs of the grizzly and pose a threat to its recovery.

With an eye toward recovery goal #1—specific protection of adequate and appropriate habitat to meet the grizzly bear's needs—we now assess what habitat may actually be available to grizzly bears in the future. Long-term viability of the remaining populations of grizzly bears is directly correlated with habitat quality and quantity. It is our firm conviction that most of the high quality habitat still available should be reserved in the form of congressionally designated wilderness or some other category that places the needs of the bears and preservation of biodiversity above those of extractive industry and development interests. With the exception of a few visionary proposals (e.g., The Northern Rockies Ecosystem Protection Act [Bader 1991]

introduced into the 103rd Congress in 1993), the numerous proposals brought before Congress for designation of new wilderness areas have failed to designate the entirety of landscape currently occupied by grizzly bears, much less the land areas necessary for range expansion as populations approach recovery. Thus, we now consider how well the areas currently reserved to the grizzly bear as a function of its threatened status (USFWS 1993) will meet its requirements for recovery. Specifically, we examine the zones designated for grizzly bear recovery to determine whether they meet the needs of viable populations of bears.

RECOVERY ZONES

A recovery zone has been defined as an area within which bears and habitat are measured to determine a population's recovery (USFWS 1993). Bears are then managed within these zones (wilderness excepted) according to agency-determined Management Situation classifications. The Management Situation criteria are a combination of perceived bear distributions and habitat quality. They are applied in the national parks, forests, and other public lands within the context of other resource uses. The criteria and the areas they designate for the bears are a source of serious ambiguity as to what is essential grizzly bear habitat. This becomes especially evident when other resource uses in multiple-use zones conflict with the preservation of the bear mandated by the National Environmental Policy Act (1969), the Endangered Species Act (1973), and the National Forest Management Act (1976).

Currently, six grizzly bear recovery zones have been delineated in the contiguous 48 states (Fig. 20.1). Four of these—the Cabinet-Yaak, the Northern Continental Divide, Selkirk Mountains, and Yellowstone recovery zones—support populations of grizzly bears that are fairly isolated geographically and demographically (a metapopulation), at least south of the Canadian border. Two other recovery zones, the North Cascades and Selway-Bitterroot, are not known to support resident grizzly bears currently, although they once did. Following studies for habitat suitability, both zones were concluded to be appropriate for reintroduction programs (Servheen *et al.* 1991).

With the exception of the Yellowstone and Selway-Bitterroot recovery zones, the zones border Canada and their grizzly bear populations are continuous into Canada. There are no persuasive data to show that the Canadian populations are any more stable than those to the south and, unfortu-

Fig. 20.1. Grizzly bear recovery zones (six) in the Northern Rockies and the North Cascades Range (USFWS 1993).

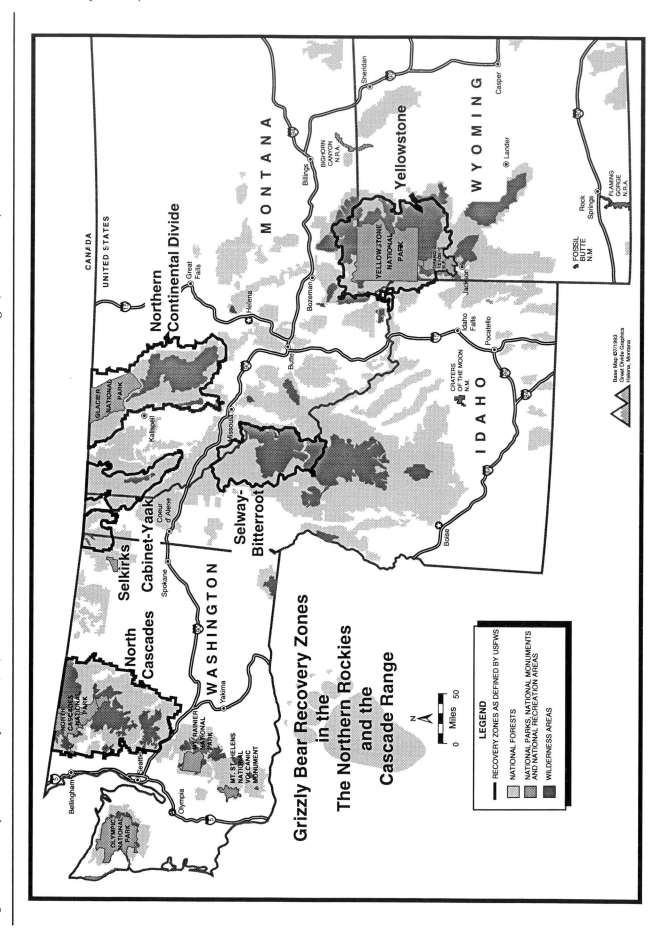

nately, Canada has made no official commitments to preserving the grizzly bear or its habitat (hunting permits are still issued). Thus, to rely on Canadian bears for dependable demographic support of the threatened U. S. populations, as does the Grizzly Bear Recovery Plan (USFWS 1993, pp. 26 and 27), is ill-advised.

An area not designated for population recovery, but under consideration by the USFWS, is the San Juan Mountains region of Colorado. This rugged area has not supported a viable population of grizzly bears for many years, and despite determined field efforts, no individuals have been officially documented since 1979. Currently, the IGBC has announced no plans for reintroducing grizzly bears into the San Juans.

Strangely, the recovery plan contains no consideration of the vast Wind River area in Wyoming as habitat where grizzly bears could be reintroduced. We believe this wildland area should be thoroughly investigated as a recovery zone since there are credible reports that members of the Yellowstone grizzly bear population are moving into the area as a result of the catastrophic Yellowstone fires of 1988.

The six recovery zones vary widely in area, estimated minimum bear populations, and mix of land management authority (Table 20.1). All lie principally within national forests, frequently encompassing substantial areas of designated and de facto wilderness. Three recovery zones—the North Cascades, the Northern Continental Divide, and the Yellowstone —include national parks within their boundaries (Fig. 20.1). Although the North Cascades, at nearly 2.6 million hectares (6.5 million ac), is the largest of the zones, it has no grizzlies. The Northern Continental Divide and Yellowstone zones, both supporting populations of bears, are the next largest. The Selkirk Mountains recovery zone is, by far, the smallest.

The Yellowstone zone contains the greatest amount of habitat specifically protected from extractive uses by inclusion within national parks (44.0%), followed by the Northern Continental Divide (16.5%), and the North Cascades (10.6%). By virtue of the Selway-Bitterroot and Frank Church-River of No Return Wilderness Areas, the Selway-Bitterroot recovery zone includes the greatest percentage of designated wilderness (49.4%) from which roading, as well as extractive industry, is prohibited (Table 20.1). Designated wilderness areas compose 36.3% of the Yellowstone, 28.3% of the North Cascades, 26.8% of the Northern Continental Divide, and only 7.5% of the Cabinet-Yaak. Because they are protected from development and motorized access, the designated wilderness areas

represent habitat bastions for the grizzly bear. They are core areas essential to long-term preservation of the grizzly.

The tiny Selkirk Mountains zone includes only 6% (16,728 hectares [41,335 ac]) of its total area in designated wilderness, and lacking national parks as well, it can provide bears only that habitat protection intrinsic to the Endangered Species Act. As we discuss later, such protection is equivocal, depending on the land management agency's (most often, the Forest Service) Management Situation assignment for the habitat and the intensity of resource use in adjacent lands.

The recovery zones originally were established either to accommodate stabilization of existing grizzly bear populations (Cabinet-Yaak, Northern Continental Divide, Selkirk Mountains, and Yellowstone) or as sites judged to be appropriate for reintroduction of grizzly bears to previously occupied habitat (North Cascades and Selway-Bitterroot). No information is provided as to how (by what criteria) these zones were determined to be appropriate. Obviously, biological, social, and financial issues are very different for each zone, as are the prospects for effecting recovery. Why six zones? Would the prospects for perpetuation of grizzly bear populations be brighter if funds and efforts were concentrated on three or four zones rather than six? We believe that they would be.

We next examine whether these six zones provide for adequate space, food, and security to support the viable metapopulations that are the ultimate goal. We present an overview of habitat characteristics; the reader is directed to the Grizzly Bear Compendium (1987) and to the individual studies cited in the text for more detailed information.

Spatial Characteristics

Although the Grizzly Bear Recovery Plan (USFWS 1993, p. 17) describes recovery zones to be areas in (presumably) larger "grizzly bear ecosystems" that contain bears, or will in the future, it later implies (p. 26) that zones are intended to accommodate the spatial needs of whatever minimum viable population is the recovery target. To determine the adequacy of the Cabinet-Yaak and Selkirk Mountains as recovery zones, the plan used 90 bears as a minimum population recovery target and average composite densities. The average densities used were one bear per 70.9 square kilometers (7090 hectares [17,519 ac]) for the Selkirk Mountains zone and, for the Cabinet-Yaak, one bear per 76 square kilometers (7600 hectares [18,780 ac]), derived from composite density figures established for bear populations in the Yellowstone, Northern

Table 20.1. Some characteristics of grizzly bear recovery zones in the lower 48 states, 1993.

Characteristic	Recovery zone					
	Cabinet-Yaak	North Cascades	N.Continental Divide	Selkirk Mtns.	Selway-Bitterroot	Yellowstone
Approx. hectares[a] (acres)	510,000 (1,260,210)	2,619,694 (6,473,264)	2,480,000 (6,128,080)	280,000 (691,880)	1,402,653 (3,465,956)	2,330,000 (5,757,430)
Estimated minimum grizzly bear population[a]	91	—	391	91	—	228
National forests	Kootenai, Lolo, Panhandle	Mount Baker, Okanogaan	Flathead, Helena, Kootenai, Lewis and Clark, Lolo	Colville, Panhandle	Bitterroot, Clearwater, Nez Perce	Beaverhead, Bridger-Teton, Custer, Gallatin, Shoshone
National parks (hectares/acres)		N. Cascades Complex (276,907/684,238)	Glacier (410,000/1,013,000)			Yellowstone (899,500/2,222,665); Grand Tetons (125,455/310,000)
Percent in national parks	0.0	10.6	16.5	0.0	0.0	44.0
Designated wilderness areas[b] (hectares/acres)	Cabinet Mtns. (38,186/94,360)	Alpine Lakes (123,596/305,407); Boulder River (19,830/49,000); Glacier Peak (233,366/576,648); Lake Chelan-Sawtooth (60,989/150,704); Henry M. Jackson (41,550/102,671); Mount Baker (47,583/117,580); Pasayten (214,427/529,850)	Bob Marshall (408,351/1,009,035); Great Bear (116,026/286,700); Lincoln-Scapegoat (97,101/239,936); Mission Mtns. (29,898/73,877); Rattlesnake (13,355/33,000)	Salmo-Priest (16,728/41,335)	Selway-Bitterroot (541,352/1,337,681); Frank Church-River of No Return[c] (150,949/372,996)	Absaroka-Beartooth[c] (233,078/575,936); Jedediah Smith[c] (23,580/58268); Lee Metcalf[c] (24,864/61,440); North Absaroka (141,861/350,538); Teton (236,936/585,468); Washakie[c] (185,183/457,588)
Percent in wilderness	7.5	28.3	26.8	6.0	49.4	36.3
Total percent of recovery zone in national parks and designated wilderness	7.5	38.9	43.3	6.0	49.4	80.3

[a] In U.S. only (U. S. Fish and Wildlife Service 1993).
[b] As of 1989 (The Wilderness Society 1989).
[c] Includes only the approximate portion within the recovery zone.

Continental Divide, and Cabinet-Yaak zones. With this mix of data, the recovery plan (pp. 26-27) concludes that the Cabinet-Yaak and the tiny Selkirk Mountains zones are adequate for population recovery.

We do not agree with the basic assumption that a population of 90 bears is an appropriate recovery goal, nor with the use of composite density values. The recovery plan (USFWS 1993, p. 26) states that these recovery zones will suffice because they are contiguous with habitat in Canada occupied by "a much larger population of grizzly bears." While convenient, this is not a scientifically sound mitigation of low population recovery goals or inadequate space when neither the bears nor their habitat requirements are accorded any special dispensation north of the border. It is especially unwise when some reports on the status of bears in Canada suggest they are being lost faster than in adjacent areas across the border in the U. S. (Horejsi 1989).

On the issue of necessary space, other researchers have argued that long-term persistence of grizzly bear populations will require far larger reserves of high quality habitat than are available in any of the recovery zones. For the most part, such arguments originate from estimates of what numbers constitute viable populations of large mammals. This approach has evolved from a more rigorous rationale than that applied in the recovery plan. Several lines of reasoning suggest that high probability of long-term persistence without loss of genetic fitness requires a population of several thousand grizzly bears. Metzgar and Bader (1992) assumed (1) an effective population (the number of individuals contributing their genes to the next generation) of 500 bears in a total population of 2000 throughout the region and (2) a population density of four bears per 259 square kilometers (100 mi^2) to conclude that approximately 129,500 square kilometers (50,000 mi^2) of high quality, secure habitat were needed to assure population recovery. Likewise, Shaffer (1992) declared that a large expansion of the presently designated recovery zones was essential to recovery of the grizzly bear in the lower 48 states.

Space necessary to ensure survival of grizzly bears in the long term is contingent on how many bears must be maintained to assure population viability. But that number is, itself, partially a function of the quality of habitat that is provided in the space. That is, the security of the habitat and the quality and stability of the food resource are important in forecasting the effects of environmental stochasticity on viability. Any evaluation of space available or necessary for population recovery, then, must include careful consideration of how well this

space meets the bears' needs for security and adequate food. A first step in obtaining this critical data is to satellite-map and describe the vegetation, ecosystem-wide (see Chapter 12). The second step is to delineate and protect corridors linking wilderness ecosystems.

Corridors

Corridors of habitat linking the major park or wilderness ecosystems should be delineated and protected so that some degree of demographic exchange between what are now "island" populations of grizzly bears is possible. To date, there are no solid data to indicate that grizzlies could or would travel relatively confined corridors. They have not shown the same propensity to wander or migrate as have wolves, cougars, or coyotes. Currently, no consensus exists among the research community as to how corridors should be located or structured, nor even whether they would function as envisioned, in grizzly population dispersal. However, if corridors prove effective "pipelines" for bear emigration and immigration, then demographic stochasticity and the decline in vigor that results from loss of genetic heterozygosity in small, isolated populations would be less of a concern over the long term (Allendorf *et al.* 1991, Nunney and Campbell 1993).

Harrison (1992), Lindenmayer and Nix (1993), and others have suggested essential components of corridor design for various species, but, as noted by Hobbs (1992), the body of data showing that such linkages actually function as conduits for wildlife or are important in their life histories is meager. He maintains that only thorough studies of real populations in real landscapes, as opposed to modeled ones, can provide a practical basis for judging the value of corridors. We agree with this assessment, but recognize that such studies could require decades, and linkages should be established immediately before expanding resource use precludes them altogether. As Hobbs (1992) remarks, protection and maintenance of existing linkages is a task more easily accomplished than attempting to retrofit them to a fractured landscape. A window of opportunity remains to preserve natural linkages in the Northern Rockies, and we must make every effort to designate these linkages and protect them.

Both American Wildlands (1993), a Montana-based environmental advocacy organization, and the USFWS (1993) have initiated studies to determine placement, structure, and habitat quality of ecosystem linkages for grizzly bears. Private lands, federal grazing allotments, developed areas, roads, and interstate highways that intervene will be serious obstacles to conceptualization and design. Even in aggregate, these may not be insurmountable problems, but their solutions will take time, possibly decades. Therefore, it is important that federal land agencies and conservation advocates continue to explore and preserve current habitat linkages between wilderness ecosystems in both the U. S. and Canada. Continued landscape fragmentation will likely make the preservation of such connections impossible in the future.

General Security Requirements

The extent to which any grizzly bear population is secure is virtually always related to human activities. Effects may be direct (human-caused mortalities) or indirect (loss or deterioration of useable habitat). It follows that requirements for security include (1) protection from purposeful takings; (2) adequate, high quality habitat remote from human activities; and (3) adequate and relatively stable food resources. When any of these requirements is not met, risk of human-related mortality increases.

Grizzly bear mortalities and their causes have been documented by Craighead *et al.* (1988b) for the Yellowstone recovery zone for the 1959-87 period. For the Yellowstone (1986-93) and the Northern Continental Divide (1980-93) recovery zones, mortalities have been documented in annual reports by Arnold Dood and Helga Pac of the Montana Department of Fish, Wildlife and Parks. Mortalities attributable to purposeful killing by humans declined substantially after sport hunting of grizzly bears was banned in Wyoming and Idaho in the late 1960s, and in Montana, with the exception of the northwestern portion, in 1975. From 1975 to 1985, sport hunting accounted for an average of 10.2 grizzly bear mortalities per year in northwestern Montana (Grizzly Bear Compendium 1987), but declined to 2.6 per year from 1986 through 1990 (Dood and Pac, Annual Reports for 1986-90). In 1991, the hunting season on grizzly bears was closed throughout the state of Montana.

Other sources of human-related mortality—viz., management control, removal to captivity (functionally, a mortality), poaching, defense of life or property, vehicular trauma—are plainly associated with the degree to which human activities encroach on grizzly bear habitat and/or with adequacy of food resources and how widely bears must range in foraging. Human activities such as real estate development, open-range grazing, timber harvest, motorized tourism, mining, and petroleum development affect all of the recovery zones to some extent. The evidence is strong that these activities, and the hundreds of miles of road that support them,

displace bears from what otherwise would be occupied habitat and fragment it into smaller, less effective parcels (Schallenberger and Jonkel 1980, Mattson *et al.* 1987, McLellan and Shackleton 1988, Aune and Kasworm 1989, Schoen and Beier 1989, Kasworm and Manley 1990, McLellan 1990, Mace and Manley 1993). Moreover, grizzly bears are more likely to suffer human-caused mortality in heavily roaded portions of their range (Schallenberger 1980, Aune and Kasworm 1989, McLellan 1989b). This tendency is a concern in the Yellowstone Ecosystem, where research indicates that when large males compete vigorously for food with females with cubs and with subadult males, the latter cohorts are forced out of the better habitat to forage in close proximity to roads and developed areas. There, the risk of death increases (Mattson *et al.* 1987).

Specific Zonal Security Assets

The six zones delineated for grizzly bear recovery (Fig. 20.1) differ greatly in space available, but also in the security provided.

Zones with National Park Cores

As discussed earlier, national parks compose portions of the Yellowstone, Northern Continental Divide, and North Cascades recovery zones, but there are none in the other three zones (Table 20.1). In national parks, bears are accorded an extra measure of protection by park management policies, and their habitat is protected from exploitation by extractive industry. However, the quality of habitat in most of the parks is diminished by tremendous tourist visitation, the usurpation of prime habitat for visitor service facilities, and the extensive roading, usually through riparian habitat, necessary to provide visitor access (Table 20.2). The parks are also crisscrossed by numerous heavily used backcountry trails with many designated campgrounds. All of these factors bring bears into close proximity with humans, often to the detriment of the bears.

The North Cascades National Park complex of 276,907 hectares (684,238 acres) is an exception among parks because roading is minimal, visitation is relatively small, and because in 1988 Congress designated 93% of the park complex as the Stephen Mather Wilderness Area (Table 20.1). The vast majority of visitors view the park complex afoot or on tour buses. This is in stark contrast to the crush of humans and vehicles that assails Yellowstone, Grand Teton, and Glacier, the other three parks included in recovery zones.

It should be noted here that the Wilderness Act of 1964 specifically directed that national parks evaluate all roadless areas within their borders that

Table 20.2. Visitor use levels and extent of roading in national parks that lie within grizzly bear recovery zones in the lower 48 states, 1992.

National park	No. of visitors, 1992	Kilometers of roading (mi)
Glacier	2,199,769	≈241 (≈150)
Grand Teton	3,012,000	≈288 (≈463)
North Cascades Complex[a]	395,476	≈92 (≈57)
Yellowstone	3,186,590	≈867 (≈539)[b]

[a] The Complex includes North Cascades National Park and the Lake Chelan and Ross Lake National Recreation Areas; 93% of the Complex was congressionally designated in 1988 as the Stephen Mather Wilderness Area.

[b] Includes 56 miles of parking areas and pullouts.

are of 2024 hectares (5000 acres) or more and prepare a proposal to Congress for official wilderness designation. Although proposals have been prepared, these have seldom been considered or authorized by Congress, and the lack of protection under tenets of the Wilderness Act has fostered continuous erosion of wilderness values. For example, in 1972 the NPS submitted a proposal to designate as wilderness 815,937 hectares (2,016,181 acres) of Yellowstone National Park's 890,328 hectares (2,200,000 acres) of roadless area. Although the proposed wilderness represented 90% of the total area of the Park, Stewart Brandborg, former Director of The Wilderness Society and an important architect of the Wilderness Act, pointed out that the 87,000 hectares (215,000 acres) omitted and reserved for development were at lower elevation and frequently in riparian zones containing critical wildlife habitat (pers. comm. 1993). However, as a result of the failure of Congress to approve the proposal, and of the NPS to pursue the issue any further, development in Yellowstone has proceeded without constraints imposed by the Wilderness Act. We can only speculate how much the wilderness qualities of Yellowstone, and of other parks that have failed to designate, have been compromised as a result. We urge park administrators to vigorously pursue wilderness designations.

Zones with Designated Wilderness Cores

Designated wilderness areas provide perhaps the greatest security to grizzly bears, superior to de facto wilderness only because provisions of the Wilderness Act preclude actions by management agencies that allow extractive commodity development and roading for motorized access. All of the recovery zones incorporate some amount of designated wilderness. The Yellowstone, with 823,067 hectares (2,033,798 acres), contains the greatest amount of designated wilderness, followed by the North Cas-

cades with 741,344 hectares (1,831,860 acres), the Selway-Bitterroot with 692,301 hectares (1,710,677 acres), and the Northern Continental Divide, with 664,731 hectares (1,642,548 acres). In all of these zones, designated wilderness represents more than 25% of the total area (Table 20.1) and forms crucial core zones of security that are more or less effective, depending on the food resource. By contrast, the Cabinet-Yaak and Selkirk Mountains, relatively small zones with only 7.5% and 6.0% of total area in designated wilderness, respectively, offer mostly lands already subjected to extractive development, grazing, and roading (particularly in the Selkirk Mountains), or in perpetual danger of such incursions. Their relatively small areas and lack of protective land designations pose a serious problem to establishing viable populations.

Zones with de Facto Wilderness

Because of the grizzly's threatened status, all habitat in recovery zones is protected, to some extent, as specified by the Endangered Species Act (1973). Habitat in private, corporate, state, or local government ownership is addressed under Section 9 of the Act, but the great majority is on federal lands administered by the Forest Service or the Bureau of Land Management and covered under Section 7. Some of the lands in recovery zones have been harvested for commodities and are roaded, often heavily. However, most zones also encompass or abut on substantial expanses of de facto wilderness extremely important to grizzly bear security and food resources, but also highly attractive to extractive industry. The miscarriage of the intent of Section 7 for delineating and preserving critical habitat as entire ecosystems has caused great concern for how de facto wilderness will be managed by the federal agencies, and what proportion will be eventually designated as wilderness. What habitat including de facto wilderness can actually be provided for population recovery?

SPECIFIC FOOD RESOURCE REQUIREMENTS

Classification of habitat in terms of bear foods and descriptions of the bear's feeding habits have been done for all of the recovery zones, but the completeness of the studies varies. Zones best characterized with respect to foods are the Yellowstone (Mealey 1980, Craighead *et al.* 1982, Knight *et al.* 1982, Mattson and Knight 1989, Mattson *et al.* 1991a; see also Chapter 10 of this publication) and the Northern Continental Divide (Mace and Jonkel 1980, Schallenberger and Jonkel 1980, Craighead *et al.* 1982, Servheen 1983, Klaver *et al.* 1986, Aune and Kasworm 1989). Information is also available for the Cabinet-Yaak (Kasworm and Manley 1988, Kasworm and Thier 1991, and Kasworm *et al.* 1991), the Selkirk Mountains (Zager 1983, Almack 1986), and the Selway-Bitterroot (Scaggs 1979). Unpublished results of an Interagency Grizzly Bear Committee survey for food resources in the Selway-Bitterroot and the North Cascades recovery zones are referenced in Servheen *et al.* (1991).

These studies show that grizzly bear foods are available in all of the recovery zones and that bears survive at some level of nutritional health in some of them, but we have little basis for comparing the relative qualities of the food resources among recovery zones. In other words, food quality, quantity, and dependability vary among zones, but how "excellent" or "poor" the overall food resource might be for support of a bear population has been little explored and no effort has been made to determine relative carrying capacities. Mattson *et al.* (1991a) provided a cursory comparison between food resources available in the Yellowstone Ecosystem and those in other North American bear ecosystems, but drew no conclusions as to overall relative quality. The only studies of which we are aware that applied a consistent quantitative methodology to compare food resources were those of Craighead *et al.* (1982) for the Scapegoat-Bob Marshall Wilderness areas and Scaggs (unpubl. manuscript) for the Selway-Bitterroot Wilderness.

As we discussed earlier (Chapters 10, 15, and 16), availability of food strongly affects bear movement and home range size. It follows that a poor food resource may cause foraging bears to range beyond their federally allotted zones into areas where their safety is compromised. Thus, until the food resources of grizzly bear habitat in the recovery zones can be compared for relative quantity and quality, the question of how many bears a zone will support has no reliable answer, and the process of calculating the size of the "all-purpose" recovery zone necessary for supporting a minimum viable population, no solid rationale.

Craighead (1980a), Craighead *et al.* (1982), and Craighead and Craighead (1991) have strongly recommended that the vegetation of the entire Northern Rockies bioregion be satellite-mapped using a regionally consistent, botanically-detailed hierarchy of vegetation and landform classifications that includes quantitative occurrence and coverage statistics for plant species. Existing data on bear feeding habits could then be applied to define quality and quantity of bear food resources from one ecosystem to another and establish reliable carrying capacities for recovery zones.

Although grizzly bears are surviving in the inhabited recovery zones, their overall nutritional status is understandably open to question. We have no information about what proportion of a population may be malnourished or may die of starvation or be predisposed to disease. Weights and measures can be used as indirect indicators of nutritional status, but influences of other habitat variables or differences in population gene pools cannot be discounted when different populations are compared. However, a study of comparative weights that did provide reasonably reliable comparisons was that of Blanchard (1987) (see also Craighead and Mitchell 1982). She found that mean weights of grizzly bears in the Yellowstone Ecosystem were consistently greater before closure of the open-pit dumps (1959-70) than they were after (1975-85). Using regression analysis to compare weights of bears before dump closure with those after, we showed that weights were significantly greater for 12 of 15 categories of age, sex, and weigh date during the preclosure period, and that rates of growth by age and within a season were greater, as well (Chapter 16). The decline in weights and growth rates after dump closure definitely was related to reduction in the foodbase because destruction of the ecocenters constituted a significant change in the habitat for the Yellowstone Ecosystem population.

The quality of the food resource governs the food-getting behavior of the bear population and, therefore, has consequences for population viability beyond that of nutritional status. Because grizzly bears forage widely in meeting nutritional needs, spatial requirements to ensure their security increase as quality and dependability of the food resource declines (Blanchard 1990, Blanchard and Knight 1991; see also Chapters 15 and 16). We showed in Chapter 11 that ecocentered bears of the Yellowstone Ecosystem, held by copious food reservoirs at the dumps, had a lower risk of mortality than did non-ecocentered bears and, in Chapter 18, that risk of mortality was associated with certain sex and age classes of bears as a function of movement tendencies that characterized the class. Also, there is good anecdotal evidence that well-fed bears are less aggressive than ones on a "survival diet" and, therefore, are at lower mortality risk.

In a similar vein, food scarcity and competition for food in the Yellowstone Ecosystem are thought to force adult female and subadult male grizzlies into close proximity of roads and developed areas where risk of death is heightened (Mattson *et al.* 1987, Knight *et al.* 1988b, Blanchard 1990, Blanchard and Knight 1991). It is not surprising, then, that occurrence of poor whitebark pine (*Pinus albicaulis*) seed crops has been shown to be directly associated with increased rates of mortality among these same two groups (Knight *et al.* 1988b; Blanchard 1990; Mattson *et al.* 1991a, 1992a).

We contend that the total food resource and its availability are the most critical characteristics for delineating critical habitat and for determining the adequacy of grizzly bear habitat in any of the recovery zones. For the wide-ranging, opportunistic grizzly, the amount of foraging space and degree of security a zone provides are intrinsic to estimating the probability of population persistence, and both are related to the quantity, quality, and diversity of the food resource. We support whole-heartedly all efforts to protect as much high-quality habitat as possible, not only for the recovery of the bear, but to preserve what remains of our nation's wilderness biodiversity. However, considering the great political and economic forces pressuring for access to our remaining wild areas (especially de facto wilderness), we fear that sufficiency of habitat is likely to be the ingredient for recovery in shortest supply. Recognizing the possibility of the worst-case scenario—that bears are relegated to surviving as separate populations in the four recovery zones now occupied—we urge that management agencies pay heed to the important role food resource plays in delineating the bears' spatial needs. We suggest they apply what we have learned about the protective properties of ecocenters to strengthen prospects for the bear's long-term survival.

In the next section, we explore current theory of population viability in large mammals. We then apply this theory in considering probabilities for long-term persistence of grizzly bears and show how judicious use of ecocenters can enhance the probabilities. Lastly, based on the Trout Creek population censuses, our most complete and detailed data, we reassess the structure of the preclosure, ecosystem-wide grizzly bear population to establish basic

parameters that, we believe, should be adopted for more rigorous viability analyses in the future.

POPULATION VIABILITY ANALYSIS AND THE ECOCENTER CONCEPT

Earlier in this chapter, we criticized the USFWS recovery plan as logically and empirically flawed. Among other things, we agreed with Shaffer (1992) that the Yellowstone recovery criteria of 15 females with cubs and 4% annual human-caused mortality are arbitrary and that the final criterion, occupancy of 16 of 18 BMUs, is potentially misleading because it presupposes habitat selection by bears, and it ignores the natural aggregating tendencies of grizzlies documented in this monograph.

In this section, we join Shaffer (1992) in promoting population viability analysis (PVA) as an objective method for setting realistic recovery targets and show how the ecocenter concept can (and, we believe, should) be incorporated into both the process of setting recovery criteria and the on-the-ground management process of meeting those targets. To provide a background for these suggestions, we first briefly review general principles of PVA and published applications of PVA to grizzly bears. For thorough reviews, we refer readers to Frankel and Soulé (1981), Shaffer and Samson (1985), Shaffer (1987, 1992), and Soulé (1987).

The Population Viability Concept

Proponents of PVA acknowledge that population extinction is often driven by deterministic forces, such as loss of habitat, environmental toxins, or the introduction of exotic diseases, predators, or competitors. However, they point out that even when such forces are absent, and hence carrying capacities remain constant, populations may still run significant risk of extinction simply as a result of normal variation in demographic parameters. Thus, if a population is very small, fluctuations in birth and death rates due entirely to the normal vagaries of chance may doom a population to extinction. Variation in year-to-year environmental conditions, both favorable and unfavorable, may have similar effects, even in large populations. These causes of variation in demographic parameters have been referred to as *demographic*, *environmental*, and *catastrophic* uncertainty (or stochasticity), respectively (Shaffer 1987). "Genetic uncertainty," the loss of genetic diversity due to chronic population and mating-system bottlenecks, may hasten the "walk" to extinction by increasing inbreeding effects and precluding genetic adaptation to environmental change (Soulé and Wilcox 1980, Frankel and Soulé 1981, Lande and Barrowclough 1987).

Population viability analysis attempts to predict, in a special limited sense, persistence times for individual populations using data in these four categories of uncertainty. The usual approach is to use population simulation software and field-derived statistics on reproduction and mortality to estimate the probability that a population with such characteristics will persist (under the status quo) for a given period into the future.

Population Viability Analysis of Grizzly Bears

The population parameters documented by Craighead *et al.* (1974) were the basis for the first computer simulation of grizzly bear population dynamics, and the same data set enabled the first real-world application of PVA to the grizzly bear (or any other species) (Shaffer 1978, 1987; Shaffer and Samson 1985). To make the preceding general comments about PVA more concrete, we review the latter three studies, and set the stage for our views about the relevance of the ecocenter concept to population viability.

Shaffer (1983) obtained mean and variance values for six demographic parameters from our study of the 1959-70 Yellowstone Ecosystem population. These parameters were litter size, annual mortality rate, annual percent of females reproducing, and annual adult, subadult, and total population sizes. The 1959-70 data were also used to estimate the effect of population density on the mean value of these six parameters (but see Chapter 9). A stochastic, discrete-time, discrete-number simulation program was then used to calculate persistence times of model populations having the demographic characteristics measured for the 1959-70 Yellowstone population. Individuals in model populations of various initial sizes (10, 15, 20, etc.) were allowed to reproduce and die according to the 1959-70 rates until the population went extinct or a predetermined period of time elapsed. Extinction was operationally defined as no remaining adult females. Fifty such simulations were run for each initial population size. Average survival time for each initial population size and the percent of populations of a specific size surviving the simulation period were calculated from the results of the 50 simulations. Shaffer and Samson (1985) later presented updated results of these simulations, some of which we reproduce in Table 20.3. Note that the simulations were not designed to consider the effects of food resource quality and stability, natural catastrophes (such as the 1988 fires), or genetic uncer-

Table 20.3. Fifty replicate simulations for each of five different initial grizzly bear population sizes to determine average time to population extinction (ATE) and the number and percent of the 50 simulations that result in extinction over a period of 100 years; adapted from simulations by Shaffer and Samson (1985) based on demographic data of Craighead *et al.* (1974).

Initial population size	Simulations in which population goes extinct		ATE
	No.	Percent	(yrs)
10	50	100	22
20	32	64	48
30	13	26	60
40	3	6	47
50	1	2	15

tainty and hence are optimistic with respect to persistence times. Note also that these are persistence times for populations initially at carrying capacity in a fixed quantity of habitat, subject only to normal annual variation in habitat quality.

For the purposes of illustration, ". . . an *arbitrary* criterion of 95% probability of persistence for 100 years . . ." was used to define a minimum viable population (Shaffer and Samson 1985, p. 147; italics ours). Only 94% of the 50 model populations with an initial size of 40 persisted for 100 years, whereas 98% of 50 populations of 50 animals did so (Table 20.3). Accordingly, the authors established 50 bears as an estimate of Minimum Viable Population size (MVP), under their specific definition of viability (95%/100 years) for populations with the demographic characteristics measured by Craighead *et al.* (1974). Of course, with a different definition of viability, the MVP estimate would be different. The authors went on to point out that only 6% of populations initially having 50 bears survived 300 years. Thus, a population size (e.g., 50) giving acceptable survival rates under one time scale of concern (100 years in this case) was totally unacceptable over a somewhat longer time scale (here, 300 years). Finally, although an initial model population of 50 had an average time to extinction of 114 years, more than half perished sooner, sometimes substantially so. Thus, *average* persistence time was a poor measure of confidence that any particular population would persist beyond the arbitrarily set criterion of 100 years.

Limitations of Viability Analyses

From this general description and the grizzly bear example, several limitations of PVA will be apparent to many readers. First, because the simulations are data-driven, they are also data-limited. Field data are measures of demographic parameters rather than true values, and these measurements are strictly applicable only to the years and population in which they are made. Normal variation in demographic parameters not captured in a field study (even a 12-year study is a short stretch of ecological time) will not be incorporated into projections, nor, obviously, can factors such as long-term food stability, genetics, and natural catastrophes, that are now excluded from simulations by design. "Errors" in estimation will be incorporated. Even in the best of circumstances, statements about persistence are probabilistic, not absolute. Thus, a population large enough to satisfy a given viability criterion ("x" probability of persisting "n" years) may, in fact, go extinct. Finally, PVA does not, at present, address habitat requirements and, therefore, is blind to steady deteriorations in habitat quantity and quality.

It is important, therefore, that the formality of the PVA process not lead us to attach to estimates of minimum viable population a precision or certainty they do not possess. Such estimates should always be treated conservatively, we believe, so that management errs on the side of population persistence rather than extinction. They should be constantly reappraised in light of new information. Population viability analysis should be regarded as a tool to identify broad but objective insights regarding population and habitat targets for recovery and maintenance. It should also encourage researchers and managers to focus their temporal and spatial scale of concern away from examining the current status of single threatened populations toward determining the long-term persistence of a series of populations distributed over broad geographic regions, the situation that exists in the contiguous 48 states today.

Population Viability Analysis and the Federal Recovery Plan

Shaffer (1992) has also applied the PVA perspective in a critique of the grizzly bear recovery plan and in his design of an alternative to that plan. We are in broad agreement with Shaffer's main conclusions, which we summarize as follows:

1. The federal recovery criterion of 15 females with cubs, although based on the MVP estimate of 50-90 bears in Shaffer (1983) and Shaffer and Samson (1985), is a misuse of their analysis. The definition of viability in these papers (95%/100 years) was chosen for illustration only and was never intended as a recommendation for management, especially for multiple populations persisting under widely varying circumstances. Considering the entire body of large vertebrate viability work, to date, the absence of genetic and catastrophic elements in their models, and a more appropriate (longer) time scale of concern, he suggests that the minimum viable population size for grizzly bears is on the order of "several thousand" bears.

2. Because this number of bears cannot be supported even in an aggregate of the six federal recovery zones, it follows that grizzly bears must be returned to a much larger portion of their former range than is currently planned.

3. For the same reason, evaluation of recovery and viability only makes sense when applied to combined populations of bears (i.e., the complex of recovery zones). A metapopulation structure will require more aggressive protection and management of the recovery zones and of habitat corridors between them than the federal plan now calls for.

4. Existing databases on bear demography should be consolidated and subjected to a rigorous, formal PVA by a multidisciplinary panel of biologists.

Whatever minimum viable population is ultimately adopted, the current field methods for determining population status and trend are inadequate and should be replaced with more rigorous estimates of abundance, natality, and mortality.

Incorporating the Ecocenter Concept in Population Viability Analysis

We have demonstrated, in the case of the centrally located Trout Creek aggregation, that ecocentered bears experienced fewer deaths from human causes because this ecocenter attracted and held aggregation members away from centers of human activity. We have also shown that ecocentered bears used the "natural" food resources as well as the food deposited at open-pit dumps, and that the ecocenter served to buffer individuals against the nutritional effects of seasonal and annual variation in these resources. As a consequence, ecocentered bears of the preclosure period had higher rates of body growth than non-ecocentered bears of the postclosure period. Although population size estimates after 1970 are less reliable than earlier ones, we demonstrated that the preclosure ecosystem population was approximately twice the size of the postclosure one. It is reasonable to suppose that the nutritional and movement-related effects of ecocenters also included reduced variation in annual rates of natality and mortality (natural and human-related). In short, logic and data suggest that ecocenters introduce a pervasive element of predictability into grizzly bear life histories. Because predictability lies at the heart of population persistence in the PVA construct, we suggest that ecocenters cannot be ignored in grizzly bear viability studies.

We see three viability-related applications of the ecocenter concept. First, degree of ecocentricity should be used to classify populations for formal PVA. Thus, existing field data on ecocentered and non-ecocentered populations should be treated separately, regardless of other similarities between populations. In particular, pre- and postclosure data for the Yellowstone populations should govern distinct sets of simulations. Similarly, to the extent that future collections of demographic data for PVAs are planned, study populations should be selected to encompass a range of ecocentricities.

Secondly, managers should recognize that a well-distributed set of ecocenters will draw and hold bears repeatedly to predictable locations. This can facilitate an accurate and noninvasive evaluation of population status with respect to recovery targets. It can prove especially valuable in making and evaluating grizzly bear transplants (discussed later).

Finally, given objective population targets for recovery, the ecocenter provides managers with a powerful tool for gaining some measure of control over data accumulation pertaining to the biological parameters governing population size and distribution. The remainder of our comments will be directed at specific applications of the ecocenter concept.

We first acknowledge that we are, in effect, proposing a program of enhancement and creation of natural and artificial ecocenters. We recognize that critics will object to any artificiality as contrary to present park policy and to recommendations of the Leopold Report (Leopold *et al.* 1963) that there be no artificial feeding of wildlife. Our response is that the latter report also advocated the control of elk numbers by shooting and live trapping and recommended "artificial" management of various kinds for other park resources. Nevertheless, we are well aware that our proposal will be extremely controversial under any circumstances. In hopes of avoiding controversy based solely upon misunderstanding, we want to be clear at the outset on four points. First, we prefer the development of natural ecocenters over artificial ecocenters whenever possible, and other things being equal. Second, when speaking of artificial ecocenters, we are not advocating the return of unsanitized open-pit refuse dumps to the Yellowstone Ecosystem. Grains, animal carcasses, and processed human food wastes are examples of ecocenter foods that can attract

bears without most of the unesthetic drawbacks inherent in the NPS's unregulated pre-1970 refuse dumps. Third, when trade-offs exist between the nutritional value (amount and type of food) and location (distance from human activity) of a proposed ecocenter, location must take precedence. The objective is not only to compensate for an adversely modified foodbase, but to draw bears away from conflicts with humans and to hold them unobtrusively within protected areas. Finally, any program of ecocenter development should be initiated as a well-planned and tightly monitored experimental program. We know too little about grizzly bears to predict the consequences of any particular regional distribution of ecocenters and the accompanying redistribution of bears.

With these comments in mind, we suggest four specific ways in which management for a network of ecocenters can facilitate recovery to population sizes and structures that meet the stricter viability criteria advocated by Shaffer (1992).

1. Ecocenter networks, by increasing per capita food resources and reducing human-caused mortality, can increase **mean** rates of natality and survival for any given density of bears inhabiting a fixed quantity of habitat; i.e., ecocenters can increase the carrying capacity of recovery areas. Other things being equal, larger initial populations have longer expected persistence times.

2. Ecocenter networks, by buffering seasonal and annual variation in other food sources, can reduce **variation** in annual rates of natality and mortality; i.e., ecocenters can reduce fluctuations in population size. Populations that fluctuate less persist longer, other things, like average carrying capacity of the habitat, being equal.

3. To the extent that transplantation is used to extend the grizzlies' current range or augment existing populations, ecocenters may be used to reduce the unwanted movement of transplanted bears outside the target area and to facilitate inclusion of these bears into the social fabric of the recipient population.

4. Ecocenters may be used to draw bears into selected, relatively "safe" linkage corridors in efforts to establish natural genetic and demographic exchange between populations in different recovery zones; i.e., ecocenters can help establish a natural metapopulation structure that might otherwise develop slowly or not at all (because without ecocenters, bears have less incentive to explore corridors) or, corridor movement may develop inappropriately (because bears might select corridors with relatively high potential for bear-human conflict).

Table 20.4. Procedure used to calculate annual point estimates of the Yellowstone Ecosystem grizzly bear population, 1960-68.

Year	I Cumulative no. of grizzlies censused at Trout Creek	II Percent of Trout Creek 1960-68 mean	III Calculated annual ecosystem population[a]
1960	98	84.4	263
1961	98	84.4	263
1962	98	84.4	263
1963	116	99.9	312
1964	128	110.2	344
1965	134	115.4	360
1966	133	114.5	357
1967	127	109.4	341
1968	113	97.3	304
Mean	116		312

[a] Calculated by multiplying decimal percents from column II times 312.

ESTABLISHING AN INTERIM RECOVERY TARGET FOR NUMBER OF ADULT FEMALES WITH CUBS

Presuming that annual mean number of adult females with cubs continues to serve as the primary indicator of population status, our objective in this section is to provide a defensible baseline which can serve, in lieu of a formal viability analysis, as an alternative to the goal of 15 females with cubs arbitrarily established by the official recovery plan (USFWS 1993). Our rationale is that the recovery target should not be less than the number of females with cubs characteristic of the relatively stable preclosure period.

To accomplish this, we will proceed as follows:

1. Establish annual point estimates of the size of the preclosure ecosystem population during 1960-68;
2. Calculate an ecosystem population age structure for 1960-68 by extrapolating the Trout Creek subpopulation age structure (Table 5.2);
3. Apply the 1960-68 Trout Creek subpopulation sex ratio (Table 5.6) to the calculated ecosystem population age structure to develop a sex-age structure for the ecosystem population; and
4. Extract from this sex-age structure the average size of the adult female segment of the 1960-68 ecosystem population.

In effect, we are assuming that the sex-age structure of the Trout Creek subpopulation, which averaged 116 bears and 37% of the total population, was our most precise data and, as such, best represented the larger ecosystem population during the same period.

Population Point Estimates

Earlier (Chapter 4), we concluded that the Craighead *et al.* (1974) upperbound assessment, which projected a starting population of 309 for 1959 and peak of 334 bears in 1967, best defined the ecosystem-wide population. Likewise, we acknowledged McCullough's mean ecosystem population estimate of 312 (McCullough 1986) to be a good approximation for the preclosure period. We will use the latter mean ecosystem population of 312 bears and the Trout Creek annual censuses (1960-68) as the bases for our computation of annual ecosystem population sizes for the same years, as follows:

1. Using the cumulative number of grizzly bears censused annually at Trout Creek (Table 5.2), we calculated a nine-year grand mean of 116 bears for the period, 1960-68 (Column I; Table 20.4).
2. Next, we calculated the percent of each annual census relative to the 1960-68 census grand mean (Column II; Table 20.4).
3. These percents were then applied to the mean ecosystem population of 312 bears to provide annual point estimates of the 1960-68 ecosystem population over the 1960-68 period (Column III; Table 20.4).

The procedure assumes that variation in Trout Creek censuses reflected ecosystem-wide trends in population size rather than, for example, annual variation in visitation to the ecocenter.

Fig. 20.2. Comparison of Trout Creek censused population to the censused and calculated ecosystem populations, 1960-68.

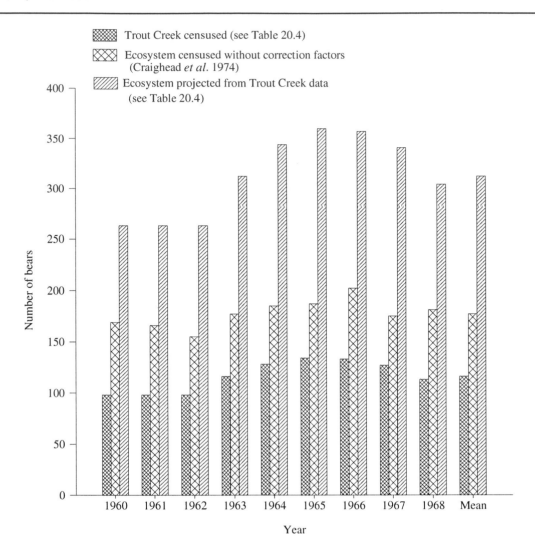

Comparison of Population Assessments

We are now in a position to compare three population assessments of the Yellowstone grizzly bears for the preclosure period (Fig. 20.2): the Trout Creek subpopulation, by direct census; the ecosystem population, by direct census, without census corrective factors; and the calculated ecosystem population derived by extrapolating the Trout Creek databases. The Trout Creek subpopulation of 116 bears represented a little over one third of the calculated mean of 312 bears for the entire Yellowstone Ecosystem. The similar trend over the 1960-68 period between the Trout Creek and extrapolated ecosystem data, on the one hand, and the direct ecosystem counts, on the other, justifies our assumption that annual variation in the Trout Creek cumulative counts reflected true variation in bear numbers.

Estimate of Ecosystem Age Structure Using the Trout Creek Database

Applying the censused age structure of the Trout Creek subpopulation (Table 5.2) to the annual point estimates of the ecosystem population (Table 20.4), we calculated annual age structures and a mean age structure for the ecosystem population (Table 20.5). For a population of 312 bears, this was 57 cubs, 43 yearlings, 33 2-year-olds, 46 subadults, and 134 adults. A comparison of this age structure, based on the Trout Creek data, with the ecosystem-wide age structures derived from our ecosystem-wide censuses (Craighead *et al.* 1974) showed close agreement (Table 20.6), indicating that the age structure of the Trout Creek subpopulation closely represented the larger ecosystem-wide population over the same period. Note, however the clear shift in the ecosystem-wide age structure toward more adults for the 1969-70 census period (Table 20.6). This was a result of manipulations and closures of ecocenters in 1969-70 that induced adults to move to Trout Creek in greater numbers than previously. This apparent change in census "efficiency" for adults was, as discussed in Chapter 4, the reason for restricting the period of this population reassessment to 1960-68.

Estimate of Ecosystem Age-Sex Structure and Mean Number of Adult Females

We applied the sex ratios observed for the Trout Creek subpopulation (Table 5.6) to develop annual age and sex structures for the extrapolated ecosystem population (Table 20.7). At Trout Creek, males predominated over females as cubs, yearlings, 2-year-olds, and 3- to 4-year-olds, but as adults, females predominated over males.

The estimated mean number of adult females in the ecosystem population from 1960 through 1968 was 76 animals (Table 20.7). On average, one-third of the adult females were with cubs each year. Thus, the adult-female-with-cubs parameter for the ecosystem population prior to dump closure was about 25. If this parameter continues in use at all, then this value, or some justifiable modification of it, should be the interim goal in achieving population recovery, rather than the questionable target of 15 females with cubs currently in use (USFWS 1993).

Furthermore, and more importantly in the long-run, the data of Tables 20.5 and 20.7, along with the individual life histories on which the analyses of Section I are based, form the appropriate empirical foundation (sex and age composition of initial model population, natality/mortality means and variances, etc.) for a formal PVA of an ecocentered grizzly bear population. Moreover, we believe that other important parameters, discussed in this chapter and elsewhere, but now excluded from formal population viability analysis by intent, must be incorporated into the modeling process to refine the assessments and to attain values that can be applied in management and in recovery planning.

MANAGING FOR GRIZZLY BEAR POPULATION RECOVERY IN THE YELLOWSTONE ECOSYSTEM

In this section, we make general and specific recommendations for grizzly bear management in the Yellowstone Ecosystem—those related to habitat requirements and those related to population dynamics. With these as background, we explore specific management actions for reestablishing a viable grizzly bear population.

Table 20.5. Age structure of the grizzly bear population in the Yellowstone Ecosystem, based on age class percents derived from bears censused at Trout Creek ecocenter, Yellowstone National Park, 1960-68.

Year	Total ecosystem population	Cubs Trout Creek observed %	Cubs Ecosystem-wide extrapolation[a] No.	Yearlings Trout Creek observed %	Yearlings Ecosystem-wide extrapolation[a] No.	2-year-olds Trout Creek observed %	2-year-olds Ecosystem-wide extrapolation[a] No.	3-4-year-olds Trout Creek observed %	3-4-year-olds Ecosystem-wide extrapolation[a] No.	Adults Trout Creek observed %	Adults Ecosystem-wide extrapolation[a] No.
1960	263	20.4	54	9.2	24	5.1	13	9.2	24	56.1	148
1961	263	18.4	48	9.2	24	11.2	30	14.3	38	46.9	123
1962	263	25.5	67	12.2	32	3.1	8	19.4	51	39.8	105
1963	312	23.3	73	18.1	56	6.9	22	12.1	38	39.7	124
1964	344	11.7	40	21.1	73	16.4	56	10.2	35	40.6	140
1965	360	22.4	81	9.7	35	15.7	56	11.9	43	40.3	145
1966	357	12.8	46	18.8	67	7.5	27	22.6	81	38.3	137
1967	341	15.0	51	13.4	46	15.0	51	16.5	56	40.2	137
1968	304	16.8	51	9.7	30	9.7	30	15.0	46	48.7	148
Mean[b]	312		56.8		43.0		32.6		45.8		134.1
Percent			18.2		13.8		10.4		14.6		43.0

[a] Calculated by multiplying the ecosystem population (column 2) by the Trout Creek (decimal) percent in an age class.
[b] Due to rounding of annual numbers in each age class to "whole bears," mean values for age classes sum to 312.3; this value is the basis for calculating percents in age classes.

Table 20.6. A comparison of ecosystem-wide age structures based on population counts and estimates.

	Ecosystem assessment 1960-68[a]	Ecosystem census 1960-68[b]	Ecosystem census 1959-70[c]	1969-70[c]
Cubs	18.2	18.6	17.6	13.1
Yearlings	13.8	13.0	12.6	12.0
Two-year-olds	10.4	10.2	9.8	8.8
Subadults[d]	14.6	14.7	14.7	14.7
Adults	43.0	43.7	45.3	51.3

[a] Database is the Trout Creek age structure (Table 5.2), applied to annual ecosystem population estimates (see Table 20.5).

[b] Database is the ecosystem-wide censuses (see Craighead *et al.* 1974), 1960-68 period.

[c] Database is the ecosystem-wide censuses (see Craighead *et al.* 1974), 1959-70 period.

[d] Three and four years of age.

Delineating the Recovery Zone

We have pointed out in the text many times that adequate habitat of sufficient area, quality, and security is essential both for recovery and for long-term survival. What then is enough habitat within the Yellowstone Ecosystem, and how do we protect, maintain, and enhance it? First, we must include within the recovery zone all grizzly bear habitat that is occupied and realistically necessary to effect recovery of the population. It would be a mistake to include areas of intense human use, even though occupied by grizzly bears; the resulting conflict would all-too-often result in the death of bears. Single-use areas, such as those supporting logging, grazing, or recreation, should be carefully evaluated for inclusion in the recovery zone. Likewise, all habitat peripheral to current recovery boundaries recently occupied as a result of emigration or range extension should be considered. The recovery boundaries must respond to the dynamics of the recovery process.

Keeping in mind the intent and clear direction of the ESA for protection of habitat critical to a threatened species, we suggest that recovery boundaries be reassessed biennially and revised, as appropriate, based on authentic bear sightings and on land-use evaluations. We recommend that de facto wilderness within the Ecosystem be officially classified as wilderness, including wildlands within Yellowstone and Grand Teton National Parks which never have been officially designated as required by the Wilderness Act.

As demonstrated in Chapter 15, the sizes of mean annual home ranges for both sexes and all age classes of the Yellowstone grizzly bear population have greatly increased over the last 20 years. Bears have moved to the peripheries of the recovery zone and beyond. Additional evidence of this expansion was provided by McDonald *et al.* (1990) in their exhaustive compilation of grizzly bear observations in the Yellowstone Ecosystem over the 1983-89 period. Figure 20.3 (adapted from Fig. 2 in McDonald *et al.* 1990) illustrates the distribution of bears for the years 1983-89, expressed as minimum convex polygons that incorporate 100% of sightings documented in those years. We have included, as well, all documented sightings for the years 1984-88 that were beyond the bounds of the polygons (adapted from Fig. 3 in McDonald *et al.* 1990). Based on these data, and on additional grizzly bear sightings reported to us by reliable sources that indicate bear activity in the Hildegard Range along the Madison River, in northeast portions of the Beartooth Mountains in the Gros Ventre Wilderness and the Wind Rivers area, and to the east and west of Jackson Hole Valley and south of Dubois in Wyoming, we believe the recovery zone requires substantial expansion to accommodate current bear distribution.

Realistically, we are aware that recovery zones cannot expand indefinitely. If the general population viability approximations of 2000 bears are correct, then we must explore ways to increase carrying capacity of the habitat already available. Maintaining and, where feasible, enhancing existing habitat quality will be as important to recovery as habitat boundary extensions. The first step toward accomplishing this is to apply the supervised approach for mapping and analyzing the vegetation using satellite multispectral imagery (Craighead *et al.* 1982, Craighead *et al.* 1988a; also see Chapter 12).

Standardizing Habitat Mapping

Managing agencies have failed to adopt or develop a detailed, systematic, standardized system for large ecosystem habitat mapping and vegetation description. It is imperative that a geographic mapping procedure, with at least the vegetation and terrain data capabilities described in Chapter 12, be adopted and employed to map the entire Yellowstone Ecosystem and the Northern Rockies bioregion, as well.

Imagery and computer capabilities for generating geographic information maps and statistical information of entire ecosystems have been greatly refined over the past decade, but ground-truthing techniques have not. A system for mapping the vegetation of large wilderness ecosystems should employ:

1. A standardized hierarchical vegetation classification based on plant succession in the region;
2. A detailed ground-truthing procedure employing standardized vegetation sampling methodology;
3. A standardized color code for vegetation and terrain features; and
4. The "supervised" approach to computerized vegetation analysis (see Chapter 12).

Standardized multispectral imagery mapping would enable resource managers to quantify and compare values of the various habitat components within the Yellowstone Grizzly Bear Recovery Zone, as well as between the other designated recovery zones. We believe this holistic approach to understanding the grizzly bears' environment is essential to any evaluation of recovery in the Yellowstone Ecosystem and to developing population viability assessment models for the future.

Unfortunately, the focus continues to be on cumulative effects analysis to define and evaluate the quality of habitat available in the six recovery zones. Recently, a habitat monitoring methodology, known as Access Management Standards, has been evoked to rate the quality of grizzly bear habitat based on the extent of human access to it. A habitat is categorized on a basis of road densities per unit of habitat.

Road management is indeed a powerful tool that can be used in conjunction with other parameters to manage grizzly bear habitat. It is difficult to understand, however, why the management agencies have not developed a standardized multispectral vegetation mapping system and applied it to

Table 20.7. Projected sex structure[a] of grizzly bear age classes in the Yellowstone Ecosystem, 1960-68.

Age class	1960		1961		1962		1963		1964		1965		1966		1967		1968		Mean number[b]		
	M	F	M	F	M	F	M	F	M	F	M	F	M	F	M	F	M	F	M	F	Total
Cubs	32	22	28	20	40	27	43	30	24	16	48	33	27	19	30	21	30	21	33.6	23.2	56.8
Yearlings	17	7	17	7	22	10	39	17	50	23	24	11	46	21	32	14	21	9	29.8	13.2	43.0
2-year-olds	9	4	21	9	6	2	15	7	39	17	39	17	19	8	35	16	21	9	22.7	9.9	32.6
3-4-year-olds	14	10	22	16	30	21	22	16	21	14	25	18	48	33	33	23	27	19	26.9	18.9	45.8
Adults	64	84	53	70	45	60	53	70	60	80	63	82	59	78	59	78	64	84	57.9	76.2	134.1
Total	136	127	141	122	143	120	173	140	194	150	199	161	199	159	189	152	163	142	170.9	141.4	312.3
Grand total	263		263		263		313		344		360		358		341		305				

[a] Derived by applying the age-specific sex ratios of the Trout Creek subpopulation (Table 5.6) to the age-class structure for the ecosystem-wide population (Table 20.5).
[b] Due to rounding of annual numbers of each sex to "whole bears," mean age-specific sex ratios may vary slightly from those for the Trout Creek subpopulation (Table 5.6).

Fig. 20.3. Sightings of grizzly bears reported from 1983 to 1989, Yellowstone Ecosystem (McDonald *et al.* 1990).

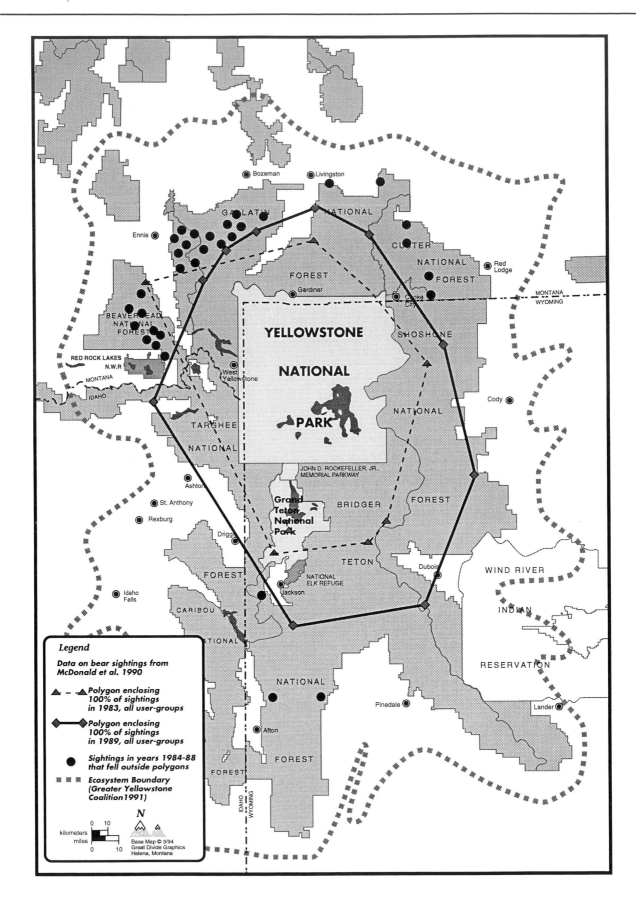

entire ecosystems. Multiple parameters such as road densities, logging operations, recreational and mining activities, grazing allotments, and numerous others can then be incorporated into the imagery base to model and evaluate grizzly bear habitat on a bioregional scale (Craighead *et al.* 1982).

Maintaining Habitat Quality

We believe maintaining the quality of existing habitat is as important as expanding the area of recovery. Much can be done to improve existing habitat; we examine some of the major steps that need more aggressive action and novel concepts that need serious consideration.

Sanitation

As we have shown, there are positive biological advantages when grizzly bears aggregate at protected ecocenters. However, when they are attracted to unsanitized food sources at designated campgrounds, recreational and hunting camp sites, commercially developed visitor centers, town garbage disposal sites, and ranch boneyards, they come in conflict with humans, and their death rate rises. It is therefore imperative that the ambitious program, initiated in the late 1960s, to sanitize the Park and adjacent areas be continued and amplified. Sanitization within the Park has been commendable, yet too many tourists are still careless with food, and bears are attracted. Intensifying patrol activities is probably not practical; however, imposing more severe fines for sanitation violations and increasing the numbers and visibility of sanitation warning signs should be effective.

Sanitation of areas outside the Park has been less successful. Adjacent towns and developed areas such as Cooke City, Gardiner, West Yellowstone, and the Flagg Ranch must develop more consistent programs for abatement. Federal programs for educating city managers and councils should be implemented, and monetary incentives developed that reward businesses and private citizens for practicing proper sanitation. It is our feeling that federal agencies must assume the lead on these initiatives and develop a mechanism for rigorous, centralized oversight.

Road Densities

Because of the broadly negative effects of roading on grizzly bears and their habitat discussed earlier in this chapter—habitat fragmentation, exclusion of bears from high quality habitat, increased risk of human-bear conflict with accompanying bear mortality—we recommend that a road density standard of one kilometer of road per 6.4 square kilometers (0.25 mi/mi^2) be established as a maximum allowable on the National Forests. This is from a third to a quarter of the allowable roading standards adopted by the National Forests within or adjoining recovery zones (USFWS 1993), standards that, in our opinion, are much too liberal. On federal or state lands that exceed the km/6.4 km^2 (0.25 mi/mi^2) standard, roads should be closed and obliterated to achieve compliance. In de facto wilderness, all types of development should be minimized. History has shown unequivocally that such encroachment has been the principal reason for deterioration of critical habitat and an important factor in the decline of grizzly bear populations. Federal standards have been too liberal.

Encroachment on grizzly bear habitat within the Yellowstone Ecosystem has been excessive on the Targhee National Forest at the southwestern border of the Park. Logging has destroyed cover and food resources, and the roading has provided easy human access.

Enhancing Natural Ecocenters

In Chapter 16, we presented evidence that natural ecocenters of a limited scope may be developing within the Yellowstone Ecosystem. It is possible that certain fishery, army cutworm, and ungulate carcass sites could be enhanced by management initiatives to increase the basic food source, concentrate its distribution, and provide a secure environment for the bears through closure or zoning the areas from visitors. Construction of natural stream barriers on Yellowstone Lake tributaries to stop and temporarily hold cutthroat trout spawning runs, and severely limiting human access during trout spawning would provide bears greater fishing opportunities. Also, annually concentrating carcasses of road-killed and winter-killed ungulates at specific remote sites available to grizzlies emerging from winter sleep would create ideal conditions for ecocenter development.

If army cutworm moth aggregations prove dependable enough to qualify as ecocenters, then ways to manage this food source effectively must be explored. At present, there are many gaps in our knowledge as to how great a role the aggregations play in the nutritional regime of grizzly bears. Detailed studies are needed to determine distribution, microhabitat requirements, and seasonal abundance of the moths within the Ecosystem. Also, quantitative data on volumes and rates of ingestion are needed for assessing the importance of this food in the overall diet of the grizzly. In short, are moths a major food source or a dietary delicacy? Because army cutworm aggregation sites have shown potential for consistently holding large numbers of bears

Encroachment on grizzly bear habitat has been excessive at certain locales within the Yellowstone Ecosystem. The result of logging practices can be seen along the south western boundary between Yellowstone National Park and Targhee National Forest. Clearcut logging has destroyed cover and food resources and the required roading has provided for easy human access.   *(Photo by Tim Crawford)*

in the isolated alpine areas of the Yellowstone Ecosystem during a portion of the visitor season, their importance as an ecological magnet may outweigh their nutritional value.

Establishing Ecocenters

We are dubious that primary ecocenters based on natural food resources (similar to those we have described for Alaska in Chapter 13) will develop within the Yellowstone Ecosystem in the foreseeable future. Spawning fish runs do not, at this time, appear to be sufficiently extensive to function as ecocenters, nor do moth aggregations or *in situ* winter-killed ungulates. The necessary nutritional base is not present or is not available in the quantity and within the time frames required. Also, if the Yellowstone population continues to gravitate to the Park periphery and beyond, access to these natural point food sources will progressively decline.

Another consideration that we believe has not been sufficiently explored is the future effect a reestablished wolf population may have on the diet of an already food-stressed grizzly bear population.

During the winter and early spring months, wolves will very likely kill many of the winter-stressed elk and bison that have, for decades, been available to grizzly bears as carcasses or as severely malnourished prey. Thus, this extremely important food source will not be available to grizzlies when they come out of winter sleep or, at best, will be greatly reduced (see Chapters 10 and 16).

Carcasses

Wild ungulate carcasses provide the most natural and, arguably, the least controversial food source that could be accumulated annually in sufficient quantity and regularity to attract and hold aggregations of bears. This was suggested by the senior author and Dr. Frank C. Craighead in the mid-1970s. The idea has been unintentionally tested in Yellowstone Park in recent years when a backcountry dump for roadkills began attracting bears in numbers (David Mattson, pers. comm. 1993). However, to our knowledge, no directed study of the phenomenon has developed, and the status of the roadkill dump is unknown.

Carcasses of winter-killed animals deposited at one or more locations and, when necessary, augmented by scheduled deposition of ungulate roadkills and control kills would constitute the ecocenter food base. There are recognized but not unsurmountable disadvantages to this approach. For example, culling ungulates within the Park would meet strong agency resistance and arouse public indignation. A comprehensive education program would have to proceed any action. Secondly, proper annual culling of the elk and bison herds will require a better understanding of the size, movements, and distribution of the herds and subherds than currently exists. Culling would be acceptable and effective only if numbers of kills could be taken and distributed so as not to destroy the basic structure of elk subpopulations. Efforts at elk herd reduction on winter range in the late 1960s caused disproportionate numbers of deaths in localized subpopulations not obvious until the main herd had dispersed to discrete summer ranges. Lastly, the program would be both time-consuming and expensive. However, we presume it would be within the capability of NPS personnel, since most of the effort would be prior to the visitor season, a "low duty" period for rangers.

Grains and Food Pellets

Large quantities of surplus grains such as corn or wheat, or of commercial food pellets will attract and hold bears for prolonged periods. These "ecocenter foods" are nutritional, relatively cheap (at least the surplus grains), and easily transported and deposited. Routinely supplying an ecocenter with these foods would be as simple and inexpensive as transporting ungulate carcasses, and perhaps more so.

Processed Human Food

The role of garbage disposal sites in the ecology of the Yellowstone grizzlies has been thoroughly documented and analyzed in this monograph. We have shown that, by functioning as demographic magnets and annually dependable sources of food, they had positive effects on the security, nutrition, social structure, and stability of the population. Therefore, establishing a modified version of the open-pit garbage dump would be a straight-forward way of creating an ecocenter. Garbage has proven nutritional qualities, is essentially cost-free, and its transport, placement, and site management are relatively simple, well-understood processes in operation every day. However, it would be necessary to develop a standardized procedure for sorting and packaging all waste foodstuffs at food services and restaurants within the Park. This processed food, free of all types of containers and inedible trash, could then be transported daily to an open-pit depository and covered lightly with soil. The soil cover is important because it forces the bears to dig and scrape for the food. This extends the feeding period and helps solidify the aggregation as a behavioral entity. Siting of ecocenters in remote, core areas of the Park is essential. For example, the upper reaches of Trout Creek or Alum Creek in Hayden Valley are sites probably as ideal for this purpose as any. The area must then be strictly zoned to exclude all visitors to provide security for the bears.

Perhaps the strongest argument for "creating" an ecocenter using processed human food is that it rightfully recognizes humans (visitors) as integral members of the Park biota, and their waste products as an ecologically acceptable food resource for grizzly bears, in lieu of the prime habitat heavily impacted and precluded by park visitors. Currently, visitors outnumber grizzlies at a ratio of approximately 10,000 to 1. Recycling processed human food to offset human ecological impacts in the Park is a concept that will become more acceptable as future park administrations face the reality of a conflicting congressional mandate to preserve, yet use the Park's resources (Craighead 1991). If desired, a strictly controlled viewing program similar to that at McNeil River, Alaska, could be designed around an ecocenter (Walker and Aumiller 1993). Such a program would offer tremendous educational, aesthetic, and scientific values. It would have few, if any, adverse affects on the aggregation of bears.

There are current examples of bears aggregating in numbers in response to garbage and to bowhead whale offal. The best known is the annual aggregation of polar bears outside the town limits of Churchill, Manitoba. Within a few years, this ecocenter has attracted observers and photographers from around the world. An even more striking example is the response of polar bears to whale offal strategically placed to lure them away from the town of Barrow, Alaska. According to Craig George (pers. comm. 1993), a wildlife biologist stationed in the town of Barrow, Alaska, disposal of whale offal resulting from the 1992 fall harvest by the resident Inupiats in an isolated area on Point Barrow (16 kilometers [10 mi] north of the town) did, in less than one year (1992), produce a small aggregation of at least 12 polar bears at the Point. A similar number were attracted to offal the following year (1993). One female from the initial aggregation established a den at the Point and emerged with two cubs. In 1993, a native hunter reported approximately 30 polar bears aggregated on a stranded bowhead whale carcass 13 kilometers (8 mi) south of the town of Barrow. At least one female denned nearby and emerged with a litter of cubs. As a result of these aggregations, according to George, a chronic problem of bears coming into the town of Barrow has been apparently alleviated. The practice of strategically placing whale offal to attract and hold polar bears will continue as an ongoing management action at Barrow, Alaska. The success of this practice from year to year, however, may, according to George, depend on the distance of pack ice from shore at the time whale offal is deposited at the point.

Multiple Food Resources

In applying the ecocenter concept, we envision creation of one large, centrally located feeding area to attract and hold approximately a third of the grizzlies inhabiting the Yellowstone Ecosystem. For this to be feasible, processed human food wastes would probably need to be augmented by additional biomass. For example, winter-kill carcasses could be gathered and deposited at the ecocenter in early spring (mid April to 31 May) and then substituted with processed human foods and carcasses of roadkills. When necessary, donated grains could be used as further supplementation. Substantial food would then be available throughout the tourist season (1 June to 30 Aug), providing all the protective advantages of the maligned garbage dumps with few, if any, of their detractions. As the bear population recovers to preclosure levels of approximately 300 individuals, other smaller ecocenters could be located so as to distribute the bears more widely, yet still contain them within the secure central core of the Ecosystem. The advantages of this centralization have been thoroughly discussed in earlier chapters.

Critics will undoubtedly argue that we have recommended a return to preclosure conditions which were considered untenable two decades ago. In essence, we have. All of our evidence indicates that viable populations of grizzly bears cannot be *realized* without a substantial increase in secure, high quality habitat in which the bear's interests are of primary concern (based on the past politics of wilderness, an unlikely possibility), a vigorous enhancement of the foodbase, or both. Artificial ecocenters are the most straight-forward solution. Arguably, grizzly bears may use wild foods more efficiently when a population is dispersed, but the sacrifice is a reduction in carrying capacity and nutritional level which translates into smaller population size and reduced probability of long-term persistence. Assessment of what is or is not acceptable in any grizzly bear recovery or management plan ultimately must be based on results of population viability analysis. A proper analysis of viability must be completed for the Yellowstone population, and we have described earlier what is necessary to accomplish this in terms of data and management actions.

Population Recovery Target

We have suggested earlier that researchers and managers focus on determining the long-term persistence of a series of isolated populations distributed over broad geographic regions, collectively referred to as a metapopulation. The Yellowstone population is but one of these. In this more limited context, we first must look at the individual populations and identify population recovery targets that are attainable within the habitat constraints of each recovery zone. These estimates, in themselves, will not be population viability goals, but rather working estimates until PVAs can be performed.

At the present time, we lack the database even to set realistic population recovery targets within the various recovery zones, let alone make population viability assessments of the overall metapopulation. However, we do have a database for the Yellowstone population that enabled us to set a realistic population recovery target that would serve until a comprehensive database can be completed and subjected to a population viability analysis (Chapter 19).

The IGBST's postclosure data set does not permit inter-period comparison of population parameters. We suggest that the data necessary for proper assessment of the population can be most quickly and accurately obtained in the future from censuses at established ecocenters. After an initial period of capture and marking (4-6 years), we believe the population could be monitored adequately without a "continuous" bear-handling program. Once a comprehensive population database has been recorded at an ecocenter, it can serve as a scientific basis for evaluating more easily obtained but less precise indicators such as "spot counts" of annual aggregation size, ecosystem-wide bear density, and numbers of females with cubs. In time, adult female death records, in conjunction with periodic ecocenter spot counts, might be the only data necessary to periodically appraise the status of the population and to recommend appropriate management actions.

MANAGING FOR GRIZZLY BEAR POPULATION RECOVERY IN THE CONTERMINOUS 48 STATES

Recent work in population viability modeling indicates that a minimum viable population of grizzly bears would be about 2000 (Soulé 1987, Harris and Allendorf 1989, Allendorf *et al.* 1991, Mace and Lande 1991, Metzgar and Bader 1992, Shaffer 1992, Nunney and Campbell 1993). These analyses, of course, do not take into account habitat quality, an important variable for which data are currently inadequate. If, however, eventual population viability analysis based on all relevant field data does indicate that a population of several thousand would be necessary for a high probability of persistence for several hundred years, then the amount and quality of secure habitat available to grizzly bears becomes a critical limiting factor.

As we discussed earlier, Metzgar and Bader (1992) estimated that a population of 2000 bears distributed at an overall density of 4 bears per 259 square kilometers (100 mi²) will require an area of secure, (presumably) high quality habitat on the order of 129,500 square kilometers (50,000 mi²). By comparison, the current recovery zones, including the North Cascades and Selway-Bitterroot Ecosystems, which are not currently inhabited by grizzly bears, comprise about 96,000 square kilometers (37,000 mi²). However, only 49% (47,117 square kilometers [18,192 mi²]) of this area is relatively protected habitat in national parks and designated wilderness and, to a large extent, its quality as grizzly bear habitat is unestablished. The majority of the remainder is in national and state forests, some of it roaded and with a history of extractive use, some of it unroaded, de facto wilderness. Little of it is designated with the needs of grizzlies as a primary concern. Large portions of the national forests surrounding and between the recovery zones must be made available for the use of grizzlies. These areas should be made secure from exploitation by virtue of wilderness designation or perhaps some other categorization (e.g., MS-1 status) which places the needs of the bears above the roading and commodity demands that dominate the broad multiple-use agenda.

Adequate undeveloped areas still exist in the Northern Rockies to support a combined population of 2000 bears. Much of this is already protected in parks and wilderness. We have shown, however, that the future of the grizzly bear is closely linked to the future of the roadless, de facto wilderness areas currently open to roading and commodity development. In and around some recovery zones, bears already inhabit substantial portions of these areas (for example, Cabinet-Yaak and Selkirk Mountains), and management strategists certainly must plan for further expansion, as recovery proceeds. Thus, we urge and support the protection of additional wildland habitat that expands and links the major park and wilderness complexes.

It is certain that habitat continuity throughout the Northern Rockies eventually will be of great importance to grizzly bears and all of the faunal species indigenous to the region. In her study of a 20,000-year vegetational and climatic history of the Pacific Northwest, Whitlock (1992) points out that linkage corridors have been, and will continue to be crucial to the movement of faunal species in response to climatic change. This might be too long-range to consider for practical purposes; however, given the growing threat of global climatic change as a result of deforestation and the "greenhouse effect," perhaps not.

We have detailed the positive effects that ecocenters could have on the recovery of the Yellowstone grizzly bear population. Our research has shown their value in (1) shifting centers of bear activity and centralizing the population in secure core areas; (2) reducing home range areas and individual ranging tendencies, and thereby, reducing risk of mortality; (3) increasing habitat carrying capacity; (4) damping annual fluctuations in population numbers; (5) damping the unpredictable nutritional stress associated with fluctuations in availability of natural foods; and (6) increasing the growth rates and size of bears in all sex and age classes. We point out here that ecocenters could be highly beneficial in stabilizing any of the extant populations.

As regards the relevance of the ecocenter concept to the management of the other threatened populations, it is noteworthy that our revised mean population estimate of 312 bears for the 1959-70 ecocentered population in the Yellowstone Ecosystem (about 20,000 square kilometers [7,800 mi²], as defined by Craighead 1980b) yields an average population density value—four bears per 259 square kilometers (100 mi²)—identical to that selected by Metzgar and Bader (1992) for calculating the spatial needs of 2000 bears. The obvious implication here is that ecocenters may well be essential to sustaining a population of 2000 bears, even given a relatively secure area of 129,500 square kilometers (50,000 mi²) and, therefore, will be a tool fundamental to the management of grizzlies throughout the Northern Rockies.

We re-emphasize that either of these actions—establishment of ecocenters or commitment of large expanses of additional habitat—pursued as prerequisites to population recovery will provoke considerable public and political controversy, with wilderness purists pitted against ecological realists and conservationists against advocates for commodity development and multiple-use. Nevertheless, as biologists, we must present our data and recommend what we believe is necessary for recovery and long-term persistence of grizzly bear populations in the lower 48 states.

REESTABLISHMENT OF GRIZZLY BEARS BY TRANSLOCATION

We have recommended what we believe is necessary for recovery based on the best biological data at our disposal. On the other hand, we have learned from experience that politics, not science, all too frequently governs resource decisions. What, then, are the obstacles to pursuing a scientific course, and what is the likelihood of attaining our goals?

To insure the persistence of the grizzly bear in the lower 48 states, it will be necessary to augment existing populations or introduce bears into large ecosystems currently devoid of grizzlies. The history of wolf introductions has made it clear that translocation of large predators should not be attempted without a comprehensive educational program to gain public consensus. Such a program should begin well in advance of the translocation, include public input forums before, during, and after the translocation, and provide a series of meetings to dispense information. The target ecosystems must be large enough to support the wide-ranging movements of grizzly bears, allowing them to establish home ranges away from potential "mortality sinks" (Chapter 11). Translocation efforts should be focused on the four largest ecosystems—Greater Yellowstone, Northern Continental Divide, Selway-Bitterroot, and North Cascades. There is no current need for translocation in the first two; among the latter, the Selway-Bitterroot would be an excellent first choice because, although it is now devoid of grizzly bears, it supported them at one time. Also, the suitability of bear habitat in this recovery zone has been more thoroughly explored than in the North Cascades (Scaggs 1979). The scientific feasibility experiment should be well advertised as such, and like wolf transplants, newly translocated grizzly bears should be destroyed if they wander into areas where they threaten people or livestock. It is important that livestock owners be promptly and fully reimbursed for any livestock losses. The research objectives of the transplant should be clearly stated, and the scientific methodology peer-reviewed by government and private-sector bear biologists.

It is not our intent to document all of the many specifics necessary for a successful transplant. Instead, we focus on two actions that our experiences with grizzly bears indicate to be important for a successful transplant in a large recovery zone uninhabited by grizzly bears. These are:

1. Satellite mapping of the vegetation of the recovery zone, including the specific release site; and
2. Establishment of a temporary ecocenter at the translocation site.

With regard to action (1), we believe it is especially important for the agencies to make thorough analyses of the target translocation environment and the entire recovery zone before grizzly bears are released—that is, habitat quality must be quantified. Construction of multispectral imagery maps supported by adequate ground truthing (Craighead *et al.* 1982, 1988a; also see Chapter 12) is the most cost-effective and precise method of accomplishing this. The procedure should be initiated simultaneously with the public education program. The resulting data would help the recovery team to make crucial decisions concerning the location of initial transplant sites and to estimate target recovery levels for the recovery zone. They would also prove invaluable for analyzing bear-habitat relationships, as the translocated grizzlies and their progeny spread throughout the zone.

Relative to action (2), we believe the establishment of a temporary ecocenter at the release site would virtually assure transplant success. We recognize, however, that this would be breaking new ground, and we know it will be controversial. However, this action is strongly justified by data presented in this monograph. The release site should be centrally located in the recovery zone and near riparian habitat, or in an open mountain basin, preferably at an elevation of 1830 to 2130 meters (6000 to 7000 ft). The translocated bears should be young animals moved in early spring to allow them to adjust to the new environment. All bears should be tatooed, color-marked, and equipped with radio collars. Ideally, there should be trail and air access to the release site, but no roads. Short-term, this may not seem practical or cost-effective, but long-term, it will greatly enhance the probability of successful research and, ultimately, a successful transplant.

A major problem with moving a grizzly bear from its home range (life range) to a release site is the tendency of the bear to return to its former range. In attempting this, it travels widely, and more so in an ecosystem-to-ecosystem transplant. The human-selected release site has no immediate biological significance to the bear unless the attraction of high quality food should prove stronger than the initial urge to emigrate. As we have shown in this monograph, fidelity to a home range is a powerful behavioral pattern. Whether transplanted bears can be seduced from the urge to return to former home ranges is unknown, and answering this question should be a focus of the transplant experiment. We believe it is possible to induce grizzly bears to establish new home ranges at or near a transplant site by creating temporary ecocenters. Caches of grain, food pellets, or animal carcasses placed at appropriate sites in the release area may attract and hold the released animals. Prior to release, food should be withheld so the bears are hungry. We would expect them to scent the nearby food sites within a few hours to a few days following release, especially if animal carcasses are used as attractants. During this period and thereafter, bears trapped at different sites will learn to know one another, to feed communally, and to acquire a taste for the basic grain or pellet attractant. If exposed to minimum disturbance, they should remain in the general area, returning periodically to the food caches. During the experiment, all feeding must be managed so the bears do not associate their food with human presence. Before grizzlies are released, a simple viewing platform could be erected to facilitate observations of the color-coded animals and the recording of their behavior.

Because the grizzly is an aggressive animal, there could be considerable risk associated with a quick, wide-dispersal release. Our experience is that grizzlies disperse rapidly from a release site, often traveling long distances and expending much energy. Without adequate nourishment in an unfamiliar environment, the released animals become hungry, aggressive, and, as a result, more dangerous. Hungry grizzly bears are dangerous. If they move to the periphery of the recovery zone, into hunter camps, or to recreation sites, they will get into trouble. A serious injury to a human or a death caused by a released grizzly would undoubtedly involve the federal government in litigation, the claimant charging "gross negligence" under the Tort Claims Act. For this reason, alone, every effort should be made to attract and hold released grizzlies at a remote release site (ecocenter) for an extended period of acclimatization.

There are other advantages to an "ecocentered release" which should be given careful consideration. By preventing dispersal until the approach of winter sleep, the released animals will tend to den near the ecocenter and return to the site in spring, thereby reducing wandering and reinforcing aggregation behavior. This will improve the chances of successful matings and provide a safe haven for another season. Researchers will then have the opportunity to gather data on mating behavior and the production of cubs and, in later years, information on weaning, sex and age composition of the expanding population, sex- and age-specific mortality, home ranges, and other important biological parameters.

Because the objective in this approach is the creation of a primary ecocenter, annual transplants should be made at the same site in the same manner, as deemed necessary. The behavior of the new recruits can be recorded as they assimilate with the original stock. If even a few bears can be induced to stay near the release site, we would expect in relatively few years the growth of a well-protected aggregation of bears exhibiting all of the ecocentric behavior of the Trout Creek or McNeil River bears.

Equally important for the future of grizzly bears, the success of the ecocentered transplants can be evaluated objectively from a strong data set. Even using radiotracking, a non-ecocentered release will be extremely difficult to evaluate and could be a complete failure because of rapid, wide dispersal of the translocated animals. Of equal concern would be equivocal results to an expensive experiment where claims could be made for or against success with no hard data supporting either contention. Finally, the ecocentered translocation experiment should be phased out gradually over a period of years, thereby enabling the researchers to document the effects of the phaseout on the population, and to take corrective management measures if necessary. As we have shown in text, a long phaseout period is essential for ecocentered bears to adjust to food withdrawal and for researchers to document range expansion, emigration, mortality, and other important biological parameters. We should not duplicate the "Yellowstone experience." Clearly, experimental translocations of grizzly bears are not to be taken lightly; they will be expensive, research intensive, and of long duration—6 to 10 years minimum. To be credible, they should not be exclusively federal "in-house" projects. We strongly recommend the establishment of a scientific oversight committee consisting of an equal number of federal, state and private-sector biological scientists whose responsibility would be periodic evaluation of the translocation project and development of management recommendations to ensure its success.

A VISION FOR THE FUTURE

There are politicians who have no use for biologists and biologists who have no use for politicians, yet infrequent but close cooperation between the two have produced the nation's major conservation achievements. It has often been a heart-rending process, but it has worked. What we, as a nation, have done in the past, we can do in the future, although the cooperative process becomes more difficult as our nation grows and our natural resources dwindle. For each step forward, we need a vision. These come from the grassroots of our society. They reveal the thinking of people on the land, people close to and in harmony with the natural world.

The Northern Rockies Ecosystem Protection Act (NREPA) may well be the conservation vision of the future; a well-considered plan that has emerged from the nation's conservation successes and failures of the past. The basic concept behind NREPA has had a long evolutionary history. It was born when the first national park was created. The concept of preserving and managing public land for the public good grew in scope and attainability with the establishment of the National Wildlife Refuge System, the National Forests, the National Wilderness System, the National Wild and Scenic Rivers System, and the Endangered Species Act—all directed toward using, yet preserving through wise management, wild America. These historic steps in the recognition and preservation of natural resources emerged spontaneously and piecemeal at the grassroots level of our society. Each step invariably led to the next. Each made portions of our natural resource heritage more secure. Most, if not all, were initially opposed by the governing status quo, but strongly advocated by a responsive public.

Ideally, these incremental steps to protect our natural resources might have led to a comprehensive, holistic policy of sustainable resource preservation and management. For many reasons, they have not been fully successful in doing this. It has taken time, experience, and acquired knowledge to recognize that the landscape of the Northern Rockies is an entity that, in the best interests of all Americans, must be understood and managed as one unit. The agencies must manage the sum of the parts, not just each piece by itself as they have done in the past. The policies of the various federal resource management agencies, although wise and visionary for their time, have proven insufficient to the task. They have been contradictory and divisive within and between agencies. Interagency cooperation has not been strong enough to rectify these differences. The result has been to fracture, rather than to unify, the management of a common resource base which underlies our economic system and is an important aspect of the American lifestyle.

Observing the cumulative impacts of extractive resource development on public forest lands in the Northern Rockies and cognizant of the developing field of ecosystem sciences, citizens in the late 1980s began a region-wide effort to draft ecosystem-based legislation. To stem the tide of habitat fragmentation, stream ensilation, and the resulting decline in biological diversity, conservationists sought to incorporate the principles of conservation biology into a multi-state legislative proposal. Grassroots activists provided detailed knowledge of local areas, assembling a proposed system of inviolate reserves and connecting corridors (Fig. 20.4, page 359). They solicited input from the members of the scientific and economic communities to refine their ideas and produce the Northern Rockies Ecosystem Protection Act. The bill has been introduced in the U. S. Congress as H.R. 852.

The Northern Rockies Ecosystem Protection Act will provide protection to large areas of unroaded de facto wilderness within the region, and it will direct that these areas and established wilderness be managed in context with the larger region, embracing human communities and extractive economies. The Act seeks to ensure ecosystem integrity and economic resource sustainability in the Northern Rocky Mountain bioregion by designating additional wilderness areas, wild and scenic rivers, biological connecting corridors, park and preserve study sites, and wildland recovery areas across a western landscape currently under federal jurisdiction. With the addition of the areas cited in the Act, the stage will be set for holistic resource management by the federal agencies.

More than just the next step in the management of our natural resources, NREPA represents the beginning of a matured conservation concept and a new conservation era, one that will require new learning, new insight, and greater cooperation among federal agencies and the public to attain common goals. It is a concept whose time has come. Like all of our major conservation movements, it will be at first, an experiment—an ongoing experiment that can result in sustainable management of extractive resources, while preserving the biological integrity of our wilderness-wildland ecosystems.

The technical science to carry out this concept is at hand. Space-age technology and the computer information revolution have made it possible to accumulate the data necessary for managing the diverse resources and economies of large biogeographic regions more readily than a single national forest was managed 15 years ago. Some of the tools at hand are refined versions of the geographic information system (GIS), satellite radio-monitoring and -tracking devices, and improved on-the-ground biological sampling techniques. The GIS is a technology that enables researchers and managers to build and collate information on a scale unimaginable just a few years ago. Spatial data, such as maps, and tabular statistical information are electronically layered to generate a database that can be video-displayed and analyzed. For example, topographic, geologic, hydrologic, and vegetation data can be digitally layered over satellite multispectral imagery, and this vast array of terrain data then computer collated and analyzed in a fraction of time and with great accuracy. To these basic terrain data can be added layers of information on land use, such as timber harvesting, road construction, recreational activities, animal censuses, and animal distribution. The technology is advancing more rapidly than its application. There is no question that our federal resource management agencies now have the technical capability of managing an entire bioregion, such as the Northern Rockies, as a single geographic unit. There is a question, however, whether the resource management agencies, our political representatives, and the Congress can make the very substantive changes that will be necessary to accept the NREPA concept, and to practice ecosystem-based management with a common, first priority goal of preserving the biological integrity of the Northern Rockies.

As individuals, most of us tend to accept, at face value, the environment we are born into, unable to envision better conditions of the past or seek brighter ones for the future. We can only vaguely imagine what changes will occur to our environment and to our intermountain society in the future. Yet we must anticipate these changes that are occurring if we are to be masters of our destiny and creators of our lifestyles, instead of victims of overburdened natural systems that must someday break from the human pressures exerted on them.

A QUESTION OF VALUES

The fundamental threat to the grizzly bear's survival is not a lack of biological knowledge, nor a lack in knowing how to apply this information. Rather, it is the threat from our politico-economic system that demands unsustainable use of our public land and water resources, whether these resources are forage, timber, fish, or minerals. Examples are legion in our conservation history because the problem is one that representative government is not good at solving. We can think of no major conservation problem that has been initially recognized and then solved by government agencies without pressure from a critical public. Potential disasters that can be foreseen clearly, but are not imminent, are allowed to fester, although the future political and economic price of avoiding them is well understood. So long as the crisis lies down the road, no sustained corrective action is taken by the agencies.

All too frequently, the portending crisis is denied and concealed by the responsible federal agencies, forced to submit to overwhelming political pressures. Our democratic processes and moral standards will be severely tested in preserving the habitat and biodiversity necessary to assure long-term survival of the grizzly bear.

The Congress and the American people have, in the past, shown great vision and leadership in creating our national parks, national forests, wildlife refuges, and wilderness areas as a means for protecting and preserving the non-extractive values that translate into beauty, science, enjoyment, philosophy, lifestyle and, for some, religion. The public lands, with their wealth of national resources, belong to all the American people, not to the citizens of any one state, nor to any one social or economic group. They are the property of the many, not the privilege of the few. The natural resources of the public lands fall into two categories of use: extractive and non-extractive. The extractive resources are timber, minerals, grazing land, gas, and oil. These cannot be used and enjoyed equally by all citizens. They have become the personal property of those few that have the power and finances to exploit them. The non-extractive ones are pristine landscapes, the mosaic of plant and animal life, the water, the soil and the basic ecological systems and biological processes. Holistically, this living entity is biodiversity, and it is absolutely essential to our future. It cannot be replicated, nor are there human-made substitutes for it. This biological web-of-life is the heritage of all Americans and it is the people's duty to preserve it.

In the ecosystem of the Northern Rockies, we find grizzly bear, grey wolf, lynx, wolverine, fisher, woodland caribou, bull trout, the great grey owl, the Harlequin duck, and many other species, holding out against the tide of extinction. These species, when endangered, are indicators of ecological malaise. Their hope, and ours, lies in regional land-use-mosaics, of protected wilderness interwoven with sustainable human settlements. This hope is severely threatened by the way in which our public land resources are manipulated by political forces. The science to manage them on a sustainable basis while protecting the region's great diversity of life must come from both the government and the private sector. Pressure to apply the science must then come from an educated public.

There is need for both economic use and preservation on our public lands; the question is,

"Where lies the balance?" Catering to special interests, with the resulting fragmentation of ecosystems, is especially unacceptable and unconscionable at a time in history when we clearly understand the necessity of preserving biological diversity and ecosystem structure and function on our public lands. We need these large areas of near-pristine environment as reservoirs of undiscovered information, as ecological benchmarks, as keys to understanding both natural and human-made changes of the future, and as sustainable economic assets, when these uses do not threaten to degrade or destroy the non-extractive values. We need them to preserve the grizzly bear.

If special interest groups and their political supporters prevail in dictating the use of public lands, it will become extremely difficult, if not impossible, to maintain the grizzly in perpetuity as a free-roaming member of the western landscape. Clearly, much more than a well-designed recovery plan is needed. Can we as a nation and as a people manage our public land resources in the future so as to provide sustainable commodity production while retaining the integrity of natural ecosystems and the basic values of our society? This will be a true test of our humanity and of the flexibility of our federal resource management agencies.

APPENDICES

Appendix A

Appendix A—Table 1. Raw and adjusted visitation at Trout Creek ecocenter by all adult male grizzly bears versus all adult females over the entire season (two-tailed Mann-Whitney test), Yellowstone National Park, 1959-70.

Population component	Raw mean visitation[a]	Adjusted mean visitation[b]
All males (*n* = 70)	0.2263	-0.0755
All females (*n* = 82)	0.2846	-0.0158
P-value	0.05	0.07

[a] Raw visitation equals fraction of complete censuses a bear was recorded as present at Trout Creek.
[b] Adjusted visitation equals raw visitation for a bear minus the mean raw visitation rate for all bears in a year.

Appendix A—Table 2. Raw and adjusted visitation at Trout Creek ecocenter by all adult male versus female grizzly bears with offspring over the entire season (two-tailed Mann-Whitney test), Yellowstone National Park, 1959-70.

Population component	Raw mean visitation[a]	Adjusted mean visitation[b]
All males (*n* = 70)	0.2151	-0.1027
All females with offspring (*n* = 52)	0.3264	0.0069
P-value	0.008	0.01

[a] Raw visitation equals fraction of complete censuses a bear was recorded as present at Trout Creek.
[b] Adjusted visitation equals raw visitation for a bear minus the mean raw visitation rate for all bears in a year.

Appendix A—Table 3. Raw and adjusted visitation at Trout Creek ecocenter by male versus female subadult grizzly bears aged 3-4 years during the breeding season (two-tailed Mann-Whitney test), Yellowstone National Park, 1959-70.

Population component	Raw mean visitation[a]	Adjusted mean visitation[b]
Males (*n* = 32)	0.2525	-0.0218
Females (*n* = 27)	0.3396	0.0245
P-value	0.20	0.47

[a] Raw visitation equals fraction of complete censuses a bear was recorded as present at Trout Creek.
[b] Adjusted visitation equals raw visitation for a bear minus the mean raw visitation rate for all bears in a year.

Appendix A—Table 4. Raw and adjusted visitation at Trout Creek ecocenter by male versus female subadult grizzly bears aged 3-4 years after the breeding season (two-tailed Mann-Whitney test), Yellowstone National Park, 1959-70.

Population component	Raw mean visitation[a]	Adjusted mean visitation[b]
Males ($n = 9$)	0.4488	0.0796
Females ($n = 13$)	0.5498	0.0381
P-value	0.59	0.73

[a] Raw visitation equals fraction of complete censuses a bear was recorded as present at Trout Creek.
[b] Adjusted visitation equals raw visitation for a bear minus the mean raw visitation rate for all bears in a year.

Appendix A—Table 5. Raw and adjusted visitation at Trout Creek ecocenter during the breeding season by 3- and 4-year-old grizzly bears versus all adult males (sexes lumped; two-tailed Mann-Whitney test), Yellowstone National Park, 1959-70.

Population component	Raw mean visitation[a]	Adjusted mean visitation[b]
3- and 4-year-olds ($n = 64$)	0.3051	0.0147
Adult males ($n = 70$)	0.2212	-0.0788
P-value	0.04	0.03

[a] Raw visitation equals fraction of complete censuses a bear was recorded as present at Trout Creek.
[b] Adjusted visitation equals raw visitation for a bear minus the mean raw visitation rate for all bears in a year.

Appendix A—Table 6. Raw and adjusted visitation at Trout Creek ecocenter after the breeding season by 3- and 4-year-old grizzly bears of both sexes versus all adult males (two-tailed Mann-Whitney test), Yellowstone National Park, 1959-70.

Population component	Raw mean visitation[a]	Adjusted mean visitation[b]
3- and 4-year-olds ($n = 34$)	0.4722	0.0945
Adult males ($n = 65$)	0.2316	-0.1534
P-value	<0.001	<0.001

[a] Raw visitation equals fraction of complete censuses a bear was recorded as present at Trout Creek.
[b] Adjusted visitation equals raw visitation for a bear minus the mean raw visitation rate for all bears in a year.

Appendix A—Table 7. Raw visitation at Trout Creek ecocenter in the breeding versus post-breeding seasons for weaned 1- and 2-year-old grizzly bears and weaned 3- and 4-year-olds (two-tailed Wilcoxon matched-pairs test), Yellowstone National Park, 1959-70.

Population component	Raw visitation[a]		P-value
	Breeding season	Postbreeding season	
1- and 2-year-olds (*n* = 38)	0.322	0.4562	0.005
3- and 4-year-olds (*n* = 38)	0.3176	0.4809	0.01

[a] Raw visitation equals fraction of complete censuses a bear was recorded as present at Trout Creek.

Appendix A—Table 8. Raw and adjusted visitation at Trout Creek ecocenter during the breeding season by weaned 1- and 2-year-old grizzly bears versus weaned 3- and 4-year-olds (sexes lumped; two-tailed Mann-Whitney test), Yellowstone National Park, 1959-70.

Population component	Raw mean visitation[a]	Adjusted mean visitation[b]
1- and 2-year-olds (*n* = 41)	0.3015	-0.0276
3- and 4-year-olds (*n* = 45)	0.2931	-0.0384
P-value	0.76	0.93

[a] Raw visitation equals fraction of complete censuses a bear was recorded as present at Trout Creek.
[b] Adjusted visitation equals raw visitation for a bear minus the mean raw visitation rate for all bears in a year.

Appendix A—Table 9. Raw and adjusted visitation at Trout Creek ecocenter after the breeding season by weaned 1- and 2-year-old grizzly bears versus weaned 3- and 4-year-olds (sexes lumped; two-tailed Mann-Whitney test) Yellowstone National Park, 1959-70.

Population component	Raw mean visitation[a]	Adjusted mean visitation[b]
1- and 2-year-olds (*n* = 32)	0.4510	-0.0405
3- and 4-year-olds (*n* = 30)	0.4363	0.0097
P-value	0.77	0.51

[a] Raw visitation equals fraction of complete censuses a bear was recorded as present at Trout Creek.
[b] Adjusted visitation equals raw visitation for a bear minus the mean raw visitation rate for all bears in a year.

Appendix B

Appendix B—Table 1. Raw and adjusted visitation by contending versus noncontending male grizzly bears during the breeding season (all years, Mann-Whitney test), Trout Creek ecocenter, Yellowstone National Park, 1959-70.

Ranking	Raw mean visitation[a]	Adjusted mean visitation[b]
Contenders[c] ($n = 14$)	0.5155	0.1620
Noncontenders ($n = 71$)	0.1480	-0.1547
P-value (two-tailed)	<0.001	<0.001

[a] Raw visitation equals fraction of complete censuses a bear was recorded as present at Trout Creek.

[b] Adjusted visitation equals raw visitation for a bear minus the mean raw visitation rate for all bears in a year.

[c] Does not include males that achieved alpha or beta status.

Appendix B—Table 2. Pair-wise comparisons (Mann-Whitney tests) of breeding season visitation by grizzly bears at Trout Creek ecocenter between categories of contending status, Yellowstone National Park, 1959-70.

Rank (n) vs. rank (n)	*P*-values[a]
Alpha (6) vs. Beta (6)	0.47
Alpha (6) vs. Other Contenders (10)	0.75
Beta (6) vs. Other Contenders (10)	0.48
Alpha + Beta (12) vs. Other Contenders (10)	0.79

[a] No adjustment for multiple comparison was made because all values substantially exceeded 0.05.

Appendix B—Table 3. Raw and adjusted visitation by grizzly bears at Trout Creek ecocenter by contending versus noncontending males in the postbreeding season (15 July-1 September; all years; Mann-Whitney test), Yellowstone National Park, 1959-70.

Ranking	Raw mean visitation[a]	Adjusted mean visitation[b]
Contenders (*n* = 14)	0.4638	-0.1017
Noncontenders (*n* = 53)	0.2416	-0.0863
P-value (two-tailed)	0.02	0.02

[a] Raw visitation equals fraction of complete censuses a bear was recorded as present at Trout Creek.
[b] Adjusted visitation equals raw visitation for a bear minus the mean raw visitation rate for all bears in a year.

Appendix B—Table 4. Comparison (Wilcoxon Matched-Pairs test) of raw visitation by grizzly bears at Trout Creek ecocenter in the breeding versus postbreeding season for noncontending and contending males, Yellowstone National Park, 1959-70.

Status	Raw visitation[a]	
	Contenders (*n* = 15)	Noncontenders (*n* = 64)
Breeding	0.4250	0.1600
Postbreeding	0.3860	0.2370
P-value[b]	0.57	0.06

[a] Raw visitation equals fraction of complete censuses a bear was recorded as present at Trout Creek.
[b] Two-tailed matched-pair test in which a male's breeding season visitation is compared against his own postbreeding visitation in the same year; a given male is used only one time in each test.

Appendix C

Appendix C—Table 1. Analysis of variance (Extra-sums-of-squares test) for 56 litter sizes recorded for 30 different adult female grizzly bears, Trout Creek ecocenter, Yellowstone National Park, 1959-70.

Source of variation	DF	SS	MS	*F*-ratio	*P*-value
Female identity	29	16.49	0.57	1.22	NS
Ecocenter status	1	0.49	0.49	1.05	NS
Female age	1	3.22	3.22	6.94	<0.02
Residual	24	11.15	0.46	—	—

Note: Young-aged (4-8 years) versus prime-aged (9-20 years) bears.

Appendix C—Table 2. Mann-Whitney comparisons (two-tailed) of mean raw and adjusted visitation at Trout Creek ecocenter during the breeding season by adult female grizzly bears with dependent yearlings versus adult females without dependent offspring, Yellowstone National Park, 1959-70.

Adult female	Raw mean visitation	Number of females (*n*)	Adjusted mean visitation	Number of females (*n*)
With yearlings	0.3122	24	0.00	24
Without offspring	0.2317	50	-0.1079	50
P-value	0.21		0.03	

Appendix C—Table 3. Mann-Whitney comparisons (two-tailed) of mean raw and adjusted visitation at Trout Creek ecocenter after the breeding season by adult female grizzly bears with dependent yearlings versus adult females without dependent offspring, Yellowstone National Park, 1959-70.

Adult female	Raw mean visitation	Number of females (*n*)	Adjusted mean visitation	Number of females (*n*)
With yearlings	0.4402	25	0.0488	20
Without offspring	0.2866	38	-0.1114	34
P-value	0.06		0.05	

Appendix C—Table 4. Mann-Whitney comparisons (two-tailed) of mean raw and adjusted visitation at Trout Creek ecocenter during the breeding season by adult female grizzly bears with dependent cubs versus adult females without dependent offspring, Yellowstone National Park, 1959-70.

Adult female	Raw mean visitation	Number of females (n)	Adjusted mean visitation	Number of females (n)
With cubs	0.2565	37	-0.0187	37
Without cubs	0.2034	46	-0.0648	46
P-value	0.40		0.35	

Appendix C—Table 5. Mann-Whitney comparisons (two-tailed) of mean raw and adjusted visitation at Trout Creek ecocenter after the breeding season by adult female grizzly bears with dependent cubs versus adult females without dependent offspring, Yellowstone National Park, 1959-70.

Adult female	Raw mean visitation	Number of females (n)	Adjusted mean visitation	Number of females (n)
With cubs	0.3950	29	0.0205	20
Without cubs	0.2608	44	-0.1710	39
P-value	0.07		0.01	

Appendix C—Table 6. Mann-Whitney comparisons (two-tailed) of mean raw and adjusted visitation at Trout Creek ecocenter for adult female grizzly bears with cubs versus those with yearlings during the breeding and postbreeding seasons, Yellowstone National Park, 1959-70.

	Mean visitation					
	Breeding			Postbreeding		
Adult female	Raw	Adjusted	Number of females (n)	Raw	Adjusted	Number of females (n)
With cubs	0.2278	-0.1181	44	0.4660	0.0098	33
With yearlings	0.3664	-0.0294	28	0.3980	-0.0651	16
P-value	0.03	0.006		0.43	0.49	

Appendix C—Table 7. Wilcoxon matched-pairs comparisons (two-tailed) of mean raw visitation at Trout Creek ecocenter during (before 15 July) and after (15 July to 1 September) the breeding season in the same year by adult female grizzly bears with cubs or yearlings, Yellowstone National Park, 1959-70.

Period	Females with cubs[a] ($n = 36$)	Females with yearlings[a] ($n = 25$)
Breeding season	0.1880	0.3820
Postbreeding season	0.4380	0.4220
P-value	<0.001	0.41

[a] A female's visitation in the breeding season is compared to her visitation in postbreeding season in the same year; a given female appears, at most, once in each test.

Appendix C—Table 8. Wilcoxon matched-pairs comparisons (two-tailed) of breeding season visitation at Trout Creek ecocenter for adult female grizzly bears when in estrus versus anestrus, Yellowstone National Park, 1959-70.

Breeding condition	Test[a]		Test[b]	
	Raw visitation ($n = 12$)	Adjusted visitation ($n = 12$)	Raw visitation ($n = 10$)	Adjusted visitation ($n = 9$)
Estrous	0.3155	0.0834	0.3052	0.0716
Anestrous	0.2520	-0.0022	0.2021	-0.0012
P-value	0.33	0.43	0.33	0.44

[a] Uses any sequence of estrus vs. anestrus.
[b] Estrous and anestrous states must precede and follow, respectively, periods of caring for offspring.

Appendix C—Table 9. Mann-Whitney comparisons (two-tailed) of mean raw and adjusted visitation at Trout Creek ecocenter for estrous versus anestrous adult female grizzly bears during the breeding and postbreeding seasons, Yellowstone National Park, 1959-70.

	Mean visitation					
	Breeding			Postbreeding		
Breeding condition	Raw	Adjusted	Number of females (n)	Raw	Adjusted	Number of females (n)
Estrous	0.2828	0.0372	31	0.3880	0.0745	26
Anestrous	0.1773	-0.0572	30	0.2032	-0.1073	24
P-value	0.01	0.01		0.02	0.03	

Appendix C—Table 10. Wilcoxon matched-pairs comparisons (two-tailed) of mean raw visitation at Trout Creek ecocenter in the breeding versus postbreeding season for estrous and anestrous adult female grizzly bears, Yellowstone National Park, 1959-70.

	Raw visitation	
Period	Estrous females[a] (n = 31)	Anestrous females[a] (n = 30)
Breeding season	0.3220	0.1340
Postbreeding season	0.4020	0.1940
P-value	0.22	0.45

[a] A female's breeding season visitation is compared against her postbreeding season visitation in the same year; a given female appears, at most, once in each test.

Literature Cited

Acevedo, W., D. Walker, L. Gaydos, and J. Wray. 1982. *Vegetation and Land Cover, Arctic National Wildlife Refuge Coastal Plain, Alaska, Map I-1443*. Miscellaneous Investigations Series. US Geological Survey, Reston, VA.

Adams, L. G., and G. A. Connery. 1983. *Buckland River Reindeer/Caribou Conflict Study*. Alaska Open file, Report 8. Bureau of Land Management, Fairbanks, AK.

Albright, H. M. 1928. *Annual Report of the Superintendent, Yellowstone National Park*. Yellowstone National Park, Mammoth WY.

Allendorf, F. W., R. B. Harris, and L. H. Metzgar. 1991. Estimation of Effective Population Size of Grizzly Bears By Computer Simulation. Pages 650-4 *in* E. C. Dudley and T. R. Dudley, eds. *The Unity of Evolutionary Biology*, Vol. II, Proceedings of the fourth international congress of systematic and evolutionary biology. Dioscorides Press, Portland, Oregon.

Almack, J. A. 1986. Grizzly Bear Habitat Use, Food Habits, and Movements in the Selkirk Mountains, Northern Idaho. Pages 150-7 *in* G. P. Contreras and K. E. Evans, eds. *Proceedings—Grizzly Bear Habitat Symposium*, GTR-INT-207. USDA Forest Service, Intermountain Research Station, Ogden, UT.

American Wildlands. 1993. *Corridors of Life, a Project to Identify, Research and Inventory Critical Biological Corridors of the Northern Rockies*. 16pp.

Arno, S. F. 1986. Whitebark pine cone crops—a diminishing source of wildlife food? *West. J. Appl. For.* 1(3):92-4.

Arno, S. F., and R. J. Hoff. 1989. *Silvics of Whitebark Pine (*Pinus albicaulis*)*. GTR-INT-253. USDA Forest Service, Intermountain Research Station, Ogden, UT. 11pp.

Aune, K. 1985. *Rocky Mountain Front Grizzly Bear Monitoring and Investigation*. Mont. Dept. of Fish, Wildl. and Parks, Helena. 139pp.

Aune, K., and B. Brannon. 1986. *Rocky Mountain Front Grizzly Bear Monitoring and Investigation*. Mont. Dept. of Fish, Wildl. and Parks, Helena. 195pp.

Aune, K., and W. Kasworm. 1989. *Final Report—East Front Grizzly Bear Study*. Montana Dept. of Fish, Wildl. and Parks, Helena. 332pp.

Aune, K., and T. Stivers. 1981. *Rocky Mountain Front Grizzly Bear Monitoring and Investigation*. Mont. Dept. of Fish, Wildl. and Parks, Helena. 50pp.

Aune, K., and T. Stivers. 1982. *Rocky Mountain Front Grizzly Bear Monitoring and Investigation*. Mont. Dept. of Fish, Wildl. and Parks, Helena. 143pp.

Aune, K., and T. Stivers. 1983. *Rocky Mountain Front Grizzly Bear Monitoring and Investigation*. Mont. Dept. of Fish, Wildl. and Parks, Helena. 180pp.

Aune, K., T. Stivers, and M. Madel. 1984. *Rocky Mountain Front Grizzly Bear Monitoring and Investigation*. Mont. Dept. of Fish, Wildl. and Parks, Helena. 239pp.

Bader, M. 1991. *The Northern Rockies Ecosystem Protection Act*. Alliance for the Wild Rockies, Missoula, MT.

Bader, M. 1992. The need for an ecosystem approach for endangered species protection. *The Public Land Law Review* 13:137-47.

Ballard, W. B., S. D. Miller, and T. H. Sparker. 1982. Home range, daily movements, and reproductive biology of brown bear in southcentral Alaska. *Can. Field-Nat.* 96(1):1-5.

Banfield, A. W. F. 1958. Distribution of the barren ground grizzly in northern Canada. *Natl. Mus. Can. Bull. No. 166* :47-59.

Barber, J. F. 1987. *Old Yellowstone Views*. Mountain Press Pub. Co., Missoula, MT. 95pp.

Barnes, V. G., and O. E. Bray. 1967. *Population Characteristics and Activities of Black Bears in Yellowstone National Park*. Final Rep. Colorado Coop. Wildl. Res. Unit, Colorado State Univ., Fort Collins. 199pp.

Barnes, V. G., Jr., and R. B. Smith. 1993. Cub adoption by brown bears, *Ursus arctos middendorffi*, on Kodiak Island, Alaska. *Can. Field-Nat.* 107(3):365-7.

503

Bartlett, R. A. 1974. *Nature's Yellowstone*, 1st ed. Univ. of New Mexico Press, Albuquerque, NM. 250pp.

Bartlett, R. A. 1985. *Yellowstone: A Wilderness Besieged*. Univ. of Arizona Press, Tucson. 437pp.

Bartos, D. L., and K. E. Gibson. 1990. Insects of Whitebark Pine With Emphasis On Mountain Pine Beetle. Pages 171-8 *in Proceedings—Symposium On Whitebark Pine Ecosystems: Ecology and Management of a High-Mountain Resource*, GTR-INT-270. USDA Forest Service, Intermountain Research Station, Ogden, UT.

Beecham, J. 1983. Population characteristics of black bears in west central Idaho. *J. Wildl. Manage.* 47(2):405-12.

Begon, M., J. L. Harper, and C. R. Townsend. 1986. *Ecology—Individuals, Populations, and Communities*. Sinauer Associates, Inc., Sunderland, MA. 876pp.

Biddle, N., ed. 1962. *The Journals of the Expedition Under the Command of Captains Lewis and Clark*, Vol. 1. The Heritage Press, New York, NY. 231pp.

Black, H. C. 1958. Black bear research in New York. *Trans. N. Amer. Wildl. Conf.* 23:443-61.

Blanchard, B. M. 1987. Size and growth patterns of the Yellowstone grizzly bear. *Int. Conf. Bear Res. and Manage.* 7:99-107.

Blanchard, B. M. 1990. Relationships Between Whitebark Pine Cone Production and Fall Grizzly Bear Movements. Pages 362-3 *in Proceedings—Symposium On Whitebark Pine Ecosystems: Ecology and Management of a High-Mountain Resource*, GTR-INT-270. USDA Forest Service, Intermountain Research Station, Ogden, UT.

Blanchard, B. M., and R. R. Knight. 1990. Reactions of grizzly bears, *Ursus arctos horribilis*, to wildfire in Yellowstone National Park, Wyoming. *Can. Field-Nat.* 104(4):592-4.

Blanchard, B. M., and R. R. Knight. 1991. Movements of Yellowstone grizzly bears. *Biol. Conserv.* 58:41-67.

Blanchard, B. M., R. R. Knight, and D. J. Mattson. 1992. Distribution of Yellowstone grizzly bears during the 1980s. *Am. Midl. Nat.* 128(2):332-3.

Bourliere *et al.* 1962. See "First World Conference on National Parks, 1964".

Bronson, F. H. 1989. *Mammalian Reproductive Biology*. Univ. of Chicago Press, Chicago. 325pp.

Brown, H. W. 1967. Trichinosis. Pages 398-400 *in* P. B. Beeson and W. McDermott, eds. *Textbook of Medicine*, 12th ed. Saunders Co., Philadelphia.

Brown, J. L. 1975. *The Evolution of Behavior*. Norton, New York. 761pp.

Buechner, H. K., F. C. Craighead, Jr., J. J. Craighead, and C. E. Cote. 1971. Satellites for research on free-roaming animals. *Bioscience* 21(24):1201-5.

Bunnell, F. L., and D. E. N. Tait. 1981. Population Dynamics of Bears—Implications. Pages 173-96 *in* C. W. Fowler and T. D. Smith, eds. *Dynamics of Large Mammal Populations*. John Wiley and Sons, New York.

Cadbury, D. A., J. G. Hawkes, and R. C. Readett. 1971. *A Computer-Mapped Flora, A Study of the County of Warwickshire*. Birmingham Natural History Society, Academic Press, New York. 768pp.

Carlson, C. E. 1978. *Noneffectiveness of Ribes Eradication As a Control of White Pine Blister Rust in Yellowstone National Park*. For. Insect and Dis. Manage. Rep. 78-18. USDA Forest Service, Northern Region, Missoula, MT. 6pp.

Chapman, J. A., J. I. Romer, and J. Stark. 1955. Ladybird beetles and army cutworm adults as food for grizzly bears in Montana. *Ecology* 36(1):156-8.

Chase, A. 1986. *Playing God in Yellowstone*. Harcourt Brace Jovanovich, San Diego. 464pp.

Cheatum, E. L. 1952. On the population dynamics of big game. *Proc. Mont. Acad. Sci.* 11:47-56.

Chittenden, H. M. 1924. *The Yellowstone National Park*, 3rd ed. J. E. Haynes, Saint Paul, MN. copyright in England. 356pp.

Christensen, N. L., J. K. Agee, P. F. Brussard, J. Hughes, D. H. Knight, G. W. Minshall, J. M. Peck, S. J. Pyne, F. J. Swanson, S. Wells, J. W. Thomas, S. E. Williams, and H. A. Wright. 1989. *Ecological Consequences of the 1988 Fires in the Greater Yellowstone Area*. Final Report of the Greater Yellowstone Postfire Ecological Assessment Workshop. 58pp.

Clements, F. E. 1916. *Plant Succession; an Analysis of the Development of Vegetation*. Carnegie Institution of Washington Publication, No. 242. Carnegie Institution of Washington, Washington, DC. 512pp.

Clutton-Brock, T. H. 1991. *The Evolution of Parental Care*. Princeton Univ. Press, Princeton. 352pp.

Cole, G. F. 1969. *Elk and the Yellowstone Ecosystem*. USDI National Park Service, Office of Natural Science Studies. Research Note, Yellowstone National Park. 14pp.

Cole, G. F. 1970. *Grizzly Bear Management in Yellowstone Park, 1970*. Pres. at 2nd Intern. Conf. on Bear Res. and Manage., Nov. 1970, Univ. of Calgary, Alberta. Research Note No. 3. Yellowstone National Park.

Cole, G. F. 1971. *Progress in Restoring a Natural Grizzly Bear Population in Yellowstone National Park*. Pres. at Amer. Assoc. for the Advance. of Sci., Symp. on Research in National Parks, Dec. 29, 1971. 21pp.

Cole, G. F. 1972a. *Interim Report On Grizzly Bear Management in Yellowstone National Park, 1972*. USDI National Park Service, Yellowstone National Park, WY. 16pp.

Cole, G. F. 1972b. Preservation and management of grizzly bears in Yellowstone National Park. *Int. Conf. Bear Res. and Manage.* 2:274-88.

Cole, G. F. 1973a. *Management Involving Grizzly Bears and Humans in Yellowstone National Park, 1970-73.* Interim Rep. USDI National Park Service. 17pp.

Cole, G. F. 1973b. *Management Involving Grizzly Bears in Yellowstone National Park, 1970-72.* Natural Resources Report, No. 7. USDI National Park Service, Mammoth, WY. 10pp.

Cole, G. F. 1974. Management involving grizzly bear and humans in Yellowstone National Park, 1970-73. *Bioscience* 24(6):335-8.

Cole, G. F. 1975. *Management Involving Grizzly Bears in Yellowstone National Park, 1970-74.* Pres. at 26th Ann. AIBS Meeting of Biological Sciences, Oregon State Univ., Corvallis. 33pp.

Cole, G. F. 1976. *Management Involving Grizzly and Black Bears in Yellowstone National Park, 1970-75.* Research Report No. 9. USDI National Park Service, Yellowstone National Park, WY. 26pp.

Colwell, R. N. 1971. *Monitoring Earth Resources from Aircraft and Spacecraft.* Special publication 275. National Aeronautics and Space Administration, Washington, DC. 170pp.

Conover, W. J. 1980. *Practical Nonparametric Statistics*, 2nd ed. Wiley, New York. 493pp.

Corliss, J. C., R. D. Pfister, R. F. Buttery, F. C. Hall, W. F. Mueggler, D. On, R. W. Phillips, W. S. Platts, and J. E. Reid. 1973. *Ecoclass—a Method for Classifying Ecosystems.* USDA Forest Service, Intermountain Forest and Range Exp. Station, Missoula, MT. 52pp.

Cowan, I. M. (Chairman), D. G. Chapman, R. S. Hoffman, D. R. McCullough, G. A. Swanson, and R. B. Weeden. 1974. *Report of Committee On the Yellowstone Grizzlies.* Natl. Acad. Sci., Washington, DC. 61pp.

Cowardin, L. M., V. Carter, F. C. Golet, and E. T. LaRee. 1979. *Classification of Wetlands and Deep Water Habitats of the United States.* FWS/OBS-79/31. USDI Fish and Wildlife Service, US GPO, Washington, DC. 103pp.

Craighead, D. J. 1995 (in prep). An integrated satellite technique to evaluate grizzly bear habitat use. Craighead Wildlife-Wildlands Institute, Missoula, MT.

Craighead, D. J., and J. J. Craighead. 1987. Tracking caribou using satellite telemetry. *Natl. Geogr. Res.* 3:462-79.

Craighead, F. C., Jr. 1976. Grizzly bear ranges and movement as determined by radiotracking. *Int. Conf. Bear Res. and Manage.* 3:97-109.

Craighead, F. C., Jr. 1979. *The Track of the Grizzly.* Sierra Club Books, San Francisco. 261pp.

Craighead, F. C., Jr., and J. J. Craighead. 1963. *Radiotracking of Grizzly Bears: Grizzly Bear Ecological Findings Obtained By Biotelemetry.* Montana Coop. Wildl. Res. Unit, Univ. of Montana, Missoula. 31pp.

Craighead, F. C., Jr., and J. J. Craighead. 1965. Tracking grizzly bears. *Bioscience* 15(2):88-92.

Craighead, F. C., Jr., and J. J. Craighead. 1972a. Data on grizzly bear denning activities and behavior obtained by using wildlife telemetry. *Int. Conf. Bear Res. and Manage.* 2:84-106.

Craighead, F. C., Jr., and J. J. Craighead. 1972b. *Grizzly Bear Prehibernation and Denning Activities As Determined By Radiotracking.* Wildlife Monogr., No. 32. The Wildl. Soc., Bethesda, MD. 35pp.

Craighead, F. C., Jr., and J. J. Craighead. 1973. *Grizzly Bear Prehibernation and Denning Activities As Determined By Radiotracking.* Montana Coop. Wildl. Res. Unit, Univ. of Montana, Missoula. 49pp.

Craighead, F. C., Jr., and J. J. Craighead. 1979. Biotelemetry research and feasibility experiments with IRLS for tracking animals by satellite. *Natl. Geogr. Soc. Res. Rep., 1970 Projects*:75-81.

Craighead, F. C., Jr., J. J. Craighead, and R. S. Davies. 1962. *Radiotracking of Grizzly Bears.* Pres. at Interdisciplinary Conference on the Use of Telemetry in Animal Behavior and Physiology in Relationship to Ecological Problems.

Craighead, F. C., Jr., J. J. Craighead, and R. S. Davies. 1963. Radiotracking of Grizzly Bears. Pages 133-48 *in* L. E. Slater, ed. *Bio-Telemetry: Proceedings of the Interdisciplinary Conference On the Use of Telemetry in Animal Behavior and Physiology in Relation to Ecological Problems.* The Macmillan Co., Pergamon Press, New York.

Craighead, F. C., Jr., J. J. Craighead, C. E. Cote, and H. K. Buechner. 1972a. Satellite and Ground Radiotracking of Elk. Pages 99-111 *in* S. R. Galler, K. Schmidt-Koenig, G. J. Jacobs, and R. E. Belleville, eds. *Animal Orientation and Navigation: A Symposium.* NASA Spec. Pub. 262. Supt. of Documents, Washington, DC.

Craighead, F. C., Jr., J. J. Craighead, C. E. Cote, and H. K. Buechner. 1978a. Satellite tracking of elk, Jackson Hole, Wyoming. *Natl. Geogr. Soc. Res. Rep., 1969 Projects*:73-88.

Craighead, F. L. 1994. *Conservation Genetics of Grizzly Bears.* Ph.D. Dissert. Montana State Univ., Bozeman. 277pp.

Craighead, J. J. 1972. *A Briefing On the Grizzly Bear Situation in Yellowstone.* Presented at Mammoth, WY, Sept. 19, 1972. Unpublished report. 11 pp.

Craighead, J. J. 1974. Determining the sex and age structure of a living population of *Ursus arctos horribilis.* First International Theriological Congress, Moscow, Russia 1:118.

Craighead, J. J. 1977. *A Delineation of Critical Grizzly Bear Habitat in the Yellowstone Region.* Montana Coop. Wildl. Res. Unit, Univ. of Montana, Missoula. 87pp.

Craighead, J. J. 1980a. A definitive system for analysis of grizzly bear habitat and other wilderness resources. *Natl. Geogr. Soc. Res. Rep., 1971 Projects*:153-62.

Craighead, J. J. 1980b. A proposed delineation of critical grizzly bear habitat in the Yellowstone region. *Int. Conf. Bear Res. and Manage.* 4:379-99.

Craighead, J. J. 1982. The future of the grizzly bear. *Wild America* 10:13-6.

Craighead, J. J. 1991. Yellowstone in Transition. Pages 27-39 *in* R. B. Keiter and M. S. Boyce, eds. *The Greater Yellowstone Ecosystem: Redefining America's Wilderness Heritage.* Yale Univ. Press, New Haven, CT.

Craighead, J. J., and D. J. Craighead. 1991. New system-techniques for ecosystem management and an application to the Yellowstone Ecosystem. *West. Wildlands* 17(1):30-9.

Craighead, J. J., and F. C. Craighead, Jr. 1967. *Management of Bears in Yellowstone National Park.* Mimeo Rep. Montana Coop. Wildl. Res. Unit, Univ. of Montana, Missoula. 113pp.

Craighead, J. J., and F. C. Craighead, Jr. 1971. Grizzly bear-man relationships in Yellowstone National Park. *Bioscience* 21(16):845-57.

Craighead, J. J., and J. A. Mitchell. 1982. Grizzly Bear. Pages 515-56 *in* J. A. Chapman and G. A. Feldhamer, eds. *Wild Mammals of North America—Biology, Management, Economics.* The Johns Hopkins Univ. Press, Baltimore.

Craighead, J. J., and R. L. Redmond. 1987. Climate and reproduction of Yellowstone grizzlies. *Nature* 327(6117):22.

Craighead, J. J., M. Hornocker, W. Woodgerd, and F. C. Craighead, Jr. 1960. Trapping, immobilizing, and color-marking grizzly bears. *Trans. N. Amer. Wildl. Conf.* 25:347-63.

Craighead, J. J., F. C. Craighead, Jr., and M. G. Hornocker. 1961. *An Ecological Study of the Grizzly Bear.* 3rd Annu. Rep. (summary of work accomplished, 1959-1961), Subproject 2.9. Montana Coop. Wildl. Res. Unit, Univ. Montana, Missoula. 59pp.

Craighead, J. J., F. C. Craighead, Jr., and M. G. Hornocker. 1964. *An Ecological Study of the Grizzly Bear.* 5th Annu. Rep. (summary of work accomplished, 1959-1963). Montana Coop. Wildl. Res. Unit, Univ. Montana, Missoula. 33pp.

Craighead, J. J., M. G. Hornocker, and F. C. Craighead, Jr. 1969. Reproductive biology of young female grizzly bears. *J. Reprod. Fertil.*(Suppl. 6):447-75.

Craighead, J. J., F. C. Craighead, Jr., and H. E. McCutchen. 1970. Age determination of grizzly bears from fourth premolar tooth sections. *J. Wildl. Manage.* 34(2):353-63.

Craighead, J. J., F. C. Craighead, Jr., J. R. Varney, and C. E. Cote. 1971. Satellite monitoring of black bear. *Bioscience* 21(24):1206-12.

Craighead, J. J., G. Atwell, and B. W. O'Gara. 1972b. *Elk Migrations in and Near Yellowstone National Park.* Wildlife Monogr., No. 29. The Wildl. Soc., Bethesda, MD. 48pp.

Craighead, J. J., F. C. Craighead, Jr., R. L. Ruff, and B. W. O'Gara. 1973a. *Home Ranges and Activity Patterns of Nonmigratory Elk of the Madison Drainage Herd As Determined By Biotelemetry.* Wildlife Monogr., No. 33. The Wildl. Soc., Bethesda, MD. 50pp.

Craighead, J. J., J. R. Varney, and F. C. Craighead, Jr. 1973b. *A Computer Analysis of the Yellowstone Grizzly Bear Population.* Montana Coop. Wildl. Res. Unit, Univ. of Montana, Missoula. 146pp.

Craighead, J. J., J. R. Varney, and F. C. Craighead, Jr. 1974. *A Population Analysis of the Yellowstone Grizzly Bears.* For. and Conserv. Exp. Sta. Bull. 40. Sch. For., Univ. Montana, Missoula. 20pp.

Craighead, J. J., F. C. Craighead, Jr., and J. Sumner. 1976. Reproductive cycles and rates in the grizzly bear, *Ursus arctos horribilis*, of the Yellowstone ecosystem. *Int. Conf. Bear Res. and Manage.* 3:337-56.

Craighead, J. J., J. R. Varney, and J. S. Sumner. 1978b. Potentialities of satellite telemetry of large mammals. *Natl. Geogr. Soc. Res. Rep., 1969 Projects*:111-4.

Craighead, J. J., J. S. Sumner, and G. B. Scaggs. 1982. *A Definitive System for Analysis of Grizzly Bear Habitat and Other Wilderness Resources.* Wildlife-Wildlands Inst. Monogr. No. 1. Univ. of Montana Foundation, Univ. of Montana, Missoula. 279pp.

Craighead, J. J., F. L. Craighead, and D. J. Craighead. 1986. Using Satellites to Evaluate Ecosystems As Grizzly Bear Habitat. Pages 101-12 *in* G. P. Contreras and K. E. Evans, eds. *Proceedings—Grizzly Bear Habitat Symposium*, GTR-INT-207. USDA Forest Service, Intermountain Research Station, Ogden, UT.

Craighead, J. J., F. L. Craighead, D. J. Craighead, and R. L. Redmond. 1988a. Mapping arctic vegetation in northwest Alaska using landsat MSS imagery. *Natl. Geogr. Res.* 4:496-527.

Craighead, J. J., K. R. Greer, R. R. Knight, and H. I. Pac. 1988b. *Grizzly Bear Mortality in the Yellowstone Ecosystem, 1959-1987.* Mont. Dept. of Fish, Wildl. and Parks, Helena. 57pp.

Crockford, J. A., F. A. Hayes, J. H. Jenkins, and S. D. Feurt. 1958. An automatic projectile type syringe. *Vet. Med.* 53(3):115-9.

Daubenmire, R. F., and J. R. Daubenmire. 1968. *Forest Vegetation of Eastern Washington and Northern Idaho.* Washington Agric. Exp. Sta. Tech. Bull. 60. 104pp.

Dean, F. C., L. M. Darling, and A. G. Lierhaus. 1986. Observations of intraspecific killing by brown bears, *Ursus arctos. Can. Field-Nat.* 100(2):208-11.

Derocher, A. E., and I. Stirling. 1994. Age-specific reproductive performance of female polar bears (*Ursus maritimus*). *J. Zool., London* 234:527-36.

Derocher, A. E., and M. Taylor. 1995 (in press). Density-dependent population regulation of polar bears. *Int. Conf. Bear Res. and Manage.* 9.

Despain, D., A. Rodman, P. Schullery, and H. Shovic. 1989. *Burned Area Survey of Yellowstone National Park: The Fires of 1988.* Division of Research and GIS Lab, Yellowstone National Park. 14pp.

Dirschl, H. G., D. L. Dabbs, and G. C. Gentle. 1974. *Landscape Classification and Plant Successional Trends in the Peace-Athabasca Delta.* Report Series 30. Canadian Wildl. Serv., Ottawa, Ontario.

Dittrich, L., and H. Kronberger. 1963. Biological-anatomical research concerning the biology of reproduction of the brown bear (*Ursus arctos* L.) and other bears in captivity. *Z. Säugetierkunde* 28(3):129.

Dood, A. R., R. D. Brannon, and R. D. Mace. 1986. *Final Programmatic Environmental Impact Statement: The Grizzly Bear in Northwestern Montana.* Mont. Dept. of Fish, Wildl. and Parks, Helena. 287pp.

Draper, N. R., and H. Smith. 1981. *Applied Regression Analysis*, 2nd ed. John Wiley, New York. 709pp.

Eberhardt, L. L. 1985. Assessing the dynamics of wild populations. *J. Wildl. Manage.* 49(4):997-1012.

Eberhardt, L. L., R. R. Knight, and B. M. Blanchard. 1986. Monitoring grizzly bear population trends. *J. Wildl. Manage.* 50(4):613-8.

Eberhardt, L. L., B. M. Blanchard, and R. R. Knight. 1994. Population trend of the Yellowstone grizzly bear as estimated from reproductive and survival rates. *Can. J. Zool.*(72):1-4.

Egbert, A. L. 1978. *The Social Behavior of Brown Bears At McNeil River, Alaska.* Ph.D. Dissert. Utah State Univ., Logan. 117pp.

Egbert, A. L., and A. W. Stokes. 1976. The social behaviour of brown bears on an Alaskan salmon stream. *Int. Conf. Bear Res. and Manage.* 3:41-56.

Elowe, K. D., and W. E. Dodge. 1989. Factors affecting black bear reproductive success and cub survival. *J. Wildl. Manage.* 53:962-8.

Erickson, A. W. 1957. Techniques for live-trapping and handling black bears. *Trans. N. Amer. Wildl. Conf.* 22:520-43.

Erickson, A. W., J. Nellor, and G. Petrides. 1964. *The Black Bear in Michigan.* Michigan State Univ. Agr. Exp. Sta., East Lansing, MI. 101pp.

First World Conference on National Parks., ed. 1964. *Proceedings of the First World Conference On National Parks, Seattle, WA, June 30-July 7, 1962.* USDI National Park Service, Seattle, WA. 471pp.

Flowerdew, J. R. 1987. *Mammals: Their Reproductive Biology and Population Ecology.* Edward Arnold, Baltimore, MD. 241pp.

Forcella, F., and T. Weaver. 1977. Biomass and productivity of the subalpine *Pinus albicaulis* -*Vaccinium scoparium* association in Montana. *Vegetatio* 33:95-106.

Frank, D. A., and S. J. McNaughton. 1992. The ecology of plants, large mammalian herbivores, and drought in Yellowstone National Park. *Ecology* 73(6):2043-58.

Frankel, O. H., and M. E. Soulé. 1981. *Conservation and Evolution.* Cambridge Univ. Press, New York. 327pp.

French, S. P., and M. G. French. 1990. Predatory behavior of grizzly bears feeding on elk calves in Yellowstone Park, 1986-88. *Int. Conf. Bear Res. and Manage.* 8:335-41.

French, S. P., M. G. French, and R. R. Knight. 1995 (in press). Grizzly bear use of army cutworm moths in the Yellowstone Ecosystem. *Int. Conf. Bear Res. and Manage.* 9.

Frison, G. C. 1978. *Prehistoric Hunters of the High Plains.* Academic Press, New York. 457pp.

Frome, M. 1984. Are biologists afraid to speak out? *Defenders* 59:40-1.

Fuller, M. R., T. E. Levanon, T. E. Strikwerda, W. S. Seegar, J. Wall, H. D. Black, F. P. Ward, P. W. Howey, and J. Partelow. 1984. Feasibility of a Bird-Borne Transmitter for Tracking Via Satellite. Pages 1-6 *in Proc. Eighth Internat. Symp. On Biotelemetry, Durbrovnik, Yugoslavia.*

Garshelis, D. L. 1995 (in press). Density-dependent population regulation of black bears. *Int. Conf. Bear Res. and Manage.* 9.

Gleason, H. A. 1926. *The Individualistic Concept of the Plant Association.* Reprinted: see Gleason, 1991.

Gleason, H. A. 1991. The Individualistic Concept of the Plant Association. Pages 98-117 *in* L. A. Real and J. H. Brown, eds. *Foundations of Ecology: Classic Papers With Commentaries.* The Univ. of Chicago Press, Chicago.

Glenn, L. P. 1980. Morphometric characteristics of brown bears on the central Alaska Peninsula. *Int. Conf. Bear Res. and Manage.* 4:313-9.

Glenn, L. P., and L. H. Miller. 1980. Seasonal movements of an Alaska Peninsula brown bear population. *Int. Conf. Bear Res. and Manage.* 4:307-12.

Glenn, L. P., J. W. Lentfer, J. B. Faro, and L. H. Miller. 1976. Reproductive biology of female brown bears (*Ursus arctos*), McNeil River, Alaska. *Int. Conf. Bear Res. and Manage.* 3:381-90.

Graves, H. S., and E. W. Nelson. 1919. *Our National Elk Herds: A Program for Conserving the Elk On National Forests About the Yellowstone National Park*. USDA Circular, No. 51. US GPO, Washington, DC. 34pp.

Greater Yellowstone Coalition. 1991. *An Environmental Profile of the Greater Yellowstone Ecosystem*. D. Glick, M. Carr, and B. Harting, eds. The Greater Yellowstone Coalition, Bozeman, MT. 132pp.

Greater Yellowstone Coordinating Committee. 1989. *The Greater Yellowstone Postfire Assessment*. S. M. Mills, ed. USDI, Washington, DC. 146pp.

Green, G. I. 1989. Dynamics of Ungulate Carcass Availability and Use By Bears On the Northern Winter Range and Firehole and Gibbon Drainages. Pages 21-32 *in* R. R. Knight, B. M. Blanchard, and D. J. Mattson. *1988 Progress Report in Yellowstone Grizzly Bear Investigations; Annual Report of the Interagency Study Team*. National Park Service.

Green, G. I., and D. J. Mattson. 1988. Dynamics of Ungulate Carcass Availability and Use By the Bears On the Northern Winter Range: 1987 Progress Report. Pages 32-50 *in* R. R. Knight, B. M. Blanchard, and D. J. Mattson. *Yellowstone Grizzly Bear Investigations: Annual Report of the Interagency Study Team, 1987*. USDI National Park Service, US GPO.

Greer, K. R. 1965. *Special Collections—Yellowstone Elk Study*. Job Completion Rep. Project W-83-R-8. Montana Fish and Game Department. 3pp.

Greer, K. R. 1966a. Fertility Rates of the Northern Yellowstone Elk Populations *in Annual Conference of the Western Association of State Fish and Game Commissions*, Conference No. 46. 10pp.

Greer, K. R. 1966b. *Special Collections—Yellowstone Elk Study*. Job Completion Rep. Project W-83-R-8. Montana Fish and Game Department. 3pp.

Greer, K. R. 1967. *Special Collections—Yellowstone Elk Study*. Job Completion Rep. Project W-83-R-10. Montana Fish and Game Department. 15pp.

Greer, K. R. 1968a. A compression method indicates fat content of elk (wapiti) femur marrows. *J. Wildl. Manage.* 32(4):747-51.

Greer, K. R. 1968b. *Special Collections—Yellowstone Elk Study*. Job Completion Rep. Project W-83-R-11. Montana Fish and Game Department. 26pp.

Gresswell, R. E., ed. 1988. *Status and Management of Interior Stocks of Cutthroat Trout*. American Fisheries Society, Bethesda, MD. 140pp.

Grimm, R. L. 1935-38. *Yellowstone Park Range Studies*. Yellowstone National Park files.

Grimm, R. L. 1939. Northern Yellowstone winter range studies. *J. Wildl. Manage.* 3(4):295-306.

Grimm, R. L. 1943-47. *Yellowstone National Park Range Studies*. Yellowstone National Park files.

Grinnell, G. B. 1876. *Report of a Reconnaissance from Carroll, Montana Territory, On the Upper Missouri, to the Yellowstone National Park, and Return, Made in the Summer of 1875*. US GPO, Washington, DC. 155pp.

Grizzly Bear Compendium. 1987. M. N. LeFranc, Jr., M. B. Moss, K. A. Patnode, and W. C. Sugg, III, eds. Interagency Grizzly Bear Committee, contracted by National Wildlife Federation, Washington, DC. 540pp.

Grodstein, F., M. B. Goldman, and D. W. Cramer. 1994. Body mass index and ovulatory infertility. *Epidemiology* 5:247-50.

Grumbine, R. E. 1994. What is ecosystem management? *Conserv. Biol.* 8(1):27-38.

Gunderson, H. L. 1976. *Mammology*. McGraw-Hill, New York. 483pp.

Gunther, K. A. 1995 (in press). Changing problems in bear management, Yellowstone National Park 20+ years after the dumps. *Int. Conf. Bear Res. and Manage.* 9.

Gunther, K. A., and R. A. Renkin. 1990. Grizzly bear predation on elk calves and other fauna of Yellowstone National Park. *Int. Conf. Bear Res. and Manage.* 8:329-34.

Haines, A. L. 1977. *The Yellowstone Story: A History of Our First National Park*, Vols. I & II. Colorado Assoc. Univ. Press, Denver. 385pp. & 543pp.

Hamer, D., S. M. Herrero, R. T. Ogilvie, T. Toth, and A. H. Marsh. 1980. *Ecological Studies of the Banff National Park Grizzly Bear, Cascade/Panther Region 1979*. Parks Canada Contract WR 48-79. 51pp.

Hamlett, G. D. 1935. Delayed implantation and discontinuous development in the mammals. *Q. Rev. Biol.* 10(4):432-47.

Hampton, H. D. 1971. *How the U.S. Cavalry Saved Our National Parks*. Indiana Univ. Press, Bloomington. 246pp.

Hansson, A. 1947. The physiology of reproduction in mink (*Mustela vison* Schreb.) with special reference to delayed implantation. *Acta Zool.* 28:1-136.

Hardin, G. 1968. The tragedy of the commons. *Science* 162:1243-8.

Harris, R. B. 1984. *Harvest Age-Structure As an Indicator of Grizzly Bear Population Status*. Masters Thesis. Univ. of Montana, Missoula, MT. 204pp.

Harris, R. B., and F. W. Allendorf. 1989. Genetically effective population size of large mammals: an assessment of estimators. *Conserv. Biol.* 3(2):181-91.

Harrison, R. L. 1992. Toward a theory of inter-refuge corridor design. *Conserv. Biol.* 6:293-5.

Harting, A., and D. Glick. 1994. *Sustaining Greater Yellowstone: A Blueprint for the Future.* Greater Yellowstone Coalition, Bozeman, MT. 222pp.

Hechtel, J. L. 1985. *Activity and Food Habits of Barren-Ground Grizzly Bears in Arctic Alaska.* Masters Thesis. Univ. of Montana, Missoula. 74pp.

Henderson, A. B. 19—. Journal of the Yellowstone Expedition of 1866 Under Captain Jeff Standifer, Also the Diaries Kept By Henderson During His Prospecting Journies in the Snake, Wind River, and Yellowstone Country During the Years 1866-1872 *in* S. Granville, ed. *Diaries of Stuart Granville and Others.* Yale Univ., New Haven. 1 microfilm reel, 35 mm.

Henry, J., and D. J. Mattson. 1988. Spring Grizzly Bear Use of Ungulate Carcasses in the Firehole River Drainage: Third Year Progress Report. Pages 51-9 *in* R. R. Knight, B. M. Blanchard, and D. J. Mattson. *Yellowstone Grizzly Bear Investigations: Annual Report of the Interagency Study Team, 1987.* USDI National Park Service, US GPO.

Herrero, S. 1972. Aspects of evolution and adaptation in American black bears (*Ursus americanus* Pallas) and brown and grizzly bears (*U. arctos* Linné) of North America. *Int. Conf. Bear Res. and Manage.* 2:221-31.

Herrero, S. 1976. Conflicts between man and grizzly bears in the national parks of North America. *Int. Conf. Bear Res. and Manage.* 3:121-45.

Hewitt, D. G. 1989. *Correcting Grizzly Bear Fecal Analysis to Observed Food Habits.* Masters Thesis. Washington State Univ., Pullman. 20pp.

Hobbs, R. J. 1992. The role of corridors in conservation: solution or bandwagon. *Trends Ecol. & Evol.* 3(11):389-91.

Hoffer, R., R. Fleming, R. Cary, P. Mroczynski, P. Krebs, D. Kreammerer, and D. Groenfeld. 1975. *Natural Resource Mapping in Mountainous Terrain By Computer Analysis of ERTS-1 Satellite Data.* Research Bulletin 919 (LARS), Information Note 061575. Purdue Univ., West Lafayette, IN.

Hogg, J. T., C. C. Hass, and D. A. Jenni. 1992. Sex-biased maternal expenditure in Rocky Mountain bighorn sheep. *Behav. Ecol. Sociobiol.* 31(4):243-51.

Horejsi, B. L. 1989. Uncontrolled land-use threatens an international grizzly bear population. *Conserv. Biol.* 3(3):219-23.

Hornocker, M. G. 1962. *Population Characteristics and Social and Reproductive Behavior of the Grizzly Bear in Yellowstone National Park.* Masters Thesis, Univ. of Montana, Missoula. 94pp.

Hoskins, W. P. 1975. *Yellowstone Lake Tributary Survey Project.* unpublished report by Interagency Grizzly Bear Study Team, Bozeman, MT. 31pp.

Houston, D. B. 1982. *The Northern Yellowstone Elk: Ecology and Management.* Macmillan Publishing Co., Inc., New York, NY. 474pp.

Husby, P., and N. McMurray. 1978. Seasonal Food Habits of Grizzly Bears (*Ursus arctos horribilis* Ord) in Northwestern Montana. Sub-Proj. No. 5. Pages 109-34 and 253-6 *in* C. Jonkel, ed. *Border Grizzly Proj. Ann. Rep. 3.* Sch. of For., Univ. of Montana, Missoula.

Hutchins, H. E., and R. M. Lanner. 1982. The central role of Clark's nutcracker in the dispersal and establishment of whitebark pine. *Oecologia* 55:192-201.

Ims, R. A. 1990. On the adaptive value of reproductive synchrony as a predator-swamping strategy. *Am. Nat.* 136:485-98.

Jonas, R. J. 1955. *A Population and Ecological Study of the Beaver (Castor canadensis) of Yellowstone National Park.* Masters Thesis. Univ. of Idaho, Moscow. 193pp.

Jonkel, C. J. 1967a. *Black Bear Population Studies.* Fed. Aid Wildl. Rest. Proj. W-98-R-1, 2, 3, 4, 5, 6 / Job B-1. Montana Dept. of Fish, Wildl. and Parks, Helena. 148pp.

Jonkel, C. J. 1967b. *The Ecology, Population Dynamics and Management of the Black Bear in the Spruce-Fir Forest of Northwestern Montana.* Ph.D. Dissert. Univ. of British Columbia, Vancouver. 170pp.

Jonkel, C. J., and I. McT. Cowan. 1971. *The Black Bear in the Spruce-Fir Forest.* Wildlife Monogr., No. 27. The Wildl. Soc., Bethesda, MD. 57pp.

Judd, S. L., R. R. Knight, and B. M. Blanchard. 1986. Denning of grizzly bears in the Yellowstone National Park area. *Int. Conf. Bear Res. and Manage.* 6:111-7.

Kasworm, W. F., and T. Manley. 1988. *Grizzly Bear and Black Bear Ecology in the Cabinet Mountains of Northwest Montana.* Mont. Dept. of Fish, Wildl. and Parks, Helena. 122pp.

Kasworm, W. F., and T. Manley. 1990. Road and trail influences on grizzly bears and black bears in northwest Montana. *Int. Conf. Bear Res. and Manage.* 8:79-84.

Kasworm, W. F., and T. J. Thier. 1991. *Cabinet-Yaak Ecosystem Grizzly Bear and Black Bear Research: 1990 Progress Report.* US Fish and Wildlife Service, Grizzly Bear Recovery Coordinator's Office. Univ. of Montana, Missoula. 53pp.

Kasworm, W. F., T. J. Thier, and C. Servheen. 1991. *Cabinet Mountains Grizzly Bear Population Augmentation: 1990 Progress Report.* US Fish and Wildlife Service, Grizzly Bear Recovery Coordinator's Office. Univ. of Montana, Missoula. 14pp.

Kay, C. E. 1985. Aspen Reproduction in the Yellowstone Park-Jackson Hole Area and Its Relationship to the Natural Regulation of Ungulates. Pages 131-60 *in* G. W. Workman, ed. *Western Elk Management: A Symposium.* Utah State Univ., Logan.

Kay, C. E. 1990. *Yellowstone's Northern Elk Herd: A Critical Evaluation of the "Natural Regulation" Paradigm.* Ph.D. Dissert. Utah State Univ., Logan. 490pp.

Kay, C. E. 1995. Browsing By Native Ungulates: Effects On Shrub and Seed Production in the Greater Yellowstone Ecosystem. Pages 310-20 *in* D. A. Roundy, D. D. McArthur, J. S. Haley, and D. K. Mann, eds. *Proceedings of the Wildland Shrub and Arid Land Symposium,* INT-GTR-315. USDA Forest Service, Denver, CO.

Kay, C. E., and F. H. Wagner. 1994. Historical Condition of Woody Vegetation On Yellowstone's Northern Range: A Critical Evaluation of the "Natural Regulation" Paradigm. Pages 151-9 *in* D. G. Despain and P. Schullery, eds. *Proceedings—Plants and Their Environment: First Biennial Scientific Conference On the Greater Yellowstone Ecosystem,* Tech. Rep. NPS/NRYELL/NRTR-93/XX. USDI-NPS Natural Resources Publication Office, Denver, CO.

Keane, R. E., and S. F. Arno. 1993. Rapid decline of whitebark pine in western Montana: Evidence from 20-year measurements. *West. J. Appl. For.* 8:44-7.

Keane, R. E., S. F. Arno, J. K. Brown, and D. F. Tomback. 1990. Simulating Disturbances and Conifer Succession in Whitebark Pine Forests. Pages 274-88 *in Proceedings—Symposium On Whitebark Pine Ecosystems: Ecology and Management of a High-Mountain Resource,* GTR-INT-270. USDA Forest Service, Intermountain Research Station, Ogden, UT.

Kilgore, B. M., C. Jones, R. Goforth, J. Beecham, J. Weigand, M. D. Strickland, F. Bunnell, and S. Herrero. 1981. *Technical Review of the Interagency Grizzly Bear Study Team and Recommendations for Future Research (Two Letters of Consultation Appended).* IGBST, Montana State Univ., Bozeman. 36pp.

King, J. M., H. C. Black, and O. H. Hewitt. 1960. Pathology, parasitology, and hematology of the black bear in New York. *N. Y. Fish Game J.* 7(2):99-111.

Kingsley, M. C. S., J. A. Nagy, and R. H. Russell. 1983. Patterns of weight gain and loss for grizzly bears in northern Canada. *Int. Conf. Bear Res. and Manage.* 5:174-8.

Kittams, W. H. 1948-58. *Northern Winter Range Studies.* Yellowstone National Park files.

Kittams, W. H. 1953. Reproduction of Yellowstone elk. *J. Wildl. Manage.* 17(2):177-84.

Kittams, W. H. 1959. Future of the Yellowstone wapiti. *Naturalist* 10(2):30-9.

Klaver, R. W., J. J. Claar, D. B. Rockwell, H. R. Mays, and C. F. Acevedo. 1986. Grizzly Bears, Insects, and People: Bear Management in the McDonald Peak Region, Montana. Pages 204-11 *in* G. P. Contreras and K. E. Evans, eds. *Proceedings—Grizzly Bear Habitat Symposium,* GTR-INT-207. USDA Forest Service, Intermountain Research Station, Ogden, UT.

Klein, D. R. 1958. *Southeast Alaska Brown Bear Studies.* Alaska Dept. of Fish and Game. Fed. Aid Wildl. Rest. Res. Job Compl. Rep. Proj. W-3-R-13. Juneau. 21pp.

Knight, R. R., and B. M. Blanchard. 1983. *Yellowstone Grizzly Bear Investigations: Annual Report of the Interagency Study Team, 1982.* USDI National Park Service, US GPO. 45pp.

Knight, R. R., and L. L. Eberhardt. 1984. Projected future abundance of the Yellowstone grizzly bear. *J. Wildl. Manage.* 48(4):1434-8.

Knight, R. R., and L. L. Eberhardt. 1985. Population dynamics of Yellowstone grizzly bears. *Ecology* 66(2):323-34.

Knight, R. R., J. Basile, K. R. Greer, S. Judd, L. Oldenburg, and L. Roop. 1975. *Annual Report—1974: Interagency Grizzly Bear Study Team.* USDI National Park Service, Washington, DC. 58pp.

Knight, R. R., J. Basile, K. R. Greer, S. Judd, L. Oldenburg, and L. Roop. 1976. *Yellowstone Grizzly Bear Investigations: Annual Report of the Interagency Study Team, 1975.* USDI National Park Service, US GPO. 46pp.

Knight, R. R., J. Basile, K. R. Greer, S. Judd, L. Oldenburg, and L. Roop. 1977. *Yellowstone Grizzly Bear Investigations: Annual Report of the Interagency Study Team, 1976.* USDI National Park Service, US GPO. 75pp.

Knight, R. R., J. Basile, K. R. Greer, S. Judd, L. Oldenburg, and L. Roop. 1978. *Yellowstone Grizzly Bear Investigations: Annual Report of the Interagency Study Team, 1977.* USDI National Park Service, US GPO. 107pp.

Knight, R. R., B. M. Blanchard, K. C. Kendall, and L. E. Oldenburg. 1980. *Yellowstone Grizzly Bear Investigations: Annual Report of the Interagency Study Team, 1978 and 1979.* USDI National Park Service, US GPO. 91pp.

Knight, R. R., B. M. Blanchard, and K. C. Kendall. 1981. *Yellowstone Grizzly Bear Investigations: Annual Report of the Interagency Study Team, 1980.* USDI National Park Service, US GPO. 55pp.

Knight, R. R., B. M. Blanchard, and K. C. Kendall. 1982. *Yellowstone Grizzly Bear Investigations: Annual Report of the Interagency Study Team, 1981.* USDI National Park Service, US GPO. 70pp.

Knight, R. (Chairman), J. Beecham, F. Bunnell, C. Servheen, and J. Swenson. 1983. *Memorandum—Interagency Grizzly Bear Steering Committee.* From Chairman, ad hoc Committee for population analysis; subject: Committee findings (4 pp.).

Knight, R. R., D. J. Mattson, and B. M. Blanchard. 1984. *Movements and Habitat Use of the Yellowstone Grizzly Bear.* USDI, Interagency Grizzly Bear Study Team, Bozeman, MT. 177pp.

Knight, R. R., B. M. Blanchard, and D. J. Mattson. 1985. *Yellowstone Grizzly Bear Investigations: Annual Report of the Interagency Study Team, 1983 and 1984.* USDI National Park Service, US GPO. 40pp.

Knight, R. R., B. M. Blanchard, and D. J. Mattson. 1986. *Yellowstone Grizzly Bear Investigations: Annual Report of the Interagency Grizzly Bear Study Team, 1985*. USDI National Park Service, US GPO. 58pp.

Knight, R. R., B. M. Blanchard, and D. J. Mattson. 1987. *Yellowstone Grizzly Bear Investigations: Annual Report of the Interagency Grizzly Bear Study Team, 1986*. USDI National Park Service, US GPO. 80pp.

Knight, R. R., J. J. Beecham, B. M. Blanchard, L. L. Eberhardt, L. H. Metzgar, C. W. Servheen, and J. Talbot. 1988a. Report of the Yellowstone Grizzly Bear Population Task Force: Equivalent Size for 45 Adult Females. Pages 153-60 *in Grizzly Bear Recovery Plan, 1993*. USFWS, Grizzly Bear Recovery Program Office, Univ. of Montana, Missoula.

Knight, R. R., B. M. Blanchard, and L. L. Eberhardt. 1988b. Mortality patterns and population sinks for Yellowstone grizzly bears, 1973-1985. *Wildl. Soc. Bull.* 16:121-5.

Knight, R. R., B. M. Blanchard, and D. J. Mattson. 1988c. *Yellowstone Grizzly Bear Investigations: Annual Report of the Interagency Study Team, 1987*. USDI National Park Service, US GPO. 33pp.

Knight, R. R., B. M. Blanchard, and D. J. Mattson. 1989. *Yellowstone Grizzly Bear Investigations: Annual Report of the Interagency Grizzly Bear Study Team, 1988*. USDI National Park Service, US GPO. 34pp.

Knight, R. R., B. M. Blanchard, and D. J. Mattson. 1990. *Yellowstone Grizzly Bear Investigations: Annual Report of the Interagency Study Team, 1989*. USDI National Park Service, US GPO. 33pp.

Knight, R. R., B. M. Blanchard, and D. J. Mattson. 1991. *Yellowstone Grizzly Bear Investigations: Annual Report of the Interagency Study Team, 1990*. USDI National Park Service, US GPO. 26pp.

Knight, R. R., B. M. Blanchard, and D. J. Mattson. 1992. *Yellowstone Grizzly Bear Investigations: Annual Report of the Interagency Study Team, 1991*. USDI National Park Service, US GPO. 31pp.

Kolenosky, G. B. 1990. Reproductive biology of black bears in east-central Ontario. *Int. Conf. Bear Res. and Manage.* 8:385-92.

Kurtén, B. 1968. *Pleistocene Mammals of Europe*. The World Naturalist. Weidenfeld and Nicolson, London. 317pp.

Lande, R., and G. F. Barrowclough. 1987. Effective Population Size, Genetic Variation, and Their Use in Population Management. Pages 87-123 *in* M. E. Soulé, ed. *Viable Populations for Conservation*. Cambridge Univ. Press, Cambridge, England.

Lavender, D. S. 1988. *The Way to the Westward Sea: Lewis and Clark Across the Continent*, 1st ed. Harper and Row, New York. 444pp.

Lentfer, J. W. 1966. *Bear Studies*. Alaska Dept. of Fish and Game. Fed. Aid Wildl. Rest. Res. Annual Rep. Vol III. Proj. W-6-R-7 and W-15-R-1. Juneau. 31pp.

Leopold, A. S., S. A. Cain, C. M. Cottam, I. N. Gabrielson, and T. L. Kimball. 1963. Wildlife management in the national parks. *Am. For.* 69(4):32-5 and 61-3. Report of Leopold Committee to Secretary, USDI.

Leopold, S. L. (Chairman), S. A. Cain, and C. E. Olmsted. 1969. *A Bear Management Policy and Program for Yellowstone National Park*. A report to the Director of National Parks by the National Sciences Advisory Committee. National Park Service. 8pp.

Lindenmayer, D. B., and H. A. Nix. 1993. Ecological principles for the design of wildlife corridors. *Conserv. Biol.* 7(3):627-30.

Luque, M. H., and A. W. Stokes. 1976. Fishing behavior of Alaska brown bear. *Int. Conf. Bear Res. and Manage.* 3:71-8.

Mace, G. M., and R. Lande. 1991. Assessing extinctions threats: toward a reevaluation of IUCN threatened species categories. *Conserv. Biol.* 5(2):148-57.

Mace, R., and C. Jonkel. 1980. Food Habits of the Grizzly Bear (Ursus arctos horribilis Ord.) in Montana. Pages 46-69 *in* C. Jonkel, ed. *Annual Report #5, Border Grizzly Project*. Univ. of Montana, Missoula, MT.

Mace, R. D., and T. L. Manley. 1993. *South Fork Flathead River Grizzly Bear Project*. Progress report for 1992. Montana Dept. of Fish, Wildl. and Parks, Helena. 34pp.

Madel, M. J. 1991. *Region Four Bear Management and Inventory: Grizzly Bear and Black Bear Species Report; Region Four Rocky Mtn. Front Grizzly Bear Management Program*. Biennial progress report—Project No. 5452. Montana Dept. of Fish, Wildl. and Parks, Helena. 51pp.

Mattson, D. J. 1995 (in prep). Sustainable grizzly bear mortality calculated from counts of unduplicated females with cubs-of-the-year. (Submitted to *Conserv. Biol.*).

Mattson, D. J., and J. J. Craighead. 1994. The Yellowstone Grizzly Bear Population: Information and Uncertainty in Endangered Species Management. Pages 101-29 *in* T. W. Clark, R. P. Reading, and A. L. Clarke, eds. *Saving Endangered Species: Professional and Organizational Lessons for Improvement*. Island Press, Washington, DC.

Mattson, D. J., and J. Henry. 1987. Spring Grizzly Bear Use of Ungulate Carcasses in the Firehole River Drainage. Pages 63-72 *in* R. R. Knight, B. M. Blanchard, and D. J. Mattson. *Yellowstone Grizzly Bear Investigations: Annual Report of the Interagency Study Team, 1986*. USDI National Park Service, US GPO.

Mattson, D. J., and C. Jonkel. 1990. Stone Pines and Bears. Pages 223-36 *in Proceedings—Symposium On Whitebark Pine Ecosystems: Ecology and Management of a High-Mountain Resource*, GTR-INT-270. USDA Forest Service, Intermountain Research Station, Ogden, UT.

Mattson, D. J., and R. R. Knight. 1989. Evaluation of Grizzly Bear Habitat Using Habitat and Cover Type Classifications. Pages 135-43 *in* D. E. Ferguson, P. Morgan, and F. D. Johnson, eds. *Proceedings—Land Classifications Based On Vegetation: Applications for Resource Management*, USFS GTR-INT-257. Washington, DC.

Mattson, D. J., and R. R. Knight. 1991. *Effects of Access On Human-Caused Mortality of Yellowstone Grizzly Bears*. Interagency Grizzly Bear Study Team. National Park Service, USDI, Bozeman, MT. 13pp.

Mattson, D. J., and R. R. Knight. 1992. Spring Bear Use of Ungulates in the Firehole River Drainage of Yellowstone National Park. Pages 5-95 through 5-120 *in* J. D. Varley and W. G. Brewster, eds. *Wolves for Yellowstone?*, A report to the US Congress, Vol. IV. Research and Analysis. National Park Service, Yellowstone National Park.

Mattson, D. J., and M. M. Reid. 1991. Conservation of the Yellowstone grizzly bear. *Conserv. Biol.* 5(3):364-72.

Mattson, D. J., R. R. Knight, and B. M. Blanchard. 1987. The effects of developments and primary roads on grizzly bear habitat use in Yellowstone National Park, Wyoming. *Int. Conf. Bear Res. and Manage.* 7:259-73.

Mattson, D. J., B. M. Blanchard, and R. R. Knight. 1991a. Food habits of Yellowstone grizzly bears, 1977-1987. *Can. J. Zool.* 69:1619-29.

Mattson, D. J., C. M. Gillin, S. A. Benson, and R. R. Knight. 1991b. Bear feeding activity at alpine insect aggregation sites in the Yellowstone ecosystem. *Can. J. Zool.* 69(9):2430-5.

Mattson, D. J., B. M. Blanchard, and R. R. Knight. 1992a. Yellowstone grizzly bear mortality, human habituation, and whitebark pine seed crops. *J. Wildl. Manage.* 56(3):432-42.

Mattson, D. J., R. R. Knight, and B. M. Blanchard. 1992b. Cannibalism and predation on black bears by grizzly bears in the Yellowstone Ecosystem. *J. Mammal.* 73(2):422-5.

Mattson, D. J., D. P. Reinhart, and B. M. Blanchard. 1994. Variation in Production and Bear Use of Whitebark Pine Seeds in the Yellowstone Area. Pages 103-18 *in* D. G. Despain and P. Schullery, eds. *Proceedings—Plants and Their Environment: First Biennial Scientific Conference On the Greater Yellowstone Ecosystem*, Tech. Rep. NPS/NRYELL/NRTR-93/XX. USDI-NPS Natural Resources Publication Office, Denver, CO.

Mattson, D. J., R. G. Wright, K. C. Kendall, and C. J. Martinka. 1995 (in press). Status and Trend of Grizzly Bear Populations in the Contiguous United States *in* E. T. LaRoe, G. S. Farris, C. E. Puckett, and P. D. Doran, eds. *Our Living Resources 1995: A Report to the Nation On the Distribution, Abundance, and Health of U.S. Plants, Animals, and Ecosystems*. USDI-National Biological Survey, Washington, DC.

Maynard Smith, J. 1982. *Evolution and the Theory of Genes*. Cambridge Univ. Press, Cambridge. 229pp.

McClintock, M. K. 1981. Social control of the ovarian cycle and the function of estrous synchrony. *Am. Zool.* 21(1):243-56.

McCullough, D. R. 1979. *The George Reserve Deer Herd: Population Ecology of a K-Selected Species*. Univ. Mich. Press, Ann Arbor. 271pp.

McCullough, D. R. 1981. Population Dynamics of the Yellowstone Grizzly Bear. Pages 173-96 *in* C. W. Fowler and T. D. Smith, eds. *Dynamics of Large Mammal Populations*. John Wiley and Sons, New York.

McCullough, D. R. 1984. Lessons from the George Reserve. Pages 211-42 *in* L. K. Halls, ed. *White-Tailed Deer: Ecology and Management*. Wildl. Manage. Instit. and Stackpole Books, Washington, DC and Harrisburg, PA.

McCullough, D. R. 1986. The Craighead's data on Yellowstone grizzly bear populations, and its relevance to current research and management. *Int. Conf. Bear Res. and Manage.* 6:21-32.

McCullough, D. R. 1990. Detecting Density Dependence: Filtering the Baby from the Bathwater. Pages 534-43 *in Trans. 55th N.A. Wildl. & Nat. Res. Conf.*

McDonald, L. L., M. D. Strickland, A. H. Wheeler, and M. A. Mullen, eds. 1990. *Sightings of Grizzly Bears in the Greater Yellowstone Ecosystem: Results of a Mail Survey Conducted 1987-1989*. Dept. of Statistics and Zoology, Univ. of Wyoming, Laramie, Wyoming. 152pp.

McLellan, B. N. 1984. *Population Parameters of the Flathead Grizzlies*. B.C. Fish and Wildl. Branch, Victoria. 28pp.

McLellan, B. N. 1989a. Dynamics of a grizzly bear population during a period of industrial resource extraction. I. Density and age-sex composition. *Can. J. Zool.* 67:1856-60.

McLellan, B. N. 1989b. Dynamics of a grizzly bear population during a period of industrial resource extraction. II. Mortality rates and causes of death. *Can. J. Zool.* 67:1861-4.

McLellan, B. N. 1989c. Dynamics of a grizzly bear population during a period of industrial resource extraction. III. Natality and rate of increase. *Can. J. Zool.* 67:1865-8.

McLellan, B. N. 1990. Relationships between human industrial activity and grizzly bears. *Int. Conf. Bear Res. and Manage.* 8:57-64.

McLellan, B. N. 1995 (in press). Density-dependent population regulation of brown bears. *Int. Conf. Bear Res. and Manage.* 9.

McLellan, B. N., and D. M. Shackleton. 1988. Grizzly bears and resource extraction industries: effects of roads on behaviour, habitat use, and demography. *J. Appl. Ecology* 25:451-60.

Meagher, M. M. 1973. *The Bison of Yellowstone National Park*, Vol. 1. National Park Service Scient. Monogr. Ser., No. 1. US GPO, Washington, DC. 161pp.

Mealey, S. P. 1975. *The Natural Food Habits of Free-Ranging Grizzly Bears in Yellowstone National Park, 1973-1974.* Masters Thesis. Montana State Univ., Bozeman. 158pp.

Mealey, S. P. 1978. *Guidelines for Management Involving Grizzly Bears in the Greater Yellowstone Area.* Interagency Grizzly Bear Study Team, Montana State Univ., Bozeman. 65pp.

Mealey, S. P. 1980. The natural food habits of grizzly bears in Yellowstone National Park, 1973-74. *Int. Conf. Bear Res. and Manage.* 4:281-92.

Merrill, E. H., R. W. M. Bramble-Brodahl, and M. S. Boyce. 1993. Estimation of green herbaceous phytomass from Landsat MSS data in Yellowstone National Park. *J. Range Manage.* 46:151-7.

Metcalf, C. L., and W. P. Flint. 1951. *Destructive and Useful Insects: Their Habits and Control,* 3rd ed. McGraw-Hill, New York. 1071pp.

Metzgar, L. H., and M. Bader. 1992. Large mammal predators in the northern Rockies: Grizzly bears and their habitat. *Northwest Environm. J.* 8(1):231-3.

Meyer, K. G., and J. P. Spencer. 1983. *Nulato Hills Digital Landcover and Terrain Data Base: User's Guide.* Remote sensing technical document. Bureau of Land Management, Anchorage, AK.

Miller, S. D. 1983. *Big Game Studies: Vol VI. Black Bear and Brown Bear.* Susitna Hydroelectric Project. Submitted to the Alaska Power Authority. Phase II Progress Rep. Alaska Dept. of Fish and Game, Juneau. 99pp.

Miller, S. D. 1987. *Big Game Studies: Vol VI. Black Bear and Brown Bear.* Susitna Hydroelectric Project. Submitted to the Alaska Power Authority. Final Report. Alaska Dept. of Fish and Game, Anchorage. 276pp.

Miller, S. D. 1990. Impact of increased bear hunting on survivorship of young bears. *Wildl. Soc. Bull.* 18(4):462-7.

Miller, S. D., and M. A. Chihuly. 1987. Characteristics of nonsport brown bear deaths in Alaska. *Int. Conf. Bear Res. and Manage.* 7:51-8.

Miller, S. D., and D. C. McAllister. 1982. *Susitna Hydroelectric Project.* Submitted to the Alaska Power Authority. Phase I Final Rep., Vol. VI. Brown bear and black bear. Alaska Dept. of Fish and Game, Juneau. 233pp.

Miller, S. D., E. F. Becker, and W. B. Ballard. 1987. Black and brown bear density estimates using modified capture-recapture techniques in Alaska. *Int. Conf. Bear Res. and Manage.* 7:23-35.

Miller, S. J., N. Barichello, and D. Tait. 1982. *The Grizzly Bears of the MacKenzie Mountains, Northwest Territories.* Completion Report. Northwest Territories Wildl. Serv., Yellowknife. 118pp.

Monnig, E., and J. Byler. 1992. *Forest Health and Ecological Integrity in the Northern Rockies.* FPM Report 92-7. USDA Forest Service, Region I, Missoula, MT. 18pp.

Mueggler, W. F., and W. P. Handl. 1974. *Mountain Grassland and Shrubland Habitat Types of Western Montana.* USDA Forest Service Intermountain For. and Range Exp. Sta. and Region I, Ogden, UT and Missoula, MT. 89pp.

Murie, A. 1940. *Ecology of the Coyote in Yellowstone.* Fauna of the National Parks of the United States, No. 4. US GPO, Washington, DC. 206pp.

Murie, A. 1944a. *The Wolves of Mount McKinley.* USDI National Park Service Fauna Series, No. 5. US GPO, Washington, DC. 238pp.

Murie, A. 1981. *The Grizzlies of Mount McKinley.* Scient. Monogr. Ser., No. 14. USDI National Park Service, US GPO, Washington, DC. 251pp.

Murie, O. J. 1944b. *Progress Report On the Yellowstone Bear Study.* USDI National Park Service. 13pp.

Nagy, J. A. 1984. Relationship of weight to chest girth in the grizzly bear. *J. Wildl. Manage.* 48(4):1439-40.

Nagy, J. A. S., and M. A. Haroldson. 1990. Comparisons of some home range and population parameters among four grizzly bear populations in Canada. *Int. Conf. Bear Res. and Manage.* 8:227-35.

Nagy, J. A., and R. H. Russell. 1978. *Ecological Studies of the Boreal Forest Grizzly Bear (Ursus arctos L.).* Annual report for 1977. Can. Wildl. Service. 72pp.

Nagy, J. A., R. H. Russell, A. M. Pearson, M. C. Kinsley, and B. C. Goski. 1983a. *Ecological Studies of the Grizzly Bear in Arctic Mountains, Northern Yukon Territory, 1972 to 1975.* Can. Wildl. Serv. 104pp.

Nagy, J. A., R. H. Russell, A. M. Pearson, M. C. Kinsley, and C. B. Larsen. 1983b. *A Study of Grizzly Bears On the Barren Grounds of Tuktoyaktuk Peninsula and Richards Island, Northwest Territories, 1974 to 1978.* Can. Wildl. Serv. 136pp.

Nelson, R. A., G. E. Folk, Jr., E. W. Pfeiffer, J. J. Craighead, C. J. Jonkel, and D. L. Steiger. 1983. Behavior, biochemistry, and hibernation in black, grizzly, and polar bears. *Int. Conf. Bear Res. and Manage.* 5:284-90.

Norusis, M. J. 1986. *SPSS/PC+ V. 2.0: For the IBM PC/XT/AT,* Vol. 2.0. SPSS, Inc., Chicago. 643pp.

Norusis, M. J. 1988. *SPSS/PC+ V. 3.0: For the IBM PC/XT/AT and PS/2.* SPSS, Inc., Chicago. 116pp.

Noss, R. F. 1990. Indicators for monitoring biodiversity: A hierarchical approach. *Conserv. Biol.* 4(4):355-64.

Noyce, K. V., and D. L. Garshelis. 1995 (in press). Body size and blood chemistry of black bears as indicators of condition and reproductive performance. *Int. Conf. Bear Res. and Manage.* 9.

Nunney, L., and K. A. Campbell. 1993. Assessing minimum viable population size: demography meets population genetics. *Trends Ecol. & Evol.* 8:234-9.

O'Neill, R. V., D. L. DeAngelis, J. B. Waide, and T. F. H. Allen. 1986. *A Hierarchical Concept of Ecosystems.* Monographs in population biology. Princeton Univ. Press, Princeton, NJ. 253pp.

Owen-Smith, R. N. 1989. *Megaherbivores: Influence of Very Large Body Size On Ecology.* Cambridge Univ. Press, New York. 369pp.

Packer, C., and A. E. Pusey. 1983. Male takeovers and female reproductive parameters: a simulation of estrous synchrony in lions (*Panthera leo*). *Anim. Behav.* 31(2):334-40.

Pallack, K. H., C. B. Nicholos, and J. E. Hines. 1990. *Statistical Inference for Capture-Recapture Experiments.* Wildlife Monogr., No. 107. The Wildl. Soc., Bethesda, MD. 97pp.

Parker, G. A. 1984. Evolutionarily Stable Strategies. Pages 30-61 *in* J. R. Krebs and N. B. Davies, eds. *Behavioral Ecology: An Evolutionary Approach.* Blackwell Press, Oxford.

Pearson, A. M. 1975. *The Northern Interior Grizzly Bear Ursus arctos L.* Canadian Wildl. Serv. Rep. Series No. 34. Ottawa 86pp.

Pease, C. M., and D. J. Mattson. 1995 (in prep.). Demography of the Yellowstone grizzly bears. (Submitted to *Ecology*).

Pelton, M. R. 1989. The Impacts of Oak Mast On Black Bears in the Southern Appalachians. Pages 7-11 *in* C. E. McGee, ed. *Proceedings Workshop, Southern Appalachian Mast Management.* USDA Forest Service and Univ. of Tennessee.

Pfister, R. D., B. L. Kovalchik, S. F. Arno, and R. C. Presby. 1977. *Forest Habitat Types of Montana.* INT-34. USDA Forest Service, Intermtn. For. and Range Exp. Sta. Rep., Ogden, UT. 174pp.

Picton, H. D. 1978. Climate and reproduction of grizzly bears in Yellowstone National Park. *Nature* 274:888-9.

Picton, H. D., and R. R. Knight. 1986. Using climate data to predict grizzly bear litter size. *Int. Conf. Bear Res. and Manage.* 6:41-4.

Picton, H. D., D. J. Mattson, B. M. Blanchard, and R. R. Knight. 1986. Climate, Carrying Capacity, and the Yellowstone Grizzly Bear. Pages 129-35 *in* G. P. Contreras and K. E. Evans, eds. *Proceedings—Grizzly Bear Habitat Symposium,* GTR-INT-207. USDA Forest Service, Intermountain Research Station, Ogden, UT.

Pritchard, G. T., and C. T. Robbins. 1990. Digestive and metabolic efficiencies of grizzly and black bears. *Can. J. Zool.* 68(8):1645-51.

Pruess, K. P. 1967. Migration of the army cutworm, *Chorizagrostis auxiliaris* (Lepidoptera: Noctuidae). I. Evidence for a migration. *Ann. Entomol. Soc. Am.* 60:910-20.

Racine, C. H. 1974. Vegetation of the Chukehi-Imuruk Area. Pages 239-79 *in* H. Melchior, ed. *Chukchi-Imuruk Biological Survey, Final Report,* vol. 2. USDI and Alaska Coop. Park Stud. Unit, Univ. of Alaska, Fairbanks.

Ramsey, M. A., and I. Stirling. 1988. Reproductive biology and ecology of female polar bears (*Ursus maritimus*). *J. Zool., London* 214:601-34.

Rausch, R. L. 1953. On the status of some Arctic mammals. *Arctic* 6:91-148.

Rausch, R. L. 1961. Notes on the black bear, *Ursus americanus* Pallus, in Alaska, with particular reference to dentition and growth. *Z. Säugetierkunde* 26(2):77-107.

Rausch, R. L. 1963. Geographic variation in size in North American brown bears, *Ursus arctos* L., as indicated by condylobasal length. *Can. J. Zool.* 41:33-45.

Rausch, R. L. 1970. Trichinosis in the Arctic. Pages 348-73 *in* S. E. Goud, ed. *Trichinosis in Man and Animals.* Charles. C. Thomas, Springfield, IL.

Rees, W. E., and M. Wackernagel. 1992. *Appropriated Carrying Capacity: Measuring the Natural Capital Requirements of the Human Economy.* Presented to the second meeting of the Internatl. Soc. for Ecolog. Econ., Stockholm, Sweden (Aug. 3-6, 1992). Univ. of British Columbia School of Community and Regional Planning, Vancouver, BC, Canada. 20pp.

Reinhart, D. P. 1988. Yellowstone Lake Tributary Study. Pages 60-9 *in* R. R. Knight, B. M. Blanchard, and D. Mattson. *Yellowstone Grizzly Bear Investigations: Annual Report of the Interagency Study Team, 1987.* USDI National Park Service, US GPO.

Reinhart, D. P. 1990a. Effects of Fire On Cutthroat Trout Spawning Streams and Associated Bear Use. Pages 19-26 *in* R. R. Knight, B. M. Blanchard, and D. Mattson. *Yellowstone Grizzly Bear Investigations: Annual Report of the Interagency Study Team, 1989.* USDI National Park Service, US GPO.

Reinhart, D. P. 1990b. *Grizzly Bear Habitat Use On Cutthroat Trout Spawning Streams in Tributaries of Yellowstone Lake.* Masters Thesis. Montana State Univ., Bozeman. 128pp.

Reinhart, D. P. 1991. The effects of fire on cutthroat spawning streams in Yellowstone Lake tributaries. *West. Wildlands* 17:34-6.

Reinhart, D., and D. Mattson. 1986. Yellowstone Lake Tributary Study. Pages 52-6 *in* R. R. Knight, B. M. Blanchard, and D. Mattson. *Yellowstone Grizzly Bear Investigations: Annual Report of the Interagency Study Team, 1985.* USDI National Park Service, US GPO.

Reinhart, D., and D. Mattson. 1987. Yellowstone Lake Tributary Study. Pages 46-62 *in* R. R. Knight, B. M. Blanchard, and D. Mattson. *Yellowstone Grizzly Bear Investigations: Annual Report of the Interagency Study Team, 1986.* USDI National Park Service, US GPO.

Reinhart, D. P., and D. J. Mattson. 1990. Bear use of cutthroat trout spawning streams in Yellowstone National Park. *Int. Conf. Bear Res. and Manage.* 8:33-350.

Reynolds, H. 1974. *North Slope Grizzly Bear Studies.* Alaska Dept. of Fish and Game. Fed. Aid Wildl. Rest. Res. Prog. Rep. Proj. W-17-6. Juneau. 27pp.

Reynolds, H. 1976. *North Slope Grizzly Bear Studies.* Alaska Fed. Aid. Wildl. Rest. Res. Final Rep. Proj. W-17-6 and 7, Juneau. 20pp.

Reynolds, H. V. 1979. *Structure, Status, Reproductive Biology, Movement, Distribution and Habitat Utilization of a Grizzly Bear Population in NPR-A (Western Brooks Range, Alaska).* Proj. 105c (Work Group 3). Alaska Dept. of Fish and Game, Juneau. 46pp.

Reynolds, H. V. 1980. *Big Game Investigations: Characteristics of Grizzly Bear Predation On Caribou in Calving Grounds of the Western Arctic Herd.* Alaska Dept. of Fish and Game. Fed. Aid Wildl. Rest. Res. Prog. Rep. Proj. W-21-2. Juneau. 12pp.

Reynolds, H. V. 1982. *Alaska Range Grizzly Bear Studies.* Alaska Dept. of Fish and Game. Fed. Aid Wildl. Rest. Res. Prog. Rep. Vol. I. Proj. W-21-2. Juneau. 12pp.

Reynolds, H. V. 1984. Grizzly Bear Population Biology in the Western Brooks Range, Alaska. Pages 2-19 *in* H. V. Reynolds and J. L. Hechtel. *Structure, Status, Reproductive Biology, Movement, Distribution, and Habitat Utilization of a Grizzly Bear Population,* Final Rep. Alaska Dept. of Fish and Game, Fed. Aid Wildl. Rest. Res., Juneau.

Reynolds, H. V. 1989. *Population Dynamics of a Hunted Grizzly Bear Population in the Northcentral Alaska Range.* Alaska Dept. of Fish and Game. Fed. Aid Wildl. Rest. Res. Prog. Rep. Proj W-23-1. Juneau. 63pp.

Reynolds, H. V. 1990. *Population Dynamics of a Hunted Grizzly Bear Population in the Northcentral Alaska Range.* Alaska Dept. of Fish and Game. Fed. Aid Wildl. Rest. Res. Prog. Rep. Proj. W-23-2 study 4.19. Juneau. 63pp.

Reynolds, H. 1993. *Evaluation of the Effects of Harvest On the Grizzly Bear Population Dynamics in the Northcentral Alaska Range.* Alaska Dept. of Fish and Game. Fed. Aid Wildl. Rest. Res. Final Rep. Proj. W-23-5. Juneau. 94pp.

Reynolds, H. V. 1995 (in press). Patterns of long-term home range use and emigration of grizzly bears in the northcentral Alaska Range. *Int. Conf. Bear Res. and Manage.* 9.

Reynolds, H. V., and J. L. Hechtel. 1980. *Big Game Investigations: Structure, Status, Reproductive Biology, Movements, Distribution and Habitat Utilization of a Grizzly Bear Population.* Alaska Dept. of Fish and Game. Fed. Aid Wildl. Rest. Res. Prog. Rep. Proj. W-17-11. Juneau. 66pp.

Reynolds, H. V., and J. L. Hechtel. 1983. *Population Structure, Reproductive Biology, and Movement Patterns of Grizzly Bears in the Northcentral Alaska Range.* Alaska Dept. of Fish and Game. Fed. Aid Wildl. Rest. Res. Prog. Rep. Proj. W-22-1. Juneau, AK. 27pp.

Reynolds, H. V., and J. L. Hechtel. 1984. *Structure, Status, Reproductive Biology, Movement, Distribution, and Habitat Utilization of a Grizzly Bear Population.* Alaska Dept. of Fish and Game. Fed. Aid Wildl. Rest. Res. Final Rep. Proj. W-21-1, W-21-2, W-22-1, W-22-2. Juneau. 29pp.

Reynolds, H. V., and J. L. Hechtel. 1986. *Population Structure, Reproductive Biology, and Movement Patterns of Grizzly Bears in the Northcentral Alaska Range.* Alaska Dept. of Fish and Game. Fed. Aid Wildl. Rest. Final Rep. Proj. W-21-2, W-22-2, W-22-3, W-22-4. Juneau. 53pp.

Reynolds, H. V., and J. L. Hechtel. 1988. *Population Dynamics of a Hunted Grizzly Bear Population in the Northcentral Alaska Range.* Alaska Dept. of Fish and Game. Fed. Aid in Wildl. Rest. Res. Prog. Rep. Project W-22-6. Juneau. 52pp.

Reynolds, H. V., J. L. Hechtel, and D. J. Reed. 1987. *Population Dynamics of a Hunted Grizzly Bear Population in the Northcentral Alaska Range.* Alaska Dept. of Fish and Game. Fed. Aid Wildl. Rest. Res. Prog. Rep. Proj. W-22-5. Juneau. 59pp.

Robbins, W. J. C., E. A. Ackerman, M. Bates, S. A. Cain, F. F. Darling, C. J. S. Durham, J. M. Fogg, T. Gill, J. M. Gillson, E. R. Hall, and C. L. Hubbs. 1963. *A Report By the Advisory Committee to the National Park Service On Research.* Natl. Aca. of Sci., National Research Council, Washington, DC. 156pp.

Rogers, E. P. 1942. *Annual Report of Superintendent, Yellowstone National Park.* US GPO, Washington, DC.

Rogers, L. 1970. Black bear of Minnesota. *Naturalist* 21(4):42-7.

Rogers, L. L. 1975. Parasites of black bears of the Lake Superior Region. *J. Wildl. Dis.* 11:189-92.

Rogers, L. 1976. Effects of mast and berry crop failures on survival, growth, and reproductive success of black bears. *Trans. N. Amer. Wildl. Conf.* 41:431-8.

Rogers, L. L. 1983. Effects of Food Supply, Predation, Cannibalism, Parasites, and Other Health Problems On Black Bear Populations. Pages 194-211 *in Symp. On Nat. Reg. Of Wildlife Populations,* Proceedings of NW Section, The Wildlife Society. For. Wildl. and Range Exp. Station, Univ. of Idaho, Moscow, ID.

Rogers, L. L. 1987. *Effects of Food Supply and Kinship On Social Behavior, Movements, and Population Growth of Black Bears in Northeastern Minnesota.* Wildlife Monogr., No. 97. The Wildl. Soc., Bethesda, MD. 72pp.

Rogers, L. L. 1989. Black Bears, People, and Garbage Dumps in Minnesota. Pages 43-6 *in* M. Bromley, ed. *Bear-People Conflicts: Proceedings of a Symposium On Management Strategies.* Northwest Territories Dept. of Renewable Resources, Yellowknife, Northwest Territories, Canada.

Rogers, L. L., and S. M. Rogers. 1976. Parasites of bears: A review. *Int. Conf. Bear Res. and Manage.* 3:411-30.

Rogers, L. L., D. W. Kuehn, A. W. Erickson, E. M. Harger, L. J. Verme, and J. J. Ozoga. 1976. Characteristics and management of black bears that feed in garbage dumps, campgrounds or residential areas. *Int. Conf. Bear Res. and Manage.* 3:169-75.

Rohlf, D. J. 1991. Six biological reasons why the endangered species act doesn't work—and what to do about it. *Conserv. Biol.* 5:273-81.

Romme, W. H. 1982. Fire and landscape diversity in subalpine forests of Yellowstone National Park. *Ecol. Monog.* 52:199-221.

Romme, W. H., and M. G. Turner. 1991. Implications of global climate change for biogeographic patterns in the Greater Yellowstone Ecosystem. *Conserv. Biol.* 5(3):373-86.

Rush, W. M., and Montana Fish and Game Commission. 1932. *Northern Yellowstone Elk Study.* Missoulian Publishing Co., Missoula, MT. 131pp.

Russell, R. H., J. W. Nolan, N. G. Woody, and G. H. Anderson. 1979. *A Study of the Grizzly Bear in Jasper National Park, 1976 to 1978.* Prep. for Parks Canada. Can. Wildl. Serv., Edmonton, Canada. 136pp.

Scaggs, G. B. 1979. *Vegetation Description of Potential Grizzly Bear Habitat in the Selway-Bitterroot Wilderness Area, Montana and Idaho.* Masters Thesis. Univ. of Montana, Missoula. 148pp.

Schallenberger, A. 1980. Review of oil and gas exploitation impacts on grizzly bears. *Int. Conf. Bear Res. and Manage.* 4:271-6.

Schallenberger, A., and C. J. Jonkel. 1980. *Rocky Mountain East Front Grizzly Studies, 1979 Ann. Rep.* Border Grizzly Project Spec. Rep. 39. Sch. of For., Univ. of Montana, Missoula. 207pp.

Scheffer, V. B. 1950. Growth layers on the teeth of Pinnipedia as an indication of age. *Science* 112(2907):309-11.

Schmidt-Nielsen, K. S. 1975. *Animal Physiology: Adaptation and Environment.* Cambridge Univ. Press, New York. 699pp.

Schoen, J. W., and L. Beier. 1986. *Brown Bear Habitat Preferences and Brown Bear Logging and Mining Relationships in Southeast Alaska.* Alaska Dept. of Fish and Game. Fed. Aid Wildl. Rest. Res. Prog. Rep. Proj. W-22-4. Juneau. 32pp.

Schoen, J. W., and LaV. R. Beier. 1989. *Brown Bear Habitat Preferences and Brown Bear Logging and Mining Relationships in Southeast Alaska.* Wildlife Research and Management Proj. No. W-23-1, Study No. 4.17, period covered 1 July 1987-30 June 1988. Juneau. 32pp.

Schoen, J. W., and LaV. R. Beier. 1990. *Brown Bear Habitat Preferences and Brown Bear Logging and Mining Relationships in Southeast Alaska.* Alaska Dept. of Fish and Game. Fed. Aid in Wildl. Rest. Res. Final Rep., Proj. W-22-1 through W-22-6 and W-23-1 through W-23-3. Juneau. 90pp.

Schwartz, C. C., and A. W. Franzmann. 1991. *Interrelationships of Black Bears to Moose and Forest Succession in the Northern Coniferous Forest.* Wildlife Monogr., No. 113. The Wildl. Soc., Bethesda, MD. 58pp.

Sellers, R. A., and L. D. Aumiller. 1995 (in press). Brown bear population parameters at McNeil River, Alaska. *Int. Conf. Bear Res. and Manage.* 9.

Servheen, C. W. 1983. Grizzly bear food habits, movements, and habitat selection in the Mission Mountains, Montana. *J. Wildl. Manage.* 47(4):1026-35.

Servheen, C., and L. C. Lee. 1979. *Mission Mountains Grizzly Bear Studies: An Interim Report, 1976-78.* Border Grizzly Proj. Montana For. Conserv. Exp. Sta., Univ. of Montana, Missoula. 299pp.

Servheen, C., A. Hamilton, R. Knight, and B. McLellan. 1991. *Report of the Technical Review Team: Evaluation of the Ability of the Bitterroot and North Cascades to Sustain Viable Grizzly Bear Populations.* Report to the Interagency Grizzly Bear Committee. 9pp.

Shaffer, M. L. 1978. *Determining Minimum Viable Population Sizes: A Case Study of the Grizzly Bear (*Ursus arctos L.*).* Ph.D. Dissert. Duke Univ., Durham, N.C. 192pp.

Shaffer, M. L. 1983. Determining minimum viable population sizes for the grizzly bear. *Int. Conf. Bear Res. and Manage.* 5:133-9.

Shaffer, M. L. 1987. Minimum Viable Populations: Coping With Uncertainty. Pages 69-86 *in* M. E. Soulé, ed. *Viable Populations for Conservation.* Cambridge Univ. Press, Cambridge, England.

Shaffer, M. L. 1992. *Keeping the Grizzly Bear in the American West.* The Wilderness Society, Washington, DC. 17pp.

Shaffer, M. L., and F. B. Samson. 1985. Population size and extinction: A note on determining critical population sizes. *Am. Nat.* 125:144-52.

Singer, F. J. 1991. The Ungulate Prey Base for Wolves in Yellowstone National Park. Pages 323-48 *in* R. B. Keiter and M. S. Boyce, eds. *The Greater Yellowstone Ecosystem: Redefining America's Wilderness Heritage.* Yale Univ. Press, New Haven, CT.

Skinner, M. P. 1925. *Bears in the Yellowstone.* A.C. McClurg and Co., Chicago. 159pp.

Skinner, M. P. 1928. The elk situation. *J. Mammal.* 9(4):309-17.

Smith, H. S. 1935. The role of biotic factors in the determination of population densities. *J. Econ. Entomol.* 28:873-98.

Sokal, R. R., and F. J. Rohlf. 1981. *Biometry: The Principles and Practice of Statistics in Biological Research*, 2nd ed. W.H. Freeman, San Francisco. 859pp.

Soulé, M. E., ed. 1987. *Viable Populations for Conservation.* Cambridge Univ. Press, Cambridge, England. 183pp.

Soulé, M. E., and B. A. Wilcox, eds. 1980. *Conservation Biology: An Evolutionary-Ecological Perspective.* Sinauer Assocs., Inc., Sunderland, MA. 395pp.

Stirling, I., A. M. Pearson, and F. L. Bunnell. 1976. Population ecology studies of polar and grizzly bears in northern Canada. *Trans. N. Amer. Wildl. Conf.* 41:421-30.

Stirling, I., W. R. Archibald, and D. DeMaster. 1977. Distribution and abundance of seals in the eastern Beaufort Sea. *J. Fish. Res. Board Can.* 34:976-88.

Stonorov, D., and A. W. Stokes. 1972. Social behavior of the Alaska brown bear. *Int. Conf. Bear Res. and Manage.* 2:232-42.

Storer, T. I., and L. P. Tevis. 1955. *California Grizzly.* Univ. Nebraska Press, Lincoln. 335pp.

Stringham, S. F. 1983. Roles of adult males in grizzly bear population biology. *Int. Conf. Bear Res. and Manage.* 5:140-51.

Stringham, S. F. 1986. Effects of climate, dump closure, and other factors on Yellowstone grizzly bear litter size. *Int. Conf. Bear Res. and Manage.* 6:226.

Stringham, S. F. 1989. Demographic Consequences of Bears Eating Garbage At Dumps: An Overview. Pages 35-42 *in* M. Bromley, ed. *Bear-People Conflicts: Proceedings of a Symposium On Management Strategies.* Northwest Territories Dept. of Renew. Resources, Yellowknife, Northwest Territories, Canada.

Stringham, S. F. 1990a. Black bear reproductive rate relative to body weight in hunted populations. *Int. Conf. Bear Res. and Manage.* 8:425-32.

Stringham, S. F. 1990b. Grizzly bear reproductive rate relative to body size. *Int. Conf. Bear Res. and Manage.* 8:433-43.

Strong, W. E. 1968. *A Trip to the Yellowstone National Park in July, August, and September, 1875*, [New Ed.]. Univ. of Oklahoma Press, Norman. 165pp.

Struhsaker, T. T., and L. Leland. 1987. Colobines: Infanticide By Adult Males. Pages 83-97 *in* B. B. Smuts, D. L. Cheney, R. M. Seyfarth, R. W. Wrangham, and T. T. Struhsaker, eds. *Primate Societies.* Univ. of Chicago Press, Chicago.

Stubbs, M. 1977. Density dependence in the life cycles of animals and its importance in K- and r-strategies. *J. Anim. Ecol.* 46:677-88.

Sumner, J. S., and J. J. Craighead. 1973. *Grizzly Bear Habitat Survey in the Scapegoat Wilderness, Montana.* Montana Coop. Wildl. Res. Unit, Univ. of Montana, Missoula. 49pp.

Tait, D. E. N. 1980. Abandonment as a reproductive tactic—the example of grizzly bears. *Am. Nat.* 115:800-8.

Talbot, S. S., C. J. Markon, and M. B. Shasby. 1984. LandSat-Facilitated Vegetation Classification of Tetlin National Wildlife Refuge, Alaska. Pages 143-51 *in* V. J. LaBau and C. L. Kerr, eds. *Inventorying Forest and Other Vegetation of the High Latitude and High Altitude Regions*, Proceedings of an international symposium. Society of American Foresters, Bethesda, MD.

Tansley, A. G. 1935. The use and abuse of vegetation concepts and terms. *Ecology* 16:287-307.

Taylor, M. 1995 (in press). Density dependent population regulation of black, grizzly, and polar bears. *Int. Conf. Bear Res. and Manage.* 9.

Thorne, E. T., R. F. Honess, and Wyoming Game and Fish Dept. 1982. *Diseases of Wildlife in Wyoming*, 2nd ed. Wyoming Game and Fish Dept. 353pp.

Thorne, E. T., M. Meagher, and R. Hillman. 1991. Brucellosis in Free-Ranging Bison: Three Perspectives. Pages 275-87 *in* R. B. Keiter and M. S. Boyce, eds. *The Greater Yellowstone Ecosystem: Redefining America's Wilderness Heritage.* Yale Univ. Press, New Haven, CT.

Tomback, D. F. 1989. The Broken Circle: Fire, Birds and Whitebark Pine. Pages 14-7 *in* T. Walsh ed. *Wilderness and Wildfire.* Misc. Publ. 50. Univ. of Montana, School of Forestry, Montana Forest and Range Experiment Station; Missoula, MT.

Trivers, R. 1974. Parent-Offspring conflict. *Am. Zool.* 14:249-64.

Trivers, R. 1985. *Social Evolution.* Benjamin/Cummings Pub. Co., Menlo Park, CA. 462pp.

Troyer, W. A., and R. J. Hensel. 1962. Cannibalism in brown bear. *Anim. Behav.* 10(3-4):231.

Troyer, W. A., and R. J. Hensel. 1964. Structure and distribution of a Kodiak bear population. *J. Wildl. Manage.* 28(4):769-72.

Troyer, W. A., and R. J. Hensel. 1969. *The Brown Bear of Kodiak Island.* USDI Bureau of Sport Fish and Wildl., Branch of Wildl. Refuges. 233pp.

Tsubota, T., Y. Takahasi, and H. Kanagawa. 1987. Changes in serum progesterone levels and growth of fetuses in Hokkaido brown bears. *Int. Conf. Bear Res. and Manage.* 7:355-8.

Tyers, D. B. 1981. *The Condition of the Northern Winter Range in Yellowstone National Park: A Discussion of the Controversy.* Ph.D. Dissert. Montana State Univ., Bozeman, MT. 170pp.

U. S. Congress. 1973. *The Endangered Species Act.* P.L. 93-205 and P.L. 94-325, as amended. US GPO, Washington, DC.

U. S. Congress. 1977. *Proposed Critical Habitat Area for Grizzly Bears.* Special hearing before subcommittee of the Committee on Appropriations. 94th Congress, Second Session. US GPO, Washington, DC. 209pp.

U. S. Congress. 1978. *Endangered Species Act Amendments of 1978, Public Law 95-632.* Nov 10, 1978. 92 Stat. 3751, Sec. 2 Section 3 of the Endangered Species Act of 1973 (16 USC 1532).

U. S. Fish and Wildlife Service. 1982. *Grizzly Bear Recovery Plan.* Fish and Wildlife Reference Service, Denver, CO. 195pp.

U. S. Fish and Wildlife Service. 1990. *Grizzly Bear Recovery Plan.* 1st review draft. U.S. Fish and Wildlife Service, Missoula, MT. 117pp.

U. S. Fish and Wildlife Service. 1992. *Grizzly Bear Recovery Plan.* 2nd review draft. Missoula, MT. 200pp.

U. S. Fish and Wildlife Service. 1993. *Grizzly Bear Recovery Plan.* Final report. Missoula, MT. 181pp.

Varney, J. R., J. J. Craighead, and J. S. Sumner. 1973. *An Evaluation of the Use of ERTS-1 Satellite Imagery for Grizzly Bear Habitat Analysis.* Prog. Rep. Montana Coop. Wildl. Res. Unit, Univ. of Montana, Missoula. 30pp.

Walker, T., and L. Aumiller. 1993. *River of Bears.* Voyageur Press, Inc, Stillwater, MN. 160pp.

Walters, J. R., and R. M. Seyfarth. 1987. Conflict and Cooperation. Pages 306-17 *in* B. B. Smuts, D. L. Cheney, R. M. Seyfarth, R. W. Wrangham, and T. T. Struhsaker, eds. *Primate Societies.* Univ. of Chicago Press, Chicago.

Wand, M., and D. Lyman. 1972a. Dangers of eating bear meat. *J. Am. Med. Assoc.* 220(2):274.

Wand, M., and D. Lyman. 1972b. Trichinosis from bear meat. *J. Am. Med. Assoc.* 220(2):245-6.

White, G. C., and R. A. Garrott. 1990. *Analysis of Wildlife Radio-Tracking Data.* Academic Press, San Diego. 383pp.

Whitlock. 1992. Vegetational and climatic history of the Pacific Northwest during the last 20,000 years: Implications for understanding present-day biodiversity. *Northwest Environm. J.* 5:5-28.

Wielgus, R. B., and F. L. Bunnell. 1994. Dynamics of a small, hunted brown bear (*Ursus arctos*) population in southwestern Alberta, Canada. *Biol. Conserv.* 67:161-6.

Wiesner, J. B., and H. F. York. 1964. National security and the nuclear-test ban. *Scient. Amer.* 211:27.

Wilson, E. O. 1975. *Sociobiology: The New Synthesis.* Belknap Press, Cambridge. 697pp.

Wimsatt, A. W. 1963. Delayed Implantation in the Ursidae, With Particular Reference to the Black Bear (*Ursus americanus* Pallas). Pages 49-76 *in* E. C. Enders, ed. *Delayed Implantation.* Univ. of Chicago Press, Chicago.

Winterberger, K. C. 1984. Landsat Data and Aerial Photographs Used in a Multiphase Sample of Vegetation and Related Resources in Interior Alaska. Pages 157-63 *in Inventorying Forest and Other Vegetation of the High Latitude and High Altitude Regions,* Proc. of internatl. symp. Society of American Foresters, Bethesda, MD.

Winters, J. B., D. E. Worley, K. R. Greer, and N. Samuel. 1971. *Sylvatic Trichiniasis in Carnivorous Mammals from Wyoming, Montana, and Idaho.* Abstract from Proc. of 46th Ann. Meet., Amer. Soc. of Parasitologists, Los Angeles, CA. August 23-27, 1971.

Worley, D. E., J. C. Fox, J. B. Winters, and K. R. Greer. 1974. Prevalence and Distribution of *Trichinella spiralis* in Carnivorous Mammals in the United States Northern Rocky Mountain Region. Pages 597-602 *in* C. W. Kim, ed. *Trichinellosis: Proc. 3rd Intern. Conf. On Trichinellosis.* Intext Educ. Publ., New York.

Wright, G. M. 1934. *Winter Range of the Northern Yellowstone Elk Herd and Suggested Program for Its Restoration.* Report to the Director. Office of National Parks, Buildings and Reservations, Washington, DC. 14pp.

Wright, G. M., and B. H. Thompson. 1935. *Wildlife Management in the National Parks.* Fauna of the National Parks, No. 2. US GPO, Washington, DC. 142pp.

Wright, G. M., J. S. Dixon, and B. H. Thompson. 1933. *A Preliminary Survey of Faunal Relations in National Parks.* Fauna of the national parks of the United States, No. 1. US GPO, Washington, DC. 157pp.

Yellowstone National Park. 1920. *[Report of Superintendent, 1920].* in National Park Service Report, 1920. pp. 197-234.

Yellowstone Nature Notes. 1920-1958. See Zaft, B. 1990.

Zaft, B., Compiler. 1990. *Bear-Related Material, Yellowstone Nature Notes—1920 to 1958.* USDI National Park Service, Yellowstone Research Library. 202pp.

Zager, P. 1983. Grizzly bears in Idaho's Selkirk Mountains: an update. *Northwest Sci.* 57(4):299-309.

Index

Note: f=figure, p=photo, t=table

Island Press Board of Directors

CHAIR
Susan E. Sechler
Executive Director, Pew Global Stewardship Initiative

VICE-CHAIR
Henry Reath
President, Collector's Reprints, Inc.

SECRETARY
Drummond Pike
President, The Tides Foundation

TREASURER
Robert E. Baensch
Senior Consultant, Baensch International Group Ltd.

Peter R. Borrelli
President, Geographic Solutions

Catherine M. Conover

Lindy Hess
Director, Radcliffe Publishing Program

Gene E. Likens
Director, The Institute of Ecosystem Studies

Jean Richardson
Director, Environmental Programs in Communities (EPIC),
University of Vermont

Charles C. Savitt
President, Center for Resource Economics/Island Press

Peter R. Stein
Managing Partner, Lyme Timber Company

Richard Trudell
Executive Director, American Indian Resources Institute

SCI QL 737 .C27 C72 1995

Craighead, John Johnson,
 1916-

The grizzly bears of
 Yellowstone